Inorganic Reactions and Methods

Volume 3

Inorganic Reactions and Methods

Distribution: VCH Verlagsgesellschaft mbH, P.O. Box 1260/1280, D-6940 Weinheim, Federal Republic of Germany
USA and Canada: VCH Publishers, Inc., 303 N.W. 12th Avenue, Deerfield Beach, FL 33442-1705, USA

Inorganic Reactions and Methods

Volume 3

The Formation of Bonds to Halogens (Part 1)

Founding Editor

J.J. Zuckerman

Editor

A.P. Hagen

Library of Congress Cataloging in Publication Data

Inorganic reactions and methods.

Includes bibliographies and indexes.
Contents: v. 1. The formation of bonds to hydrogen—
pt. 2, v. 2. The formation of the bond to hydrogen—
—v. 15. Electron-transfer and electrochemical
reactions; photochemical and other energized reactions.
1. Chemical reaction, Conditions and laws of—
Collected works. 2. Chemistry, Inorganic—Synthesis—
Collected works. I. Zuckerman, Jerry J.
QD501.I623 1987 541.3'9 85-15627
ISBN 0-89573-250-5 (set)

Printed in the United States of America.

ISBN 0-89573-253-X VCH Publishers
ISBN 3-527-26261-X VCH Verlagsgesellschaft

Contents of Volume 3

2.2. Formation of Halogen–Halogen Bonds

2.3. Formation of Bonds between Halogens and Group-VB (O, S, Se, Te, Po) Elements

2.3. Formation of Bonds between Halogens and Group-VB (O, S, Se, Te, Po) Elements

2.4. Formation of Bonds between Halogens and Group-VB (N, P, As, Sb, Bi) Elements

2.5. The Formation of the Halogen–Group-IVB (C, Si, Ge, Sn, Pb) Element Bond

How to Use this Book

1. Organization of Subject Matter

1.1. Logic of Subdivision and Add-On Chapters

This volume is part of a series that describes all of inorganic reaction chemistry. The contents are subdivided systematically and so are the contents of the entire series: Using the periodic system as a correlative device, it is shown how bonds between pairs of elements can be made. Treatment begins with hydrogen making a bond to itself in H_2 and proceeds according to the periodic table with the bonds formed by hydrogen to the halogens, the groups headed by oxygen, nitrogen, carbon, boron, beryllium and lithium, to the transition and inner-transition metals and to the members of group zero. Next it is considered how the halogens form bonds among themselves and then to the elements of the main groups VI to I, the transition and inner-transition metals and the zero-group gases. The process repeats itself with descriptions of the members of each successive periodic group making bonds to all the remaining elements not yet treated until group zero is reached. At this point all actual as well as possible combinations have been covered.

The focus is on the primary formation of bonds, not on subsequent reactions of the products to form other bonds. These latter reactions are covered at the places where the formation of those bonds is described. Reactions in which atoms merely change their oxidation states are not included, nor are reactions in which the same pairs of elements come together again in the product (for example, in metatheses or redistributions). Physical and spectroscopic properties or structural details of the products are not covered by the reaction volumes which are concerned with synthetic utility based on yield, economy of ingredients, purity of product, specificity, etc. The preparation of short-lived transient species is not described.

While in principle the systematization described above could suffice to deal with all the relevant material, there are other topics that inorganic chemists customarily identify as being useful in organizing reaction information and that do not fit into the scheme. These topics are the subject of eight additional chapters constituting the last four books of the series. These chapters are systematic only within their own confines. Their inclusion is based on the best judgment of the Editorial Advisory Board as to what would be most useful currently as well as effective in guiding the future of inorganic reaction chemistry.

1.2. Use of Decimal Section Numbers

The organization of the material is readily apparent through the use of numbers and headings. Chapters are broken down into divisions, sections and subsections, which have short descriptive headings and are numbered according to the following scheme:

1. Major Heading
1.1. Chapter Heading
1.1.1. Division Heading
1.1.1.1. Section Heading
1.1.1.1.1. Subsection Heading

Further subdivision of a five-digit "slice" utilizes lower-case Roman numerals in parentheses: (i), (ii), (iii), etc. It is often found that as a consequence of the organization, cognate material is located in different chapters but in similarly numbered pieces, i.e., in parallel sections. Section numbers, rather than page numbers, are the key by which the material is accessed through the various indexes.

1.3. Building of Headings

1.3.1. Headings Forming Part of a Sentence

Most headings are sentence-fragment phrases which constitute sentences when combined. Usually a period signifies the end of a combined sentence. In order to reconstitute the context in which a heading is to be read, superior-rank titles are printed as running heads on each page. When the sentences are put together from their constituent parts, they describe the contents of the piece at hand. For an example, see 2.3 below.

1.3.2. Headings Forming Part of an Enumeration

For some material it is not useful to construct title sentences as described above. In these cases hierarchical lists, in which the topics are enumerated, are more appropriate. To inform the reader fully about the nature of the material being described, the headings of connected sections that are superior in hierarchy always occur as running heads at the top of each page.

2. Access and Reference Tools

2.1. Plan of the Entire Series (Front Endpaper)

Printed on the inside of the front cover is a list, compiled from all 18 reaction volumes, of the major and chapter headings, that is, all headings that are preceded by a one- or two-digit decimal section number. This list shows in which volumes the headings occur and highlights the contents of the volume that is at hand by means of a gray tint.

2.2. Contents of the Volume at Hand

All the headings, down to the title of the smallest decimal-numbered subsection, are listed in the detailed table of contents of each volume. For each heading the table of contents shows the decimal section number by which it is preceded and the number of the page on which it is found. Beside the decimal section numbers, successive indentations reveal the hierarchy of the sections and thereby facilitate the comprehension of the phrase (or of the enumerative sequence) to which the headings of hierarchically successive sections combine. To reconstitute the context in which the heading of a section must be read to become meaningful, relevant headings of sections superior in hierarchy are repeated at the top of every page of the table of contents. The repetitive occurrences of these headings is indicated by the fact that position and page numbers are omitted.

2.3. Running Heads

In order to indicate the hierarchical position of a section, the top of every page of text shows the headings of up to three connected sections that are superior in hierarchy. These running heads provide the context within which the title of the section under discussion becomes meaningful. As an example, the page of Volume 1 on which section 1.4.9.1.3 "in the Production of Methanol" starts, carries the running heads:

1.4. The Formation of Bonds between Hydrogen and O,S,Se,Te,Po
1.4.9. by Industrial Processes
1.4.9.1. Involving Oxygen Compounds

whereby the phrase "in the Production of Methanol" is put into its proper perspective.

2.4. List of Abbreviations

Preceding the indexes there is a list of those abbreviations that are frequently used in the text of the volume at hand or in companion volumes. This list varies somewhat in length from volume to volume; that is, it becomes more comprehensive as new volumes are published.

Abbreviations that are used incidentally or have no general applicability are not included in the list but are explained at the place of occurrence in the text.

2.5. Author Index

The author index is compiled by computer from the lists of references. Thus it tells whose publications are cited and in that respect is comprehensive. It is not a list of authors, beyond those cited in the references, whose results are reported in the text. However, as the references cited are leading ones, consulting them, along with the use of appropriate works of the secondary literature, will rapidly lead to the complete literature related to any particular subject covered.

Each entry in the author index refers the user to the appropriate section number.

2.6. Compound Index

The compound index lists individual, fully specified compositions of matter that are mentioned in the text. It is an index of empirical formulas, ordered according to the following system: the elements within a given formula occur in alphabetical sequence except for C, or C and H if present, which always come first. Thus, the empirical formula

for $Ti(SO_4)_2$ is O_8S_2Ti
$BH_3 \cdot NH_3$ BH_6N
Be_2CO_3 CBe_2O_3
$CsHBr_2$ Br_2CsH
$Al(HCO_3)_3$ $C_3H_3AlO_9$

The formulas themselves are ordered alphanumerically without exception; that is, the formulas listed above follow each other in the sequence BH_6N, Br_2CsH, CBe_2O_3, $C_3H_3AlO_9$, O_8S_2Ti.

A compound index constructed by these principles tells whether a given compound is present. It cannot provide information about compound classes, for example, all aluminum derivatives or all compounds containing phosphorus.

In order to open this route of access as well, the compound index is augmented by successively permuted versions of all empirical formulas. Thus the number of appearances that an empirical formula makes in the compound index is equal to the number of elements it contains. As an example, $C_3H_3AlO_9$, mentioned above, will appear as such and, at the appropriate positions in the alphanumeric sequence, as $H_3AlO_9{*}C_3$, $AlO_9{*}C_3H_3$ and $O_9{*}C_3H_3Al$. The asterisk identifies a permuted formula and allows the original formula to be reconstructed by shifting to the front the elements that follow the asterisk.

Each nonpermuted formula is followed by linerarized structural formulas that indicate how the elements are combined in groups. They reveal the connectivity of the compounds underlying each empirical formula and serve to distinguish substances which are identical in composition but differ in the arrangement of elements (isomers). As an example, the empirical formula $C_4H_{10}O$ might be followed by the linearized structural formulas $(CH_3CH_2)_2O$, $CH_3(CH_2)_2OCH_3$, $(CH_3)_2CHOCH_3$, $CH_3(CH_2)_3OH$, $(CH_3)_2CHCH_2OH$ and $CH_3CH_2(CH_3)CHOH$ to identify the various ethers and alcohols that have the element count $C_4H_{10}O$.

Each linearized structural formula is followed in a third column by keywords describing the context in which it is discussed and by the number(s) of the section(s) in which it occurs.

2.7. Subject Index

The subject index provides access to the text by way of methods, techniques, reaction types, apparatus, effects and other phenomena. Also, it lists compound classes such as organotin compounds or rare-earth hydrides which cannot be expressed by the empirical formulas of the compound index.

For multiple entries, additional keywords indicate contexts and thereby avoid the retrieval of information that is irrelevant to the user's need.

Again, section numbers are used to direct the reader to those positions in the book where substantial information is to be found.

2.8. Periodic Table (Back Endpaper)

Reference to periodic groups avoids cumbersome enumerations. Section headings in the series employ the nomenclature.

Unfortunately, however, there is at the present time no general agreement on group designations. In fact, the scheme that is most widely used (combining a group number with the letters A and B) is accompanied by two mutually contradictory interpretations. Thus, titanium may be a group IVA or group IVB element depending on the school to which one adheres or the part of the world in which one resides.

In order to clarify the situation for the purposes of the series, a suitable labeled periodic table is printed on the inside back cover of each volume. All references to periodic group designations in the series refer to this scheme.

Preface to the Series

Inorganic Reactions and Methods constitutes a closed-end series of books designed to present the state of the art of synthetic inorganic chemistry in an unprecedented manner. So far, access to knowledge in inorganic chemistry has been provided almost exclusively using the elements or classes of compounds as starting points. In the first 18 volumes of **Inorganic Reactions and Methods**, it is bond formation and type of reaction that form the basis of classification.

This new route of access has required new approaches. Rather than sewing together a collection of review articles, a framework has had to be designed that reflects the creative potential of the science and is hoped to stimulate its further development by identifying areas of research that are most likely to be fruitful.

The reaction volumes describe methods by which bonds between the elements can be formed. The work opens with hydrogen making a bond to itself in H_2 and proceeds through the formation of bonds between hydrogen and the halogens, the groups headed by oxygen, nitrogen, carbon, boron, beryllium and lithium to the formation of bonds between hydrogen and the transition and inner-transition metals and elements of group zero. This pattern is repeated across the periodic system until all possible combinations of the elements have been treated. This plan allows most reaction topics to be included in the sequence where appropriate. Reaction types that do not arise from the systematics of the plan are brought together in the concluding chapters on oxidative addition and reductive elimination, insertions and their reverse, electron transfer and electrochemistry, photochemical and other energized reactions, oligomerization and polymerization, inorganic and bioinorganic catalysis and the formation of intercalation compounds and ceramics.

The project has engaged a large number of the most able inorganic chemists as Editorial Advisors creating overall policy, as Editorial Consultants designing detailed plans for the subsections of the work, and as authors whose expertise has been crucial for the quality of the treatment. The conception of the series and the details of its technical realization were the subject of careful planning for several years. The distinguished chemists who form the Editorial Advisory Board have devoted themselves to this exercise, reflecting the great importance of the project.

It was a consequence of the systematics of the overall plan that publication of a volume had to await delivery of its very last contribution. Thus was the defect side of the genius of the system revealed, as the excruciating process of extracting the rate-limiting manuscripts began. Intense editorial effort was required in order to bring forth the work in a timely way. The production process had to be designed so that the insertion of new material was possible up to the very last stage, enabling authors to update their pieces with the latest developments. The publisher supported the cost of a computerized bibliographic search of the literature and a second one for updating.

Each contribution has been subjected to an intensive process of scientific and linguistic editing in order to homogenize the numerous individual pieces, as well as to provide the highest practicable density of information. This had several important consequences. First, virtually all semblances of the authors' individual styles have been excised. Second, it was learned during the editorial process that greater economy of language could be achieved by dropping conventionally employed modifiers (such as very) and eliminating italics used for emphasis, quotation marks around nonquoted words, or parentheses around phrases, the result being a gain in clarity and readability. Because the series focuses on the chemistry rather than the chemical literature, the need to tell who has reported what, how and when can be considered of secondary importance. This has made it possible to bring all sentences describing experiments into the present tense. Information on who published what is still to be found in the reference lists. A further consequence is that authors have been burdened neither with identifying leading practitioners, nor with attributing priority for discovery, a job that taxes even the talents of professional historians of science. The authors' task then devolved to one of describing inorganic chemical reactions, with emphasis on synthetic utility, yield, economy, availability of starting materials, purity of product, specificity, side reactions, etc.

The elimination of the names of people from the text is by far the most controversial feature. Chemistry is plagued by the use of nondescriptive names in place of more expository terms. We have everything from Abegg's rule, Adkin's catalyst, Admiralty brass, Alfven number, the Amadori rearrangement and Adurssov oxidation to the Zdanovskii law, Zeeman effect, Zincke cleavage and Zinin reduction. Even well-practiced chemists cannot define these terms precisely except for their own areas of specialty, and no single source exists to serve as a guide. Despite these arguments, the attempt to replace names of people by more descriptive phrases was met in many cases by a warmly negative reaction by our colleague authors, notwithstanding the obvious improvements wrought in terms of lucidity, freedom from obscurity and obfuscation and, especially, ease of access to information by the outsider or student.

Further steps toward universality are taken by the replacement of element and compound names wherever possible by symbols and formulas, and by adding to data in older units their recalculated SI equivalents. The usefulness of the reference sections has been increased by giving journal-title abbreviations according to the *Chemical Abstracts Service Source Index*, by listing in each reference all of its authors and by accompanying references to patents and journals that may be difficult to access by their *Chemical Abstracts* citations. Mathematical signs and common abbreviations are employed to help condense prose and a glossary of the latter is provided in each volume. Dangerous or potentially dangerous procedures are highlighted in safety notes printed in boldface type.

The organization of the material should become readily apparent from an examination of the headings listed in the table of contents. Combining the words constituting the headings, starting with the major heading (one digit) and continuing through the major chapter heading (two digits), division heading (three digits), section heading (four digits) to the subsection heading (five digits), reveals at once the subject of a "slice" of the plan. Each slice is a self-contained unit. It includes its own list of references and provides definitions of unusual terms that may be used in it. The reader, therefore, through the table of contents alone, can in most instances quickly reach the desired material and derive the information wanted.

In addition there is for each volume an author index (derived from the lists of references) and a subject index that lists compound classes, methods, techniques, apparatus, effects and other phenomena. An index of empirical formulas is also provided. Here in each formula the element symbols are arranged in alphabetical order except that C, or C and H if present, always come first. Moreover, each empirical formula is permuted successively. Each permuted formula is placed in its alphabetical position and cross referenced to the original formula. Therefore, the number of appearances that an empirical formula makes in the index equals the number of its elements. By this procedure all compounds containing a given element come together in one place in the index. Each original empirical formula is followed by a linearized structural formula and keywords describing the context in which the compound is discussed. All indexes refer the user to subsection rather than page number.

Because the choice of designations of groups in the periodic table is currently in a state of flux, it was decided to conform to the practice of several leading inorganic texts. To avoid confusion an appropriately labeled periodic table is printed on the back endpaper.

From the nature of the work it is obvious that probably not more than two persons will ever read it entire: myself and the publisher's copy editor, Dr. Lindsay S. Ardwin. She, as well as Ms. Mary C. Stradner, Production Manager of VCH Publishers, are to be thanked for their unflagging devotion to the highest editorial standards. The original conception for this series was the brainchild of Dr. Hans F. Ebel, Director of the Editorial Department of VCH Verlagsgesellschaft in Weinheim, Federal Republic of Germany, who also played midwife at the birth of the plan of these reaction volumes with my former mentor, Professor Alan G. MacDiarmid of the University of Pennsylvania, and me in attendance, during the Anaheim, California, American Chemical Society Meeting in the Spring of 1978. Much of what has finally emerged is the product of the inventiveness and imagination of Professor Helmut Grünewald, President of VCH Verlagsgesellschaft. It is a pleasure to acknowledge that I have learned much from him during the course of our association. Ms. Nancy L. Burnett is to be thanked for typing everything that had to do with the series from its inception to this time. Directing an operation of this magnitude without her help would have been unimaginable. My wife Rose stood by with good cheer while two rooms of our home filled up with 10,000 manuscript pages, their copies and attendant correspondence.

Finally, and most important, an enormous debt of gratitude toward all our authors is to be recorded. These experts were asked to prepare brief summaries of their knowledge, ordered in logical sequence by our plan. In addition, they often involved themselves in improving the original conception by recommending further refinements and elaborations. The plan of the work as it is being published can truly be said to be the product of the labors of the advisors and consultants on the editorial side as well as the many, many authors who were able to augment more general knowledge with their own detailed information and ideas. Because of the unusually strict requirements of the series, authors had not only to compose their pieces to fit within narrowly constrained limits of space, format and scope, but after delivery to a short deadline were expected to stand by while an intrusive editorial process homogenized their own prose styles out of existence and shrank the length of their expositions. These long-suffering colleagues had then to endure the wait for the very last manuscript scheduled for their volume to be delivered so that their work could be published, often after a further diligent search of the literature to insure that the latest discoveries were being cited and that claims for facts now proved false were eliminated. To these co-workers (270 for the reaction volumes alone), from whom so much was demanded but who continued to place their knowledge and talents unstintingly at the disposal of the project, we dedicate this series.

J. J. ZUCKERMAN
Norman, Oklahoma
July 4, 1985

The scientific community is appreciative of the JJZ vision for a systematic inorganic chemistry. Many of the contributions had been edited prior to his death; therefore, his precise syntax will remain an important part of the series. Dr. Lindsay S. Ardwin, the copy editor, deserves special thanks for her tireless efforts to maintain the high standards of previous volumes.

A.P. HAGEN
Norman, Oklahoma
October 10, 1989

Editorial Consultants to the Series

Contributors to Volume 3

Professor J. M. Bellama
Department of Chemistry
University of Maryland
College Park, MD 20742
(*Sections* 2.5.6–2.5.9)

Professor Dr. H. J. Breunig
Anorganische Chemie
Universität Bremen
D-2800 Bremen 33, BRD
(*Sections* 2.4.7–2.4.9, 2.4.11–2.4.13)

Dr. K. O. Christe
Rockwell International Corp.
6633 Canoga Avenue
Canoga Park, CA 91304
(*Section* 2.2.7)

Dr. M. Delmas
Laboratoire des Organometalliques
Université de Droit
13397 Marseille Cedex 13
(*Sections* 2.5.2–2.5.5)

Professor D. D. Desmarteau
Department of Chemistry
Clemson University
Clemson, SC 29634
(*Sections* 2.2.2.1, 2.3.3–2.3.5)

Professor J. E. Drake
Department of Chemistry
University of Windsor
Windsor, Ontario N9B 3P4
(*Sections* 2.5.10, 2.5.11)

Professor M. Fild
Lehrstuhl B. für
Anorganische Chemie
Technische Universität
D-3300 Braunschweig, BRD
(*Sections* 2.4.5.6, 2.4.10)

Professor R. Filler
Department of Chemistry
Illinois Institute of Technology
Chicago, IL 60616
(*Section* 2.2.5)

Professor Dr. I. Haiduc
Chemistry Department
Universitatea Babes-Bolyai
R-3400 Cluj Napoca, Romania
(*Sections* 2.3.9, 2.3.10)

Mr. W. D. Lee
Department of Chemistry
The University of Oklahoma
Norman, OK 73019
(*Sections* 2.2.2.5, 2.2.4.4, 2.2.6.3, 2.3.13, 2.4.14, 2.5.17)

Professor J. C. Maire
Laboratoire des Organometalliques
Université de Droit
13397 Marseille Cedex 13
(*Sections* 2.5.2–2.5.4)

Professor Dr. R. Mews
Anorganische Chemie
Universität Bremen
D-2800 Bremen 33, BRD
(*Sections* 2.3.2, 2.3.6–2.3.8, 2.3.11, 2.3.12)

Professor J. B. Milne
Department of Chemistry
University of Ottawa
Ottawa, Ontario K1N 9B4
(*Sections* 2.2.2.2–2.2.2.4)

Dr. B. B. Randolph
Lubricating Specialties
2400 Michelson Suite B
Irvine, CA 92617
(*Sections* 2.2.2.1, 2.3.3–2.3.5)

Professor G.-V. Röschenthaler
Anorganische Chemie
Universität Bremen
D-2800 Bremen 33, BRD
(*Sections* 2.4.2–2.4.5)

Professor Jacob Shamir
Department of Chemistry
The Hebrew University
Jerusalem, Israel
(*Sections* 2.2.3, 2.2.4–2.2.4.3, 2.2.6–2.2.6.2)

Professor C. H. van Dyke
Department of Chemistry
Carnegie–Mellan University
Pittsburgh, PA 15213
(*Sections* 2.5.12–2.5.16)

2. Formation of Bonds to Halogens (Part 1)

2.1. Introduction

Bonds to halogens are nearly as ubiquitous as those to hydrogen and, like the latter require two volumes of our series to describe their formation. The elaboration begins in Volume 3 with the halogens themselves, along with the interhalogens as cationic, anionic and neutral species. Volume 3 also includes the formation of halogen bonds to the elements of main groups VIB (O, S, Se, Te, Po), VB (N, P, As, Sb, Bi) and IVB (C, Si, Ge, Sn, Pb). The bonds to the remainder of the groups and elements are then covered in Volume 4.

(J. J. ZUCKERMAN)

2.2. Formation of Halogen–Halogen Bonds

2.2.1. Introduction

The elementary homonuclear dihalogens are produced by chemical or electrochemical oxidation of halide salts or of hydrogen halides by chemical reduction of halates and hypohalites or by the thermal dissociation of the halides of certain elements in higher oxidation states. Formation of elemental fluorine is a special case and proceeds only by anodic oxidation of fluorides or, chemically, from high oxidation state fluorides.

(J. J. ZUCKERMAN)

2.2.2. Preparation of the Elemental Halogens

2.2.2.1. by Anodic Oxidation

2.2.2.1.1. of Fluoride Ion.

The anodic oxidation of fluoride ion is the only practical method available for the formation of elemental F_2. Fluorine can be generated from high oxidation state fluorides such as XeF_6 [1], AuF_5 [2], $[NF_4][BF_4]$[3], and $K_2[NiF_6]\cdot KF$[4], although prior availability of F_2 is required for their preparation. A true chemical synthesis of F_2 from K_2MnF_6 and SbF_5 has been achieved[5]. These fluorides are readily prepared from $SbCl_5$, $KMnO_4$, and HF.

The preparation of F_2 is carried out industrially and in the laboratory using solutions of KF in anhyd HF as the electrolyte[6–12]. Owing to the high oxidation potential of F^- ($E \simeq 2.8$ V) and the properties of F_2 itself, no alternative electrolyte containing fluoride is known. Some fused-salt mixtures may be utilized as electrolytes, but these offer no advantage in terms of cost or ease of operation. The electrode reactions are:

$$2\ F^- \rightarrow F_2 + 2\ e^- \text{ (anode)} \tag{a}$$

$$2\ [HF_2]^- + 2\ e^- \rightarrow H_2 + 4\ F^- \text{ (cathode)} \tag{b}$$

A diagram of the components of a fluorine cell is shown in Fig. 1.

The electrolytic production of F_2 is straightforward in principle, but there are difficult problems in practice[6,7]: the high vapor pressure of HF, the corrosion of the cell materials and the mixing of H_2 and F_2 gases formed during electrolysis and cell voltages. The vapor pressure of the HF is controlled effectively by temperature (T) and electrolyte composition. By holding the electrolyte composition at ca. KF·2 HF, the electrolyte (mp) is kept sufficiently low (near 70°C) and the partial pressure of HF at this T is then <15 torr (<2000 Pa). Using a composition near KF·HF raises the mp to ca. 240°C. Here the partial pressure of HF is low, but the mp changes rapidly with composition, and this electrolyte is now seldom used.

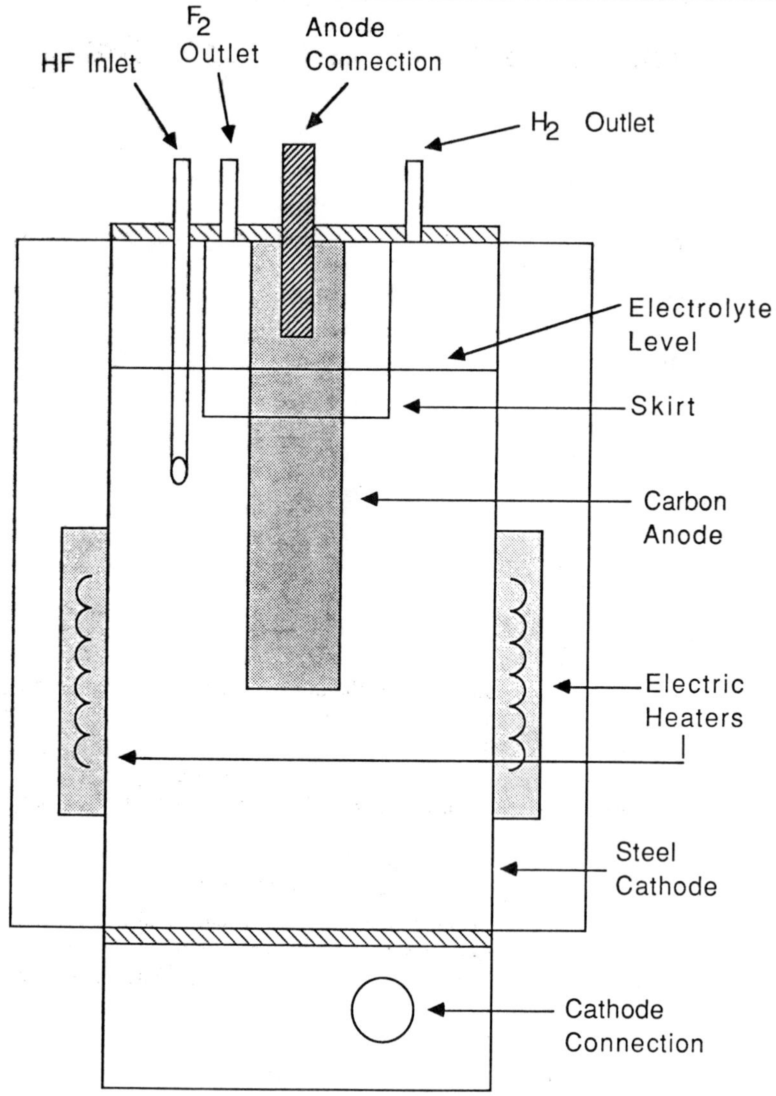

Figure 1. Arrangement of Components in a Fluorine Cell.

The control of the corrosion caused by HF and F_2 is important in obtaining pure F_2 and for the long-term operation of the cell. Currently, most cells are constructed of mild steel with hard carbon anodes. The skirt (see Fig. 1) needed to separate the anode and cathode near the upper portion of the anode is usually constructed of corrosion-resistant Ni–Cu alloy. Gasket and insulating materials vary widely but polytetrafluoroethylene or an elastomer such as neoprene rubber can be used. Connection to the anode is critical, and this is done almost exclusively with Cu metal. However, depending on details of the cell and how it is operated, this is usually the most critical part.

The mixing of H_2 and F_2 during cell operation is prevented by the skirt. This device need only penetrate a few cm below the surface of the electrolyte, because the contact angle between the amorphous carbon anode and the electrolyte is $\sim 160°$, and F_2 is evolved only at the electrode–electrolyte–gas space interface, provided the voltage of the cell is kept within limits.

The proper cell voltage depends on many factors but is determined by the current density at which polarization problems occur, the current density at which H_2 bubbles enter the F_2 compartment and vice versa, the current density at which the anode T becomes too high and the current density allowed by the electrical contact to the anode. Generally, for Cu–carbon contacts using hard (ungraphitized) carbon, the working cell voltage is nominally 8–12 V and 10–6000 A with current densities as high as 1663 A/m^2.

With pure, nearly anhyd HF and relatively high-purity KF, the production of F_2 by anodic oxidation of F^- can be carried out with current efficiencies of 90–95 % in a continuous process. The energy efficiency of fluorine cells, however, is low. The F_2 thus formed is contaminated with HF, which is removed by NaF scrubbers. The byproduct H_2 is scrubbed with caustic soda. After removal of the HF from the F_2, the main impurities are CF_4, O_2 and N_2. If wet HF is used, significant amounts of OF_2 may also be present.

Cells can produce g to kg amounts of F_2 per h, depending on design and size. Industrial production is done with banks of cells of modest size at 4,000–6,000 A. A problem with all cells, depending mostly on the type of carbon anode but also on H_2O in the electrolyte, excessively high current density, poor anode contacts and improper conditions of T and electrolyte concentrations, is polarization, which causes the cell voltage to increase with a loss of current. Polarized cells can usually be corrected by stopping the process and applying a high voltage for 3–5 min, followed by a cooling period of a few min. The addition of small amounts of LiF to the electrolyte is also used to prevent polarization of the anode.

Fluorine gas can be prepared from anhydrous HF–NaF solutions by photooxidation at F_2-doped n-type TiO_2 electrodes[13].

(D. D. DESMARTEAU, B. B. RANDOLPH)

1. K. Seppelt, H. H. Rupp, *Z. Anorg. Allg. Chem.*, *409*, 331 (1974).
2. L. Stein, in *Halogen Chemistry*, V. Gutmann, ed., Vol. 1, Academic Press, New York, 1967, p. 133.
3. K. O. Christe, R. D. Wilson, C. J. Schack, *Inorg. Chem.*, *16*, 937 (1977).
4. L. B. Asprey, *J. Fluorine Chem.*, *7*, 359 (1976).
5. K. O. Christe, *Inorg. Chem.*, *25*, 3721 (1986).
6. A. J. Rudge, *The Manufacture and Use of Fluorine and Its Compounds*, Oxford Univ. Press, Oxford, 1962.
7. G. H. Cady, in *Fluorine Chemistry*, J. H. Simons, ed., Vol. 1, Academic Press, New York, 1950, p. 293.
8. J. H. Simons, *Inorg. Synth.*, *1*, 138 (1939).
9. G. H. Cady, *Inorg. Synth*, *1*, 142 (1939).
10. H. B. Neumarks, M. Siegmund, in *Kirk–Othmer Encyclopaedia of Chemical Technology*, 2nd ed., Wiley, New York, Vol. 9, 1966, p. 506.
11. *Ind. Eng. Chem.*, *39*, 243 (1947).
12. A. J. Woytek, in *Kirk–Othmer Encyclopaedia of Chemical Technology*, 3rd ed., Wiley, New York, Vol. 10, 1980, p. 1030.
13. C. M. Wang, Q. C. Mir, S. Maleknia, T. E. Mallouk, *J. Am. Chem. Soc.*, *110*, 3710 (1988).

2.2.2.1.2. of Chloride Ion.

The formation of Cl_2 by the anodic oxidation of Cl^- is the major process[1,2] for the manufacture of Cl_2. Unlike F_2, Cl_2 continues to be produced by chemical oxidation, mainly[3] with O_2. Environmental concerns may lead to an increase in chemical oxidation methods for Cl_2 formation[4].

The standard electrode potentials for aq Cl^-/Cl_2 are[5]:

$$Cl_{2(g)} + 2\,e^- \rightarrow 2\,Cl^- \qquad 1.35828\ \text{V} \tag{a}$$

$$Cl_{2(aq)} + 2\,e^- \rightarrow 2\,Cl^- \qquad 1.396\ \text{V} \tag{b}$$

$$Cl^-_{3(aq)} + 2\,e^- \rightarrow 3\,Cl^- \qquad 1.4152\ \text{V} \tag{c}$$

The equilibrium constant for $[Cl_3]^-$(aq) are:

$$Cl_{2(g)} + Cl^- \rightarrow [Cl_3]^- \qquad 0.119\ \text{atm}^{-1} \tag{d}$$

$$Cl_{2(aq)} + Cl^- \rightarrow [Cl_3]^- \qquad 0.215\ \text{kg/mol} \tag{e}$$

Thus, $[Cl_3]^-$ is less important than $[Br_3]^-$ or $[I_3]^-$. Owing to the commercial importance of Cl_2, the kinetics of electrochemical behavior of Cl^- is extensively studied[6] and leads to a mechanism at aged graphite electrodes:

$$Cl^-_{(aq)} - e^- \rightarrow Cl_{(ads)} \tag{f}$$

$$Cl_{(ads)} + Cl^-_{(aq)} - e^- \rightarrow Cl_{2(aq)} \tag{g}$$

with both Eqs. (f) and (g) being equally rate controlling[7].

The production of Cl_2 is carried out in H_2O using two primary cell types, the Hg cell and the diaphragm cell. In the former, of which there are many types, the electrochemical reactions are:

Anode $$Cl^- \rightarrow \tfrac{1}{2}\,Cl_2 + e^- \tag{h}$$

Cathode $$Na^+ + Hg + e^- \rightarrow Na(Hg) \tag{i}$$

Overall $$NaCl + Hg \rightarrow Na(Hg) + \tfrac{1}{2}\,Cl_2 \tag{j}$$

The Na(Hg) is treated with H_2O to make NaOH, H_2 and Hg. The Hg is then recycled as part of flowing cathode system in juxtaposition to fixed carbon anodes. The Na(Hg)-H_2O reaction is carried out in a decomposer, which serves as a short-circuited cell consisting of a Na(Hg) anode and graphite cathode. The Hg cell principle is shown schematically in Fig. 1.

Mercury-chlorine cells use NaCl(aq) or brine as a source of Cl^-, although KCl is also used. The brine is saturated with NaCl at 50–60°C and, depending on the source of NaCl, is purified before it is acidified with HCl to a pH of 2.5–5 for use in the cell. The brine is then electrolyzed at ca. 4.0 V in banks of cells having capacities of 60,000–180,000 A and current efficiencies of 94–96% based on NaCl. The gaseous Cl_2 is appropriately purified, depending on its intended use. The Na formed on the surface of the Hg is absorbed to form the Na(Hg) amalgam and is later converted to NaOH or caustic H_2 gas in the decomposer. The caustic solution is normally 50% by wt. The spent brine contains some HOCl, and this is converted to Cl_2 and removed along with dissolved Cl_2 by acidifying the anolyte with HCl. The brine is then recycled after appropriate purification and resaturation with NaCl.

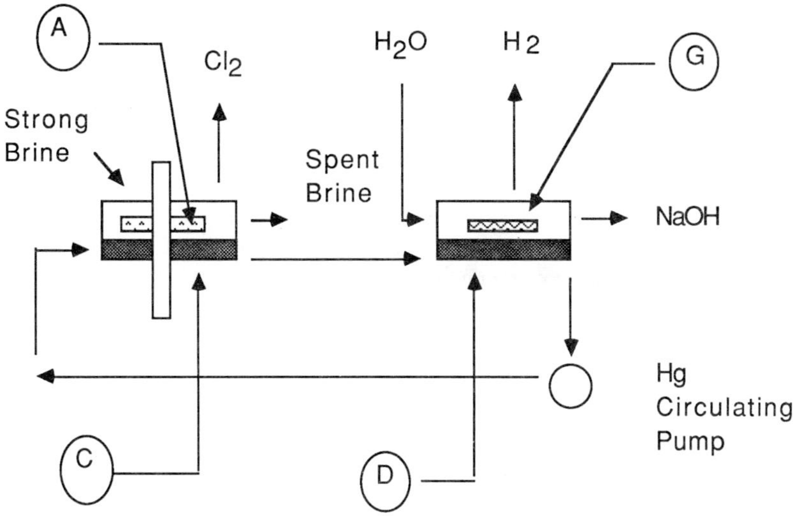

ELECTROLYZER

A = Graphite Anode
C = Hg Cathode

DECOMPOSER

G = Graphite Grids
D = Amalgam Anode

Figure 1. The Mercury Cell Principle.

The principal competing reactions in the process are:

$$H_2O \rightarrow H_2 + O^{\cdot} \text{ (electrolysis)} \tag{k}$$

$$C + 2\,O^{\cdot} \rightarrow CO_2 \text{ (anode)} \tag{l}$$

$$Cl_2 + H_2O \text{ (brine)} \rightarrow HCl + HOCl \tag{m}$$

$$Na(Hg) + \tfrac{1}{2}\,Cl_2 \rightarrow NaCl + Hg \tag{n}$$

$$Na(Hg) + HCl \rightarrow NaCl + \tfrac{1}{2}\,H_2 + Hg \tag{o}$$

Depending on conditions, H_2 formed along with Cl_2 is a problem and must be monitored closely in the continuous operation of Hg cells. Newer cell designs employ metal anodes of coated Ti. These anodes have longer lifetimes than those of carbon and do not require adjustment to maintain the appropriate gap between anode and cathode as required in more conventional cells.

Diaphragm cells account for ca. 70% of all Cl_2 produced, and many different designs are in use. Most use graphite anodes and steel-screen cathodes. The anode reaction is the same as the Hg cell, but the cathode reaction is:

$$2\,H_2O \rightarrow H_2 + 2\,OH + 2\,e^- \tag{p}$$

and the overall reaction is:

$$2\ NaCl + H_2O \rightarrow 2\ NaOH + Cl_2 + H_2 \qquad (q)$$

The classic diaphragm material is asbestos, which is placed on the cathode support as a slurry. The pressure between anode and cathode separated by the diaphragm is maintained at ca. 5 in. WG, and the pH of the anolyte between 3 and 4. Under these conditions the cell can be kept in equilibrium to produce 97% Cl_2 at the anode, nearly 100% H_2 at the cathode and 11–12% NaOH in the catholyte effluent. A schematic diaphragm cell is shown in Fig. 2. Banks of commercial cells operate at ~4 V d.c. and 10,000–50,000 A, depending on size and design. Current efficiencies are near 96 %. The

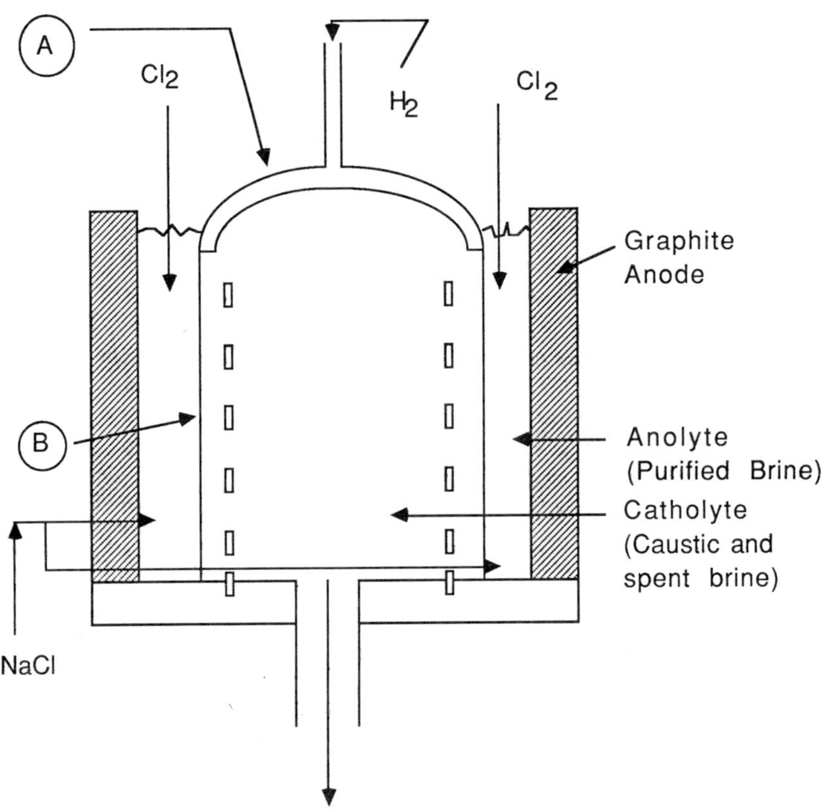

Figure 2. Schematic Diaphragm Cell.

feed brine is saturated with NaCl at 70°C (26.6%) and the spent cell liquor contains ≈16% NaCl. The Cl_2 formed exits as a hot gas and is scrubbed with H_2O to cool it, followed by drying with H_2SO_4. The main impurities are O_2, N_2, CO_2 and some chlorinated hydrocarbons. The Cl_2 is further purified by fractionation and then liquefied for storage and shipping.

Newer diaphragm cells replace asbestos diaphragms with semipermiable membranes[8]. New fluorinated polymer membranes are especially promising. A membrane allows much simpler design and operation of cells. Diaphragm and membrane cells are compared in Fig. 3.

Another electrochemical method uses a diaphragm cell but employs conc HCl (muriatic acid) as the Cl^- source[1–3]. Hydrogen and Cl_2 of high purity are produced by the process, and the HCl feed can be the byproduct of an organic chlorination process carried out nearby. The cells are different in design and operation from cells using NaCl. The anode and cathode are both graphite, and a cell consists of sandwiches of vertical graphite plates and diaphragms with electrical connections to the two end plates only. All internal plates are then bipolar, acting as cathode on one side and anode on the other. A typical commercial assembly consists of 40 electrolytic cells, with a single cell voltage of 2.3 V and a total voltage to the electrolyzer of 93 V at 1600 A.

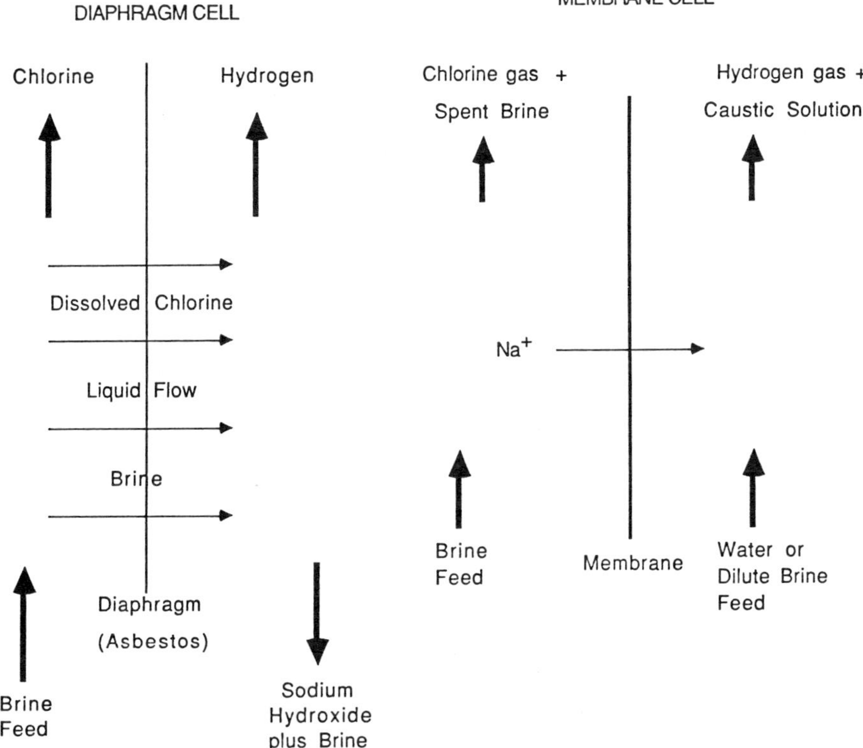

Figure 3. Schematic Comparison of Diaphragm and Membrane Cells.

Two other indirect processes based on HCl are the electrolyses[1] of $CuCl_2$ and $NiCl_2$, whose reactions are:

$$2\,CuCl_2 \xrightarrow{\text{electrolysis}} 2\,CuCl + Cl_2 \qquad (r)$$

$$2\,CuCl + 2\,HCl + \tfrac{1}{2}\,O_2 \rightarrow 2\,CuCl_2 + H_2O \qquad (s)$$

$$2\,HCl + \tfrac{1}{2}\,O_2 \rightarrow Cl_2 + H_2O\ \text{(net)} \qquad (t)$$

and:

$$NiCl_2 \xrightarrow{\text{electrolysis}} Ni + Cl_2 \qquad (u)$$

$$Ni + 2\,HCl \rightarrow H_2 + NiCl_2 \qquad (v)$$

$$2\,HCl \rightarrow H_2 + Cl_2\ \text{(net)} \qquad (w)$$

These two processes are not competitive for commercial Cl_2 production. Other electrolytic processes involving aqueous solutions of other metals and fused metal chlorides can be used[9] to produce Cl_2. Of these, none is a commercial source of Cl_2, but some serve to produce certain metals, such as Be and Na.

The electrochemical generation of Cl_2 is important[10]. The processes involve the use of metal[11] or metal oxide[12,13] electrodes. Chlorine is generated by the visible light-driven photoxidation of Cl^- at $MoSe_2$ and MoS_2 electrodes[14]. These are not alternatives for the commercial production of Cl_2.

(D. D. DESMARTEAU, B. B. RANDOLPH)

1. J. S. Sconce, *Chlorine. Its Manufacture, Properties and Uses*, ACS Monograph Series, No. 154, Reinhold, New York, 1962.
2. L. Mond, *J. Soc. Chem. Ind. (London)*, 713 (1896).
3. J. J. Leddy, T. C. Jones Jr., B. S. Lowry, F. W. Spiller, R. E. Wing, C. D. Binger, in *Kirk–Othmer Encyclopaedia of Chemical Technology*, 3rd ed., Wiley, New York, Vol. 1, 1978, p. 799.
4. L. J. Goldwater, *Sci. Am., 224*, 15 (1971).
5. A. Cerquetti, P. Longhi, T. Mussini, G. Natta, *J. Electroanal. Chem., 20*, 411 (1969).
6. T. Mussini, G. Faita, in *Encyclopedia of Electrochemistry of the Elements*, A. J. Bard, ed., Vol. 1, Marcel Dekker, New York, 1973, p. 1.
7. L. J. J. Janssen, J. G. Hoogland, *Electrochim. Acta, 15*, 941 (1970).
8. D. J. Vaughan, *DuPont Innovation, 4*, 10 (1973).
9. *Gmelins Handbuch der Anorganischen Chemie*, 8 Auflage, Teil A, Chlor, 1968, p. 2.
10. W. J. Vining, T. J. Meyer, *J. Electroanal. Chem. Interfacial Electrochim., 195*, 183 (1985).
11. M. Hara, K. Asami, K. Hashimoto, T. Masumoto, *Electrochim. Acta, 31*, 481 (1987).
12. R. Boggio, A. Carugati, G. Lodi, S. Trasatti, *J. Appl. Electrochem., 15*, 335 (1985).
13. V. Connsoni, S. Trasatli, F. Pollack, W. E. O'Grady, *J. Electroanal. Chem. Interfacial Electrochem., 228*, 393 (1987).
14. C. P. Kubiak, L. F. Schneeymeyer, M. S. Wrighton, *J. Am. Chem. Soc., 102*, 6898 (1980).

2.2.2.1.3. of Bromide Ion.

The anodic oxidation of Br^- to form Br_2 is important only in an historical sense. There have, however, been some studies recently on the photooxidation of Br^-.[1,2] Bromine is produced today[3–5] almost exclusively by chemical oxidation of Br^- with Cl_2. Bromine, first obtained commercially by chemical oxidation with MnO_2, was produced by electrochemical oxidation for a short period before the use of Cl_2 became widespread.

The standard electrode potentials for the Br_2/Br^- couple are[6,7]:

$$Br_{2(l)} + 2\ e^- \rightarrow 2\ Br^- \quad 1.0152\ V \quad (a)$$

$$Br_{2(aq)} + 2\ e^- \rightarrow 2\ Br^- \quad 1.0874\ V \quad (b)$$

$$[Br_3]^-_{(aq)} + 2\ e^- \rightarrow 3\ Br^- \quad 1.0503\ V \quad (c)$$

For the Br_2/Br^- couple at a Pt electrode, the following scheme accounts for the kinetic data[8]:

$$Br_2 + e^- \rightarrow [Br_2]^-_{(ads)} \quad (d)$$

$$[Br_2]^-_{(ads)} \rightarrow Br^- + Br_{(ads)} \text{ (fast equilibrium)} \quad (e)$$

$$Br_{(ads)} + e^- \rightarrow Br^- \quad (f)$$

$$2\ Br_{(ads)} \rightarrow Br_2 \quad (g)$$

The anodic oxidation of Br^- is easier than that of Cl^-, and Br_2 can be obtained by electrolysis of a mixture of Cl^- and Br^-. In one process, anode and cathode are carbon sticks arranged in clusters and separated by a porous clay diaphragm. The potential drop for each cell is 3.4 V, and current densities are near 115 A/m^2. The brine feed is obtained as byproduct from the extraction of KCl from carnallites ($K_2MgCl_4 \cdot 6\ H_2O$). The brine is split into catholyte and anolyte streams. The cell reactions are:

$$2\ Br^- \rightarrow Br_2 + 2\ e^- \text{ (anode)} \quad (h)$$

$$Mg^{2+} + 2\ e^- + 2\ H_2O \rightarrow Mg(OH)_2 + H_2 \text{ (cathode)} \quad (i)$$

$$MgBr_2 + 2\ H_2O \rightarrow Br_2 + Mg(OH)_2 + H_2 \text{ (net)} \quad (j)$$

After exiting the cell, the anolyte is stripped of Br_2 by heating and then returned to the cathode compartment, where the Mg^{2+} is removed as $Mg(OH)_2$. Current efficiencies are low (68–70%) and clogging of the diaphragm by $Mg(OH)_2$ is a problem.

Another process employs bipolar carbon electrodes. The $Mg(OH)_2$ is deposited on the cathode side of the electrode and, after a time, the direction of the current is reversed to remove it. It floats to the surface of the brine and is removed from the exiting cell liquor by filtration. This liquor is then stripped of Br_2 by heating. This process avoids the problem of $Mg(OH)_2$ clogging the cell diaphragms, but the low current efficiency (40–50%) makes the process uncompetitive. The energy costs of producing Br_2 electrolytically are ca. 1.5 times those of producing Cl_2 electrolytically and then using the Cl_2 to oxidize Br^-. Therefore, electrolytic processes are abandoned in favor of chemical oxidation by Cl_2 in industrial processes.

Some Br_2 is produced electrolytically for the production of bromates[9]. In alkaline solution (pH 8–9) the Br_2 formed at the anode reacts to form $[OBr]^-$, which disproportionates to $[BrO_3]^-$ and Br^- under the reaction conditions.

(D. D. DESMARTEAU, B. B. RANDOLPH)

1. C. P. Kubiak, L. F. Schneeymeyer, M. S. Wrighton, *J. Am. Chem. Soc., 102*, 6898 (1980).
2. C. Levy-Clement, A. Heller, W. A. Bonner, B. A. Parkinson, *J. Electrochem. Soc., 129*, 1701 (1982).
3. F. Yaron, in *Bromine and Its Compounds*, Z. E. Jolles, ed., Academic Press, New York, 1966, p. 3.
4. C. J. Mayac, in *Kirk–Othmer Encyclopaedia of Chemical Technology*, 3rd ed., Wiley, New York, Vol. 13, 1981, p. 649.

5. A. J. Downs, C. J. Adams, in *Comprehensive Inorganic Chemistry*, A. F. Trotman-Dickenson, ed., Vol. 2, Pergamon Press, Oxford, 1973, p. 1124.
6. W. M. Latimer, *Oxidation Potentials*, Prentice-Hall, Englewood Cliffs, NJ, 1952, p. 60.
7. T. Mussini, G. Faita, *Ric. Sci.*, *36*, 175 (1966).
8. T. Mussini, G. Faita, in *Encyclopedia of Electrochemistry of the Elements*, Vol. 1, A. J. Bard, ed., Marcell Dekker, New York, 1973, p. 78.
9. C. A. Hampel, ed., *Encyclopedia of Electrochemistry*, Reinhold, New York, 1964, p. 127.

2.2.2.1.4. of Iodide Ion.

Iodide is easily oxidized to I_2. The standard potentials in H_2O are [1,2]:

$$I_2(s) + 2\,e^- \rightarrow 2\,I^- \quad 0.536\ V \tag{a}$$

$$I_2(aq) + 2\,e^- \rightarrow 2\,I^- \quad 0.621\ V \tag{b}$$

$$[I_3]^- + 2\,e^- \rightarrow 3\,I^- \quad 0.536\ V \tag{c}$$

However, the anodic oxidation of I^- does not compete with chemical oxidation methods of preparing I_2 owing to low concentrations of I^- in natural brines[3].

Oxidation of I^- at Pt electrodes[2] occurs at the Pt surface, but there is disagreement regarding which step is rate determining in the electrode process[4]:

$$I^- \rightarrow I(ads) + e^- \tag{d}$$

$$I(ads) + I^- \rightarrow I_2 + e^- \tag{e}$$

$$2\,I(ads) \rightarrow I_2 \tag{f}$$

$$I_2 + I^- \rightarrow [I_3]^- \tag{g}$$

Reactions (d), (f) and (g) constitute the electrode process and the exchange between the first layer and the solution is determined by Eq. (e).

Additional procedures include the photooxidation of HI[5], a 200% yield of I_2 from I^- using an electrosynthetic scheme in which a common reactant is oxidized to the same product in both anodic and cathodic cell compartments[6], oxidation of I^- at Pt electrodes[7], and a rotating-disk electrode generation of I_2 from conc I_2/I^- solutions[8].

(D. D. DESMARTEAU, B. B. RANDOLPH)

1. W. M. Latimer, *Oxidation Potentials*, 2nd ed., Prentice-Hall, New York, NY, 1952, p 64.
2. P. G. Desideri, L. Lepri, D. Heimler, in *Encyclopedia of Electrochemistry of the Elements*, A. J. Bard, ed., Vol. 1, Marcel Dekker, New York, 1973, p. 91.
3. A. J. Downs, C. J. Adams, in *Comprehensive Inorganic Chemistry*, A. F. Trotman-Dickenson, ed., Vol. 2, Pergamon Press, Oxford, 1973, p. 1140.
4. L. M. Dané, L. J. J. Janssen, J. G. Hoogland, *Electrochim. Acta*, *13*, 507 (1968).
5. C. Levy-Clement, A. Heller, W. A. Bonner, B. A. Parkinson, *J. Electrochem. Soc.*, *129*, 1701 (1982).
6. R. J. H. Chan, C. Ueda, T. Kuwana, *J. Am. Chem. Soc.*, *105*, 3713 (1983).
7. S. Swathirajan, S. Bruckenstein, *J. Electroanal. Chem. Interfacial Electrochem.*, *143*, 167 (1983).
8. X. Liao, K. Tanno, F. Kurosawa, *J. Electroanal. Chem. Interfacial Electrochem.*, *239*, 149 (1988).

2.2.2.2. by Chemical Oxidation

Commercial production of halogens was by chemical oxidation of halide[1–6]. While Br_2 and I_2 are still produced in this way, chemical oxidation to make Cl_2 is supplanted by electrolytic oxidation (§2.2.2.1). Commercially produced halogen is employed in the

TABLE 1. REDUCTION POTENTIALS FOR Cl, Br AND I[9]

Reaction	Electrode potential (V)		
	Cl	Br	I
$X_2 + 2\ e^- = 2\ X^-$	1.3583	1.065	0.535
$2\ HOX + 2\ H^+ + 2\ e^- = X_2 + H_2O$	1.63	1.59	1.45
$HOX + H^+ + 2\ e^- = X^- + H_2O$	1.49	1.33	0.99
$2\ XO_3^- + 12\ H^+ + 10\ e^- = X_2 + 6\ H_2O$	1.47	1.52	1.195
$XO_3^- + 6\ H^+ + 6\ e^- = X^- + 3\ H_2O$	1.45	1.44	1.085

laboratory, but when the commercial product is not available, chemical methods are used for Cl_2, Br_2 and I_2 synthesis[7,8]. As shown in Table 1, the ease of oxidation of halide ion follows the order I > Br > Cl, and the heavier the halogen, a wider range of oxidizing agents may be employed in principle for the preparation of the elemental halogen.

The early work on the preparation of elemental halogens is reviewed[1–3]. This survey stresses work done since the last review in 1973.

(J. B. MILNE)

1. *Gmelins Handbuch der Anorganischen Chemie*, 8 Auflage: '*Chlor*', System-nummer 6, Teil A; Verlag Chemie, Weinheim/Bergstr., 1968; '*Brom*', System-nummer 7, Verlag Chemie, Berlin, 1931; '*Jod*', System-nummer 8, Verlag Chemie, Berlin, 1933.
2. J. W. Mellor, *A Comprehensive Treatise on Inorganic and Theoretical Chemistry*, Vol. II, Longmans, Green, London, 1922; Supplement II, Part I, Longmans, Green, London, 1956.
3. A. J. Downs, C. J. Adams, in '*Comprehensive Inorganic Chemistry*', A. F. Trotman-Dickenson, ed., Pergamon Press, Oxford, 1973.
4. *Kirk-Othmer Encyclopedia of Chemical Technology*, 3rd ed., Vols. 1, 4, 13, Wiley, New York, 1978.
5. J. S. Robinson, ed., *Chlorine Production Processes: Recent and Energy-Saving Developments*, Chemical Technology Review No. 185, Noyes Data, Park Ridge, NJ, 1981.
6. E. Booth, *Chem. Ind.* (*London*), 838 (1978); 52 (1979).
7. G. Brauer, *Handbook of Preparative Inorganic Chemistry*, Academic Press, New York, 1963.
8. H. Guerin, *Sci. Prog. Decouv.*, 37 (1970); *Chem. Abstr.*, *74*, 14620 (1971).
9. R. C. Weast, ed., *Handbook of Chemistry and Physics*, 6th ed., CRC Press, Boca Raton, FL, 1986.

2.2.2.2.1. of Fluoride.

All commercial fluorine is produced by electrolysis (§2.2.2.1). The first example of the chemical synthesis of F_2 from materials not requiring F_2 for their preparation depends upon the displacement of an unstable fluoride from its complex by a strong electron-pair acceptor acid[1]:

$$2\ K[MnO_4] + 2\ KF + 10\ HF + 3\ H_2O_2 \xrightarrow{50\%\ aq\ HF} 2\ K_2[MnF_6] + 8\ H_2O + 3\ O_2 \quad (a)$$

$$SbCl_5 + 5\ HF \rightarrow SbF_5 + 5\ HCl \quad (b)$$

$$K_2[MnF_6] + 2\ SbF_5 \rightarrow 2\ K[SbF_6] + MnF_3 + \tfrac{1}{2}\ F_2 \quad (c)$$

(J. B. MILNE)

1. K. O. Christe, *Inorg. Chem.*, *25*, 3721 (1986).

2.2.2.2.2. of Chloride.

The Cl_2 of commerce is produced by the electrolysis of NaCl, KCl or HCl. Only three chemical oxidation processes are commercially available today. One, known since 1868[1], employs the reaction of HCl with O_2:

$$4\ HCl + O_2 \xrightarrow[CuCl_2]{430°–475°C} 2\ Cl_2 + 2\ H_2O \qquad (a)$$

and is used to produce small quantities of Cl_2. Many catalysts are available for this reaction, which is exothermic only to the extent[2] of 59 kJ/mol and, because of an unfavorable entropy change, must be run at relatively low T. Catalysts such as MnO_2[3], various oxides and chlorides of Fe, Mg, Cu and Cd[4], perlite and obsidian[5], Cr_2O_3, UO_3, CeO_2 and V_2O_5[1] are suggested. For the $CuCl_2$-catalyzed reaction, the sequence involves an equilibrium between Cu(I) and Cu(II) species[6]:

$$2\ CuCl_2 \rightarrow 2\ CuCl + Cl_2 \qquad (b)$$

$$4\ CuCl + O_2 \rightarrow 2\ Cu_2OCl_2 \qquad (c)$$

$$Cu_2OCl_2 + 2\ HCl \rightarrow 2\ CuCl_2 + H_2O \qquad (d)$$

Another process[1,6,7,8] also uses HCl and O_2 to produce Cl_2. The catalyst is a solution of oxides of nitrogen in H_2SO_4. The process operates at low T relative to that in Eq. (a), but allowance must be made for the corrosive nature of the materials used[6]. The sequence is:

$$NO + NO_2 + 2\ H_2SO_4 \rightarrow 2\ NO[HSO_4] + H_2O \qquad (e)$$

$$2\ NO[HSO_4] + 2\ HCl \rightarrow 2\ NOCl + 2\ H_2SO_4 \qquad (f)$$

$$2\ NOCl + O_2 \rightarrow 2\ NO_2 + Cl_2 \qquad (g)$$

$$NO_2 + 2\ HCl \rightarrow Cl_2 + NO + H_2O \qquad (h)$$

$$2\ NO + O_2 \rightarrow 2\ NO_2 \qquad (i)$$

A solution of NO_2 in 75% H_2SO_4 at 60°C converts 97% of the NOCl to Cl_2, which may be washed with 75% H_2SO_4 to give[9] effluent gas with 94 vol% Cl_2 and 6 vol% HCl.

Chlorine can be produced[1] from KCl and HNO_3. Production in this way is economical only because there is a market for the byproduct, KNO_3. The reaction sequence is related to Eqs. (e)–(i), but H_2SO_4 is not used:

$$3\ KCl + 4\ HNO_3 \xrightarrow{10°C} 3\ KNO_3 + NOCl + Cl_2 + 2\ H_2O \qquad (j)$$

$$2\ NOCl + 4\ HNO_3 \rightarrow 6\ NO_2 + Cl_2 + 2\ H_2O \qquad (k)$$

$$3\ NO_2 + H_2O \rightarrow NO + 2\ HNO_3 \qquad (l)$$

$$2\ NO + O_2 \rightarrow 2\ NO_2 \qquad (m)$$

Similar processes are patented[10,11]. Related processes using NaCl in place of KCl are not commercially viable because demand for $NaNO_3$ is low[1].

Other processes are suggested for bulk production of Cl_2 from chlorides by chemical oxidation. Nitrogen dioxide-O_2 mixtures are used[1]:

$$2\ NO_2 + 2\ KCl + O_2 \rightarrow 2\ KNO_3 + Cl_2 \tag{n}$$

$$2\ NO_2 + 2\ NH_4Cl + O_2 \rightarrow 2\ NH_4NO_3 + Cl_2 \tag{o}$$

These oxidations take place at relatively low T (~200°C). Oxidation with O_2 alone requires higher T and a starting chloride that forms a stable oxide or oxo-anion, e.g.:

$$4\ FeCl_3 + 3\ O_2 \xrightarrow{700°-900°C} 2\ Fe_2O_3 + 6\ Cl_2 \tag{p}$$

$$4\ NaFeCl_4 + 3\ O_2 \xrightarrow{440°-650°C} 2\ Fe_2O_3 + 4\ NaCl + 6\ Cl_2 \tag{q}$$

$$SiO_2 \cdot Al_2O_3 + 4\ NaCl + O_2 \xrightarrow{850°C} 2\ Cl_2 + Na_4[SiAl_2O_7] \tag{r}$$

Sulfur trioxide can be used as the oxidizing agent[1]:

$$NaCl + SO_3 \rightarrow NaSO_3Cl \tag{s}$$

$$3\ NaSO_3Cl \xrightarrow{225°C} Na_2S_2O_7 + NaCl + SO_2 + Cl_2 \tag{t}$$

$$2\ HCl + 2\ SO_3 \xrightarrow[HgCl_2]{160°C} H_2SO_4 + SO_2 + Cl_2 \tag{u}$$

With NH_4Cl as starting material, Cl_2 is produced catalytically in a modification of Eq. (a)[12].

$$O_2 + 2\ NH_4Cl \xrightarrow[MgO]{600°-700°C} 2\ NH_3 + H_2O + Cl_2 \tag{v}$$

Catalysts similar to those for Eq. (a) are also used[1,4]. Both O_3 and H_2O_2 also oxidize[1] HCl:

$$O_3 + 6\ HCl \xrightarrow{catal} 3\ H_2O + 3\ Cl_2 \tag{w}$$

$$2\ HCl + H_2O_2 \xrightarrow{27°-60°C} 2\ H_2O + Cl_2 \tag{x}$$

The disadvantage of both these processes is the cost of the oxidant.

Chlorine used for industrially important organic chlorinations is produced from HCl and O_2 without separation of the Cl_2 [2]. These oxychlorinations are used for the chlorination of ethylene and benzene:

$$4\ HCl + O_2 + 2\ C_2H_4 \xrightarrow[catal]{250°-300°C} 2\ ClCH_2CH_2Cl + 2\ H_2O \tag{y}$$

Hydrogen chloride is often a byproduct of other processes carried out in the plant, such as the cracking of $ClCH_2CH_2Cl$:

$$ClCH_2CH_2Cl \xrightarrow[catal]{500°C} CH_2CHCl + HCl \tag{z}$$

On a laboratory scale, where convenience often outweighs expense, Cl_2 is most conveniently generated[13] from conc HCl and $KMnO_4$:

$$2\ KMnO_4 + 16\ HCl \rightarrow 2\ KCl + 2\ MnCl_2 + 5\ Cl_2 + 8\ H_2O \qquad \text{(aa)}$$

or MnO_2:

$$MnO_2 + 4\ HCl \rightarrow MnCl_2 + 2\ H_2O + Cl_2 \qquad \text{(ab)}$$

(J. B. MILNE)

1. J. S. Robinson, ed., *Chlorine Production Processes: Recent and Energy-Saving Developments*, Chemical Technology Review No. 185, Noyes Data, Park Ridge, NJ, 1981.
2. H. Guerin, *Sci. Prog. Decouv.*, 37 (1970); *Chem. Abstr.*, *74*, 14620 (1971).
3. U. Sh. Rysaev, A. M. Potapov, Ev. V. P. Vasil, I. I. Laktionov, F. Sh. Rysaev, USSR Pat. 3,432,371 (1982); *Chem. Abstr. 100*, 9,423 (1984).
4. G. Laplace, G. Diprose, Ger. Pat. 2,048,753 (1971); *Chem. Abstr. 75*, 7990 (1971).
5. M. S. Salakhov, M. M. Guseinov, Ch. A. Chalabiev, I. R. Akhverdiev USSR Pat. 487,018 (1975); *Chem. Abstr. 84*, 61,993 (1976).
6. *Kirk-Othmer Encyclopedia of Chemical Technology*, 3rd ed., Vols. 1, 4, 13, Wiley, New York, 1978.
7. J. T. Reding, B. P. Shepherd, *US Environmental Protection Agency Rep.* PB-241-927 (1975); *Chem. Abstr.*, *83*, 182011 (1975).
8. L. W. Bostwick, *Chem. Eng. News*, *83*, 86 (1976).
9. M. Mokuji, Jpn. Pat. 70-09244 (1970); *Chem. Abstr. 73*, 47158 (1970).
10. T. Takashi, Jpn. Pat. 1,936,223 (1970); *Chem. Abstr*, *72*, 91813 (1970).
11. H. A. DeVries, R. Goettsch, A. H. DeRooij, Netherl. Pat. 70-06500 (1971); *Chem. Abstr. 76*, 61428 (1972).
12. F. K. Mikhailov, A. Z. Khomenko, *Mezhdunar. Simp. Sots. Stran. Sodovoi Prom.* (*Mater*), 90 (1971); *Chem. Abstr.*, *80*, 5289 (1974).
13. G. Brauer, *Handbook of Preparative Inorganic Chemistry*, Academic Press, New York, 1963.

2.2.2.2.3. of Bromide.

Commercial production of Br_2 is all done by Cl_2 oxidation of bromide[1]:

$$Cl_2 + 2\ Br^- \rightarrow 2\ Cl^- + Br_2 \qquad \text{(a)}$$

For dilute solutions, the Br_2 is swept out with air and concentrated by trapping in aq Na_2CO_3, where disproportionation takes place:

$$3\ Br_2 + 6\ OH^- \rightarrow BrO_3^- + 5\ Br^- + 3\ H_2O \qquad \text{(b)}$$

from which Br_2 is swept after acidification with H_2SO_4. The Br_2, alternatively, may be concentrated by reduction with SO_2 in solution:

$$Br_2 + SO_2 + 2\ H_2O \rightarrow 2\ HBr + H_2SO_4 \qquad \text{(c)}$$

and steam swept from the solution after reoxidation with Cl_2. A patented process[2] involves trapping Br_2 in saturated aq NH_4Br to a concentration of 100 g Br_2/L from which it is distilled:

$$Br_2 + Br^- \rightarrow [Br_3]^- \qquad \text{(d)}$$

Bromine can also be produced by electrolysis[1] as well as by a modification of Eq. (a), §2.2.2.2.2[3,4]:

$$4\ HBr + O_2 \xrightarrow[\text{catal}]{} 2\ Br_2 + 2\ H_2O \qquad \text{(e)}$$

but these procedures are not commercial. Oxidation of HBr by O_2, catalyzed by oxides of nitrogen in H_2SO_4, parallels Eqs. (e)–(i), §2.2.2.2.2, for Cl_2 production[5]. Oxidation of Br^--containing brines with O_3 gives Br_2 in 93% yield[6,7]:

$$O_3 + 6\ HBr \xrightarrow[pH\ 2-3]{25°C} 3\ Br_2 + 3\ H_2O \tag{f}$$

On a laboratory scale, Br_2 is produced conveniently by oxidation of Br^- with $KMnO_4$[8]:

$$2\ MnO_4^- + 10\ Br^- + 16\ H^+ \rightarrow 2\ Mn^{2+} + 8\ H_2O + 5\ Br_2 \tag{g}$$

or $K_2Cr_2O_7$[9]:

$$Cr_2O_7^{2-} + 6\ Br^- + 14\ H^+ \rightarrow 2\ Cr^{3+} + 7\ H_2O + 3\ Br_2 \tag{h}$$

(J. B. MILNE)

1. *Kirk-Othmer Encyclopedia of Chemical Technology*, 3rd ed., Vol. 4, Wiley, New York, 1978.
2. P. Teissedre, M. Viard, Fr. Pat. 2,511,356 (1981); *Chem. Abstr., 99,* 7,697 (1983).
3. J. S. Robinson, ed., *Chlorine Production Processes: Recent and Energy-Saving Developments*, Chemical Technology Review No. 185, Noyes Data Corp., Park Ridge, NJ, 1981.
4. G. Eisenhauer, R. Strang, A. Arno, Eur. Pat. Appl. EP 179,163 (1986); *Chem. Abstr., 104,* 227167 (1986).
5. C. P. Van-Dijk, US Pat. 4,027,000 (1977); *Chem. Abstr. 87,* 41295 (1977).
6. V. M. Kaut, S. L. Gobov, L. L. Chusova, *Vopr. Khim. Khim. Tekhnol. 70,* 88 (1983); *Chem. Abstr., 99,* 214857 (1983).
7. L. L. Chusova, S. L. Gobov, V. M. Kaut, N. V. Girgilzhiinik, T. A. Serdechnaya, USSR SU 1,139,700 (1985); *Chem. Abstr., 102,* 169139 (1985).
8. J. S. V. Reddering, *Specktrum, 17,* 23 (1979); *Chem. Abstr., 91,* 55519 (1979).
9. G. Brauer, *Handbook of Preparative Inorganic Chemistry*, Academic Press, New York, 1963.

2.2.2.2.4. of Iodide.

The generation of I_2 by chemical oxidation is accomplished by a wider range of oxidants than for Br_2 and Cl_2. Relatively weak oxidizing agents, such as Cu^{2+}, Fe^{3+}, Sb(V) and $[NO_2]^-$, can oxidize I^- to I_2 quantitatively. Because of the ease of volumetric titration of I_2, iodometric analysis of such oxidizing agents is widely applied[1].

The I_2 produced commercially is made by chemical oxidations, numerous[2,3] in the past but more restricted today.

One procedure for I_2 production is analogous to that used for Br_2. The brine is oxidized with Cl_2 and the I_2 is air swept from the solution and concentrated by reduction to HI with SO_2. The HI is then reoxidized with Cl_2, filtered and further purified by melting under conc H_2SO_4:

$$2\ I^- + Cl_2 \rightarrow I_2 + 2\ Cl^- \tag{a}$$

$$I_2 + SO_2 + 2\ H_2O \rightarrow 2\ HI + H_2SO_4 \tag{b}$$

Another method of concentration of I_2 in the Cl_2-oxidized brine is by passing the solution over Cu bales, whereby insoluble CuI is formed. Subsequent agitation of the bales separates the CuI, which is filtered and dried. Iodine may be recovered[2] from CuI by heating at 350°–375°C with NaOH to produce NaI, leaching out the NaI and reoxidizing with Cl_2:

$$I_2 + 2\ Cu \rightarrow 2\ CuI \tag{c}$$

$$2\ CuI + 2\ NaOH \xrightarrow{350°-375°C} 2\ NaI + Cu + CuO + H_2O \tag{d}$$

Iodine can be concentrated by ion exchange. The I_2 in the form of polyiodide is adsorbed onto the anion-exchange resin and then eluted with NaOH followed by NaCl. The resultant solution is acidified and oxidized to recover I_2:

$$3\ I_3^- + 6\ [OH]^- \rightarrow [IO_3]^- + 8\ I^- + 3\ H_2O \qquad (e)$$

$$2\ [IO_3]^- + 16\ I^- + 3\ Cl_2 + 12\ H^+ \rightarrow 9\ I_2 + 6\ Cl^- + 6\ H_2O \qquad (f)$$

Another industrial process uses a prior concentration of iodide in brine by precipitation as insoluble AgI. Treatment of the AgI with scrap iron gives FeI_2:

$$2\ AgI + Fe \rightarrow FeI_2 + 2\ Ag \qquad (g)$$

The silver is separated, dissolved in nitric acid and recycled. The FeI_2 solution is treated with Cl_2 and the I_2 separated.

Other chemical oxidation methods are of commercial interest. A process similar to those for Cl_2 (§2.2.2.2.2), using O_2, O_3, HNO_3, NO or NO_2 to oxidize HI and employing various catalysts, is patented[4]. Iodine may be adsorbed onto an anion exchanger in the chloride form after oxidation of I^- by $NaNO_2$ at pH 1.5–2.0[5].

$$2\ HNO_2 + 2\ H^+ + 2\ I^- \rightarrow I_2 + 2\ NO + 2\ H_2O \qquad (h)$$

The polyhalide is eluted with 2 N Na_2SO_3 and $NaNO_3$ and reoxidized after acidification. Iodine produced by oxidation of I^- in brine with Cl_2 may be adsorbed on activated carbon as a concentration step. Treatment with SO_2 yields HI, which is then reoxidized with Cl_2[6]. Solvent extraction with tri-n-butylphosphate can also be used as a concentration step after oxidation with H_2O_2 of I^--containing solutions acidified with H_2SO_4[7]:

$$H_2O_2 + 2\ I^- + 2\ H^+ \rightarrow I_2 + 2\ H_2O \qquad (i)$$

Ferrous sulfate has been suggested as a catalyst for this oxidation, which is regarded as less caustic than those reactions commonly used[8].

Preparation of I_2 by oxidation of iodide on a laboratory scale may be accomplished conveniently with many common oxidizing agents, e.g., MnO_2, MnO_4^-, NO_2^-, Cl_2, $Cr_2O_7^{2-}$ or Cu^{2+}. Elemental I_2 may be removed from the mixtures by filtration with further purification by sublimation or by steam distillation[9].

(J. B. MILNE)

1. A. I. Vogel, *A Textbook of Quantitative Inorganic Analysis*, 3rd ed., Longmans, Green, London, 1961, pp. 343, 383.
2. J. W. Mellor, *A Comprehensive Treatise on Inorganic and Theoretical Chemistry*, Vol. II, Longmans, Green, London, 1922 ; Supplement II, Part I, Longmans, Green, London, 1956.
3. E. Booth, *Chem. Ind.* (*London*), 838 (1978); 52 (1979).
4. *Kirk-Othmer Encyclopedia of Chemical Technology*, 3rd ed., Vol 13, Wiley, New York, 1978.
5. P. K. Pazvantov, *Khim. Ind.* (*Sofia*), *46*, 226 (1974).
6. I. Kubinski, I. Lipinska, Pol. Pat. 63,328 (1971); *Chem. Abstr.*, *76*, 115,604 (1972).
7. S. T. Takezhanov, G. L. Pashkov, A. S. Kulenov, K. Z. Kuanysheva, USSR Pat. 300,058 (1974); *Chem. Abstr.*, *82*, 158,270 (1975).
8. H. H. Weetall, W. Hertl, *Inorg. Chim. Acta*, *104*, 119 (1985).
9. G. Brauer, *Handbook of Preparative Inorganic Chemistry*, Academic Press, New York, 1963.

2.2.2.3. by Reduction.

Consideration of the reduction potentials for halate and hypohalite species shown in Table 1, §2.2.2.2, shows these to be relatively strong oxidizing agents and likely sources of halogen synthesis by chemical reduction. This preparative route, however, is less attractive than halide oxidation. The salts of halogens in positive oxidation states are more expensive than the halides, and the oxoanions are slow to be reduced. The ClO_3^- and ClO_4^- anions exemplify this inert character. Moreover, the reduction potential of the halate ions to halogen is similar to that to halide, and selective reduction to halogen is more difficult to achieve than selective oxidation of halide.

The only industrially important process for making halogen by reduction is that for I_2, but chemical oxidation methods account for most production now[1]. In this process[2], $NaIO_3$ is extracted from the caliche (Chile saltpeter), where it is concentrated in the mother liquors on crystallization of the $NaNO_3$. Part of the $NaIO_3$ solution is treated with stoichiometric $NaHSO_3$ to reduce the iodate to NaI and then added to stoichiometric aq $NaIO_3$ to produce I_2:

$$2\ NaIO_3 + 6\ NaHSO_3 \rightarrow 3\ I_2 + 3\ Na_2SO_4 + 3\ H_2SO_4 \tag{a}$$

$$5\ NaI + NaIO_3 + 3\ H_2SO_4 \rightarrow 3\ I_2 + 3\ Na_2SO_4 + 3\ H_2O \tag{b}$$

Iodine is filtered and purified by sublimation[3].

Reductions of species with halogen in positive oxidation states to the element are important in analysis[4]:

$$2\ [IO_4]^- + 16\ H^+ + 14\ I^- \rightarrow 8\ I_2 + 8\ H_2O \tag{c}$$

$$[XO_3]^- + 6\ I^- + 6\ H^+ \rightarrow 3\ I_2 + X^- + 3\ H_2O \tag{d}$$

where X = Cl, Br;

$$2\ [XO_3]^- + 10\ X^- + 12\ H^+ \rightarrow 6\ X_2 + 6\ H_2O \tag{e}$$

where X = Br, I; and:

$$HClO + 2\ I^- + H^+ \rightarrow I_2 + Cl^- + H_2O \tag{f}$$

These reactions, however, are rarely used for synthesis.

Other commercially important reductions are those where controlled Cl_2 production is required, e.g., in the sterilization of swimming-pool water. A mixture of $NaClO_3$ with oxalic acid, tartaric acid or NaCl, and $KHSO_4$ slowly generates Cl_2 as well as small amounts of ClO_2 on contact with H_2O[5]:

$$5\ H_2C_2O_4 + 2\ ClO_3^- + 2\ H^+ \rightarrow Cl_2 + 10\ CO_2 + 6\ H_2O \tag{g}$$

$$C_4H_6O_6 + 2\ ClO_3^- + 2\ H^+ \rightarrow Cl_2 + 4\ CO_2 + 4\ H_2O \tag{h}$$

$$2\ NaClO_3 + 10\ NaCl + 12\ H^+ \rightarrow 6\ Cl_2 + 6\ H_2O + 12\ Na^+ \tag{i}$$

Trichloroisocyanuric acid or $Ca(OCl)_2$ on contact NaCl, $NaHSO_4$ and H_2O yield Cl_2 under vacuum. Chlorine generation ceases when the vacuum is removed[6,7]:

$$Ca(OCl)_2 + 2\ NaCl + 4\ H^+ \rightarrow 2\ Cl_2 + Ca^{2+} + 2\ Na^+ + 2\ H_2O \qquad (j)$$

$$\text{Trichloroisocyanuric acid} + 3\ NaCl + 3\ H^+ \rightarrow \text{Cyanuric acid} + 3\ Na^+ + 3\ Cl_2 \qquad (k)$$

Trichloroisocyanuric acid   Cyanuric acid

(J. B. MILNE)

1. E. Booth, *Chem. Ind. (London)*, 838 (1978); 52 (1979).
2. *Kirk-Othmer Encyclopedia of Chemical Technology*, 3rd ed., Vols. 1, 4, 13, Wiley, New York, 1978.
3. J. L. Tonelli, E. M. Cortes, *Minerales*, *38*, 39 (1983); *Chem. Abstr.*, *101*, 173866 (1984).
4. A. I. Vogel, *A Textbook of Quantitative Inorganic Analysis*, 3rd Edition, Longmans, Green, London, 1961, pp. 343, 383.
5. P. G. Bjorklund, K. G. Karsson, Ger. Pat. 2,425,319 (1974); *Chem. Abstr.*, *82*, 102,997 (1975).
6. W. B. Murray, Ger. Pat. 2,351,441 (1974); *Chem. Abstr.*, *82*, 61,232 (1975).
7. E. M. Smolin, L. Rapoport, in '*Heterocyclic Compounds*', Vol. 13, E. M. Smolin, L. Rapoport, eds., Interscience, New York, 1959, p. 391.

2.2.2.4. by Dissociation of Halides.

Elements with more than one positive oxidation state form halides, which undergo thermal dissociation to give elemental halogen, e.g.:

$$AuCl_3 \xrightarrow{250^\circ C} AuCl + Cl_2 \qquad (a)$$

$$TlCl_3 \xrightarrow{40^\circ C} TlCl + Cl_2 \qquad (b)$$

$$PbCl_4 \rightarrow PbCl_2 + Cl_2 \qquad (c)$$

$$SbCl_5 \xrightarrow{140^\circ C} SbCl_3 + Cl_2 \qquad (d)$$

$$2\ SeBr_4 \xrightarrow{70^\circ\text{–}80^\circ C} Se_2Br_2 + 3\ Br_2 \qquad (e)$$

$$ClF_5 \xrightarrow{165^\circ C} ClF_3 + F_2 \qquad (f)$$

$$2\ VI_3 \xrightarrow{280^\circ C} 2\ VI_2 + I_2 \qquad (g)$$

Complex halides can also be decomposed thermally to produce halogen, e.g.:

$$4\ SOCl_2 \xrightarrow{400^\circ C} S_2Cl_2 + 2\ SO_2 + 3\ Cl_2 \qquad (h)$$

$$2\ VOBr_2 \xrightarrow{350^\circ C} 2\ VOBr + Br_2 \qquad (i)$$

Whereas none of these processes is practicable for halogen synthesis, dissociation of a halide is important in industrial production. The decomposition of $CuCl_2$ is a step in Eq. (a), §2.2.2.2.2[1]:

$$2\ CuCl_2 \xrightarrow{480°C} 2\ CuCl + Cl_2 \tag{j}$$

In the oxidation of HCl by SO_3, the thermal dissociation of SO_2Cl_2 is a step[2]:

$$SO_2Cl_2 \xrightarrow{100°C} SO_2 + Cl_2 \tag{k}$$

In the sequence[1] Eqs. (e)–(i), §2.2.2.2.2, and other processes, thermal dissociation of NOCl and NO_2Cl occurs:

$$2\ NOCl \rightarrow 2\ NO + Cl_2 \tag{l}$$

$$2\ NO_2Cl \rightarrow 2\ NO_2 + Cl_2 \tag{m}$$

(J. B. MILNE)

1. *Kirk-Othmer Encyclopedia of Chemical Technology*, 3rd ed. Vols. 1, 4, 13, Wiley, New York, 1978.
2. J. S. Robinson, ed., *Chlorine Production Processes*: *Recent and Energy-Saving Developments*, Chemical Technology Review No. 185, Noyes Data Corp., Park Ridge, NJ, 1981.

2.2.2.5. Diastatine Synthesis.

There is little evidence for the existence of diastatine (At_2).

Zero-state astatine, identified by its volatility and extractability into non-polar organic solvents, is characterized by irreproducible behavior and probably consists of a mixture of components[1,2]. Halogens in their zero valence states are reactive, and their reactions often involve a dissociative preequilibrium:

$$X_2 \rightarrow 2\ X \tag{a}$$

or:

$$X_2 \rightarrow X^+ + X^- \tag{b}$$

or:

$$X_2 + H_2O \rightarrow HOX + H^+ + X^- \tag{c}$$

Astatine is studied as 10^{-9}M concentrations, at which the equilibria (a)–(c) will be shifted toward the right. The result is increased reactivity toward both major and minor components of the reaction system, which accounts for low survival of At_2.

A radiogaschromatographic[3,4] study has claimed the identification of At_2. However, since the species persists in the presence of xs I_2, the identification must be considered questionable. Formation of At_2^+ has been observed in a plasma ion source mass spectrometer, presumably as the result of the reaction[5]

$$At + At^+ \rightarrow At_2^+ \tag{d}$$

(W. D. LEE)

1. G. L. Johnson, R. F. Leininger, E. Segre, *J. Chem. Phys.*, *17*, 1 (1949).
2. E. H. Appelman, *J. Am. Chem. Soc.*, *83*, 805 (1961).

3. K. Otozai, N. Takahashi, *Radiochim. Acta*, *31*, 201 (1982).
4. N. Takahashi, K. Otozai, *J. Radioanal. Nucl. Chem.*, *103*, 1 (1986).
5. N. A. Golovkov, I. I. Gromova, M. Janicki, Yu. V. Norseev, V. G. Sandukovskii, L. Vasaros, *Radiochem. Radioanal. Lett.*, *44*, 67 (1980).

2.2.3. Preparation of Cationic Polyhalogens (Homonuclear and Heteronuclear)

Polyhalogen cations composed solely of halogens can be stabilized in the solid by complex counteranions, usually derived from electron-pair acceptor acids. These cations also exist in strongly acidic solutions either obtained by dissolving the solid in suitable solvents or prepared directly in solution from elemental halogens or interhalogens.

The heteronuclear cations are mostly monomers, $[XY_{n-1}]^+$, where the more electropositive central halogen in a positive oxidation state is bonded to one or several more electronegative halogens in a uninegative oxidation state. These heteronuclear cations are derivatives of the neutral interhalogens, XY_n, in which the central halogen, X, is in the same oxidation state as it is in the cation, releasing a halide anion. Such compounds are synthesized from a neutral interhalogen, XY_n, with an electron-pair acceptor acid halide, MY_m, accepting the halide anion transferred from XY_n. These neutral interhalogens are amphoteric and undergo autoioniation:

$$2\,XY_n \rightleftharpoons [XY_{n-1}]^+ + [XY_{n+1}]^- \qquad \text{(a)}$$

Therefore, the cations $[XY_4]^+$ and $[XY_6]^+$ are known only with Y = F, as no interhalogens are available in which the central halogen is in the +V and +VII oxidation state except the halogen fluorides, e.g., ClF_5 or IF_7.

Oxidation of a neutral halogen or interhalogen gives homonuclear cations. Oxidations are the sole route to some heteronuclear cations, such as $[ClF_6]^+$ or $[BrF_6]^+$, whose parent compounds, the neutral ClF_7 or BrF_7, are unknown.

These compounds are reactive and hydrolyze and, therefore, must be handled excluding moisture. Precautions with the halogen fluorides require the use of chemically resistant apparatus as used in fluorine chemistry[1], including vacuum lines and apparatus constructed of suitable metals, such as Ni, Cu or Ni–Cu alloys and working at moderate pressure and T. Such fluoroplastics as polytetrafluoroethylene or polychlorotrifluoroethylene and related copolymers can also be used.

The safety precautions as used in fluorine chemistry, such as working in well-ventilated hoods, chemical disposal of unreacted chemicals, etc., have to be exercised.

(J. SHAMIR)

1. J. H. Canterford, T. A. O'Donnell, in *Technique of Inorganic Chemistry*, H. B. Jonassen, A. Weissberger, eds., Vol. VII, Interscience, New York 1968, p. 273.

2.2.3.1. by Halide-Anion Transfer in Acid–Base Reactions.

A neutral interhalogen, XY_n, is reacted with an electron-pair acceptor acid halide, MY_m, to give a halide-anion transfer:

$$XY_n + MY_m \rightarrow [XY_{n-1}]^+[MY_{m+1}]^- \qquad \text{(a)}$$

TABLE 1. SOLIDS DERIVED FROM NEUTRAL INTERHALOGENS EXIST WITH THE FOLLOWING LISTED COUNTERANIONS.[a]

Cation	Counteranions
$[ClF_2]^+$	$[BF_4]^-$, $[PF_6]^-$, $[AsF_6]^-$, $[SbF_6]^-$, $[Sb_2F_{11}]^-$, $[Sb_4F_{21}]^-$, $[PtF_6]^-$, $[RuF_6]^-$[4], $[SnF_6]^{2-}$
$[BrF_2]^+$	$[AsF_6]^-$, $[SbF_6]^-$, $[Sb_3F_{16}]^-$, $[BiF_6]^-$, $[TaF_6]^-$, $[NbF_6]^-$, $[GeF_6]^{2-}$, $[SnF_6]^{2-}$, $[AuF_4]^-$, $[SO_3F]^-$, $[PtF_6]^{2-}$, $[PdF_4]^-$
$[IF_2]^+$	$[BF_4]^-$, $[AsF_6]^-$, $[SbF_6]^-$
$[ICl_2]^+$	$[AlCl_4]^-$, $[SbCl_6]^-$, $[SbF_6]^-$[6], $[Sb_2F_{11}]^-$, $[SO_3Cl]^-$
$[Cl_2F]^+$	$[BF_4]^-$, $[AsF_6]^-$
$[I_2Cl]^+$	$[AlCl_4]^-$, $[SbCl_6]^-$
$[ClF_4]^+$	$[AsF_6]^-$, $[SbF_6]^-$, $[Sb_2F_{11}]^-$, $[Sb_4F_{21}]^-$, $[PtF_6]^-$
$[BrF_4]^+$,	$[AsF_6]^-$, $[Sb_2F_{11}]^-$, $[SnF_6]^{2-}$
$[IF_4]^+$	$[SbF_6]^-$, $[Sb_2F_{11}]^-$, $[SO_3F]^-$, $[SnF_6]^{2-}$, $[PtF_6]^-$
$[IF_6]^+$	$[AsF_6]^-$, $[SbF_6]^-$, $[Sb_2F_{11}]^-$
$[I_3Cl_2]^+$	$[SbCl_6]^-$[7,9]
$[I_3Br_2]^+$	$[SbCl_6]^-$[9]
$[ClF_6]^+$	$[PtF_6]^-$[8]

[a] The data are from refs. 1–3 and 5 except for the compounds having a different reference number cited.

When n = 1, no monoatomic cation is formed, a multimolecular reaction taking place instead:

$$2\ XY + MY_m \rightarrow [X_2Y]^+[MY_{m+1}]^- \quad \text{(b)}$$

or

$$3\ XY + MY_m \rightarrow [X_3Y_2]^+[MY_{m+1}]^- \quad \text{(c)}$$

Such reactions take place by direct interaction of the reagents, or in a solution of one of the reagents in xs or in another unreacting solvent such as anhyd HF. For some, the electron-pair acceptor acid, MF_m, can be formed in situ prior to salt formation by reacting a derivative of M such as a lower oxidation fluoride, other halides, oxides, or the element itself with xs halogen fluoride.

The more volatile reactant should be in xs and then pumped off after completion of the reaction, leaving a pure solid product. Similarly, when the preparation is done in solution the solid product is left behind after the solvent and xs reagent are pumped off.

Table 1 summarizes the solid products obtained by this method.

(J. SHAMIR)

1. J. Shamir, *Struct. Bonding, 37*, 141 (1979).
2. R. J. Gillespie, M. J. Morton, *Q. Rev. Chem. Soc., 25*, 553 (1971).
3. A. J. Downs, C. J. Adams, in *Comprehensive Inorganic Chemistry*, A. F. Trotman-Dickenson, ed., Vol. 2, Pergamon Press, Oxford, 1973, p. 1340.
4. R. C. Burns, T. O. O'Donnell, *J. Inorg. Nucl. Chem., 42*, 1613 (1980).
5. B. D. Stepin, *Russ. Chem. Rev., 56*, 726 (1987).
6. T. Birchall, R. D. Meyers, *Inorg. Chem., 20*, 2207 (1981).
7. N. Thorup, J. Shamir, *Inorg. Nucl. Chem. Lett., 17*, 193 (1981).
8. F. Q. Roberto, US 3709748; *Chem. Abstr., 78*, 99980 (1973).
9. S. Pohl, W. Saak, *Z. Naturforsch., Teil. B, 36B*, 283 (1981).

2.2.3.2. by Oxidation of Elemental Halogens or Interhalogens.

This is the best method for homonuclear polyhalogen cations as well as for some heteronuclear ones.

In these preparations an electron-pair acceptor acid forming the counteranion is required to stabilize the solid salts, and a strongly acidic environment is necessary for stabilization in solution.

TABLE 1. CATIONIC POLYHALOGENS PREPARED BY THE OXIDATIVE METHOD[a]

Cl_3^+	Solid: $Cl_2 + 2\,[Cl_2F][AsF_6] \xrightarrow{-78°C} 2\,[Cl_3][AsF_6] + F_2$
	$Cl_2 + ClF + AsF_5 \xrightarrow{-78°C} [Cl_3][AsF_6]$
	$Cl_2 + ClF + SbF_5 \xrightarrow{\text{in HF}} [Cl_3][SbF_6]$
Br_3^+	Solid: $Br_2 + BrF_3$ (or BrF_5) $+ AsF_5 \xrightarrow{-50°C} [Br_3][AsF_6]$
	$\frac{3}{2}\,Br_2 + [O_2][AsF_6] \rightarrow [Br_3][AsF_6] + O_2$
	$4\,BrSO_3F + Au \xrightarrow{60°C} [Br_3][Au(SO_3F)_4]$ [4]
	Also formed in HSO_3F solution or in super acids using $S_2O_6F_2$ as oxidizer
I_3^+	Solid: $I_2 + ICl + AlCl_3 \rightarrow [I_3][AlCl_4]$
	$3\,I_2 + 3\,AsF_5 \xrightarrow{\text{in } AsF_3} 2\,[I_3][AsF_6] + AsF_3$
	$3\,I_2 + 3\,SbF_5 \xrightarrow{\text{in } SO_2} 2\,[I_3][SbF_6] + SbF_3$
	$3\,I_2 + S_2O_6F_2 \rightarrow 2\,[I_3][SO_3F]$
	Also formed in H_2SO_4, oleum or HSO_3F solutions using HIO_3, I_2O_5, KIO_3, $IOSO_4$ or $S_2O_6F_2$ as oxidizers
I_5^+	Solid: $2\,I_2 + ICl + AlCl_3 \rightarrow [I_5][AlCl_4]$
	Also formed in H_2SO_4 or HSO_3F solutions using HIO_3 and $S_2O_6F_2$ as oxidizers
Br_2^+	Solid: $9\,Br_2 + 2\,BrF_5 + 30\,SbF_5 \rightarrow 10\,[Br_2][Sb_3F_{16}]$
	$2\,Br_2 + S_2O_6F_2 + 10\,SbF_5 \xrightarrow[\text{xs}SbF_5]{25°C} 2\,[Br_2][Sb_3F_{16}]$ [5]
	Also formed in super-acids, using $S_2O_6F_2$ as oxidizer
I_2^+	Solid: $I_2 + \text{xs}SbF_5 \rightarrow [I_2][Sb_2F_{11}]$
	$I_2 + 2\,SbF_5$ (or TaF_5) $\xrightarrow{IF_5} [I_2][Sb_2F_{11}]$ (or $[I_2][Ta_2F_{11}]$)
	$2\,I_2 + 5\,SbF_5 \xrightarrow{SO_2} 2\,[I_2][Sb_2F_{11}] + SbF_3$
	$2\,I_2 + S_2O_6F_2 + 8\,SbF_5 \xrightarrow[\text{xs}SbF_5]{50°C} 2\,[I_2][Sb_2F_{11}]$ [5]
	Also formed in HSO_3F, SbF_5, IF_5 or $H_2S_2O_7$ solutions and in oleum, using HIO_3, $S_2O_6F_2$ and $K_2S_2O_8$ as oxidizers

I_4^{2+}	In HSO_3F solution at low T
$[ICl_2]^+$	Solid: $I_2 + 2\ SbCl_5 + xsCl_2 \rightarrow [ICl_2][SbCl_6]$ $I_2 + 2\ AlCl_3 + Cl_2 \xrightarrow{-70°C} [ICl_2][AlCl_4]$ $IOSO_2F + Cl_2 \rightarrow [ICl_2][SO_3F]$ $[IBr_2][SO_3F] + Cl_2 \rightarrow [ICl_2][SO_3F]$ Also formed in H_2SO_4 or $H_2S_2O_7$ solution using HIO_3 as oxidizer
$[IBr_2]^+$	Solid: $IOSO_2F + Br_2 \rightarrow [IBr_2][SO_3F]$ $[I_3][SO_3F] + Br_2 \rightarrow [IBr_2][SO_3F]$ $ISO_3CF_3 + Br_2 \rightarrow [IBr_2][SO_3CF_3]$ Also formed in solutions of H_2SO_4 or $H_2S_2O_7$ using HIO_3 as oxidizer
$[I_2Cl]^+$	Solid: $I_2 + 2\ SbCl_5 \rightarrow [I_2Cl][SbCl_6] + SbCl_3$ $I_2 + SbCl_5 + xsCl_2 \xrightarrow{-78°C} [I_2Cl][SbCl_6]$ $IOSO_2F + ICl \rightarrow [I_2Cl][SO_3F]$ $[I_3][SO_3F] + [ICl_2][SO_3F] \rightarrow 2\ [I_2Cl][SO_3F]$ Also formed in solutions of H_2SO_4, oleum or HSO_3F using HIO_3 as oxidizer
$[I_2Br]^+$	Solid: $IOSO_2F + IBr \rightarrow [I_2Br][SO_3F]$ $[I_3][SO_3F] + [IBr_2][SO_3F] \rightarrow 2\ [I_2Br][SO_3F]$ Also formed in H_2SO_4 solution using HIO_3 as oxidizer.
$[BrICl]^+$	Solid: $IBr + SbCl_5 + xsCl_2 \rightarrow [BrICl][SbCl_6]$ $[ICl_2][SO_3F] + [IBr_2][SO_3F] \rightarrow 2\ [BrICl][SO_3F]$ Also formed in H_2SO_4 and HSO_3F solutions.
$[I_3Cl_2]^+$	Solid: $I_2 + SbCl_5 + xsCl_2 \rightarrow [I_3Cl_2][SbCl_6]$[6] $I_2 + SbCl_5 \rightarrow [I_3Cl_2][SbCl_6]$
$[ClF_6]^+$	Solid: $ClF_5 + PtF_6 \xrightarrow{UV} [ClF_6][PtF_6]$ $FClO_3 + PtF_6 \rightarrow [ClF_6][PtF_6]$
$[BrF_6]^+$	Solid: $BrF_5 + [KrF]^+$ (or $[Kr_2F_3]^+$)$[AsF_6] \rightarrow [BrF_6][AsF_6] + Kr$ $BrF_5 + [KrF]^+$ (or $[Kr_2F_3]^+$)$[SbF_6] \rightarrow [BrF_6][Sb_2F_{11}] + Kr$
$[IF_6]^+$	Solid: $IF_5 + [KrF][Sb_2F_{11}] \rightarrow [IF_6][Sb_2F_{11}] + Kr$ $IF_5 + AuF_5 + SbF_5 \rightarrow [IF_6][SbF_6] + AuF_3$ $IF_5 + K[AuF_6] + 2\ SbF_5 \rightarrow [IF_6][SbF_6] + K[SbF_6] + AuF_3$ $3\ IF_5 + 4\ [O_2][AuF_6] \rightarrow 3\ [IF_6][AuF_6] + 4\ O_2 + AuF_3$

[a] The data are from refs. 1–3, except for the compounds having a different reference number cited.

Strong oxidizers, such as a more electronegative halogen or an interhalogen composed of a more electronegative halogen than the one to be oxidized are required. Another polyhalogen cation, disulfuryl fluoride, iodic acid and cationic species such as O_2^+, $[KrF]^+$ and $[Kr_2F_3]^+$ can also be used as oxidizers. Electron-pair acceptor acids, such as AsF_5, SbF_5 and $SbCl_5$, can also act as the oxidizer in addition to serving as a halide acceptor when used in xs.

It is easier to prepare the lower oxidation state cations, X_3^+, than the higher ones, X_2^+. As a result, all the halogens except fluorine form X_3^+ cations, whereas only Br_2^+, I_2^+

and I_4^{2+} are known in the X_2^+ series. The order of electrophilicity $Br_2^+ > Br_3^+ > I_2^+ > I_3^+$ is indicated by the acid strengths of the media that stabilize these cations.

The Br_2^+ ion can exist in solution of a super acid of SbF_5–HSO_3F–3 SO_3 only, whereas Br_3^+ is formed in solution of super acid, but the solid salt can be dissolved in fluorosulfuric acid. These cationic species are formed by oxidizing Br_2 with stoichiometric disulfuryl fluoride.

The I_2^+ ion can exist in less acidic solutions, such as fluorosulfuric acid or in 44.9 and 65% oleum. Iodine compounds such as ICl, KI, IBr, $I(py)_2NO_3$, $IOSO_2F$, $IOSO_2Cl$, IO_2CCH_3, and N-iodo organic compounds can be oxidized to I_2^+, using oxidizers such as iodic acid, disulfuryl fluoride or an xs of the oxidizing acidic solvents themselves. Finally, the I_3^+ can be stabilized in weaker acids, such as H_2SO_4, and can be prepared by oxidizing not only I_2 but even iodosyl compounds or N-iodo organics. The concentration ratio of the halogen to the oxidizer must be controlled, because the product obtained depends on that ratio, e.g.:

$$2\,X_2 + S_2O_6F_2 \rightarrow 2\,X_2^+ + 2\,[SO_3F]^- \qquad \text{(a)}$$

$$3\,X_2 + S_2O_6F_2 \rightarrow 2\,X_3^+ + 2\,[SO_3F]^- \qquad \text{(b)}$$

Some heteronuclear polyhalogen cations can be prepared only by the oxidative method, because their parent interhalogen compounds are not known and they therefore cannot be prepared by using the halide transfer method, e.g., $[IBr_2]^+$, $[BrICl]^+$, $[ClF_6]^+$ and $[BrF_6]^+$. The latter two form when oxidizing fluorides, such as PtF_6 and Kr fluoride salts, are used.

Table 1 summarizes the oxidative reactions forming solids as well as known stable solutions.

(J. SHAMIR)

1. J. Shamir, *Struct. Bonding, 37*, 141 (1979).
2. R. J. Gillespie, M. J. Morton, *Q. Rev. Chem. Soc., 25*, 553 (1971).
3. A. J. Downs, C. J. Adams, in *Comprehensive Inorganic Chemistry*, A. F. Trotman-Dickenson, ed., Vol. 2, Pergamon Press, Oxford, 1973, p. 1340.
4. K. C. Lee, F. Aubke, *Inorg. Chem.*, 19, 119 (1980).
5. W. W. Wilson, R. C. Thompson, F. Aubke, *Inorg. Chem., 19*, 1489 (1980).
6. N. Thorup, J. Shamir, *Inorg. Nucl. Chem. Lett., 17*, 193 (1981).

2.2.4. Preparation of Neutral Interhalogens

Interhalogens, XY_n, are known[1–3] in which n = 1, 3, 5 and 7. All XY_n compounds in which n > 1 are actually halogen fluorides, XF_n, except the dimer, I_2Cl_6. These are reactive materials that hydrolyze easily, and, therefore moisture must be excluded of; e.g., **halogen fluorides must be handled as described in §2.2.3.**

Interhalogens are prepared directly from the elements. Some are prepared from equilibria of higher oxidation state halogen fluorides with the same halogen to form a lower oxidation state compound. Other fluorinating agents can be used instead of F_2 to prepare halogen fluorides. Fluorination of halides instead of elemental halogen is sometimes preferable, leading to purer products. Impurities such as HF, halogen, other halogen fluorides and oxyhalogen can be removed by reacting the material, first with NaF to form the nonvolatile $Na[HF_2]$, then removing HF and fractionally distilling or subliming, either by trap to trap or low T distillation for final purification.

(J. SHAMIR)

1. A. J. Downs, C. J. Adams, in *Comprehensive Inorganic Chemistry*, A. F. Trotman-Dickenson, ed., Vol. 2, Pergamon Press, Oxford, 1973, p. 1476.
2. L. Stein, in *Halogen Chemistry*, V. Gutmann, ed., Vol. 1, Academic Press, New York, p. 133.
3. H. Meinert, *Z. Chem.*, *7*, 41 (1967).

2.2.4.1. from the Elements

The halogens react with each other forming binary interhalogens. Because these are equilibria, the working conditions, such as relative concentrations, T and pressures, determine the products[1–3].

(J. SHAMIR)

1. A. J. Downs, C. J. Adams, in *Comprehensive Inorganic Chemistry*, A. F. Trotman-Dickenson, ed., Vol. 2, Pergamon Press, Oxford, 1973, p. 1476.
2. L. Stein, in *Halogen Chemistry*, V. Gutmann, ed., Vol. 1, Academic Press, New York, p. 133.
3. H. Meinert, *Z. Chem.*, *7*, 41 (1967).

2.2.4.1.1. by Fluorination of Cl_2, Br_2 and I_2.

This important method has been reviewed[1–3].

The three stable chlorine fluorides can be prepared by direct fluorination of Cl_2. The products obtained depend on the concentrations, T and pressure.

Chlorine monofluoride is obtained by heating a 1:1 gas mixture at 220°–250°C:

$$Cl_2 + F_2 \xrightarrow{220°–250°C} 2\ ClF \qquad (a)$$

Heating a mixture of $Cl_2 + 3\ F_2$ results in chlorine trifluoride:

$$Cl_2 + 3\ F_2 \xrightarrow{200°–300°C} 2\ ClF_3 \qquad (b)$$

Chlorine pentafluoride can be prepared by direct interaction only under high pressure, with 1:14 xs F_2:

$$Cl_2 + 5\ F_2 \xrightarrow[250\ atm]{350°C} 2\ ClF_5 \qquad (c)$$

The use of NiF_2 as catalyst results in high yields even at lower T and pressure[4].

Fluorination of Br_2 results in two stable compounds: BrF_3, which is unstable and disproportionates to Br_2, and BrF_5. Fluorination of Br_2 with BrF_3 or BrF_5 produces BrF in larger concentration at higher T:

$$Br_2 + BrF_3 \rightleftharpoons 3\ BrF \qquad (d)$$

or

$$2\ Br_2 + BrF_5 \rightleftharpoons 5\ BrF \qquad (e)$$

which disproportionates at RT to Br_2 and a higher bromine fluoride.

The reaction between F_2 and Br_2 at RT forms BrF_3:

$$Br_2 + 3\ F_2 \rightarrow 2\ BrF_3 \quad \text{(f)}$$

Equation (f) can be used to determine the purity of F_2 gas when a known vol of gas is reacted with a known amount of Br_2 dissolved in BrF_3. A red solution forms and becomes discolored at the end point, becoming slightly yellowish. At low T BrF is obtained for reagent use[5].

Bromine pentafluoride is prepared from the elements at ≥ 150°C:

$$Br_2 + 5\ F_2 \xrightarrow{\geq 150°C} 2\ BrF_5 \quad \text{(g)}$$

or by fluorination of BrF_3:

$$BrF_3 + F_2 \xrightarrow{200°C} BrF_5 \quad \text{(h)}$$

Instead of using F_2, bromine trifluoride can also be prepared by passing chlorine trifluoride into liq Br_2 at 0°C:

$$Br_2 + 2\ ClF_3 \xrightarrow{0°C} 2\ BrF_3 + Cl_2 \quad \text{(i)}$$

Only the higher iodine fluorides, namely the penta- and heptafluorides, are stable. Iodine monofluoride is unstable and cannot be isolated at RT because it disproportionates to I_2 and IF_5. It can be prepared by fluorination of I_2 at −78°C with IF_3:

$$I_2 + IF_3 \xrightarrow[-78°C]{\text{in } CCl_3F} 3\ IF \quad \text{(j)}$$

with AgF:

$$I_2 + AgF \xrightarrow{-10° \text{ to } 30°C} IF + AgI \quad \text{(k)}$$

being swept away with N_2 and trapped at −78°C, or from the elements at low T[5].

Iodine trifluoride, which is stable only ≤ −35°C, when it starts to disproportionate to I_2 and IF_5, is prepared by direct interaction of the elements at low T:

$$I_2 + 3\ F_2 \xrightarrow[-78°C]{\text{in } CCl_3F} 2\ IF_3 \quad \text{(l)}$$

or by fluorination with XeF_2:

$$I_2 + 3\ XeF_2 \rightarrow 2\ IF_3 + 3\ Xe \quad \text{(m)}$$

Iodine pentafluoride can be prepared from the elements at RT:

$$I_2 + 5\ F_2 \rightarrow 2\ IF_5 \quad \text{(n)}$$

Fluorination can also take place by using ClF_3, BrF_3, AgF or RuF_5. It is also possible to fluorinate iodine pentoxide or metal iodates with F_2, ClF_3 and BrF_3 or SF_4.

Iodine heptafluoride is formed by fluorination at higher T:

$$I_2 + 7\ F_2 \xrightarrow{250°\text{–}270°C} 2\ IF_7 \quad \text{(o)}$$

$$IF_5 + F_2 \xrightarrow{150°C} IF_7 \quad \text{(p)}$$

(J. SHAMIR)

1. A. J. Downs, C. J. Adams, in *Comprehensive Inorganic Chemistry*, A. F. Trotman-Dickenson, ed., Vol. 2, Pergamon Press, Oxford, 1973, p. 1476.
2. L. Stein, in *Halogen Chemistry*, V. Gutmann, ed., Vol. 1, Academic Press, New York, p. 133.
3. H. Meinert, *Z. Chem.*, *7*, 41 (1967).
4. A. Smalc, B. Zemva, J. Slivnik, K. Lutar, *J. Fluorine Chem.*, *17*, 381 (1981).
5. S. Rozen, M. Brand, *J. Org. Chem.*, *50*, 3248 (1985).

2.2.4.1.2. by Chlorination of Br_2 and I_2.

Bromine chloride cannot be isolated because it decomposes to the elements. The direct reaction can take place in the gas phase or in CCl_4 or H_2O solution at RT.

Chlorination of I_2 can result in two compounds. Iodine monochloride can be prepared from equimol quantities of the elements, either directly or in solution:

$$I_2 + Cl_2 \rightarrow 2\ ICl \tag{a}$$

Reaction with xs liq Cl_2 at −78°C results in the dimeric iodine trichloride:

$$I_2 + 3\ Cl_2 \xrightarrow{-78^\circ C} I_2Cl_6 \tag{b}$$

(J. SHAMIR)

1. A. J. Downs, C. J. Adams, in *Comprehensive Inorganic Chemistry*, A. F. Trotman-Dickenson, ed., Vol. 2, Pergamon Press, Oxford, 1973, p. 1476.
2. H. Meinert, *Z. Chem.*, *7*, 41 (1967).

2.2.4.1.3. by Bromination of I_2.

Direct reaction of Br_2 and I_2 under reflux results in solid IBr:

$$I_2 + Br_2 \xrightarrow{\text{reflux}} 2\ IBr \tag{a}$$

At 25°C, 8.8% is dissociated to the elements[1–2].

(J. SHAMIR)

1. A. J. Downs, C. J. Adams, in *Comprehensive Inorganic Chemistry*, A. F. Trotman-Dickenson, ed., Vol. 2, Pergamon Press, Oxford, 1973, p. 1476.
2. H. Meinert, *Z. Chem.*, *7*, 41 (1967).

2.2.4.2. from the Oxidation of Anionic Halides.

Chlorine pentafluoride can be prepared at normal pressures by fluorinating halides or fluorohalides:

$$MCl + 3\ F_2 \xrightarrow{100^\circ\text{–}300^\circ C} MF + ClF_5 \quad (50\text{–}70\%) \tag{a}$$

where M is an alkali metal. Higher yields are obtained by:

$$MClF_4 + F_2 \xrightarrow{80^\circ\text{–}150^\circ C} MF + ClF_5 \quad (90\%) \tag{b}$$

The product of fluorination of a halide salt is purer than that obtained from an element because the element cannot always be purified and dried as easily, resulting in products containing mostly a halogen oxide and HF, whereas the halide can be obtained pure.

The following reactions also produce halogen fluorides:

$$KBr + 3\ F_2 \xrightarrow{25°C} KF + \underset{(50\%)}{BrF_5} \quad \text{(c)}$$

$$HI \text{ or } MI + F_2 \rightarrow MF + IF_5 \quad \text{(d)}$$

$$KI + 4\ F_2 \rightarrow KF + IF_7 \quad \text{(e)}$$

Variations are:

$$AgF + I_2 \rightarrow AgI + IF \quad \text{(f)}$$

$$5\ AgF + 3\ I_2 \rightarrow 5\ AgI + IF_5 \quad \text{(g)}$$

Reactions in aq acid at RT can form iodine chlorides such as:

$$I^- + Cl^- \rightarrow ICl + 2\ e^- \quad \text{(h)}$$

in which $KMnO_4$, KIO_3 or chlorine water serves as oxidizer.

A similar reaction is:

$$I_2 + 5\ Cl^- + [ClO_3]^- + 6\ H^+ \rightarrow I_2Cl_6 + 3\ H_2O \quad \text{(i)}$$

(J. SHAMIR)

1. A. J. Downs, C. J. Adams, in *Comprehensive Inorganic Chemistry*, A. F. Trotman-Dickenson, ed., Vol. 2, Pergamon Press, Oxford, 1973, p. 1476.
2. H. Meinert, *Z. Chem.*, *7*, 41 (1967).

2.2.4.3. from the Oxidation of Organoiodides

These compounds are derivatives of interhalogens in which a terminal halogen is substituted by an organic part. They can be prepared by halogenation of organoiodides.

(J. SHAMIR)

2.2.4.3.1. by F_2 or Cl_2.

Perfluoroalkyl iodides, R_fI ($R_f = C_2F_5$, n-C_3F_7, n-C_4F_9), can be fluorinated by F_2 to form R_fIF_2.

Substituted iodobenzenes can be chlorinated by passing Cl_2 into $CHCl_3$ solutions to form:

$$ArI + Cl_2 \xrightarrow{CHCl_3} ArICl_2 \quad \text{(a)}$$

(J. SHAMIR)

1. A. J. Downs, C. J. Adams, in *Comprehensive Inorganic Chemistry*, A. F. Trotman-Dickenson, ed., Vol. 2, Pergamon Press, Oxford, 1973, p. 1564.

2.2.4.3.2. by Oxidizing Fluorinating Agents (Excluding F_2).

Perfluoroalkyl iododifluorides can be prepared[1] with halogen fluorides instead of using F_2:

$$R_fI + XF_n \rightarrow R_fIF_2 \tag{a}$$

When using xs halogen fluoride, the iodotetrafluorides are obtained[1]:

$$R_fI + ClF_3\ (BrF_3, BrF_5) \rightarrow R_fIF_4 \tag{b}$$

(J. SHAMIR)

1. A. J. Downs, C. J. Adams, in *Comprehensive Inorganic Chemistry*, A. F. Trotman-Dickenson, ed., Vol. 2, Pergamon Press, Oxford, 1973, p. 1576.

2.2.4.4. Preparation of Neutral and Cationic Interhalogen Compounds Containing Astatine

The interhalogens AtI, AtBr, and AtCl can be prepared in the vapor phase from astatine with I_2, Br_2, and Cl_2, respectively[1–3]. Positive identification of the products has been obtained by mass spectrometry[1]. Excess Br_2 and Cl_2 may be distilled away in vacuo at −78°C. Excess I_2 may be removed by fractional vacuum distillation at RT[2]. Introduction of iodine, bromine or chlorine vapors to the He carrier gas during mass separation of astatine produces mass lines attributed to the cations $[AtI]^+$, $[AtBr]^+$ and $[AtCl]^+$ respectively in a plasma ion source mass spectrometer[4]. This is taken as evidence of the initial production of the neutral molecules in the plasma ion source. In H_2O AtI is formed by the reaction of astatine with a mixture of I_2 and I^-, whereas AtBr is formed[5] by the reaction with a mixture of IBr, I_2, and Br^-. The I^- and Br^- ions are used to control the redox potentials of the mixtures. Depending on the I^- and Br^- concentrations, various amounts of $[AtI_2]^-$ are also formed (see §2.2.6.3).

Astatine and I_2 can be extracted together from H_2O into $CHCl_3$. The astatine can then be cocrystallized[6] with the I_2 presumably as AtI. The AtI is also prepared from astatine present in an irradiated Bi_2O_3 target by dissolution in aq $HClO_4$ containing I_2 and precipitation of the Bi as its phosphate.[7]

Ion exchange of the product formed by oxidation of astatine with $[Cr_2O_7]^{2-}$ in HNO_3 indicate that addition of Cl^- converts an originally formed cationic species to neutral molecules, which could be AtCl or OAtCl[8].

Efforts to prepare and identify an astatine fluoride have failed. Nonvolatile products may result from reaction of an initially formed fluoride with the walls of the glass apparatus. Synthesis of astatine fluorides should be possible, however, in liq halogen fluoride solvents, as in the characterization of fluorides of Rn[9].

(W. D. LEE)

1. E. H. Appelman, E. N. Sloth, M. H. Studier, *Inorg. Chem.*, *5*, 766 (1966).
2. G.-J. Meyer, K. Rössler, *Radiochem. Radioanal. Lett.*, *25*, 377 (1976).
3. J. Merinis, Y. Legoux, G. Bouissières, *Radiochem. Radioanal. Lett.*, *11*, 59 (1972).
4. N. A. Golovkov, I. I. Gromova, M. Janicki, Yu. V. Norseev, V. G. Sandukovskii, L. Vasaros, *Radiochem. Radioanal. Lett.*, *44*, 67 (1980).
5. E. H. Appelman, *J. Phys. Chem.*, *65*, 325 (1961).
6. A. H. W. Aten Jr., J. G. van Raaphorst, G. Nooteboom, G. Blasse, *J. Inorg. Nucl. Chem.*, *15*, 198 (1960).

7. A. H. W. Aten Jr., T. Doorgeest, U. Hollstein, P. H. Moeken, *Analyst (London)*, *77*, 774 (1952).
8. Yu. V. Norseev, V. A. Khalkin, *J. Inorg. Nucl. Chem.*, *30*, 3239 (1968).
9. L. Stein, *Science*, *168*, 362 (1970).

2.2.5. Preparation of Aryliodo Fluorides from Aryliodo Oxides

2.2.5.1. from Aryliodoso and Aryliodo Compounds and HF.

Iodosobenzene (C_6H_5IO), an unstable compound that is oxidized to iodoxybenzene ($C_6H_5IO_2$), can be regarded as the anhydride of the hypothetical acid $C_6H_5I(OH)_2$. Derivatives of this acid are known, including iodobenzene dichloride, iodobenzene difluoride and iodosobenzene diacetate, -bis-trifluoroacetate and dibenzoate. Iodosobenzene is obtained by alkaline hydrolysis of iodobenzene dichloride[1,2], which is prepared[3] from iodobenzene with Cl_2:

$$C_6H_5I + Cl_2 \xrightarrow{CHCl_3} C_6H_5ICl_2 \qquad (a)$$

$$C_6H_5ICl_2 + 2\ NaOH \rightarrow C_6H_5IO + 2\ NaCl + H_2O \qquad (b)$$

p-Iodosotoluene is obtained[4] in 85% yield by treating p-iodotoluene with aq NaOH. Iodoso compounds can also be prepared by electrophilic aromatic substitution, possibly involving the iodoxyl ion, $[IO]^+$, e.g.:

$$3\ I_2O_5 + 2\ I_2 + 5\ H_2SO_4 \rightarrow 5\ (IO)_2SO_4 + 5\ H_2O \qquad (c)$$

$$2\ C_6H_6 + (IO)_2SO_4 \rightarrow 2\ C_6H_5IO + H_2SO_4 \qquad (d)$$

iodosobenzene can be converted[5] to phenyliodo difluoride with HF. p-Iodosotoluene in glacial acetic acid treated with 46% aq HF yields[4] p-tolyliodo difluoride:

$$CH_3\text{—}C_6H_4\text{—}I{=}O + 2\ HF \rightarrow CH_3\text{—}C_6H_4\text{—}IF_2 + H_2O \qquad (e)$$

A one-step reaction of HgO and aq HF with aryliodo dichlorides in CH_2Cl_2 provides a simple and rapid preparation of aryliodo difluorides[6]:

$$ArICl_2 + HgO + 2\ HF \rightarrow ArIF_2 + HgCl_2 + H_2O \qquad (f)$$

($Ar = C_6H_5$, p-ClC_6H_4, p-MeC_6H_4, p-$NO_2C_6H_4$)

p-(Iodophenyl)acetic acid difluoride is readily prepared from p-aminophenylacetic acid[7]:

$$p\text{-}H_2NC_6H_4CH_2CO_2H \xrightarrow[\text{KI, Cl}_2,\ \text{HF, HgO}]{\text{HONO }(0^\circ)} p\text{-}F_2IC_6H_4CH_2CO_2H \qquad (g)$$

(R. FILLER)

1. *Org. Syn.*, Coll. Vol. III, Wiley, New York, 1955, p. 483.
2. *Org. Syn.*, Coll. Vol. V, Wiley, New York, 1973, p. 658.
3. *Org. Syn.*, Coll. Vol. III, Wiley, New York, 1955, p. 482.
4. B. S. Garvey, L. F. Halley, C. F. H. Allen, *J. Am. Chem. Soc.*, **59**, 1827 (1937).

5. O. Dimroth, W. Bockemüller, *Chem. Ber.*, *64B*, 522 (1931).
6. W. Carpenter, *J. Org. Chem.*, *31*, 2688 (1966).
7. T. B. Patrick, J. J. Scheibel, W. E. Hall, Y. H. Lee, *J. Org. Chem.*, *45*, 4492 (1980).

2.2.5.2. from Aryliodo Compounds and SF_4.

The reaction of aryliodoso compounds or aryliodo-bis-trifluoroacetates with SF_4 in CH_2Cl_2 at $-20°C$ gives aryliodo difluorides in 59–100% yield[1]. Pentafluoroethyliodo difluoride, $CF_3CF_2IF_2$, is also prepared by this route:

$$ArI{=}O + SF_4 \xrightarrow[-20°C]{CH_2Cl_2} ArIF_2 \qquad \text{(a)}$$

or

$$ArI(O_2CCF_3)_2$$

where Ar = C_6H_5, p-MeC_6H_4, o-$NO_2C_6H_4$, m-FC_6H_4, p-FC_6H_4, C_6F_5, β-pyridyl).

(R. FILLER)

1. V. V. Lyalin, V. V. Orda, L. A. Alekseeva, L. M. Yagupol'skii, *Russ. J. Org. Chem.* (*Engl. Transl.*), 329 (1970).

2.2.5.3. by Electrochemical Fluorination of Aryliodo Compounds.

Phenyliodo difluoride (iodobenzene difluoride) is prepared by electrochemical fluorination, using AgF in CH_3CN with a Pt electrode[1]:

$$C_6H_5I + AgF \rightarrow C_6H_5IF_2 \qquad \text{(a)}$$

(R. FILLER)

1. H. Schmidt, H. Meinert, *Angew. Chem.*, *72*, 109 (1960).

2.2.5.4. from Aryliodo Compounds and XeF_2.

Aryliodo compounds react at RT with XeF_2 in the presence of a trace of anhyd HF to give aryliodo difluorides in 95% yield[1]:

$$XC_6H_4I + XeF_2 \xrightarrow[CH_2Cl_2]{\text{trace HF}} XC_6H_4IF_2 + Xe \qquad \text{(a)}$$

where (X = H, m-OMe, p-OMe, m-Cl, p-Cl, m-NO_2. The m-chloro compound is the most stable of these products. Polymer-supported aryliodo difluorides have been obtained by reaction of XeF_2 with iodinated "pop-corn" polystyrene[2–5] and cross-linked polystyrene[6].

(R. FILLER)

1. M. Zupan, A. Pollak, *J. Fluorine Chem.*, 7, 445 (1976).
2. M. Zupan, A. Pollak, *J. Chem. Soc. Chem. Commun.*, 715 (1975).

3. M. Zupan, *Coll. Czech. Chem. Commun.*, *42*, 266 (1977).
4. A. Gregorcic, M. Zupan, *Bull. Chem. Soc. Jpn.*, *50*, 517 (1976).
5. A. Gregorcic, M. Zupan, *J. Chem. Soc. Perkin Trans. 1*, 1446 (1977).
6. B. Šket, M. Zupan, P. Zupet, *Tetrahedron*, *40*, 1603 (1984).

2.2.5.5. from Aryliodo Compounds and F_2.

Aryliodo difluorides are prepared[1] in 70–90% yields by oxidative liquid-phase direct fluorination of aryliodo compounds at −78°C under N_2:

$$XC_6H_4I + F_2/N_2 \xrightarrow[-78°C]{CFCl_3} XC_6H_4IF_2 \qquad (a)$$

where X = H, o-F, o-Cl, m-Cl, p-Cl, o-Me, p-Me, o-CF_3, o-OMe.

Similarly, pentafluorophenyliodo difluoride and -tetrafluoride have also been prepared[2]:

$$C_5F_5I + F_2 \rightarrow C_6F_5IF_2 \xrightarrow{F_2} C_6F_5IF_4 \qquad (b)$$

(R. FILLER)

1. I. Ruppert, *J. Fluorine Chem.*, *15*, 173 (1980).
2. R. Kasemann, G. Klein, D. Naumann, *J. Fluorine Chem.*, *29*, 99 (1985).

2.2.6. Preparation of Anionic Polyhalides (Homonuclear and Heteronuclear)

Polyhalide anions can be prepared[1–6], like the polyhalogen cations, based on the amphoteric properties of neutral interhalogens:

$$2\,XY_n \rightleftharpoons [XY_{n-1}]^+ + [XY_{n+1}]^- \qquad (a)$$

and in case of n = 1:

$$3\,XY \rightleftharpoons [X_2Y]^+ + [XY_2]^- \qquad (b)$$

or

$$4\,XY \rightleftharpoons [X_3Y_2]^+ + [XY_2]^- \qquad (c)$$

Interhalogens react with electron-pair donor bases, such as metal halides, transferring the halide anion to be accepted by the neutral interhalogen and forming the related polyhalide anion:

$$XY_n + M^+Y^- \rightarrow M^+[XY_{n+1}]^- \qquad (d)$$

However, polyhalide anions are prepared by other methods although their parent interhalogens are not isolated.

In heteronuclear polyhalide anions, the different halogens give $[X_mY_nZ_p]^-$, where m + n + p is an odd number. The most electropositive halogen is at the center of the anion. Different preparative methods result in the same product, as long as isomers are not possible. For example, in

$$\left.\begin{array}{l} KI + Cl_2 \\ KCl + ICl \\ K[ICl_4] + KI \end{array}\right\} \rightarrow K[ICl_2] \qquad \text{(e)}$$

only the dichloroiodate is formed, with iodine at the center. Larger countercations e.g., the larger alkali metals, are required to stabilize such salts, whereas smaller cations stabilize hydrates such as $NH_4I_3 \cdot 3\ H_2O$, $NaI_3 \cdot NaI_5 \cdot 4\ H_2O$ and $NaI \cdot NaI_3 \cdot 6\ H_2O$. The hydrate complexes the cation and increases its size, as in $LiI_4 \cdot H_2O$, which is actually $[Li(H_2O)_4]_2^+ I_8^{2-}$. In addition, $[PCl_4]^+$, nitrosyl $[NO]^+$ and nitryl $[NO_2]^+$ species can serve as countercations[7,8]. Some bulky organic cations, such as $[R_4N]^+$, $[Ph_4As]^+$ and $[R_3S]^+$ can also serve as countercations for polyhalide anions not derived from halogen fluorides. These latter would destroy the organics, eliminating their possible use as countercations. Small inorganic multivalent cations can be increased in size by complexation, e.g., with $[Ni(NH_3)_4]^{2+}$ or $[Co(NH_3)_6]^{3+}$, thus rendering them possible countercations.

The stability of the polyhalide anions decreases with decrease in size $I_3^- > Br_3^- > Cl_3^-$. Similarly, the symmetric heteronuclear polyhalides are more stable than the nonsymmetric ones: $I_3^- > [IBr_2]^- > [ICl_2]^- > [I_2Br]^- > Br_3^- > [BrCl_2]^- > [Br_2Cl]^-$. Sometimes the products include as an impurity xs of the simple halide.

(J. SHAMIR)

1. J. Shamir, *Israel J. Chem.*, *17*, 37 (1978).
2. A. I. Popov, in *Halogen Chemistry*, Vol. 1, V. Gutmann, ed., Academic Press, New York 1967, p. 225.
3. A. A. Opalovskii, *Russ. Chem. Rev.* (*Engl. Transl.*), *36*, 711 (1967).
4. H. Meinert, *Z. Chem.*, *7*, 41 (1967).
5. A. J. Downs, C. J. Adams, in *Comprehensive Inorganic Chemistry*, A. F. Trotman-Dickenson, ed., Vol. 2., Pergamon Press, Oxford, 1973, p. 1534.
6. B. D. Stepin, S. B. Stepina, *Russ. Chem. Rev. (Engl. Transl.)*, *55*, 812 (1986).
7. R. Rufaelof, J. Shamir, *Spectrochim. Acta*, *30A*, 1305 (1974).
8. J. Shamir, S. Schneider, A. Bino, S. Cohen, *Inorg. Chim. Acta*, *114*, 35 (1986).

2.2.6.1. from Elemental Halogens with Halide Anion.

Reacting a halogen with its own halide is the only method for homonuclear polyhalide anions, e.g.:

$$KI + I_2 \rightarrow KI_3 \qquad \text{(a)}$$

All the halogens but fluorine form such trihalide anions, X_3^-.

Homonuclear anions with more than three atoms are known only for iodine, namely I_5^-, I_7^-, I_9^- and I_8^{2-}. The structures of these polyiodides differ from other similar polyhalides, in which the most electropositive halogen is at the center, being in a high positive oxidation state, and the surrounding halogens are uninegative. The polyiodide structures are chains linked together by alternating I_2 molecules and iodide anions.

Reacting an elemental halogen with a different halide can form heteronuclear trihalide anions, such as XY_2^- [1–6].

A heteronuclear pentahalide anion can also be formed by reacting Cl_2 with a heteronuclear polyhalide anion, e.g.:

$$[ICl_2]^- + Cl_2 \rightarrow [ICl_4]^- \quad \text{(b)}$$

Fluorination of chlorides or bromides in a flow system forms the halogen fluoride anions, e.g.:

$$MCl + F_2 \xrightarrow{90^\circ\text{–}250^\circ C} M[ClF_4] \quad \text{(c)}$$

$$MBr + F_2 \rightarrow M[BrF_4] \quad \text{(d)}$$

$$MI + F_2 \rightarrow M[IF_6] \text{ or } M[IF_4] \quad \text{(e)}$$

$$NiCl_2 + F_2 \xrightarrow[300\text{ torr}]{150\text{–}200^\circ C} Ni[ClF_2]_2 \text{ or } Ni[ClF_4]_2 \quad \text{(f)}$$

In a static system the reaction:

$$MCl + F_2 \xrightarrow{300^\circ C} M[ClF_4] \quad \text{(g)}$$

can take place.

(J. SHAMIR)

1. J. Shamir, *Israel J. Chem.*, *17*, 37 (1978).
2. A. I. Popov, in *Halogen Chemistry*, Vol. 1, V. Gutmann, ed., Academic Press, New York 1967, p. 225.
3. A. A. Opalovskii, *Russ. Chem. Rev.* (*Engl. Transl.*), *36*, 711 (1967).
4. H. Meinert, *Z. Chem.*, *7*, 41 (1967).
5. A. J. Downs, C. J. Adams, in *Comprehensive Inorganic Chemistry*, A. F. Trotman-Dickenson, ed., Vol. 2., Pergamon Press, Oxford, 1973, p. 1534.
6. H. Zimmer, M. Jayquant, A. Amer, B. S. Ault, *Z. Naturforsch., Teil B*, *38B*, 103 (1983).
7. B. D. Stepin, S. B. Stepina, *Russ. Chem Rev.* (*Eng. Transl.*), *55* 812 (1986).

2.2.6.2. from Interhalogen Compounds with Halide Anion.

The most pronounced are reactions of interhalogens (except ClF_5 and IF_7) acting as halide acceptors with electron-pair donor bases, such as group-IA halides, or nitrosyl and nitryl halides serving as halide donors forming polyhalide anions:

$$MY + XY_n \rightarrow M[XY_{n+1}] \quad \text{(a)}$$

When performed in xs interhalogen as a solvent, the xs is pumped off after completion, leaving the solid product. A second noninteracting solvent, such as anhyd HF or CH_3CN, CCl_4 or dichloroethane, facilitates the process. Mixtures of halogen fluorides can be used for this purpose, e.g.:

$$ClF_3 + BrF_5 \rightleftharpoons [ClF_2]^+ + [BrF_6]^- \quad \text{(b)}$$

forms with RbF or CsF the $[BrF_6]^-$ salts, whereas with KF the product $K[ClF_4]$ is formed. Similarly:

$$ClF_3 + IF_5 \leftrightarrows [ClF_2]^+ + [IF_6]^- \quad \text{(c)}$$

forms the $K[IF_6]$ salt.

Although the bromine and iodine monofluorides are unstable, the related bromo and iodo difluoride anions can be prepared. Distilling equimolar Br_2 and BrF_3 on CsF:

$$3\ CsF + (Br_2 + BrF_3) \rightarrow 3\ Cs[BrF_2] \quad \text{(d)}$$
$$\rightleftharpoons 3\ BrF$$

However, the $[IF_2]^-$ anion is prepared only by exchange of a different polyhalide anion with a fluoride:

$$[Et_4N][ICl_2] + 2\ AgF \xrightarrow{CH_3CN} [Et_4N][IF_2] + 2\ AgCl \quad \text{(e)}$$

The $[IF_4]^-$ anion is prepared from the slightly stable IF_3, reacting directly on a fluoride anion at low T:

$$MF + IF_3 \xrightarrow[CH_3CN]{-50^\circ C} M[IF_4] \quad \text{(f)}$$

where M = K, Rb, Cs. With xs fluoride, a multivalent polyhalide anion is formed:

$$3\ CsF + IF_3 \xrightarrow[CH_3CN]{-40^\circ C} Cs_3[IF_6] \quad \text{(g)}$$

The $[IF_4]^-$ anion can also be prepared by reduction of IF_5 with an iodide anion:

$$5\ MI + 4\ IF_5 \rightarrow 5\ M[IF_4] + 2\ I_2 \quad \text{(h)}$$

The few interhalogens not containing fluorine, which are all diatomic except for I_2Cl_6, can be used to prepare numerous heteronuclear polyhalides. These form trihalide anions when reacting with a halide. Pentahalide anions can be formed when reacting with an already existing polyhalide anion. The anions can be prepared by:

$$Y^- + XY \rightarrow [XY_2]^- \quad \text{(i)}$$

$$X^- + XY \rightarrow [X_2Y]^- \quad \text{(j)}$$

$$X^- + YZ \rightarrow [XYZ]^- \quad \text{(k)}$$

$$X_3^- + XY \rightarrow [X_4Y]^- \quad \text{(l)}$$

$$[XY_2]^- + XY \rightarrow [X_2Y_3]^- \quad \text{(m)}$$

$$[XY_2]^- + XZ \rightarrow [X_2Y_2Z]^- \quad \text{(n)}$$

$$[XY_2] + YZ \rightarrow [XY_3Z]^- \quad \text{(o)}$$

$$[XYZ]^- + XY \rightarrow [X_2Y_2Z]^- \quad \text{(p)}$$

Iodine trichloride can also form pentahalide anions when reacting with a chloride anion to form $[ICl_4]^-$, with a bromine to form $[IBrCl_3]^-$, with a fluoride to form $[ICl_3F]^-$ and with an iodide to form $[I_2Cl_3]^-$ polyhalide anions.

Table 1 summarizes the solid polyhalide anions.

(J. SHAMIR)

1. J. Shamir, *Israel J. Chem.*, *17*, 37 (1978).
2. A. I. Popov, in *Halogen Chemistry*, Vol. 1, V. Gutmann, ed., Academic Press, New York 1967, p. 225.

TABLE 1. POLYHALIDE ANIONS, $[X_mY_nZ]^-$

Trihalides	Pentahalides	Heptahalides	Nonahalide	Other
I_3^-	I_5^-	I_7^-	I_9^-	I_8^{2-}
$[I_2Br]^-$	$[I_4Cl]^-$	$[I_6Br]^-$		
$[I_2Cl]^-$	$[I_4Br]^-$	$[IF_6]^-$		
$[IBr_2]^-$	$[I_2Br_3]^-$	$[Br_6Cl]^-$		
$[ICl_2]^-$	$[I_2Br_2Cl]^-$	$[BrF_6]^-$		
$[IBrCl]^-$	$[I_2BrCl_2]^-$			
$[IBrF]^-$	$[I_2Cl_3]^-$			
$[IF_2]^-$	$[IBrCl_3]^-$			
$[Br_3]^-$	$[ICl_4]^-$			
$[Br_2Cl]^-$	$[ICl_3F]^-$			
$[BrCl_2]^-$	$[IF_4]^-$			
$[BrF_2]^-$	$[BrF_4]^-$			
Cl_3^-	$[ClF_4]^-$			
$[ClF_2]^-$				

3. A. A. Opalovskii, *Russ. Chem. Rev. (Engl. Transl.)*, *36*, 711 (1967).
4. H. Meinert, *Z. Chem.*, *7*, 41 (1967).
5. A. J. Downs, C. J. Adams, in *Comprehensive Inorganic Chemistry*, A. F. Trotman-Dickenson, ed., Vol. 2., Pergamon Press, Oxford, 1973, p. 1534.
6. A. I. Popov, R. E. Buckles, *Inorg. Synth.*, *5*, 167 (1957).

2.2.6.3. Preparation of Anionic Polyhalides Containing Astatine.

Reactions of AtI with I^-, Br^-, and Cl^-, and reactions of AtBr with Br^- and Cl^- produce mixed polyhalide ions via the equilibria[1]:

$$AtI + I^- \rightleftharpoons [AtI_2]^- \tag{a}$$

$$AtI + Br^- \rightleftharpoons [AtIBr]^- \tag{b}$$

$$AtI + Cl^- \rightleftharpoons [AtICl]^- \tag{c}$$

$$AtBr + Br^- \rightleftharpoons [AtBr_2]^- \tag{d}$$

$$AtBr + Cl^- \rightleftharpoons [AtBrCl]^- \tag{e}$$

At low I^- concentration, $[AtBr_2]^-$ and $[AtCl_2]^-$ can be formed:

$$[AtIBr]^- + Br^- \rightleftharpoons [AtBr_2]^- + I^- \tag{f}$$

$$[AtICl]^- + Cl^- \rightleftharpoons [AtCl_2]^- + I^- \tag{g}$$

One or more anionic chlorocomplexes form[2,3] after oxidation of astatine with Cl_2, $[Cr_2O_7]^{2-}$, or hot $[S_2O_8]^{2-}$, followed by treatment with HCl. These complexes may be $[AtCl_2]^-$ and/or $[AtOCl_2]^-$. Bromo- and iodo- complexes can be formed by treating the oxidized astatine with HBr or HI[4]. On the basis of electromigration studies, it has been suggested $[AtX_2]^-$ anions are formed after oxidation with hot $[Cr_2O_7]^{2-}$, while $[AtOX_2]^-$ anions are formed after oxidation with hot $[S_2O_8]^{2-}$ or with XeF_4[4]. The complexes hydrolyze in alkaline solution[4].

(W. D. LEE)

1. E. H. Appelman, *J. Phys. Chem.*, *65*, 325 (1961).
2. H. M. Neumann, *J. Inorg. Nucl. Chem.*, *4*, 349 (1957).
3. Yu. V. Norseev, V. A. Khalkin, *J. Inorg. Nucl. Chem.*, *30*, 3239 (1968).
4. I. Dreyer, R. Dreyer, V. A. Khalkin, M. Milanov, *Radiochem. Radioanal. Lett.*, *40*, 145 (1979).

2.2.7. Preparation of Halogen Oxyfluorides

Halogen oxyfluorides are prepared by forming either halogen–fluorine or halogen–oxygen bonds. However, because this section deals exlusively with the formation of the halogen–halogen bond, preparative methods for halogen oxyfluorides based on the formation of halogen–oxygen bonds are not included here.

(K. O. CHRISTE)

2.2.7.1. from Halogen Oxides with F_2.

The interaction of F_2 with chlorine oxides, such as ClO_2[1–5], Cl_2O_6[6–7], or Cl_2O_7[8], produces $FClO_2$. Quantitative yields of $FClO_2$ are obtained from ClO_2 under moderate conditions. Inert solvents, such as $CFCl_3$, or diluents, such as air or N_2, are recommended because of **the explosive nature of the chlorine oxides**. For the higher chlorine oxides, the need for elevated T and the decreased yields of $FClO_2$ suggest that the primary step involves thermal decomposition to ClO_2, which is then fluorinated. Therefore, all these reactions involve the step:

$$2\ ClO_2 + F_2 \rightarrow 2\ FClO_2 \qquad \text{(a)}$$

In view of the shock sensitivity of chlorine oxides, none of the above methods is recommended for the large-scale production of $FClO_2$, and necessary safety precautions must be used. The recommended method for $FClO_2$ is from $NaClO_3$ with ClF_3 (see §2.2.7.6).

The low-T (−78°C) fluorination of Cl_2O with F_2 produces[9] ClF_3O. Depending on conditions, the byproducts can be either ClF or ClF_3:

$$2\ F_2 + Cl_2O \rightarrow ClF_3O + ClF \qquad \text{(b)}$$

$$3\ F_2 + Cl_2O \rightarrow ClF_3O + ClF_3 \qquad \text{(c)}$$

When no catalyst is used, or if KF and NaF are present as catalysts, ClF is the main byproduct. When the more basic alkali-metal fluorides RbF and CsF are used, ClF_3 is the coproduct. The formation of ClF_3 rather than ClF is associated with the more ready formation of $[ClF_2]^-$ intermediates with RbF and CsF. Yields of ClF_3O from Cl_2O vary and are affected by the alkali fluoride present. Yields of $>40\%$ are obtained and reach $>80\%$ using NaF or CsF. Because NaF does not form an adduct with ClF_3O[10], stabilization of the product by complex formation does not influence the ClF_3O yields.

Owing to unpredictable explosions experienced with liq Cl_2O, attempts are made to circumvent the Cl_2O isolation step, and the crude Cl_2O, still absorbed on the Hg(II) salts, is fluorinated directly. Again, ClF_3O is formed, but its yield is too low to make this synthetic route attractive.

The fluorination of solid Cl_2O to ClF_3O proceeds at −196°C provided the F_2 is suitably activated by such methods as glow discharge. Unactivated F_2 does not interact with Cl_2O at $-196°C$. The low yield of ClF_3O (1–2%) makes this modification impractical.

Because of the shock sensitivity of Cl_2O, its fluorination reaction is not the preferred method for the preparation of ClF_3O. **Replacement of Cl_2O by the more stable $ClONO_2$ results in a safer process** (see §2.2.7.5.).

By analogy with ClO_2, BrO_2 is fluorinated by F_2 to give $FBrO_2$ in high yield[11]:

$$2\ BrO_2 + F_2 \rightarrow 2\ FBrO_2 \tag{d}$$

However, direct reaction with F_2 is not practical because even at $-78°C$ the reaction is vigorous, resulting in **spontaneous decomposition of the BrO_2 and explosions.** Liquid Cl_2 or perfluoropentane is used as a diluent to moderate the reaction: BrO_2 is not soluble in these solvents and must be suspended. Again, **this method is not recommended for larger scale preparations of $FBrO_2$ and appropriate safety precautions must be taken.**

The fluorination of I_2O_5 with F_2 in anhyd HF may[12] result in FIO_2, but this claim is refuted by subsequent study[13], which shows that anhyd HF quantitatively converts I_2O_5 or FIO_2 to IF_5.

(K. O. CHRISTE)

1. H. Schmitz, H. J. Schumacher, *Z. Anorg. Allg. Chem., 249*, 238 (1942).
2. P. J. Aymonino, J. E. Sicre, H. J. Schumacher, *J. Chem. Phys., 22*, 756 (1954).
3. J. E. Sicre, H. J. Schumacher, *Z. Anorg. Allg. Chem., 286*, 232 (1956).
4. M. Schmeisser, F. L. Ebenhöch, *Angew Chem., 66*, 230 (1954).
5. W. Kwasnick, in *Handbook of Preparative Inorganic Chemistry*, G. Brauer, ed., 2nd ed., Vol. 1, Academic Press, New York, 1963, p. 1.
6. A. J. Arvia, P. J. Aymonino, *Spectrochim. Acta, 19*, 1449 (1963).
7. A. J. Arvia, W. H. Basualdo, H. J. Schumacher, *Z. Anorg. Allg. Chem., 286*, 58 (1956).
8. R. V. Figini, E. Goloccia, H. J. Schumacher, *Z. Phys. Chem. (Frankfurt am Main) (N.S.), 14*, 32 (1958).
9. D. Pilipovich, C. B. Lindahl, C. J. Schack, R. D. Wilson, K. O. Christe, *Inorg. Chem., 11*, 2189 (1972).
10. K. O. Christe, C. J. Schack, D. Pilipovich, *Inorg. Chem., 11*, 2205 (1972).
11. M. Schmeisser, E. Pammer, *Angew, Chem., 67*, 156 (1955).
12. M. Schmeisser, K. Brändle, *Adv. Inorg. Chem. Radiochem., 5*, 41 (1963).
13. H. Selig, U. El-Gad, *J. Inorg. Nucl. Chem., 35*, 3517 (1973).

2.2.7.2. from Halogen Oxides with Fluorinating Agents Other than F_2.

In the $FClO_2$ synthesis from ClO_2 and F_2, the latter can be replaced by other fluorinating agents: the passage of ClO_2, diluted by N_2, over AgF_2[1] or CoF_3[2] at RT produces pure $FClO_2$ in high yield:

$$AgF_2 + ClO_2 \rightarrow AgF + FClO_2 \tag{a}$$

The consumption of AgF_2 can be followed by the color change of AgF_2 (dark brown) to AgF (yellow).

Halogen fluorides can also fluorinate ClO_2 to $FClO_2$: e.g., the passage of ClO_2 through liq BrF_3 at 30°C proceeds[1]:

$$6\ ClO_2 + 2\ BrF_3 \rightarrow 6\ FClO_2 + Br_2 \tag{b}$$

Risk of explosion in the ClO_2-AgF_2 reaction is reduced when ClO_2 is replaced by the less dangerous Cl_2O. The yield of $FClO_2$ is 35%[3,4]. Similarly, Cl_2O can be fluorinated at −78°C with either ClF[5]:

$$2\ Cl_2O + ClF \rightarrow FClO_2 + 2\ Cl_2 \qquad (c)$$

or ClF_3O[6]:

$$Cl_2O + ClF_3O \rightarrow FClO_2 + 2\ ClF \qquad (d)$$

In these two reactions the unstable FClO molecule is formed as an intermediate but disproportionates to yield $FClO_2$ and ClF (see §2.2.7.4.).

Fluorination of Cl_2O_6 with BrF_3 or BrF_5[7], HF[8] and FNO_2[1] also yields principally $FClO_2$. The latter reaction is carried out in $CFCl_3$ at 0°C:

$$Cl_2O_6 + FNO_2 \rightarrow FClO_2 + [NO_2][ClO_4] \qquad (e)$$

These reactions are carried out at low T at which decomposition of Cl_2O_6 to $2\ ClO_2 + O_2$ can be excluded. They can be rationalized[2] by polarization of Cl_2O_6 to $[ClO_2]^+[ClO_4]^-$.

$$[ClO_2][ClO_4] + FNO_2 \rightarrow [NO_2][ClO_4] + FClO_2 \qquad (f)$$

The ionic structure of solid Cl_2O_6 is confirmed by vibrational spectroscopy[9].

Fluorination of BrO_2 to $FBrO_2$ is achieved[2,10] using BrF_5:

$$10\ BrO_2 + 2\ BrF_5 \rightarrow 10\ FBrO_2 + Br_2 \qquad (g)$$

The reaction is carried out at −55°C in liq BrF_5, and the $FBrO_2$ is separated from the Br_2 byproduct and xs BrF_5 by vacuum fractionation. This reaction can be simplified further by preparing[2,10,11] the BrO_2 in situ by passing O_3 through Br_2 in BrF_5:

$$BrF_5 + 2\ Br_2 + 10\ O_3 \rightarrow 5\ FBrO_2 + 10\ O_2 \qquad (h)$$

When I_2O_5 is dissolved in boiling IF_5, white hygroscopic needles of IF_3O separate on cooling[12–14]:

$$I_2O_5 + 3\ IF_5 \rightarrow 5\ IF_3O \qquad (i)$$

A modification of this reaction is used to prepare[15] KIF_4O:

$$5\ KF + I_2O_5 + 3\ IF_5 \rightarrow 5\ KIF_4O \qquad (j)$$

(K. O. CHRISTE)

1. M. Schmeisser, K. Fink, *Angew. Chem.*, *69*, 780 (1957).
2. M. Schmeisser, K. Brändle, *Adv. Inorg. Chem. Radiochem.*, *5*, 41 (1963).
3. Y. Macheteau, J. Gillardeau, *Bull. Soc. Chim. Fr.*, 4075 (1967).
4. J. Gillardeau, Y. Macheteau, Fr. Pat. 1,527,112 (1968), *Chem. Abstr.*, *71*, 23,395 (1969).
5. K. O. Christe, *Inorg. Chem.*, *11*, 1220 (1972).
6. C. J. Schack, C. B. Lindahl, D. Pilipovich, K. O. Christe, *Inorg. Chem.*, *11*, 2201 (1972).
7. R. Weiss, Ph.D. Thesis, Techn. Univ., Aachen, Germany (1959).
8. M. Schmeisser, *Angew. Chem.*, *67*, 493 (1955).
9. A. C. Pavia, J. L. Pascal, A. Potier, *C.R. Hebd. Seances. Acad. Sci., Ser. C*, *272*, 1492 (1971).
10. M. Schmeisser, E. Pammer, *Angew. Chem.*, *69*, 781 (1957).
11. M. Schmeisser, L. Taglinger, *Chem. Ber.*, *94*, 1533 (1961).
12. E. E. Aynsley, R. Nichols, P. L. Robinson, *J. Chem. Soc.* 623 (1953).
13. E. E. Aynsley, *J. Chem. Soc.*, 2425 (1958).

14. E. E. Aynsley, M. L. Hair, *J. Chem. Soc.*, 3747 (1958).
15. K. O. Christe, R. D. Wilson, E. C. Curtis, W. Kuhlmann, W. Sawodny, *Inorg. Chem.*, *17*, 533 (1978).

2.2.7.3. from Halogen Oxyfluorides with Fluorinating Agents.

The fluorination of a chlorine oxyfluoride to one of higher oxidation state is difficult owing to the scarcity of stable low (+III) and high (+VII) oxidation state oxyfluorides. One example[1,2] is the fluorination of $FClO_2$ by the strong oxidizer, PtF_6:

$$2\ FClO_2 + 2\ PtF_6 \rightarrow [ClF_2O_2][PtF_6] + [ClO_2][PtF_6] \qquad (a)$$

Several side reactions compete, and the yield of $[ClF_2O_2]^+$ varies with changes in conditions. The $[ClF_2O_2][PtF_6]$ can be converted to ClF_3O_2 by displacement using FNO_2:

$$[ClF_2O_2][PtF_6] + FNO_2 \rightarrow [NO_2][PtF_6] + ClF_3O_2 \qquad (b)$$

Oxygen and fluorine ligands scramble in the synthesis of ClF_3O from mixtures of chlorine-, fluorine-, and oxygen-containing starting materials using W-photolysis[3–5]. A claim[5] for the formation of ClF_5O in the UV-photolysis of the ClF_5–OF_2 system cannot be confirmed[6].

Kinetic study of the photolyses of the ClF_3–O_2 and Cl_2–F_2–O_2 systems[7] shows that contrary to the original report[3], the rate of ClF_3O formation is the same for both systems, increasing with O_2 concentration, and independent of irradiation time. Furthermore, the rate of ClF_3O formation is proportional to the intensity of the 184.7 nm band of the Hg spectrum indicating that the dissociation of O_2 to two ground-state, ^3P, oxygen atoms is the primary photochemical process. The mechanism requires photochemical dissociation of ClF_3 as well:

$$O_2 \xrightarrow{h\nu(184.7\ nm)} O + O \qquad (c)$$

$$ClF_3 \xrightarrow{h\nu(200\text{–}350\ nm)} ClF_2 + F \qquad (d)$$

$$O + ClF_2 \rightarrow ClF_2O \qquad (e)$$

$$ClF_2O + F_2 \rightarrow ClF_3O + F \qquad (f)$$

In the photolysis of ClF_3 under similar conditions, a photochemical steady state is quickly achieved where $(F_2) = (ClF) = \alpha(ClF_3)$, and α has a value of ≈ 1 at low and of ≈ 3 at high pressures. These results, together with the known photochemical decomposition[8] of OF_2, explain why ClF_3O can be generated by the photolysis of so many different starting materials, including the halogen oxyfluorides $FClO_2$, $FClO_3$ and IF_5O.

For bromine oxyfluorides, fluorinations use the powerful oxidizers PtF_6 and KrF_2. With PtF_6[9]:

$$2\ FBrO_2 + 5\ PtF_6 \rightarrow 2\ [BrF_2O][PtF_6] + [O_2][PtF_6] + 2\ PtF_5 \qquad (g)$$

$$PtF_5 + FBrO_2 \rightarrow [BrO_2][PtF_6] \qquad (h)$$

$$2\ PtF_6 + 2\ FBrO_2 \rightarrow 2\ [BrO_2][PtF_6] + F_2 \qquad (i)$$

but there is no evidence for $[BrO_2F_2]^+$. Similarly, KrF_2 with either $FBrO_2$, BrF_3O or $FBrO_3$[10–12] does not produce any novel bromine(+VII) compounds, but proceeds:

$$2\ FBrO_2 + 2\ KrF_2 \rightarrow 2\ BrF_3O + 2\ Kr + O_2 \qquad (j)$$

$$2\ BrF_3O + 2\ KrF_2 \rightarrow 2\ BrF_5 + 2\ Kr + O_2 \qquad (k)$$

Iodine(+V) oxyfluorides or oxide are fluorinated by BrF_5 to yield $FBrO_2$ and IF_5:

$$2\ I_2O_5 + 5\ BrF_5 \rightarrow 5\ FBrO_2 + 4\ IF_5 \qquad (l)$$

$$FIO_2 + BrF_5 \rightarrow FBrO_2 + IF_5 \qquad (m)$$

$$2\ IF_3O + BrF_5 \rightarrow FBrO_2 + 2\ IF_5 \qquad (n)$$

but heptavalent IF_3O_2 or its AsF_5 or SbF_5 adducts are fluorinated to give IF_5O[11]:

$$2\ IF_3O_2 + BrF_5 \rightarrow 2\ IF_5O + FBrO_2 \qquad (o)$$

$$IF_3O_2 \cdot AsF_5 + BrF_5 \rightarrow IF_5O + [BrF_2O][AsF_6] \qquad (p)$$

The addition of HF to IF_3O_2 forms OIF_4OH reversibly[13,14]:

$$IF_3O_2 + HF \underset{+SO_3}{\overset{+HF}{\rightleftharpoons}} OIF_4OH \qquad (q)$$

and is the addition of HF across an X=O double bond to form a new X—F bond.

(K. O. CHRISTE)

1. K. O. Christe, *Inorg. Chem., 12*, 1580 (1973).
2. K. O. Christe, R. D. Wilson, *Inorg. Chem., 12*, 1356 (1973.
3. D. Pilipovich, H. H. Rogers, R. D. Wilson, *Inorg Chem., 11*, 2192 (1972).
4. R. Bougon, J. Isabey, P. Plurien, *C.R. Hebd. Seances Acad. Sci., Ser. C, 271*, 1366 (1970).
5. K. Züchner, O. Glemser, *Angew. Chem., Int. Ed. Engl., 11*, 1094 (1972).
6. K. O. Christe, C. J. Schack, *Adv. Inorg. Chem. Radiochem., 18*, 319 (1976).
7. A. E. Axworthy, K. H. Mueller, R. D. Wilson, *Photochemistry of Interhalogen Compounds of Interest as Rocket Propellants*, Final Report, Contract No. AFOSR-TR-73-2183 (1973).
8. R. Gatti, E. Staricco, J. E. Sicre, H. J. Schumacher, *Z. Phys. Chem. (Frankfurt am Main) (N.S.), 23*, 164 (1960).
9. M. Adelheim, E. Jacob, *Angew. Chem., Int. Ed. Engl., 16*, 461 (1977).
10. R. J. Gillspie, P. Spekkens, *J. Chem. Soc., Dalton Trans.*, 1539 (1977).
11. R. J. Gillespie, P. H. Spekkens, *Isr. J. Chem., 17*, 11 (1978).
12. R. Bougon, T. Bui Huy, P. Charpin, R. J. Gillspie, P. H. Spekkens, *J. Chem. Soc., Dalton Trans.*, 6 (1979).
13. A. Engelbrecht, P. Peterfy, *Angew. Chem., Int. Ed. Engl., 8*, 768 (1969).
14. A. Engelbrecht, P. Peterfy, E. Schandara, *Z. Anorg. Allg. Chem., 384*, 202 (1971).

2.2.7.4. by Disproportionation of Halogen Oxyfluorides.

Thermally unstable FClO disproportionates at RT[1,2]:

$$2\ FClO \rightarrow FClO_2 + ClF \qquad (a)$$

This reaction explains the formation of $FClO_2$ and ClF where FClO is the expected product (see §2.2.7.2.).

Iodine trifluoride oxide is stable at RT but at 100°C undergoes a reversible change[3–5] to IF_5 and FIO_2:

$$2\ IF_3O \rightleftharpoons IF_5 + FIO_2 \qquad (b)$$

This reaction is also involved in the thermal and photochemical decomposition[6,7] of IF_3O_2:

$$2\ IF_3O_2 \rightarrow 2\ IF_3O + O_2 \tag{c}$$

$$2\ IF_3O \rightarrow IF_5 + FIO_2 \tag{d}$$

(K. O. CHRISTE)

1. K. O. Christe, C. J. Schack, *Adv. Inorg. Chem. Radiochem.*, *18*, 319 (1976).
2. T. D. Cooper, F. N. Dost, C. H. Wang, *J. Inorg. Nucl. Chem.*, *34*, 3564 (1972).
3. E. E. Aynsley, R. Nichols, P. L. Robinson, *J. Chem. Soc.*, *623* (1953).
4. E. E. Aynsley, *J. Chem. Soc.*, *2425* (1958).
5. E. E. Aynsley, M. L. Hair, *J. Chem. Soc.*, *3747* (1958).
6. A. Englebrecht, P. Peterfy, *Angew. Chem., Int. Ed. Engl.*, *8*, 768 (1969).
7. A. Engelbrecht, P. Peterfy, E. Schandara, *Z. Anorg. Allg. Chem.*, *384*, 202 (1971).

2.2.7.5. from Positive Halogen Compounds with Fluorinating Agents.

For the synthesis of ClF_3O, fluorination of $ClONO_2$ by F_2 at $-35°C$[1] is best:

$$2\ F_2 + ClONO_2 \rightarrow ClF_3O + FNO_2 \tag{a}$$

Contrary to the chlorine oxides, $ClONO_2$ is not shock sensitive. Other advantages include (a) less F_2 is required than in the fluorination of Cl_2O, which yields ClF_3 as coproduct; (b) volatility differences of the products FNO_2 and ClF_3O ($\Delta T_{bp} \approx 100°C$) permit separation by fractional condensation; (c) $ClONO_2$ can be prepared more conveniently; and (d) yields of ClF_3O are higher.

In the fluorination of $ClONO_2$, side reactions compete with the fluorination step. These arise from thermal decomposition of the starting material owing to inefficient removal of the heat of reaction. Hence, the rate of the competing reactions is affected by T. At $\gtrsim$RT, decomposition of the hypochlorite is favored and little or no ClF_3O is formed, resulting in rapid, uncontrolled reactions. Thermal decomposition preceding the fluorination step yields only intermediates incapable of producing ClF_3O. Thus, in order to maximize fluorination, long times at $T < 0°C$ are indicated.

Similarly, $ClOSO_2F$ interacts[2] with ClF_3O:

$$2\ ClOSO_2F + ClF_3O \rightarrow FClO_2 + 2\ ClF + S_2O_5F_2 \tag{b}$$

and

$$ClOSO_2F + ClF_3 \rightarrow FClO_2 + ClF + SO_2F_2 \tag{c}$$

These reactions can be rationalized in terms of a reduction of ClF_3O to the unstable FClO, which decomposes to $FClO_2$ and ClF.

(K. O. CHRISTE)

1. D. Pilipovich, C. B. Lindahl, C. J. Schack, R. D. Wilson, K. O. Christe, *Inorg. Chem.*, 11, 2189 (1972).
2. C. J. Schack, C. B. Lindahl, D. Pilipovich, K. O. Christe, *Inorg. Chem.*, *11*, 2201 (1972).

2.2.7.6. from Halogen Oxyacids and Their Salts with Fluorinating Agents.

Whereas fluorination of halogen oxyacids results in the formation of the corresponding fluorooxy compounds:

$$HOClO_3 + F_2 \rightarrow FOClO_3 + HF \qquad (a)$$

fluorination of the salts of these halogen oxyacids is a more useful method for halogen oxyfluorides. The halogen oxyfluoride product depends on the starting material and the conditions. Thus the fluorination of perchlorates is a high yield synthesis of perchloryl fluoride. Heating $KClO_4$ to 70°–120°C in xs SbF_5 produces $FClO_3$ in 50% yield[2]. The yield of $FClO_3$ can be increased to 90% and the T lowered to 20°–50°C, when a mixture of HF–SbF_5 is used[3,4]. Slightly lower yields are obtained when the HF solvent is replaced by AsF_3, IF_5, or BrF_5.

Most of the commercial processes are based on[5] $HOSO_2F$. Evolution of $FClO_3$ starts at 50°C and goes to completion at 85°–110°C. The yields of $FClO_3$ vary from 50 to 80%[5–8] and, if necessary, the $HOSO_2F$ can be regenerated[5]. The reaction can be carried out in glass apparatus. Such additives as 5–25% SbF_3 to the $HOSO_2F$ increase the yield of $FClO_3$ to 90% but hinder regeneration of $HOSO_2F$[9]. Addition of HF–BF_3 increases the $FClO_3$ yield to 85% but requires elevated pressure. Zinc, Al, Ag, and Pb fluorides decrease the yield of $FClO_3$.

Highest yields of perchloryl fluoride (97%) are achieved with fluorosulfonic acid and SbF_5 (superacid) as fluorinating medium. Potassium, Na, Li, Mg, Ba, Ca, and Ag perchlorates and perchloric acid itself undergo the reaction. Commercial reagents are used, and their additional purification is not necessary; unlike previous methods, this preparation of perchloryl fluoride can be carried out at RT. At 100°–135°C reaction time is 1–10 min, which allows the process to be carried out continuously in a packed column. The purity of product obtained after the usual purification reaches ≥98%; air and CO_2 are present as trace impurities[10].

The reaction mechanism between $[ClO_4]^-$ and superacids is not established[3,4,11,15]. Based on present understanding of superacid chemistry[16,18] and of complex formation of $FClO_3$, $[ClO_3]^+$ is not an intermediate. Furthermore, the high yields of $FClO_3$ (≲97%) would be surprising in view of the instability of $[ClO_3]^+$. Other mechanisms involving protonated perchloric acid[19] are more plausible:

$$4\,HF + 2\,SbF_5 \rightarrow 2\,[H_2F]^+ + 2\,[SbF_6]^- \qquad (b)$$

$$2\,[H_2F]^+ + [ClO_4]^- \rightarrow [H_2OClO_3]^+ + 2\,HF \qquad (c)$$

$$[H_2OClO_3]^+ + HF \rightarrow FClO_3 + [H_3O]^+ \qquad (d)$$

$$[ClO_4]^- + 3\,HF + 2\,SbF_5 \rightarrow FClO_3 + [H_3O]^+ + 2\,[SbF_6]^- \qquad (e)$$

Other syntheses of $FClO_3$ from metal perchlorates include the electrolysis of a saturated solution of $NaClO_4$ in anhyd HF, with a current efficiency of 10%[20], and the fluorination of NO_2ClO_4 by ClF_3 at RT, which results in the formation of $FClO_3$ and smaller amounts of $FClO_2$, ClO_2 and $ClNO_2$[21]. The corresponding reaction of $KClO_4$ with BrF_3 yields $FClO_2$ in 97% yield[21]:

$$3\,KClO_4 + 5\,BrF_3 \rightarrow 3\,KBrF_4 + Br_2 + 3\,O_2 + 3\,FClO_2 \qquad (f)$$

If the $MClO_4$ starting material is replaced by $MClO_3$, the main fluorination product is $FClO_2$, e.g., the reaction of $NaClO_3$ with equimolar ClF_3 produces $FClO_2$ in high yield[22]. This method is based on the reaction of ClF_3 gas with $KClO_3$ to give $FClO_2$ in high yield.[20–23] The substitution of $KClO_3$ by $NaClO_3$ is significant because the product NaF does not form an adduct with ClF_3, whereas KF does. This decrease by 60% the amount of ClF_3 required. By analogy with the $KClO_3$ reaction with BrF_3[24], the idealized stoichiometry is:

$$6\ NaClO_3 + 4\ ClF_3 \rightarrow 6\ NaF + 2\ Cl_2 + 3\ O_2 + 6\ FClO_2 \qquad (g)$$

The use of ClF_3 in slight xs avoids the formation of shock-sensitive chlorine oxides. The $KClO_3$ reaction with BrF_3[24]:

$$6\ KClO_3 + 10\ BrF_3 \rightarrow 6\ KBrF_4 + 2\ Br_2 + 3\ O_2 + 6\ FClO_2 \qquad (h)$$

also produces $FClO_2$ in high yield, but it is difficult to obtain pure, colorless $FClO_2$ by this method.

The action of F_2 on $KClO_3$ is not synthetically useful for $FClO_2$ owing to the large amounts of $FClO_3$ always formed[20,25–27].

Low T fluorination of $NaClO_2$ with F_2 produces $FClO_2$ as the main product, however, small amounts of ClF_3O are also obtained in addition to ClF_3, ClF and Cl_2[28].

Fluorination of alkali-metal perbromates with HF and electron-pair acceptor acids is analogous to that of the perchlorates and produces $FBrO_3$ in high yield. The reactions are carried out in HF using SbF_5[29], AsF_5, BrF_5 or $[BrF_6][AsF_6]$[30]:

$$KBrO_4 + 2\ AsF_5 + 3\ HF \rightarrow FBrO_3 + [H_3O][AsF_6] + KAsF_6 \qquad (i)$$

$$2\ KBrO_4 + BrF_5 + 2\ HF \rightarrow 2\ FBrO_3 + FBrO_2 + 2\ KHF_2 \qquad (j)$$

$$2\ KBrO_4 + 2\ [BrF_6][AsF_6] \rightarrow 2\ FBrO_3 + 2\ BrF_5 + O_2 + 2\ KAsF_6 \qquad (k)$$

In the absence of HF, $CsBrO_4$ reacts[31] with BrF_5 and F_2 at RT to produce solid $CsBrF_4O$, with $FBrO_3$ and $FBrO_2$ as the volatile products. Potassium perbromate is less reactive than the Cs salt and requires prolonged heating at 80°C to achieve conversion to $KBrF_4O$. In the absence of F_2, the conversion of $CsBrO_4$ to $CsBrF_4O$ is slow, even at 80°C, and is not catalyzed by HF.

The reaction of $KBrO_3$ with BrF_5 is complex. According to the original report[32] $KBrO_3$ reacts with BrF_5 at −50°C:

$$2\ KBrO_3 + 2\ BrF_5 \rightarrow 2\ KBrF_4 + 2\ FBrO_2 + O_2 \qquad (l)$$

Subsequent work[33] shows that $KBrO_3$ reacts with equimolar BrF_5 at RT:

$$2\ KBrO_3 + 2\ BrF_5 \rightarrow 2\ KBrO_2F_2 + 2\ BrF_3 + O_2 \qquad (m)$$

slowly when xs BrF_5 is used; 1:2.5 $KBrO_3$ and BrF_5 does not react at RT$FBrO_2$[34], but a rapid reaction occurs when catalytic HF is added. The solid product of this reaction consists of $KBrO_2F_2$ and $KBrF_4O$; $FBrO_2$ is the volatile product. The formation[35] of $KBrF_4O$ is shown by using xs BrF_5 and F_2 at 80°C, which results in quantitative formation of $KBrF_4O$. This reaction can be difficult to duplicate, however, and can result in the formation[31] of $KBrF_4$. The $KBrO_3$ reaction with BrF_5 is further modified[34] by reacting $KBrF_6$ with $KBrO_3$ in CH_3CN:

$$KBrF_6 + KBrO_3 \xrightarrow{CH_3CN} KBrO_2F_2 + KBrF_4O \qquad (n)$$

The $KBrF_4O$ produced is soluble in CH_3CN, whereas $KBrO_2F_2$ is not, and the two products can be separated by extraction with CH_3CN.

The reactions of BrF_5 with $[BrO_3]^-$ or $[BrO_4]^-$ involve oxygen-fluorine exchange. From the quantitative yields of $[BrF_4O]^-$, a free-radical mechanism involving the addition of oxygen atoms to bromine fluorides is unlikely. The mechanism involves addition of BrF_5 or $[BrF_6]^-$ across a Br=O double bond of $[BrO_4]^-$ of $[BrO_3]^-$, followed by $FBrO_3$ or $FBrO_2$ elimination with $[BrF_4O]^-$ formation.

Periodates e.g., $Ba_3H_4(IO_6)_2$, are fluorinated, by HSO_3F to tetrafluoroorthoperiodic acid[36]:

$$Ba_3H_4(IO_6)_2 \xrightarrow{HSO_3F} HOIF_4O \tag{o}$$

The same product is also formed from $HOIO_3$ or $NaIO_4$ in anhyd HF [37]. When $CsIO_4$ is repeatedly treated with anhyd HF, and the solvent is pumped away, the less volatile acid, $HOIF_4O$, displaces HF from the $CsHF_2$ with quantitative formation of $CsIF_4O_2$ (cis:trans $\approx 2:1$)[38]. This is a convenient synthesis of $[IF_4O_2]^-$ salts:

$$CsIO_4 + 6\ HF \xrightarrow{HF} CsHF_2 + HOIF_4O + 2\ H_2O \tag{p}$$

$$CsHF_2 + HOIF_4O \xrightarrow{HF} CsIF_4O_2 + 2\ HF \tag{q}$$

With F_2 at 60°C, $CsIO_4$ is converted[38] mainly to $CsIF_8$ and $CsIF_6$, but the products also contain lesser amounts of $CsIF_4O$ and cis- and trans-$CsIF_4O_2$. With xs ClF_5 at RT, $CsIO_4$ is slowly converted to $CsIF_8$, trans-$CsIF_4O_2$ and some $CsIF_4O$. With the more reactive fluorinating agent, ClF_3, complete conversion of $CsIO_4$ is obtained at RT:

$$3\ CsIO_4 + 11\ ClF_3 \rightarrow 6\ FClO_2 + 3\ ClF + 2\ CsClF_4 + CsIF_6 \cdot 2\ IF_5 \tag{r}$$

With BrF_5, the main reaction is:

$$CsIO_4 + 2\ BrF_5 \rightarrow CsIF_4O_2 + 2\ BrF_3O \tag{s}$$

This reaction is analogous to that[39] for KIO_4 and IF_5, i.e.:

$$KIO_4 + 2\ IF_5 \rightarrow KIF_4O_2 + 2\ IF_3O \tag{t}$$

and produces trans-$[IF_4O_2]^-$. Under the conditions (120°C, vacuum) used for the removal of xs IF_5, IF_3O disproportionates:

$$2\ IF_3O \rightarrow FIO_2 + IF_5 \tag{u}$$

resulting in FIO_2 and KIF_4O_2 as the final products. Compared to the IF_5 reaction, the BrF_5 synthesis offers the advantage that the BrF_3O and BrF_3 byproducts are volatile and can be pumped off easily. However the resulting solid product is contaminated with $CsBrF_4$.

Iodates are fluorinated but, whereas the fluorination of HIO_3 in aq HF forms the $[IO_2F_2]^-$ anion[40], solutions of $NaIO_3$ in anhyd HF contain IF_5[41]. With IF_5, MIO_3 or KIO_2F_2 produces the corresponding $[IF_4O]^-$ salts[42]:

$$MIO_3 + 2\ MF + 2\ IF_5 \rightarrow 3\ MIF_4O \qquad (M = Cs, K) \tag{v}$$

$$KIO_2F_2 + KF + IF_5 \rightarrow 2\ KIF_4O \tag{w}$$

(K. O. CHRISTE)

1. G. H. Rohrback, G. H. Cady, *J. Am. Chem. Soc.*, *69*, 677 (1947).
2. A. Engelbrecht, US Pat. 2,942,947 (1960); *Chem. Abstr.*, *54*, 21,681 (1960).
3. C. A. Wamser, W. B. Fox, D. Gould, B. Sukornick, *Inorg. Chem.*, *7*, 1933 (1968).
4. C. A. Wamser, B. Sukornick, W. B. Fox, D. Gould, *Inorg. Synth.*, *14*, 29 (1973).
5. G. Barth-Wehrenalp, *J. Inorg. Nucl. Chem.*, *2*, 266 (1956).
6. G. Barth-Wehrenalp, US Pat. 2,942,948 (1960), *Chem. Abstr.*, *54*, 21,681c (1960).
7. W. Lalande, US Pat. 2,982,617 (1961), *Chem. Abstr.*, *55*, 16,926 (1961).
8. W. Kwasnick, in *Handbook of Preparative Inorganic Chemistry*, G. Brauer, ed., 2nd ed., Vol. 1, Academic Press, New York, 1963, p. 1.
9. D. Dess, US Pat. 2,982,618 (1961); *Chem. Abstr.*, *55*, 16,927 (1961).
10. G. Barth-Wehrenalp, H. Mandell, US Pat. 2,942,949 (1960), *Chem. Abstr. 54*, 21,681 (1960).
11. J. Barr, R. J. Gillespie, R. C. Thompson, *Inorg. Chem.*, *3*, 1149 (1964).
12. G. Barth-Wehrenalp, *J. Inorg. Nucl. Chem.*, *4*, 374 (1957).
13. K. Lang, Ph.D. Thesis, Univ. Munich, Germany, 1956.
14. A. A. Woolf, *J. Inorg. Nucl. Chem.*, *3*, 250 (1956).
15. M. Schmeisser, K. Brändle, *Adv. Inorg. Chem. Radiochem.*, *5*, 41 (1963).
16. K. O. Christe, C. J. Schack, R. D. Wilson, *Inorg. Chem.*, *14*, 2224 (1975).
17. R. J. Gillespie, *Acc. Chem. Res.*, *1*, 202 (1968).
18. G. A. Olah, A. M. White, D. H. O'Brien, *Chem. Rev.*, *70*, 561 (1970).
19. K. Lang, Diplom. Theses, Univ. Munich, Germany (1955).
20. A. Engelbrecht, H. Atzwanger, *Monatsh. Chem.*, *83*, 1087 (1952); *J. Inorg. Nucl. Chem.*, *2*, 348 (1956).
21. A. W. Beardell, C. J. Grelecki, US Pat. 3,404,958 (1968), *Chem. Abstr.*, *69*, 108,231 (1968).
22. K. O. Christe, R. D. Wilson, C. J. Schack, *Inorg. Nucl. Chem. Lett.*, *11*, 161 (1975).
23. D. F. Smith, G. M. Begun, W. H. Fletcher, *Spectrochim. Acta*, *20*, 1763 (1964).
24. A. A. Woolf, *J. Chem. Soc.*, 4113 (1954).
25. H. Bode, E. Klesper, *Angew. Chem.*, *66*, 605 (1954).
26. A. Engelbrecht, *Angew. Chem.*, *66*, 442 (1954).
27. J. E. Sicre, H. J. Schumacher, *Angew. Chem.*, *69*, 226 (1957).
28. D. Pilipovich, C. B. Lindahl, C. J. Schack, R. D. Wilson, K. O. Christe, *Inorg. Chem.*, *11*, 2189 (1972).
29. E. H. Appelman, M. H. Studier, *J. Am. Chem. Soc.*, *91*, 4561 (1969).
30. P. Spekkens, Ph.D. Thesis, McMaster Univ., Hamilton, Ontario, Canada, 1977.
31. K. O. Christe, R. D. Wilson, E. C. Curtis, W. Kuhlmann, W. Sawodny, *Inorg. Chem.*, *17*, 533 (1978).
32. M. Schmeisser, E. Pammer, *Angew. Chem.*, *69*, 781 (1957).
33. R. Bougon, G. Tantot, *C.R. Hebd. Seances Acad. Sci., Ser. C*, *281*, 271 (1975).
34. R. J. Gillespie, P. Spekkens, *J. Chem. Soc., Dalton Trans.*, 2391 (1976).
35. R. Bougon, T. Bui Huy, P. Charpin, G. Tantot, *C.R. Hebd. Seances Acad. Sci., Ser. C*, *283*, 71 (1976).
36. A. Engelbrecht, P. Peterfy, *Angew. Chem., Int. Ed. Engl.*, *8*, 768 (1969).
37. A. Engelbrecht, P. Peterfy, E. Schandara, *Z. Anorg. Allg. Chem.*, *384*, 202 (1971).
38. H. Selig, U. Elgad, *J. Inorg. Nucl. Chem. Suppl.*, 91 (1976).
39. K. O. Christe, R. D. Wilson, C. J. Schack, *Inorg. Chem.*, *20*, 2104 (1981).
40. R. J. Gillespie, J. P. Krasznai, *Inorg. Chem.*, *16*, 1384 (1977).
41. J. B. Milne, D. Moffett, *Inorg. Chem.*, *14*, 1077 (1975).
42. H. Selig, U. El-Gad, *J. Inorg. Nucl. Chem.*, *35*, 3517 (1973).
43. J. B. Milne, D. M. Moffett, *Inorg. Chem.*, *15*, 2165 (1976).

2.2.7.7. by Elimination.

Elimination of halogen oxyfluorides is involved in the CsF-catalyzed $FClO_3$ elimination of fluorocarbon perchlorates[1]:

$$R_fCF_2OClO_3 \xrightarrow{CsF} R_fCFO + FClO_3 \qquad (a)$$

which proceeds at 60°C by attack of the CF_2 carbon atom by the fluoride anion, followed by an interval nucleophilic displacement reaction and $FClO_3$ elimination:

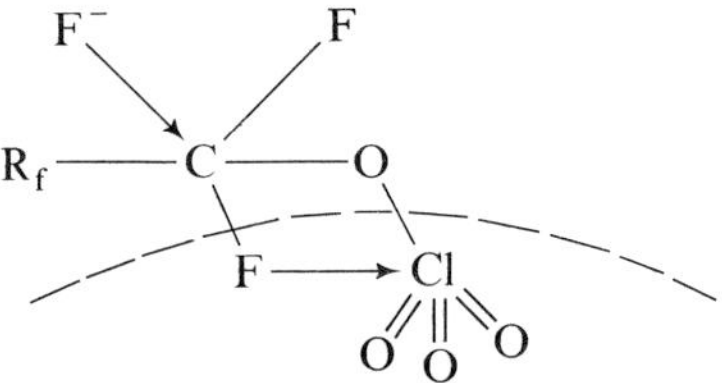

The reactions of BrF_5 with $[BrO_3]^-$ or $[BrO_4]^-$ (see §2.2.7.6.) may also involve similar intermediates that decompose with $FBrO_2$ or $FBrO_3$ elimination, but these intermediates are not isolated.

(K. O. CHRISTE)

1. C. J. Schack, D. Pilipovich, J. F. Hon, *Inorg. Chem., 12*, 897 (1973).

2.3. Formation of Bonds between Halogens and Group-VB (O, S, Se, Te, Po) Elements

2.3.1. Introduction

Formation of the halides of the group-VIB elements results from direct combination with the halogens themselves or with hydrogen or other inorganic or organic halides. The process also can proceed by cleavage of a group-VIB hydride or other compound containing bonds to an element drawn from the main groups headed by carbon, oxygen, nitrogen, etc. Further oxidative additions to low oxidation state group-VIB compounds takes place with the halogens, hydrogen or other inorganic or organic halides. Metatheses exchange one halide for another.

(J. ZUCKERMAN)

2.3.2. from the Elements

2.3.2.1. from Halogenation by Elemental Halogens

2.3.2.1.1. to Give the Oxygen–Halogen Bond.

The known, binary oxygen–fluorine species are O_nF_2 (n = 2–6) and O_nF (n = 1–4), but reliable evidence exists only for[1–3] OF_2, OF, O_2F_2, O_2F and O_4F_2. Except OF_2, all species are unstable at RT, but the radicals OF and O_2F and the neutral peroxides O_2F_2 and O_4F_2 are characterized by spectroscopy and decomposition reactions.

Because of the instability of the O—F bond, no oxygen fluorides are formed in thermal reactions. When an O_2-F_2 mixture is heated to 240°–500°C in a flow system, however, and the reaction mixture is quenched to −196°C, $[O_2F]^{\cdot}$ radicals are obtained[4].

For OF_2 the best method of preparation is the cleavage of the O—H bond by F_2 (see §2.3.3.), but it is also produced in silent electrical discharges from an O_2-F_2 mixture[5]. Usually, under these and similar conditions (glow discharge), the O—O bond remains intact and $O_{2n}F_2$ species are formed. The reaction of O_2 with F_2 occurs on passing an electrical discharge through an O_2-F_2 mixture at low pressure, with the discharge vessel maintained at low T:[6]

$$O_2 + F_2 \xrightarrow{\text{discharge}} O_2F_2 + O_4F_2 + O_2F \tag{a}$$

In addition, O_4F_2 is formed[7], which is in equilibrium with the $[O_2F]^{\circ}$ radical[8–10]. The existence of O_3F_2[11], O_5F_2 and O_6F_2[12] is questionable[2]. The O_2F_2-O_4F_2 ratio is T dependent, as a result of O_2F, being more thermally stable[13–15].

Similarly, radiolysis of liq O_2-F_2 mixtures at 77 K with 3 MeV radiation forms

O_2F_2 and O_4F_2, with small amounts of higher oxides[16]. Ultraviolet irradiation of liq O_2-F_2 forms O_2F_2[17–19]:

$$O_2 + F_2 \xrightarrow[-196°C]{UV} O_2F_2 \tag{b}$$

In a vol of 100 mL, >1 mol of O_2F_2 can be prepared. This reaction is also useful for purifying FO_2. Neither Cl_2 nor Br_2 react with O_2, but at 325°C and with a partial pressure of O_2 of 1.2×10^8 Pa (1200 atm), I_2 yields 2.3% diiodine pentoxide[21].

The action of oxygen atoms on Cl_2 gives ClO and ClO_2 as primary products, in a reaction of known kinetics[22], Cl_2O_6 is also formed[23], but chlorine oxides are much better prepared by less direct methods or from O_3 and Cl_2 (see §2.3.2.5.). Whereas BrO_2 is isolated at low T after mixtures of Br_2 vapor and O_2 are subjected to electrical discharge[21], other methods are more useful for O—Br bond formation.

(R. MEWS)

1. E. A. V. Ebsworth, J. A. Connor, J. J. Turner, in *Comprehensive Inorganic Chemistry*, A. F. Trotman-Dickenson, ed., Vol. 2, Pergamon Press, Oxford, 1973, Ch. 22, p. 685.
2. I. J. Solomon, in *Kirk–Othmer's Encyclopedia of Chemical Technology*, Vol. 10, Wiley, New York, 1980, p. 733.
3. *Gmelin, Handbuch der Anorganischen Chemie, Fluorine, Suppl. Vol 2*, Springer Verlag, Berlin, 1980, p. 139.
4. V. I. Vedeneev, Yu. M. Gershenzon, A. P. Dement'ev, A. P. Nalbandyan, O. M. Sarkisov, *Izv. Akad. Nauk SSSR, Ser. Khim.*, 1438 (1970).
5. I. V. Nikitin, A. V. Dudin, *Izv. Akad. Nauk SSSR, Ser. Khim.*, 2127 (1977).
6. O. Ruff, W. Menzel, *Z. Anorg. Allg. Chem.*, *211*, 204 (1933); *217*, 85 (1934).
7. I. V. Nikitin, V. Ya. Rosolovskii, *Izv. Akad. Nauk SSR, Ser. Khim.*, 266 (1968).
8. A. D. Kirshenbaum, J. G. Aston, A. V. Grosse, *US Dept. Comm. Office Tech. Serv. D. B. Rept.*, *149*, 443 (1961); *Chem. Abstr. 56*, 13,754 (1962).
9. A. D. Kirshenbaum, A. V. Grosse, *J. Am. Chem. Soc.*, *81* 1277 (1959).
10. A. G. Streng, *Can. J. Chem.*, *44*, 1476 (1966).
11. R. A. Hemstreet, A. H. Taylor, *US Pat.* 3,437,582 (1969); *Chem. Abstr. 71*, 5018 (1969).
12. A. G. Streng, A. V. Grosse, *J. Am. Chem. Soc.*, *88*, 169 (1966).
13. I. V. Nikitin, A. V. Dudin, V. Ya. Rosolovskii, *Izv. Akad. Nauk. SSSR, Ser. Khim.*, 269 (1973).
14. P. P. Chegodaev, V. I. Tupikov, E. C. Strukov, *Russ. J. Phys. Chem. (Engl. Transl.) 47*, 1315 (1973).
15. P. P. Chegodaev, V. I. Tupikov, *Dokl. Akad. Nauk SSSR, 210*, 647 (1973).
16. C. T. Goetschel, V. A. Campanile, C. D. Wagner, J. N. Wilson, *J. Am. Chem. Soc.*, *91*, 4702 (1969); *Am. Chem. Soc., Div. Fuel Chem., Prepr. 12*, 101 (1968); *Chem. Abstr. 71*, 118,324 (1969).
17. S. Aoyama, S. Sakuraba, *J. Chem. Soc. Jpn.*, *59*, 1321 (1938).
18. A. Kirshenbaum, A. Grosse, J. Astor, *J. Am. Chem. Soc.*, *81*, 6398 (1959).
19. A. Smalc, K. Lutar, J. Slivnik, *J. Fluorine Chem.*, *6*, 287 (1975).
20. J. B. Levy, B. K. W. Copeland, *J. Phys. Chem.*, *72*, 3168 (1968).
21. Ref. 1, p. 1225.
22. H. Niki, B. Weinstock, *J. Chem. Phys.*, *47*, 3249 (1967).
23. Ref. 1, p. 1362.

2.3.2.1.2. to Give the Sulfur–Halogen Bond.

Reviews of the sulfur halides[1] and sulfur fluorides are available[2–4]. Therefore, only the newer findings are discussed here.

The neutral binary fluorides FSSF, SSF_2, SF_2, S_2F_4, SF_4, SF_6 and S_2F_{10} are known. Besides these species, stable salts with $[SF_3]^+$ or $[SF_5]^-$ ions are isolated. Short-lived radical species ($[SF]^\cdot$, $[SF_3]^\cdot$ or $[SF_5]^\cdot$) are detected in the thermolysis or photolysis of SF_4, SF_2 or SF_5X (X = Cl, Br).

The fluorides can be obtained from direct fluorination of elemental sulfur, but the most favorable product is the hexafluoride[5,6], SF_6, prepared by burning sulfur in F_2:

$$S_{(liq)} + xs\ F_2 \xrightarrow[\text{flow}]{150^\circ\text{–}200^\circ C} SF_6 \qquad \text{(a)}$$

$$S_8 + xs\ F_2 \xrightarrow[\text{static}]{80^\circ\text{–}150^\circ C} SF_6 \qquad \text{(b)}$$

Either in a flow system at 150°–200°C or in static systems at 80°–150°C yields up to 99% are achieved. Minor impurities are either decomposed by thermolysis (e.g., S_2F_2, S_2F_{10}) or hydrolysis (e.g., SF_4). The preparation of SF_6 is reviewed[7]. The first step of the reaction between sulfur and F_2 forms fluorosulfanes, and successive F_2 addition, rearrangements and decompositions give the higher fluorinated species:

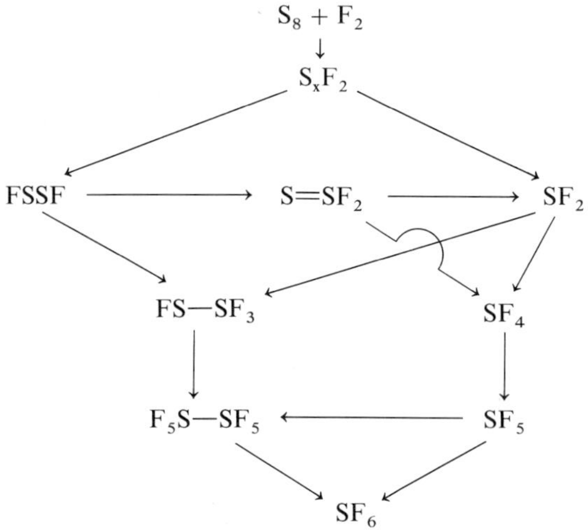

With xs F_2, the final step is the formation of SF_6.

The reaction scheme shows some of the possible pathways; all of the intermediates are characterized, some are prepared by other routes. From the direct fluorination, it is possible to isolate—besides SF_6—S_2F_2, S_2F_4, SF_4 and S_2F_{10}, but often metathesis (see §2.3.10.) is the better synthetic approach to the lower fluorides[2]. Under special conditions, these intermediate fluorides can be formed in good yield. In a metal vacuum apparatus from sulfur vapor and F_2, S_2F_4 is obtained[8]:

$$S_{x(g)} + F_2 \rightarrow FSSF_3 + (S_2F_2, SF_4, SF_6) \qquad \text{(c)}$$

At −78°C the impurities given are separated from S_2F_4, which is isolated in a pure state in yields ≤50%.

Sulfur tetrafluoride is important for inorganic and organic syntheses. The yields in its preparation from the elements depend on the conditions (T, P, concentrations, catalysis); low T favors the formation of SF_4; e.g., 1:3 F_2:N_2 is led over sulfur films (condensed on glass surfaces) at $-75°C$[9]:

$$S_{8\,(film)} + 1{:}3\ F_2\text{–}N_2 \xrightarrow{-75°C} \underset{(40\%)}{SF_4} \tag{d}$$

$$S_8 + 1{:}3\text{–}1{:}1\ F_2\text{–}N_2 \xrightarrow[-78°C]{CFCl_3} \underset{(70\%)}{SF_4} \tag{e}$$

On a small scale, SF_4 is also obtained[10] from fluorination of S_8 suspensions in $CFCl_3$. More drastic conditions can be applied if a large xs of sulfur is present, e.g., 83% yields are obtained[11] by blowing F_2 over the surface of molten sulfur at 120°–150°C, by bubbling F_2 through sulfur at 220°C[12] (97%) (this yield is not reproducible[13]). For large-scale preparation, fluorination of sulfur dissolved in sulfur chlorides is the best method[13,14]:

$$S_xCl_2 + 2(x-1)\ F_2 \xrightarrow{0°\text{–}30°C} (x-1)\ SF_4 + SCl_2 \tag{f}$$

$$SCl_2 + 1/8(x-1)\ S_8 \rightarrow S_xCl_2 \tag{g}$$

The first step is the oxidative fluorination of sulfur chlorides; SF_4 is obtained in 98% purity.

In the electrochemical fluorination (80°C; S_8:KF:HF = 1:8:16 or S_8:NH_4F:HF = 1:2:6) SF_4 is only a byproduct, SF_6 being the main product[15].

Fluorination of sulfur–boron mixtures leads to salts with the trifluorosulfur cation[16,17]:

$$S_8 - B + xs\ F_2 - N_2 \rightarrow [SF_3][BF_4] \tag{h}$$

Owing to cation formation further oxidation of the sulfur is prevented. From this salt pure SF_4 is obtained by displacement. If no such stabilization is provided, xs F_2 gives[7] SF_6. Traces of the intermediate fluoride, $F_5S\text{–}SF_5$, are present ($\leq 0.1\%$). **This extremely toxic compound** is separated from SF_6 by distillation[18,19]. A more direct route to S_2F_{10} is by reduction[20] of SF_5Cl.

Sulfur oxyfluorides are formed[21,22] if sulfur is burned simultaneously in F_2 and O_2:

$$S_{(liq)} + 0.34{:}1\ O_2\text{–}F_2 \xrightarrow{308°C} \underset{(40.6\%)}{OSF_4} \tag{i}$$

$$S_{(liq)} + 3.5{:}1\ O_2\text{–}F_2 \xrightarrow{170°\text{–}180°C} \underset{(70\%)}{SO_2F_2} + \underset{(29.6\%)}{SF_4} \tag{j}$$

The known binary sulfur chlorides are S_2Cl_2, SCl_2, SCl_4 and the chlorosulfanes S_xCl_2. Mixed halides, e.g., S_2ClF_3 S_2ClF, SCl_3F, SCl_2F_2 and $SClF_3$ (§2.3.10.1 and §2.3.9.3.1), are confirmed by spectroscopic methods. Only S_2Cl_2 is stable at RT; the S—Cl bond in higher oxidation states is stabilized either by cation formation (e.g., in $[SCl_3]^+$) or by introduction of electron-withdrawing ligands (e.g., SF_5Cl).

By direct chlorination S_2Cl_2, SCl_2, SCl_4 and, in the presence of electron-pair acceptor acids, the $[SCl_3]^+$ cation is formed, the chlorosulfanes S_xCl_2 are synthesized from S_2Cl_2 and sulfanes.

When Cl_2 is passed into solid or molten sulfur or into a suspension of sulfur in S_2Cl_2, disulfur dichloride is produced[23,24] primarily:

$$S_8 + Cl_2 \xrightarrow{RT-300^\circ C} S_2Cl_2 \xrightarrow[Cl_2]{} SCl_2 \xrightarrow[-78^\circ C]{liq\ Cl_2} SCl_4 \quad (k)$$

From S_2Cl_2 and liq Cl_2 in a closed system, red SCl_2 is formed. Freshly distilled S_2Cl_2 requires a longer induction period. If the S_2Cl_2 already contains some SCl_2 or if the S_2Cl_2 is aged, the reaction starts immediately; catalysts [e.g., ICl_3, $SnCl_4$, $SbCl_5$, Fe(III) salts][25,26] also accelerate the reaction.

At -78°C a mixture of SCl_2 and stoichiometric Cl_2 will solidify to give[27,28] SCl_4. The compound is stable only at low T in a sealed tube. According to vibrational spectra $[SCl_3]^+$ and Cl^- ions are present in the solid[29]. Therefore, trichlorosulfur salts are formed directly from the elements in the presence of electron-pair acceptor acids[30–34], e.g.:

$$\tfrac{2}{8} S_8 + 7\,Cl_2 + 2\,I_2 \rightarrow 2\,[SCl_3][ICl_4] \quad (l)$$

$$\tfrac{1}{8} S_8 + 2\,Cl_2 + AlCl_3 \rightarrow [SCl_3][AlCl_4] \quad (m)$$

$$\tfrac{1}{8} S_8 + xs\ Cl_2 + 2\,AsF_3 \rightarrow [SCl_3][AsF_6] + AsCl_3 \quad (n)$$

The Al salt is indefinitely stable at RT. The last two salts are also obtained by electrooxidation of sulfur in molten[35] $AlCl_3$–NaCl.

In the sulfur–bromine system only S_nBr_2(n = 1, 2 and X) exist, but not the tetra- or hexabromide. Stabilization of the S—Br bond in higher oxidation states is possible, similar to the Cl system, through ion formation (as in $[SBr_3]^+$) or by introducing electronegative groups (as in SF_5Br). Direct combination of the elements at RT forms S_2Br_2[36]:

$$\tfrac{1}{4} S_8 + Br_2 \rightarrow S_2Br_2 \quad (o)$$

The garnet red, oily liquid is distillable under reduced pressure; at >100°C decomposition into the elements occurs.

The other bromides mentioned are obtained only by metathetical reactions (see §2.3.10.2. and §2.3.10.3.).

The tribromosulfur (IV) cation is prepared quantitatively from the elements in liq SO_2 with[37] AsF_5 or SbF_5:

$$\tfrac{1}{4} S_8 + 3\,Br_2 + 2\,AsF_5 \rightarrow 2\,[SBr_3][AsF_6] + AsF_3 \quad (p)$$

Sulfur and I_2 do not react, the only existing binary sulfur iodide, S_2I_2, being formed by metathesis from S_2Cl_2 (See §2.3.10.1. and §2.3.10.2). In the presence of electron-pair acceptor acids cationic species with S—I bonds are isolated[38]:

$$\tfrac{1}{4} S_8 + 2\,I_2 + 3\,AsF_5 \xrightarrow{SO_2} [S_2I_4][AsF_6]_2 + AsF_3 \quad (q)$$

Salts with the $[S_7I]^+$-cation are obtained either by oxidizing sulfur with iodine cations or I_2 with sulfur cations[39,40]:

$$\left.\begin{array}{l}[I_3][AsF_6] + S_8 \\ [I_2][Sb_2F_{11}] + S_8 \\ [S_{19}][AsF_6] + I_2\end{array}\right\} \rightarrow [S_7I][MF_6] \quad (r)$$

Changing the stoichiometry or the solvent in Eq. (q) leads to solids containing the $[S_7I]^+$ cation[41–43]:

$$4\ S_8 + 2\ I_2 + 9\ AsF_5 \xrightarrow{SO_2} (S_7I)_4S_4(AsF_6)_6 + 3\ AsF_3 \qquad (s)$$

$$\tfrac{28}{8}\ S_8 + 3\ I_2 + 10\ SbF_5 \xrightarrow{AsF_3} 2\ [(S_7I)_2I][SbF_6]_3 + (SbF_3)_3SbF_5 \qquad (t)$$

In the latter sulfur iodine cation two $[S_7I]^+$ are linked by I^+.

Reviews of the S—Cl[44,45], S—Br[46,47] and S—I bonds[48,49] are available.

(R. MEWS)

1. *Gmelin, Handbuch der Anorganischen Chemie, Schwefel, Erg. V Band. 2, Sulfur Halides*, Springer Verlag, Berlin, 1978.
2. F. Seel, *Adv. Inorg. Chem. Radiochem.*, 16, 297 (1974).
3. S. P. v. Halasz, O. Glemser, in *Sulfur in Organic and Inorganic Chemistry*, Vol. 1 A. Senning, ed., Marcel Dekker, New York, 1971, p. 207.
4. J. M. Shreeve, in ref. 3, Vol. 4, 1982, p. 131.
5. H. Moissan, *Ann. Chim. (Paris)*, *24*, 239 (1891).
6. H. Moissan, P. Lebeau, *C.R. Hebd. Seances Acad. Sci.*, *130*, 865, 984, 1436 (1900).
7. Ref. 1, p. 99
8. A. Haas, H. Willner, *Z. Anorg. Allg. Chem.*, *462*, 57 (1980).
9. F. Brown, P. L. Robinson, *J. Chem. Soc.*, 3147 (1955).
10. D. Naumann, D. K. Padma, *Z. Anorg. Allg. Chem.*, *401*, 53 (1973).
11. Y. Tanaka, T. Watanabe, H. Nakayama, Showa Denko K. K., Jpn. Kokai 75-3399 6 (1973/75); *Chem. Abstr.*, *83*, 62,982 (1975).
12. S. Kleinberg, J. F. Tompkins, US Pat. 3,399,036 (1968); *Chem. Abstr.*, *69*, 78,864 (1968).
13. W. Becher, J. Massonne, *Chem.-Ztg.*, *98*, 117 (1974).
14. W. Becher, J. Massonne, W. Pohlmeyer, Ger. Off. 2,217,971 (1973); *Chem. Abstr.*, *80*, 38,921 (1974).
15. A. Engelbrecht, E. Mayer, C. Pupp, *Monatsh. Chem.*, *95*, 633 (1964).
16. N. Bartlett, P. L. Robinson, *Proc. Chem. Soc.*, 230 (1957).
17. N. Bartlett, P. L. Robinson, *J. Chem. Soc.*, 3417 (1961).
18. K. G. Denbigh, R. Whytlaw-Gray, *J. Chem. Soc.*, 1346 (1934).
19. B. Cohen, A. G. MacDiarmid, *Inorg. Chem.*, *1*, 754 (1962).
20. H. L. Roberts, *J. Chem. Soc.*, 3183 (1962).
21. H. Wada, E. Uefuki, Jpn. Pat., 7,208,444 (1972); *Chem. Abstr.*, *79*, 33,233 (1973).
22. H. Wada, E. Uefuki, Jpn. Pat. 7,208,443 (1972); *Chem. Abstr.*, *78*, 149,417 (1973).
23. *Gmelin, Handbuch der Anorganischen Chemie, Schwefel, Vol. B3*, Verlag Chemie, Weinheim, 1963, p. 1761; ref. 1, p. 239.
24. P. Macaluso, in *Kirk–Othmer Encyclopedia of Chemical Technology*, 2nd ed., Vol. 19, Wiley, New York, 1969, p. 371, 390.
25. Ref. 23, p. 1780; ref. 1, p. 221.
26. G. Riesz, in *Ullmanns Encyclopadie der Technischen Chemie*, W. Forst, ed., 3rd ed., Vol. 15, Urban and Schwarzenberg, Berlin, 1964, p. 491.
27. Ref. 23, p. 1789; ref. 1, p. 272.
28. F. Feher, in *Handbook of Preparative Inorganic Chemistry* G. Brauer, ed., 2nd ed., Vol. 1, Academic Press, New York, 1960, p. 376.
29. M. Feuerhahn, R. Minkwitz, *Z. Anorg. Allg. Chem.*, *426*, 247 (1976).
30. G. Mamantov, R. Marassi, F. W. Poulsen, S. E. Springer, J. P. Wiaux, R. Huglen, N. R. Smyrl, *J. Inorg. Nucl. Chem.*, *41*, 260 (1979).
31. R. Hulme, US Pat. 4,172,115 (1979); *Chem. Abstr. 92*, 25,131 (1980).
32. Ref. 1, p. 290.
33. A. Finch, P. N. Gates, T. H. Page, *Inorg. Chim. Acta*, *25*, L49 (1977).
34. R. Minkwitz, K. Jänichem, H. Prenzel, V. Wötfel, *Z. Naturforsch. Teil B*, *40*, 53 (1985) and ref. therein.

35. R. Marassi, G. Mamantov, M. Matsunaga, S. E. Springer, J. P. Wiaux, *J. Electrochem. Soc.*, *126*, 231 (1979).
36. Ref. 23, pp. 239, 1865.
37. J. Passmore, E. K. Richardson, P. Taylor, *Inorg. Chem.*, *17*, 1681 (1978).
38. J. Passmore, G. Sutherland, T. Whidden, P. S. White, *J. Chem. Soc., Chem. Commun.*, 289 (1980).
39. J. Passmore, P. Taylor, T. K. Whidden, P. S. White, *J. Chem. Soc., Chem. Commun.*, 689 (1976).
40. J. Passmore, G. Sutherland, P. Taylor, T. Whidden, P. S. White, *Inorg. Chem.*, *20*, 3839 (1981).
41. J. Passmore, G. Sutherland, P. S. White, *J. Chem. Soc., Chem. Commun.*, 330 (1980).
42. J. Passmore, G. Sutherland, P. S. White, *J. Chem. Soc., Chem. Commun.*, 901 (1979).
43. J. Passmore, G. Sutherland, P. S. White, *Inorg. Chem.*, *21*, 2717 (1982).
44. C. R. Russ, I. B. Douglass, in *Sulfur in Organic and Inorganic Chemistry*, Vol. 1., A. Senning, ed., Marcel Dekker, New York, 1971, p. 239.
45. W. R. Hardstaff, R. F. Laugler, in *Sulfur in Organic and Inorganic Chemistry*, Vol. 4, A. Senning, ed., Marcel Dekker, New York, 1982, p. 193.
46. P. S. Magee, in *Sulfur in Organic and Inorganic Chemistry* Vol. 1, A. Senning, ed., Marcel Dekker, New York, 1971, p. 261.
47. P. S. Magee, in *Sulfur in Organic and Inorganic Chemistry* Vol. 4, A. Senning, ed., Marcel Dekker, New York, 1982, p. 283.
48. J. P. Danehy, in *Sulfur in Organic and Inorganic Chemistry* Vol. 1, A. Senning, ed., Marcel Dekker, New York, 1971, p. 327.
49. L. Field, C. M. Lukehart, in *Sulfur in Organic and Inorganic Chemistry* Vol. 4, A. Senning, ed., Marcel Dekker, New York, 1982, p. 327.

2.3.2.1.3. to Give the Se–Halogen Bond.

In Se—F chemistry SeF_6 and SeF_4 are known[1–4] along with the lower fluorides, $SeSeF_2$, FSeSeF and SeF_2. The latter two are obtained[5] together with SeF_4 when dilute F_2 acts on hot Se. Polyselenides, FSe_nF (n = 3 or 4), are also present:

$$\text{Se} + 1{:}200\text{–}1{:}1000\ F_2\text{–Ar} \xrightarrow{\approx 200^\circ\text{C}} \text{FSeSeF} + SeF_2 + SeF_4 \xrightarrow{\text{UV}} SeSeF_2 \qquad \text{(a)}$$

(UV photolysis arrow leads from FSeSeF down to $SeSeF_2$.)

The unstable lower fluorides are trapped at low T and characterized by IR spectroscopy; FSeSeF is partially isomerized by UV photolysis to give[5] $SeSeF_2$.

When 1:1 F_2-N_2 is passed over Se at 0°C, SeF_4 is obtained[6,7]. The white, hygroscopic tetrafluoride, mp −9.5°C is separated from minor amounts of SeF_6 by vacuum techniques:

$$\text{Se} + F_2\text{-}N_2 \xrightarrow{0^\circ\text{C}} SeF_4 \qquad \text{(b)}$$

$$\text{Se} + F_2 \xrightarrow{250^\circ\text{–}300^\circ\text{C}} SeF_6 \qquad \text{(c)}$$

Liquid Se and F_2 react at 250°–300°C to form SeF_6. The SeF_4, also present in the resulting gas, is condensed and returned to the reaction vessel to form more SeF_6[8,9]. Impurities in the hexafluoride, [mp −34.6°C, 1500 mm (2×10^5 Pa); sublimation without melting at −46.6°C/atm (1.01×10^5 Pa)[10]] are removed by aq KOH because SeF_6 is inert toward H_2O like its sulfur analogue.

In the Se—Cl system are Se_2Cl_2 and $SeCl_4$[10,11], whereas $SeCl_2$ might be present in equilibrium with the other chlorides in solution[12] or in the vapor over solid $SeCl_4$[13]. The monohalide is made directly from the stoichiometric elements or, better, by adding the halogen to a suspension of Se in CS_2 until dissolution of the element is complete and precipitation of the tetrahalide begins[14]. Vibrational spectra show only the product; no $SeSeCl_2$ can be detected[15]:

$$2\ Se + Cl_2 \xrightarrow{CS_2} ClSeSeCl \xrightarrow{3\ Cl_2} 2\ SeCl_4 \qquad (d)$$

Excess Cl_2 gives $SeCl_4$ in almost quantitative yield and high purity[16,17] if moisture is excluded. The solid consists, like the analogous sulfur and Te compounds, of $[SeCl_3]^+$ and Cl^- ions[18].

In the vapor, $SeCl_4$ is almost completely dissociated into Cl_2 and lower chlorides[19].

The $[SeCl_3]^+$ cation[20], which is formed as the hexafluoroarsenate by chlorinating Se in $AsCl_3$ solution and adding AsF_3, is stable[21].

$$Se + 3\ Cl_2 + 2\ AsF_3 \rightarrow [SeCl_3][AsF_6] + AsCl_3 \qquad (e)$$

The stable bromides Se_2Br_2 and $SeBr_4$, are formed from the elements at RT[22,23]; $SeBr_2$ also may be present[24–26].

The existence of Se_2Br_2 and $SeBr_4$ is confirmed in the solid[27,28]. The phase relations and crystal structures in the Se–Br system shows that two thermodynamically stable, congruently melting phases exist, α-Se_2Br_2, mp +5°C and α-$SeBr_4$, mp 123°C[29]. Two other phases, β-Se_2Br_2 and β-$SeBr_4$ are metastable and are irreversibly transformed to the α-modifications by annealing[29]:

$$2\ Se + Br_2 \xrightarrow{RT} \alpha\text{-}Se_2Br_2 \xrightarrow{Br_2} \beta\text{-}SeBr_4 \rightarrow \alpha\text{-}SeBr_4 \qquad (f)$$

The color of freshly prepared $SeBr_4$ is orange–red owing to the presence of β-$SeBr_4$ which transforms at RT in a few days to the stable, brown–red–black α-$SeBr_4$. Both modifications contain tetrameric, cubane-like $(SeBr_4)_4$ molecules, in which $[SeBr_3]^+$ units are linked together by Br^- bridges.

Salts with the $[SeBr_3]^+$ and probably the $[Se_2Br_5]^+$ cation are obtained from the elements in the presence of electron-pair acceptor of acids or by bromination of the $[Se_4]^{2+}$ cation[30]:

$$Se + xs\ Br_2 + xs\ AsF_5 \rightarrow [SeBr_3][AsF_6] \qquad (g)$$

$$[Se_4][MF_6]_2 + Br_2 \rightarrow 2\ [SeBr_3][MF_6] + 2\ SeBr_4 \qquad (h)$$

No stable neutral Se iodides are known. The Se iodides Se_3I_2, SeI_2 and SeI_4 (?), may be present in solutions of Se and I_2 in CS_2[31], but no compound formation is detected in the solid[23]. With electron-pair acceptor fluoro acids, however, stabilization of the (+4) oxidation state is also observed[33]:

$$12\ Se + I_2 + 3\ AsF_5 \xrightarrow{SO_2} 2\ [Se_6I][AsF_6] + AsF_3 \qquad (i)^{32}$$

$$2\ Se + 3\ I_2 + 3\ AsF_5 \rightarrow 2\ [SeI_3][AsF_6] + AsF_3 \qquad (j)^{33}$$

(R. MEWS)

1. *Gmelin, Handbuch der Anorganischen Chemie, Selen, Erg. Band. A3* Springer Verlag, Berlin, 1981.
2. *Gmelin, Handbook of Inorganic Chemistry, Selenium, Suppl. Vol. B2,* Springer Verlag, Berlin, 1984.

3. S. N. Chizhevskaya, N. Kh. Abrikosov, B. B. Azizova, *Izv. Akad. Nauk SSSR, Neorgan. Mater. 9*, 218 (1973).
4. J. W. George, *Prog. Inorg. Chem. 2*, 23 (1960).
5. A. Haas, H. Willner, *Z. Anorg. Allg. Chem., 454*, 17 (1979); *Ber. Bunsenges. Phys. Chem., 82*, 24 (1978).
6. E. E. Aynsley, R. D. Peacock, P. L. Robinson, *J. Chem. Soc.*, 1231 (1952).
7. C. Dagron, *Hebd. Seances Acad. Sci., 244*, 1779 (1957).
8. D. M. Yost, W. H. Claussen, *J. Am. Chem. Soc., 55*, 885 (1933).
9. A. A. Banks, Br. Pat. 713,119 (1954); *Chem. Abstr., 49*, 7204 (1955).
10. K. W. Bagnall *The Chemistry of Se, Te, and Po*, Elsevier, Amsterdam, 1966; also, in *Comprehensive Inorganic Chemistry*, Vol. 2, A. F. Trotman-Dickenson, ed., Pergamon Press, Oxford, 1973, p. 935.
11. Ref. 1, p. 239, ref. 2, p. 134.
12. M. Lundkvist, M. Lellep, *Acta Chem. Scand., 22*, 291 (1968).
13. D. M. Yost, C. E. Kircher, *J. Am. Chem. Soc., 52*, 4680 (1930).
14. H. Stammreich, R. Forneris, *Spectrochim. Acta, 8*, 46 (1956).
15. P. J. Hendra, P. J. D. Park, *J. Chem. Soc.*, 908 (1968).
16. Z. Vavrin, R. Resl, *Chem. Zvesti, 8*, 241 (1954); *Chem. Abstr. 49*, 7430 (1955).
17. A. I. Alekperov, M. A. Babaeva, *Azerb. Khim. Zh.*, 170 (1967); *Chem. Abstr., 69*, 46,380 (1968).
18. C. B. Shoemaker, S. C. Abrahams, *Acta Crystallogr. 18*, 296 (1965).
19. K. W. Bagnall, in *Comprehensive Inorganic Chemistry*, Vol. 2, A. F. Trotman-Dickenson, ed., Pergamon Press, Oxford, 1973, p. 958.
20. Ref. 2, p. 174.
21. L. Kolditz, W. Schäfer, *Z. Chem., 1*, 124 (1961).
22. R. Schneider, *Pogg. Ann., 129*, 453 (1886).
23. Ref. 1, p. 241, ref. 2, p. 222, 228.
24. N. W. Tideswell, J. D. McCullough, *J. Am. Chem. Soc., 78*, 3026 (1956).
25. N. Katsaros, J. W. George, *J. Chem. Soc., Chem. Commun.*, 662 (1968).
26. K. Hogberg, M. Lundqvist, *Acta Chem. Scand., 24*, 255 (1970).
27. G. V. Golubkova, E. S. Petrov, A. N. Kanev, *Izv. Sib. Otd. Akad. Nauk SSSR, Ser. Khim. Nauk*, 51 (1976); *Chem. Abstr., 84*, 173,074 (1976).
28. G. V. Golubkova, E. S. Petrov, A. N. Kanev, *Izv. Sib. Otd. Akad. Nauk SSSR, Ser. Khim. Nauk*, 39 (1975); *Chem. Abstr., 84*, 22,743 (1976).
29. P. Born, R. Kniep, D. Mootz, *Z. Anorg. Allg. Chem., 451*, 12 (1979).
30. W. V. F. Brooks, J. Passmore, E. K. Richardson, *Can. J. Chem., 57*, 3230 (1979).
31. A. F. Kapustinskii, Yu.M. Golutvin, *J. Gen. Chem. USSR, (Engl. Transl.), 17*, 2010 (1947).
32. W. A. Shantha Nandana, J. Passmore, P. S. White, *J. Chem. Soc., Chem. Comm.*, 526 (1983).
33. J. Passmore, P. Taylor, *J. Chem. Soc., Dalton Trans.*, 804 (1976).

2.3.2.1.4. to Give the Te–Halogen Bond.

Reviews of the inorganic chemistry of Te[1], especially the halide chemistry[2,3] are available.

In the Te–F system only TeF_4 and TeF_6 are stable species; Te_2F_{10} does not exist but TeF_2 can be stabilized as a thiourea complex[4]. At 0°C, TeF_4 is formed from the elements[5]:

$$Te + F_2 \xrightarrow{0°C} TeF_4 \quad \text{(a)}$$

$$Te + F_2 \xrightarrow[N_2]{150°C} TeF_6 \quad \text{(b)}$$

$$Te + F_2 \xrightarrow[N_2(O_2)]{40°-80°C} F_5TeOTeF_5, Te_3F_{14}O_2 \quad \text{(c)}$$

At 150°C TeF_6 is the only product[6]. Because commercial F_2 contains O_2, $F_5TeOTeF_5$ and $F_5TeOTeF_4OTeF_5$ are also formed at low T, but at higher T the Te—O bond cleaves[5–7]. The yield of $F_5TeOTeF_5$ is higher (40–50%) if the fluorination is performed in the presence of $CaCl_2 \cdot x\,H_2O$ powder, or if an F_2-O_2 mixture is passed over 1:1 Te-TeO_2. In earlier reports $(F_5Te)_2O$ is described as Te_2F_{10} which is not known[2,8,9].

Te reacts with Cl_2 to yield[10], besides $TeCl_4$, the subhalides Te_2Cl and Te_3Cl_2; Te_2Cl_2 and $TeCl_2$ are unstable, existing only in the gas phase. The Te—Cl bond in hexavalent Te is stabilized if strong electron-withdrawing groups are present (e.g., in TeF_5Cl), just as for sulfur and Se.

If 1:1 Te and Cl_2 is heated to 300°C and the reaction mixture is cooled by 3°C/min, besides, Te_3Cl_2 and $TeCl_4$, Te_2Cl is also formed[11]. If 1:2 Te-Cl_2 is heated in a silica vessel to 200°-220°C and the mixture is homogenized at 300°-350°C, after quenching and tempering at 180°C, only $TeCl_4$ and Te_3Cl_2 are formed[12,13]:

$$Te + Cl_2 \xrightarrow{300°C} Te_3Cl_2 + TeCl_4 + Te_2Cl \qquad (d)$$

$$Te + 2\,Cl_2 \xrightarrow{200°–220°C} Te_3Cl_2 + TeCl_4 \qquad (e)$$

These results and the Te-Cl_2 phase diagram show that only Te_3Cl_2 and $TeCl_4$ are stable species[10,12–14], and that Te_2Cl is metastable[12]. In flow systems from powdered or granulated Te and O_2-free Cl_2 with exclusion of moisture, pure $TeCl_4$ is obtained[2,15]:

$$Te + xs\;Cl_2 \xrightarrow{175°–400°C} TeCl_4 \qquad (f)$$

which can be purified by distillation at >390°C.

The reaction is faster if Cl_2 is passed through a suspension of Te in an organic solvent [ROH, $(CH_3)_2CO$, CH_3COOH][15]. Because heat is evolved, cooling to 25°-30°C is necessary[16]. Pure $TeCl_4$ forms in HCl by anodic dissolution (via Cl_2 produced in situ) of Te at 20°-50°C[17]. Chlorination of Te in $AsCl_3$-AsF_3 leads to $[TeCl_3]^+[AsF_6]^-$[18].

In the Te-Br system, Te_2Br and $TeBr_4$ exist in the solid, whereas TeBr and $TeBr_2$ are known only as gases[17]. If 1:1 Te and Br_2 is heated in a closed vessel to 220°-260°C (Br_2 is slowly distilled onto heated Te), Te_2Br and $TeBr_4$ are isolated[12]:

$$Te + Br_2 \xrightarrow{220°–260°C} Te_2Br + TeBr_4 \qquad (g)$$

$$Te + xs\;Br_2 \rightarrow TeBr_4 \qquad (h)$$

With xs Br_2, $TeBr_4$ is formed either in static or flow systems at 0°-360°C[3,15,20]. The orange crystals can be purified by sublimation [380-400°C/150 torr (2×10^4 Pa)][20]. In the presence of AsF_5, $[TeBr_3][AsF_6]$ is obtained[21].

$$Te + xs\;Br_2 + xs\;AsF_5 \xrightarrow[-AsF_3]{SO_2} [TeBr_3][AsF_6] \qquad (i)$$

Besides TeI_4, the subiodides Te_2I, α-TeI and β-TeI can be prepared from elemental Te and I_2. The latter are formed under hydrothermal conditions (Table 1[22]).

The Te_2I and β-TeI subhalides form dark, metallic needles, whereas β-TeI forms dark, metallic, compact crystals[11].

In the vapor above solid Te in the Te-I_2 system, TeI_2 is detected at >270°C[22]:

$$Te + I_2 \rightarrow TeI_4 \rightleftharpoons TeI_2 + I_2 \qquad (j)$$

$$Te + xs\;I_2 \xrightarrow{230°–300°C} TeI_4 \qquad (k)$$

TABLE 1. HYDROTHERMAL GROWING CONDITIONS FOR Te SUBHALIDES[11]

Subhalide[a]	Te_2I	β-TeI	α-TeI
Starting material, Te:I	2.5:1	2.5:1	2.5:1
Solvent	10 mMHI	10 mMHI	10 mMHI
Degree of filling (%)	65	65	65
T (°C)	280 → 265	198 → 192	195
Time (days)	10	7	8

[a] The autoclave is kept in a 45° position.

In closed ampules at 230°–300°C with a slight I_2 xs, black, crystalline TeI_4 is formed[22–24]; in SO_2 soln $[TeI_3][AsF_6]$ forms[25].

(R. MEWS)

1. A. Engelbrecht, F. Sladky, *MTP Int. Rev. Inorg. Chem., Ser. Two, 3*, 137 (1975).
2. Gmelin *Handbuch der Anorganischen Chemie*, Tellurium, Suppl. Vol. 2, Chlorides, Fluorides, Springer Verlag, Berlin, 1977.
3. Gmelin *Handbuch der Anorganischen Chemie*, Tellurium, Suppl. Vol. 3, Bromides, Iodides, Springer Verlag, Berlin, 1978.
4. D. Foss, S. Hauge, *Acta Chem. Scand., 15*, 1523 (1961).
5. R. Campbell, P. L. Robinson, *J. Chem. Soc.*, 3454 (1956).
6. Ref. 2, p. 19.
7. W. D. English, J. W. Dale, *J. Chem. Soc.*, 2498 (1953).
8. H. Bürger, *Z. Anorg. Allg. Chem., 360*, 97 (1968).
9. P. M. Watkins, *J. Chem. Educ., 51*, 520 (1974).
10. Ref. 1, p. 67 ff.
11. R. Kniep, D. Mootz, A. Rabenau, *Z. Anorg. Allg. Chem., 422*, 17 (1976).
12. A. Rabenau, H. Rau, G. Rosenstein, *Angew. Chem., Int. Ed. Engl., 9*, 802 (1970).
13. A. Rabenau, H. Rau, *Z. Anorg. Allg. Chem., 395*, 273 (1973).
14. S. A. Ivashin, E. S. Petrov, *Izv. Sib. Otd. Akad. Nauk SSSR Ser. Khim. Nauk*, 48 (1970); *Chem. Abstr., 76*, 145,518 (1972).
15. H. Oppermann, G. Stöver, E. Wolf, *Z. Anorg. Allg. Chem., 410*, 179 (1974).
16. A. I. Alekperov, M. A. Babaeva, *Azerb. Khim. Zh.*, 170 (1967); *Chem. Abstr., 69*, 46,380 (1968).
17. A. Y. Ryazanov, A. I. Volfson, G. D. Chigrinova, USSR Pat. 138,922 (1961); *Chem. Abstr., 56*, 99,743 (1962).
18. L. Kolditz, W. Schäfer, *Z. Anorg. Allg. Chem., 315*, 35 (1962).
19. Ref. 3, p. 1.
20. G. M. Serebrennikova, L. A. Sazikova, B. D. Stepin, *Russ. J. Inorg. Chem. (Engl. Transl.), 12*, 716 (1967).
21. W. F. V. Brooks, J. Passmore, E. K. Richardson, *Can. J. Chem., 57*, 3230 (1979).
22. H. Oppermann, G. Stöver, E. Wolf, *Z. Anorg. Allg. Chem., 419*, 200 (1976).
23. H. Oppermann, G. Kunze, E. Wolf, *Z. Anorg. Allg. Chem., 432*, 182 (1977).
24. V. V. Safonov, O. V. Lemeshko, G. B. Korshunov, *Russ. J. Inorg. Chem. (Engl. Transl.), 16*, 1217 (1971).
25. J. Passmore, J. Sutherland, P. S. White, *Can. J. Chem., 59*, 2876 (1981).

2.3.2.1.5. to Give the Po–Halogen Bond.

The chemistry of the Po halides is similar to that of Te, but the bipositive derivatives are more stable, the quadripositive halides less stable than their Te analogues and the subhalides are not known. The halides are covalent, volatile and hydrolyzable[1].

Probably PoF_6 is formed from F_2 and ^{208}Po plated on Pt:

$$Po + xs\ F_2 \rightarrow PoF_6 \quad (a)$$

The product is stable in the vapor, but on cooling a nonvolatile compound is formed, resulting from the radiation-induced decomposition of PoF_6. This material is assumed[2] to be PoF_4.

The metal reacts with dry halogens (Cl_2, Br_2) at 200°–250°C to give the tetrahalides:

$$Po + 2\ Cl_2 \xrightarrow{200°C} PoCl_4 \quad (b)^{3-5}$$

$$Po + 2\ Br_2 \xrightarrow[1\ h,\ 250°C]{200\ mm\ (2.7 \times 10^4\ Pa)} PoBr_4 \quad (c)^{4,6}$$

$$Po + 2\ Br_2 \xrightarrow[250°C,\ 5\ min]{N_2} PoBr_4 \quad (d)^{7}$$

The bright-yellow $PoCl_4$ and the bright-red $PoBr_4$ decompose to give the dihalides (ruby-red $PoCl_2$ and $PoBr_2$). Black PoI_4 is difficult to obtain because of its degradation to PoI_2 and I_2:

$$Po + 2\ I_2 \xrightarrow[1\ mm]{40°C} PoI_4 \quad (e)$$

Polonium does not react with I_2 in CCl_4 but does in C_6H_6.

Vapor pressure measurements show that gaseous PoI_4 is dissociated partially to PoI_2 and I_2; in the PoI_4–I_2 system the existence of PoI_6 is also assumed[9].

(R. MEWS)

1. K. W. Bagnall, *The Chemistry of Se, Te and Po*, Elsevier, Amsterdam, 1966; also, in *Comprehensive Inorganic Chemistry*, A. F. Trotman-Dickenson, ed., Pergamon Press, Oxford, 1973, p. 958.
2. B. Weinstock, C. L. Chernick, *J. Am. Chem. Soc.*, *82*, 4116 (1960).
3. E. F. Joy, *US Atomic Energy Commission Rep. M-4123* (1947; declassified 1955).
4. J. J. Burbage, *Rec. Chem. Prog.*, *14*, 157 (1953).
5. K. W. Bagnall, W. R. M. D'Eye, J. W. Freeman, *J. Chem. Soc.*, 2320 (1955).
6. E. F. Joy, *Chem. Eng. News*, *32*, 4848 (1954).
7. K. W. Bagnall, W. R. M. D'Eye, J. H. Freeman, *J. Chem. Soc.*, 3959 (1955).
8. K. W. Bagnall, W. R. M. D'Eye, J. H. Freeman, *J. Chem. Soc.*, 3385 (1956).
9. A. S. Abakumov, M. L. Malyshev, *Radiokhimiya*, *18*, 894 (1976); *Chem. Abstr.*, *86*, 60,689 (1977).

2.3.2.2. from Halogenation by Hydrogen Halides.

Because O, S, Se and Te are nonmetallic no reaction occurs with hydrogen halides. Only Po dissolves in HX (X = F, Cl, Br, I); in HF solutions either complex fluoropolonites are present[1], or PoF_4 may be the product[2]:

$$Po + 4\ HF \rightarrow PoF_4(?) + 2\ H_2 \quad (a)$$

$$Po + 4\ HCl \rightarrow PoCl_4 + 2\ H_2 \quad (b)$$

$$Po + 4\ HBr \rightarrow PoBr_4 + 2\ H_2 \quad (c)$$

By evaporation of the solutions obtained by dissolving metallic Po in HCl, $PoCl_4$ and $PoBr_4$ are isolated[3–6].

(R. MEWS)

1. H. V. Moyer, in *Polonium, US Atomic Energy Commission* Rep. TID-5221, H. V. Moyer, ed., Oak Ridge, TN, (1956); *Chem. Abstr., 50,* 13,590 (1956).
2. E. H. Belcher, *Proc. R. Soc. (London) Ser. A,216,* 90 (1953).
3. K. W. Bagnall, R. W. M. D'Eye, J. H. Freeman, *J. Chem. Soc.,* 2320 (1955).
4. J. J. Burbage, *Rec. Chem. Prog., 14,* 157 (1953).
5. E. F. Joy, *US Atomic Energy Commission* Rep. M-4123 (1947; declassified 1955).
6. K. W. Bagnall, R. W. M. D'Eye, J. H. Freeman, *J. Chem. Soc.,* 3385 (1956).

2.3.2.3. from Halogenation by Oxidizing Halides

2.3.2.3.1. to give the Oxygen–Halogen Bond.

Owing to the lability of oxygen–halogen bonds, the possibilities for their formation from the element and oxidizing halides are limited. Possible oxidizers are the noble gas fluorides, halogen fluorides and OF_2 itself. The low-T discharge or photolysis of liq and gas mixtures of OF_2 and O_2 forms[1,2] O_2F_2.

(R. MEWS)

1. S. Aoyama, S. Sakuraba, *J. Chem. Soc. Jpn., 62,* 208 (1941).
2. A. G. Streng, L. V. Streng, *Inorg. Nucl. Chem. Lett., 2,* 107 (1966).

2.3.2.3.2. to Give the Sulfur–Halogen Bond.

The oxidation of sulfur by halogen-containing compounds instead of the halogens themselves is applied to the preparation of sulfur fluorides. Few examples of S–Cl or S—Br bond formation by means of metal or nonmetal halides are known.

Because the action of F_2 is difficult to control, routes to the lower sulfur fluorides from fluorides with different oxidizing properties are developed. As oxidative fluorinating agents, besides metal fluorides nitrogen, oxygen, halogen and noble-gas fluorides are also used.

Metal fluorides, such as AgF, Hg_2F_2 and HgF_2, are only weakly oxidizing. With gaseous or molten sulfur, FSSF, SSF_2 and SF_4 are formed[1–4]:

$$AgF + S_{(melt)} \xrightarrow[vac]{125^\circ C} FSSF\ (+SSF_2 + SF_4) \tag{a}$$

$$AgF\text{–}S_8 \xrightarrow[vac]{100^\circ\text{–}170^\circ C} FSSF + \cdots \tag{b}$$

Most conveniently, FSSF is prepared by adding AgF in small portions to molten sulfur at 125°C[3] or by heating AgF–S_8 slowly to 100°C and then to 170°C[4]. In the first procedure glass vessels, and in the second a Ni–Cu (Monel) alloy apparatus, are used. These reactions are performed in vacuum; the gaseous products are pumped off and condensed at −196°C. Separation of the sulfur fluorides by low-T distillation techniques is possible. If the crude mixture is kept at −25°C for some time, isomerization of FSSF to SSF_2 occurs[4,5]. Under special conditions the fluorosulfanes FS_3F and FS_4F are formed[6–8] from gaseous sulfur and AgF.

According to IR[8], NMR[9] and mass spectrometric[8] investigations, SF_2 is also present among the products from sulfur and AgF, but it cannot be isolated pure. In all reactions that lead to SF_2, S_2F_4 is also present, the first intermediate in the disproportionation of SF_2[1,8]:

$$2\ SF_2 \rightarrow FSSF_3 \rightarrow \text{dec} \tag{c}$$

This disproportionation also explains the presence of SF_4 among the reaction products[4,5,9,10–13]. The tetrafluoride is formed when high oxidation state transition-metal fluorides are used as fluorinating agents[14,15], e.g., CoF_3. Under more drastic conditions (250°C, rotating cylinder) SF_6 is the main product[16].

Under controlled conditions, VF_5[17] and UF_6[18] will react with sulfur to give SF_4 in high purity:

$$\tfrac{1}{8}\ S_8 + 4\ VF_5 \xrightarrow{-196°-25°C} SF_4 + 4\ VF_4 \tag{d}$$

$$S_{x(vap)} + 2\ UF_6 \xrightarrow[vac]{300°C} SF_4 + 2\ UF_4 \tag{e}$$

$$S_{x(liq)} + 2\ UF_6 \xrightarrow{150°C} SF_4 + 2\ UF_4 \tag{f}$$

Halogen fluorides and sulfur give good yields of SF_4 e.g., with IF_5[19–21] and BrF_3[22]. Under mild conditions no SF_6 is observed. The former reaction is suggested[21] for the preparation of isotopically enriched $^{35}SF_4$. At 200°C SF_6 is formed in quantitatively[23,24]:

$$\tfrac{5}{8}\ S_8 + 4\ IF_5 \xrightarrow[12\ h]{100°-300°C} 5\ SF_4 + 2\ I_2\ (84\%) \tag{g}$$

$$\tfrac{3}{8}\ S_8 + 4\ BrF_3 \rightarrow 3\ SF_4 + 2\ Br_2 \tag{h}$$

$$\tfrac{1}{x} S_x + 2\ BrF_3 \xrightarrow{200°C} SF_6 + Br_2 \tag{i}$$

$$\tfrac{1}{8}\ S_8 + 4\ ClF \xrightarrow{RT} SF_4 + 2\ Cl_2 \tag{j}$$

$$\tfrac{7}{8}\ S_8 + 4\ ClF_3\text{-}N_2 \xrightarrow[120°-160°C]{RT\ or} 3\ SF_4 + 2\ S_2Cl_2 \tag{k}$$

$$S_8 + xs\ ClF_3 \xrightarrow[N_2]{<100°C} SF_4, SF_6, SF_5Cl, Cl_2 \tag{l}$$

In static systems at RT SF_4 is formed from powdered sulfur and ClF[25,26]. Similar results are obtained in flow systems with ClF_3 at RT with powdered[27] or liq sulfur[28] at 120°-160°C. At 105°C with ClF_3 S—F and S—Cl bonds also form, along with SF_4, SF_6 and SF_5Cl in 15-20% yield[29].

Sulfur reacts with group-VIB and -VB fluorine compounds:

$$\tfrac{1}{8}\ S_8 + 2\ CF_3OF \xrightarrow{RT} SF_4 + 2\ OCF_2 \tag{m}$$

$$\tfrac{1}{8}\ S_8 + NF_3 \rightarrow SSF_2, SF_4, SF_6, NSF \tag{n}$$

$$\tfrac{1}{8}\ S_8 + N_2F_4 \rightarrow SF_5NF_2 \tag{o}$$

Besides several side products, CF_3OF forms SF_4 at RT[30], whereas with NF_3[31] and N_2F_4[32] S—F and N—S bonds also form.

The reaction of sulfur with oxidizing fluorides has advantages, but for the preparation of sulfur fluorides usually the direct fluorination (see §2.3.2.1.) or metathetical reactions (see §2.3.10.1.) are better approaches.

The same is true for sulfur chlorides, where only the direct syntheses from the elements is a useful route. Chlorination of sulfur at 250°–450°C with $AlCl_3$ gives[33] S_2Cl_2; ICl_3 reacts simultaneously as chlorinating agent and as electron-pair acceptor acid[34]:

$$\frac{3}{8}\,S_8 + 7\,ICl_3 \rightarrow 3\,[SCl_3][ICl_4] + 2\,I_2 \qquad \text{(p)}$$

The reaction of sulfur with N,N-dichlorosulfonylamines forms S—N and S—Cl bonds[34]:

$$RNX_2 + \frac{1}{8}\,S_8 \rightarrow RNSX_2\ (+RNSNR + X_2) \qquad \text{(q)}^{37}$$

where $X = Cl$, $R = R'SO_2$; $X = Cl, Br$, $R = CF_3, C_2F_5$. Better results are obtained when sulfur[35,36] chlorides are used instead of sulfur.

(R. MEWS)

1. F. Seel, *Adv. Inorg. Chem. Radiochem., 16*, 297 (1974).
2. F. Seel, R. Budenz, *Chimia, 17*, 355 (1963).
3. F. Seel, R. Budenz, D. Werner, *Chem. Ber., 97*, 1369 (1964).
4. R. D. Brown, G. P. Pez, *Austr. J. Chem., 20*, 2305 (1967).
5. R. D. Brown, G. P. Pez, M. F. O'Dwyer, *Austr. J. Chem., 18*, 627 (1965).
6. F. Seel, R. Budenz, W. Gombler, H. Seitter, *Z. Anorg. Allg. Chem., 380*, 262 (1971).
7. K.-P. Wanczek, C. Bliefert, R. Budenz, *Z. Naturforsch., TeilA, 30*, 1156 (1975).
8. F. Seel, R. Budenz, K.-P. Wanczek, *Chem. Ber., 103*, 3946 (1970).
9. F. Seel, E. Heinrich, W. Gombler, R. Budenz, *Chimia, 23*, 73 (1969).
10. F. Seel, R. Budenze, *Chem. Ber., 98*, 251 (1965).
11. R. L. Kuczkowski, *J. Am. Chem. Soc., 86*, 3617 (1964).
12. D. K. Padma, S. R. Satyanarayana, *J. Inorg. Nucl. Chem., 28*, 2432 (1966).
13. B. Meyer, T. V. Oommen, B. Gotthardt, T. R. Hooper, *Inorg. Chem., 10*, 1632 (1971).
14. J. Fischer, W. Jaenckner, *Angew. Chem., 42*, 810 (1929).
15. W. Schmidt, *Monatsh. Chem., 85*, 452 (1953).
16. J. Nakamura, T. Takeuchi, H. Tomioka, Jpn. Kokai 76–25497 (1976); *Chem. Abstr., 85*, 80,444 (1976).
17. J. A. Canterford, T. A. O'Donnell, *Inorg. Chem., 6*, 541 (1967).
18. J. Aubert, B. Cochet-Muchy, J. P. Cuer, Fr. Pat. 1,586,833 (1970); *Chem. Abstr., 74*, 143,947 (1971).
19. L. R. Belohlav, E. T. McBee, *Ind. Eng. Chem., 51*, 1102 (1959).
20. C. W. Tullock, F. C. Fawcett, W. C. Smith, D. D. Coffman, *J. Am. Chem. Soc., 82*, 539 (1960).
21. G. A. Kolta, G. Webb, J. M. Winfield, *J. Fluorine Chem., 19*, 89 (1981/82).
22. M. J. Nichols, Ph.D. Diss., Durham Univ., 1958.
23. H. Purchelt, B. R. Sabels, T. C. Hoernig, *Geochim. Cosmochim. Acta, 35*, 625 (1971).
24. J. A. Kolta, J. Webb, J. Winfield, *J. Fluor. Chem., 19*, 89 (1981).
25. J. J. Pitts, A. W. Jache US Pat. 3,373,000 (1968); *Chem. Abstr., 68*, 88,678 (1968).
26. J. J. Pitts, A. W. Jache, *Inorg. Chem., 7*, 1661 (1968).
27. F. Nyman, H. L. Roberts, *J. Chem. Soc.*, 3180 (1962).
28. W. Becher, J. Massonne, *Chem.-Ztg., 98*, 117 (1974).
29. H. L. Roberts, Br. Pat. 883,673 (1959); *Chem. Abstr., 56*, 11,217 (1962).
30. R. S. Porter, G. H. Cady, *J. Am. Chem. Soc., 79*, 2625 (1957).
31. O. Glemser, U. Biermann, J. Knaack, *Chem. Ber., 98*, 446 (1965).
32. A. L. Logothetis, G. N. Sausen, R. J. Shozda, *Inorg. Chem., 2*, 173 (1963).
33. P. Hagenmuller, J. Rouxel, J. David, A. Colin, B. L. Neindre, *Z. Anorg. Allg. Chem., 323*, (1963).
34. A. Finch, P. N. Gates, T. H. Page, *Inorg. Chim. Acta, 25*, L49 (1977).
35. A. V. Kirsanov, E. S. Leochenko, L. M. Markovski, in *I.U.P.A.C. 9, Organic Sulfur Chemistry*, R. K. A. Freidlina, A. E. Skorova, eds., Pergamon Press, Oxford, 1981, p. 109. Oxidative aminations of sulfur and sulfur-halogen compounds are reviewed.

36. J. S. Borovikove, E. S. Lerchenko, E. I. Borovik, E. A. Darmokhval, *J. Org. Chem. USSR*, *20*, 190 (1984).
37. M. Geisel, Ph.D. Diss, Göttingen 1984.

2.3.2.3.3. to Give the Se–Halogen Bond.

Selenium is oxidized by halogen halides, oxygen, sulfur, nitrogen halides, PCl_5 and transition-metal halides. Fluorinating agents, such as halogen fluorides or metal fluorides[1], form SeF_4 e.g.:

$$Se + 4\ ClF \xrightarrow{RT} SeF_4 + 2\ Cl_2 \qquad (a)$$

$$3\ Se + 4\ ClF_3 \xrightarrow[90°C]{SeF_4} 3\ SeF_4 + 2\ Cl_2 \qquad (b)$$

whereas ClF reacts with powdered Se at RT to form SeF_4 and Cl_2; no SeF_6 is observed[2]. The same products are also produced with xs ClF_3, either neat[3] or in SeF_4[4] or HF [5,6] as solvents. In the fluorination with NO_2F [7], $NO_2F \cdot 5\ HF$ [8], NOF[9] and NOF · 3 HF [8], SeF_4 or its fluoro complexes are the products:

$$4\ NO_2F + Se \rightarrow 4\ NO_2 + SeF_4 \xrightarrow{NO_2F} NO_2SeF_5 \qquad (c)$$

$$NOF \cdot 3\ HF + Se \rightarrow SeF_4 \cdot x\ NOF \qquad (d)$$

The tetrafluoride is also obtained[10] by heating Se with AgF:

$$5\ Se + 4\ AgF \rightarrow SeF_4 + 2\ Ag_2Se_2 \qquad (e)$$

From molten ICl and Se, $SeCl_4$ is formed[11]:

$$Se + 4\ ICl \rightarrow SeCl_4 + 2\ I_2 \qquad (f)$$

from IBr, $SeBr_4$[12]:

$$Se + 4\ IBr \rightarrow SeBr_4 + 2\ Br_2 \qquad (g)$$

Sulfur chlorides oxidize Se to $SeCl_4$ or Se_2Cl_2[13–15]:

$$Se + xs\ SCl_2 \rightarrow SeCl_4 + S_2Cl_2 \qquad (h)$$

$$2\ Se + S_2Cl_2 \xrightarrow{600°C} Se_2Cl + \tfrac{2}{x} S_x \qquad (i)$$

$$Se + xs\ SO_2Cl_2 \xrightarrow[-SO_2]{SbCl_5} [SeCl_3][SbCl_6] \qquad (j)$$

With SCl_2, Se gives $SeCl_4$ and S_2Cl_2 exothermically[13]. With S_2Cl_2, reaction occurs at >600°C. Oxidative chlorination with SO_2Cl_2 in the presence of $SbCl_5$ gives trichloroselenium hexachloroantimonate[15], but the oxidizing agent may also be $SbCl_5$ because PCl_5, too, reacts slowly to form[16] $PCl_5 \cdot SeCl_4$. The equilibrium[17]:

$$Se + SeX_2 \rightleftharpoons Se_2X_2 \qquad (k)$$

(X = Cl, Br) may be understood as an oxidation of the element by the dihalides. Chlorination by metal halides, such as $HgCl_2$ or Hg_2Cl_2, also forms Se_2Cl_2[14]. With $[NO][SbCl_6]$ in liq SO_2 $[Se_9Cl]^+[SbCl_6]^-$ is obtained[18].

Aminoselenium chlorides are obtained from Se and dichloroamines[18]:

$$RNCl_2 + Se \rightarrow RNSeCl_2 \qquad (l)$$

($R = C_6F_5SO_2, C_6F_5CO$) or N-chlorocarboxamides:

$$2\ RCONHCl + Se \rightarrow (RCONH)_2SeCl_2 \qquad (m)$$

(R = aryl[21,22], Me[22]).

$$Na[RSO_2NCl] + Se \xrightarrow[20°C,\ 2d]{ClCH_2CH_2Cl} Na_2[(RSO_2N)_2SeCl_2] \qquad (n)$$

where $R = C_6H_5$, p-$CH_3C_6H_4$[23]. Bromoamines form Se_2Br_2[24]:

$$2\ (CF_3)_2NBr + 2\ Se \rightarrow Se_2Br_2 + [(CF_3)_2N]_2 \qquad (o)$$

$$CuBr_2 + 3\ Se \rightarrow Se_2Br_2 + CuSe \qquad (p)^{25}$$

Iodine salts also oxidize Se[26]:

$$2\ [I_2][Sb_2F_{11}] + 2\ Se \xrightarrow[RT]{SO_2} Se_2I_4(Sb_2F_{11})_2 \qquad (q)$$

(R. MEWS)

1. A. Engelbrecht, W. Sladky, *Adv. Inorg. Chem. Radiochem.*, *24*, 189 (1981).
2. J. J. Pitts, A. W. Jache, *Inorg. Chem.*, *7*, 1661 (1968).
3. W. Hückel, *Nachr. Akad. Wiss. Göttingen, Math. Physik., Kl.*, Nr. 1, 36 (146/48); *0Chem. Abstr.*, *43*, 6793 (1949).
4. G. A. Olah, M. Nojima, I. Kerekes, *J. Am. Chem. Soc.*, *96*, 925 (1974).
5. A. F. Clifford, H. C. Beachell, W. M. Jack, *J. Inorg. Nucl. Chem.*, *5*, 57 (1957).
6. A. F. Clifford, A. G. Morris, *J. Inorg. Nucl. Chem.*, *5*, 71 (1957).
7. E. E. Aynsley, G. Hetherington, P. L. Robinson, *J. Chem. Soc.*, 1119 (1954).
8. F. Seel, W. Birnkraut, D. Werner, *Chem. Ber.*, *95*, 1264 (1962).
9. F. Seel, H. Massat, *Z. Anorg. Allg. Chem.*, *280*, 186 (1955).
10. O. Glemser, F. Meyer, A. Haas, *Naturwissenschaften*, *52*, 130 (1965).
11. V. Gutmann, *Z. Anorg. Allg. Chem.*, *264*, 169 (1951).
12. V. Gutmann, *Monatsh. Chem.*, *82*, 280 (1951).
13. R. C. Paul, S. Singh, *J. Indian Chem. Soc.*, *35*, 909 (1958).
14. H. H. Anderson, C. Steinbrecher, N 70-8511 (1958) 1; *Nucl. Sci. Abstr.*, *12*, 15,326 (1958).
15. J. R. Masaguer, *An. R. Soc. Esp. Fis. Quim.*, *Ser. B53*, 518 (1957).
16. V. Lenher, C. H. Kao, *J. Am. Chem. Soc.*, *48*, 1550 (1926).
17. M. Lundkvist, M. Lellep, *Acta Chem. Scand.*, *24*, 255 (1970).
18. K. Faggiani, R. J. Gillespie, J. W. Kolis, K. C. Mathotra, *J. Chem. Soc. Chem. Commun*, 591 (1987).
19. A. V. Zibarev, G. N. Dolenko, S. A. Krupoder, L. N. Mazalov, O. Kh. Poleshchuk, G. G. Furin, G. G. Yakobson, *J. Org. Chem. USSR (Engl. Transl.)*, *16*, 347 (1980).
20. J. S. Thrasher, K. Seppelt, *Z. Anorg. Allg. Chem.*, *507*, 7 (1983).
21. N. Ya. Derkach, T. V. Lyapina, E. S. Levchenko, *J. Org. Chem. (Engl. Transl.)*, *10*, 145 (1974).
22. N. Ya. Derkach, T. V. Lyapina, E. S. Levchenko, *J. Org. Chem. (Engl. Transl.)*, *14*, 256 (1978).
23. G. G. Barashenkov, N. Ya. Derkack, *J. Org. Chem. (Engl. Transl.)*, *22*, 1189 (1986).
24. R. C. Dobbie, H. J. Emeleus, *J. Chem. Soc. Suppl. No 1*, 5894 (1964).
25. A. A. Babitsyna, T. A. Emel'yanova, V. T. Kalinnikov, *Russ. J. Inorg. Chem.*, *25*, 288 (1980).
26. W. A. Shantha Nandana, J. Passmore, P. S. White, C. M. Wong, *J. Chem. Soc. Chem. Comm.*, 1098 (1982).

2.3.2.3.4. to Give the Te–Halogen Bond.

Tellurium behaves toward oxidizing agents as its lighter homologues do, and the same reagents are used. However, because this element is more easily oxidized, reaction occurs even with phosphorus bromides and iodides. If ClF_3 is dropped on finely divided Te in HF at $-78°C$, TeF_6 is formed[1]:

$$Te + 2\ ClF_3 \xrightarrow[-78°C]{HF} TeF_6 + Cl_2 \quad (a)$$

whereas ClF oxidizes the element to give either TeF_4:

$$Te + 4\ ClF \xrightarrow[\text{Ni-Cu alloy autoclave}]{200°C} TeF_4 + 2\ Cl_2 \quad (b)$$

or TeF_6 depending on the stoichiometry[2]:

$$Te + 6\ ClF \xrightarrow[\text{Ni-Cu, autoclave}]{200°C} TeF_6 + 3\ Cl_2 \quad (c)$$

With Te and TeF_6 disproportionation takes place[8] slowly to give TeF_4 in poor yield:

$$Te + 2\ TeF_6 \xrightarrow[100\ h]{180°C} 3\ TeF_4 \quad (d)$$

Only CoF_3 gives the hexavalent fluoride[3]. Fluoro complexes or TeF_4 are formed by other oxidizing nonmetal fluorides, (e.g., NOF[4], NO_2F[5]):

$$Te + 5\ NOF \cdot 3\ HF \xrightarrow{20°\text{–}120°C} [NO][TeF_5] + 4\ NO + 15\ HF \quad (e)^4$$

$$Te + 5\ NO_2F \xrightarrow[\text{exothermic}]{} [NO_2][TeF_5] + 4\ NO_2 \quad (f)^5$$

and metal fluorides (e.g., PbF_2[6], CuF_2[7], FeF_3[7]):

$$2\ CuF_2 + Te \xrightarrow{800°C} 2\ Cu + TeF_4 \quad (68\%) \quad (g)^7$$

$$4\ FeF_3 + Te \xrightarrow{800°C} 4\ FeF_2 + TeF_4 \quad (74\%) \quad (h)^7$$

Metal fluorides require high T. With FeF_3 the yield of TeF_4 is higher, but the product is contaminated by FeF_2. Under milder conditions fluorides are formed from tellurides[9]:

$$CuTe + 2\ CuF_2 \xrightarrow{195°C} 3\ Cu + TeF_4 \quad (i)$$

The preparation of $TeCl_4$ from ICl and Te can be used as synthetic method for the tetrachloride[10]:

$$4\ ICl + Te \rightarrow TeCl_4 + 2\ I_2 \quad (j)$$

If ICl (mp 27.2°C) is slowly added to finely powdered Te, conversion is quantitative. With sulfur chlorides (SCl_2 [11], SO_2Cl_2 [12]) $TeCl_4$ is formed:

$$Te + 4\ SCl_2 \rightarrow TeCl_4 + 2\ S_2Cl_2 \quad (k)^{11}$$

$$Te + 2\ SO_2Cl_2 \xrightarrow{SbCl_5} [TeCl_3][SbCl_6] + 2\ SO_2 \quad (l)^{12}$$

PCl_5 [13] also gives the tetrachloride: The chlorinating power of SO_2Cl_2 is enhanced by strong electron-pair acceptor acids. The acid itself can be the oxidizing agent. Because $TeCl_2$ exists only in the vapor[14], in the solid a mixture of Te and $TeCl_4$ is present[15], earlier claims for the formation of $TeCl_2$ from the Te and metal chlorides[16] are incorrect.

An active bromination agent for Te is IBr, the molten reagent (mp 41°C) is added slowly to powdered Te[17]:

$$4\ IBr + Te \rightarrow TeBr_4 + 2\ I_2 \quad (m)$$

The reaction mixture is extracted with CCl_4, leaving $TeBr_4$, as a residue.

Bromination also occurs[13] with PBr_5 at 100°C; with phosphorus iodides TeI_4 is formed[17]:

$$Te + 2\ P_2I_4 \xrightarrow{100°C} TeI_4 + P_4 \quad (n)$$

$$3\ Te + 4\ PI_3 \xrightarrow{CS_2} 3\ TeI_4 + P_4 \quad (o)$$

(R. MEWS)

1. A. F. Clifford, A. G. Morris, *J. Inorg. Nucl. Chem.*, *5*, 71 (1957).
2. J. J. Pitts, A. W. Jache, US Pat. 3,373,000 (1968); *Chem. Abstr.*, *68*, 88,678 (1968).
3. H. R. Gwinn, Y-568 (1950) 1/7; *Nucl. Sci. Abstr. 4*, 2641 (1950).
4. F. Seel, W. Birnkraut, D. Werner, *Chem. Ber.*, *95*, 1264 (1962); *Angew Chem.* 73, 806 (1961).
5. E. E. Aynsley, G. Hetherington, P. L. Robinson, *J. Chem. Soc.*, 1119 (1954).
6. E. Montiguie, *Ann. Pharm. Fr.*, *1*, 107 (1943).
7. J. H. Moss, R. Ottie, J. B. Wilford, *J. Fluorine Chem.*, *3*, 317 (1973/74).
8. J. H. Junkins, H. A. Bernhardt, E. J. Barber, *J. Am. Chem. Soc.*, *74*, 5749 (1952).
9. M. J. Nichols, Ph.D. Diss., Univ. Durham, 1958.
10. V. Gutmann, *Z. Anorg. Allg. Chem.*, *264*, 169 (1959).
11. R. C. Paul, G. D. Singh, *J. Indian Chem. Soc.*, *35*, 909 (1958)
12. J. R. Masaguer, *An. R. Soc. Esp. Fis. Quim., Ser. B*, *53*, 518 (1957); *Chem. Abstr.*, *55*, 12131 (1961).
13. E. Montiguie, *Bull. Soc. Chim. Fr.*, 180 (1948).
14. S. A. Ivashin, E. S. Petrov, *Izv. Sibirsk Otd. Akad. Nauk SSSR Ser. Khim. Nauk*, *5*, 48 (1970); *Chem. Abstr. 74*, 57,890 (1971).
15. G. C. Christensen, J. Alstadt, *Radiochem. Radioanal. Lett.*, *13*, 227 (1973).
16. H. H. Anderson, L. Steinbrecher, NYO-8511, 1 (1958); *Nucl. Sci. Abstr.*, *12*, 15,326 (1958).
17. V. Gutmann, *Monatsh. Chem.*, *82*, 280 (1951).

2.3.2.4. from Halogenation by Organic Halides.

Bonds to carbon and to halogen are formed only rarely from organic halides and group-VIB elements. Only halogenation is observed, giving products obtainable also directly from the elements.

Carbon tetrabromide brominates sulfur at 120°–160°C, giving[1] S_2Br_2, Br_2 and CS_2. In the irradiation ($\lambda = 253.7$ nm) of sulfur in n-propylbromide, S_2Br_2 is formed[2]. From

Se and CCl_4 in a high-frequency discharge also CCl_3SeCl is observed[3] besides $SeCl_2$, CSe_2, C_2Cl_6, etc.; $SeCCl_2$ may be an intermediate.

Toward Te, carbon chlorides react only as chlorinating agents[4], but Te—C and halogen bonds are formed[5] simultaneously from diphenyliodonium chloride and Te:

$$[Ph_2I]Cl + Te \rightarrow Ph_2TeCl_2 + \cdots \quad (a)$$

Diorganotellurium diiodides are produced in yields of $\leq 50\%$ from Te and organic iodides[6]:

$$2\ RI + Te \rightarrow R_2TeI_2 \quad (b)$$

($R = CH_3$, C_2H_5, $C_6H_5CH_2$, C_6F_5)[7–11], whereas Te and $BrCH_2CH_2I$ give a mixed halide[2]:

$$Te + 2\ BrCH_2CH_2I \rightarrow TeBr_2I_2 + 2\ C_2H_4 \quad (c)$$

Organic halides $X(CH_2)_nX$ combine with Te to give telluracycloalkane 1,1-dihalides[12–14], if n = 4 or 5, but dihalides with n = 2, or 3 give indefinite products[12]. The reactivity of the dihalides decreases with decreasing atomic mass of the halogen.

Reviews of the organic chemistry of Se[15] and Te[6] are available.

(R. MEWS)

1. H. V. A. Briscoe, J. B. Peel, J. R. Rowlands, *J. Chem. Soc.*, 1766 (1929).
2. M. Elbanowski, *Poznan Tow. Pryj. Nauk, Pr. Kom. Mat. Przyr. Pr. Chem.*, *12*, 95 (1971); *Chem. Abstr.*, *75*, 43,039 (1971).
3. R. Steudel, *Z. Naturforsch., Teil B*, *23*, 1163 (1968); *Tetrahedron Lett.*, 1845 (1967).
4. E. Montignie, *Bull. Soc. Chim. F.*, 180 (1948).
5. R. B. Sandin, F. T. McClure, F. Irwin, *J. Am. Chem. Soc.*, *61*, 2944 (1939).
6. K. J. Irgolic in *The Organic Chemistry of Tellurium*, Gordon and Breach, New York, 1974, p. 143 ff.
7. S. C. Cohen, M. L. N. Reddy, A. G. Massey, *J. Organomet. Chem.*, *11*, 563 (1968).
8. E. Demarcay, *Bull. Soc. Chim. Fr.*, 99 (1883).
9. A. Scott, *Proc. Chem. Soc. (London)*, *20*, 156 (1904).
10. R. H. Vernon, *J. Chem. Soc.*, 86 (1920).
11. R. H. Vernon, *J. Chem. Soc.*, 687 (1921).
12. M. V. Farrar, J. M. Gulland, *J. Chem. Soc.*, 12 (1945).
13. G. T. Morgan, H. Burgess, *J. Chem. Soc.*, 32 (1928).
14. G. T. Morgan, F. H. Burstall, *J. Chem. Soc.*, 180 (1931).
15. K. J. Irgolic, M. V. Kudchadker, in *Selenium*, R. A. Zingaro, W. C. Cooper, eds., Van Nostrand-Reinhold, New York, 1974, p. 408 ff.

2.3.2.5. from Oxidation of Elemental Halogens by Ozone.

No products with O—F bonds are isolated[1,2] from the reaction of O_3 and F_2 at 0°–20°C, although $[OF]^{\cdot}$ radicals are formed in the initial step[2,3]:

$$O_3 + F_2 \rightarrow [OF]^{\cdot} + O_2 \rightarrow \tfrac{1}{2}\,F_2 + \tfrac{1}{2}\,O_2 \quad (a)$$

At −150°C from the photolysis (365 nm) of 2:1, O_3 and F_2, O_2F_2 is prepared, but no oxygen fluorides is found at −78°C[4]. Formation of O_2F_2 might result from the dimerization of $[OF]^{\cdot}$ radicals[4] or from fluorine addition to O_2 (see §2.3.2.1.1.). Chlorine and O_3 give[5] Cl_2O_6:

$$Cl_2 + 6\ O_3 \xrightarrow{620\ nm} Cl_2O_6 \rightarrow Cl_2O_7 \quad (b)$$

which decomposes to form[6,7] Cl_2O_7.

In contradiction to this, Cl_2O_7 is found as the only Cl—O product in the light-induced reaction[8]:

$$Cl_2 + 7\,O_3 \xrightarrow[365\text{ nm}]{-10.5°\text{ to }0°C} Cl_2O_7 + 7\,O_2 \qquad \text{(c)}$$

Bromine and O_3 form BrO_2 at low T[9]:

$$Br_2 + 4\,O_3 \xrightarrow[CFCl_3]{-50°C} 2\,BrO_2 + 4\,O_2 \qquad \text{(d)}$$

At RT, higher oxides form, which might be BrO_3[10] or Br_3O_8[11]. Iodine and O_3 give I_3O_9 independent of the conditions[12–15]:

$$\tfrac{3}{2}\,I_2 + 9\,O_3 \rightarrow I_3O_9 + 9\,O_2 \qquad \text{(e)}$$

(R. MEWS)

1. E. Briner, R. Tolun, *Helv. Chim. Acta.*, *31*, 937 (1948).
2. E. H. Staricco, J. E. Sicre, H. J. Schumacher, *Z. Physik. Chem.*, *31*, 385 (1962).
3. H. G. Wagner, C. Zetsch, J. Warnatz, *Ber. Bunsenges. Phys. Chem.*, *76*, 526 (1972).
4. A. D. Kirshenbaum, *Inorg. Nucl. Chem. Lett.*, *1*, 121 (1965).
5. M. Bodenstein, P. Hartek, E. Padelt, *Z. Anorg. Allg. Chem.*, *147*, 233 (1925).
6. A. J. Arvia, W. H. Basualdo, H. J. Schumacher, *Z. Anorg. Allg. Chem.*, *286*, 58 (1956).
7. H. J. Schumacher, *Z. Physik. Chem. (Leipzig)*, *13*, 353 (1957).
8. R. W. Davidson, D. G. Williams, *J. Phys. Chem.* *77*, 2515 (1973).
9. M. Schmeisser, K. Jörger, *Angew. Chem.*, *71*, 523 (1959).
10. A. Pflugmacher, H. J. Raben, H. Dahman, *Z. Anorg. Allg. Chem.*, *279*, 313 (1959).
11. A. J. Arvia, P. J. Aymonino, H. J. Schumacher, *Z. Anorg. Allg. Chem.*, *298*, 1 (1959).
12. R. K. Ball, J. R. Partington, *J. Chem. Soc.*, 1258 (1935).
13. F. Fichter, F. Rohner, *Chem. Ber.*, *42*, 4093 (1909).
14. T. Kikindai, *C. R. Hebd. Seances Acad. Sci.*, *240*, 1102 (1955).
15. M. Schmeisser, K. Brändle, *Adv. Inor. Chem. Radiochem.*, *5*, 41 (1963).

2.3.3. from Cleavage of the Group VIB–Hydrogen Bond (Excluding Polonium)

2.3.3.1. by Halogens.

The cleavage of group VIB—H bonds by halogens to form a VIB-halogen bond is limited to compounds of oxygen and sulfur. Compounds containing Se—H and Te—H bonds are easily oxidized, and simple substitution by halogen is rare.

The cleavage of O—H bonds by halogens is important for the synthesis of hypohalites and fluoroxy derivatives. With H_2O, halogens react to form HOX[1]:

$$H_2O + X_2 \rightarrow HOX + HX \qquad \text{(a)}$$

(X = Cl, Br, I). However, the sat HOX conc is only 0.030 M for Cl_2 and decreases for Br_2 and I_2. With F_2, HOF is not present in aq F_2 but can be prepared by a low-T flow reaction of F_2 with ice[2].

Reaction of X_2 with aq $[OH]^-$ yields nearly stable $[OCl]^-$ at 22°C and $[OBr]^-$ solutions at <0°C, but $[OI]^-$ is unstable under all conditions:

$$2\,[HO]^- + X_2 \rightarrow [OX]^- + X^- + H_2O \qquad \text{(b)}$$

$$3\,[OX]^- \rightarrow [XO_3]^- + 2\,X^- \qquad \text{(c)}$$

(X = Cl, Br). Reaction of F_2 with 2% aq NaOH yields[3] OF_2. Small amounts of OF_2 can also be obtained[4] in excellent yield from F_2 with H_2O absorbed on xs KF:

$$KF{\cdot}H_2O + KF + F_2 \xrightarrow{KF} 2\,KF{\cdot}HF + OF_2 \qquad \text{(d)}$$

The preparation of other O—F compounds is achieved with F_2 and alcohols and acids. For alcohols, the yields are low and the O—F bond may be formed not directly, by cleavage of O—H bonds, but from an intermediate ketone or acid fluoride. For example, fluorination of CH_3OH in a Cu tube at 350°C produces CF_3OF in excellent yield, but COF_2 is probably produced first and subsequently fluorinated to CF_3OF. Preparative reactions are found with $HONO_2$, $HOClO_3$, $HOC(CF_3)_3$ and CF_3CO_2H to form in each case the **explosive fluoroxy compound**.

$$HONO_{2(l)} + F_2 \xrightarrow{25°C} \underset{90\%}{FONO_2} \qquad \text{(e)}^5$$

$$HOClO_{3(l)}70\% + F_2 \xrightarrow{25°C} \underset{90\%}{FOClO_3} \qquad \text{(f)}^5$$

$$CF_3CO_2H_{(g)} + F_2 \xrightarrow{25°C} \underset{25\%}{CF_3CO_2F} \qquad \text{(g)}^{5,6}$$

$$HOC(CF_3)_3 + F_2 \xrightarrow[CsF]{-78°C} \underset{98\%}{FOC(CF_3)_3} \qquad \text{(h)}^7$$

$$CH_3CO_2H + F_2 \xrightarrow{K[O_2CH_3]} CH_3CO_2F \qquad \text{(i)}^8$$

Compounds containing the -OOF group are also formed from hydroperoxides with F_2 over CsF:

$$ROOH + F_2 \xrightarrow[-78°C]{CsF} \underset{35\%}{ROOF} \qquad \text{(j)}$$

where R = CF_3[9], SF_5[10].

Reaction of S—H bonds with halogens leads to formation of S—X bonds (X = F, Cl, Br), but simple substitution of H by X is rare; e.g., F_2 and H_2S form SF_6 and HF, whereas Cl_2 and H_2S form S_2Cl_2 and HCl, followed by further reaction of S_2Cl_2 with H_2S to form polysulfanes[11]. Simple substitution of hydrogen by halogen seems to be limited to compounds where the hydrogen is made more acidic by the presence of electron-withdrawing groups, and then only with X = Cl or Br:

$$CF_3C(O)SH + X_2 \xrightarrow{-78^\circ C} \underset{98\%}{CF_3C(O)SX} + HX \qquad (k)^{12}$$

$$CF_3SSH + Cl_2 \xrightarrow{-78^\circ C} \underset{64\%}{CF_3SSCl} + HCl \qquad (l)^{13}$$

(D. D. DESMARTEAU, B. B. RANDOLPH)

1. F. A. Cotton, G. Wilkinson, *Advanced Inorganic Chemistry*, 4th ed., Wiley, New York, 1980, p. 556.
2. E. H. Appelman, *Acc. Chem. Res.*, *6*, 113 (1973).
3. Yost, D. M., *Inorg. Synth.*, *1*, 109 (1934).
4. A. H. Borning, K. E. Pullen, *Inorg. Chem.*, *8*, 1791 (1969).
5. M. Lustig, J. M. Shreeve, *Adv. Fluorine Chem.*, *7*, 175 (1973).
6. A. Sekiya, D. D. DesMarteau, *Inorg. Chem.*, *19*, 1328 (1980). The reaction CF_3CO_2H with F_2 over CsF produces CF_3CO_2F in high yield, but it cannot be isolated because it is subsequently fluorinated again to form $CF_3CF(OF)_2$.
7. D. D. DesMarteau, unpublished results.
8. E. H. Appelman, M. H. Mendelsohn, H. Kim, *J. Am. Chem. Soc.*, *107*, 6515 (1985).
9. D. D. DesMarteau, *Inorg. Chem. 11*, 193 (1972).
10. D. D. DesMarteau, R. M. Hammaker, *Israel J. Chem.*, *17*, 103 (1978).
11. K. W. C. Burton, P. Machmer, in *Inorganic Sulfur Chemistry*, G. Nickless, ed., Elsevier, New York, 1968, p. 339.
12. W. V. Rochat, G. L. Gard, *J. Org. Chem.*, *34*, 4173 (1969).
13. W. Gambler, *Z. Anorg. Allg. Chem.*, *416*, 235 (1975).

2.3.3.2. by Oxidizing Halides.

Oxidizing halides group VIB–halogen bonds cleave VIB–hydrogen bonds to form O—X bonds (X = Cl, Br, I). As for the related reaction with elemental halogens, this reaction can form S—X bonds, but is of little use for Se—X or Te—X bonds.

The most important reaction is that of ClF with acidic hydrogen or oxygen[1–5]. Reactions are carried out at lower T, depending on the stability of OCl compound desired, in reactors resistant to HF:

$$ROH + ClF \xrightarrow{-196^\circ - 22^\circ C} ROCl + HF \qquad (a)$$

The HF product can be distilled or absorbed on NaF (see Table 1). The scope of this reaction is not completely known. Many strong acids react in high yield, but $HOPOF_2$ and $(HO)_2SO_2$ do not lead to the expected products.

Other halogen fluorides also react with acids to yield the corresponding halogen derivatives. This reaction may be more general than is currently known; e.g., Br(III) compounds form in excellent yield:

$$3\ F_5SeOH + BrF_3 \rightarrow Br(OSeF_5)_3 + 3\ HF \qquad (b)^{14}$$

$$3\ HOSO_2F + BrF_3 \rightarrow Br(OsO_2F)_3 + 3\ HF \qquad (c)^{15}$$

TABLE 1. REACTIONS OF ClF WITH ROH TO PRODUCE ROCl

ROH	T (°C)	Product (yield in %)	Ref.
$HONO_2$	−78	$ClONO_2$ (90)	6
$(CF_3)_3COH$	22	$(CF_3)_3COCl$ (~100)	7
CF_3CO_2H	−111 to −78	CF_3CO_2Cl (~100)	8,9
CF_3SO_2OH	−111 to −78	CF_3SO_2OCl (~100)	8,10
CF_3OOH	−111	CF_3OOCl (95)	11
F_5SeOH	22	F_5SeOCl (92)	12
cis,trans-$TeF_4(OH)_2$	−78	cis,trans-$TeF_4(OCl)_2$ (~100)	13

The pseudointerhalogen compound, $ClOSO_2F$, is sometimes better than ClF; e.g., CF_3CO_2Cl can be prepared in excellent yield[16]:

$$CF_3CO_2H + ClOSO_2F \rightarrow CF_3CO_2Cl + HOSO_2F \tag{d}$$

The advantage of Eq. (d) over that with ClF is that the $HOSO_2F$ is of lower volatility than HF, and the reaction may be carried out in borosilicate glass. A related reaction yields $I(OSO_2CF_3)_3$ from $I(OSO_2F)_3$, which is generated[17] in situ from I_2 and $S_2O_6F_2$:

$$I_2 + 2\ S_2O_6F_2 + 6\ HOSO_2CF_3 \xrightarrow{22°C} 2\ I(OSO_2CF_3)_3 + 6\ HOSO_2F \tag{e}$$

Organic compounds containing O—X bonds (X = Cl, Br) are prepared in solution from $[OX]^-$ and HOX, e.g., with t-butanol[18]:

$$(CH_3)_3COH + HOBr \xrightarrow[22°C]{H_2O} H_2O + (CH_3)_3COBr \tag{f}$$

$$(CH_3)_3COH + [OCl]^- \xrightarrow[20°C]{[OH]^-} (CH_3)_3COCl + [OH]^- \tag{g}$$

In Eq. (g) the $[OCl]^-$ is generated by passing Cl_2 into aq alkaline $(CH_3)_3COH$.

Finally, certain N-bromo compounds, expecially N-bromoacetamide (NBA) and N-bromosuccinimide (NBS) form HOBr and ROBr in aq soln[19,20]:

$$NBA + CH_3OH \xrightarrow{CH_3OH} CH_3OBr + \text{acetamide} \tag{h}$$

$$NBS + H_2O \xrightarrow{H_2O} HOBr + \text{succinimide} \tag{i}$$

(D. D. DESMARTEAU, B. B. RANDOLPH)

1. F. Aubke, D. D. DesMarteau, *Fluorine Chem. Rev.*, *8*, 73 (1977).
2. C. J. Schack, K. O. Christe, *Israel J. Chem.*, *17*, 20 (1978).
3. F. M. Mukhametshin, *Russ. Chem. Rev.* (*Engl. Transl*), *49*, 668 (1980).
4. F. Haspel-Hentrich, J. M. Shreeve, *Inorg. Synth.*, *24*, 58 (1988).
5. J. M. Shreeve, *Adv. Inorg. Chem. Radiochem.*, *26*, 119 (1983).
6. C. J. Schack, *Inorg. Chem.*, *6*, 1938 (1967).
7. D. E. Young, L. R. Anderson, D. E. Gould, W. B. Fox, *J. Am. Chem. Soc.*, *92*, 2313 (1970).
8. D. D. DesMarteau, *J. Am. Chem. Soc.*, *100*, 340 (1978).
9. I. Tari, D. D. DesMarteau, *Inorg. Chem.*, *11*, 3205 (1979).
10. Y. Katsuhara, R. M. Hammaker, D. D DesMarteau, *Inorg. Chem.*, *19*, 607 (1980).

11. C. T. Ratcliffe, C. V. Hardin, L. R. Anderson, W. B. Fox, *J. Am. Chem. Soc.*, *93*, 3886 (1971).
12. P. Huppmann, D. Lentz, K. Seppelt, *Z. Anorg. Allg. Ch9em.*, *472*, 26 (1981).
13. B. Potter, D. Lentz, H. Pritzkow, K. Seppelt, *Angew. Chem., Int. Ed. Engl.*, *20*, 1036 (1981).
14. K. Seppelt, *Chem. Ber.*, *106*, 157 (1973).
15. K. Seppelt, private communication.
16. C. J. Schack, K. O. Christe, *J. Fluorine Chem.*, *12*, 325 (1978).
17. J. R. Dalziel, F. Aubke, *Inorg. Chem.*, *12*, 2707 (1973).
18. C. Walling, A. Padwa, *J. Org. Chem.*, *27*, 2976 (1962).
19. M. Fieser, L. F. Fieser, *Reagents for Organic Synthesis*, Vol. 4, Wiley, New York, (1974), p. 49.
20. H. Rapaport, C. H. Lovell, H. R. Reist, M. E. Warren Jr., *J. Am. Chem. Soc.*, *89*, 1942 (1967).

2.3.4. from Cleavage of the Group VIB–Carbon bond

2.3.4.1. by Halogens.

The cleavage of group VIB–carbon bonds by halogens is rarely used to form group VIB–halogen bonds. Although many such bonds can be cleaved by halogen, the compounds derived are often more readily obtained by other routes. However, certain sulfur–fluorine compounds are exceptions.

Cleavage of S—C and Se—C bonds by X_2 (X = F, Cl, Br) proceeds in high yield, e.g.:

$$CS_2 \xrightarrow[F_2]{200°–300°C} CF_4 + 2\, SF_6 \qquad (a)^{1}$$

$$KSCN + 6\, F_2 \xrightarrow{-78°C} CF_3NF_2 + SF_6 + KF \qquad (b)^{2}$$

$$CS_2 + 3\, Cl_2 \xrightarrow{\Delta} CCl_4 + S_2Cl_2 \qquad (c)^{3}$$

$$SeCS + 4\, Cl_2 \xrightarrow{CCl_4} SeCl_4 + ClSCCl_3 \qquad (d)^{4}$$

$$RSeCN + Br_2 \rightarrow RSeBr + BrCN \qquad (e)^{5}$$

Only Eq. (e) is used in the laboratory, but Eq. (b) is an excellent source of the difluoroamine.

In fluorine chemistry, compounds containing S—C bonds are fluorinated. Some of these reactions are the only route to certain S—F bonds. However, higher yields of most are obtained using oxidizing fluorides, such as AgF_2 or CoF_3, or by electrochemical fluorination (see §2.3.4.2.).

(D. D. DESMARTEAU, B. B. RANDOLPH)

1. B. Cochet-Muchy, J. P. Cuer, Fr. Pat. 1,445,502 (1965/66), *Chem. Abstr.*, *66*, 47,928 (1967).
2. J. K. Ruff, *J. Org. Chem.*, *32*, 1675 (1967).
3. *Gmelins Handbuch der Anorganischen Chemie*, Syst. Nr. 9, Erg. Band. 2, Springer-Verlag, Berlin, 1978, p. 239.
4. H. V. A. Brisioe, J. B. Pell, P. L. Robinson, *J. Chem. Soc.*, 1048 (1929).
5. K. W. Bagnall, *The Chemistry of Selenium, Tellenium and Polonium*, Elsevier, New York, 1966, p. 164.

2.3.4.2. by Oxidizing Halides or Electrochemical Fluorination.

Cleavage of group VIB-carbon bonds by oxidizing halides to form group-VIB halides is important only for sulfur fluorides. Many reactions of S-, Se- and Te-carbon compounds with oxidizing halides lead to novel halogen derivatives, but these rarely involve cleavage of the group VIB-carbon bonds. Instead, they involve oxidative addition of halogen and cleavage of group VIB—VIB bonds[1–3].

The oxidizing halides used to fluorinate C—S bonds are AgF_2, CoF_3 and the high-valent Ni fluoride formed on the surface the Ni anode in the electrochemical fluorination process. Examples of the three fluorination types are given in Table 1. Yields are low, but several products are not obtained by other means. Weaker C—Se and C—Te bonds render similar reactions with these materials unfavorable.

TABLE 1. FLUORINATION BY CLEAVAGE OF C—S BONDS

Compound	Method	Products	Ref.
CS_2	CoF_3, 250°C	CF_3SF_5 (40%), CF_4, SF_6	4
COS	CoF_3, 200°C	SF_6, COF_2	5
CS_2	AgF_2	CF_3SF_3, SF_4, SF_6	6
$(CF_3S)_2C{=}S$	AgF_2	CF_3SF_3, CF_3SF_5, $(CF_3S)_2$	6
CS_2	ECF[a]	CF_3SF_5 (90%)	7
$(n\text{-}C_4H_9)_2S$	ECF[a]	$n\text{-}C_4F_9SF_5$, $(CF_2)_4$, $(n\text{-}C_4F_9)_2SF_4$	8
$CH_3S(CH_2)_2SCH_3$	ECF[a]	CF_3SF_5, $C_2F_5SF_4CF_3$	9

[a] ECF = electrochemical fluorination.

(D. D. DESMARTEAU, B. B. RANDOLPH)

1. *Gmelins Handbuch der Anorganischen Chemie*, Band 2, Teil 2, Verlag Chemie, Weinheim, 1973.
2. R. D. Dresdner, T. R. Hooper, *Fluorine Chem. Rev.*, *4*, 1 (1969).
3. R. E. Banks, R. N. Haszeldine, in *The Chemistry of Organic Sulfur Compounds*, N. Kharasch, C. Y. Meyers, eds., Vol. 2, Pergamon Press, Oxford, 1966, p. 137.
4. G. A. Silvey, G. H. Cady, *J. Am. Chem. Soc.*, *72*, 3624 (1950).
5. G. A. Silvey, G. H. Cady, *J. Am. Chem. Soc.*, *74*, 5792 (1952).
6. W. A. Sheppard, *J. Am. Chem. Soc.*, *84*, 3058 (1962).
7. A. F. Clifford, H. F. El-Shamy, H. S. Emeléus, R. N. Haszeldine, *J. Chem. Soc.*, 2372 (1953).
8. F. W. Hoffman, T. C. Simmons, R. B. Beck, H. V. Holler, T. Katz, R. J. Koshar, E. R. Larson, J. I. Mulvaney, F. E. Rogers, B. Singleton, R. S. Sparks, *J. Am. Chem. Soc.*, *79*, 3424 (1957).
9. R. D. Dresdner, J. A. Young, *J. Am. Chem. Soc.*, *81*, 574 (1959).

2.3.5. from Cleavage of the Group VIB–Group IVB Element Bond

2.3.5.1. by Halogens.

Cleavage of group IVB—O bonds by halogens is not known to form O—X bonds. Except for F_2, the reverse reaction may be more favorable. Although certain O—F compounds may be produced by fluorination of SiF_3OR or other Si-O bonds, no successful synthesis by this route is known.

The cleavage of group IVB—S, —Se and —Te bonds by halogen is facile. Most of these reactions are with Si—S bonds, as the number of compounds with Si—Se and Si—Te bonds is small, and group IVB-S, -Se, and -Te bonded compounds are rare[1–8]. Cleavage of Si—S bonds occurs, e.g.:

$$R_3SiSH + 3\,Br_2 \rightarrow 2\,R_3SiBr + S_2Br_2 + 2\,HBr \quad (a)^5$$

where R = alkyl;

$$n\text{-}C_4H_9SSi(CH_3)_3 + Br_2 \xrightarrow{-70°C} (CH_3)_3SiBr + n\text{-}C_4H_9SBr \quad (b)^6$$

$$n\text{-}C_4H_9SBr \xrightarrow{Br_2} (n\text{-}C_4H_9S)_2$$

Cleavage of $R_3SiSSiR_3$ by Cl_2, Br_2, I_2 is exothermic, and only Si—I and —S bonds are formed[8] with I_2. With $R_3SiSeSiR_3$, Br_2 gives only R_3SiBr and Se[9]. Cleavage of Ge—S bonds in $(CH_3C_6H_4S)_4Ge$ by Br_2 results in $(CH_3C_6H_4S)_3GeBr$ and the disulfide[10].

These examples indicate the lack of useful group VIB—halogen bond formation by group VIB-IVB cleavage by halogens.

(D. D. DESMARTEAU, B. B. RANDOLPH)

1. S. N. Borisov, M. G. Voronkov, E. Ya. Lukevits, *Organosilicon Derivatives of Phosphorus and Sulfur*, Plenum Press, New York, 1971, p. 157.
2. D. Brandes, *Organomet. Chem. Rev. 7*, 257 (1979).
3. C. H. Van Dyke, in *Organometallic Compounds of the Group-IV Elements*, A. G. MacDiarmid, ed., Part I, Marcel Dekker, New York, 1972.
4. H. C. Clark, R. J. Puddeplatt, S. E. Cook, F. W. Frey, H. Shapiro, in *Organometallic Compounds of Group-IV Elements*, A. G. MacDiarmid, ed., Part II, Marcel Dekker, New York, 1972.
5. Ref. 2, p. 270.
6. E. W. Abel, *J. Chem. Soc.*, 5975 (1964).
7. L. Wolinski, H. Tieckelmann, H. W. Post, *J. Org. Chem., 16*, 1138 (1951).
8. Ref. 2, p. 337.
9. Ref. 2, p. 366.
10. H. S. Backer, F. J. Strenstra, *Recl. Trav. Chim. Pays-Bas, 52*, 1033 (1933).

2.3.5.2. by Oxidizing Halides.

Cleavage of VIB-IVB bonds by oxidizing halides is little explored, given that only Si—O bonds will likely lead to useful syntheses; e.g., IF_5 acts on $CH_3OSi(CH_3)_3$[1]:

$$IF_5 + CH_3OSi(CH_3)_3 \xrightarrow{20°C} CH_3OIF_4 + FSi(CH_3)_3 \quad (a)$$

With Si—O bonds, therefore, as in $(CH_3)_3SiOR$, such interhalogens as ClF should form ROCl and $(CH_3)_3SiF$. The use of such reagents as F_3SiOR should also prove satisfactory, but the fluorinated materials are more difficult to obtain. However, SiF_3OCH_3 is prepared easily, and SiF_3Cl and SiF_3Br are easily prepared from this[2]. The latter may be used to prepare $ROSiF_3$ derivatives.

(D. D. DESMARTEAU, B. B. RANDOLPH)

1. G. Oates, J. M. Winfield, *Inorg. Nucl. Chem. Lett., 8*, 1093 (1972).
2. W. Airey, G. M. Sheldrick, *J. Inorg. Nucl. Chem., 32*, 1827 (1970).

2.3.6. from Cleavage of the Group VIB–Oxygen Bond

2.3.6.1. by Halogens.

2.3.6.1.1. to Give the Sulfur–Halogen Bond.

The cleavage of S—O bonds by elemental halogen is not useful for the formation of sulfur–halogen bonds. This bond, especially in the oxides, is stable, being cleaved only by F_2 under drastic conditions. At 200°C, SO_2 with xs F_2 gives[1,2] O_2SF_2 along with OSF_2, OSF_4 and SF_5OF:

$$SO_2 + \text{xs } F_2 \rightarrow O_2SF_2, OSF_2, OSF_4, SF_5OF \xrightarrow{650°C} SF_6 + O_2 \qquad (a)$$

At higher T, SF_6 is isolated in good yields[2–4]; SF_6 is formed by direct fluorination or by disproportionation[5] of OSF_4 formed as an intermediate:

$$2\ OSF_4 \xrightarrow{\Delta} SF_6 + O_2SF_2 \qquad (b)$$

Catalytic fluorination[6] of OSF_2 in the presence of CsF yields SF_6 as a decomposition product, from the intermediate $[OSF_5]^-$. Other anions, e.g., $[S_2O_3]^{2-}$, react similarly under mild conditions[7]:

$$Na_2S_2O_3 + F_2 \xrightarrow{-80°–15°C} \underset{(30\%)}{SF_6} + \underset{(8\%)}{OSF_2} + \underset{(62\%)}{O_2SF_2} \qquad (c)$$

Besides other compounds, SF_6 is formed[8] from the electrolysis of $[NH_4]_2SO_4$-KF-HF melts. Traces are also observed in the electrochemical fluorination[9] of SO_2Cl_2 or $SOCl_2$.

Similar reactions are not known with the other halogens; the formation of $OSCl_2$ from SO_2 is possible only if the reaction with Cl_2 is carried out in the presence of an oxygen acceptor:

$$SO_2 + CO + Cl_2 \xrightarrow{200–300°C} OSCl_2 + CO_2 \qquad (d)^{10}$$

$$SO_2 + SCl_2 + Cl_2 \xrightarrow{175–210°C} OSCl_2 + S_xCl_2 \qquad (e)^{11,12}$$

$$SO_3 + S_2Cl_2 + Cl_2 \xrightarrow[\text{activated carbon}]{180°C} OSCl_2 \qquad (f)^{13}$$

(R. MEWS)

1. A. Vanderchmitt, *Rep. CEA-R-4613* (1974); *Nucl. Sci. Abstr., 31*, 8,027 (1975); *Chem. Abstr., 83*, 21,249 (1975).
2. cf. F. B. Dudley, G. H. Cady, D. F. Eggers, *J. Am. Chem. Soc., 78*, 1533 (1956).
3. W. Kwasnik, in *Handbook of Preparative Inorganic Chemistry*, G. Brauer, ed., Vol. 1, 2nd ed, Academic Press, NY, 1975, p. 170.
4. H. Jonas, O. Ruff, W. Kwasnik, J. Söll, R. Zimmermann, H. Glissmann, *Angew. Chem., 61*, 32 (1949).
5. W. J. Middleton, US Pat. 2,912,307 (1959); *Chem. Abstr., 54*, 3890 (1960).
6. J. K. Ruff, M. Lustig, *Inorg. Chem., 3*, 917 (1964).
7. M. Picon, L. Domange, *C. R. Hebd. Seances Acad. Sci.*, 236, 704 (1953).
8. A. Engelbrecht, E. Mayer, C. Pupp, *Monatsh. Chem., 95*, 633 (1964).

9. S. Nagase, T. Abe, H. Baba, *Bull. Chem. Soc. Jpn.*, *42*, 2062 (1969).
10. E. F. Fricke, US Pat. 2,471,946 (1947); *Chem. Abstr.*, *43*, 7201 (1949).
11. A. Pechukas, US Pat. 2,431,823 (1947); *Chem. Abstr.*, *42*, 2409 (1948).
12. G. M. Strongin, A. N. Bodrova, Yu. M. Al'tshuler, *Vestn. Tekhn. Ekon. Inf. Nauchno.-Issled. Inst. Tekhn-Ekon. Issled. Gos. Kom. Sov. Min. SSSR po. Khim*, *9*, 25 (1961); *Ref. Zh. Khim.*, 12,k77 (1962); *Chem. Abstr.*, *58*, 5,285 (1963).
13. H. Jones, P. Lueg, Ger. Pat. 939,571 (1956); *Chem. Abstr.*, *52*, 14,113 (1958).

2.3.6.1.2. to Give the Se–Halogen Bond.

The Se—O bond is more readily cleaved by F_2 than that of the lighter homologue. Passage of F_2 diluted with N_2 over dry SeO_2 yields the oxyfluoride in $\leq 80\%$ yield of the dioxide used[1]:

$$SeO_2 + F_2 \rightarrow OSeF_2 + \tfrac{1}{2} O_2 \qquad \text{(a)}$$

The reaction is carried out in glass. Products resulting from Se—O bond cleavage and oxidative addition of fluorine are also formed[2–5]:

$$SeO_2 + \text{xs } F_2 \xrightarrow{60^\circ\text{–}90^\circ C} F_5SeOF + F_5SeOOSeF_5 + SeF_6 \qquad \text{(b)}$$

Low T ($\approx 40^\circ C$) favors $SeOF_2$, and small amounts of SeF_4 are also observed. Further oxidation and addition to the Se=O bond needs xs F_2 and higher T. Using a flow system and an AgF_2 catalyst, good yields of the peroxide can be obtained:

$$SeO_2 + F_2\text{-}N_2 \xrightarrow[AgF_2]{110^\circ C} SeF_6 + \underset{(12\%)}{Se_2O_2F_{10}} \qquad \text{(c)}$$

The Se—O bond in the peroxide is cleaved by F_2 at 70°C to give SeF_6 (78%) and SeF_5OF (18%)[2].

In the fluorination of SeO_2 SeF_5OSeF_5 is also formed[3,4]:

$$SeO_2 + F_2 \rightarrow F_5SeOOSeF_5 + F_5SeOSeF_5 \qquad \text{(d)}$$

Thermal decomposition of the peroxide gives SeF_6, O_2 and SeF_4; $F_5SeOSeF_5$ may form from decomposition of the peroxide and recombination of $[SeF_5O]^{\cdot}$ and $[SeF_5]^{\cdot}$ radicals[3,4].

Reaction occurs even more readily with anions; $KSeO_2F$ gives[6] SeF_5OF at $-87^\circ C$:

$$KSeO_2F + F_2 \rightarrow KF + \underset{(14\%)}{SeF_5OF} + \underset{(16\%)}{SeF_4(OF)_2} + SeF_6 + O_2 \qquad \text{(e)}$$

The only known group-VIB bis(hypofluorite), $SeF_4(OF)_2$, is also isolated.

(R. MEWS)

1. E. E. Aynsley, R. D. Peacock, P. L. Robinson, *J. Chem. Soc.*, 1231 (1952).
2. G. Mitra, G. H. Cady, *J. Am. Chem. Soc.*, *81*, 2646 (1959).
3. W. L. Reichert, G. H. Cady, *US Nat. Tech. Inf. Serv. AD Rep.* No. 750282 (1972), *Chem. Abstr.*, *78*, 79,126 (1973).
4. W. L. Reichert, G. H. Cady, *Inorg. Chem.*, *12*, 769 (1973).
5. G. Mitra, *J. Indian Chem. Soc.*, *37*, 804 (1960).
6. J. E. Smith, G. H. Cady, *Inorg. Chem.*, *9*, 1293 (1970).

2.3.6.1.3. to Give the Te–Halogen Bond.

The oxides TeO_3 and TeO_2 are almost quantitatively converted[1] to TeF_6 by F_2:

$$TeO_3 + F_2\text{-}N_2 \xrightarrow{100°-250°C} TeF_6 \qquad (a)$$

$$TeO_2 + F_2\text{-}N_2 \xrightarrow{100°-200°C} TeF_6 \qquad (b)$$

In the last reaction traces of Te_2F_{10} (really $F_5TeOTeF_5$)[2] are also detected. The fluorinations are performed in Ni vessels, and the products are collected in borosilicate glass traps at −180°C. If a Te-TeO_2 mixture is fluorinated by F_2 diluted with O_2 at 50–60°C in a flow system, the yield of $F_5TeOTeF_5$ is increased to 40–45%; $Te_3F_{14}O_2$ in ≈5% yield is also isolated:

$$1{:}1\ Te\text{-}TeO_2 + 1{:}2.5\ F_2/O_2 \xrightarrow{50°-60°C} F_5TeOTeF_5 + Te_3O_2F_{14} \qquad (c)$$

The Te—O bond is also cleaved by Cl_2; e.g., $TeCl_4$ is formed[3] by heating TeO_2 with Cl_2 in CCl_4 to 100°C:

$$TeO_2 + Cl_2 \xrightarrow[100°C]{CCl_4} TeCl_4 + O_2 \qquad (d)$$

(R. MEWS)

1. R. Campbell, P. L. Robinson, *J. Chem. Soc.*, 3454 (1956).
2. P. M. Watkins, *J. Chem. Educ.*, *51*, 520 (1974).
3. E. Montignie, *Bull. Soc. Chim. Fr.*, 180 (1948).

2.3.6.2. by Hydrogen Halides

2.3.6.2.1. to Give the Sulfur-Halogen Bond.

Fluorosulfuric acid is formed from aq HF and H_2SO_4 at RT[1]:

$$H_2SO_4 + HF \rightleftharpoons HSO_3F + H_2O \qquad (a)$$

The HSO_3F conc in this equilibrium depends on the H_2O content. With K_2SO_4 reaction occurs according to[2]:

$$4\ HF + K_2SO_4 \rightleftharpoons HSO_3F + 2\ K^+ + [H_3O]^+ + 3\ F^- \qquad (b)$$

In organosulfur chemistry cleavage of the S—O bond by hydrogen halides is a more useful synthetic approach[3]. Sulfinate esters and HCl form an equilibrium containing small amounts of sulfinyl chloride[3]:

$$CH_3\text{—}\overset{\overset{\displaystyle O}{\|}}{S}\text{—}OCH_3 + HCl \rightleftharpoons CH_3\overset{\overset{\displaystyle O}{\|}}{S}Cl + CH_3OH, \qquad (c)$$

while sulfoxides yield unstable dichlorides (chlorosulfonium chlorides, $[R_2SCl]Cl$[4,5] or dibromides (bromosulfonium chlorides, $[ArRSBr]Br$)[6].

Sulfonic bromides and acid anhydrides are cleaved by HX, e.g.[7]:

$$\mathrm{ArS(=O)_2{-}O{-}S(=O)_2Ar + HBr \xrightarrow[70^\circ\text{–}75^\circ C]{AcOH} ArSO_2Br + ArSO_3H + AsSSAr} \quad \underset{(79\%)}{} \qquad \text{(d)}$$

where Ar = p-IC_6H_4.

Sulfenyl bromides can be prepared from sulfonic acids in cold acetic acid, but from hot solutions only the disulfide is isolated[8]:

$$\mathrm{ArSO_3H + 3\ HBr \xrightarrow{AcOH} ArSBr + 2\ H_2O + Br_2} \qquad \text{(e)}$$

Owing to the instability of the S—I bond, no stable product is isolated from bond cleavage of organic sulfites by HI, although OSI_2 may be an intermediate[9].

(R. MEWS)

1. W. Traube, E. Reubke, *Chem. Ber.*, *54*, 1618 (1921).
2. K. Fredenhagen, H. Fredenhagen, *Z. Anorg. Allg. Chem.*, *243*, 39 (1939).
3. R. V. Norton, G. M. Beverly, I. B. Douglass, *J. Org. Chem.*, *32*, 3645 (1967).
4. F. G. Bordwell, B. M. Pitt, *J. Am. Chem. Soc.*, *77*, 572 (1955).
5. W. R. Hardstaff, R. F. Langler, in *Sulfur in Organic and Inorganic Chemistry*, Vol. 4, A Senning, ed., Marcel Dekker, New York, 1982 p. 200.
6. T. Zincke, A. Dahn, *Chem. Ber.*, *45*, 3457 (1912).
7. V. O. Lukashevich, *Dokl. Akad, Nauk SSSR*, *103*, 627 (1955).
8. K. Fries, G. Schürmann, *Chem. Ber.*, *47*, 1195 (1914).
9. M. Davis, H. Szkuta, A. J. Krubsack, in *Mechanisms of Reactions of Sulfur Compounds*, Vol. 5, N. Kharasch, ed., Intra Science Research Foundation, Santa Monica, CA, 1970, p. 1ff.

2.3.6.2.2. to Give the Se–Halogen Bond.

This approach to Se—halogen bond formation is used in inorganic and organic Se chemistry; e.g., HF converts SeO_2 completely to $OSeF_2$ from 4:1 HF:SeO_2[1]:

$$\mathrm{SeO_2 + 4\ HF \rightarrow OSeF_2 + [H_3O][HF_2]} \qquad \text{(a)}$$

Selenic acid dissolves in anhyd HF to give colorless solutions; in the H_2SeO_4:HF conc range 1:3.8–1:35.4, formation of F_5SeOH and two other products (one is $HSeO_3F$) is observed by ^{19}F-NMR spectroscopy[2]. If the resulting H_2O is removed by HSO_3F, good yields in $HOSeF_5$ are obtained[3]:

$$\mathrm{H_2SeO_4 + 3\ HSO_3F + 2\ HF \rightarrow HOSeF_5 + 2\ H_2SO_4} \qquad \text{(b)}$$

This substance is also obtained from SeO_2F_2 and HF at RT[4]:

$$\mathrm{3\ SeO_2F_2 + 4\ HF \rightarrow 2\ HOSeF_4 + H_2SeO_4} \qquad \text{(c)}$$

Anhydrous HCl forms an adduct[5,6], $SeO_2 \cdot 2\ HCl$, which can be dehydrated to $SeOCl_2$:

$$\mathrm{SeO_2 + 2\ HCl \rightarrow SeO_2 \cdot 2\ HCl \xrightarrow[-H_2O]{} OSeCl_2} \qquad \text{(d)}$$

In H_2O the following equilibrium is observed[6,7]:

$$H_2SeO_3 + 2\ HCl \rightleftharpoons OSeCl_2 + H_2O \tag{e}$$

$$OSeCl_2 + 3\ HCl \rightleftharpoons [H_3O]^+ + [SeCl_5]^- \tag{f}$$

At higher HCl concentrations (15.18M, 0°C), the intermediate $OSeCl_2$ forms the pentachloroselenate ion[8]. When HCl is bubbled through an ice-cold, saturated solution containing 2:1 NH_4Cl:SeO_2 until saturation, crystals of bright yellow $(NH_4)_2SeCl_6$ deposit on standing[9].

With HBr and SeO_2, $OSeBr_2$ is the product from solutions containing 1:7–8 SeO_2 and HBr. If the amount of HBr is raised, $[SeBr_6]^{2-}$ forms[10]. From these solutions pure $SeBr_4$ is isolated by extraction with organic solvents[11,12]:

$$SeO_2 + 2\ HBr \xrightarrow{-H_2O} OSeBr_2 \xrightarrow{+2\ HBr,\ -H_2O} SeBr_4 \xrightarrow{+2\ Br^-} [SeBr_6]^{2-} \tag{g}$$

Similar reactions are possible with organoseleniums. Diorganoselenium oxides and their hydrates react with hydrohalic acids in H_2O or organic solvents to give diorganoselenium dihalides[13,14]:

$$Me_2SeO + 2\ HX \rightarrow Me_2SeX_2 + H_2O \tag{h}$$

(X = Cl[15], Br[16])

$$Ar_2SeO + 2\ HX \rightarrow Ar_2SeX_2 \tag{i}$$

Selenones are reduced to give the dihalides[17]:

$$R_2SeO_2 + 4\ HCl \rightarrow R_2SeCl_2 + Cl_2 + 2\ H_2O \tag{j}$$

and seleninyl halides are formed[18] from selenic acids RSeOH; conc HBr solutions transform selenic acids into their tribromides[19]:

$$RSeOOH + 3\ HBr \rightarrow RSeBr_3 + 2\ H_2O \tag{k}$$

Organoselenium iodides are better prepared by oxidative addition or metathesis.

(R. MEWS)

1. H. Selig, U. El-Gad, *J. Inorg. Nucl. Chem., H. H. Hyman Mem. Vol.*, 233 (1976).
2. U. El-Gad, H. Selig, *Inorg. Chem., 14*, 140 (1975).
3. K. Seppelt, *Angew. Chem., Int. Ed. Engl., 11*, 630 (1972).
4. K. Seppelt, *Z. Anorg. Allg. Chem., 428*, 35 (1977).
5. G. B. L. Smith, J. Jackson, *Inorg. Synth., 3*, 130 (1950).
6. F. Feher, in *Handbook of Preparative Inorganic Chemistry*, G. Brauer, ed., 2nd ed., Academic Press, New York, 1975, p. 429.
7. A. K. Babko, T. T. Mityureva, *Russ. J. Inorg. Chem. (Engl. Transl.), 6*, 213 (1961).
8. J. Milne, P. La Haie, *Inorg. Chem., 18*, 3180 (1979).
9. P. La Haie, J. Milne, *Inorg. Chem., 18*, 632 (1979).
10. L. Futekov, H. Specker, *Z. Anal. Chem.*, 276, 41 (1975).
11. E. Bock, R. Bock, Ger. Pat. 937,644 (1956); *Chem. Abstr., 52*, 19,042 (1958).
12. Yu. V. Migalina, V. I. Staninets, V. G. Leudel, I. M. Balog, V. A. Palyulin, A. S. Koz'min, N. S. Zefirov, *Chem. Heterocycl. Compounds USSR, 13*, 49 (1977).
13. A. Rheinboldt, in *Methoden der Organischen Chemie-Houben Weyl*, Vol. IX *S, Se, Te Verbindungen*, E. Müller, O. Bayer, H. Meerwein, K. Ziegler, eds., Georg Thieme, Stuttgart, 1955. pp. 1010, 1026.
14. K. J. Irgolic, M. V. Kudchadker in *Selenium*, R. A. Zingaro, C. W. Cooper, eds., Van Nostrand-Reinhold, New York, 1974, p. 408 ff.

15. R. Paetzold, U. Lindner, G. Bochmann, P. Reich, *Z. Anorg. Allg. Chem.*, *352*, 295 (1967).
16. Ref. 12, p. 1013.
17. Ref. 12, p. 1033.
18. H. Rheinboldt, E. Giesbrecht, *Chem. Ber.*, *89*, 631 (1956).
19. Ref. 12, p. 1133.

2.3.6.2.3. to Give the Te–Halogen Bond.

The Te—O bond is cleaved by hydrogen halides. If TeO_2 is dissolved in 40% aq HF in the presence of alkali fluorides, pentafluorotellurates (−1) are formed[1–5] which give TeF_4 on heating[5]:

$$TeO_2 + MF + 4\ HF \rightarrow M^+[TeF_5]^- + 2\ H_2O \tag{a}$$

$$TeO_2 + C_5H_5N + 4\ HF \rightarrow [C_5H_5NH][TeF_5]^- + 2\ H_2O \tag{b}$$

However, if TeO_2 is heated 1:1:4 with $KHCO_3$ and 26 M HF with little H_2O a clear solution results and, after cooling, large crystals of $KTeF_4OH$ separate[6]:

$$KHCO_3 + TeO_2 + 4\ HF \rightarrow KTeF_4OH + 2\ H_2O + CO_2 \tag{c}$$

If no cation is present the free acid $H_2Te_2O_3F_4$ is isolated[7]:

$$2\ TeO_2 + 4\ HF \rightarrow HOTeF_2{-}O{-}TeF_2OH + H_2O \tag{d}$$

In the solvolysis of orthotelluric acid, $Te(OH)_6$, in HF[8–14] ≤4 OH groups are replaced by fluorine. The stereochemical route of the solvolysis can be followed by ^{125}Te-NMR spectroscopy. Most of the intermediate species are isolated[13]; e.g., cis-$(HO)_4TeF_2$ is the final product of the solvolysis of $(HO)_6Te$ in 48% aq HF. Further fluorination only occurs in anhyd HF with the formation of 1:1 sym- and asym-$(HO)_3TeF_3$. Only the latter reacts further to give trans-$(HO)_2TeF_4$. Only cis-$(HO)_2TeF_4$, a hydrolysis product of TeF_6, gives TeF_5OH and traces of TeF_6 with anhyd HF. These results are summarized[12]:

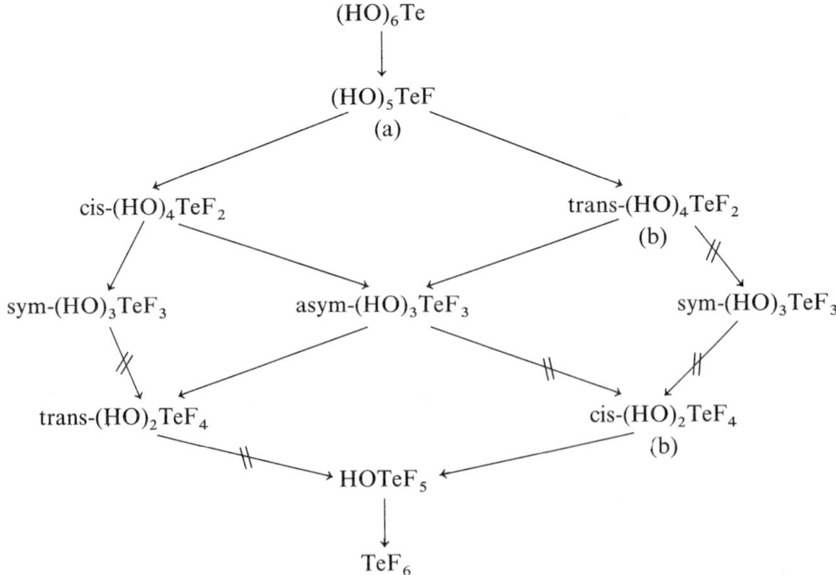

[(a) is not observed in 48% aq HF; (b) is the hydrolysis product of TeF_6].

Under special conditions pure trans-$(HO)_2TeF_4$ is obtained by dissolving $Te(OH)_6$ in 48% aq HF. After addition of pyridine [$(HO)_6Te$:HF:PY = 1:10:3] the solution is evaporated and the pyridinium salt isolated in 90% yield. Pure trans-$(HO)_2TeF_4$ is sublimed from a solution of the trans salt in 96% H_2SO_4 at 130°–150°C in vacuo[12]; and $F_5Te(OH)$ is prepared from orthotelluric acid and SO_3-stabilized HF, HSO_3F:

$$Te(OH)_6 + 5\ FSO_3H \rightarrow 5\ H_2SO_4 + HOTeF_5 \qquad (e)$$

Because the resulting H_2O is eliminated by the reagent, it is possible to replace all five OH groups[14].

Gaseous HCl reacts with powdered and preheated (300°C) TeO_2 at RT to give $TeCl_4$[15]:

$$TeO_2 + 4\ HCl \rightarrow TeCl_4 + 2\ H_2O \qquad (f)$$

$$TeCl_4 + 2\ HCl + nH_2O \rightarrow TeCl_4 \cdot 2\ HCl \cdot n\ H_2O \qquad (g)$$

which associates HCl in the presence of traces of H_2O. The same product is obtained with conc aq HCl. In these solutions H_2TeCl_6 is present[15,16].

Numerous hexachlorotellurates(IV) are prepared by dissolving TeO_2 in hot conc HCl and adding the metal chloride[17]:

$$TeO_2 + 4\ HCl \xrightarrow[-2H_2O]{} TeCl_4 \xrightarrow{2\ MCl} M_2TeCl_6 \qquad (h)$$

Alkyl and arylammonium hexachlorotellurates, as well as salts of heterocyclic nitrogen bases, are prepared by this method[17,18].

If H_2TeO_3 is dissolved in 0.5 M HCl ($\leq$10M Cl^-) at 18°C, $[TeCl_6]^{2-}$ forms[19] along with oxyanions, such as $[TeOCl_4]^{2-}$, $[TeOCl_3]^-$, as well as $TeOCl_2$ and TeO(OH)Cl; the latter two may be formulated as chlorohydroxo complexes[20].

The oxide TeO_2, prepared from Te and conc HNO_3, gives $TeBr_4$ in aq HBr[21]:

$$TeO_2 + 4\ HBr \rightleftarrows TeBr_4 + 2\ H_2O \qquad (6)$$

The reaction is reversible and hydrolysis occurs at pH 4.5. Pure $TeBr_4$ is isolated after adding conc H_2SO_4.

Organotellurium dihalides can be prepared similarly. In conc hydrohalic acids, tellurides are oxidized to telluroxides, which then combine with the acids forming the dichlorides[22–27] or dibromides[27]:

$$R_2Te + \tfrac{1}{2}\ O_2 + 2\ HCl \xrightarrow[-H_2O]{HCl\text{—}ether} R_2TeCl_2 \qquad (j)$$

$$R_2Te + \tfrac{1}{2}\ O_2 + 2\ HBr \xrightarrow[-H_2O]{HBr\text{—}ether} TeBr_2 \qquad (k)$$

In organohydroxytelluriums, the OH group may also be exchanged via HI[28,29].

(R. MEWS)

1. H. L. Wells, J. M. Willis, *Am. J. Sci.*, *12*, 190 (1901).
2. N. N. Greenwood, A. C. Sarma, B. P. Straughan, *J. Chem. Soc.*, *A*, 1446 (1966).
3. T. C. Gibb, R. Greatrex, N. N. Greenwood, A. C. Sarma, *J. Chem. Soc.*, *A*, 212 (1970).
4. J. B. Milne, D. Moffett, *Inorg. Chem.*, *12*, 2240 (1973).
5. J. Carre, P. Germain, J. Thorey, G. J. Perachon, *J. Fluorine Chem.*, *3*, 241 (1986).

6. J. B. Milne, D. Moffett, *Inorg. Chem.*, *13*, 2750 (1974).
7. J. C. Jumas, M. Maurin, E. Phillippot, *J. Fluorine Chem.*, *8*, 329 (1976).
8. A. Engelbrecht, F. Sladky, *Monatsh. Chem.*, *96*, 159 (1965).
9. L. Kolditz, I. Fitz, *Z. Anorg. Allg. Chem.*, *349*, 175 (1967).
10. G. W. Fraser, G. D. Meikle, *J. Chem. Soc., Chem. Commun.*, 624 (1974).
11. U. Elgad, H. Selig, *Inorg. Chem.*, *14*, 140 (1975).
12. W. Tötsch, F. Sladky, *J. Chem. Soc., Chem. Commun.*, 927 (1980).
13. W. Tötsch, P. Peringer, F. Sladky, *J. Chem. Soc., Chem. Commun.*, 841 (1981).
14. A. Engelbrecht, F. Sladky, *Angew. Chem., Int. Ed. Engl.*, *3*, 383 (1964).
15. P. Khodadad, *Ann. Chim. (Paris)*, *10*, 83 (1965).
16. D. M. Adams, D. M. Morris, *J. Chem. Soc.*, *A*, 2067 (1967).
17. *Gmelins, Handbuch der Anorganischen Chemie, Te, Erg. Band. B 2*, Springer Verlag, Berlin, 1977, p. 134 ff.
18. E. R. Clark, W. R. McWhinnie, J. Mallaki, N. S. Dance, C. H. W. Jones, *Inorg. Chim, Acta*, *41*, 279 (1980).
19. B. I. Nabivanets, E. E. Kapantsyan, *Russ. J. Inorg. Chem. (Engl. Transl.)*, *13*, 946 (1968).
20. E. Petkowa, K. Vassilev, *Dokl. Bulg. Akad. Nauk*, *21*, 1173 (1968); *Chem. Abstr.*, *70*, 72,434 (1969).
21. P. Dupuy, *C.R. Hebd. Seances Acad. Sci.*, *237*, 718 (1953).
22. K. Lederer, *Justus Liebigs Ann. Chem.*, *391*, 326 (1912).
23. M. De Moura Campos, E. L. Suranyi, J. De Andrade Jr., N. Petragnani, *Tetrahedron*, *20*, 2797 (1964).
24. H. K. Livingstone, R. Krosec, *J. Polym. Sci., Part B*, *9*, 95 (1971).
25. K. J. Irgolic, in *The Organic Chemistry of Tellurium*, Gordon and Breach, New York, 1974, p. 177 ff. Methods for the preparation of organic dihalides are compared.
26. H. D. K. Drew, *J. Chem. Soc.*, 560 (1929).
27. M. P. Balfe, C. A. Chaplin, H. Phillips, *J. Chem. Soc.*, 341 (1938).
28. R. H. Vernon, *J. Chem. Soc.*, *117*, 86 (1920); *117*, 889 (1920).
29. F. L. Gilbert, T. M. Lowry, *J. Chem. Soc.*, 3179 (1928).

2.3.6.2.4. to Give the Po–Halogen Bond.

The white solid obtained by treating Po(IV) hydroxide with dilute HF may be PoF_4 or a basic compound[1]:

$$PoO_2 + 4\ HF \rightarrow PoF_4(?) + 2\ H_2O \tag{a}$$

By heating the dioxide in dry HCl gas, $PoCl_4$ is obtained[2]:

$$PoO_2 + 4\ HCl \rightarrow PoCl_4 + 2\ H_2O \tag{b}$$

The $PoCl_4$ product is hydroscopic, like $TeCl_4$, and is rapidly hydrolyzed by moist air to give a white solid[3] that is mixture of a basic chloride and a hydroxide or hydrated oxide[2]. Evaporation of HCl solutions of $PoCl_4$ gives an oxychloride[4]. These solutions contain $[PoCl_6]^{2-}$ which may be precipitated as greenish Cs hexachloropolonite[2,5].

The behavior of PoO_2 toward HBr is analogous:

$$PoO_2 + 4\ HBr \xrightarrow[-2\,H_2O]{} PoBr_4 \xrightarrow{\text{HBr soln}} [PoBr_6]^{2-} \tag{c}$$

$$[PoBr_6]^{2-} \xrightarrow{-30^\circ C} H_2PoBr_6(?) \qquad [PoBr_6]^{2-} \xrightarrow{2\,CsBr} Cs_2PoBr_6$$

The tetrabromide is prepared from dry HBr or from solutions[6]. Cooling of dil aq HBr to −30°C precipitates blackish-brown complex hexabromopolonous acid[6], whereas addition of CsBr yields[6] dark red Cs_2PoBr_6.

With HI, a black addition compound, $PoO_2 \cdot 2\,HI$, is formed in the cold. Heating PoO_2 and dry HI gas to 200°C gives PoI_4. The behavior in solution is similar to that of the other tetrahalides[7].

(R. MEWS)

1. K. W. Bagnall, J. H. Freeman, D. S. Robertson, P. S. Robinson, M. A. A. Stewart, *UK Atomic Energy Authority, unclassified report C/R* 2566 (1958).
2. K. W. Bagnall, R. W. M. D'Eye, J. H. Freeman, *J. Chem. Soc.*, 2320 (1955).
3. E. F. Joy, *US Atomic Energy Commission Rep. M-4123* (1947), declassified (1955).
4. G. Boussiéres, *Bull. Soc. Chim. Fr.* 536 (1952).
5. E. Staritzky, *US Atomic Energy Commission Rep. LA 1286* (1951).
6. K. W. Bagnall, R. W. M. D'Eye, J. H. Freeman, *J. Chem. Soc.*, 3959 (1955).
7. K. W. Bagnall, *The Chemistry of Selenium, Tellurium, and Polonium*, Elsevier, Amsterdam, 1966.

2.3.6.3. by Organic Halides

2.3.6.3.1. to Give the Sulfur–Halogen Bond.

Haloalkenes are inert toward SO_2, but they add to the S—O bond in SO_3. In a subsequent step, halogen is transferred from carbon to sulfur. No sulfur—halogen bond forms with bromides or iodides.

The photolysis of 1,1-difluoroalkenes and SO_2 in CCl_2F_2 gives sulfinic fluorides[1,2]:

$$RR'C{=}CF_2 + SO_2 \xrightarrow[-25^\circ \text{ to } -35^\circ C]{UV} RR'C(SOF)(COF) \quad (a)$$

where R, R^1 = CF_3, F; F, F; F, Cl; Cl, Cl.

Fluoroalkenes add SO_3 to give β-sultones[3,4]:

$$RFC{=}CF_2 + SO_3 \rightarrow F{-}\underset{|}{\overset{F}{C}}{-}\underset{|}{\overset{R}{C}}{-}F \text{ (ring: } C{-}O{-}SO_2{-}C\text{)} \quad (b)$$

$$\xrightarrow{\text{base}} O{=}C(F){-}C(R)(F){-}SO_2F \qquad \xrightarrow{\text{ROH}} ROC({=}O){-}C(R)(F){-}SO_2F$$

Ring opening is catalyzed by t-amines, ethers, etc.[4,5]. The fluorocarbonyl is attacked by nucleophiles, such as amines, or alcohols[5,6], while the FSO_2 group remains intact;

SO_2 reacts with CCl_4 and other chlorocarbons only in the presence of catalysts ($AlCl_3$) at $>150°C$[7]. With phosgene, similar drastic conditions are necessary[8].

$$SO_2 + CCl_4 \xrightarrow{150°C} OSCl_2 + COCl_2 \quad (c)$$

$$SO_2 + COCl_2 \xrightarrow{AlCl_3} SOCl_2 + CO_2 \quad (d)$$

$$SO_2 + 2\ COCl_2 \xrightarrow[\text{activated carbon}]{300°C} SCl_4 + 2\ CO_2 \quad (e)$$

With carbon catalysts at 300°C, oxygen is completely exchanged[8].

At RT, $HCCl_3$[3] and CCl_4[3,9,10] are attacked by SO_3 with partial O-Cl exchange, and bromocarbons reduce the trioxide[11]:

$$HCCl_3 + SO_3 \rightarrow CO + HSO_3Cl + S_2O_5Cl_2 \quad (f)$$

$$CCl_4 + n\ SO_3 \rightarrow S_nO_{3n-1}Cl_2 + OCCl_2 \quad (g)$$

With exclusion of moisture from the reaction with CCl_4, phosgene and polysulfuryl chlorides, $S_nO_{3n-1}Cl_2$ (n-5) are formed[9,10]. With increasing n stability decreases, and earlier attempts describe[3] only $S_2O_5Cl_2$. This method is also useful for $S_2O_5Cl_2$ and $S_3O_8Cl_2$[3,10]. Simultaneous C—S and C—Cl bond formation can occur, e.g.[12]:

$$ClCN + SO_3 \rightarrow ClSO_2NCO + ClSO_2OSO_2NCO \quad (h)$$

Organic sulfonyl chlorides can be prepared in 80–90% yield by heating of their salts with benzotrihalides (or derivatives)[13]:

$$C_6H_5CCl_3 + RSO_3Na \rightarrow RSO_2Cl + C_6H_5COCl + NaCl \quad (i)$$

$$C_6H_5CCl_3 + 2\ RSO_2Na \rightarrow 2\ RSO_2Cl + C_6H_5COONa + NaCl \quad (j)$$

$$(CH_2SO_3H)_2 + 2\ COCl_2 \rightarrow (CH_2SO_2Cl)_2 + 2\ CO_2 + 2\ HCl \quad (k)$$

Sulfonic acids are transformed to the chlorides by phosgene in toluene quantitatively[14].

(R. MEWS)

1. D. Sianesi, G. C. Bernardi, G. Moggi, *Tetrahedron Lett.*, 1313 (1970).
2. G. C. Bernardi, G. Moggi, D. Sianesi, *Ann. Chim. (Rome)*, *62*, 95 (1972).
3. E. E. Gilbert, *Chem. Rev.*, *62*, 549 (1962).
4. I. L. Knunyants, G. A. Sokol'skii, *Angew. Chem., Int. End. Engl.*, *11*, 583 (1972). Reviews the chemistry of the sultones.
5. *Gmelins Handbuch der Anorganischen Chemie, Perfluorohalogenoorgano Compounds*, Vol. 2 (1973), pp. 106 ff, 116, 139, 144.
6. N. P. Aktaev, G. A. Sokol'skii, I. L. Knunyants, *Izv. Akad. Nauk SSR, Ser. Khim.*, 2530 (1975).
7. C. E. Frank, P. T. Hallowell, C. W. Theobald, G. T. Vaala, *Ind. Eng. Chem.*, *41*, 2061 (1949).
8. H. Hecht, R. Greese, G. Jander, *Z. Anorg. Allg. Chem.*, *269*, 262 (1952).
9. R. J. Gillespie, E. A. Robinson, *Can. J. Chem.*, *39*, 2189 (1961).
10. H. A. Lehmann, G. Ladwig, *Z. Anorg. Allg. Chem.*, *284*, 1 (1956).
11. G. Siegemund, *Angew. Chem., Int. Ed. Engl.*, *12*, 918 (1973).
12. R. Graf, *Chem. Ber.*, *89*, 1071 (1956).
13. M. Quadflieg, in *Houben-Weyl Methoden der Organischen Chemie*, Vol. IX, S, Se, Te Compounds, E. Müller, O. Bayer, H. Meerwein, K. Ziegler, eds., Georg Thieme Verlag, Stuttgart, 1955, pp. 397, 567.
14. Ref. 13, p. 391.

2.3.6.3.2. to Give the Se–Halogen Bond.

Selenium-halogen bond formation by this approach is limited. The Se—O bond in selenous acid ester is cleaved by acetyl chloride:

$$OSe(OEt)_2 + CH_3C(O)Cl \rightarrow OSe\begin{matrix} \diagup Cl \\ \diagdown OEt \end{matrix} + CH_3C(O)OEt \qquad (a)$$

Refluxing a 1:2 mixture 4 h at 80°-90°C forms ethoxyseleninyl chloride. From a 1:4 ratio, $SeCl_4$ forms along with a complex red liquid of indefinite composition[1].

Selenium trioxide reacts violently with CCl_4[2]; with $(C_6H_5)_3CCl$ explosions occur, at RT. At −80°C in liq SO_2 addition to the oxide with formation of $[(C_6H_5)_3C]^+[SeO_3Cl]^-$ is observed[3].

(R. MEWS)

1. R. C. Mehrotra, S. N. Mathur, *Indian J. Chem.*, *5*, 375 (1967).
2. K. N. Mochalov, S. N. Kondrat'ev, G. I. Blagoveshchenskaya, E. E. Sidorova, *Dokl. Akad. Nauk SSSR*, *167*, 361 (1966).
3. R. C. Paul, R. D. Sharma, K. C. Mehrotra, *Indian J. Chem.*, 12, 320 (1974).

2.3.6.3.3. to Give the Te– and Po–Halogen Bond.

Pure $TeCl_4$ is obtained from TeO_2 and CCl_4 at 500°C [1,2]:

$$TeO_2 + CCl_4\text{-}CO_2 \xrightarrow[-COCl_2]{500°C} TeCl_4 + Te_6O_{11}Cl_2 \qquad (a)$$

in a borosilicate tube; CCl_4 (kept at 40°-50°C) is transferred into the reaction zone by a stream of CO_2, and $TeCl_4$ is collected in the colder parts of the tube at 150°C, while $Te_6O_{11}Cl_2$, which is formed from $TeCl_4$ and TeO_2, condenses at 280°C.

On warming, benzyl chloride also acts on TeO_2 to give[3] $TeCl_4$.

Similarly to $TeCl_4$, $PoCl_4$ is produced by heating PoO_2 with CCl_4 vapor[4,5]:

$$PoO_2 + 2\ CCl_4 \xrightarrow{200°C} PoCl_4 + 2\ COCl_2 \qquad (b)$$

(R. MEWS)

1. P. Khodadad, *Ann. Chim. (Paris)*, *10*, 83 (1965).
2. P. Khodadad, *Bull. Soc. Chim. Fr.*, 468 (1965).
3. C. H. Fischer, A. Eisner, *J. Org. Chem.*, *6*, 169 (1941).
4. J. J. Burbage, *Rec. Chem. Prog.*, *14*, 157 (1953).
5. E. F. Joy, *US Atomic Energy Commission Rep. M-4123* (1947); declassified 1955.

2.3.6.4. by Metal Halides

2.3.6.4.1. to Give the Sulfur–Halogen Bond.

The S—O bond in SO_2, used as solvent in halogen chemistry, is resistant to attack by fluorides. Only the reactive VF_5 is solvolyzed[1]:

$$SO_2 + VF_5 \rightarrow OSF_2 + OVF_3 \qquad (a)$$

with ionic fluorides formation of fluorosulfinates occurs (see §2.3.7.4).

Sulfur trioxide undergoes similar reactions; the primary step is the formation of fluorosulfates which decompose at elevated T to give SO_2F_2:

$$2\ SO_3 + BaF_2(Ba[SiF_6]_2) \xrightarrow{500^\circ C} BaSO_4 + SO_2F_2(+2\ SiF_4) \quad \text{(b)}$$

$$2\ SO_3 + ZnF_2 \xrightarrow{200^\circ C} Zn(SO_3F)_2 \xrightarrow{500^\circ C} ZnSO_4 + SO_2F_2 \quad \text{(c)}$$

The reaction of $BaF_2[Ba(SiF_6)_2]$ with SO_3 vapor at 500°C is suggested[2] for the preparation of chlorine-free SO_2F_2; $Zn(SO_3F)_2$, formed at 200°C incompletely, decomposes at 500°C. At the same T, HgF_2 also gives[3] SO_2F_2. In contrast, transition-metal pentafluorides, such as VF_5, TaF_5 or NbF_5 [1], produce $S_2O_5F_2$, e.g.:

$$VF_5 + 2\ SO_3 \rightarrow S_2O_5F_2 + VOF_3 \quad \text{(d)}$$

The S—O single bond in sulfonic acid anhydrides is cleaved by nucleophilic attack by F^-. Pyrosulfuryl chloride yields SO_2ClF (and some S_2O_5ClF) on treatment with AgF[4]:

$$ClSO_2OSO_2Cl + AgF \rightarrow ClSO_2F + AgOSO_2Cl \quad \text{(e)}$$

$$(RSO_2)_2O + NaF \rightarrow RSO_2F + NaOSO_2R \quad \text{(f)}$$

Organosulfonyl fluorides are obtained[5] in yields better than 80% by heating the anhydrides with NaF.

From SO_2 and a eutectic mixture of $AlCl_3$-NaCl is produced[6] $OSCl_2$, but for syntheses of thionylchloride exchange by nonmetal halides (see §2.3.6.5.1.) is preferred. However, solvolysis of transition-metal chlorides is a useful route to transition-metal oxychlorides[7,8] and bromides[8]:

$$2\ VCl_4 + SO_2 \rightarrow OSCl_2 + VCl_3 + VOCl_3 \quad \text{(g)}$$

$$WCl_6 + SO_2 \xrightarrow{60^\circ-70^\circ C} OSCl_2 + OWCl_4 \quad \text{(h)}$$

$$NbBr_5 + SO_2 \xrightarrow{70^\circ-80^\circ C} OSBr_2 + ONbBr_3 \quad \text{(i)}$$

Although UCl_4 does not react, heating a mixture of UCl_4 and UCl_5 with SO_2 to 80°–90° deposits UO_2Cl_2 and a solution of $OSCl_2$ is formed. The reaction occurs in two steps, UCl_5 disproportionating to give UCl_4 and the unstable UCl_6, the latter being solvolysed[8].

Similar solvolyses are observed with dimethylsulfoxide (DMSO), with $NbCl_5$ the oxychloride and $(CH_3)_2SCl_2$ is formed[9]:

$$(CH_3)_2SO + NbCl_5 \rightarrow (CH_3)_2SCl_2 + NbOCl_3 \quad \text{(j)}$$

Toward halides SO_3 is more reactive; even NaCl forms pyrosulfuryl chloride under mild conditions:

$$7\ SO_3 + 2\ NaCl \xrightarrow[8\ \text{days}]{40^\circ C} S_2O_5Cl_2 + Na_2S_5O_{16} \quad \text{(k)}$$

The reaction with NaCl is complete after 1 week at 45°C[10]. Small amounts of SO_2Cl_2 also form.

At higher T the S—Cl bond is cleaved[11]:

$$2\ SO_3 + NaCl \xrightarrow{500°-600°C} Na_2SO_4 + SO_2 + Cl_2 \qquad (l)$$

With more covalent metal chlorides, such $GeCl_4$, $NbCl_5$ or $TaCl_5$[12], the first step is insertion into the metal–Cl bond to form the chlorosulfate group, then decomposition to give $S_2O_5Cl_2$ or SO_2Cl_2 occurs, e.g.:

$$2\ NbCl_5 + 14\ SO_3 \xrightarrow[\text{reflux 40 h}]{SO_2Cl_2} Nb_2O(SO_4)_4 + 5\ S_2O_5Cl_2 \qquad (m)$$

Chloride ion as a nucleophile cleaves S—O—S bridges; organosulfonyl chlorides are prepared in excellent yield from anhydrides and $ZnCl_2$[5], $AlCl_3$[5] or NaCl[13], e.g.:

$$(CF_3CFClSO_2)_2O + NaCl \xrightarrow{-40°C} CF_3CFClSO_2Cl + Na[CF_3CFClSO_3] \qquad (n)$$

$$2\ (RSO_2)O + ZnCl_2 \rightarrow 2\ RSO_2Cl + Zn(OSO_2R)_2 \qquad (o)$$

Ionic bromides or iodides such as KBr or KI, react with SO_3, to give Br_2 and I_2, no S—halogen bond being formed[14] (at lower T KSO_3Br is described; cf. §2.3.7.4.)

(R. MEWS)

1. H. C. Clark, H. J. Emeléus, *J. Chem. Soc.*, 190 (1958).
2. H. K. H. Lam, H. T. Fullam, US Pat. 3,403,144 (1968); *Chem. Abstr.*, *69*, 108,225 (1968).
3. E. L. Muetterties, D. D. Coffman, *J. Am. Chem. Soc.*, *80*, 5914 (1958).
4. R. C. Brastead, in *Comprehensive Inorganic Chemistry*, Vol. 8, Van Nostrand, Princeton NJ, 1961, p. 107.
5. G. Van Dyke-Tiers, *J. Org. Chem.*, *28*, 1244 (1963).
6. C. S. Sherer, Ph.D. *Diss., Univ. Alabama*, 1967; *Diss. Abstr.*, *B28*, 1843 (1967); *Chem. Abstr.*, *68*, 74,761 (1968).
7. J. E. Drake, J. Vekris, J. S. Wood, *J. Inorg. Nucl. Chem.*, *30*, 3380 (1968).
8. W. Behne, G. Jander, H. Hecht, *Z. Anorg. Allg. Chem.*, *269*, 249 (1952).
9. D. B. Copley, A. Thompson, *J. Less-Common Met.*, *6*, 407 (1960).
10. E. Pruskaric, R. de Jaeger, J. Heubel, *Rev. Chim. Miner.*, *12*, 374 (1975).
11. R. Brännland, *Tek. Tidskr.*, *85*, 629 (1955); *Chem. Abstr.*, *49*, 16,367 (1955).
12. E. Hayek, K. Hinterauer, *Monatsh. Chem.*, *82*, 205 (1951).
13. G. A. Sokol'skii, L. I. Ragulin, G. P. Ovsyannikov, I. L. Knunyants, *Acad. Sci. USSR Bull. Chem. Sci.*, 1174 (1971).
14. M. Fioretti, U. Croatto, *Gazz. Chim. Ital.*, *70*, 850 (1940).

2.3.6.4.2. to Give the Se–Halogen Bond.

Fluorination of SeO_2 to give $F_5SeOOSeF_5$ and $F_5SeOSeF_5$ is catalyzed by AgF_2, the mechanism involving intermediate attack of the metal halide[1,2].

If SeO_3 is heated with alkali-metal fluorides without a solvent, SeO_2F_2 is formed. Byproducts and conditions depend on the metal ion[3–5]:

$$2\ NaF + 3\ SeO_3 \rightarrow Na_2Se_2O_7 + SeO_2F_2 \qquad (a)$$

$$KF + SeO_3 \xrightarrow{120°C} KSeO_3F + K_2SeO_4 + SeO_2F_2 \qquad (b)$$

$$CsF + SeO_3 \xrightarrow{80°C} CsSeO_3F + SeO_2F_2 \qquad (c)$$

e.g., $AlCl_3$ reacts under dry N_2 with SeO_2 in a salt melt from equimol LiCl–NaCl–KCl at 500°C:

$$3\ SeO_2 + 2\ AlCl_3 \rightarrow 3\ SeOCl_2 + Al_2O_3 \qquad \text{(d)}$$

After 3 h, $SeOCl_2$ is prepared in 74% yield.

Metal chlorides and bromides are oxidized by SeO_3 to form selenates, SeO_2, and elemental halogen[7]; e.g.:

$$2\ KBr + 2\ SeO_3 \rightarrow K_2SeO_4 + SeO_2 + Br_2 \qquad \text{(e)}$$

(R. MEWS)

1. G. Mitra, G. H. Cady, *J. Am. Chem. Soc.*, *81*, 2646 (1959).
2. W. L. Reichert, G. H. Cady, *Inorg. Chem.*, *12*, 769 (1973).
3. D. Kempe, G. Schmitt, *Z. Chem.*, *5*, 427 (1965).
4. K. Dostal, M. Cernik, *Z. Chem.*, *6*, 424 (1966).
5. A. J. Edwards, M. A. Mouty, R. D. Peacock, *J. Chem. Soc.*, *A*, 557 (1967).
6. R. S. Drago, K. W. Whitten, *Inorg. Chem.*, *5*, 677 (1966).
7. J. Frkan, *Chem. Zvesti*, *12*, 330 (1958); *Chem. Abstr.*, *52*, 16110 (1958).

2.3.6.4.3. to Give the Te–Halogen Bond.

Fluorination of TeO_2 by metal fluorides affords[1] TeF_4, e.g.:

$$2\ CuF_2 + 3\ TeO_2 \xrightarrow{370°\text{–}480°C} TeF_4 + 2\ CuTeO_3 \left(\xrightarrow{500°\text{–}950°C} Cu_2O + CuO + TeO_2 \right) \qquad \text{(a)}$$

$$4\ FeF_3 + 3\ TeO_2 \xrightarrow[90\ min]{700°C} 3\ TeF_4 + 2\ Fe_2O_3 \qquad \text{(b)}$$

The latter is the most satisfactory, with yields $\leq 77\%$. Because the metal oxides might be transferred back to the fluorides via HF, HF is the indirect source of fluorine in these reactions.

The tetrachloride forms[2] from TeO_2 and $FeCl_3$, and TeI_4 is prepared from the oxide and AlI_3 at 230°C under reduced pressure[3].

(R. MEWS)

1. J. H. Moss, R. Ottie, J. B. Wilford, *J. Fluorine Chem.*, *3*, 317 (1973/74).
2. E. Montignie, *Bull. Soc. Chim. Fr.*, 175 (1946).
3. M. Chaigneau, *Bull. Soc. Chim. Fr.*, 886 (1957).

2.3.6.5. by Nonmetal Halides

2.3.6.5.1. to Give the Sulfur–Halogen Bond.

The formation of the Sulfur—halogen bond from oxo species yields the Cl derivatives, whereas S—F or S—Br bonds are better obtained by metathesis from the chlorides. Most reactions are performed with the oxides (SO_2, SO_3), but their derivatives (the related acids, anions, esters, imino derivatives, etc.) follow similar routes. The reactions of SO_2 are important because of the importance of liq SO_2 as a solvent in fluorine chemistry. Electron-pair acceptor acids are involved; therefore, it is useful to know under what conditions the solvent is attacked. Because the Sb halides behave like their lighter homologous, they are discussed in this section.

Sulfur dioxide reacts with XeF_2 and halogen fluorides with oxidative addition without S—O bond cleavage with SF_4 dismutation taking place under mild conditions[1,2]:

$$SF_4 + SO_2 \xrightarrow[50°-150°C]{8\ h} 2\ OSF_2 \quad (a)$$

With SF_6 the S—O bond is cleaved only under drastic conditions; e.g., the laser-induced reaction of SF_6 gives[3] OSF_2, SF_4 and SO_2F_2:

$$SO_2 + SF_6 \xrightarrow{CO_2\ laser} OSF_2, SF_4, SO_2F_2 \quad (b)$$

The primary step is the decomposition of SF_6 to give SF_4 and F_2 (via $SF_5\cdot$ and $F\cdot$) followed by dismutation and oxidative addition.

If 1:1 or 2:1 NF_3:SO_2 are heated to 440°C for 16 h, SF_6 is formed in 50–90% yield[4]:

$$2\ NF_3 + SO_2 \rightarrow SF_6 + 2\ NO \quad (c)$$

If OSF_2 is the primary product, it is deoxygenated and fluorinated by nitrogen fluorides, as it is known from the photochemical reaction of OSF_2 and N_2F_4[5]. Nitrosyl fluoride, formed from NOX (X = Cl, BF_4) in the presence of fluoride ion, attacks SO_2[6]:

$$5\ [(CH_3)_4N]F + 2\ NOX + 4\ SO_2 \rightarrow 5\ [(CH_3)_4N]^+ + 2\ X^- + N_2 + 3\ [SO_3F]^- + OSF_2 \quad (d)$$

Even after heating for 2 h at 500°C[7] PF_5 does not react with SO_2; only in the CO_2-laser-induced reaction are OSF_2 and OPF_3 formed[8,9]:

$$SO_2 + PF_5 \xrightarrow{CO_2\ laser} OSF_2 + OPF_3 \quad (e)$$

neither do AsF_5, SbF_5 and BF_3 fluorinate SO_2 under normal conditions (with SbF_5 a stable adduct, F_5SbOSO, is formed[10]).

Because of the electron-pair acceptor acidity of SO_3 and its tendency for coordination expansion, the trioxide is more reactive toward halides than SO_2. The primary step is an insertion into the E—X bond:

$$-E(X)_2 + SO_3 \rightarrow X{-}S(=O)_2{-}O{-}E(X) \xrightarrow{SO_3} X{-}S(=O)_2{-}O{-}S(=O)_2{-}X + 1/n({-}O{-}E(O{-}){-})_n$$

$$X{-}S(=O)_2{-}O{-}E(X) \rightarrow X{-}S(=O)_2{-}X + 1/n({-}O{-}E(O{-}){-})_n \quad (f)$$

This scheme is followed with IF_5 where $S_2O_5F_2$ is formed in good yield[11]. With XeF_6 more extensive O—S bond cleavage is observed, for on heating OSF_4 is also formed[12].

Oxygen difluoride fluorinates SO_3 at 350°-500°C to form SF_6 along with other products[13]. This deoxygenating fluorination results from the instability of initially formed S—O—F species. Similar decompositions are observed for SF_5—OF under these conditions; e.g., SF_6 is observed when the hypofluorite is heated at >200°C [14] or irradiated with UV light[14–19]. The same pathway is followed by CF_3OF with SO_3 at 203°-234°C in Al vessels[20,21]; the addition product is formed primarily:

$$CF_3OF + SO_3 \rightarrow CF_3OOSO_2F(+S_2O_6F_2) \xrightarrow{\Delta} SO_2F_2 + COF_2 + O_2 \quad \text{(g)}$$

At higher T these decompose with O—S bond cleavage[22].

With SF_4, SO_3 reacts at 50°-150°C to give a mixture of oxyfluorides[1,2,23]:

$$SF_4 + SO_3 \rightarrow OSF_2 + S_2O_5F_2(+SO_2F_2) \quad \text{(h)}$$

The OSF_2 and $S_2O_5F_2$ are also formed from 2:1 HSO_3F: SF_4:

$$2\ HOSO_2F + SF_4 \rightarrow FSO_2OSO_2F + OSF_2 + 2\ HF \quad \text{(i)}$$

whereas with sulfates and sulfites O_2SF_2 and OSF_2 are obtained[2]:

$$Na_2SO_3 + SF_4(1{:}5) \xrightarrow{350°C} 2\ NaF + 2\ OSF_2 \quad \text{(j)}$$

$$CaSO_4 + SF_4(1{:}5) \xrightarrow{350°C} CaF_2 + OSF_2 + O_2SF_2 \quad \text{(k)}$$

Formation of O_2SF_2 from SO_3 and SF_6 is observed[14] on heating to 250°C for 24 h:

$$SF_6 + 2\ SO_3 \rightarrow 3\ SO_2F_2 \quad \text{(l)}$$

Pentafluoroselenic acid, $HOSeF_5$, is a strong fluorinating agent. Even H_2SO_4 is transformed into fluorosulfuric acid[25]:

$$HOSeF_5 + H_2SO_4 \rightarrow SeO_2F_2 + HSO_3F \quad \text{(m)}$$

$$2\ HOSeF_5 + 4\ SO_3 \rightarrow O_2SeF_2 + 2\ HSO_3F + F_5SeOSO_2OSO_2F \quad \text{(n)}$$

With SO_3 in fluorosulfuric acid, however, only addition products are isolated.

Under pressure NF_3 and SO_3 form SO_2F_2 and SF_6, but with xs NF_3(NF_3:SO_3 > 2:1) SF_6 is obtained exclusively[4]:

$$3\ SO_3 + 2\ NF_3 \xrightarrow{230°-440°C} 3\ SO_2F_2 + NO + NO_2 \xrightarrow{\text{xs } NF_3} SF_6 + NO, NO_2 \quad \text{(o)}$$

Although SO_3 inserts into the As—F bonds of AsF_3 to give complicated structures[26] (see §2.3.7.5); with SbF_3 the simple insertion product, $Sb(OSO_2F)_3$, is formed[27,28], which decomposes reversibly. If SbF_5 is added to refluxing SO_3, however, oxyfluorides result[29] in addition to the simple addition product, $SbF_5 \cdot SO_3$:

$$SO_3 + SbF_5 \rightarrow SbF_5 \cdot SO_3 + SO_2F_2, S_2O_5F_2, S_3O_8F_2, S_4O_{11}F_2, S_5O_{14}F_2, S_6O_{17}F_2 \quad \text{(p)}$$

If liq SO_3 is saturated with BF_3, a colorless oil separates after addition of sulfuric acid (70%)[30]. According to NMR measurements this oil contains[31] $S_2O_5F_2$, $S_3O_8F_2$, $S_4O_{11}F_2$, $S_5O_{14}F_2$, and $S_6O_{17}F_2$.

Aside from oxide chemistry, only few examples of O—F bond exchanges by nonmetal fluorides are known. One of the S—O single bonds in dialkylsulfites, $(RO)_2SO$, is readily cleaved by SF_4 to give sulfinic alkoxides in 90% yields[32]:

$$(RO)_2S{=}O + SF_4 \rightarrow ROS(O)F + OSF_2 + RF \quad (q)$$

Trifluorosulfonium salts, e.g. $[SF_3][BF_4]$, convert sulfoxides to the corresponding monofluorocations[33]:

$$PhS(O)CF_3 + [SF_3][BF_4] \xrightarrow[25°C/30\ min]{pentane} [PhS(F)CF_3][BF_4] + OSF_2 \quad (r)$$

Cleavage of the S—O bond in oxides by nonmetal chlorides is an important synthetic method, e.g., SO_2 with S_2Cl_2 at 150°–400°C in the presence of Cl_2 and activated carbon as catalyst to give $OSCl_2$[34,35]:

$$S_2Cl_2 + 2\ SO_2 + 3\ Cl_2 \xrightarrow{150°–400°C} 4\ OSCl_2 \quad (s)$$

Whereas PCl_3 is inert toward SO_2, PCl_5 reacts vigorously at RT[36,37]:

$$SO_2 + PCl_5 \rightarrow OSCl_2 + OPCl_3 \quad (t)$$

at −50°C several days are required for completion[36].

For laboratory-scale preparations, dry SO_2 is blown over solid PCl_5. Completion is reached after the PCl_5 dissolves. Moisture must be excluded. Yields of ≤50% can be obtained[38]. Arsenic and Sb halides are not attacked by SO_2[36,38]

At RT $SiCl_4$ does not react with SO_2; BCl_3 is solvolyzed by SO_2 over several years to give $OSCl_2$ and B_2O_3[38]. This reaction is catalyzed by Cl^- through intermediate formation of $[ClSO_2]^-$, which is attacked more easily by BCl_3[39]:

$$3\ SO_2 + 2\ BCl_3 \xrightarrow[2–7\ days,\ RT]{Cl^-} B_2O_3 + 3\ OSCl_2 \quad (u)$$

Sulfur trioxide with SCl_2 gives[40] $S_2O_5Cl_2$, or with other nonmetal halides (e.g., PCl_5, $OPCl_3$, $SiCl_4$, $HSiCl_3$[41]) also the pyrosulfurylchloride, or SO_2Cl_2 (with BCl_3[42]).

The Sb chlorides, $SbCl_3$[43] and $SbCl_5$[44,45], exchange as do the metal halides. In the first step SO_3 is inserted into the Sb—Cl bond to form a chlorosulfate group, which decomposes to give $S_2O_5Cl_2$ or SO_2Cl_2:

$$SbCl_3 + xs\ SO_3 \xrightarrow{50°–60°C} S_2O_5Cl_2 + [Sb(SO_4)Cl] \quad (v)$$

$$2\ SbCl_5 + xs\ SO_3 \rightarrow 2\ SbCl_4[(SO_3)_xCl] \rightarrow (Cl_4Sb)_2SO_4 + S_nO_{3n-1}Cl_2 \quad (w)$$

The SO_3-delivering reagents, such as sulfates react analogously. Sulfuric acid itself reacts with PCl_5 to give HSO_3Cl. With xs H_2SO_4, only HCl and SO_3 (or $H_2S_2O_7$) are formed[46]. The attack of PBr_5 on SO_2 is slower than with PCl_5[36], but if the electron-pair acceptor acidity of the brominating agent is enhanced, as when SO_2 is introduced into a cooled mixture of 1:1 PCl_3:Br_2, $OSBr_2$ is formed in good yield[47]:

$$SO_2 + PCl_3Br_2 \rightarrow \underset{(80–90\%)}{OSBr_2} + OPCl_3 \quad (x)$$

Also, BBr_3 reacts vigorously with SO_2[38]:

$$3\ SO_2 + 2\ BBr_3 \rightarrow 3\ OSBr_2 + B_2O_3 \quad (y)$$

whereas from heating with BI_3 in a heavy-walled glass tube only B_2O_3, sulfur and I_2 result[38].

Mixed Sb halides react with SO_3 to form sulfurylchlorofluoride and sulfurylbromofluoride[28]:

$$SO_3 + SbF_3Cl_2 \xrightarrow{180°C} S_2O_5F_2 + SO_2ClF \quad (z)$$

$$SO_3 + SbF_3\text{-}Br_2 \rightarrow SO_2BrF \quad (aa)$$

Exchange of oxygen for halogen is a useful method in organosulfur and in S—N chemistry. Chlorosulfonium cations are formed[48] from sulfoxides and $OSCl_2$:

$$RR'SO + OSCl_2 \rightarrow [RS(Cl)R']Cl + SO_2 \quad (ab)$$

Sulfinyl chlorides form from sulfinic acids and thionyl chloride[49,50]:

$$RSO_2H + OSCl_2 \rightarrow RSOCl + HCl + SO_2 \quad (ac)$$

whereas sulfonic acids, RSO_3H, and their salts $[RSO_3]M$ react with such inorganic halides as FSO_3H, $ClSO_3H$, $OSCl_2$, PCl_5 and $OPCl_3$ to give the corresponding sulfonyl halides[51], e.g.:

$$RSO_2OH + PCl_5 \rightarrow RSO_2Cl + OPCl_3 + HCl \quad (ad)$$

$$[Fe(C_5H_4SO_3^-)_2]Pb^{2+} + OPCl_3 \rightarrow Fe(C_5H_4SO_2Cl)_2 + PbPO_3Cl \quad (ae)$$

Sulfinyl bromides[52] and sulfonyl bromides[53,54] can be prepared similarly:

$$Cl_3CSO_2H + OSBr_2 \xrightarrow[71\%]{0°\text{–}20°C} Cl_3CSOBr + SO_2 + HBr \quad (af)$$

$$Cl_3CSO_2K + OSBr_2 \xrightarrow[25°C, 48\%]{CCl_4} Cl_3CSOBr + SO_2 + KBr \quad (ag)$$

Purity is critical. Impure reactants give different results. Organosulfonyl bromides are prepared as are the chlorides[53,54]:

$$RSO_3K + PBr_5 \rightarrow RSO_2Br + KBr + OPBr_3 \quad (ah)$$

but this method is seldom used.

In the reaction of thionylimides, RNSO, with $OSCl_2$ the S—N linkage remains intact, and sulfur dichlorideimides are formed[55]:

$$ClSO_2NSO + OSCl_2 \rightarrow ClSO_2NSCl_2 + SO_2 \quad (ai)$$

$$HNSO + OSCl_2 \rightarrow NSCl + SO_2 + HCl \quad (aj)$$

Likewise, HNSO reacts with $OSCl_2$ to give[56] NSCl and SO_2. Similar results are obtained with PCl_5:

$$RNSO + PCl_5 \rightarrow RNSCl_2 + OPCl_3 \quad \text{(ak)}$$

where R = ClC_6H_4[57], FSO_2[58].

Sulfamic acid gives $ClSO_2NPCl_3$, which is an intermediate in the formation of $(NSOCl)_3$[59], and mixed systems such as $(NSOCl)_x(NPCl_2)_{3-x}$[60]:

$$H_3NSO_3 + PCl_5 \rightarrow ClSO_2NPCl_3 + OPCl_3 + 3\,HCl \quad \text{(al)}$$

Similar exchanges with phosphorus pentahalides are found for cyclic systems[61]:

$$[(NSOF)_2NSO]^- + PX_5 \rightarrow (NSOF)_2NSX + OPX_3 + X^- \quad \text{(am)}$$

where X = F, Cl.

(R. MEWS)

1. A. L. Oppegard, W. C. Smith, E. L. Muetterties, V. A. Engelhardt, *J. Am. Chem. Soc.*, *82*, 3835 (1960).
2. W. C. Smith, US Pat. 3,026,179 (1962), *Chem. Abstr.*, *57*, 13,411 (1962).
3. V. D. Klimov, V. A. Kuz'menko, V. A. Legasov, *Russ. J. Inorg. Chem. (Engl. Transl.)*, *21*, 1156 (1976).
4. O. Glemser, U. Biermann, *Chem. Ber.*, *100*, 1184 (1967).
5. M. Lustig, C. L. Bumgardner, J. K. Ruff, *Inorg. Chem.*, *3*, 917 (1964).
6. F. Seel, H. Meier, *Z. Anorg. Allg. Chem.*, *274*, 197 (1953).
7. A. P. Hagen, B. W. Callaway, *Inorg. Chem.*, *17*, 554 (1978).
8. A. V. Eletskii, V. D. Klimov, V. A. Legasov, *High Energy Chem. USSR (Engl. Transl.)*, *10*, 110 (1976).
9. V. D. Klimov, V. A. Kuz'menko, V. A. Legasov, *Russ. J. Phys. Chem. (Engl. Trans.)*, *51*, 559 (1977).
10. E. E. Aynsley, R. D. Peacock, P. L. Robinson, *Chem. Ind. (London)*, 1117 (1951).
11. W. Schmidt, *Monatsh. Chem.*, *85*, 452 (1954).
12. G. R. Zeilenga, Ph.D. Diss., Purdue Univ., 1967; *Diss. Abstr.*, *B28*, 4484 (1968); *Chem. Abstr.*, *69*, 48,858 (1968).
13. A. Engelbrecht, E. Nachbaur, C. Pupp. *Monatsh. Chem.*, *95*, 219 (1964).
14. S. M. Williamson, G. H. Cady, *Inorg. Chem.*, *1*, 673 (1962).
15. C. I. Merrill, G. H. Cady, *J. Am. Chem. Soc.*, *83*, 298 (1961).
16. W. H. Hale, S. M. Williamson, *Inorg. Chem.*, *4*, 1342 (1965).
17. R. Czerepinski, G. H. Cady, *J. Am. Chem. Soc.*, *90*, 3954 (1968).
18. G. Pass, H. L. Roberts, *Inorg. Chem.*, *2*, 1016 (1963).
19. B. W. Tattershall, G. H. Cady, *J. Inorg. Nucl. Chem.*, *29*, 3003 (1967).
20. W. P. van Meter, G. H. Cady, *J. Am. Chem. Soc.*, *82*, 6005 (1960).
21. M. Lustig, J. M. Shreeve, *Adv. Fluorine Chem.*, *7*, 175 (1973).
22. J. Czarnowski, E. Castellano, H. J. Schumacher, *Z. Phys. Chem. (Frankfurt am Main)*, *65*, 225 (1969).
23. H. C. Clark, *Chem. Rev.*, *58*, 869 (1958).
24. J. R. Case, F. Nyman, *Nature (London)*, *193*, 473 (1962).
25. K. Seppelt, *Chem. Ber.*, *105*, 3131 (1972).
26. *Gmelins Handbuch der Anorganischen Chemie, Sulfur, Suppl.* Vol. 3 *Sulfuroxides*, Springer Verlag, Berlin, 1980, p. 300 ff.
27. E. L. Muetterties, D. D. Coffman, *J. Am. Chem. Soc.*, *80*, 5914 (1958).
28. E. Hayek, A. Czaloun, B. Krismer, *Monatsh. Chem.*, *87*, 741 (1956).
29. R. J. Gillespie, R. A. Rothenbury, *Can. J. Chem.*, *42*, 416 (1964).
30. H. A. Lehmann, L. Kolditz, *Z. Anorg. Allg. Chem.*, *272*, 73 (1953).
31. R. J. Gillespie, J. v. Oubridge, E. A. Robinson, *Proc. Chem. Soc. (London)*, 428 (1961).
32. J. I. Darragh, A. M. Noble, D. W. A. Sharp, D. W. Walker, J. M. Winfield, *Inorg. Nucl. Chem. Lett.*, *4*, 517 (1968).

33. L. M. Yagupol'skii, N. V. Kondratenko, G. N. Timofeeva, *Zh. Org. Khim*, *20*, 115 (1984).
34. F. C. Trager, US Pat. 2,779,663 (1957); *Chem. Abstr. 51*, 8396 (1957).
35. A. Pechukas, US Pat. 2,431,823 (1947); *Chem. Abstr.*, *42*, 2409 (1948).
36. W. Behne, G. Jander, H. Hecht, *Z. Anorg. Allg. Chem.*, *269*, 249 (1952).
37. F. Feher, in *Handbook of Preparative Inorganic Chemistry*, G. Brauer, ed., Academic Press, Vol. 1, 2nd ed, NY, 1975 p. 382.
38. H. Hecht, R. Greese, G. Jander, *Z. Anorg. Allg. Chem.*, *269*, 262 (1952).
39. A. B. Burg, E. R. Birnbaum, *J. Inorg. Nucl. Chem.*, *7*, 146 (1958).
40. M. Sveda, US Pat. 2,530,410 (1950); *Chem. Abstr. 45*, 4898 (1951).
41. *Gmelins, Handbuch der Anorganischen Chemie, Sulfur, B2*, Verlag Chemie, Weinheim, 1960; p. 1821.
42. Ref. 41, p. 1807.
43. H. A. Lehmann, L. Riesel, *Z. Anorg. Allg. Chem.*, *371*, 281 (1969).
44. H. A. Lehmann, L. Riesel, *Z. Anorg. Allg. Chem.*, *371*, 274 (1969).
45. R. Appel, *Z. Anorg. Allg. Chem.*, *285*, 114 (1956).
46. Ref. 41, p. 1835.
47. H. Dohse, R. Mühdemann, Ger. Pat. 665,061 (1938), *Chem. Abstr.*, *33*, 1891 (1939).
48. W. R. Hardstaff, R. F. Langler, in *Sulfur in Organic and Inorganic Chemistry*, Vol. 4, A. Senning, ed., Marcel Dekker, New York, 1982, p. 200.
49. J. L. Kice, G. Guaraldi, *J. Org. Chem.*, *31*, 3568 (1966).
50. J. L. Kice, G. Guaraldi, C. G. Venier, *J. Org. Chem.*, *31*, 3561 (1966).
51. I. Haiduc, K. J. Wynne, in *Methodicum Chimicum*, Vol. 7, H. Zimmer, K. Niedenzu, eds., Georg Thieme Verlag, Stuttgart, 1976, p. 707ff.
52. A. Senning, S. Kaae, C. Jacobsen, P. Kelly, *Acta Chem. Scand.*, *22*, 3256 (1968).
53. W. H. Hunter, B. E. Sorensen, *J. Am. Chem. Soc.*, *54*, 3364 (1932).
54. A. H. Kohlhase, *J. Am. Chem. Soc.*, *54*, 2441 (1932).
55. H. W. Roesky, *Angew. Chem.*, *Int. Ed. Engl.*, *7*, 63 (1968).
56. R. L. Dekock, M. S. Haddad, *Inorg. Chem.*, *16*, 216 (1977).
57. E. S. Levchenko, I. E. Sheinman, *J. Gen. Chem. USSR. (Engl. Transl.)*, *36*, 446 (1966).
58. H. W. Roesky, *Angew. Chem.*, *Int. Ed. Engl.*, *6*, 711 (1967).
59. A. V. Kirsanov, *Izv. Akad. Nauk SSR, Otd. Khim. Nauk*, 426 (1950); *J. Gen. Chem. USSR (Engl. Transl.)*, *22*, 93 (1952).
60. *Gmelins Handbuch der Anorganischen Chemie, Sulfur-Nitrogen Compounds I*, Springer Verlag, Berlin, 1977, p. 28ff.
61. D. L. Wagner, H. Wagner, O. Glemser, *Chem. Ber.*, *108*, 2469 (1975).

2.3.6.5.2. to Give the Se–Halogen Bond.

The Se—O bond is cleaved more easily by nonmetal halides than is the S—O bond. The dioxide, e.g., is quantitatively oxidized by XeF_2 vapor to give[1] SeF_6:

$$SeO_2 + 3\ XeF_2 \rightarrow SeF_6 + O_2 + 3\ Xe \qquad (a)$$

One of the Se—O bonds is preserved when the reaction with ClF is carried out under controlled conditions[2]:

$$SeO_2 + 2\ ClF \rightarrow OSeF_2 + (Cl_2O) \qquad (b)$$

$$SeO_2 + 4\ ClF \rightarrow SeF_4 + (2\ Cl_2O) \qquad (c)$$

From 1:3.5 SeO_2:ClF, $OSeF_2$ can be obtained in 95% yield; at 1:6, SeF_4 is formed along with other products. Similar results are found[3] with SeO_2 and ClF_3. Bromine trifluoride converts SeO_2 to SeF_6[4]:

$$SeO_2 + 2\ BrF_3 \rightarrow SeF_6 + O_2 + Br_2 \qquad (d)$$

The fluorination of SeO_2 with SF_4 under autogeneous P gives[5] SeF_4:

$$SeO_2 + 2\ SF_4 \xrightarrow{100°-250°C} SeF_4 + 2\ OSF_2 \quad (e)$$

$$SeO_2 + SeF_4 \rightarrow 2\ OSeF_2 \quad (f)$$

From xs SeO_2 with SeF_4, seleninyl fluoride is prepared[6,7].

Nitryl fluoride gives SeF_4 at RT, too, but it combines with xs NO_2F to form white, crystalline NO_2SeF_5[8]:

$$SeO_2 + 5\ NO_2F \rightarrow NO_2SeF_5 + 4\ NO_2 + O_2 \quad (g)$$

and SeO_3 reacts with SeF_4[9,10]:

$$SeO_3 + SeF_4 \rightarrow SeO_2F_2 + SeOF_2 \quad (h)$$

If xs SeO_3 is present a second product is formed that is either the unstable adduct, $SeF_4 \cdot SeO_3$ (actually $[SeF_3][SeO_3F]$) or $Se_2O_5F_2$. A plausible mechanism is:

$$SeO_3 + SeF_4 \rightleftharpoons [SeF_3][SeO_3F] \rightarrow OSeF_2 + O_2SeF_2 \quad (i)$$

$$SeO_3 + SeOF_2 \rightarrow SeO_2 + SeO_2F_2 \quad (j)$$

$$SeO_3 + SeO_2F_2 \rightarrow FSeO_2OSeO_2F \quad (k)$$

From SeO_3 and HSO_3F is prepared SeO_2F_2 (cf. §2.3.6.2.2)[11,12];

$$HSO_3F + SeO_3 \rightarrow HSeO_3F + SO_3 \quad (l)$$

$$HSeO_3F + HSO_3F \rightarrow SeO_2F_2 + H_2SO_4 \quad (m)$$

On heating, a vigorous reaction takes place. Byproducts are $OSeF_2$, $S_2O_5F_2$ and $S_3O_8F_2$[11]. Refluxing $BaSeO_4$ with xs FSO_3H also produces SeO_2F_2 in excellent yield[12]. Reaction of SeO_2F_2 with HSO_3F containing KHF_2 or HF gives $HOSeF_5$[14–16], but HF may be the halogenating agent (cf. §2.3.6.2.2); SeO_2F_2 itself acts as a fluorinating agent towards selenates. If the compound is condensed onto K_2SeO_4 at $-60°C$ and the mixture warmed, colorless crystalline $KSeO_3F$ is formed[17]:

$$K_2SeO_4 + SeO_2F_2 \rightarrow 2\ KSeO_3F \quad (n)$$

$$MHSeO_4 + SeO_2F_2 \rightarrow MSeO_3F + HOSeO_2F \quad (o)$$

Hydrogen selenates behave similarly[18].

If 1.5 to 2-fold xs PF_5 is distilled onto SeO_3 in liq SO_2, a colorless solution results at RT. Spectroscopic investigation shows[19] the presence of $Se_2O_2F_2$, $Se_2O_5F_2$ and OPF_3:

$$x\ SeO_3 + PF_5 \rightarrow Se_xO_{3x-1}F_2 + OPF_3\ (x = 1, 2) \quad (p)$$

After several days all the $Se_2O_5F_2$ is converted to SeO_2F_2. Higher Se oxyfluorides, $Se_xO_{3x-1}F_2$ (x = 1–3), are formed[20] according to NMR spectroscopy from 1:1–6:1 SeO_3: AsF_3 in SO_2. The degree of branching and oligomerization depends on reactant ratio and time. At RT, withtout a solvent, xs AsF_3 (10–20%) forms a clear solution from which SeO_2F_2[19,21] and SeF_6[21] are slowly evolved. After several days, crystalline As(III)FSe(VI)O_4 precipitates[19,21,22]. Refluxing the mixture is a route to SeO_2F_2[21]:

$$SeO_3 + xs\ AsF_3 \rightarrow SeO_2F_2 + (AsOF)_x \quad (q)$$

Similar results are obtained[23] with SbF_3.

Also, KBF_4 behaves like KF; BF_3 is split off and SeO_2F_2 is formed[16]:

$$2\ KBF_4 + 2\ SeO_3 \rightarrow 2\ BF_3 + SeO_2F_2 + K_2SeO_4 \qquad (r)$$

Fluoroesters of selenious and selenic acid are prepared[24,25] by mixing 1:1 $(RO)_2SeO:SeOF_2$:

$$(RO)_2SeO + SeOF_2 \rightarrow 2\ FSe(O)OR \qquad (s)$$

or $(RO)_2SeO_2$ and SeO_2F_2[26]:

$$(RO)_2SeO_2 + SeO_2F_2 \rightarrow 2\ FSeO_2OR \qquad (t)$$

Ammonium aminoselenate is fluorinated by SeO_2F_2, as well[27]:

$$[NH_4][NH_2SO_3] + 2\ SeO_2F_2 \rightarrow [NH_4][N(SeO_2F)_2] + HF + HSeO_3F \qquad (u)$$

Either neat[28] or in SO_2Cl_2, SeO_2 reacts with $OSCl_2$ to give $OSeCl_2$:

$$SeO_2 + 2\ OSCl_2 \rightarrow SeCl_4 + 2\ SO_2 \qquad (v)$$

From SeO_2 and $SeCl_4$, $OSeCl_2$ also forms[29]:

$$SeO_2 + SeCl_4 \rightarrow 2\ OSeCl_2 \qquad (w)$$

With SeO_3 and $SeCl_4$ in $OPCl_3$ solution, a white solid is formed[30] and Cl_2 evolves. A reaction mechanism similar to that with SeF_4 is accepted[10]:

$$\begin{array}{c} SeO_3 + SeCl_4 \rightarrow \{SeO_2Cl_2\} + OSeCl_2 \rightarrow SeO_2 + Cl_2 \\ \downarrow \\ SeO_2 + Cl_2 \end{array} \qquad (x)$$

From SeO_3 and PCl_3 in SO_2Cl_2 soln, $OSeCl_2$ forms[31]:

$$SeO_3 + 2\ PCl_3 \xrightarrow{SO_2Cl_2} OSeCl_2 + 2\ OPCl_3 \qquad (y)$$

Mixing 1:2 Se_2Br_2: SeO_2, with 3 mol Cl_2, gives, after warming to RT over several h, SeOBrCl in 13% yield[32]:

$$Se_2Br_2 + 2\ SeO_2 + 3\ Cl_2 \rightarrow 2\ SeOBrCl + 2\ SeOCl_2 \qquad (z)$$

Antimony pentachloride[33] and selenium tetrahalides cleave the Se—O bond in alkoxides. Redistributions of ligands, as in the fluorine system, are observed:

$$3\ (MeO)_4Se + SeX_4 \rightarrow 4\ (MeO)_3SeX \qquad (aa)^{34}$$

$$(EtO)_2SeO + SeX_4 \rightarrow 2\ EtOSe(O)X \qquad (ab)^{32}$$

where X = Cl, Br.

(R. MEWS)

1. N. N. Aleinikov, D. N. Sokolov, L. K. Gombeva, B. L. Korsunskii, F. I. Dubovitskii, *Acad. Sci. USSR Bull. Chem. Sci. (Engl. Trans.)*, 2552 (1973).
2. C. Lau, J. Passmore, *J. Fluorine Chem.*, *6*, 77 (1973).
3. W. Hückel, *Nachr. Akad. Wiss. Göttingen, Math. Phys. Kl.*, 36 (1946); *Chem. Abstr.*, *43*, 6793 (1949).
4. H. J. Emeléus, A. A. Woolf, *J. Chem. Soc.*, 164 (1950).

5. A. G. Oppegard, W. C. Smith, E. L. Muetterties, V. A. Engelhardt, *J. Am. Chem. Soc.*, *82*, 3835 (1960).
6. R. D. Peacock, *J. Chem. Soc.*, 3617 (1953).
7. L. E. Alexander, I. R. Beattie, *J. Chem. Soc., Dalton Trans.*, 1745 (1972).
8. E. E. Aynsley, G. Hetherington, P. L. Robinson, *J. Chem. Soc.*, 1119 (1954).
9. T. Birchall, K. J. Gillespie, S. C. Vekris, *Can. J. Chem.*, *43*, 1672 (1965).
10. A. J. Edwards, M. A. Mouty, R. D. Peacock, *J. Chem. Soc., A*, 557 (1967).
11. R. C. Paul, S. K. Sharma, R. D. Sharma, K. K. Paul, *J. Inorg. Nucl. Chem.*, *33*, 2905 (1971).
12. K. Dostal, M. Cernik, *Z. Chem.*, *6*, 424 (1966).
13. A. Engelbrecht, B. Stoll, *Z. Anorg. Allg. Chem.*, *292*, 20 (1957).
14. K. Seppelt, *Chem. Ber.*, *105*, 2431 (1972).
15. K. Seppelt, *Angew. Chem., Int. Ed. Engl.*, *11*, 723 (1972).
16. K. Seppelt, *Chem. Ber.*, *105*, 3131 (1972).
17. P. Martin, A. Scholer, E. Class, *Chimia*, *21*, 162 (1967).
18. M. Cernik, K. Dostal, *Z. Anorg. Allg. Chem.*, *425*, 37 (1976).
19. J. Touzin, L. Mitacek, *Z. Chem.*, *20*, 32 (1980).
20. J. Touzin, L. Meznik, L. Mitacek, *Coll. Czech. Chem. Commun.*, *44*, 1530 (1979).
21. H.-G. Jerschkewitz, *Angew. Chem.*, *69*, 562 (1957).
22. J. Touzin, L. Mitacek, *Coll. Czech. Chem. Commun.*, *44*, 2743 (1979).
23. J. Touzin, L. Mitacek, *Coll. Czech. Chem. Commun.*, *44*, 2751 (1979).
24. R. Paetzold, K. Aurich, *Z. Chem.*, *2*, 60 (1962).
25. R. Paetzold, K. Aurich, *Z. Anorg. Allg. Chem.*, *317*, 151 (1962).
26. R. Paetzold, R. Kurze, G. Engelhardt, *Z. Anorg. Allg. Chem.*, *353*, 62 (1967).
27. A. Ruzicka, *Z. Chem.*, *19*, 75 (1979).
28. R. Pactzold, K. Auria, *Z. Anorg. Allg. Chem.*, *315*, 72 (1962).
29. V. V. Safonov, E. A. Fedorov, *Russ. J. Inorg. Chem.*, *23*, 1097 (1978).
30. T. Birchall, R. J. Gillespie, S. L. Vekris, *Can. J. Chem.*, *43*, 1672 (1965).
31. E. Class, *Experientia*, *22*, 133 (1966).
32. N. N. Yarovenko, M. A. Raksha, G. B. Gazieva, *J. Gen. Chem. USSR (Engl. Transl.)*, *31*, 3737 (1961).
33. R. C. Paul, K. K. Bhasin, R. D. Sharma, *Indian J. Chem.*, *13*, 723 (1975).
34. M. Reichenbaecher, R. Paetzold, *Z. Anorg. Allg. Chem.*, *400*, 176 (1973).

2.3.6.5.3. to Give the Te–Halogen and Po–Halogen Bond.

The Te—O bond in Te oxides cleaves and is used in synthesis as in Se chemistry. Halogen fluorides and TeO_2 give a mixture of oxygen-free species:

$$TeO_2 + ClF_3 \rightarrow TeF_6 + (F_5Te)_2O \qquad (a)^1$$

$$TeO_2 + ClF \rightarrow TeF_5Cl \qquad (b)^2$$

$$TeO_2 + BrF_5 \rightarrow TeF_6 + (TeF_5)_2O \qquad (c)^1$$

$$TeO_2 + BrF_3 \xrightarrow{100^\circ\text{–}450^\circ C} TeF_4 + TeF_6 + (TeF_5)_2O + F_5TeOTeF_4OTeF_5 \qquad (d)^3$$

but in some cases hexavalent species containing the F_5TeO group are isolated.

The action SF_4[4] or SeF_4[5,6] on TeO_2 gives TeF_4:

$$TeO_2 + 2\ SF_4 \xrightarrow[150^\circ C]{12\ h} TeF_4 + 2\ SOF_2 \qquad (e)$$

$$TeO_2 + 2\ SeF_4 \rightarrow TeF_4 + 2\ SeOF_2 \qquad (f)$$

whereas heating with $M^+[TeF_5]^-$ in the presence of metal fluorides[7] leads to salts containing the $[OTeF_4]^{2-}$ anion:

$$TeO_2 + MTeF_5 + 3\ MF \rightarrow 2\ M_2TeOF_4 \qquad (g)$$

Heating the components in a Pt boat to 550°C yields Cs_2OTeF_4 and K_2OTeF_4[7]. Reaction (e) must be carried out in a stainless-steel autoclave under pressure. Tellurium oxides, tellurates or other compounds containing oxygen are fluorinated by this method; TeF_4 remains as a crystalline residue and is purified by sublimation. In method (f), SeF_4 is condensed[4] onto powdered TeO_2 and the mixture is heated to 80°C, where the reaction starts. After refluxing for 15 min, the volatiles ($SeOF_2$, xs SeF_4) are removed at 100°C and 10^{-2} torr (1 Pa)[5,6]. If the fluorination is carried out in the presence of metal fluorides, pentafluorotellurates are obtained directly[8]; NO_2F itself acts as deoxygenating and fluoride ion delivering agent[9,10]:

$$5\ NO_2F + TeO_2 \rightarrow (NO_2)TeF_5 + 4\ NO_2 + O_2 \tag{h}$$

After warming, the exothermic reaction starts, keeping the mixture liquid. After cooling, colorless crystals separate.

Pentafluorotelluric acid, $HOTeF_5$, (mp 39.1°C, bp 59.7°C) is obtained in high yield from BaH_4TeO_6 and $HOSO_2F$[11–14]:

$$BaH_4TeO_6 + 7\ HOSO_2F \rightarrow HOTeF_5 + Ba(SO_3F)_2 + 5\ H_2SO_4 \tag{i}$$

Aside from F_5TeOH, two other volatile compounds containing the pentafluorooxo group [F_5TeOSO_3H and $(F_5TeO)_2SO_2$] and small amounts of TeF_6 are formed depending on the H_2O content of the tellurate. Best results are obtained if dried BaH_4TeO_6 (180°C) is added to vacuum-distilled, SO_3-free FSO_3H at −20°C. The mixture is heated with vigorous stirring, and pure TeF_5OH distills. It is also possible to prepare TeF_5OH by dissolving 1:12 $Te(OH)_6$ in HSO_3F[15]. Pentafluorotelluric acid is used in synthetic chemistry[16,17].

Few examples of O–Cl or O–Br exchange by nonmetal halides are known, e.g., TeO_2 with $OSCl_2$ gives[18] $TeCl_4$:

$$TeO_2 + 2\ SOCl_2 \xrightarrow{\Delta} TeCl_4 + 2\ SO_2 \tag{j}$$

$$11\ TeO_2 + TeCl_4 \rightarrow 2\ Te_6O_{11}Cl_2 \tag{k}$$

$$TeO_2 + TeCl_4 \rightarrow 2\ TeOCl_2 \tag{l}$$

In the $TeCl_4$–TeO_2 system, $Te_6Cl_2O_{11}$ and two eutectics are found[19]; in the gas phase above this system, $OTeCl_2$ is present according to tensimetric measurements and transport reactions[20]. The TeO_2–$TeBr_4$ system is similar with $Te_6Br_2O_{11}$ formed[21,22]. The oxide TeO_2 reacts exothermally with BBr_3; by heating the compounds for 5 h to 110°–120°C, $TeBr_4$ is produced[23]:

$$3\ TeO_2 + 4\ BBr_3 \rightarrow 3\ TeBr_4 + 2\ B_2O_3 \tag{m}$$

In the TeO_2–TeI_4 system, $Te_6O_{11}I_2$ is not confirmed[24], but from the transport behavior of TeO_2 with I_2 and TeI_4, the existence of $TeOI_2$ is derived[25]:

From PoO_2 and inorganic chlorides $PoCl_4$ is formed[26], e.g.:

$$PoO_2 + 2\ OSCl_2 \rightarrow PoCl_4 + 2\ SO_2 \tag{n}$$

$$PoO_2 + 2\ PCl_5 \rightarrow PoCl_4 + 2\ OPCl_3 \tag{o}$$

(R. MEWS)

1. D. R. Vissers, M. J. Steindler, US At. Energy Commiss. access. No. 43,164, Rept. ANL-7142 (1966); *Chem. Abstr.*, *66*, 71,964 (1967).
2. C. Lau, J. Passmore, *Inorg. Chem.*, *13*, 2278 (1974).
3. T. Sakurai, *J. Nucl. Sci. Technol.*, *10*, 130 (1973).
4. W. C. Smith, US Pat. 3,026,179 (1962), *Chem. Absr.*, *57*, 13,411 (1962).
5. R. Campbell, P. L. Robinson, *J. Chem. Soc.*, 785, (1956).
6. C. Dragon, *C.R. Hebd. Seances Acad. Sci.*, *255*, 122 (1962).
7. J. B. Milne, D. Mofett, *Inorg. Chem.*, *12*, 2240 (1973).
8. A. J. Edwards, M. A. Mouty, R. D. Peacock, A. J. Suddens, *J. Chem. Soc.*, 4087 (1964).
9. E. E. Aynsley, G. Hetherington, P. L. Robinson, *J. Chem. Soc.*, 1119 (1954).
10. G. Hetherington, P. L. Robinson, in *Recent Aspects of the Inorganic Chemistry of Nitrogen*, Chem. Soc., London, Spec. Publ. No. 10, 1957, p. 23.
11. H. Bürger, *Z. Anorg. Allg. Chem.*, *360*, 97 (1968).
12. A. Engelbrecht, W. Loreck, W. Nehoda, *Z. Anorg. Allg. Chem.*, *360*, 88 (1968).
13. A. Engelbrecht, F. Sladky, *Angew. Chem.*, *Int. Ed. Engl.*, *3*, 383 (1964).
14. A. Engelbrecht, F. Sladky, *Monatsh. Chem.*, *96*, 159 (1965).
15. K. Seppelt, D. Nöthe, *Inorg. Chem.*, *12*, 2727 (1973).
16. A. Engelbrecht, F. Sladky, *Adv. Inorg. Chem. Radiochem.*, *24*, 189 (1981).
17. K. Seppelt, *Angew. Chem.*, *Int. Ed. Engl.*, *21*, 877 (1982).
18. V. Lenher, *J. Am. Chem. Soc.*, *30*, 737 (1908).
19. H. Oppermann, G. Kunze, W. Reichelt, *Z. Anorg. Allg. Chem.*, *429*, 18 (1977).
20. H. Oppermann, G. Kunze, cited in 19.
21. P. Khodadad, *Ann. Chim. (Paris)*, *10*, 83 (1965).
22. V. V. Safonov, N. S. Nikulenko, M. G. Varfolomeev, V. A. Grinko, V. I. Ksenzenko, *Russ. J. Inorg. Chem.* (*Engl. Transl.*) *20*, 1370 (1975).
23. P. M. Druce, M. F. Lappert, *J. Chem. Soc.*, *A*, 3595 (1971).
24. H. Oppermann, G. Kunze, E. Wolf, G. A. Kokovin, I. M. Sichova, G. E. Osipova, *Z. Anorg. Allg. Chem.*, *461*, 165 (1980).
25. K. W. Bagnall, R. W. M. D'Eye, J. H. Freeman, *J. Chem. Soc.*, 2320 (1955).

2.3.7 by Halogenation of Group-VIB Oxides and Oxoanions

The action of halogens or interhalogens on group-IVB–group-VIB oxoanions or oxo-species in the presence of metal fluorides will give ROF or ROCl[1]. In this section only the reactions of group-VIB species are described.

(R. MEWS)

1. J. M. Shreeve, *Adv. Inorg. Chem. Radiochem.*, *26*, 119 (1983).

2.3.7.1 with Elemental Halogens

2.3.7.1.1. to Give the Sulfur–Halogen Bond.

Only F_2 adds to SO_3; SO_2 is also attacked by Cl_2, whereas the less oxidizing halogens, Br_2 and I_2 react only with anions or lower oxides (Br_2 only). Halogen addition to sulfur oxides can be photolytically or thermally induced or catalyzed by metal fluorides because anionic intermediates are more readily attacked:

$$\gt S{=}O + F^- \rightarrow \left[\gt S(F){-}O \right]^- \xrightarrow[-X^-]{+X_2} F{-}\underset{(\mathbf{A})}{\overset{|}{\underset{|}{S}}}{-}O{-}X \rightarrow (F)(X)\gt S{=}O \qquad \text{(a)}$$

If sulfur is not in the 6+ oxidation state, the hypohalites (**A**) will oxidatively rearrange to the final products, as in the fluorination of OSF_2 and OSF_4.

The photochemically induced addition of F_2 to SO_3 starts with the formation of $[F]^{\cdot}$ radicals[1]:

$$F_2 \xrightarrow{h\nu} 2\,[F]^{\cdot} \tag{b}$$

$$F^{\cdot} + SO_3 \rightarrow [FSO_3]^{\cdot} \rightarrow \tfrac{1}{2}\,FSO_2OOSO_2F \xrightarrow{F^{\cdot}} FSO_2OF \tag{c}$$

With xs F_2, the initially formed peroxodisulfuryldifluoride, $S_2O_6F_2$, gives the hypofluorite, FSO_3F.

Heating 2:1 SO_3:F_2 gives peroxodisulfuryldifluoride, $S_2O_6F_2$, in high yield[2] with some $S_2O_5F_2$ and traces of FSO_3F[2]. The peroxide is prepared >90% yield on a large scale by flowing F_2 and xs SO_3 over an AgF_2 catalyst at 160°C[3,4]:

$$2\,SO_2 + F_2 \xrightarrow[\text{autoclave}]{170^\circ C} FSO_2OOSO_2F + S_2O_5F_2\ (+\,FSO_3F) \tag{d}$$

$$\text{xs } SO_3 + F_2 \xrightarrow[160^\circ C]{\text{flow over } AgF_2} FSO_2OOSO_2F\ (+\,FSO_3F) \tag{e}$$

Caution should be exercised because even with xs SO_3 small amounts of the explosive hypofluorite, FSO_3F, are formed[5]. The relative amounts of $S_2O_6F_2$, FSO_2OF and degradation products, such as SO_2F_2, depend on conditions. Low T favors the formation of $S_2O_6F_2$, but at >170°C, large amounts of FSO_3F are observed. With xs SO_3, even at high T, $S_2O_6F_2$ is formed[3–8]. The hypofluorite itself is obtained in high yield[6] at 220°C with xs F_2:

$$SO_3 + \text{xs } F_2 \xrightarrow{AgF_2} FSO_2OF \tag{f}$$

Catalysis by AgF_2 arises from the intermediate formation of fluorosulfate anion[9,10]:

$$AgF_2 + 2\,SO_3 \rightarrow Ag(SO_3F)_2 \xrightarrow[-AgF_2]{F_2} [FSO_3]^{\cdot} \tag{g}$$

$$[FSO_3]^{\cdot} \rightarrow S_2O_6F_2 \qquad [FSO_3]^{\cdot} \xrightarrow{\frac{1}{2}F_2} FSO_2OF$$

The sulfate ion itself adds F_2; the resulting fluoroxysulfates are powerful oxidants and fluorinating agents[11,12]:

$$M_2SO_4 + F_2 \xrightarrow{0^\circ\text{–}4^\circ C} MF + MSO_4F \tag{h}$$

where M = Rb, Cs.

The resulting yellowish-white salts are stable, when stored at −10°C.

Hypofluorite formation from S(VI) oxyanions is general; e.g., from OSF_4 and F_2 pentafluorosulfurhypofluorite is formed quantitatively via the $[SF_5O]^-$ anion:

$$OSF_4 + F_2 \xrightarrow{AgF_2} SF_5OF \qquad \text{(i)}^{13\text{–}15}$$

$$OSF_4 + CsF \rightarrow Cs[OSF_5] \xrightarrow{F_2} CsF + SF_5OF \tag{j}$$

Fluorination of SO_2 quickly yields sulfuryl fluoride:

$$SO_2 + F_2 \xrightarrow[200°C]{AgF_2} SO_2F_2 \quad (k)^{13}$$

$$SO_2 + F_2 \xrightarrow{500°C} SO_2F_2 \quad (l)^{16,17}$$

With xs F_2 formation of SF_6 starts at 200°C, and OSF_2, OSF_4 and SF_5OF are also observed.

Electrochemical fluorination of SO_2 in liq HF gives SO_2F_2 in 90% yield[18].

From SO_2 and Cl_2, SO_2Cl_2 is formed in excellent yield under mild conditions if catalysts (e.g., activated charcoal) are present, or if Cl radicals are produced photochemically[19]:

$$SO_2 + Cl_2 \xrightarrow[\text{or } h\nu]{\text{catal}} SO_2Cl_2 \quad (m)$$

In the presence of metal fluorides, SO_2FCl results from the chlorination of $[FSO_2]^-$[20]:

$$MF + SO_2 \rightarrow M[SO_2F] \xrightarrow{Cl_2} MCl + SO_2FCl \quad (n)$$

If SO_2, Cl_2 and HF are passed over a BaF_2 catalyst at 125°–400°C, SO_2F_2 is formed in >80% yields[21]. Similarly, sulfonyl chlorides form from organic sulfinate salts:

$$M[RSO_2] + X_2 \rightarrow RSO_2X + MX \quad (o)^{22-24}$$

where X = Cl, Br, I;

$$R_fSO_2\text{—}ZnX + Cl_2 \rightarrow R_fSO_2Cl + ZnXCl \quad (p)^{25,26}$$

where R_f = perfluoroalkyl;

$$R\text{—}\overset{\overset{\displaystyle O}{\|}}{S}\text{—}OR + X_2 \rightarrow RSO_2X + RX \quad (q)^{27}$$

where X = Cl, Br.

Even Br_2 and I_2 oxidize these salts to give the corresponding sulfonyl halides in good yield. Stabilities decrease from Cl to I, aryl derivatives being more stable than alkylsulfonyl halides. Sulfonyl halides are also obtained from sulfinate esters.

Addition of Br_2 to SO is a purely inorganic example for an S—Br bond formation from sulfur oxides[28]:

$$SO + Br_2 \rightarrow OSBr_2 \quad (r)$$

(R. MEWS)

1. M. Gambaruto, J. E. Sicre, H. J. Schumacher, *J. Fluorine Chem.*, *5*, 175 (1979).
2. J. K. Ruff, R. F. Merritt, *Inorg. Chem.*, *7*, 1219 (1968).
3. F. B. Dudley, G. H. Cady, *J. Am. Chem. Soc.*, *79*, 513 (1957).
4. J. M. Shreeve, G. H. Cady, *Inorg. Synth.*, *7*, 124 (1963).
5. G. H. Cady, *Inorg. Synth.*, *11*, 155 (1968).
6. F. B. Dudley, G. H. Cady, D. F. Eggers, *J. Am. Chem. Soc.*, *78*, 290 (1956).
7. M. Lustig, J. M. Shreeve, *Adv. Fluorine Chem.*, *7*, 175 (1973).
8. R. A. De Marco, J. M. Shreeve, *Adv. Inorg. Chem. Radiochem.*, *16*, 109 (1974).

9. G. H. Cady, *Intra-Sci. Chem. Rep.*, *5*; *Chem. Abstr.*, *74*, 150433 (1971).
10. F. B. Dudley, *J. Chem. Soc.*, 3407 (1963).
11. E. H. Appelman, L. J. Basile, R. C. Thompson, *J. Am. Chem. Soc.*, *101*, 3384 (1979).
12. M. V. Steele, P. A. G. O'Hare, E. H. Appelman, *Inorg. Chem.*, *20*, 1022 (1981).
13. F. B. Dudley, G. H. Cady, D. F. Eggers, *J. Am. Chem. Soc.*, *78*, 1553 (1956).
14. J. K. Ruff, M. Lustig, *Inorg. Chem.*, *3*, 1422 (1964).
15. J. K. Ruff, *Inorg. Synth.*, *11*, 131 (1968).
16. A. Vanderchmitt, Rep. CEA-R-4613, 1974; *Chem. Abstr.*, *83*, 21,249 (1975).
17. *Gmelins Handbuch der Anorganischen Chemie*, *Sulfur*, Suppl. Vol. 3 (*Sulfur Oxides*), Springer Verlag Berlin 1980, p. 177 ff.
18. S. Nagase, T. Abe, H. Baba, *Bull. Chem. Soc. Jpn.*, *42*, 2062 (1969).
19. Ref. 17, p. 178.
20. F. Seel *Inorg. Synth.*, *9*, 111 (1967).
21. E. S. Jones, M. A. Robinson, R. E. Eibeck, US Pat. 4,003,984 (1977); *Chem. Abstr.*, *80*, 108,788 (1977), Ger. Pat. 2,643,521 (1977); *Chem. Abstr.*, *87*, 55,235 (1977); cf. D. M. Cook, US Pat. 4,102,987 (1978); *Chem. Abstr.*, *90*, 124,053 (1979).
22. P. S. Magee, in *Sulfur in Organic and Inorganic Chemistry*, A. Senning, ed., Marcel Dekker, New York, Vol. 1, 1971 p. 306; Vol. 4, 1982 p. 314.
23. J. P. Danehy, in *Sulfur in Organic and Inorganic Chemistry*, A. Senning, ed., Vol. 1 Marcel Dekker, New York, 1971, p. 336.
24. L. Field, C. M. Lukehart, in *Sulfur in Organic and Inorganic Chemistry*, A. Senning, ed., Vol. 4, Marcel Dekker, New York, 1982, pp. 356, 359.
25. A. Commeyras, H. Blancou, P. Moreau, Ger. Offen. 2,708,751 (1977); *Chem. Abstr.*, *87*, 183970 (1977).
26. A. Commeyras, H. Blancou, A. Lantz, Ger. Offen. 2,756,169 (1978); *Chem. Abstr.*, *89*, 108161 (1978).
27. I. B. Douglass, *J. Org. Chem.*, *30*, 633 (1965).
28. P. W. Schenk, *Z. Anorg. Allg. Chem.*, *223*, 385 (1937).

2.3.7.1.2. to Give the Se–, Te–, Po–Halogen Bond.

Elemental halogens attack the chalcogen–oxygen bond in di- or trioxides. Finely divided SeO_2 and F_2 leads only to the formation of SeO_2F_2 [1]:

$$SeO_2 + F_2 \rightarrow SeO_2F_2 \qquad \text{(a)}$$

Products resulting from Se—O bond cleavage, e.g., SeF_6 or SeF_5OF (see §2.3.6.1.2), are dominant. The synthesis of the hypofluorite is improved by starting from $K[SeOF_3]$ or $Cs[OSeF_5]$:

$$KSeOF_3 \xrightarrow{F_2} KSeOF_5 \xrightarrow[-KF]{F_2} F_5SeOF \qquad \text{(b)}^{2}$$

$$CsOSeF_5 \xrightarrow[-CsF]{F_2} F_5SeOF \qquad \text{(c)}^{3}$$

$$Hg(OSeF_5)_2 \xrightarrow[-HgF_2]{F_2} 2\ F_5SeOF \qquad \text{(d)}^{3}$$

Fluorination of the mercurial is the best method, only small amounts of SeF_6 and $F_5SeOOSeF_5$ being formed[3].

Fluorine addition to the fluoroselenite anion gives bis(fluoroxy)selenium tetrafluoride in $\leq 50\%$ yield[4]:

$$K[SeO_2F] + 3\ F_2 \rightarrow KF + SeF_4(OF)_2 \qquad \text{(e)}$$

The higher homologues of F_2 will not add to Se oxides or oxoanions; halogenation of TeO_2 leads to TeO bond cleavage (see §2.3.6.1.3).

(R. MEWS)

1. G. Mitra, *J. Indian Chem. Soc.*, *37*, 804 (1960).
2. J. E. Smith, G. H. Cady, *Inorg. Chem.*, *9*, 1442 (1970).
3. K. Seppelt, *Chem. Ber.*, *106*, 157 (1973).
4. J. E. Smith, G. H. Cady, *Inorg. Chem.*, *9*, 1293 (1970).

2.3.7.2. with Halogen Halides

2.3.7.2.1. to Give the Sulfur–Halogen Bond.

Halogen fluorides and chlorides add to SO_3 and SO_2. If SO_3 and ClF are warmed slowly to RT, $ClOSO_2F$ is formed in 90% yield[1,2] with some SO_2F_2 and Cl_2:

$$SO_3 + ClF \rightarrow ClOSO_2F \quad (a)$$

Solution of SO_3 in BrF_3 occurs without evolution of O_2 or formation of Br_2, $[BrF_2]^+[SO_3F]^-$ being formed[3,4]. With BrF_5 the same product is obtained besides several others[4].

If SO_3 is distilled into ICl at 0°C, ISO_3Cl is produced in high yield[5,6]; the products from ICl_3 depend on the stoichiometry[5,7]:

$$ICl + SO_3 \xrightarrow{0°C} IOSO_2Cl \quad (b)$$

$$ICl_3 + SO_3 \xrightarrow[CCl_4]{0\text{–}10°C} [ICl_2][SO_3Cl] \quad (c)$$

$$ICl_3 + 3\ SO_3 \rightarrow I(SO_3Cl)_3 \quad (d)$$

The pentafluorosulfuroxo anion reacts with ClF to give the hypochlorite in high yield[8–12]:

$$CsF + OSF_4 \rightarrow CsOSF_5 \xrightarrow[-CsF]{ClF} SF_5OCl \quad (e)$$

If SO_2 is slowly warmed from −78°C to RT, SO_2ClF is formed in 99% yield[13] plus small amounts of SO_2F_2 and Cl_2:

$$SO_2 + ClF \xrightarrow[\text{autoclave}]{-78°C \text{ to RT}} SO_2ClF \quad (f)$$

whereas BrF_3 reacts uncontrollably with SO_2 under various conditions. From stoichiometric mixtures of Br_2 and BrF_3, sulfurylbromofluoride is formed quantitatively[14]:

$$BrF_3 + Br_2 + 3\ SO_2 \rightarrow SO_2BrF \quad (g)$$

but IF_5 does not attack[14,15] SO_2.

S-Perfluoroalkylsulfinate salts and ClF give the corresponding sulfonyl chlorides[16]:

$$Cs[CF_3SO_2] + ClF \rightarrow CF_3SO_2Cl + CsF \quad (h)$$

(R. MEWS)

1. C. J. Schack, R. D. Wilson, *Inorg. Chem.*, *9*, 311 (1970).
2. C. V. Hardin, C. T. Ratcliffe, L. R. Anderson, W. B. Fox, *Inorg. Chem.*, *9*, 1938 (1970).
3. A. A. Woolf, *J. Fluorine Chem.*, *1*, 127 (1971).
4. M. Gross, H. Meinert, R. A. Grimmer, *Z. Chem.*, *10*, 441 (1970).
5. R. C. Paul, A. Arora, K. C. Malhotra, *J. Inorg. Nucl Chem.*, *33*, 991 (1971).
6. C. Tiar, R. Mercier, M. Camelot, *C.R. Hebd. Seances Acad. Sci.*, *268*, 1825 (1969).
7. R. C. Paul, C. L. Arora, K. C. Malhotra, *Indian J. Chem.*, *9*, 473 (1971).
8. D. E. Gould, L. R. Anderson, D. E. Young, W. B. Fox, *J. Am. Chem. Soc.*, *91*, 1310 (1964).
9. D. E. Gould, L. R. Anderson, W. B. Fox, Ger. Offen. 1,928,539 (1969); *Chem. Abstr.*, *72*, 54743 (1970).
10. D. E. Young, D. E. Gould, L. R. Anderson, W. B. Fox, Ger. Offen. 1,958,346 (1971); *Chem. Abstr.*, *74*, 53036 (1971).
11. C. J. Schack, R. D. Wilson, J. S. Muirhead, S. N. Conz, *J. Am. Chem. Soc.*, *91*, 2907 (1969).
12. C. J. Schack, R. D. Wilson, Ger. Offen. 1,949,658 (1970); *Chem. Abstr.*, *73*, 16,921 (1970).
13. C. J. Schack, R. D. Wilson, *Inorg. Chem.*, *9*, 311 (1970).
14. H. Jonas, *Z. Anorg. Allg. Chem.*, *265*, 273 (1951).
15. E. E. Ainsley, R. Nichols, P. L. Robinson, *J. Chem. Soc.*, 623 (1953).
16. A. Majid, J. M. Shreeve, *Inorg. Chem.*, *13*, 2710 (1974).

2.3.7.2.2. to Give the Se–, Te–, Po–Halogen Bond.

Pentafluoroselenium hypochlorite[1] and its Te analogue[2,3] are prepared in excellent yield from the corresponding mercurial and ClF:

$$Hg(OSeF_5)_2 + 2\ ClF \rightarrow HgF_2 + 2\ F_5SeOCl \quad (a)$$

$$Hg(OTeF_5)_2 + 2\ ClF \rightarrow HgF_2 + 2\ F_5TeOCl \quad (b)$$

Because BrF and IF are unstable, the analogous hypobromites and iodites are prepared by other routes[1,2]:

$$BrF_3 + 3\ SeF_5OH \rightarrow 3\ HF + (F_5SeO)_3Br \quad (c)$$

$$(F_5SeO)_3Br + Br_2 \rightarrow 3\ F_5SeOBr \quad (d)$$

$$F_5SeOCl + ICl \rightarrow F_5SeOI + Cl_2 \quad (e)$$

$$3\ F_5SeOCl + ICl_3 \rightarrow I(OSeF_5)_3 + 3\ Cl_2 \quad (f)$$

Yields in these reactions are good (70–90%), but the iodo derivatives decompose at RT.

The pentafluorotellurate group behaves similarly[2]:

$$2\ F_5TeOCl + Br_2 \rightarrow 2\ F_5TeOBr + Cl_2 \quad (g)$$

$$F_5TeOCl + ICl \rightarrow F_5TeOI, I(OTeF_5)_3, ICl_3, Cl_2 \quad (h)$$

$$3\ F_5TeOCl + ICl_3 \rightarrow I(OTeF_5)_3 + 3\ Cl_2 \quad (i)$$

The hypoiodite decomposes to give[2] I_2 and $I(OTeF_5)_3$.

(R. MEWS)

1. K. Seppelt, *Chem. Ber.*, *106*, 157 (1973).
2. K. Seppelt, *Chem. Ber.*, *106*, 1920 (1973).
3. K. Seppelt, D. Nöthe, *Inorg. Chem.*, *12*, 2727 (1973).

2.3.7.3. with Nonmetal Halides

2.3.7.3.1. to Give the Sulfur–Halogen Bond.

This synthetic approach is limited to fluorides and chlorides because only they form stable bonds to high oxidation state sulfur. The formation of the bromosulfate from $[SO_3Br]^-$ is observed.

Fluorinations by XeF_2 are catalyzed by electron-pair acceptor acids; thus the reaction with SO_3 occurs below RT[1]:

$$3\ XeF_2 + 6\ SO_3 \xrightarrow{RT} 3\ FSO_2OOSO_2F + 3\ Xe \tag{a}$$

$$XeF_6 + 6\ SO_3 \xrightarrow{70°C} 3\ FSO_2OOSO_2F + Xe \tag{b}$$

In the latter, $[XeF_5]^+[SO_3F]^-$ may be the primary product[2].

From SO_3 and oxygen fluorides or hypofluorites fluorosulfonyl peroxides form.

With O_2F_2, SO_3 reacts explosively, and in such solvents as SO_2F_2 or $C_2F_4Cl_2$, 1:1 $S_2O_6F_2$:SO_4F_2 form[3]:

$$SO_3 + 2\ O_2F_2 \xrightarrow[-80°C]{solvent} FSO_2OOSO_2F + FSO_2OOF \tag{c}$$

Irradiation of gaseous SO_3 and OF_2 ($\lambda > 350$ nm) will give sulfurylfluoride fluoroperoxide[4–6], FSO_2OOF:

$$SO_3 + OF_2 \xrightarrow{h\nu} FSO_2OOF\ (+\ S_2O_5F_2) \tag{d}$$

Under these conditions the peroxide forms[5] only with xs OF_2; unfiltered UV light decomposes the product[6].

Similar pathways are followed by hypofluorites, R—OF (R = SF_5 [7], FSO_2 [8–10], CF_3 [11–13]):

$$SF_5OF + SO_3 \rightarrow SF_5OOSO_2F \tag{e}$$

$$FSO_2OF + SO_3 \rightarrow FSO_2OOSO_2F \tag{f}$$

$$CF_3OF + SO_3 \xrightarrow[autoclave]{245°–260°C} CF_3OOSO_2F\ (+\ SO_2F_2, COF_2, O_2) \tag{g}$$

$$CF_3OF + SO_3 \xrightarrow{flow} S_2O_6F_2 + CF_3OOSO_2F \tag{h}$$

Product distribution in Eq. (h) depends on conditions[11–13]; in flow systems CF_3OF acts mainly as a fluorine source[11,12]. The addition of hypofluorites across the S=O double bond of OSF_4 is related, e.g.:

$$SF_5OF + OSF_4 \xrightarrow[AgF_2\ catal]{N_2,\ 225°C} SF_5OOSF_5 \tag{i}$$ [14]

A hydroxylamine derivative results[15,16] from N_2F_4 and SO_3:

$$N_2F_4 + SO_3 \xrightarrow[h\nu]{\lambda = 253.7\ nm} FSO_2ONF_2 \tag{j}$$

By insertion of SO_3 into group VIB- and VB-halogen bonds, mainly ionic species with halosulfate ions are formed, e.g.:

$$SeX_4 + SO_3 \rightarrow [SeX_3][SO_3X] \ (X = F, Cl) \qquad (k)^{17-22}$$

$$TeCl_4 + SO_3 \rightarrow [TeCl_3][SO_3Cl] \qquad (l)^{19-22}$$

$$NO_nF + SO_3 \rightarrow [NO_n][SO_3F] \ (n = 1, 2) \qquad (m)^{23}$$

$$NO_nCl + SO_3 \rightarrow [NO_n][SO_3Cl], [NO_n][S_2O_6Cl] \ (n = 1, 2) \qquad (n)^{24}$$

$$OPX_3 + SO_3 \rightleftarrows [OPX_2][SO_3X] \ (X = Cl, Br) \qquad (o)^{24,25}$$

The reaction of AsF_3 and SO_3 produces[26,27] $2\,AsF_3 \cdot 3\,SO_3$, and by NMR measurements three different species ($AsF_3 \cdot 3\,SO_3$, $2\,AsF_3 \cdot 3\,SO_3$ and $3\,AsF_3 \cdot 3\,SO_3$) are detected bonded with bridging fluorosulfate groups[28]. However, in liq SO_2 the AsF_3-SO_3[29] and SbF_3-SO_3[30] systems by different spectroscopic methods show only simple insertion products:

$$EF_3 + n\,SO_3 \xrightarrow{SO_2} EF_{3-n}(OSO_2F)_n \qquad (p)$$

where E = As, Sb; n = 1-3;

$$SbF_3 + SO_3 \underset{\Delta}{\overset{120°C}{\rightleftarrows}} Sb(OSO_2F)_3 \qquad (q)^{27,31}$$

Similar patterns are found for SbX_5 and the trichlorides:

$$SbF_5 + SO_3 \rightarrow SbF_4(OSO_2F) \qquad (r)^{32}$$

$$SbCl_5 + SO_3 \rightarrow SbCl_5 \cdot n\,SO_3 \qquad (s)$$

$$ECl_3 + SO_3 \rightarrow ECl_{3-n}(SO_3Cl)_n \qquad (t)$$

where E = As, n = 1, 3[33,35,36]; E = Sb, n = 1, 2[34].

Group-IVB halides add across the S—O bond of SO_3 to give esters of halosulfonic acids, e.g.:

$$R_f{-}C(=O)F + SO_3 \rightarrow R_f{-}C(=O){-}OSO_2F \qquad (u)^{38}$$

where R_f = polyfluoroalkyl group;

$$CF_3CF{=}CF_2 + SO_3 \xrightarrow{BrF_3} CF_2{=}CF{-}CF_2OSO_2F + F_3C(F)\overline{C{-}C}(F)F \ (\text{ring closed by } O_2S{-}O) \qquad (v)^{37}$$

$$\text{hexafluorocyclobutene } C_4F_6 \xrightarrow[100°C]{SO_3} C_4F_5(OSO_2F) + C_4F_4(OSO_2F)_2 + C_4F_3(OSO_2F)_3 \qquad (w)^{39}$$

Similar insertions are possible with organosilyl halides resulting in halosulfonic acid esters[39]:

$$Me_3SiCl + SO_3 \xrightarrow{-30^\circ C} Me_3SiOSO_2Cl \quad (x)$$

Halosulfonic acids are prepared in quantitative yield from SO_3 and a slight xs of HX:

$$HX + SO_3 \rightarrow HSO_3X \quad (y)$$

where X = F[41,42,43], Cl[41,44], Br[45].

Addition of xs HBr to SO_3 in liq SO_2 at $-35^\circ C$ followed by solvent stripping at the same T gives HSO_3Br, mp -8 to $-6^\circ C$. Below $-30^\circ C$ HSO_3Br is stable for a day; at $0^\circ C$ fast decomposition takes place[41].

Fluorination of SO_2 by XeF_2 is catalyzed by electron-pair acceptor acids, such as BF_3 or HF (via $[XeF]^+$)[1], or by halide ions ($[XSO_2]^-$ being formed as an intermediate)[46]:

$$XeF_2 + SO_2 \rightarrow Xe + SO_2F_2 \quad (z)$$

Dioxylgenyl salts[47] and oxygen fluorides act as fluorine sources; SO_2F_2 is observed in all reactions[3,6,48–51]:

$$[O_2][SbF_6] + SO_2 \rightarrow SO_2F_2 \quad (aa)$$

$$OF_2 + SO_2 \xrightarrow[h\nu]{\lambda = 365\ nm} SO_2F_2 + S_2O_5F_2 \quad (ab)$$

$$O_2F_2 + SO_2 \xrightarrow{-160^\circ C} SO_2F_2, FSO_2OF, FSO_2OOF, FSO_2OSO_2F \quad (ac)$$

$$SO_2 + O_4F_2 \xrightarrow[CF_3Cl]{-183^\circ C} SO_2F_2, FSO_2OOF, S_2O_5F_2 \quad (ad)$$

At 300°–500°C SO_2 reacts with OF_2 to give[48] SO_2F_2 and OSF_2, besides SO_2F_2, FSO_2OSO_2F is also formed by photolysis[6]. Addition of O_2F_2[3,49–51] and O_4F_2[51] to SO_2 occurs at low T.

Group-VIB hypofluorites add to SO_2, whereas perfluoroalkyl derivatives give numerous products:

$$SF_5OF + SO_2 \xrightarrow{50^\circ C} F_5SOSO_2F \quad (ae)^{52}$$

$$SF_5OOSF_5 + SO_2 \xrightarrow{225^\circ C} F_5SOSO_2F, (F_5SO)_2SO_2, SO_2F_2, SF_5OSF_5, OSF_4, SF_6 \quad (af)^{7,53}$$

$$FSO_2OF + SO_2 \rightarrow FSO_2OSO_2F \quad (ag)^{10,54,55}$$

$$CF_2(OF)_2 + SO_2 \xrightarrow{60^\circ C} SO_2F_2 + COF_2 \quad (ah)^{56}$$

$$\xrightarrow{UV} S_2O_5F_2 + COF_2$$

$$CF_3OF + SO_2 \xrightarrow{170^\circ–185^\circ C} \quad (ai)^{11}$$

$$CF_4, SO_2F_2, S_2O_5F_2, CF_3OSO_2F, (CF_3O)_2SO_2, CF_3OSO_2OSO_2F, (CF_3OSO_2)O$$

$$CF_3OOF + SO_2 \rightarrow CF_3OSO_2F + CF_3OSO_2OSO_2F \quad (aj)^{57}$$

Numerous products are also formed from CF_3OF and gaseous SO_2 at 170°–180°C, or with liq SO_2 in an autoclave at 90°C[52].

Hypochlorites add directly to SO_2 at RT or below, to give excellent yields of the corresponding chlorosulfates:

$$SF_5OCl + SO_2 \rightarrow SF_5OSO_2Cl \qquad (ak)^{58}$$

$$FSO_2OCl + SO_2 \rightarrow FSO_2OSO_2Cl \qquad (al)^{59}$$

$$R_fSO_2OCl + SO_2 \rightarrow R_fSO_2OSO_2Cl \qquad (am)^{60}$$

$$R_fOCl + SO_2 \rightarrow R_fOSO_2Cl \qquad (an)^{61}$$

where $R_f = CF_3, C_2F_5, (CF_3)_2CF, (CF_3)_2CH, (CF_3)_3C, CF_3(CH_3)CH, CF_3CH_2$.

Nitrogen fluorides mainly act as fluorine sources:

$$\left.\begin{array}{l} cis\text{-}N_2F_2 + SO_2 \xrightarrow{100°C} \\ N_2F_2 + SO_2 \xrightarrow{h\nu} \end{array}\right\} SO_2F_2 + N_2 + S_2O_5F_2, OSF_2, OSF_4, N_2O \qquad (ao)^{62,63}$$

$$N_2F_4 + 2\ SO_2 \xrightarrow{CO_2\ laser} 2\ SO_2F_2 + N_2 \qquad (ap)^{64}$$

but under less drastic conditions, simultaneous formation of S—N and S—F bonds is observed[16,65]:

$$N_2F_4 + SO_2 \rightarrow FSO_2NF_2 \qquad (aq)$$

$$F_5SNF_2 + SO_2 \xrightarrow[90\ h]{75°C} FSO_2NF_2 + SF_4 \qquad (ar)^{66}$$

Nitrosyl fluoride adds to SO_2 giving[67,68] unstable FSO_2NO:

$$SO_2 + NOF \rightarrow FSO_2NO \qquad (as)$$

The decomposition pressure of nitroso sulfuryl fluoride at −18°C is 760 torr (1.01×10^5 Pa).

Simultaneous C—S and S—Cl bond formation is observed in sulfochlorination[69,70]:

$$RH + SO_2 + Cl_2 \xrightarrow[-HCl]{h\nu} RSO_2Cl \qquad (at)$$

while aromatic sulfonyl chlorides and bromides are prepared[71,72] from diazonium salts and SO_2 in the presence of $CuCl_2$:

$$[RN_2]X + SO_2 \xrightarrow[-N_2]{CuCl_2} \underset{(40\text{–}90\%)}{RSO_2X} \qquad (au)$$

(R. MEWS)

1. N. Bartlett, F. O. Sladky, *J. Chem. Soc., Chem. Commun.*,1046 (1968).
2. M. Eisenberg, D. D. DesMarteau, *Inorg. Chem., 11*, 2641 (1972).
3. I. J. Solomon, A. J. Kacmarek, J. M. McDonough, *J. Chem. Eng. Data, 13*, 529 (1968).

4. R. Gatti, E. H. Staricco, J. E. Sicre, H. J. Schumacher, *Z. Phys. Chem. (Frankfurt am Main)*, *36*, 211 (1963); *Angew. Chem. Int. Ed. Engl.*, *2*, 149 (1963); *An. Asoc. Quim. Argentina*, *52*, 167 (1964); *Chem. Abstr.*, *64*, 2902 (1966).
5. M. Gambaruto, J. E. Sicre, H. J. Schumacher, *J. Fluorine Chem.*, *5*, 175 (1975).
6. G. Franz, F. Neumayr, *Inorg. Chem.*, *3*, 921 (1964).
7. C. I. Merrill, G. H. Cady, *J. Am. Chem. Soc.*, *85*, 909 (1963).
8. F. B. Dudley, G. H. Cady, *J. Am. Chem. Soc.*, *79*, 513 (1957).
9. J. Czarnowski, E. Castellano, H. J. Schumacher, *Z. Phys. Chem. (Frankfurt am Main)*, *57*, 249 (1968).
10. W. H. Basualdo, H. J. Schumacher, *Z. Phys. Chem. (Frankfurt am Main)*, *47*, 57 (1965).
11. W. P. van Meter, G. H. Cady, *J. Am. Chem. Soc.*, *82*, 6005 (1960).
12. M. Lustig, J. M. Shreeve, *Adv. Fluorine Chem.*, *7*, 175, 182, 184 (1973).
13. J. Czarnowski, E. Castellano, H. J. Schumacher, *Z. Phys. Chem. (Frankfurt am Main)*, *65*, 225 (1969).
14. C. I. Merrill, G. H. Cady, *J. Am. Chem. Soc.*, *83*, 298 (1961).
15. M. Lustig, C. L. Bumgardner, J. K. Ruff, *Inorg. Chem.*, *3*, 917 (1964).
16. G. W. Fraser, J. M. Shreeve, M. Lustig, C. L. Bumgardner, *Inorg. Synth.*, *12*, 299 (1970).
17. R. J. Gillespie, W. A. Whitla, *Can. J. Chem.*, *47*, 4153 (1969).
18. R. D. Peacock, *J. Chem. Soc.*, 3617 (1953).
19. H. Gerding, *Recl. Trav. Chim. Pays-Bas*, *75*, 589 (1956).
20. H. Gerding, D. J. Stufkens, H. Gijben, *Recl. Trav. Chim. Pays-Bas*, *89*, 619 (1970).
21. R. C. Paul, C. L. Arora, K. C. Malhotra, *Indian J. Chem.*, *9*, 473 (1971).
22. R. C. Paul, K. K. Paul, K. C. Malhotra, *Aust. J. Chem.*, *22*, 847 (1969).
23. *Gmelins Handbuch der Anorganischen Chemie*, *Sulfur*, Suppl. Vol. 3, Springer Verlag, Berlin, 1980, p. 289ff.
24. J. Devynck, *Ann. Chim. (Paris)*, *7*, 321 (1972).
25. R. C. Paul, S. K. Vasisht, *J. Indian Chem. Soc.*, *43*, 141 (1966).
26. A. Engelbrecht, A. Aignesberger, E. Hayek, *Monatsh. Chem.*, *86*, 470 (1955).
27. E. L. Muetterties, D. D. Coffman, *J. Am. Chem. Soc.*, *80*, 5914 (1958).
28. R. J. Gillispie, J. V. Oubridge, *Proc. Chem. Soc.*, 308 (1960).
29. J. Touzin, L. Mitacek, *Coll. Czech. Chem. Commun.*, *44*, 1525 (1979).
30. J. Touzin, L. Mitacek, *Coll. Czech. Chem. Commun.*, *44*, 2751 (1979).
31. E. Hayek, A. Czaloun, B. Krismer, *Monatsh. Chem.*, *87*, 741 (1956).
32. R. J. Gillespie, R. A. Rothenbury, *Can. J. Chem.*, *42*, 416 (1964).
33. C. L. Arora, R. Kumar, B. B. Sandhir, *Ann. Chim. (Paris)*, *2*, 219 (1977).
34. L. Riesel, H. A. Lehmann, *Z. Anorg. Allg. Chem.*, *371*, 289 (1969).
35. H. A. Lehmann, L. Reisel, *Z. Anorg. Allg. Chem.*, *371*, 281 (1969).
36. C. L. Arora, B. B. Sandhir, *Ann. Chim. (Paris)*, *1*, 301 (1976).
37. J. Touzin, M. Jaros, *Z. Chem.*, *11*, 469 (1971).
38. C. G. Krespan, D. C. England, *J. Org. Chem.*, *40*, 2937 (1975); *J. Am. Chem. Soc.*, *103*, 5598 (1981).
39. B. E. Smart, *J. Org. Chem.*, *41*, 2353 (1976).
40. M. Schmidt, H. Schmidbaur, *Angew. Chem.*, *70*, 657 (1958); *Chem. Ber.*, *95*, 47 (1962).
41. Ref. 23, p. 291.
42. A. W. Jache, *Adv. Inorg. Chem. Radiochem.*, *16*, 177 (1974).
43. W. R. Wheatley, D. E. Treadway, R. G. Toennies, US Pat. 3,957,959 (1976), *Chem. Abstr.*, *85*, 80,438 (1976).
44. T. Kuboyama, H. Nagata, Jpn. Pat. 76-10, 840 (1960); *Chem. Abstr.*, *85*, 126,731 (1976).
45. M. Schmidt, G. Talsky, *Z. Anorg. Allg. Chem.*, *303*, 210 (1960).
46. I. L. Wilson, *J. Fluorine Chem.* *5*, 13 (1975).
47. L. Stein, F. A. Hohorst, *Inorg. Nucl. Chem.*, *H. H. Hyman Memor. Vol.*, 73 (1976).
48. A. Engelbrecht, E. Nachbaur, C. Pupp, *Monatsh. Chem.*, *95*, 219 (1964).
49. I. J. Solomon, *U.S. Govt. Res. Develop. Rep.*, *41*, 44 (1966); *Chem. Abstr.*, *66*, 101,245 (1967).
50. I. J. Solomon, A. J. Kacmarek, J. Raney, *Inorg. Chem.*, *7*, 1221 (1968).
51. I. J. Solomon, A. J. Kacmarek, *J. Fluorine. Chem.*, *1*, 255 (1971).
52. G. Pass, H. L. Roberts, *Inorg. Chem.*, *2*, 1016 (1963).
53. G. Pass, *J. Chem. Soc.*, 6047 (1963).
54. J. E. Roberts, G. H. Cady, *J. Am. Chem. Soc.*, *82*, 354 (1960).
55. E. Castellano, H. J. Schumacher, *Z. Phys. Chem.*, *(Frankfurt am Main)*, *54*, 77 (1967).

56. F. A. Hohorst, J. M. Shreeve, *Inorg. Chem.*, *7*, 624 (1968).
57. R. A. DeMarco, W. B. Fox, *Inorg. Nucl. Chem. Lett.*, *10*, 695 (1974).
58. D. E. Young, L. R. Anderson, D. E. Gould, W. B. Fox, *J. Am. Chem. Soc.*, *92*, 2313 (1970), *Tetrahedron Lett.*, 723 (1969).
59. A. V. Fokin, A. D. Nikolaeva, Yu. N. Studnev, A. N. Rapkin, N. A. Proshin, L. D. Kuznetsova, *Acad. Sci. USSR Bull. Chem. Sci.*, 915 (1975).
60. M. Lustig, *Inorg. Chem.*, *4*, 104 (1965).
61. D. E. Young, L. R. Anderson, B. W. Fox, US Pat. 3,654,335 (1972); *Chem. Abstr.*, *77*, 4,893 (1972); US Pat. 3,681,423 (1972); *Chem. Abstr.*, *77*, 151,464 (1972).
62. H. W. Roesky, O. Glemser, D. Bormann, *Chem. Ber.*, *99*, 1589 (1966).
63. M. Lustig, *Inorg. Nucl. Chem. Lett.*, *5*, 723 (1969).
64. V. V. Gorlevskii, A. N. Oraevskii, A. V. Pankratov, A. N. Skachkov, V. M. Shabarskin, *High Energy Chem. USSR* (*Engl. Transl.*), *10*, 389 (1976).
65. M. Lustig, C. L. Bumgardner, F. A. Johnson, J. K. Ruff, *Inorg. Chem.*, *3*, 1165 (1964).
66. J. L. Boivin, *Can. J. Chem.*, *42*, 2744 (1964).
67. G. Balz, E. Mailänder, *Z. Anorg. Allg. Chem.*, *217*, 161 (1934).
68. F. Seel, H. Massat, *Z. Anorg. Allg. Chem.*, *280*, 186 (1955).
69. C. F. Reed, US Pat. 2,046,090 (1936); *Chem. Abstr.*, *30*, 5,593 (1936).
70. H. Eckoldt, in *Houben-Weyl, Methoden der Organischen Chemie, S, Se, Te Verbindungen*, Vol. IX; E. Müller, O. Bayer, H. Meeswein, K. Ziegler, eds., Georg Thieme Verlag, Stuttgart, 1955, p. 411.
71. e.g., N. B. Chapman, K. Clarke, S. N. Sawhney, *J. Chem. Soc.*, *C*, 518 (1968).
72. I. Haiduc, K. J. Wynne, in *Methodicum Chimicum*, H. Zimmer, K. Niedenzu, eds., Georg Thieme Verlag, Stuttgart, 1976, p. 704.

2.3.7.3.2. to Give the Se– and Te–Halogen Bonds.

The element-oxygen bond in the higher S—O homologues is labile, and simple additions are frequent only with SeO_3. For SeO_2, only insertion into the Se—F bond of SeO_2F_2 is known[1]:

$$SeO_2 + SeO_2F_2 \rightarrow FSeO_2OSeOF \quad (a)$$

However, TeO_3 and TeO_2 react with Te—O bond cleavage. From the oxoanion, $[OTeF_5]^-$, and FSO_2OF, F_5TeOF is prepared[2]:

$$Cs[OTeF_5] + FSO_2OF \rightarrow Cs[OSO_2F] + F_5TeOF \quad (b)$$

On heating, SeO_3 reacts vigorously with HSO_3F; $HSeO_3F$ is formed in the first step[3]:

$$HSO_3F + SeO_3 \rightarrow HSeO_3F + SO_3 \quad (c)$$

$$HSeO_3F + HSO_3F \rightarrow [H_2SeO_3F]^+ + [SO_3F]^- \rightarrow SeO_2F_2 + H_2SO_4 \quad (d)$$

Byproducts[3] are $OSeF_2$, $S_2O_5F_2$ and $S_3O_8F_2$. Treatment of $(SeO_3)_4$ with xs SeO_2F_2 gives $Se_2O_5F_2$ via polyselenonyl fluorides[4]:

$$(SeO_3)_4 + SeO_2F_2 \xrightarrow{120°C} FSeO_2(OSeO_2)_4F \quad (e)$$

$$(SeO_3)_4 + FSeO_2(OSeO_2)_nF \rightarrow FSeO_2(OSeO)_{n+4}F \quad (f)$$

with $n \leq 4$:

$$FSeO_2(OSeO_2)_nF + \text{xs } SeO_2F_2 \rightarrow FSeO_2OSeO_2F \quad (g)$$

The primary product of SeO_3 with SeF_4 is $[SeF_3][SeO_3F]$, which decomposes to give[5,6] $OSeF_2$ and O_2SeF_2. Similarly, SeO_3 abstracts Cl^- from chlorides of S, Se and Te, e.g.:

$$S_4N_3Cl + SeO_3 \xrightarrow[-40°C]{SO_2} [S_4N_3][SeO_3Cl] \qquad (h)^7$$

$$2n\ ECl_4 + 2n\ SeO_3 \xrightarrow[-40°C]{SO_2} \left[\begin{array}{cccccccc} & O & & Cl & O & & Cl & \\ & & Se & & & Se & & \\ Cl & O & & O & Cl & O & & O \\ & E & & & & E & & \\ Cl & & Cl & & Cl & & Cl & \end{array}\right]_n \qquad (i)^7$$

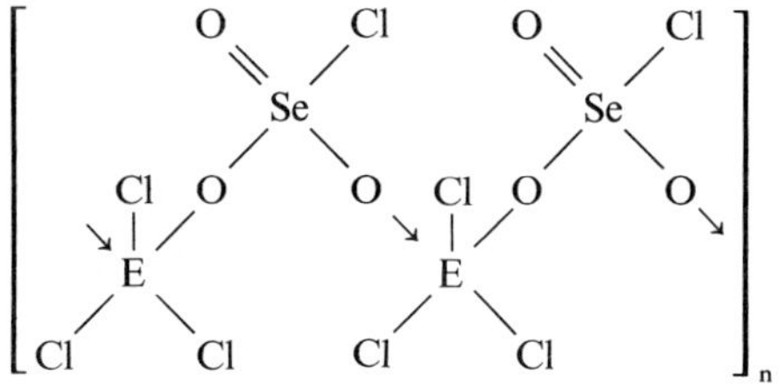

where E = Se, Te.

Mixing SeO_3 and S_4N_3Cl in liq. SO_2 at −40°C yields the pale yellow chloroselenate, which, like other chloroselenates, is not stable above 0°C. The 1:1 adducts of $SeCl_4$ and $TeCl_4$ show similar stabilities, decomposing above 0°C. With PCl_5 at −20°C in CCl_4, white, crystalline $PCl_4(OSeO_2Cl)$ is formed[7], and conductometric titrations indicate that $AsCl_3$[8], $SbCl_3$[8,9] and $SbCl_5$[8,9] form chloroselenates and dichloroselenates similarly, e.g.:

$$AsCl_3 + SeO_3 \rightarrow Cl_2AsOSeO_2Cl \qquad (j)$$

$$AsCl_3 + 2\ SeO_3 \rightarrow ClAs(OSeO_2Cl)_2 \qquad (k)$$

$$2\ SbF_3 + (n+1)SeO_3 \xrightarrow[10\text{–}12\ \text{days}]{RT,\ SO_2} [SbF_2]_2[Se_nO_{3n+1}] + SeO_2F_2 \qquad (l)$$

with $n \geq 1$.

With AsF_3 O—Se bond cleavage occurs (see §2.3.6.5.), and shaking SeO_3 and SbF_3 for 10–12 days in a sealed vessel in SO_2 forms SeO_2F_2 and solids containing $[SeO_3F]^-$ and $[Se_nO_{3n+1}]^{2-}$ anions and $[SbF_2]^+$ cations[10].

Insertion of SeO_3 into the group IVB—Cl bond[11,12] also occurs:

$$R_3MCl + SeO_3 \rightarrow R_3MOSeO_2Cl \qquad (m)$$

(M = Si, Sn). **These reactions must be carried out at low T and under anhydrous conditions to avoid uncontrolled reactions. At RT Ph_3CCl and SeO_3 explode.** At −80°C in liq SO_2, however, the reddish chloroselenate, $[Ph_3C]^+[SeO_3Cl]^-$, is obtained[7]:

$$Ph_3CCl + SeO_3 \xrightarrow[SO_2]{-80°C} [Ph_3C][SeO_3Cl] \qquad (n)$$

At RT, from SeO_3 and anhydr HF, fluoroselenic acid is formed as a clear, viscous, hygroscopic liquid[13]:

$$SeO_3 + HX \rightarrow HSeO_3X \qquad (o)$$

where X = F, Cl).

The reaction with HCl is more vigorous[11,12]; from direct action at −30°C, Cl_2 is formed[12]. However, condensing HCl on freshly prepared SeO_3 at −186°C, warming slowly to $-80°C$ and removing xs HCl under vacuum, gives[12] $HSeO_3Cl$. The acid is also formed in liq SO_2 at $-80°C$[12].

(R. MEWS)

1. K. Dostal, *MTP Int. Rev., Inorg. Chem., Ser. Two, 3*, 107 (1975).
2. C. J. Schack, W. W. Wilson, K. O. Christe, *Inorg. Chem., 22*, 18 (1983).
3. R. C. Paul, S. K. Sharma, R. D. Sharma, K. K. Paul, K. C. Malhotra, *J. Inorg. Nucl. Chem., 33*, 2905 (1971).
4. K. Dostal, *Folia Fac. Sci. Nat., Univ. Purkyn, Brun., 18*, 1 (1977); *Chem. Abstr., 90*, 196,975 (1979).
5. T. Birchall, R. J. Gillespie, S. L. Vekris, *Can. J. Chem., 43*, 1672 (1965).
6. A. J. Edwards, M. A. Mouty, R. D. Peacock, *J. Chem. Soc., A*, 557 (1967).
7. R. C. Paul, R. D. Sharma, K. C. Malhortra, *Indian J. Chem., 12*, 320 (1974).
8. J. Touzin, M. Jaros, *Z. Chem., 11*, 469 (1971).
9. J. Touzin, P. Bauer, M. Jaros, *Coll. Czech. Chem. Commun., 41*, 2997 (1976).
10. J. Touzin, L. Mitacek, *Coll. Czech. Chem. Commun., 44*, 2751 (1979).
11. M. Schmidt, P. Bornmann, I. Wilhelm, *Angew. Chem., Int. Ed. Engl., 2*, 691 (1963).
12. M. Schmidt, I. Wilhelm, *Chem. Ber., 97*, 876 (1964).
13. H. Bartels, E. Class, *Helv. Chim. Acta, 45*, 179 (1962).

2.3.7.4. with Metal Halides

2.3.7.4.1. to Give the Sulfur–Halogen Bond.

Sulfur dioxide and, especially, the trioxide exhibit electron-pair acceptor acidity; e.g., halogen ions add to give halosulfinates and halosulfates. Similarly, the pentafluoroxysulfate is formed[1,2] from OSF_4 and CsF:

$$O{=}SF_4 + CsF \rightarrow Cs[OSF_5] \qquad \text{(a)}$$

Salts of the lighter homologues of Cs are less stable.

Metal fluorosulfates can be prepared[3–5] from anhydr metal fluorides and xs SO_3, and fluorodisulfates result[6] from mixing SO_3 with less than stoichiometric amounts of MSO_3F:

$$NaF + SO_{3(g)} \xrightarrow{200°C} Na[SO_3F] \qquad \text{(b)}^7$$

$$KF + \text{xs } SO_{3(g)} \xrightarrow{RT} K[(SO_3)_nF] \xrightarrow{vac} K[S_2O_6F] \xrightarrow[vac]{35°C} K[SO_3F] \qquad \text{(c)}^8$$

$$CsF + SO_{3(g)} \xrightarrow{100°C} CsSO_3F \qquad \text{(d)}^9$$

Between 50 and 60°C, $TlSO_3F$[7], $AgSO_3F$[10] and $Ag[SO_3F]_2$[11] form from the corresponding fluorides. Only the higher alkaline-earth metals form pure fluorosulfates[12]:

$$MF_2 + 2\ SO_3 \xrightarrow{200°C} M[SO_3F]_2 \qquad \text{(e)}$$

(M = Ca, Sr, Ba), whereas BeF_2[10], MgF_2[10], PbF_2[10], ZnF_2[10], NiF_2[7] and CuF_2[7] give impure products.

Insertion of SO_3 into transition-metal fluorine bonds occurs, e.g., of CrF_5[13], CrO_2F_2[14], UF_6[15,16], but fluorination to give $S_2O_6F_2$ is also observed:

$$CrF_5 + 5\ SO_3 \xrightarrow{55°C} Cr(SO_3F)_3 + S_2O_6F_2 \tag{f}$$

$$CrO_2F_2 + 2\ SO_3 \xrightarrow[\text{liq } SO_2]{RT} CrO_2(OSO_2F)_2 \tag{g}$$

$$2\ UF_6 + 8\ SO_3 \xrightarrow{RT} 2\ UF_2(SO_3F)_3 + S_2O_6F_2 \tag{h}$$

$$2\ UF_6 + 10\ SO_3 \xrightarrow[-50°C]{CCl_3F} 2\ UF(SO_3F)_4 + S_2O_6F_2 \tag{i}$$

$$\downarrow 80\text{–}100°C$$

$$UF_2(SO_3F)_3,\ UF_3(SO_3F)_2$$

Action of SO_3 on chlorides and bromides has to be carefully controlled, because, as with iodides, oxidation to elemental halogen is observed[17]:

$$NaCl + SO_3 \xrightarrow{-10°C} Na[SO_3Cl] \qquad (j)^{18}$$

$$\downarrow SO_2(liq)$$

$$Na[S_3O_9Cl] \qquad (k)^{19}$$

$$KCl + SO_3 \xrightarrow[SO_2(liq)]{-10°C} K[S_2O_6Cl] \qquad (l)^{19}$$

$$NH_4Cl + SO_3 \xrightarrow[SO_2(liq)]{-10°C} [NH_4][S_2O_6Cl] \qquad (m)^{19}$$

Chlorosulfates of Al [e.g., $AlCl_2(SO_3Cl)$[20], $Al(SO_3Cl)_3$[19–21], $MAl(SO_3Cl)_4$[22]] and Ga [$Ga(SO_3Cl)_3$[21], $MGa(SO_3Cl)_4$[22]] are obtained similarly.

The preparation of $K[SO_3Br]$ is effected[23] by gradual combination of KBr and SO_3 solutions in SO_2:

$$KBr\text{-}SO_2 + xs\ SO_3\text{-}SO_2 \xrightarrow[SO_2]{-30°C} K[SO_3Br] \tag{n}$$

The acceptor strength of SO_2 is lower than that of SO_3; only from the fluorides of the heavier alkali metals are pure fluorosulfinates formed[24,25]:

$$KF + SO_2 \xrightarrow[5\ days]{RT} K[SO_2F] \tag{o}$$

$$MF + SO_2 \xrightarrow{5\text{–}6\ days} M[SO_2F] \tag{p}$$

where M = Rb, Cs. Even after 7 weeks, only 2.3% conversion to $Na[SO_2F]$ is obtained with NaF, and LiF and CaF_2 are inert.

Tetraalkylammonium halides behave similarly to metal halides[24,25]; e.g., $[Me_4N]SO_2F$ is stable up to 140°C, and the Cl, Br and I derivatives decompose at 88, 41 and 12.4°C, respectively[24]. Halosulfinate ions in solution are characterized spectroscopically[26–28], and iodosulfinates by x-ray[29]:

$$R^+I^- + SO_2 \xrightarrow{CH_3CN} R^+[ISO_2]^- \quad (q)$$

where R = $Ph_3P(CH_2C_6H_5)$, Rb(18-crown-6)[29]. Similar interactions are found in organometallic iodo complexes such as $Pt(CH_3)(PPh_3)_2I \cdot SO_2$[30] and $Cu_mI_m(PR_3)_n \cdot x\ SO_2$[31].

(R. MEWS)

1. M. Lustig, J. K. Ruff, *Inorg. Chem., 6,* 2115 (1967).
2. K. O. Christe, C. J. Schack, D. Pilipovich, E. C. Curtis, W. Sawodny, *Inorg. Chem., 12,* 620 (1973).
3. *Gmelins Handbuch der Anorganischen Chemie, Sulfur, Suppl. Vol. 3,* Springer Verlag, Berlin, 1980, p. 306.
4. W. V. Rochat, G. L. Gard, *Inorg. Chem., 8,* 158 (1969).
5. R. E. Noftle, G. H. Cady, *J. Inorg. Nucl. Chem., 29,* 969 (1967).
6. P. Vast, R. Heubel, *C.R. Hebd. Seances Acad. Sci., C, 267,* 236 (1968).
7. F. B. Dudley, *J. Chem. Soc.* 3407 (1963).
8. H. A. Lehmann, L. Kolditz, *Z. Anorg. Allg. Chem., 272,* 69 (1953).
9. J. K. Ruff, M. Lustig, *Inorg. Chem., 3,* 1422 (1964).
10. E. Hayek, A. Czaloun, B. Krismer, *Monatsh. Chem., 89,* 741 (1956).
11. P. C. Leung, F. Aubke, *Inorg. Chem., 17,* 1765 (1978).
12. E. L. Muetterties, D. D. Coffman, *J. Am. Chem. Soc., 80,* 5914 (1958).
13. S. D. Brown, G. L. Gard, *Inorg. Nucl. Chem. Lett., 11,* 19 (1975).
14. S. D. Brown, P. J. Green, G. L. Gard, *J. Fluorine Chem., 5,* 203 (1975).
15. W. W. Wilson, C. Naulin, R. Bougon, *Inorg. Chem., 16,* 2252 (1977).
16. J. P. Masson, C. Naulin, P. Charpin, R. Bougon, *Inorg. Chem., 17,* 1858 (1978).
17. Ref. 3, p. 307.
18. E. Puskaric, R. de Jaeger, J. Heubel, *Rev. Chim. Miner., 12,* 374 (1975).
19. G. H. Weinreich, *Bull. Soc. Chim. Fr.,* 2820 (1963).
20. C. L. Arora, R. Kumar, S. S. Bhardwaj, *Ann. Chim. (Paris), 2,* 279 (1977).
21. B. Vandorpe, M. Drache, B. Dubois, *C.R. Hebd. Seances Acad. Sci., Ser. C, 271,* 1076 (1970).
22. B. Vandorpe, M. Drache, *Bull. Soc. Chim. Fr.,* 2978 (1971).
23. S. Noel, M. Wartel, J. Heubel, *C.R. Hebd. Seances Acad Sci., Ser. C, 264,* 446 (1967).
24. F. Seel, L. Riehl, *Z. Anorg. Allg. Chem., 282,* 293 (1955).
25. F. Seel, H. Jonas, L. Riehl, J. Langer, *Angew. Chem., 67,* 32 (1955).
26. E. J. Woodhause, T. H. Norris, *Inorg. Chem., 10,* 614 (1971).
27. A. Salama, S. B. Salama, M. Sobeir, S. Wasif, *J. Chem. Soc., A,* 1112 (1971).
28. D. F. Burow, *Inorg. Chem., 11,* 573 (1972).
29. P. G. Eller, G. J. Kubas, *Inorg. Chem., 17,* 894 (1978).
30. M. Snow, J. A. Ibers, *Inorg. Chem., 12,* 224 (1973).
31. P. G. Eller, G. J. Kubas, R. R. Ryan, *Inorg. Chem., 16,* 2454 (1977).

2.3.7.4.2. to Give Se– and Te–Halogen Bond.

Metal halides add to SeO_3, SeO_2 and TeO_2, whereas halotellurates are not obtained by direct combination. From SeO_3 and alkali fluorides monofluoroselenates(VI) are formed. The reactions are carried out in liq SO_2 at −25°C with an xs of ~15% SeO_3. After 20 h the salts can be isolated[1] by evaporating SO_2 and xs SeO_3:

$$SeO_3 + MF \xrightarrow[-25°C]{SO_2} M[SeO_3F] \quad (a)$$

where M = Li, Na, K, Rb, Cs.

The $KSeO_3F$ product is also prepared from HF solutions[2]. Heating SeO_3 with KF (120°C) or CsF (80°C) with solvents gives[1] SeO_2F_2 plus M_2SeO_4 and $MSeO_3F$.

Fusing SeO_2 together with alkali halides 1:1 and 2:1 gives:

$$MX + SeO_2 \rightarrow M[SeO_2X] \qquad (b)^{3,4}$$

where M = K, Cs if X = F and M = Li, Na if X = Cl, Br;

$$MX + 2\ SeO_2 \rightarrow M[XSe_2O_4] \qquad (c)^{3}$$

where M = K, Cs if X = Cl, Br.

Iodides are oxidized but LiF and NaCl do not interact. The $KSeO_2F$ product can also be prepared from dimethylsulfoxide (DMSO) solution[5]; $MSeO_2F$ are made in liq SO_2(M = Cs[6]) or even from H_2O (M = Cs, Me_4N, Et_4N)[7,8]. The $[SeO_2F]^-$ ion is an intermediate in the AgF–AgF_2-catalyzed fluorination of SeO_2 to give[9] SeO_2F_2. Similar to SeO_2, $OSeF_2$ adds F^- to form salts containing the $[OSeF_3]^-$ anion[10]:

$$KF + OSeF_2 \rightarrow K[OSeF_3] \qquad (d)$$

(mp 138°C). The salt is stable to ≈400°C. Fluorotellurates, $MTeO_3F$ (M = Na, NH_4), are not well characterized. Compounds of approximately this formula are obtained by thermal condensation of $Te(OH)_6 \cdot NaF$ at 100°C or $Te(OH)_6 \cdot 1.5\ NH_4F$ at 180°C; from $Te(OH)_6 \cdot 2\ KF$ above 100°C $K_2TeO_3F_2$ is obtained[11].

In contrast to SeO_2, 2:1 RbF and CsF add to TeO_2[12]:

$$2\ MF + TeO_2 \xrightarrow{800°C} M_2[TeO_2F_2] \qquad (e)^{12}$$

where M = Rb, Cs;

$$MF + TeO_2 \rightarrow M[TeO_2F] \qquad (f)^{7}$$

where M = Na, K. Heating the components under N_2 to 800°C in a Pt boat results in a colorless melt, which gives colorless crystals on cooling[12]. These salts cannot be obtained from aq solution.

(R. MEWS)

1. A. J. Edwards, M. A. Mouty, R. D. Peacock, *J. Chem. Soc., A*, 557 (1967).
2. K. Dostal, M. Cernik, *Z. Chem.*, *6*, 424 (1966).
3. R. Paetzold, K. Aurich, *Z. Chem.*, *6*, 265 (1965).
4. R. Paetzold, K. Aurich. *Z. Anorg. Allg. Chem.*, *335*, 281 (1965).
5. R. J. Gillespie, P. Spekkens, J. O. Milne, D. Moffett, *J. Fluorine Chem.*, *7*, 43 (1976).
6. F. Seel, D. Gölitz, *Z. Anorg. Allg. Chem.*, *327*, 28 (1964).
7. J. Milne, *Inorg. Chem.*, *17*, 3592 (1978).
8. *Gmelins Handbuch der Anorganischen Chemie, Suppl.*, Vol. B1, *Se*, Springer Verlag, Heidelberg, 1981, p. 228.
9. G. Mitra, *J. Indian Chem. Soc.*, *37*, 804 (1960).
10. R. Paetzold, K. Aurich, *Z. Anorg. Allg. Chem.*, *348*, 94 (1966).
11. L. Kolditz, I. Fitz, *Z. Anorg. Allg. Chem.*, *349*, 184 (1967).
12. J. B. Milne, D. Mofett, *Inorg. Chem.*, *12*, 2240 (1973).

2.3.8. from Cleavage of the Group-VIB–Group VIB Element Bond

2.3.8.1. by Elemental Halogens

2.3.8.1.1. to Give the Oxygen–Halogen Bond.

The most stable compounds containing O—O single bonds are peroxides with fluorinated ligands[1], $R_fO—OR_f$, e.g., $FSO_2O—OSO_2F$. The low O—O bond energy (92.05 kJ/mole) and the stability of the $[SO_3F]^{\cdot}$ radical contribute to its reactivity and versatility. In a flow system at 250°C with $F_2—N_2$ the **highly explosive hypofluorite** (§2.3.7.1.1) is formed quantitatively[2,3]:

$$FSO_2OOSO_2F + F_2\text{-}N_2 \xrightarrow{250°C} 2\ FSO_2OF \qquad \text{(a)}$$

The reaction can also be induced photolytically[4]:

$$FSO_2OOSO_2F + F_2 \xrightarrow{h\nu} 2\ FSO_2OF \qquad \text{(b)}$$

with Cl_2 reaction occurs only with difficulty; at 25°C it goes to completion only after several weeks and at 125°C after several days[5,6]:

$$Cl_2 + S_2O_6F_2 \xrightarrow[5\ days]{125°C} 2\ ClOSO_2F \qquad \text{(c)}$$

$$Br_2 + 3\ S_2O_6F_2 \xrightarrow{25°C} 2\ Br(OSO_2F)_3 \qquad \text{(d)}^7$$

$$Br(OSO_2F)_3 + Br_2 \xrightarrow{25°C} 3\ Br(OSO_2F) \qquad \text{(e)}^{7-9}$$

$$I_2 + S_2O_6F_2 \xrightarrow{25°C} I[I(OSO_2F)_2] \qquad \text{(f)}^{10-12}$$

$$I_2 + 3\ S_2O_6F_2 \rightarrow 2\ I(OSO_2F)_3 \qquad \text{(g)}^{7,11}$$

whereas Br_2 and I_2 combine 1:1 or 1:3 with the radical according to the stoichiometry.

Similarly, hypofluorites are formed from other peroxides; e.g., the O—O bond in bis(pentafluoroselenium)peroxide is cleaved at 70°C in a flow system[13]:

$$F_5SeOOSeF_5 + F_2 \rightarrow 2\ F_5SeOF\ (+SeF_6) \qquad \text{(h)}$$

$$F(O)CO\text{-}OC(O)F + F_2 \xrightarrow{h\nu} 2\ FC(O)OF \qquad \text{(i)}$$

Fluoroformylhypofluorite is prepared[14] by the borosilicate-filtered irradiation of a 1:3.5 peroxide and F_2. No reaction of the latter peroxides with other halogens is known.

(R. MEWS)

1. R. A. de Marco, J. M. Shreeve, *Adv. Inorg. Chem. Radiochem.*, *16*, 109 (1974).
2. F. B. Dudley, G. H. Cady, *J. Am. Chem. Soc.*, *79*, 513 (1957).
3. J. E. Roberts, G. H. Cady, *J. Am. Chem. Soc.*, *81*, 4166 (1959).
4. R. Gatti, J. E. Sicre, H. J. Schumacher, *Z. Phys. Chem. (Frankfurt am Main)*, *40*, 127 (1964).
5. W. P. Gilbreath, G. H. Cady, *Inorg. Chem.*, *2*, 496 (1963).
6. E. J. Vasini, H. J. Schumacher, *Z. Phys. Chim. (Frankfurt am Main)*, *65*, 238 (1969).

7. J. E. Roberts, G. H. Cady, *J. Am. Chem. Soc.*, *82*, 352 (1960).
8. F. Aubke, R. J. Gillespie, *Inorg. Chem.*, *7*, 599 (1968).
9. R. J. Gillespie, M. J. Morton, *Inorg. Chem.*, *11*, 591 (1972).
10. F. Aubke, G. H. Cady, *Inorg. Chem.*, *4*, 269 (1965).
11. C. Chung, G. H. Cady, *Inorg. Chem.*, *11*, 2528 (1972).
12. M. J. Collins, G. Dénès, R. J. Gillespie, *J. Chem. Soc. Chem. Comm.*, 1296 (1984).
13. W. L. Reichert, G. H. Cady, *Inorg. Chem.*, *12*, 709 (1973).
14. R. Cauble, G. H. Cady, *J. Am. Chem. Soc.*, *89*, 5161 (1967).

2.3.8.1.2. to Give the Sulfur–Halogen Bond.

The synthesis of sulfur halides from elemental sulfur and halogens involves several S—S bond cleavage steps, as discussed in §2.3.2.2.:

$$S_8 \xrightarrow{X_2} XSSX \xrightarrow{X_2} SX_2 \rightarrow \cdots \quad \text{(a)}$$

In the fluorination of sulfur, one of these intermediates can be S_2F_{10}, which is further attacked under mild conditions to give SF_6. For the formation of SF_5X (X = Cl, Br) heating to 150°-200°C is necessary:

$$S_2F_{10} + Cl_2 \xrightarrow[\text{microwave}]{150°C} 2\ SF_5Cl \quad (100\%) \quad \text{(b)}^{1-4}$$

$$S_2F_{10} + Br_2 \xrightarrow{150°C} 2\ SF_5Br \quad (80\%) \quad \text{(c)}^{1,5}$$

The reactions are carried out in borosilicate- or Ni–Cu alloy vessels, in static or flow systems. Pentafluorosulfur chloride is formed quantitatively, but even with a fourfold xs of Br_2, formation of SF_5Br is incomplete. A better approach to SF_5X are oxidative additions to SF_4 (see §2.3.9.), which is more readily available and less hazardous to handle.

The S—S bond cleavage of disulfides by Cl_2 gives sulfenyl chlorides[6–11], where with F_2 bond cleavage and oxidative addition are observed simultaneously[12]:

$$CF_3SSCF_3 + F_2 \xrightarrow[38\ h]{-120°C} 2\ CF_3SF_3 \quad (90\%) \quad \text{(d)}$$

$$RSSR + Cl_2 \rightarrow 2\ RSCl \xrightarrow{Cl_2} RSCl_3 \quad \text{(e)}$$

$$RSSR + Br_2 \rightarrow 2\ RSBr \quad \text{(f)}$$

Chlorinolysis of aromatic and aliphatic disulfides in aprotic media furnishes 2 equiv of sulfenyl chloride; xs Cl_2 at low T leads[13] to unstable $RSCl_3$. Bromides, obtained in the same way, are unstable[14,15].

S,S′-Perfluoroalkyl disulfides are attacked by Cl_2 at higher T and P or under UV-irradiation, e.g.:

$$CF_3S{-}SCF_3 + Cl_2 \xrightarrow{UV} 2\ CF_3SCl \quad \text{(g)}^{16}$$

$$R_fSSR_f + Cl_2 \xrightarrow[\text{and/or UV}]{80°-200°C} 2\ R_fSCl \quad \text{(h)}^{17}$$

Related is the S—S bond cleavage in thiocyanogen:

$$(SCN)_2 + Cl_2 \rightarrow 2\ ClSCN \qquad (i)^{18,19}$$

$$(SCN)_2 + Br_2 \rightleftharpoons 2\ BrSCN \qquad (j)^{20}$$

$$(SCN)_2 + I_2 \rightleftharpoons 2\ ISCN \rightarrow 2,4,6\text{-}N_3C_3(SI)_3 \qquad (k)^{21,22}$$

which reacts even with I_2 to form an appreciable equilibrium concentration of ISCN, with 1,3,5-triazine-2,4,6-trisulfenyl iodide being formed by its trimerization.

Sulfonyl halides are isolated from the cleavage of the S—S bond in thiosulfonate esters[23] and salts[24], e.g.:

$$CH_3SO_2SCH_3 + Cl_2 \rightarrow CH_3SO_2Cl + CH_3SCl_3 \qquad (l)$$

$$RBr + Na_2S_2O_3 \xrightarrow{50\%\ CH_3OH} Na[RSSO_3] + NaBr \qquad (m)$$

$$\xrightarrow[\text{3. } +Cl_2]{\text{1. strip } CH_3OH \text{ 2. add ice} + AcOH} \underset{\text{(mainly)}}{RSO_2Br + RSO_2Cl} \qquad (n)$$

The last synthesis is a sequence of S—S bond cleavage, oxidative chlorination, followed by hydrolysis and metathetical exchange with bromide ion.

(R. MEWS)

1. B. Cohen, A. G. MacDiarmid, *Inorg. Chem.*, *4*, 1782 (1965).
2. F. A. Cotton, J. W. George, *Proc. Chem. Soc. (London)*, 317 (1959).
3. B. Cohen, A. G. MacDiarmid, *Chem. Ind. (London)*, 1866 (1962).
4. H. J. Emeléus, B. Tittle, *J. Chem. Soc.* 1644 (1963).
5. T. A. Kovacina, A. D. Berry, W. B. Fox, *J. Fluorine Chem.*, *7*, 430 (1976).
6. L. Field, in *Organic Chemistry of Sulfur*, S. Oae, ed., Plenum Press, New York, 1977, p. 303.
7. C. R. Russ, I. B. Douglass, in *Sulfur in Organic and Inorganic Chemistry*, A. Senning, ed., Vol. 1, Marcel Dekker, New York, 1971, p. 239.
8. W. R. Hardstaff, R. F. Langler, in *Sulfur in Organic and Inorganic Chemistry*, A. Senning, ed., Vol. 4, Marcel Dekker, New York, 1982, p. 193.
9. M. L. Kee, I. B. Douglass, *Org. Prep. Proc.*, *2*, 235 (1970).
10. E. Kühle, *Synthesis*, 561 (1970).
11. A. Schöberl, A. Wagner, in *Houben-Weyl, Methoden der Organischen Chemie*, Vol. IX, *S, Se, Te Verbindungen*, E. Müller, O. Bayer, H. Meerwein, K. Ziegler, eds., Georg Thieme Verlag, Stuttgart, 1955, p. 269ff.
12. R. W. Braun, A. H. Cowley, M. C. Cushner, R. J. Lagow, *Inorg. Chem.*, *17*, 1679 (1978).
13. E.g., K. R. Brower, I. B. Douglass, *J. Am. Chem. Soc.*, *73*, 5787 (1951).
14. P. S. Magee, in *Sulfur in Organic and Inorganic Chemistry*, Vol. 1, A. Senning, ed., Marcel Dekker, New York, 1971, p. 261.
15. P. S. Magee, in *Sulfur in Organic and Inorganic Chemistry*, Vol. 4, A. Senning, ed., Marcel Dekker, New York, 1982, p. 283.
16. R. N. Haszeldine, J. M. Kidd, *J. Chem. Soc.*, 3219 (1953).
17. *Gmelins Handbook of Inorganic Chemistry*, *Perfluorohalogenoorgano Compounds of the Main-Group Elements*, Vol. 9/1, Verlag Chemie, Weinheim, 1973, p. 158ff.
18. C. Raby, J. Claude, J. Buxeraud, F. Moreau, *Bull. Soc. Pharm. (Bordeaux)*, *114*, 147 (1975); *Chem. Abstr.*, *86*, 65,108 (1977).
19. R. G. R. Bacon, R. G. Guy, *J. Chem. Soc.*, 318 (196).
20. M. J. Nelson, A. D. E. Pullin, *J. Chem. Soc.*, 604 (1960).
21. J. C. Hinshaw, *Tetrahedron Lett.*, 3567 (1972).
22. R. C. Cambie, H. H. Lee, P. S. Rutledge, P. D. Woodgate, *J. Chem. Soc.*, *Perkin Trans. 1*, 757 (1979).

23. I. B. Douglass, C. E. Osborne, *J. Am. Chem. Soc.*, *75*, 4582 (1953).
24. C. Ziegler, J. M. Sprague, *J. Org. Chem.*, *16*, 621 (1951).

2.3.8.1.3. to Give the Se– and Te–Halogen Bond.

Seleninyl chlorides and bromides are prepared by cleavage of the Se—Se bond in diselenides, but the fluorides and iodides are rather unstable. With Cl_2 and Br_2 xs elemental halogen must be avoided to prevent trihalide formation[1,2]:

$$R_2Se_2 + X_2 \rightarrow 2\ RSeX \tag{a}$$

e.g., R, X = CF_3, Cl[3,4]; CF_3, Br[4,5]; CHF_2, Cl[6]; $BrCF_2CF_2$, Cl[7]; CH_3; Cl; C_2H_5, Cl[3]; C_2F_5, Cl[8]; C_2H_5, Br[9]; p-$NO_2C_6H_4$, Br[10]. Whereas halogenations of alkyl diselenides are carried out under controlled conditions (low T, solvents), perfluoroalkyl derivatives and halogens are reacted neat. Seleninyl iodides are only stable when R is extremely bulky.

$$\left[2,4,6\text{-}((CH_3)_3C)_3C_6H_2\text{—Se—} \right]_2 + I_2 \rightarrow 2\ 2,4,6\text{-}((CH_3)_3C)_3C_6H_2\text{—SeI} \tag{b}$$

The Te—Te bond in organic ditellurides is also cleaved by halogens (Cl_2, Br_2, I_2), but the expected tellurenyl halides are unstable and react further, as observed in Se chemistry, to give the trihalides[12–14]:

$$RTe\text{—}TeR + X_2 \rightarrow 2\ [RTeX] \tag{c}$$

$$[RTeX] \xrightarrow{X_2} 2\ RTeX_3 \qquad [RTeX] \rightarrow R_2TeX_2 + Te$$

Aryltellurenyl halides form from 1:1 diarylditellurides and halogen[15,16]. By using a nonpolar solvent in which the products are insoluble, further oxidation to Te(IV) derivatives is avoided. The tellurenyl halides decompose[17] to give diorganotellurium dihalides and elemental Te. The stability of RTeX decreases in the order I > Br > Cl and is also dependent on the aryl substituent. With ligands containing S or Se as donor atoms, tellurenyl halides can be stabilized as PhTeX·L (X = Cl, Br; L = thiourea, selenourea, etc.)[18], and β-naphthyltellurenyl iodide forms an adduct[19] with Ph_3P.

(R. MEWS)

1. H. Rheinboldt, in *Houben-Weyl Methoden der Organischen Chemie* Vol. IX *(S, Se, Te Compounds)*, E. Müller, O. Bayer, H. Meerwein, K. Ziegler, eds., Georg Thieme Verlag, Stuttgart, 1955, p. 1161ff.
2. K. J. Irgolic, M. K. Kudchadker, in *Selenium*, R. A. Zingaro, C. W. Cooper, eds., Van Nostrand, New York, 1974, p. 415.
3. R. Paetzold, E. Wolfram, *Z. Anorg. Allg. Chem.*, *352*, 167 (1967).
4. N. N. Yarovenkov, V. N. Shemanina, G. B. Gazieva, *J. Gen. Chem. USSR (Engl. Transl.)*, *29*, 924 (1959).
5. J. W. Dale, H. J. Emeléus, R. N. Haszeldine, *J. Chem. Soc.*, 2939 (1958).
6. N. N. Yarovenko, M. A. Raksha, *J. Gen. Chem. USSR (Engl. Transl.)*, *30*, 4027 (1960).
7. N. N. Yarovenko, M. A. Raksha, V. N. Shemanina, *J. Gen. Chem. USSR (Engl. Transl.)*, *30*, 4032 (1960).

8. N. Wellcman, H. Regev, *J. Chem. Soc.*, 7511 (1965).
9. G. Bergson, G. Nordström, *Ark. Kemi, 17*, 569 (1961); *Chem. Abstr., 56*, 4606 (1962).
10. Ref. 1, p. 1164.
11. W.-W. duMont, S. Kubiniok, H.-J. von Schuering, U. Peters, *Angew. Chem. Int. Ed. Engl., 26*, 780 (1987).
12. Ref. 1, p. 1161.
13. K. J. Irgolic, in *The Organic Chemistry of Tellurium*, Gordon and Breach, London, 1974, p. 78.
14. K. J. Wynne, P. S. Pearson, *Inorg. Chem., 9*, 106 (1970).
15. G. Vincentini, E. Giesbrecht, C. M. R. Pitombo, *Chem. Ber., 92*, 40 (1959).
16. P. Schulz, G. Klar, *Z. Naturforsch., Teil B., 30*, 40 (1975); *30*, 43 (1975).
17. W. L. Dorn, A. Knoechel, P. Schulz, G. Klar, *Z. Naturforsch., Teil B., 31*, 1043 (1976).
18. S. Hange, O. Vikane, *Acta Chem., Scand., 27*, 3596 (1973).
19. N. Peturagnani, M. de Moura Campos, *Tetrahedron., 21*, 13 (1965).

2.3.8.2. by Hydrogen Halides.

No examples of this type of reaction are known.

(R. MEWS)

2.3.8.3. from Cleavage of the Group VIB–Element Bond by Other Halides

2.3.8.3.1. to Give the Oxygen–Halogen Bond.

Only the extremely oxidizing peroxide $FSO_2O{-}OSO_2F$, converts halogen compounds to the fluorosulfato-halogens; $ClOSO_2F$ can form in the reaction with metal chlorides[1–3] and $BrOSO_2F$ forms with CF_3Br[4]. Complex anions are isolated:

$$KBr + 2\ S_2O_6F_2 \xrightarrow{50°C} K[Br(OSO_2F)_4] \qquad (a)^{2}$$

$$KBrO_3 + 2\ S_2O_6F_2 \xrightarrow{0°C} K[Br(OSO_2F)_4] + 1.5\ O_2 \qquad (b)^{6,7}$$

$$CsBr + S_2O_6F_2 \xrightarrow{Br_2} Cs[Br(OSO_2F)_2] \qquad (c)^{8}$$

$$2\ ICl + 3\ S_2O_6F_2 \rightarrow 2\ I(OSO_2F)_3 + Cl_2 \qquad (d)^{9}$$

$$KI + 2\ S_2O_6F_2 \rightarrow K[I(OSO_2F)_4] \qquad (e)^{5}$$

$$KICl_4 + 2\ S_2O_6F_2 \xrightarrow[24\,h]{70°C} K[I(OSO_2F)_4] + 2\ Cl_2 \qquad (f)^{6}$$

(R. MEWS)

1. G. C. Kleinkopf, J. M. Shreeve, *Inorg. Chem., 3*, 607 (1964).
2. R. F. Noftle, G. H. Cady, *J. Inorg. Nucl. Chem., 29*, 969 (1967).
3. P. A. Yeats, B. L. Poh, B. F. E. Ford, J. R. Sams, F. Aubke, *J. Chem. Soc. A*, 2188 (1970).
4. C. T. Ratcliff, J. M. Shreeve, *Inorg. Chem., 3*, 631 (1964).
5. M. Lustig, G. H. Cady, *Inorg. Chem., 1*, 714 (1962).
6. H. H. Carter, S. P. L. Jones, F. Aubke, *Inorg. Chem., 9*, 2485 (1970).
7. P. A. Yeats, B. Landa, J. R. Sams, F. Aubke, *Inorg. Chem., 15*, 1452 (1976).
8. C. Chung, G. H. Cady, *Z. Anorg. Allg. Chem., 385*, 18 (1971).
9. J. M. Shreeve, G. H. Cady, *J. Am. Chem. Soc., 83*, 4521 (1961).

2.3.8.3.2. to Give the Sulfur–Halogen Bond.

Few compounds containing S—S bonds are used to synthesize sulfur–halogen derivatives. The cleavage of the S—S bond in disulfides is used to prepare S—F and S—Cl compounds from S_2F_{10} and fluorine-containing radicals (e.g., $[NF_2]^\cdot$[1–3], $[COF]^\cdot$[4], $[CF_3NO]^\cdot$[5], $[(CF_3)_2NO]^{\cdot\cdot}$[5]); SF_6 is formed as a byproduct.

From S_2F_{10} and ICl at 140°C SF_5Cl and I_2 are formed, $[SF_5]^\cdot$ radicals are intermediates:

$$S_2F_{10} + 2\ ICl \xrightarrow{140°C} 2\ SF_5Cl + I_2 \tag{a}$$

$$S_2F_{10} + BCl_3 \xrightarrow{100°C} SF_5Cl + Cl_2, SO_2, SOF_2, SiF_4, BF_3 \tag{b}$$

$$S_2F_{10} + Al_2Cl_6 \rightarrow SF_6, Cl_2 \tag{c}$$

With BCl_3, SF_5Cl and numerous decomposition products are found[7], from Al_2Cl_6 only SF_6 and Cl_2 are detected[7].

Action of chlorine fluorides on perfluoroalkyl disulfides cleaves S—S bonds and causes oxidative additions[8,9]:

$$CF_3SSCF_3 + 2\ ClF_3 \xrightarrow[CF_2Cl_2]{-78°C} 2\ CF_3SF_3 + Cl_2 \tag{d}$$

$$(R_f)_2S_2 + ClF \xrightarrow{25°C} R_fSF_4Cl + R_fSF_5 \tag{e}$$

Addition of aliquots of ClF_3 to CF_3SSCF_3 in CF_2Cl_2 at −78°C produces CF_3SF_3[8]; with ClF at RT S(VI) derivatives are formed[9]. Higher metal fluorides, such as AgF_2[10] or CoF_3[11], give sulfur trifluorides, R_fSF_3, or pentafluorides, R_fSF_5, depending on the conditions, whereas from perfluoroaryl disulfides under moderate conditions sulfenyl fluorides are formed[12]:

$$(4\text{-}R_fC_6F_4S)_2 + AgF_2 \xrightarrow[CFCl_2CF_2Cl]{20°C,\ 10\ h} 4\text{-}R_fC_6F_4\text{—}SF \tag{f}$$

where if R_f = F, yield is 63; if $R_f = CF_3$ yield is 70%;

$$(C_6F_5S)_2 + AgF_2 \xrightarrow[CFCl_2CF_2Cl]{50°C\ 5\ h} 2\ C_6F_5SF_3 \tag{g}$$ [14,15]

$$(n\text{-}C_4H_9S)_2 + AgF_2 \rightarrow 2\ (n\text{-}C_4H_8F)SF_3 \tag{h}$$ [14]

Higher T gives only the trifluoride[12,13]. The sulfenyl fluorides are unstable; e.g., from $(C_6H_5S)_2$ and AgF_2 only the trifluoride[13,14] and the pentafluoride[13,16] are isolated. In this reaction and in the formation of thianthrene from $(C_6H_5S)_2$ and SF_4[17], the sulfenyl fluoride is an intermediate:

$$C_6H_5SSC_6H_5 + SF_4 \xrightarrow{liq\ SF_4} 2\ C_6H_5SF + SF_2 \tag{i}$$

↓

[thianthrene structure] + 2 HF

Sulfenyl chlorides are most commonly produced by chlorinolysis of S—S (or C—S) bonds with either Cl_2 (see §2.3.8.1.) or chlorinating agents[18,19], such as SO_2Cl_2, CH_3SCl_3, N-chloro succinimide $(CH_3CO)_2NCl$, e.g.:

$$RSSR + SO_2Cl_2 \xrightarrow[-15°C]{CCl_4} 2\ RSCl + SO_2 \qquad (j)$$

These exothermic reactions are best performed in a solvent at low T; otherwise the alkyl chain is also chlorinated[20–22]. Perfluoroalkyl disulfides do not react under these conditions.

Simultaneous cleavage and formation of S—S bonds occurs in the preparation of sulfinyl chlorides from sulfinic acid thioesters[23]:

$$CH_3\overset{O}{\overset{\|}{S}}SCH_3 + CH_3SCl \rightarrow CH_3\overset{O}{\overset{\|}{S}}Cl + (CH_3S)_2 \qquad (k)$$

(R. MEWS)

1. E. C. Stump, C. D. Padgett, W. S. Brey, *Inorg. Chem., 2*, 648 (1963).
2. G. H. Cady, D. F. Eggers, B. Tittle, *Proc. Chem. Soc. (London)*, 65 (1963).
3. L. Boivin, *Can. J. Chem., 42*, 2744 (1964).
4. R. Czerepinski, G. H. Cady, *J. Am. Chem. Soc., 90*, 3954 (1968).
5. M. D. Vorob'ev, A. S. Filatov, M. A. Englin, *J. Org. Chem. USSR (Engl. Transl.), 9*, 326 (1973).
6. M. D. Vorob'ev, A. S. Filatov, M. A. Englin, *J. Gen. Chem. USSR (Engl. Transl.), 44*, 2677 (1974).
7. B. Cohen, A. G. MacDiarmid, *Inorg. Chem., 4*, 1782 (1965).
8. G. H. Sprenger, A. H. Cowley, *J. Fluorine Chem. 7*, 333 (1976).
9. T. Abe, J. M. Shreeve, *J. Fluorine Chem., 3*, 187 (1973/74).
10. E. W. Lawless, L. D. Harman, *Inorg. Chem., 7*, 391 (1968).
11. G. A. R. Brandt, H. J. Emeléus, R. N. Haszeldine, *J. Chem. Soc.*, 2198 (1952).
12. G. G. Furin, T. V. Terent'eva, G. G. Yakobson, *Izv. Sibirsk. Otd. Nauk SSR, Ser. Khim Nauk, 14*, 78 (1972); *Chem. Abstr., 78*, 83,964 (1973).
13. W. A. Sheppard, *J. Am. Chem. Soc., 82*, 4751 (1960).
14. W. A. Sheppard, *J. Am. Chem. Soc., 84*, 3058 (1962).
15. W. A. Sheppard, S. S. Foster, *J. Fluorine Chem. 2*, 53 (1972).
16. W. A. Sheppard, *J. Am. Chem. Soc., 84*, 3064 (1962).
17. F. Seel, R. Budenz, R. D. Flaccus, R. Staab, *J. Fluorine Chem., 12*, 437 (1979).
18. C. R. Russ, I. B. Douglass, in *Sulfur in Organic and Inorganic Chemistry*, A. Senning, ed., Vol. 1, Marcel Dekker, New York, 1971, p. 243.
19. W. R. Hardstaff, R. F. Langer, in *Sulfur in Organic and Inorganic Chemistry*, A. Senning, ed., Vol. 4, Marcel Dekker, New York, 1982, p. 195.
20. A. Brintzinger, K. Pfannstiel, H. Koddebusch, K. Kling, *Chem. Ber., 83*, 87 (1950).
21. I. B. Douglass, K. R. Brower, F. T. Martin, *J. Am. Chem. Soc., 74*, 5770 (1952).
22. H. Brintzinger, H. Koddebusch, K. E. Kling, G. Jung, *Chem. Ber., 85*, 455 (1952).
23. I. B. Douglass, D. A. Koop, *J. Org. Chem., 27*, 1398 (1962).

2.3.8.3.3. to Give the Se–Halogen Bond.

Cleavage of the Se—Se bond in diselenides by halides resembles the corresponding reactions of disulfides. Selenyl fluorides formed in the first step with halogen fluorides (e.g., BrF_3[1], ClF[2,3]) are immediately oxidized to Se(IV):

$$CF_3SeSeCF_3 \xrightarrow[CFCl_3,\ 78°\text{–}0°C]{BrF_3} 2\ \{CF_3SeF\} \rightarrow 2\ CF_3SeF_3 \qquad (a)$$

$$C_2F_5SeSeC_2F_5 \xrightarrow[-196°\text{ to }0°C]{ClF} 2\ \{C_2F_5SeF\} \rightarrow 2\ C_2F_5SeF_3 \rightarrow 2\ C_2F_5SeF_4Cl \qquad (b)$$

further addition gives Se(VI) derivatives (see §2.3.9.3.2).

Also, from perfluoroaryl diselenides and AgF_2, only Se(IV) derivatives can be isolated[4,5]:

$$(4\text{-}RC_6F_4Se)_2 + AgF_2 \xrightarrow{CFCl_2CF_2Cl} 2\,[4\text{-}RC_6F_4SeF] \rightarrow 2\ 4\text{-}RC_6F_4SeF_3 \quad \text{(c)}$$

where R = F, CF_3;

$$(RSe)_2 + 6\,AgF_2 \xrightarrow{CFCl_2CF_2Cl} 2\,RSeF_3 + 6\,AgF \quad \text{(d)}$$

where R = C_6H_5, 4-$CH_3C_6H_4$, 4-FC_6H_4, 2-$C_2H_5C_6H_4$, 2-$NO_2C_6H_4$[6].

Stoichiometric SO_2Cl_2[7–10] cleaves the Se—Se bond, but xs halogenating agent forms trihalides also. With $OSCl_2$ no further oxidation is observed:

$$RSeSeR + SO_2Cl_2 \rightarrow 2\,RSeCl + SO_2 \xrightarrow{xs\ SO_2Cl_2} RSeCl_3 \quad \text{(e)}$$

$$2\,RSeSeR + OSCl_2 \rightarrow 4\,RSeCl + SO_2 + \tfrac{1}{8}\,S_8 \quad \text{(f)}$$

Disproportionation occurs in the $(CF_3Se)_2$–CF_3SeCl_3 system at RT[11]:

$$CF_3SeSeCF_3 + CF_3SeCl_3 \rightarrow 3\,CF_3SeCl \quad \text{(g)}$$

With N,N′-dichloroamides the Se—Se bond is cleaved with simultaneous formation of Se—Cl and Se—N bonds[12]:

$$R_2Se_2 + R'CONCl_2 \xrightarrow[CCl_4]{5^\circ\text{–}10^\circ C} RSe(Cl){=}N\overset{\overset{O}{\|}}{C}R' \quad \text{(h)}$$

(R. MEWS)

1. E. Lehmann, *J. Chem. Res.*, *1*, 42 (1978).
2. D. D. Desjardins, C. Lau, J. Passmore, *Inorg. Nucl. Chem. Lett.*, *9*, 1037 (1973).
3. C. Lau, J. Passmore, *J. Fluorine Chem.*, *7*, 261 (1976).
4. G. G. Furin, T. V. Terent'eva, G. G. Yakobson, *Izv. Sib. Otd. Akad. Nauk SSR, Ser. Khim. Nauk*, 78 (1972); *Chem. Abstr.*, *78*, 83,964 (1973).
5. G. G. Furin, O. I. Andreevskaya, G. G. Yakobson, *Izv. Sib. Otd. Akad. Nauk SSR, Ser. Khim Nauk*, 141 (1981); *Chem. Abstr.*, *95*, 132,412 (1981).
6. W. M. Maxwell, K. J. Wynne, *Inorg. Chem.*, *20*, 1707 (1981).
7. H. Rheinboldt, in *Houben-Weyl Methoden der Organischen Chemie Vol. IX, S, Se, Te compounds*, E. Müller, O. Bayer, H. Meerwein, K. Ziegler, eds., Georg Thieme Verlag, Stuttgart, 1955, pp. 1161, 1162.
8. S. Keimatsu, I. Satoda, *J. Pharm, Soc. Jpn.*, *55*, 233 (1935); *Chem. Abstr.*, *31*, 6661 (1937).
9. R. Paetzold, D. Knaust, *Z. Chem.*, *10*, 269 (1970).
10. O. Behagel, H. Seibert, *Chem. Ber.*, *66*, 714 (1933).
11. J. W. Dale, H. J. Emeléus, R. N. Haszeldine, *J. Chem. Soc.*, 2939 (1958).
12. N. Ya. Derkach, T. V. Lyapina, E. S. Levchenko, *J. Org. Chem. USSR (Engl. Transl.)*, *16*, 31 (1980).

2.3.8.3.4. to Give the Te–Halogen Bond.

Perfluoroalkyl ditellurides[1] are difficult to obtain; therefore, little of their chemistry is known. Reactions follow routes similar to those for the lighter homologues:

$$F_5C_2TeTeC_2F_5 + 6\,ClF \xrightarrow{-78^\circ C} 2\,F_5C_2TeF_3 + 3\,Cl_2 \quad \text{(a)}$$

Besides the trifluoride, small amounts of trans-$F_5C_2TeClF_4$ and TeF_5Cl are found[2].

The Te—Te bond in dioxytellurides is cleaved by S, Se and Te halides (e.g., SF_4[3], SO_2Cl_2, Ph_2SeBr_2, Ph_2SeCl_2, $TeCl_4$, $TeBr_4$, TeI_4, $ArTeCl_3$[4,5]) to give diaryl dihalides:

$$(RC_6H_4)_2Te + SF_4 \rightarrow (RC_6H_4)_2TeF_2 \quad \text{(b)}$$

where R = H, 4-OCH_3, n = 1, 2;

$$2\ ArTe{-}TeAr + 2\ ArTeCl_3 \rightarrow 6\ [ArTeCl] \rightarrow 3\ Ar_2TeCl_2 + 3\ Te \quad \text{(c)}$$

$$ArTe{-}TeAr + SO_2Cl_2 \xrightarrow[-SO_2]{} 2\ [ArTeCl] \rightarrow Ar_2TeCl_2 + Te \quad \text{(d)}$$

With SF_4 the difluorides $ArTeF_2$ are prepared from Ar_2Te_n (n = 1, 2) independently of n. Tellurenyl halides are formed from cleavages; similar observations are made with organic halides:

$$ArTe{-}TeAr + BrCH_2CH_2Br \xrightarrow[-C_2H_4]{} Ar_2TeBr_2 + Te \quad \text{(e)}$$

$$2\ ArTe{-}TeAr + 2\ CH_3I \rightarrow 2\ ArTeCH_3 + 2\ [ArTeI] \quad \text{(f)}$$

$$ArTeCH_3 \xrightarrow{CH_3I} [(CH_3)_2TeAr]I \qquad [ArTeI] \rightarrow Ar_2TeI_2 + Te$$

Organic 1,2-dibromides act as brominating agents[4]. With CH_3I, the primary products react further to give Te iodides and diaryltellurium diiodides[6,7].

Similar results are obtained with diazonium chlorides[8]; e.g., treating Ph_2Te_2 in Me_2CO containing $Cu(OAc)_2$ with $[ArN_2]^+Cl^-$ in aq HCl affords $ArTeCl_2Ph$ (Ar = F-, 4-OCH_3-, H-C_6H_4) in 83–86% yield.

(R. MEWS)

1. H. L. Paige, J. Passmore, *Inorg. Nucl. Chem. Lett.*, *9*, 277 (1973).
2. C. D. Desjardins, C. Lau, J. Passmore, *Inorg. Nucl. Chem. Lett.*, *10*, 151 (1974).
3. I. D. Sadekov, A. Ya. Bushkov, L. N. Markovskii, V. I. Minkin, *J. Gen. Chem. USSR (Engl. Transl.)*, *46*, 1617 (1976).
4. N. Petragnani, M. de Moura Campos, *Chem. Ber.*, *94*, 1759 (1961).
5. W. L. Dorn, A. Knoechel, P. Schulz, G. Klar, *Z. Naturforsch., Teil B*, *31*, 1043 (1976).
6. L. Reichel, E. Kirschbaum, *Ann. Chem.*, *523*, 211 (1936).
7. G. T. Morgan, H. D. K. Drew, *J. Chem. Soc.*, 2307 (1925).
8. I. D. Sadekov, A. A. Maksimenko, *J. Org. Chem. USSR (Engl. Transl.)*, *14*, 2411 (1979).

2.3.9. from Cleavage of the Group VIB–Nitrogen Bond

2.3.9.1. by Halogens

This reaction type is not suited for the formation of bonds between Group VIB elements and halogens, as only accidental examples can be given.

The O—N bond is not cleaved by halogens. In the reaction of NO_2 with atomic fluorine, a compound containing an O—F bond is formed[1] but only the π bond is cleaved in the process and the reactions can be best described as an addition:

$$NO_2 + F_{at} \rightarrow FON{=}O \quad \text{(a)}$$

Chlorine with NO gives[2,3] the addition compound ClNO without cleaving the O—N bond.

The S—N bond is resistant to attack by halogens, although cleavage may occur in the sulfur nitrides; however, these are often accompanied by molecular rearrangements and other transformations and no straightforward mechanisms for the formation of sulfur–halogen products can be suggested. For example, thiazyl trifluoride, NSF_3, is cleaved by F_2 at $-196°C$ to give SF_6, but an S—N—S compound is also formed[4]:

$$F_3S{\equiv}N + F_2 \xrightarrow{-196°C} SF_6 + F_5S{=}N{-}SF_5 + N_2 \qquad \text{(b)}$$

Tetrasulfur tetranitride is cleaved by F_2 diluted with He at $-78°C$ to give trifluorocyclotrithiazene, a cyclic trimer $(FSN)_3$, probably via[5] monomeric FSN:

$$3\ S_4N_4 + 6\ F_2 \rightarrow 12\ FSN \rightarrow 4\ (FSN)_3 \qquad \text{(c)}$$

The intermediate $F_2S_4N_4$ can be identified[6].

The stability of S—N bonds toward F_2 is illustrated by the photochemical reaction of $CF_3CF_2{-}N{=}SF_2$, which gives only fluorination products $CF_3CF_2{-}N{=}SF_4$, $CF_3CF_2{-}NFSF_5$, $(CF_3CF_2N)_2SF_2$ and $(CF_3CF_2NF)_2SF_4$, but no S—N bond cleavage occurs[7].

Elemental chlorine reacts with tetrasulfur tetranitride at RT to form trichlorocyclotrithiazene in high yield[8–11]; an intermediate addition compound, $Cl_2S_4N_4$, can be isolated if the reaction is carried out[12–15] at $-60°C$; probably the precursor of the cyclic trimer is the Cl—S≡N monomer[16]:

$$S_4N_4 + Cl_2 \xrightarrow{-60°C} Cl_2S_4N_4 \xrightarrow[RT]{+Cl_2} ClSN \rightarrow (ClSN)_3 \qquad \text{(d)}$$

The chlorination of the $S_4N_4 \cdot SbCl_5$ adduct also gives[17] a cyclic compound containing an S—Cl bond, $[S_3N_2Cl_6]^-$:

$$S_4N_4 \cdot SbCl_5 + Cl_2 \rightarrow [S_3N_2Cl]^+[SbCl_6] \qquad \text{(e)}$$

This five-membered ring, in turn, can be chlorinated (as Cl^- salt) to give trichlorocyclotrithiazene[18]:

$$[S_3N_2Cl]^+Cl + Cl_2 \rightarrow Cl{-}S{\equiv}N \rightarrow (ClSN)_3 \qquad \text{(f)}$$

Both S—N and S—S bond cleavage and recombination of the monomeric units are involved; the complete stoichiometry is not established, and the mechanism is unknown.

A linear compound, $S_3N_2O_2$[19], when chlorinated, undergoes complex transformations involving SN cleavage and recombination of the Cl—S≡N units, with formation of an oxo derivative of trichlorocyclotrithiazene[20]:

$$O{=}S{=}N{-}S{-}N{=}S{=}O + Cl_2 \rightarrow \text{cyclo-}[OS(Cl){=}N{-}S(Cl){=}N{-}S(Cl){=}N{-}] \qquad (g)$$

The reaction of tetrasulfur tetranitride with Br_2 is influenced by the conditions. For example, solid S_4N_4 reacts with Br_2 vapor to give a partially brominated, highly conducting polymer[21–24], $(Br_{0.4}SN)_x$, whereas in CS_2, a cyclic compound, CS_2NBr_2X ($X = Br^-$ or$[Br_3]^-$), is formed with the participation of the solvent, in addition to $[S_4N_3]^+Br^-$ and $[S_4N_3]^+Br_3^-$[25,26]:

$$S_4N_4 + Br_2 + CS_2 \rightarrow \left[\text{BrS-}\overset{}{C}\langle S^+{=}N{-}S{-}N{=}\rangle\right]X \qquad (h)$$

In CCl_4 and CS_2 the compounds $(Br_{0.33}SN)_x$, $[S_4N_3]^+Br^-$, $[S_4N_3]^+Br_3^-$ and $S_3N_2Br_2$ can also be obtained[27].

With liquid Br_2 the product is only $S_4N_3Br_3$, containing no S—Br bonds[28]. Mass spectrometry demonstrates the formation of BrSN intermediates in the reaction of S_4N_4 with bromine[29].

The chlorination of the eight-membered P—N—S ring in 1,3-$(Ph_2PN)_2(SN)_2$ yields a six-membered ring derivative via elimination of ClSN from $(Ph_2PN)_2(SNCl)_2$ by cleavage of an SN bond[30]:

$$\text{cyclo-}[Ph_2P{=}N{-}PPh_2{=}N{-}S{-}N{=}S{=}N{-}] \xrightarrow{+Cl_2} \text{cyclo-}[Ph_2P{=}N{-}PPh_2{=}N{-}S(Cl){=}N{-}S(Cl){=}N{-}] \xrightarrow{-ClSN} \text{cyclo-}[Ph_2P{=}N{-}PPh_2{=}N{-}S(Cl){=}N{-}] \qquad (i)$$

Thionylaniline, C_6H_5NSO, is cleaved by Br_2 to form thionyl bromide, $SOBr_2$, in low yield[31].

Nitrososulfuryl fluoride is cleaved by Cl_2 and Br_2 to form mixed sulfuryl halides[32]:

$$ONSO_2F + X_2 \rightarrow XSO_2F + NOX \qquad (j)$$

($X = Cl$, Br).

A reaction involving selective S—N bond cleavage in S—N compounds is illustrated by the following[33] (R = *p*-ClC_6H_4, *p*-$O_2N—C_6H_4$, *p*-MeC_6H_4):

$$R{-}\underset{\underset{N-SO_2Ph}{\|}}{S}{-}NClBu\text{-}t + Cl_2 \rightarrow R{-}\underset{\underset{N-SO_2Ph}{\|}}{S}{-}Cl \quad + \text{other products} \qquad (k)$$

Sulfenyl bromides[34] and iodides[35] can be prepared in high yields (90% and 50–84%, respectively) by cleavage of sulfonyl hydrazides with Br_2 in HCl_3 (or NaBr–$NaBrO_3$ mixture) and I_2 (dissolved in aq KI):

$$RSO_2NHNH_2 + 2\ X_2 \rightarrow RSO_2X + N_2 + 3\ HX \tag{l}$$

(X = Br, I).

This converts sulfonyl chlorides into bromides and iodides.

The chemistry of Se—N compounds is less investigated. Pure tetraselenium tetranitride, Se_4N_4, reacts explosively with dry Cl_2, but when Cl_2 is diluted with CO_2, $SeCl_4$ and SeO_2 can be isolated as products[36]. When Se_4N_4 reacts with Cl_2 in $HCCl_3$, the product[37] is also S_eCl_4. Bromine vapor diluted with CO_2 cleaves Se_4N_4 to form $[NH_4]_2[SeBr_6]$, but I_2 does not react[37] with Se_4N_4.

Tellurium nitride, Te_3N_4, explodes[38] with Cl_2 and Br_2, but when suspended in ether the chlorination with gaseous Cl_2 produces $TeCl_4$ and $[NH_4]_2[TeCl_6]$; Br_2 reacts similarly[39].

Tellurium–N bonds in diaryltellurium(IV) imines are cleaved by Cl_2 to form diaryltellurium dichlorides[40]:

$$R_2Te{=}NR' + Cl_2 \rightarrow R_2TeCl_2 + \text{unknown byproducts} \tag{m}$$

(I. HAIDUC)

1. D. M. Fasano, N. S. Nogar, *J. Chem. Phys.*, *78*, 6688 (1983).
2. D. R. Herschbach, H. S. Johnston, K. S. Pitzer, R. E. Powell, *J. Chem. Phys.*, *25*, 736 (1956).
3. W. C. Nottingham, J. R. Sutter, *Int. J. Chem. Kinet.*, *18*, 1289 (1986).
4. A. Waterfeld, R. Mews, *Angew. Chem., Int. Ed. Engl.*, *21*, 354 (1982).
5. N. J. Maraschin, R. L. Lagow, *J. Am. Chem. Soc.*, *94*, 8601 (1972).
6. I. Ruppert, *J. Fluorine Chem.*, *20*, 241 (1982).
7. I. Stahl, R. Mews, O. Glemser, *Angew. Chem., Int. Ed. Engl.*, *19*, 408 (1980).
8. E. Demarçay, *C. R. Hebd. Seances. Acad. Sci.*, *91*, 854, 1066 (1880).
9. W. Muthmann, E. Seitter, *Chem. Ber.*, *30*, 627 (1897).
10. A. Mouwsen, *Chem. Ber.*, *64*, 2311 (1931); *65*, 1724 (1932).
11. H. Schröder, O. Glemser, *Z. Anorg. Allg. Chem.*, *298*, 78 (1959).
12. J. Nelson, H. G. Heal, *Inorg. Nucl. Chem. Lett.*, *6*, 429 (1970).
13. H. Vincent, Y. Monteil, *Synth. React. Inorg. Metal.-Org. Chem.*, *8*, 51 (1978).
14. L. Zborilova, P. Gebauer, *Z. Chem.*, *19*, 32 (1979).
15. L. Zborilova, P. Gebauer, *Z. Anorg. Allg. Chem.*, *448*, 5 (1979).
16. M. Geohring, *Ergebnisse und Probleme der Chemie der Schwefel–Stickstoffverbindungen*, Akademie Verlag, Berlin, 1957, p. 64.
17. W. Isenberg, N. K. Homsy, J. Anhaus, H. W. Roesky, G. M. Sheldrick, *Z. Naturforsch., Sect. B*, *38*, 808 (1983).
18. W. L. Jolly, K. D. Maguire, *Inorg. Synth.*, *9*, 102 (1967).
19. J. Weiss, *Z. Naturforsch.*, *16b*, 477 (1961).
20. D. Schläfer, M. Becke-Goehring, *Z. Anorg. Allg. Chem.*, *362*, 1 (1968).
21. G. B. Street, R. L. Bingham, J. I. Crowley, J. Kuyper, *J. Chem. Soc., Chem. Commun.*, *1977*, 464.
22. M. Akhtar, C. K. Chiang, A. J. Heeger, A. G. MacDiarmid, *J. Chem. Soc., Chem. Commun.*, *1977*, 846.
23. M. Akhtar, C. K. Chiang, A. J. Heeger, J. Milliken, A. G. MacDiarmid, *Inorg. Chem.*, *17*, 1539 (1978).
24. G. B. Street, T. C. Clarke, G. Wolmershäuser, *Org. Coat. Plast. Chem.*, *38*, 623 (1978).
25. G. Wolmershäuser, G. B. Street, R. D. Smith, *Inorg. Chem.*, *18*, 383 (1979).
26. G. Wolmershäauser, C. Krüger, Y. H. Tsai, *Chem. Ber.*, *115*, 1126 (1982).
27. H. Vincent, Y. Monteil, M. R. Berthet, *Synth. React. Inorg. Metal.-Org. Chem.*, *10*, 99 (1980).
28. G. Wolmershäuser, G. B. Street, *Inorg. Chem.*, *17*, 1685 (1978).

29. R. D. Smith, G. B. Street, *Inorg. Chem.*, *17*, 938 (1978).
30. N. Burford, T. Chivers, M. N. S. Rao, J. F. Richardson, *Inorg. Chem.*, *23*, 1946 (1984).
31. A. Michaelis, *Chem. Ber.*, *24*, 745 (1891).
32. F. Seel, H. Massat, *Z. Anorg. Allg. Chem.*, *280*, 186 (1955).
33. T. N. Dubinina, L. N. Markovskii, E. S. Levchenko, *Russ. J. Org. Chem.*, *13*, 1497 (1977).
34. A. C. Poshkus, J. E. Herwith, F. A. Magnota, *J. Org. Chem.*, *28*, 2766 (1963).
35. L. M. Litvinenko, V. A. Dadali, V. A. Savelova, T. I. Krichevtsova, *J. Gen. Chem. USSR*, *34*, 3780 (1964).
36. W. Strecker, H. E. Schwarzkopf, *Z. Anorg. Allg. Chem.*, *221*, 194 (1935).
37. W. Strecker, C. Claus, *Chem. Ber.*, *56*, 367 (1923).
38. W. Strecker, W. Ebert, *Chem. Ber.*, *58*, 2534 (1925).
39. W. Strecker, C. Mahr, *Z. Anorg. Allg. Chem.*, *221*, 200 (1935).
40. V. I. Naddaka, V. P. Garkin, V. I. Minkin, *Z. Org. Khim.*, *16*, 1619 (1980).

2.3.9.2. by Hydrogen Halides.

The S–N bond is expected to be sensitive to hydrogen halides, but the reactivity and the products strongly depends upon the oxidation states of sulfur and the reaction conditions. Few such reactions are described, and the reason seems obvious: S—N compounds are formed from S–halogen reagents, and the reverse reaction, with regeneration of the halide, is not an attractive one.

Sulfur(II)-nitrogen compounds, i.e., sulfenamides, $RSNR'_2$, are cleaved by gaseous or conc. HCl to form sulfenyl chlorides[1,2]:

$$RSNR'_2 + 2\ HCl \rightarrow RSCl + R'_2NH \cdot HCl \qquad (a)$$

Sulfenyl isocyanates are cleaved by hydrogen halides to form sulfenyl halides[3]:

$$C_6F_5SNCO + HX \rightarrow C_6F_5SX + HNCO \qquad (b)$$

Sulfur(II) chloride results from the cleavage of N-tetraorganyl thioamides[4]:

$$R_2NSNR_2 + 4\ HCl \rightarrow SCl_2 + 2\ [R_2NH_2]^+Cl^- \qquad (c)$$

Bis(benzenesulfene)imides, $(PhS)_2NR$, are cleaved by 3N hydrochloric acid to form primary alkylamine hydrochlorides, $RNH_2 \cdot HCl$, but the fate of the sulfur component is not established[45]. With gaseous HCl the cleavage may occur to form the sulfenyl chloride, PhSCl.

The cleavage of sulfenamides with hydrofluoric acid gives unexpected products. Heated at 90°C for 2 h with hydrofluoric acid, CF_3SNEt_2 gives $(CF_3S)_2$ and a residue containing 1:3 CF_3SO_2F and $(CF_3S)_2$. The reaction of Cl_3CSNEt_2 with HF at −78°C produces 65% $FCl_2CS—Cl$, 18% Cl_3CSCl, 11% ClF_2CSCl and 3% Cl_3CF. From $PhSNEt_2$ and HF a 31% yield of Ph_2S_2 is obtained, and $EtSNEt_2$ with HF at −30°C yields 52% $(EtS)_2$ and 33% $EtSO_2SEt$. In all these reactions, intermediate formation of sulfenyl fluorides, RS—F, by S—N bond cleavage is assumed, but further secondary reactions lead to the products reported[6].

Thionylaniline in hydrocarbon solution is cleaved by dry HCl gas at −10°C to form thionyl chloride (in 94% yield) and aniline hydrochloride; at 60–80°C the reverse reaction takes place[7]:

$$C_6H_5NSO + 3\ HCl \xrightarrow{-10°C} OSCl_2 + C_6H_5NH_2 \cdot HCl \qquad (d)$$

Tetrasulfur tetranitride is reported to form a $S_4N_4 \cdot 2\ HF$ adduct[8], but complete cleavage with HF may occur under more drastic conditions[9]:

$$S_4N_4 + HF \rightarrow SF_6 + SF_4 + O_2SF_2 + NF_3 \quad \text{(e)}$$

The source of oxygnen in the formation of O_2SF_2 is probably the wall of the reaction vessel. In the presence of Cu(II) oxide at 100°C, dry HF cleaves S_4N_4 to form OSF_2 quantitatively[10].

Other hydrogen halides react with tetrasulfur tetranitride to form ionic products, $[S_4N_3]^+ X^-$ (X = Cl[11,12], Br, I[12]), that do not contain direct (covalent) sulfur–halogen bonds.

Gaseous HBr cleaves S_4N_3Cl and $S_3N_2Cl_2$ at RT and $S_6N_4Cl_2$ (at 0°C) to form S_2Br_2, in addition to S_4N_3Br and other products[13].

Sulfonyl fluorides can be prepared in 53–78% yields by treating sulfonyl amides with sodium nitrite in anhyd HF[14,15]:

$$RSO_2NH_2 + NaNO_2 + 2\ HF \rightarrow RSO_2F + N_2 + NaF + H_2O \quad \text{(f)}$$

This reaction may have preparative value for the synthesis of arylsulfonyl fluorides from available chlorides via sulfonylamides.

Tetraselenium tetranitride, Se_4N_4, explodes with dry gaseous HCl, but when the latter is diluted with CO_2 the products are $SeCl_4$ and SeO_2[16].

Diaryltellurium imined are readily cleaved by HCl to form the corresponding dihalides[17]:

$$R_2Te{=}NR' + 3\ HCl \rightarrow R_2TeCl_2 + R'NH_2 \cdot HCl \quad \text{(g)}$$

Tellurium nitride, Te_3N_4, reacts with HBr to form[18] $[NH_4]_2[TeBr_6]$, but the reaction with HCl occurs explosively[19].

(I. HAIDUC)

1. T. Zincke, *Justis Liebigs Ann. Chem.*, *400*, 1 (1913).
2. T. Zincke, H. Röse, *Justus Liebigs Ann. Chem.*, *406*, 103 (1914).
3. R. J. Neil, M. E. Peach, *J. Fluorine Chem.*, *1*, 257 (1972).
4. H. Lecher, *Chem. Ber.*, *58*, 421 (1925).
5. T. Mukaiyama, T. Taguchi, *Tetrahedron Lett.*, *1970*, 3411.
6. I. L. Knunyants, I. N. Rozhkov, *Izvest. Akad. Nauk SSSR, Ser. Khim.*, *1970*, 2264.
7. P. Carré, D. Lieberman, *Bull. Soc. Chim. Fr.*, *6*, 579 (1939).
8. S. P. S. Jadon, *Curr. Sci.* (*India*), *55*, 781 (1986).
9. A. Engelbrecht, E. Mayer, C. Pupp, *Monatsch. Chem.*, *95*, 633 (1964).
10. O. Ruff, C. Thiel, *Chem. Ber.*, *38*, 549 (1905).
11. A. G. MacDiarmid, *Nature* (*London*), *164*, 1131 (1949).
12. A. G. MacDiarmid, *J. Am. Chem. Soc.*, *78*, 3871 (1956).
13. H. Vincent, Y. Monteil, M. P. Berthet, *Z. Anorg. Allg. Chem.*, *471*, 233 (1980).
14. R. L. Fern, C. A. Vanderwerf, *J. Am. Chem. Soc.*, *72*, 4809 (1950).
15. B. J. Halperin, M. Krska, E. Levy, C. A. Vanderwerf, *J. Am. Chem. Soc.*, *73*, 1857 (1951).
16. W. Strecker, H. E. Schwarzkopf, *Z. Anorg. Allg. Chem.*, *221*, 194 (1935).
17. V. I. Naddake, V. P. Garkin, V. I. Minkin, *Z. Org. Khim.*, *16*, 1619 (1980).
18. W. Strecker, C. Mahr, *Z. Anorg. Allg. Chem.*, *221*, 200 (1935).
19. W. Strecker, W. Ebert, *Chem. Ber.*, *58*, 2534 (1925).

2.3.9.3. by Other Halides.

Tetrasulfur tetranitride reacts with numerous metal and nonmetal halides, to give a variety of compounds containing sulfur–halogen bonds, following ring cleavage. The nature of the products depends upon the reaction conditions, and different products can be obtained with the same reagent by changing the T, molar ratio, solvents, etc. For example, mild fluorination of tetrasulfur tetranitride with silver(II) fluoride in boiling CCl_4 yields the cyclic addition product $(FSN)_4$, which preserves the eight-membered ring[1–3]; with a higher reagent ratio the ring is completely cleaved to FSN[1,4] and F_3SN[5,6]. The same reaction can also produce[1,7] SN_2F_2, a compound containing S—F bond:

$$S_4N_4 + 4\ AgF_2 \rightarrow (FSN)_4 + 4\ AgF \quad \text{(a)}$$

$$S_4N_4 + 12\ AgF_2 \rightarrow 4\ F_3SN + 12\ AgF \quad \text{(b)}$$

At 20°C, after 12 h, the product[8] is $S_3N_2F_2$; SN_2F_2 can also be cleaved by AgF_2 at RT to form[9] F_3SN.

The reactions of tetrasulfur tetranitride with SeF_4, SF_4 and IF_5 yield[10] F_3SN and FSN.

Thionyl chloride and bromide cleave a sulfur–nitrogen bond in N-tetraalkylthioylamides, with formation of amidosulfurous acid halides (thionyl amide halides) (X = Cl[11], Br[12]):

$$R_2N\underset{\underset{O}{\|}}{S}NR_2 + SOX_2 \rightarrow 2\ R_2N\underset{\underset{O}{\|}}{S}X \quad \text{(c)}$$

Methylsulfonyl amide reacts with chlorosulfonic acid to form methylsulfonyl chloride[13]:

$$MeSO_2NH_2 + ClSO_3H \rightarrow MeSO_2Cl + H_2NSO_3H \quad \text{(d)}$$

Sulfur—chlorine bonds are formed[14] with N-substituted sulfonamides with PF_5:

$$RSO_2NHR' + PCl_5 \rightarrow R{-}\overset{\overset{NR'}{\|}}{\underset{\underset{O}{\|}}{S}}{-}Cl + HCl + OPCl_3 \quad \text{(e)}$$

with preservation of the S—N bond, suggesting that S(VI)—N bonds are not sensitive to this reagent. Unsubstituted arylsulfonamides undergo condensation with PCl_5 without cleavage of the S—N bond and formation of sulfonamides[15,16]:

$$RSO_2{-}NH_2 + PCl_5 \rightarrow RSO_2{-}N{=}PCl_3 + 2\ HCl \quad \text{(f)}$$

On heating, however, atomatic sulfonamides can be converted by PCl_5 into the corresponding sulfonyl chlorides[17]:

$$RSO_2{-}NH_2 + PCl_5 \xrightarrow{-120°C} RSO_2{-}Cl + OPCl_3 + HCl + H_2O \quad \text{(g)}$$

Tetrasulfur tetranitride is cleaved by various metal and nonmetal chlorides. The reaction with sulfuryl chloride produces the cyclic trimer[18], $(ClSN)_3$, via intermediate ClSN; the stoichiometry of this reaction cannot be described fully because not all the products are identified. The cleavage of S_4N_4 by iron(III) chloride in thionyl chloride yields only the five-membered ring derivative[19], $[ClS_4N_3]^+[FeCl_4]^-$:

$$S_4N_4 + FeCl_3 \xrightarrow{OSCl_2} \left[ClS_4N_3\right]^{+} [FeCl_4] \qquad (h)$$

The chloride of this cyclic cation, $[ClS_4N_3]^+Cl^-$, is cleaved by sulfuryl chloride, followed by trimerization, to yield trichlorocyclotrithiazene, sulfur(II) chloride and sulfur dioxide[20]:

$$3\ [ClS_4N_3]^+Cl^- + 3\ O_2SCl_2 \rightarrow (ClSN)_3 + 3\ SCl_2 + 3\ SO_2 \qquad (i)$$

The trimer can in turn be cleaved[21,22] by sulfur(II) chloride in the presence of $AlCl_3$:

$$(ClSN)_3 + 3\ SCl_2 + 3\ AlCl_3 \xrightarrow{OSCl_2} 3\ [N(SCl)_2][AlCl_4] \qquad (j)$$

The reaction of tetrasulfur tetranitride with PCl_3 is complex and occurs with S—N bond cleavage, but the product is a cyclic compound, formed in low yield[23]:

$$S_4N_4 + PCl_3 \rightarrow ClS(N)_3(PCl_2)_2 \text{ (six-membered ring)} \qquad (k)$$

Derivatives of the same six-membered ring, containing PClPh or PPh_2 units, are formed in the reactions of S_4N_4 with $PhPCl_2$ and Ph_2PCl, respectively[23,24].

The reactions of S_4N_4 with PCl_5 in liquid SO_2 is complex and yields simple adducts and a mixture of chlorosulfanes and chlorophosphazenes, but $SbCl_5$, $AlCl_3$, AsF_5 and SbF_5 produce[25,26] $[S_4N_4]^{2+}$ salts.

A ring contraction reaction with elimination of ClSN occurs on treating $Ph_4P_2N_4S_2$ with sulfuryl chloride[27]:

$$Ph_4P_2N_4S_2 + SO_2Cl_2 \xrightarrow{-\,ClSN} ClS(N)_3(PPh_2)_2 + \text{other products} \qquad (l)$$

The reaction of $S({=}NSiMe_3)_2$ with PCl_5 involves complex transformations, including some S—N bond cleavages, with the recombination of the building units into a six-membered ring derivative[28]:

$$Me_3SiN{=}S{=}NSiMe_3 + PCl_5 \xrightarrow{-\,Me_3SiCl} ClS(N)_3(PCl_2)_2 \qquad (m)$$

Some decomposition reactions of sulfur–nitrogen–fluorine compounds occur with intramolecular S—N bond cleavage[29,30]:

$$O{=}S(NSF_2)(NFS_2) \xrightarrow{RT,\ -FSN} O{=}S(F)(NSF_2) \xrightarrow{-FSN} O{=}S(F)(F) \qquad (n)$$

$$OSNSF_5 \rightarrow N{\equiv}SF_3 + OSF_2 + SO_2 + F_5SNSNSF_5 \qquad (o)$$

Somewhat similar reactions of Se—N compounds lead to formation of selenium(IV) fluoride and chloride and thiazyl trifluoride[31]:

$$2\ Cl_2Se{=}NSF_5 \xrightarrow{RT} SeF_4 + SeCl_4 + 2\ NSF_3 \qquad (p)$$

Diaryltellurium(IV) imines are cleaved[32] by acetyl chloride to form dichlorides, R_2TeCl_2.

The examples cited show that the formation of sulfur–halogen bonds by cleavages of S—N bonds with either halogens, hydrogen halides or other metal and nonmetal halides are not systematically used for synthetic purposes. However, much interesting chemistry can be anticipated in this little explored area. Reactions of this type can be expected to have preparative value in converting available S—Cl compounds to other S fluorides, bromides or iodides via intermediate S—N compounds, or in the synthesis of mixed N—S—X compounds, by selective cleavage of just one S—N bond in compounds containing N—S—N groups.

(I. HAIDUC)

1. O. Glemser, H. Schröder, H. Haeseler, *Z. Anorg. Allg. Chem., 279*, 28 (1955).
2. O. Glemser, *Prep. Inorg. React., 1*, 227 (1964).
3. O. Glemser, R. Mews, *Angew. Chem., Int. Ed. Engl. 19*, 883 (1980).
4. O. Glemser, H. Schröder, H. Haeseler, *Naturwissenschaften, 42*, 44 (1955).
5. O. Glemser, H. Richert, *Z. Anorg. Allg. Chem., 307*, 313 (1961).
6. O. Glemser, H. Meyer, A. Haas, *Chem. Ber., 97*, 1704 (1964).
7. H. Schröder, O. Glemser, *Z. Anorg. Allg. Chem., 298*, 78 (1959).
8. O. Glemser, E. Wyszomirski, *Angew. Chem., 69*, 534 (1957).
9. O. Glemser, H. Schröder, *Z. Anorg. Allg. Chem., 284*, 97 (1956).
10. B. Cohen, T. R. Hooper, D. Hugill, R. D. Peacock, *Nature (London), 207*, 748 (1965).
11. A. Dorlars, in *Houben-Weyl Methoden der Organischen Chemie*, Band XI/2, 734 (1958).
12. A. Senning, S. Kaae, C. Jacobsen, P. Kelly, *Acta Chem. Scand., 22*, 3256 (1968).
13. I. G. Schroetter, *Ger. Pat.*, 634,687 (1933); *C. Zbl., II*, 3447 (1936).
14. J. Braun, K. Weissbach, *Chem. Ber., 63*, 2836 (1930).
15. A. V. Kirsanov, *J. Gen. Chem. USSR, 22*, 269 (1952).
16. A. V. Kirsanov, V. I. Sevchenko, *J. Gen. Chem. USSR., 24*, 474 (1954).
17. R. Delaby, J. V. Harispe, J. Paris, *Bull. Soc. Chim. Fr., 1945*, 954.
18. G. G. Alange, A. J. Banister, and B. Bell, *J. Chem. Soc., Dalton Trans., 1972*, 2399.
19. H. M. M. Shearer, A. J. Banister, J. Halfpenny, G. Whitehead, *Polyhedron, 2*, 149 (1983).
20. G. G. Alange, A. J. Banister, B. Bell, *J. Chem. Soc., Dalton Trans., 1972*, 2399.
21. O. Glemser, B. Krebs, J. Wegener, E. Kindler, *Angew. Chem., Int. Ed. Engl., 8*, 598 (1969).
22. G. G. Alange, A. J. Banister, P. J. Dainty, *Inorg. Nucl. Chem. Lett., 15*, 175 (1979).
23. T. Chivers, M. N. S. Rao, *Inorg. Chem., 23*, 3605 (1984).
24. T. Chivers, M. N. S. Rao, J. F. Richardson, *J. Chem. Soc., Chem. Commun., 1982*, 982.
25. R. J. Gillespie, J. P. Kent, J. F. Sawyer, D. R. Slim, J. D. Tyrer, *Inorg. Chem., 20*, 3799 (1981).
26. J. Bojes, T. Chivers, A. W. Cordes, G. Maclean, R. T. Oakley, *Inorg. Chem., 20*, 16 (1981).
27. N. Burford, T. Chivers, M. N. S. Rao, J. F. Richardson, *Inorg. Chem., 23*, 1946 (1984).

28. S. Pohl, O. Petersen, H. W. Roesky, *Chem. Ber.*, *112*, 1545 (1979).
29. I. Stahl, R. Mews, O. Glemser, *Chem. Ber.*, *113*, 2430 (1980).
30. J. S. Trasher, G. A. Iannaccone, N. S. Hosmare, D. E. Maurer, A. F. Clifford, *J. Fluorine Chem.*, *18*, 537 (1981).
31. J. S. Trasher, K. Seppelt, *Z. Anorg. Allg. Chem.*, *507*, 7 (1983).
32. V. I. Naddke, V. P. Garkin, V. I. Minkin, *Z. Org. Khim.*, *16*, 1619 (1980).

2.3.10. from Cleavage of the Group VIB–Group VB Bonds

2.3.10.1. by Halogens.

The formation of bonds between halogens and group-VIB elements (O, S, Se, Te, Po) by cleavage of their bonds to group-VB elements, As, Sb, Bi) is limited. The O—P, O—As, O—Sb and O—Bi bonds react with halogens only in the presence of reducing agents (to fix the oxygen), leading to group-VB halides, and no oxygen—halogen bond can be formed by such reactions.

The bonds of S, Se and Te with P, As, Sb, or Bi are sensitive to at least some halogens, to creating halogeno derivatives of group-VIB elements. These reactions are accompanied by the formation of group-VB halide elements as well.

Tetraphosphorus trisulfide, P_4S_3, is attacked by Cl_2, but the products are not identified[1]; this sulfide also reacts with I_2 to form a compound in which the halogen is attached to phosphorus, rather than sulfur, e.g., $I_2P_4S_3$[2], another product being[3] P_4S_7.

A patent[4] describes the reaction of P_4S_{10} with Cl_2 in the presence of alcohols, using toluene or xylene as solvent, with formation of $(RO)_2P(S)Cl$.

Other P—S compounds, e.g., trialkyl thiphosphates, react with Cl_2 to form sulfur-chlorine bonds, i.e., dialkylphosphorylsulfenyl chlorides, but without cleavage of the P—S bond[5]:

$$(RO)_3PS + Cl_2 \rightarrow \left[(RO)_2\overset{+}{\underset{\displaystyle S{-}Cl}{\overset{|}{P}}}{-}O{-}R\right]Cl \rightarrow (RO)_2\underset{\displaystyle S{-}Cl}{\overset{|}{P}}O + RCl \tag{a}$$

Tetraphosphorus decasulfide reacts[6] with Br_2 in $HCCl_3$ to give unidentified orange crystals that eliminate Br_2; no reaction takes place with I_2.

Chlorine reacts with As_2S_3 with formation[7] of 2 $AsCl_3 \cdot 3\ SCl_2$, but this claim has not been confirmed by later authors[8]. This reaction produces $AsCl_3$ and elemental sulfur[9]. At RT As_2S_3 is attacked by Br_2 in an exothermic reaction, to form[10] S_2Br_2 and $AsBr_3$; I_2 forms only elemental sulfur and AsI_3. Another arsenic sulfide, described as As_2S_2 (As_4S_4), reacts[7] with Cl_2 to form S_2Cl_2.

In the reaction of Sb_2S_3 with F_2 SbF_3 is isolated but the fate of sulfur is not established[12]. Antimony(III) sulfide is chlorinated by gaseous Cl_2 on heating, to form[13] $SbCl_3$ and SCl_2, and $SbCl_5 \cdot SCl_4$ is a product of this reaction[14,15]. Perhaps the latter is $[SCl_3]^+[SbCl_6]^-$. With bromine at RT, Sb_2S_3 forms[10] S_2Br_2, $SbBr_3$ and SbSBr: with I_2 the products are[16] SbI_3 and SbSI.

The Sb–S melt (1:2 ratio) containing antimony selenides, treated with dry Cl_2 forms[15] $SbCl_5 \cdot SeCl_4$.

(I. HAIDUC)

1. G. Lemoine, *C. R. Hebd. Seances Acad. Sci., 96*, 1630 (1883); *Chem. Ber., 16*, 1672 (1883).
2. L. Wolter, *Chem. Zg., 31*, 640 (1907).
3. R. D. Thompson, C. J. Wilkins, *J. Inorg. Nucl. Chem., 3*, 187 (1956/1957).
4. K. Imamura, S. Nabekawa, K. Takeuchi, T. Tsuchiya, *Japan. Kokai*, 77,116,422 (1977); *Chem. Abstr., 88*, 50,278j.
5. J. Michalski, A. Skowronska, *Chem. Ind. (London), 1958*, 1199.
6. R. F. Hunter, *Chem. News, 131*, 38 (1925).
7. H. Rose, *Poggendorff Ann., 42*, 517 (1837).
8. J. R. Partington, *J. Chem. Soc., 1929*, 2577.
9. L. F. Nilson, *J. Prakt. Chem., 12*, 330 (1875).
10. E. Montignie, *Bull. Soc. Chim. France, 9*, 654 (1942).
11. R. Schneider, *J. Prakt. Chem., 36*, 498 (1887).
12. H. Moissan, *Ann. Chim. Phys., 24*, 262 (1891).
13. H. Rose, *Poggendorff Ann.*, 3, 445 (1825).
14. H. Rose, *Poggendorff Ann., 42*, 534 (1837).
15. R. Weber, *Poggendorff Ann., 125*, 81 (1867).
16. R. Schneider, *Poggendorff Ann., 110*, 150 (1860).

2.3.10.2. by Hydrogen Halides

The reactions of group VIB–group VB compounds with hydrogen halides are not suitable for preparing group VIB halides. Liquid HCl with As_2S_3 forms[1] $AsCl_3$ and H_2S and Sb_2S_3 reacts with HCl gas at RT to form[2] $SbCl_3$ and H_2S.

In the presence of quaternary phosphonium halides, P_4S_{10} reacts with HCl and HBr to form dihalogenodithiophosphate salts and hydrogen sulfide[3], without sulfur—halogen bond formation:

$$P_4S_{10} + 4\ HCl + 4\ [Ph_3PMe]Cl \rightarrow 4\ [Ph_3PMe][Cl_2PS_2] + 2\ H_2S \qquad (a)$$

$$P_4S_{10} + 4\ HBr + 4\ [Ph_4P]Br^- \rightarrow 4\ [Ph_4P][Br_2PS_2] + 2\ H_2S \qquad (b)$$

Antimony(III) sulfide in similar reactions forms complex salts containing the $[Sb_2Br_8]^{2-}$ anion and hydrogen sulfide[4].

(I. HAIDUC)

1. G. Jander, H. Schmidt, *Wien. Chem. Z., 46*, 49 (1943).
2. C. Tookey, *J. Chem. Soc., 15*, 463 (1862).
3. U. Müller, A. T. Mohammed, *Z. Anorg. Allg. Chem., 514*, 164.
4. A. T. Mohammed, U. Müller, *Z. Naturforch., Teil B, 40*, 562 (1985).

2.3.10.3. by Other Halides

Tetraphosphorus decasulfide, P_4S_{10}, reacts with thionyl chloride, $OSCl_2$, in a sealed tube at 100–150°C, to form S_2Cl_2, $SPCl_3$, SO_2 and elemental sulfur[1,2], but with PCl_5 $SPCl_3$ is obtained, and no sulfur chloride is formed[23]. Similarly, only $SPBr_3$ is obtained with PBr_5[24].

Other halogenating agents also form only thiophosphoryl halides in reactions with P_4S_{10}; P_4S_{10} reacts with anhyd $FeCl_3$ to form $SPCl_3$ (and $FeCl_2$ and FeS_2)[25], and with BiF_3[26] and PbF_2[27] to form SPF_3.

In the reaction of As_2S_3 with $OSCl_2$ the products are[8] S_2Cl_2, $AsCl_3$ and SO_2:

$$As_2S_3 + 6\ OSCl_2 \rightarrow 3\ S_2Cl_2 + 2\ AsCl_3 + 3\ SO_2 \tag{a}$$

Phosphorus pentachloride chlorinates the selenide P_2Se_5 to form Se_2Cl_2, in addition to other products[29].

It can be seen from the reference list that most reactions cited in this section represent very old chemistry and perhaps some of these would deserve reinvestigation by modern means.

(I. HAIDUC)

1. H. Prinz, *Mitt. Chem. Labor. Univ. Jena, 1879*, 82.
2. H. Prinz, *Justus Liebigs Ann. Chem., 223*, 355 (1884).
3. R. Weber, *J. Prakt. Chem., 77*, 65 (1859).
4. H. S. Broth, C. A. Seabright, *Inorg. Synth., 2*, 153 (1946).
5. E. Glatzel, *Chem. Ber., 23*, 37 (1890).
6. T. E. Thorpe, J. W. Rodger, *J. Chem. Soc., 53*, 766 (1888).
7. H. H. Anderson, *J. Am. Chem. Soc., 72*, 2761 (1950).
8. H. B. North, C. B. Conover, *J. Am. Chem. Soc., 37*, 2486.
9. E. Beaudrimont, *Ann. Chim. Phys., 2*, 9 (1864).

2.3.11. from Further Oxidative Addition of Low Oxidation State Group-VIB Compounds

2.3.11.1. with Elemental Halogens

2.3.11.1.1. to Give the Sulfur–Halogen Bond.

Although F_2 oxidizes low valent, group-VIB compounds to give +6 and, under mild conditions, +4 oxidation state derivatives, few examples for the oxidative addition of Cl_2 and Br_2 are known in the chemistry of sulfur.

Elemental sulfur and F_2 react to give lower fluorides, but these are oxidized by xs F_2. Quantitative formation of SF_6 from SF_4 occurs at 25°–250°C[1]:

$$SF_4 + F_2 \xrightarrow{25°\text{–}250°C} SF_6 \tag{a}$$

Kinetic investigations suggest that $[SF_5]^{\bullet}$ radicals are formed in the first step[2].

Electrochemical fluorination of SF_4 in HF gives varying yields of SF_6, depending on the conditions[3,4]. Formation of $[SF_5]^{\bullet}$ radicals is confirmed by interaction of atomic fluorine with SF_4 in an Ar matrix[5]. Sulfur hexafluoride is observed, along with other products, when sulfur-containing compounds, e.g., sulfides[6], MeSCN[7], CS_2[8], COS[9], NSF_3[10], or organic sulfides, are subjected to F_2 under drastic conditions.

Organic sulfur compounds are prepared by electrochemical fluorination[11]. In addition to oxidative addition of F_2 to the sulfur, H–F exchange at carbon and fragmentations are observed. Acyclic and cyclic perfluoroalkyl sulfur(VI) derivatives can be prepared[12,13] (only S—F products are given):

$$CS_2 + [F_2] \rightarrow \underset{(>90\%)}{CF_3SF_5}, SF_6, \underset{(0.5\%)}{CF_2(SF_5)_2}, \underset{(0.5\%)}{CF_2(SF_3)_2} \qquad (b)^{14}$$

$$(n\text{-}C_4H_9)_2S + [F_2] \rightarrow \underset{(\approx 14\%)}{n\text{-}C_4F_9SF_5}, \underset{(10\%)}{(n\text{-}C_4F_9)_2SF_4}, \underset{(2.8\%)}{\overline{CF_2(CF_2)_3SF_4}} \qquad (c)^{15}$$

$$CH_3SC_2H_4SCH_3 + [F_2] \rightarrow$$
$$\underset{(21\%)}{CF_3SF_5}, \underset{(8.5\%)}{C_2F_5SF_4CF_3}, \underset{(2\%)}{(C_2F_5)_2SF_4}, \underset{(1\%)}{CF_3SF_4C_2F_4SF_5}, \underset{(0.7\%)}{(C_2F_4SF_4)_2} \qquad (d)^{16}$$

$$O(C_2H_4)_2S + [F_2] \rightarrow \underset{(\approx 4\%)}{C_2F_5SF_5}, \underset{(\approx 4\%)}{C_2F_5OC_2F_4\text{-}SF_5}, \underset{(\approx 17\%)}{\overline{C_2F_4OC_2F_4SF_4}} \qquad (e)^{17}$$

$$HSCH_2COOH + [F_2] \rightarrow \underset{(3\%)}{SF_3CF_2COOH}, \underset{(2\%)}{SF_5CF_2COF} \qquad (f)^{17}$$

Acyclic and cyclic thiols and dithiols produce perfluorinated and partially fluorinated S(VI) derivatives[18–20].

Because even electrochemical fluorination gives compounds in low yield other mild direct fluorinations are used:

$$CS_2 + F_2\text{-}N_2 \xrightarrow{50°C} SF_5CF_2SF_3, SF_5CF_2SF_5, F_3CSF_5 \qquad (g)^{21}$$

$$CS_2 + F_2\text{-}He \xrightarrow{-120°C} \underset{(60\%)}{F_3SCF_2SF_3} \qquad (h)^{22}$$

$$CS_2 + F_2 \xrightarrow[-95°C]{7 \times 10^2 - 4 \times 10^3\ Pa} \underset{(68\%)}{F_3SCF_2SF_3} \qquad (i)^{23}$$

$$CF_3SSCF_3 + F_2\text{-}He \xrightarrow{-120°C} \underset{(90\%)}{2\ F_3CSF_3} \qquad (j)^{24}$$

$$CF_3SCF_3 + F_2 \xrightarrow[C_2F_6]{-78°C} F_3CSF_2CF_3 \qquad (k)^{25}$$

The fluorination with He diluted F_2 in inert solvents at lower T is even more promising because the heat of reaction is dissipated:

$$(O_2N\text{-}C_6H_3(NO_2)\text{-}S)_2 + F_2\text{-}He \xrightarrow[0°\text{–}5°C]{HF} 2\ O_2N\text{-}C_6H_3(NO_2)\text{-}SF_3 \qquad (l)$$

$$Ph_2S + F_2\text{-}He \xrightarrow[-80°C]{CFCl_3} \underset{(91\%)}{Ph_2SF_2} \qquad (m)^{27}$$

This method can be applied to organometallic derivatives, e.g.:

$$Hg(SCF_3)_2 + F_2\text{-}He \xrightarrow[-110°C]{CCl_3F} Hg(SF_2CF_3)_2 \qquad (n)^{28}$$

and strained heterocycles:

$$F_2C\langle S, S\rangle CF_2 + F_2\text{-}N_2 \xrightarrow[-45°C]{CCl_3F} F_2C\langle SF_2, S\rangle CF_2 \qquad (o)^{29}$$

$$\downarrow F_2\text{-}N_2,\ MeCN,\ -45°\text{–}0°C$$

$$F_2C\langle SF_2, SF_2\rangle CF_2$$

$$F_2C\langle S, SO_2\rangle CF_2 + F_2\text{-}N_2 \xrightarrow[0°C]{MeCN} F_2C\langle SF_2, SO_2\rangle CF_2 \qquad (p)^{30}$$

$$Cl_2C\langle S, SO_2\rangle CCl_2 + F_2\text{-}N_2 \xrightarrow[-25°C]{CCl_3F} Cl_2C\langle SF_2, SO_2\rangle CCl_2 \qquad (q)^{29,30}$$

This method also works for the higher homologs[26], Se and Te. Under these conditions oxidation to hexavalent chalcogenoranes, e.g.:

$$Ph_2SF_2 + F_2\text{-}He \xrightarrow[-78°C]{CFCl_3} Ph_2SF_4 \qquad (r)^{31}$$

is slow. For the oxidation of OSF_2 with F_2, photolytic initiation or higher T[32] (and catalysts[33,34]) are necessary:

$$OSF_2 + F_2 \xrightarrow{\Delta} OSF_4 \qquad (s)^{32-35}$$

$$OSF_2 + F_2 \xrightarrow{h\nu} OSF_4 + O_2SF_2 \qquad (t)^{36,37}$$

when fluorine is replaced by organic groups, oxidation of the sulfur occurs at low T.

$$R_2SO + F_2\text{-}He \xrightarrow[-80°C]{CFCl_3} R_2SOF_2 \ (80\text{–}90\%) \qquad (u)^{38}$$

where R = C_6H_5, p-FC_6H_4;

$$ArSOF + F_2\text{-}He \xrightarrow[-78°C]{CFCl_3} ArSOF_3 \qquad (v)^{39}$$

where if Ar = C_6H_5, the yield is 38%, and if Ar = p-FC_6H_4 the yield is 22%.

Elemental fluorine also adds to low-valent cyclic and acyclic sulfur–nitrogen compounds; e.g., direct fluorination of S_4N_4 cleaves the ring system to smaller units[40], but fluorination with dil F_2 (F_2:He = 1:100) at low T gives the cyclic thiazenes $S_4N_4F_4$ (12%) and $S_3N_3F_3$ (81%)[41]. Even milder is fluorination in the liquid phase with dil F_2 (F_2:He = 1:5)[42]:

$$S_4N_4 + F_2\text{–}N_2 \rightarrow NSF + NSF_3 + SF_5NSF_2 + \text{decomp. products} \qquad (w)^{40}$$

$$S_4N_4 + F_2\text{–He} \rightarrow S_4N_4F_4 + S_3N_3F_3 \qquad (x)^{41}$$

$$S_4N_4 + F_2\text{–He} \xrightarrow{CFCl_3} S_4N_4F_2 + S_4N_4F_4 + S_3N_3F_3 \qquad (y)^{42}$$

Monitoring the reaction by ^{19}F NMR, it is possible to get the primary fluorination product, $S_4N_4F_2$, quantitatively together with traces[42] of $S_4N_4F_4$ and $S_3N_3F_3$. Thionylimides and diimides, the aza analogues of SO_2, add F_2 under more drastic conditions[43–45]:

$$RNSO + F_2 \xrightarrow{h\nu} RNSOF_2 \qquad (z)$$

where R = FSO_2[43], CF_3SO_2[44], CH_3SO_2[45];

$$RNSNR + F_2 \xrightarrow{h\nu} RN{=}SF_2{=}NR \qquad (aa)$$

where R = FSO_2[46], CF_3SO_2[44]. Direct fluorination of sulfur difluorideimides results in oxidative addition and simultaneous S—N bond cleavage:

$$R_fN{=}SF_2 + F_2 \rightarrow R_fNSF_2NR_f + R_fNFSF_5 + R_fNF_2 + SF_6 \qquad (ab)$$

where (R_f = CF_3, C_2F_5[47], FSO_2[48]). If the UV-initiated, gas-phase fluorination is monitored by IR spectroscopy, the primary addition product R_fNSF_4 can be isolated in 30% yield[49]:

$$C_2F_5NSF_2 + F_2 \xrightarrow{h\nu} [C_2F_5NSF_3]^{\cdot} \rightarrow C_2F_5NSF_4$$

$$[C_2F_5NSF_3]^{\cdot} \xrightarrow{-SF_4} C_2F_5NSF_2NC_2F_5 \xrightarrow{F_2} C_2F_5NSF_4NC_2F_5 \xrightarrow[-C_2F_5NF_2]{F_2} C_2F_5NFSF_5$$

$$C_2F_5NSF_4 \xrightarrow{F_2} C_2F_5NFSF_5 \qquad (ac)^{49}$$

but the difluoride diimide is formed simultaneously. Addition of further F_2 is slower.

Fluorination of the S≡N triple bond in NSF_3 at low T gives, along with SF_6 and N_2, the unexpected SF_5NSF_4:

$$2\ NSF_3 + \tfrac{3}{2}\ F_2 \rightarrow SF_5NSF_4 + \tfrac{1}{2}\ N_2 \qquad (ad)^{50}$$

1,3-Fluorine addition is observed with sulfenyl isocyanates[51]:

$$CF_3SNCO + F_2\text{-}N_2 \xrightarrow{-78°C} FC(O)NSFCF_3 \quad (72\%) \tag{ae}$$

Oxidative addition and Si—N bond cleavage is found with silylamino sulfenyl derivatives at −65°C or at −80°C [52]:

$$CF_3SN(SiMe_3)_2 + X_2 \rightarrow XNSXCF_3 + 2\ Me_3SiX \tag{af}$$

where X = F, Cl.

Under matrix conditions the oxidative addition of F_2 is possible even to SCl_2:

$$SCl_2\text{-}Ar + F_2\text{-}Ar \rightarrow SF_2Cl_2 \qquad \text{(ag)}^{53}$$

Oxidative addition of chlorine to low-valent sulfur compounds is rare, but SCl_2 is formed[54,55] from S_2Cl_2 and Cl_2:

$$S_2Cl_2 + Cl_2 \rightleftharpoons 2\ SCl_2 \tag{ah}$$

The dichloride must be stabilized by PCl_3 or PCl_5. From SCl_2 and stoichiometric Cl_2 colorless SCl_4 is formed at −78°C; at −30°C it decomposes to the starting materials:

$$SCl_2 + Cl_2 \underset{-30°C}{\overset{-78°C}{\rightleftharpoons}} SCl_4 \qquad \text{(ai)}^{56}$$

Solid SCl_4 consists of $[SCl_3]^+$ cations bridged by Cl^- as concluded from vibrational spectroscopy. If SCl_2 chlorination is carried out in the presence of lone-pair acceptor acids, trichlorosulfonium salts are formed quantitatively[58]. These $[SCl_3]^+$ salts are indefinitely stable at RT. The chloride ion donor ability of SCl_4 and the stabilizing effect of salt formation is shown by Eq. (ai)–a(l).

$$SCl_2 + Cl_2 \xrightarrow{AsF_3} [SCl_3][AsF_6] \qquad \text{(aj)}^{57,58}$$

$$SCl_2 + Cl_2 \xrightarrow{AlCl_3} [SCl_3][AlCl_4] \qquad \text{(ak)}^{59,60}$$

$$2\ SCl_2 + I_2 + 5\ Cl_2 \rightarrow 2\ [SCl_3][ICl_4] \qquad \text{(al)}^{61}$$

Chlorine will add to the C=S double bond to give sulfenyl derivatives:

$$CS_2 + 6\ Cl_2 \rightarrow SCl_2 + Cl_3CSCl \qquad \text{(am)}^{62}$$

$$CS_2 + 2\ Cl_2 \rightarrow Cl_2C(SCl)_2 \qquad \text{(an)}^{65}$$

$$RR'C{=}S + Cl_2 \rightarrow RR'CClSCl \qquad \text{(ao)}^{62,64}$$

Alkylsulfenylchlorides are further oxidized to S(IV) species:

$$RSCl + Cl_2 \rightarrow RSCl_3 \qquad \text{(ap)}^{65,66}$$

$$R_2S + Cl_2 \rightarrow R_2SCl_2 \qquad \text{(aq)}^{67,68}$$

These unstable solids are ionic, too. With BCl_3 the stable organochlorosulfonium salts $[RSCl_2][BCl_4]$ and $[R_2SCl][BCl_4]$ are formed[69]. The trifluoromethyl dichlorosulfonium salt decomposes at RT[70]:

$$2\ CF_3SCl + Cl_2 + 3\ AsF_5 \xrightarrow{SO_2} 2\ [CF_3SCl_2][AsF_6] + AsF_3 \tag{ar}$$

In trialkylsulfonium derivatives only ionic halogen is present (these compounds are not discussed). Oxochlorosulfuranes can also be isolated; e.g., the chlorination of $[(CF_3)_2CHO]_2S$ gives a stable chlorosulfurane[71]:

$$3\,[(CF_3)_2CHO]_2S + 2\,Cl_2 \rightarrow [(CF_3)_2CHO]_3SCl + SCl_2 \qquad \text{(as)}$$

Especially stable species are formed when the sulfur is in ring systems e.g.,

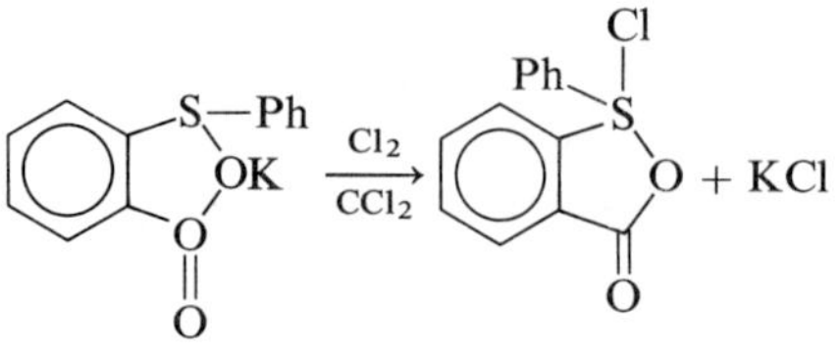

$$\text{(at)}^{72}$$

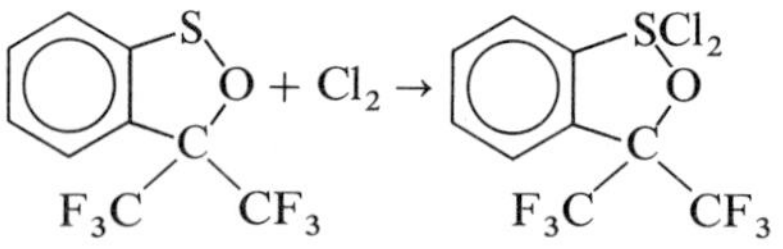

$$\text{(au)}^{73}$$

While cations resists oxidation, even S(IV) anions are oxidized by Cl_2, e.g.:

$$SF_4 + CsF + Cl_2 \rightarrow Cs[SF_5] + Cl_2 \rightarrow CsCl + SF_5Cl \qquad \text{(av)}^{74}$$

$$CF_3SF_3 + CsF + Cl_2 \rightarrow Cs[CF_3SF_4] + Cl_2 \rightarrow CsCl + CF_3SF_4Cl \qquad \text{(aw)}^{75}$$

$$R_fNSO + CsF + Cl_2 \rightarrow Cs[R_fNSOF] + Cl_2 \rightarrow R_fNSOClF + CsCl \qquad \text{(ax)}$$

where R_f = perfluoroalkyl[76], -aryl[77], SF_5[78].

Addition of Cl_2 to sulfur-nitrogen compounds is possible in special systems:

$$R\overset{O}{\overset{\|}{S}}NHR' + Cl_2 \rightarrow R\overset{O}{\overset{\|}{S}}(NR)Cl + HCl \qquad \text{(ay)}^{79}$$

$$Na[(PhSO_2N)_2S(OMe)] + Cl_2 \rightarrow Na[(PhSO_2N)_3SCl] \qquad \text{(az)}^{80}$$

$$(CF_3)_2C{=}N{-}S{-}F + Cl_2 \rightarrow (CF_3)_2CClNSFCl \qquad \text{(ba)}^{81}$$

$$(CF_3)_2C{=}NSR + Cl_2 \xrightarrow[\text{1 day}]{CCl_4/20^\circ C} \underset{(80\text{–}92\%)}{(CF_3)_2CClNSRCl} \qquad \text{(bb)}^{82}$$

where R = Cl, RC_6H_4.

$$S(NSO)_2 + \text{xs } Cl_2 \rightarrow (NSCl)_n(NSOCl)_{3-n} \qquad \text{(bc)}^{83,84}$$

with (n = 0–2);

$$S_4N_4 + Cl_2 \rightarrow S_4N_4Cl_2 \xrightarrow{Cl_2} \{S_4N_4Cl_4\} \rightarrow S_3N_3Cl_3 + \{NSCl\} \qquad \text{(bd)}^{85,86}$$

With S_4N_4, reactions analogous[41] to those with F_2[41] form $S_4N_4Cl_2$[85,86]. Further chlorination cleaves the ring system[87]. A rare example of an oxidative addition to a cation is found with $[NS_2]^+$:

$$[NS_2][AsF_6] + X_2 \rightarrow [N(SX)_2][AsF_5] \qquad \text{(be)}^{88}$$

where (X = Cl, Br). The dibromo cation is unstable and loses halogen under vacuum. the tribromo analogue of the $[SCl_3]^+$ cation is obtained similarly:

$$S_2Br_2 + Br_2 + AlBr_3 \xrightarrow{CH_2Cl_2} [SBr_3][AlBr_4] \qquad (bf)^{89}$$

Chlorobromo sulfonium ions ($[SCl_2Br]^+$, $[SClBr_2]^+$)[89] are also prepared by this method. From cryoscopic and conductance measurements, the formation of the $[S_2Cl_2Br]^+$ cation is deduced[90]:

$$2\ S_2Cl_2 + Br_2 + 6\ H_2S_2O_7 \rightarrow 2\ [S_2Cl_2Br]^+ + SO_2 + 2\ [HS_3O_{10}]^- + 5\ H_2SO_4 \qquad (bg)$$

Similar to the sulfenyl chlorides, the bromides are formed by addition of Br_2 across the C=S double bond:

$$XYC{=}S + Br_2 \rightarrow XYBrCSBr \qquad (bh)$$

where X = Y = F[91], CF_3[92], CF_3S[93]; X = F, Y = Cl[91], SCF_3[93]. Bromosulfonium bromides are formed by the addition to organic sulfides, e.g.[94,95]; pentacoordination is preserved only in special cases, e.g.:

$$(H_3C)_2S + Br_2 \xrightarrow[0°C]{CHCl_3} [(H_3C)_2SBr]Br \qquad (bi)^{94,95}$$

$$(H_3C)_3C{-}C_6H_4{-}S{-}C_6H_3(KOC(CF_3)_2){-}C(CH_3)_3 + Br_2 \rightarrow$$

$$(H_3C)_3C{-}C_6H_3{<}[C(CF_3)_2{-}O{-}S(Br)]{-}C_6H_4{-}C(CH_3)_3 \qquad (bj)^{96}$$

In the presence of electron pair acceptor acids salts are formed[97]:

$$2\ CF_3SBr + Br_2 + 3\ SbF_5 \xrightarrow{SO_2} 2\ [CF_3SBr_2][SbF_6] + SbF_3 \qquad (bk)$$

and S_4N_4 is brominated to $(NSBr)_x$ of unknown structure[98]. Bromine also adds to the $(NS)_x$ polymer[99,100] enhancing superconducting properties [e.g., in $(NSBr_{0.4})_x$[101]] Bromine adds to the $[N_2S_3]^{+}$ radical cation quantitatively:

$$2\ [N_2S_3]^{+\cdot}[AsF_6]^- + Br_2 \xrightarrow{SO_2} 2\ [N_2S_3Br][AsF_6] \qquad (bl)^{102}$$

Halogenation of phosphorus-containing S—N heterocycles gives ring species with S—Br bonds:

$$Ph_2P(N{-}S{-}N)_2(S{-}S)PPh_2 + Br_2 \xrightarrow[23°C]{CH_2Cl_2} Ph_2P(NS(Br)N)_2PPh_2 \qquad (bm)^{103}$$

and S—I bonds:

$$[Ph_2P(N)_2S]_2\text{-type ring } (Ph_2P{=}N{-}S{-}N{=}PPh_2 \text{ bridged via } N, \text{ S}\cdots\text{S}) + I_2 \xrightarrow{CH_2Cl_2} 2\ (Ph_2P)_2N_3S{-}I \qquad (bn)^{103}$$

(R. MEWS)

1. F. A. Cotton, J. W. George, *J. Inorg. Nucl. Chem.*, *7*, 397 (1958).
2. A. C. Gonzalez, H. J. Schumacher, *Z. Naturforsch., Teil B. 36*, 1381 (1981).
3. E. H. Man, US Pat. 2,904,476 (1959); *Chem. Abstr.*, *54*, 117 (1960).
4. J. Nakamura, S. Sato, Japan Kokai 73-16899 (1973); *Chem. Abstr.*, *79*, 55572 (1973).
5. R. R. Smardzewski, W. B. Fox, *J. Chem. Phys.*, *67*, 2309 (1977).
6. *Gmelins Handbook of Inorganic Chemistry, Sulfur, Suppl. 2, Sulfur Halides*, Springer Verlag, Berlin, 1978, p. 102 ff.
7. J. K. Ruff, *J. Org. Chem.*, *32*, 1675 (1967).
8. B. Cochet-Muchy, J. P. Cuer Fr. Pat. 1,445,502 (1965/66); *Chem. Abstr. 66*, 47,928 (1967).
9. A. Haas, H. Willner, *Z. Anorg. Allg. Chem.*, *462*, 57 (1980).
10. P. A. G. O'Hare, W. N. Hubbard, O. Glemser, J. Wegener, *J. Chem. Thermodyn.*, *2*, 71 (1970).
11. J. H. Simons, *Fluorine Chem.*, J. H. Simons, ed., Academic Press, NY, Vol. 1, 1950), p. 414.
12. *Gmelins, Handbook of Inorganic Chemistry, Perfluoroorgano Compounds of the Main Group Elements*, Vol. 2, Verlag Chemie, Weinheim, 1973, p. 167 ff.
13. S. Nagase, *Fluorine Chem. Rev.*, *1*, 77 (1967).
14. A. F. Clifford, H. K. El-Shamy, H. J. Emeléus, R. N. Haszeldine, *J. Chem. Soc.*, 2372 (1953).
15. F. W. Hoffmann, T. C. Simmons, R. B. Beck, H. V. Holler, T. Katz, R. J. Koshar, E. R. Larsen, J. E. Mulvaney, F. E. Rogers, B. Singleton, R. S. Sparks, *J. Am. Chem. Soc.*, *79*, 3424 (1957).
16. R. D. Dresdner, J. A. Young, *J. Am. Chem. Soc.*, *81*, 574 (1959).
17. R. N. Haszeldine, F. Nyman, *J. Chem. Soc.*, 2684 (1956).
18. T. Abe, S. Nagase, H. Baba, *Bull. Chem. Soc. Jpn.*, *46*, 3845 (1973).
19. H. Baba, K. Kodaira, S. Nagase, T. Abe, *Bull. Chem. Soc. Jpn.*, *50*, 2809 (1977).
20. H. Baba, K. Kodaira, S. Nagase, T. Abe, *Bull. Chem. Soc. Jpn.*, *51*, 1891 (1978).
21. E. A. Tyczkowski, L. A. Bigelow, *J. Am. Chem. Soc.*, *75*, 3523 (1953).
22. L. A. Shimp, R. J. Lagow, *Inorg. Chem.*, *11*, 2974 (1977).
23. A. Waterfeld, R. Mews, *J. Fluorine Chem.*, *23*, 325 (1983).
24. R. W. Braun, A. H. Cowley, M. C. Cushner, R. J. Lagow, *Inorg. Chem.*, *17*, 1679 (1978).
25. E. W. Lawless, *Inorg. Chem.*, *9*, 2796 (197).
26. D. L. Chamberlain, Jr., N. Kharasch, *J. Am. Chem. Soc.*, *77*, 1041 (1955).
27. I. Ruppert, *Chem. Ber.*, *112*, 3023 (1979).
28. R. J. Lagow, personal communication, 1982.
29. W. Sundermeyer, M. Witz, *J. Fluorine Chem.*, *26*, 359 (1984).
30. R. Schork, W. Sundermeyer, *Chem. Ber.*, *118*, 1415 (1985).
31. I. Ruppert, *J. Fluorine Chem.*, *13*, 81 (1979).
32. H. Jonas, *Z. Anorg. Allg. Chem.*, *265*, 273 (1951).
33. F. B. Dudley, G. H. Cady, D. F. Eggers Jr., *J. Am. Chem. Soc.*, *78*, 1553 (1958).
34. J. K. Ruff, M. Lustig, *Inorg. Chem.*, *3*, 1422 (1964).
35. J. K. Ruff, *Inorg. Syntheses*, *11*, 131 (1968).
36. C. Vallana, E. Castellano, H. J. Schumacher, *Z. Phys. Chem. (Frankfurt-am-Main)*, *42*, 260 (1964).
37. E. Castellano, H. J. Schumacher, *Z. Phys. Chem. (Frankfurt-am-Main)*, *40*, 51 (1964).
38. I. Ruppert, *Angew. Chem., Int. Ed. Engl.*, *18*, 878 (1979).
39. I. Ruppert, *Chem. Ber.*, *113*, 1047 (1980).
40. B. Cohen, T. R. Hooper, R. D. Peacock, *J. Chem. Soc., Chem. Commun.* 32 (1966); *Nature (London)*, *207*, 748 (1965).
41. N. J. Maraschin, R. J. Lagow, *J. Am. Chem. Soc.*, *94*, 8601 (1972).

42. I. Ruppert, *J. Fluorine Chem.*, *20*, 241 (1982).
43. H. W. Roesky, D. P. Babb, *Inorg. Chem.*, *8*, 1733 (1969).
44. H. W. Roesky, G. Holtschneider, *Z. Anorg. Allg. Chem.*, *378*, 168 (1970).
45. H. W. Roesky, *Inorg. Nucl. Chem. Lett.*, *6*, 759 (1970).
46. H. W. Roesky, D. P. Babb, *Angew. Chem.*, *Int. Ed. Engl.*, *8*, 510 (1969).
47. M. Lustig, J. K. Ruff, *Inorg. Chem.*, *4*, 1444 (1965).
48. H. W. Roesky, *Angew. Chem.*, *Int. Ed. Engl.*, *7*, 630 (1968).
49. I. Stahl, R. Mews, O. Glemser, *Angew. Chem.*, *Int. Ed. Engl.*, *19*, 408 (1980).
50. A. Waterfeld, R. Mews, *Angew. Chem.*, *Int. Ed. Engl.*, *21*, 354 (1982).
51. D. Bielefeld, A. Haas, *Chem. Ber.*, *116*, 1257 (1983).
52. A. Haas, R. Walz, *Chem. Ber.*, *118*, 3248 (1985).
53. R. Minkwitz, M. Nass, J. Sawatzki, *J. Fluorine Chem.*, *31*, 175 (1986).
54. Ref. 6, p. 221 ff.
55. F. Feher, in *Handbook of Preparative Inorganic Chemistry*, G. Brauer, ed., Vol. 1, Academic Press, New York, 1975, p. 371.
56. Ref. 6, p. 386.
57. L. Kolditz, W. Schäfer, *Z. Chem.*, *1*, 124 (1961); *Z. Anorg. Allg. Chem.*, *315*, 35 (1962).
58. L. Kolditz, T. Moya, M. Calov, *Z. Chem.*, *24*, 51 (1984).
59. H. E. Doorenbos, J. C. Evans, R. O. Kagel, *J. Phys. Chem.*, *74*, 3385 (1970).
60. G. Mamantov, R. Marassi, F. W. Poulsen, S. E. Springer, J. P. Wiaux, R. Huglen, N. R. Smyrl, *J. Inorg. Nucl. Chem.*, *41*, 260 (1979).
61. A. Finch, P. N. Gates, T. H. Page, K. B. Dillon, T. C. Waddington, *J. Chem. Soc.*, *Dalton Trans.*, 2401 (1980).
62. E. Kühle, *Synthesis*, 561 (1970).
63. H. M. Pitt, H. Bender, US Pat. 3,331,872 (1967); *Chem. Abstr.*, *67*, 63,784 (1967).
64. A. Haas, U. Niemann, *Adv. Inorg. Chem. Radiochem.*, *18*, 143 (1976).
65. I. B. Douglass, R. V. Norton, R. L. Weichman, R. B. Clarkson, *J. Org. Chem.*, *34*, 1803 (1969).
66. E. N. Givens, H. Kwart, *J. Am. Chem. Soc.*, *90*, 386 (1968).
67. G. E. Wilson, M. M. Y. Chang, *J. Am. Chem. Soc.*, *96*, 7533 (1974).
68. N. C. Baenziger, R. E. Buckles, R. J. Mauer, T. D. Simpson, *J. Am. Chem. Soc.*, *91*, 5749 (1969).
69. M. E. Peach, *Can. J. Chem.*, *47*, 1675 (1969).
70. R. Minkwitz, M. Nass, A. Radúnz, H. Preut, *Z. Naturforsch.*, *Teil B*, *40*, 1123 (1985).
71. G. V. Röschenthaler, *Angew. Chem.*, *Int. Ed. Engl.*, *16*, 862 (1977).
72. P. Livant, J. C. Martin, *J. Am. Chem. Soc.*, *99*, 5761 (1977).
73. L. D. Martin, E. F. Perozzi, J. C. Martin, *J. Am. Chem. Soc.*, *101*, 3595 (1979).
74. C. W. Tullock, D. D. Coffman, E. L. Muetterties, *J. Am. Chem. Soc.*, *86*, 357 (1964).
75. J. I. Darragh, G. Haran, D. W. A. Sharp, *J. Chem. Soc.*, *Dalton Trans.*, 2289 (1973).
76. R. Mews, P. Kricke, I. Stahl, *Z. Naturforsch.*, *Tei B*, *36*, 1093 (1981).
77. T. Abe, J. M. Shreeve, *Inorg. Chem.*, *19*, 3063 (1980).
78. P. Kricke, I. Stahl, R. Mews, O. Glemser, *Chem. Ber.*, *114*, 3467 (1981).
79. E. U. Jonsson, C. C. Bacon, C. R. Johnson, *J. Am. Chem. Soc.*, *93*, 5306 (1971).
80. H. W. Roesky, W. Schmieder, W. Isenberg, W. S Sheldrick, G. M. Sheldrick, *Chem. Ber.*, *115*, 2714 (1982).
81. J. Varwig, R. Mews, *J. Chem. Res.*, (M), *1*, 2744 (1977); (S), *1*, 245 (1977).
82. Yu. G. Shermolovich, V. S. Talanov, V. V. Pirozhenko, L. N. Markorskii, *J. Org. Chem. USSR*, *18*, 2240 (1982).
83. D. Schläfer, M. Becke-Goehring, *Z. Anorg. Allg. Chem.*, *362*, 1 (1968).
84. J. Weiss, R. Mews, O. Glemser, *J. Inorg. Nucl. Chem.*, *Suppl.* 213 (1976).
85. L. Zbořilová, P. Gebauer, *Z. Chem.*, *19*, 32 (1979).
86. L. Zbořilová, P. Gebauer, *Z. Anorg. Allg. Chem.*, *448*, 5 (1979).
87. A. Meuwsen, *Ber. Deutsch. Chem. Ges.*, *64*, 2311 (1931).
88. W. V. F. Brooks, G. K. MacLean, J. Passmore, P. S. White, C. M. Wong, *J. Chem. Soc.*, *Dalton Trans.*, 1961 (1983).
89. H. F. Askew, P. N. Gates, *J. Chem. Res.*, (S), 116 (1980).
90. A. Bali, K. C. Malhotra, *Aust. J. Chem.*, *28*, 983 (1975).
91. N. N. Yarovenko, A. S. Vasil'eva, *J. Gen. Chem. USSR* (*Engl. Transl.*), *29*, 3754 (1959).
92. W. J. Middleton, W. H. Sharkey, *J. Org. Chem.*, *30*, 1384 (1965).
93. A. Haas, W. Klug, *Chem. Ber.*, *101*, 2609 (1968).
94. F. Boberg, G. Winter, G. R. Schultze, *Chem. Ber.*, *89*, 1160 (1956).

95. For further examples cf. P. S. Magee, in *Sulfur in Inorganic and Organic Chemistry*, A. Senning, ed., Marcel Dekker, New York, Vol. 1, 261 (1971); Vol. 4, 283 (1982).
96. J. C. Martin, E. F. Perozzi, *J. Am. Chem. Soc.*, *96*, 3155 (1974).
97. R. Minkwitz, R. Lekies, H. Preut, *Z. Naturforsch., Teil B*, *42*, 1227 (1987).
98. A. Clever, W. Muthmann, *Ber. Deutsch. Chem. Ges.*, *24*, 340 (1896).
99. C. Bernard, A. Herold, M. Le Laurain, G. Robert, *C.R. Hebd. Seances Acad. Sci., Ser. C*, *283*, 625 (1976).
100. M. Akhtar, C. K. Chiang, A. J. Heeger, A. G. MacDiarmid, *J. Chem. Soc., Chem. Commun.*, 846 (1977).
101. J. F. Kwak, R. L. Greene, W. W. Fuller, *Phys. Rev. B, Condens. Matter*, *20*, 2658 (1979).
102. D. K. Padma, R. Mews, *Z. Naturforsch., Teil B 42*, 699 (1984).
103. N. Burford, T. Chivers, M. N. S. Rao, J. F. Richardson, *Inorg. Chem.*, *23*, 1946 (1984).

2.3.11.1.2. to Give the Se–Halogen Bond.

Oxidative addition of F_2 to SeF_4 gives[1,2] SeF_6, as in the sulfur system:

$$SeF_4 + F_2 \rightarrow SeF_6 \qquad (a)^{1,2}$$

From $OSeF_2$ and F_2 in the presence of KF, the pentafluorohypofluorite is formed:

$$OSeF_2 + F_2 \xrightarrow{KF} F_5SeOF \qquad (b)^{3,4}$$

$$OSeF_2 + F_2 + HF \rightarrow F_5SeOH \qquad (c)^{5}$$

The last reaction can only be controlled with difficulty and explosions can occur[5]. The oxidative addition of F_2 to Se_2Cl_2[6], $SeOCl_2$[7] and SeO_2[7–10] leads to Se—Cl and Se—O bond cleavage and is discussed in (§2.3.10, §2.3.6 and §2.3.7). Uranium selenides, USe_n ($n = 1 - 3$)[11], and Li_2Se[12] give SeF_4 and SeF_6, depending on the conditions. Dialkyl- and diarylselenium difluorides are prepared in the low-T liquid-phase fluorination:

$$RR'Se + F_2 \rightarrow RR'SeF_2 \qquad (d)^{13}$$

where if $R = Ph$ and $R' = CH_3$, yield is (53%), and if $R = R' = Ph$, yield is 86%. Perfluoroethylselenium trifluoride adds F_2 in the presence of CsF to give the pentafluoride[14]:

$$C_2F_5SeF_3 \cdot CsF + F_2 \xrightarrow[-CsF]{} C_2F_5SeF_5 \qquad (e)^{14}$$

whereas CF_3SeF_5 is formed from the trifluoride and liq F_2 in low yield[15]:

$$CF_3SeF_3 + F_2 \xrightarrow{-196°C} CF_3SeF_5\ (+CF_4, SeF_6, SeF_4) \qquad (f)^{15}$$

The unstable $(CF_3)_2SeF_4$ is also formed under these conditions[15]:

$$(CF_3)_2SeF_2 + F_2 \xrightarrow[100h]{-196°C} (CF_3)_2SeF_4(?) \qquad (g)^{15}$$

Selenium(II) halides are unstable, but anionic and cationic derivatives form from oxidative addition, e.g.:

$$[SeCN]^- + Br_2 \rightarrow [SeBr_2CN]^- \qquad (h)^{16}$$

$$Cl_2Se_2 + Cl_2 \xrightarrow[HSO_3Cl]{H_2S_2O_7} [Cl_3Se_2]^+ \qquad (i)^{17,18}$$

$$Cl_2Se_2 + Br_2 \xrightarrow[HSO_3Cl]{H_2S_2O_7} [Cl_2BrSe_2]^+ \qquad (j)^{17,18}$$

$$Br_2Se_2 + Br_2 \xrightarrow{H_2S_2O_7} [Br_3Se_2]^+ \qquad (k)^{17}$$

Alkylammonium dibromocyanoselenates(II) are prepared from selenocyanates and Br_2, whereas chlorination is achieved[16] by O_2SCl_2. In $H_2S_2O_7$ or HSO_3Cl containing Cl_2 or Br_2, Se_2X_2 forms the cations $[Cl_3Se_2]^+$, $[Cl_2BrSe_2]^+$ and $[Br_3Se_2]^+$, which disproportionate[17,18] to $[SeX_3]^+$ and Se. Inorganic and organic Se(IV)-Cl and -Br derivatives are more stable than their sulfur(IV) analogues, whereas S(II) species are easier isolated. Diselenium dichlorides[19] and dibromides[20] are oxidized to the tetrahalides. The dihalides exist in the gas phase[22] or in equilibrium in solution[23]:

$$Se_2X_2 + X_2 \rightarrow [2\ SeX_2] \xrightarrow{2\ X_2} 2\ SeX_4 \qquad (l)^{19,20}$$

$$Se_2X_2 + 3\ Y_2 \xrightarrow{CS_2} 2\ SeXY_3 \qquad (m)^{21}$$

where if X = Cl, Y = Br; if X = Br, Y = Cl. In the chlorination of organoselenium derivatives, aside from oxidative addition, C—Se bond cleavage is often observed, e.g.:

$$CSe_2 + Cl_2 \rightarrow CCl_3SeCl + SeCl_4 \qquad (n)^{24}$$

$$RSeCN + Cl_2 \rightarrow RSeCl_3 + ClCN \qquad (o)^{25}$$

$$RSeSeR + X_2 \rightarrow [RSeX] \xrightarrow{X_2} RSeX_3 \qquad (p)$$

where X = Cl, Br[27,28];

$$RSeR' + X_2 \rightarrow RSeX_2R' \qquad (q)^{28,29}$$

where X = Cl, Br, I.

Selenium trihalides; $RSeX_3$ (X = Cl, Br), form directly from diselenides and elemental halogen, the expected seleninyl halides are isolated only with difficulty. The diorganylselenium dihalides, $RSeX_2R$, (X = Cl, Br, I), are prepared from the diorganylselenides. Chlorine, Br_2 or I_2, in Et_2O, CCl_4 or $CHCl_3$, converts the selenides quantitatively into the dihalides. The reaction can be used to purify the selenides via their dihalides, which are reduced back to the parent selenides[29].

$$(CF_3)_2Se + Cl_2 \rightarrow CF_3SeCl_3 + CF_3Cl \qquad (r)^{30}$$

In $(CF_3)_2Se$ one Se—C bond is cleaved and only the trihalide is isolated[30]; aryltrifluoromethylselenium dichlorides are formed:

$$CF_3SePh + Cl_2 \rightarrow CF_3SeCl_2Ph \qquad (s)^{31}$$

Chlorine, but not Br_2, forms a trichloroderivative[32]:

$$\mathrm{RC}(\!=\!\mathrm{O}\rightarrow\mathrm{SeCl})\mathrm{CH}\!=\!\mathrm{C}(\mathrm{Me})\text{—} + \mathrm{Cl_2} \xrightarrow[0^\circ\mathrm{C}]{\mathrm{CH_2Cl_2}} \mathrm{RC}(\!=\!\mathrm{O}\rightarrow\mathrm{SeCl_3})\mathrm{CH}\!=\!\mathrm{C}(\mathrm{Me})\text{—} \quad \text{(t)}$$

where R = p-$MeOC_6H_4$.

The structure of $[Se_4I_4]^{2+}$ from ^{77}Se NMR is $[I_2SeSeSeSeI_2]^{2+}$.

$$[\mathrm{Se_4}][\mathrm{AsF_6}]_2 + 2\,\mathrm{I_2} \xrightarrow{\mathrm{SO_2}} [\mathrm{Se_4I_4}][\mathrm{AsF_6}]_2 \quad \text{(u)}^{33}$$

(R. MEWS)

1. A. A. Banks, Br. Pat. 713,119 (1954).
2. J. Bousquet, J. Carre, P. Claudy, M. Kollmannsberger, J. Thourey, P. Barberi, *J. Calorim. Anal. Therm.* (*Prepr.*), *8*, I-83 (1977).
3. J. E. Smith, G. H. Cady, *Inorg. Chem.*, *9*, 1442 (1970).
4. K. Seppelt, *Chem. Ber.*, *106*, 157 (1973).
5. K. Seppelt, *Angew. Chem., Int. Ed. Engl.*, *11*, 630 (1972).
6. P. C. Goggin, *J. Inorg. Nucl. Chem.*, *28*, 661 (1966).
7. G. Mitra, G. H. Cady, *J. Am. Chem. Soc.*, *81*, 2646 (1959).
8. W. L. Reichert, G. H. Cady, *Inorg. Chem.*, *12*, 769 (1973).
9. G. Mitra, *J. Indian Chem. Soc.*, *37*, 804 (1960).
10. J. E. Smith, G. H. Cady, *Inorg. Chem.*, *9*, 1293 (1970).
11. P. Khodadad, *Bull. Soc. Chim. Fr.*, 133 (1961).
12. M. Ader, *J. Chem. Thermodyn.*, *6*, 587 (1974).
13. I. Ruppert, *Chem. Ber.*, *112*, 3023 (1979).
14. C. Lau, J. Passmore, *J. Fluorine Chem.*, *7*, 261 (1970).
15. A. Haas, H.-U. Weiler, *Chem. Ber.*, *118*, 943 (1985).
16. K. J. Wynne, J. Golden, *Inorg. Chem.*, *13*, 185 (1974).
17. A. Bali, K. C. Malhotra, *J. Inorg. Nucl. Chem.*, *39*, 957 (1977).
18. R. C. Paul, D. Konwer, J. K. Puri, *Indian J. Chem.*, *A21*, 81 (1982).
19. F. Fehér in *Handbook of Preparative Inorganic Chemistry*, G. Brauer, ed., Vol. 1, Academic Press, New York, 1975, p. 424.
20. *Gmelin's Handbook of Inorganic Chemistry*, *Selenium*, Vol. B2, Springer Verlag, Berlin, 1984, p. 238ff.
21. G. C. Hayward, P. J. Hendra, *J. Chem. Soc.*, *A*, 643 (1967).
22. M. Lundkvist, M. Lellep, *Acta Chem. Scand.*, *22*, 291 (1968).
23. G. V. Golubkova, E. S. Petrov, A. N. Kanev, *Izv. Sib., Otd. Akad. Nauk SSSR, Ser. Khim. Nauk*, 51 (1976); *Chem. Abstr.*, *84*, 173,074 (1976).
24. D. J. G. Ives, R. W. Pittman, W. Wardlaw, *J. Chem. Soc.*, 1080 (1947).
25. H. Rheinboldt, in *Houben-Weyl Methoden der Organischen Chemie*, Vol. IX, *S, Se and Te Verbindungen*, E. Müller, O. Bayer, H. Meerwein, K. Ziegler, eds., 4th ed., Georg Thieme Verlag, Stuttgart, 1955, p. 42.
26. Ref. 25, pp. 1130, 1132.
27. K. J. Wynne, J. W. George, *J. Am. Chem. Soc.*, *91*, 1649 (1969).
28. Ref. 25, p. 1011ff.
29. K. J. Irgolic, M. V. Kudchadker, in *Selenium*, R. A. Zingaro, W. C. Cooper, eds., Van Nostrand-Reinhold, New York, 1974, pp. 497ff,
30. J. W. Dale, H. J. Emeléus, R. N. Haszeldine, *J. Chem. Soc.*, 2939 (1958).
31. L. M. Yagupol'skii, V. G. Voloshchuk, *J. Gen. Chem. USSR* (*Engl. Transl.*), *37*, 1543 (1967).
32. M. R. Detty, H. R. Luss, J. M. McKelvey, S. M. Geer, *J. Org. Chem.*, *51*, 1692 (1986).
33. M. M. Carnell, F. Grein, M. Murchie, J. Passmore, C.-M. Wong, *J. Chem. Soc. Chem. Commun.*, 225 (1986).

2.3.11.1.3 to Give the Te– and Po–Halogen Bond.

Lower halides of Te are oxidized by F_2 at high T to give[1] TeF_6. Under more controlled conditions (dil F_2 in N_2, 25°C, flow system), one of the halogen bonds in TeX_4 (X = Cl, Br) is preserved, but amounts of TeF_6 are formed:

$$TeX_4 + \tfrac{5}{2} F_2 \rightarrow TeF_5X + \tfrac{3}{2} X_2 \qquad (a)^2$$

$$2\ TeF_4 + F_2 + Br_2 \rightarrow \underset{(83\%)}{2\ TeF_5Br} \qquad (b)^3$$

A convenient synthesis for TeF_5Br is the simultaneous action of F_2 and Br_2 on TeF_4.

In the fluorine-oxygen system several F_5TeO– derivatives were characterized:

$$Te(OTeF_5)_4 + F_2 \rightarrow cis\text{-}F_2Te(OTeF_5)_4 + trans\text{-}F_2Te(OTeF_5)_4 + FTe(OTeF_5)_5 \qquad (c)^{4,5}$$

Direct fluorination of diorganoselenides is possible in the liquid phase at low T, e.g.[6,7]:

$$RR'Te + F_2\text{–}Ar \xrightarrow[-78°C]{CFCl_3} RR'TeF_2 \qquad (d)$$

where if R = R' = CH_3, yield is 22%[6], = Ph, yield is 94%[6]; = CF_3[7]; if R = CH_3 and R' = Ph, yield is 53%[6]. Chlorine oxidizes $TeCl_2$[8], and $TeBr_2$[9], to give the tetrahalides (cf. §2.3.2.1.4, about the existence of the dihalides):

$$TeCl_2 + Cl_2 \rightarrow TeCl_4 \qquad (e)^9$$

$$TeBr_2 + Cl_2 \rightarrow TeCl_4 + Br_2 \qquad (f)^{10}$$

Liquid Br_2 oxidizes the dihalides to give $TeBr_4$[9] and $TeCl_2Br_2$[8]:

$$TeBr_2 + Br_2 \rightarrow TeBr_4 \qquad (g)^{10}$$

$$TeCl_2 + Br_2 \rightarrow TeCl_2Br_2 \qquad (h)^9$$

Ether solutions of $TeBr_2$ and I_2 give garnate-red $TeBr_2I_2$[9]:

$$TeBr_2 + I_2 \rightarrow TeBr_2I_2 \qquad (i)^{10}$$

From metal tellurides and Cl_2, $TeCl_4$ forms[11] and $PoCl_2Br_2$, $PoCl_2I_2$ and $PoBr_2I_2$ are obtained similarly[12,13].

Telluride dihalides are prepared from tellurides and halogens in ether, $CHCl_3$ or CCl_4:

$$R_2A + X_2 \rightarrow R_2TeX_2 \qquad (j)$$

where if A = Te, R = alkyl, aryl; X = Cl, Br, I_2[14,15]; if A = Po, R = p-$CH_3C_6H_4$, X = Br, I[16];

$$(CF_3)_2Te + X_2 \xrightarrow{CFCl_3} (CF_3)_2TeX_2 \xrightarrow{X_2} CF_3TeX_3 + CF_3X \qquad (k)^{17}$$

where X = Cl, Br. The chlorides and bromides are isolated in high yields, but with xs halogen C—Te bond cleavage is observed[17].

The trihalides are isolated similarly from ditellurides and elemental halogen[18,19]:

$$RTeTeR + 3\ X_2 \xrightarrow{CCl_4 \text{ or } CHCl_3} 2\ RTeCl_3 \qquad (l)$$

where X = Cl, Br, I. The expected tellurenyl halides are not stable, being oxidized further immediately. Further additions are known:

$$\mathrm{RC(O{-}Te{\leftarrow}O{=}CAr)(=CH{-}C{=}CH{-})} + X_2 \rightarrow \mathrm{RC(O{-}TeX{\leftarrow}O{=}CAr)(=CH{-}C{=}CH{-})} \qquad (m)^{20}$$

$$\mathrm{H_2C(CH_2{-}TeX)(C(Ar){=}O{\rightarrow})} + X_2 \rightarrow \mathrm{H_2C(CH_2{-}TeX_3)(C(Ar){=}O)} \qquad (n)^{21}$$

where X = Cl, Br.

(R. MEWS)

1. *Gmelins Handbuch der Anorganischen Chemie, Te*, Suppl. Band. 2, Springer Verlag, Berlin, 1977, p. 19.
2. G. W. Fraser, R. D. Peacock, P. M. Watkins, *J. Chem. Soc., Chem. Commun.*, 1257 (1968).
3. L. Lawlor, J. Passmore, *Inorg. Chem., 18*, 2921 (1979).
4. D. Lentz, H. Pritzkow, K. Seppelt, *Angew. Chem., Int. Ed. Engl., 16*, 729 (1977).
5. D. Lentz, H. Pritzkow, K. Seppelt, *Inorg. Chem., 17*, 1926 (1978).
6. I. Ruppert, *Chem. Ber., 112*, 3023 (1979).
7. D. Naumann, S. Herberg, *J. Fluorine Chem. 19*, 205 (1982).
8. G. Klein, D. Naumann, *J. Fluorine Chem., 30*, 259 (1985).
9. E. E. Aynsley, *J. Chem. Soc.*, 3016 (1953).
10. E. E. Aynsley, R. H. Watson, *J. Chem. Soc.*, 2603 (1955).
11. E. Montiguie, *Bull. Soc. Chim. Fr.*, 748 (1947).
12. K. W. Bagnall, R. W. M. D'Eye, J. H. Freeman, *J. Chem. Soc.*, 3959 (1955).
13. K. W. Bagnall, R. W. M. D'Eye, J. H. Freeman, *J. Chem. Soc.*, 3385 (1956).
14. K. J. Irgolic, in *The Organic Chemistry of Tellurium*, Gordon and Breach, New York, 1974, pp. 143, 280ff.
15. H. Rheinboldt, in *Houben-Weyl Methoden der Organischen Chemie*, Vol. IX, *S, Se und Te Verbindungen*, E. Müller, ed., Georg Thieme Verlag, Stuttgart, 1955, pp. 1064, 1065.
16. V. D. Nefedov, V. E. Zhuravlev, M. A. Toropova, S. A. Grachev, A. V. Levchenko, *J. Gen. Chem. USSR (Engl. Transl.), 35*, 1440 (1965).
17. S. Herberg, D. Naumann, *Z. Anorg. Allg. Chem., 494*, 151 (1982).
18. Ref. 13, p. 78.
19. Ref. 14, p. 1156.
20. M. R. Detty, H. R. Luss, *J. Org. Chem., 48*, 5149 (1983).
21. M. R. Detty, H. R. Luss, M. McKelvey, S. M. Geer, *J. Org. Chem., 51*, 1692 (1986).

2.3.11.2. with Hydrogen Halides.

No examples of this type of reaction are known.

(R. MEWS)

2.3.11.3. with Nonmetal Halides

2.3.11.3.1. to Give the Sulfur–Halogen Bond.

Nonmetal halides, e.g., noble-gas fluorides, interhalogens, oxygen fluorides, hypohalites, nitrogen halides and group-VB pentahalides undergo oxidative addition to group-VIB compounds. The most useful reagents for oxidative fluorinations are ClF and CF_3OF, but the synthetic potential of noble-gas fluorides, has not yet been realized.

In group-VIB chemistry KrF_2 is not used as a fluorinating agent, and although XeF_6 and XeF_4 are more powerful fluorinating agents than XeF_2 (e.g., SF_4 is oxidized by XeF_4 at 23°C[1]), the latter is safer to handle and its oxidizing power can be enhanced by electron-pair acceptor acid catalysts, e.g., BF_3 or HF. These polarize the Xe—F bond, $F—\overset{\delta+}{Xe}\cdots\overset{\delta-}{F}\cdots HF$, leading to stronger electrophilicity[2].

From arylsulfides, difluorosulfuranes are formed; e.g.:

$$PhSR + XeF_2 \rightarrow PhSF_2R + Xe \qquad (a)^3$$

where R = Ph, R_3C;

$$p\text{-}RC_6H_4SCF_3 + XeF_2 \xrightarrow{HF} p\text{-}RC_6H_4SF_2CF_3 + Xe \qquad (b)^4$$

where R = H, Cl, NO_2;

$$(H_3C)_2S + XeF_2 \xrightarrow{-Xe,\ -HF} FCH_2SCH_3 \rightarrow [(H_3C)_2SCH_2SCH_3][F(HF)_n] \qquad (c)^5$$

$$(H_3C)_2S + XeF_2 \xrightarrow[-23°C]{HF} [(H_3C)_2SF][F(HF)_n] \qquad (d)^5$$

In liq HF the fluorination of $(CH_3)_2S$ gives fluorosulfonium ions. With BF_3 or AsF_5, stable tetrafluoroborates or hexafluoroarsenates are formed. Neutral difluorosulfuranes decompose to give fluoroalkylsulfides as subsequent products (Eq. c).

In alkylsulfides with substituents having α-hydrogen atoms α-fluorination is observed along with olefin formation in some cases. Bisaryl- or aryltrifluoromethylsulfides give the desired products in quantitative yield in the presence of HF as catalyst[4].

In sulfur–nitrogen chemistry, XeF_2 is a useful reagent. Because of its low polarity, only little S—N bond cleavage occurs, in contrast to the halogen fluorides. Without catalysts, S(II) is oxidized to S(IV). With BF_3, S(VI) species are also formed[6–8], but carbon–sulfur bonds are attacked, too, e.g.:

$$(CF_3)_2C{=}NSX + XeF_2 \rightarrow (F_3C)_2\overset{F}{\overset{|}{C}}{-}N{=}S\begin{matrix}\diagup F\\ \diagdown X\end{matrix} + Xe \qquad (e)^{6,7}$$

where X = F[6], Cl[6], NCO[7], i-C_3F_7[8], $CCl(CF_3)_2$[8];

$$(CF_3)_2C{=}NSN{=}C(CF_3)_2 + XeF_2 \rightarrow (CF_3)_2\overset{F}{C}N{=}S{=}N\overset{F}{C}(CF_3)_2 \qquad (f)^6$$

$$(CF_3)_2C{=}NSN{=}C(CF_3)_2 + 2\ XeF_2 \xrightarrow{BF_3} (CF_3)_2\overset{F}{C}N{=}SF_2{=}N\overset{F}{C}(CF_3)_2 \qquad (g)^6$$

$$CF_3N{=}S{=}O + XeF_2 \rightarrow CF_3NSOF_2 + Xe \qquad (h)^9$$

$$(CF_3)_2C{=}N\overset{O}{\overset{\|}{S}}CF_3 + XeF_2 \rightarrow (CF_3)_2\overset{F}{C}N{=}\overset{O}{\overset{\|}{S}}\begin{matrix}\diagup F\\ \diagdown CF_3\end{matrix} + (CF_3)_2CFNSOF_2 + Xe + CF_4 \qquad (i)^{10}$$

The $[NS_2]^+$ cation can be fluorinated:

$$[NS_2][AsF_6] + 2\ XeF_2 \xrightarrow[\substack{1.\ 0°C,\ 0.5\ h \\ 2.\ 20°C,\ 12\ h}]{\text{liq } SO_2F_2} [N(SF_2)_2][AsF_6] + 2\ Xe \qquad (j)^{11}$$

$$CF_3CN_3S_2 + XeF_2 \rightarrow CF_3CN_3S_2F_2 \qquad (k)^{12}$$

Sulfites are oxidized to give fluorosulfates:

$$(RO)_2SO + XeF_2 \xrightarrow[3\ h]{-196°C \text{ to } +20°C} \underset{(60\%)}{ROSO_2F} + RF + Xe \qquad (l)^{13}$$

$$(RO)_2SO + 2\ ClF \xrightarrow[\text{then } +20°C/12\ h]{-196°C \text{ to } -78°C/2\text{-}3\ h} RSO_2F + RF + Cl_2 \qquad (m)^{13}$$

where R = CF_3CH_2 (98–100%), $(CF_3)_2CH$ (80%), C_2H_5 (45%).

Because of their oxidizing power, halogen fluorides are used for oxidative additions. With ClF, fluorination or chlorofluorination takes place and bromofluorination with 1:1 BrF_3:Br_2. However, IF_5 acts only as a fluorinating agent.

Metal sulfides are burned quantitatively by BrF_3 to give SF_6 and metal fluorides[14]. With sulfur chlorides and bromides, halogen exchange is the first step, oxidative additions than follows[15]. The products obtained from SF_4 and ClF, depending on the T and mol ratio, are[16–22] either SF_6 or SF_5Cl:

$$SF_4 + ClF \xrightarrow{(CsF)} SF_5Cl \qquad (n)^{16-18,21,22}$$

$$SF_4 + 2\ ClF \rightarrow SF_6 + Cl_2 \qquad (o)^{19}$$

$$2\ SF_4 + ClF_3 \rightarrow SF_6 + SF_5Cl \qquad (p)^{16,17}$$

$$2\ SF_4 + 2\ ClF \xrightarrow{\text{CW-laser}} S_2F_{10} + Cl_2 \qquad (q)^{23}$$

With CsF, high yields of SF_5Cl are obtained. Almost 1:1 SF_6 and SF_5Cl are isolated when 2:1 SF_4 and ClF_3 are heated to 120°–180°C in Cu vessels. The laser-induced process yields[23] S_2F_{10} and Cl_2. However, BrF_3 does not react until 300°C [24], whereas a BrF_3–Br_2 mixture gives SF_5Br in 39%[25,26] or <10%[27] yield:

$$SF_4 + BrF_5\text{–}Br_2\ 10{:}1 \xrightarrow[5\ days]{100°C} SF_5Br \qquad (r)$$

The reaction is catalyzed by CsF, with $Cs[SF_5]$ as an intermediate[28]. When OSF_2 is fluorinated by ClF[29] or BrF_5[30], OSF_4 is formed in good yield:

$$OSF_2 + 2\ ClF \rightarrow OSF_4 + Cl_2 \qquad (s)$$

$$OSF_2 + BrF_5 \rightarrow OSF_4 + BrF_3 \qquad (t)$$

In Eq. (p) Cl_2 is removed from OSF_4 with Hg at RT. These reactions can be applied if no F_2 is available. No mixed halide is detected among the products. Under matrix conditions chlorofluoro sulfuranes are obtained[31]:

$$SCl_2 + ClF \rightarrow SCl_3F, SCl_2F_2, SClF_3, SF_4 \qquad (u)$$

In the presence of electron pair acceptor acids, ClF acts as a chlorinating agent:

$$CF_3SCl + AsF_3 + ClF \rightarrow [CF_3SCl_2][AsF_6] \qquad (v)^{32}$$

$$OSF_2 + [Cl_2F]X \rightarrow [OSClF_2]X + ClF \qquad (w)$$

where X = AsF_6, PF_6, SbF_6, Sb_2F_{11}. Mass spectra of these salts are consistent with initial dissociation into $OSClF_3$ and the corresponding pentafluoride[34].

Halogen halides in perfluoroorganosulfur chemistry add across the C=S double bond to form sulfenyl halides[35,36]:

$$RR'C{=}S + ClF \rightarrow RR'C(F)SCl \qquad (x)^{36}$$

where if R = F and R' = NCS the yield is 25%; if R' is CF_3S the yield is 30%; if R = R' = CF_3S the yield is 32%. At low T ClF is a mild and selective fluorinating agent in organosulfur and perfluoroorganosulfur chemistry. With perfluorinated sulfides at low T, S(II) is oxidized to S(IV) derivatives; at RT S(VI) species form exclusively, e.g.[37–42]:

$$ClF + \overline{CF_2{-}CF_2{-}CF_2{-}CF_2{-}S} \rightarrow \overline{CF_2{-}CF_2{-}CF_2{-}CF_2{-}SF_2} + \overline{CF_2{-}CF_2{-}CF_2{-}CF_2{-}SF_4} \qquad (y)$$

>4:1	$\xrightarrow[8\ h]{-40°C}$	(74%)	(13.7%)
>4:1	$\xrightarrow[12\ h]{25°C}$	(7.8%)	(63.2%)

$$ClF + CF_3{-}S{-}CF_3 \rightarrow CF_3SF_2CF_3 + CF_3{-}SF_4{-}CF_3 \qquad (z)$$

>4:1	$\xrightarrow[12\ h]{-78°C}$	(>99%)	(0%)
>4:1	$\xrightarrow[12\ h]{25°}$	(0%)	(42%)

In the same way $CF_3SF_2R_f$ (R_f = C_2F_5, C_3F_7)[39], is prepared[41] under mild conditions (−78°C, CF_2Cl_2 as a solvent), and similar results are obtained with[43] ClF_3. **The reactions of organic sulfides with ClF or ClF_3 are potentially hazardous, and violent explosions occur**[41]. Fluorination of $F_2C{=}S$ or tetrafluorodithietane with IF_5 at 150°–200°C forms the S(IV) derivative CF_3SF_3 in 65% and 55% yield, respectively[44]. The dithietane structure is not degraded with ClF under mild conditions[45]; S(VI) derivatives are isolated:

$$S(CF_2)_2S \xrightarrow[0°C,\ 12\ h]{xs\ ClF} Cl_2F_2S(CF_2)_2SCl_2F_2\ 58\% \xrightarrow[25°C,\ 12\ h]{xs\ ClF} F_4S(CF_2)_2SF_4 \qquad (aa)$$

Alkylperfluoroalkylsulfides are oxidized to S(VI) derivatives[46,47], perfluorosulfoxides give oxydifluorides[48]:

$$CF_3SMe + xs\ ClF \rightarrow CF_3SF_4Me \tag{ab}$$

$$(CF_3S)_2CH_2 + xs\ ClF \rightarrow (CF_3SF_4)_2CH_2 + CF_3SF_4CH_2F \tag{ac}$$

$$CF_3SEt + xs\ ClF \rightarrow trans\text{-}CF_3SF_4Et \tag{ad}$$

$$R_fR_f'S{=}O + 2\ ClF \xrightarrow{-78^\circ C} R_fR_f'S(O)F_2 + Cl_2 \tag{ae}$$

where $R_f = R_f' = CF_3, C_2F_5$; or $R_f = CF_3$, $R_f' = C_2F_5$. Oxidative additions of ClF to sulfenylchlorides[49], disulfides[49] (with primary cleavage of the S—S bond (see §2.3.8) and sulfur trifluorides[50] forms perfluoroalkylsulfur tetrafluoride chlorides:

$$R_fSCl + ClF \xrightarrow{25^\circ C} R_fSF_4Cl \tag{af}$$

where $R_f = CF_3, C_2F_5, n\text{-}C_3F_7, n\text{-}C_4F_9$[49];

$$R_fSSR_f' + ClF \rightarrow R_fSF_4Cl + R_f'SF_4Cl \tag{ag}$$

where ($R = R_f' = CF_3$; $R_f = CF_3$, $R_f' = C_2F_5$[50];

$$CF_2(SF_3)_2 + ClF \rightarrow CF_2(SF_4Cl)_2 \tag{ah}$$

Difluoropersulfuranes are prepared from the corresponding sulfuranes via BrF_3[51,52]:

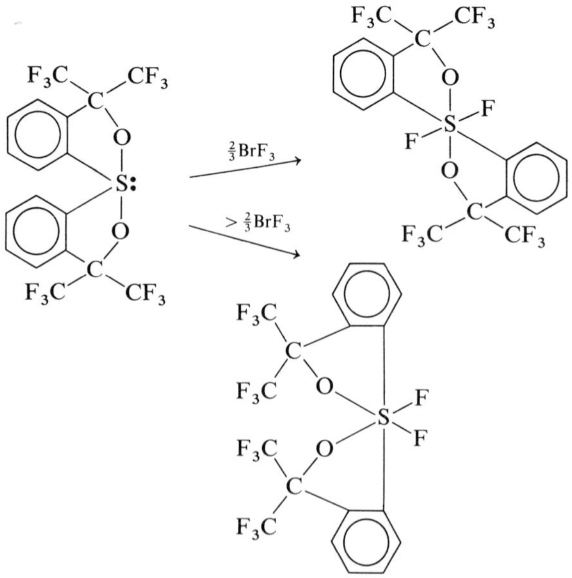

$\frac{2}{3}BrF_3$ → ; $> \frac{2}{3}BrF_3$ → (ai)

Halogen fluorides are not useful reagents in S—N chemistry, because of the polar nature of the X—F bond, S—N bond-cleavage occurs[53–56], e.g.:

$$R_fN{=}SF_2 + 2\ ClF \rightarrow R_fNCl_2 + SF_4 \qquad \text{(aj)}$$

$$FSO_2NSF_2 + 2\ ClF \rightarrow FSO_2NCl_2 + SF_4 \qquad \text{(ak)}$$

$$S_4N_4 + ClF \xrightarrow[2.\ 20°C,\ 10\ h]{1.\ -78°C,\ 4\ h} S_4N_4Cl_2 \xrightarrow{ClF} NSCl + (NSCl)_3$$

$$\xrightarrow{ClF} NSF + ClNSF_2 \xrightarrow{ClF} [Cl_2NSF_3] \qquad \text{(al)}$$

$$\xrightarrow{ClF} \tfrac{1}{2}\ N_2 + \tfrac{3}{4}\ Cl_2 + SF_4$$

The $S_4N_4 \cdot S_4N_4F_4$ product resulting from IF_5 and S_4N_4[57] is $S_4N_4F_2$[58]:

$$2\ S_4N_4 + xs\ IF_5 \rightarrow S_4N_4 \cdot S_4N_4F_4 = 2\ S_4N_4F_2 \qquad \text{(am)}$$

Oxygen fluorides and hypofluorites R_fOF ($R_f = SF_5$, SeF_5FSO_2, CF_3) also add to low valent S compounds. With OF_2, oxygen and fluorine bonds are formed simultaneously, whereas O_2F_2 only acts as fluorinating agent. The gas-phase reaction (10 min, 450°C) of 2:1 OF_2 and SF_4 gives O_2SF_2, SF_6, OSF_4 and O_2[59]:

$$xs\ OF_2 + SF_4 \xrightarrow{450°C} SF_6 + O_2SF_2 + OSF_4 \qquad \text{(an)}$$

whereas in the photolysis in CF_3Cl at -80°C SF_5OF is formed. The ESR spectra show $[SF_5]^{\cdot}$, $[SF_5O]^{\cdot}$ and $[SF_5OSF_4]^{\cdot}$ radicals as intermediates[60]:

$$OF_2 + SF_4 \xrightarrow[-80°C]{h\nu} SF_5OF \qquad \text{(ao)}$$

With xs SF_4, SF_5OSF_5 is isolated in 20% yield[61]:

$$OF_2 + xs\ SF_4 \xrightarrow[-78°C]{h\nu} SF_5OSF_5 \qquad \text{(ap)}$$

However, **O_2F_2 and SF_4 in ClO_3F react violently even at < -140°C** to give SF_6 and O_2; an intense violet intermediate (O_2SF_6?) is observed[62]:

$$O_2F_2 + SF_4 \xrightarrow[ClO_3F]{-143°C} SF_6 + O_2 \qquad \text{(aq)}$$

Hypofluorites like SF_5OF or SeF_5OF also oxidize SF_4:

$$SF_5OF + SF_4 \xrightarrow{75°C} SF_5OSF_5 + SF_5OOSF_5 + SF_5OSF_4OSF_5 \qquad \text{(ar)}^{63,64}$$

$$SF_4 + SF_5OF \xrightarrow[Ni\ reactor]{140°C} SF_5OSF_5 + SF_5OOSF_5 + OSF_4, SF_6 \qquad \text{(as)}^{65}$$

$$SF_4 + SeF_5OF \xrightarrow[10\ days]{20°C} \underset{(15\%)}{SeF_5OSF_5} + SO_2F_2 + OSF_2 + OSF_4 + SeF_6 \qquad \text{(at)}^{66}$$

When SF_4 reacts with SF_5OF in the presence of O_2, OSF-containing species can be isolated[63].

Similar additions occur with FSO_2OF[67], FSO_2OCl[68] and CF_3SO_2OCl[67], whereas with FSO_2OBr[67] only decomposition products are observed[67], e.g.:

$$SF_4 + FSO_2OF \xrightarrow[3\,h]{70°C} FSO_2OSF_5 \qquad (au)^{67}$$

$$FSO_2OCl + SF_4 \xrightarrow{0°C,\ 1.5\,h} SF_4(Cl)OSO_2F \quad (86\%\ cis,\ 14\%\ trans) \qquad (av)^{68}$$

$$F_3CSO_2OCl + SF_4 \xrightarrow{-111\ to\ 22°} cis\text{-}SF_4(Cl)OSO_2CF_3 \qquad (aw)^{69}$$

$$RSO_2OCl + SF_4 \xrightarrow{-90°\ to\ -40°C} [SF_3][OSO_2R] \qquad (ax)^{68}$$

where R = F, CF_3. Liquid (10 h, 75°C) CF_3OF reacts with SF_4 to give CF_3OSF_5 in 98% yield:

$$CF_3OF + SF_4 \xrightarrow{75°C} CF_3OSF_5 \qquad (ay)^{63}$$

$$CF_3OF + SF_4 \xrightarrow{h\nu} CF_3OSF_5 + cis\text{-}(CF_3O)_2SF_4 + SiF_4, COF_2, SO_2F_2, OSF_2, CF_3OOCF_3 \qquad (az)^{70}$$

According to EPR, a radical mechanism is followed[60]. If O_2 is present, a complicated mixture results[63].

Other hypofluorites react similarly, e.g.:

$$FC(O)OF + SF_4 \xrightarrow{h\nu} FC(O)OSF_5 + CO_2, OCF_2, SiF_4, SF_6, SO_2F_2, OSF_2 \qquad (ba)^{71}$$

$$F_2C(OF)_2 + SF_4 \rightarrow SF_6 + COF_2 + \tfrac{1}{2}O_2 \qquad (bb)^{72}$$

$$CF_3OOF + SF_4 \rightarrow CF_3OSF_5 + CF_3OSF_4OSF_5 \qquad (bc)^{73}$$

Similar oxidative additions with OSF_2 occur:

$$O_2F_2 + OSF_2 \rightarrow OSF_4 + O_2 \qquad (bd)^{74}$$

In the irradiation with CF_3OF and FSO_2OF the $[OSF_3]^\bullet$ radical is an intermediate[75], but no stable oxo derivative of OSF_4 is produced. Only CF_3OF is used for oxidative fluorinations in organosulfur chemistry. Dialkyl- and diarysulfides react at −78°C to give difluorosulfuranes:

$$\text{R(H)C}\langle\rangle\text{S} + xs\ CF_3OF \xrightarrow[CH_2Cl_2]{-78°C} \text{R(H)C}\langle\rangle\text{SF}_2 + OCF_2 \qquad (be)^{75}$$

where R = H, CH_3. Tetrahydrothiophene, n-Pr_2S, Ph_2S and perfluorodiphenylsulfide are converted to difluorosulfuranes which are characterized by ^1H- and ^{19}F-NMR spectroscopy[76] and by hydrolysis to the corresponding sulfoxides.

Tetrafluorosulfuranes are formed when xs CF_3OF reacts with cyclic sulfides, e.g.:

$$\text{R(H)C}[(CH_2)_n][(CH_2)_m]\text{S} + xs\ CF_3OF \xrightarrow{-78°C} \text{R(H)C}[(CH_2)_n][(CH_2)_m]\text{SF}_4 + OCF_2 \qquad (bf)^{77}$$

Similarly, Ph_2S gives Ph_2SF_4; however, from n-Pr_2S and CF_3OF, n-$Pr_2S(OCF_3)F_3$ was obtained[77].

With hypochlorites usually only S—O bonds are formed[78]. From CF_3SCl and $(CF_3)_3COCl$, a chlorosulfurane is isolated[79]:

$$CF_3SCl + 2\,(CF_3)_3COCl \rightarrow CF_3S[OC(CF_3)_3]_2Cl + Cl_2 \qquad \text{(bg)}$$

Whether the S—Cl bond is preserved during the reaction is not known. With cyclic systems, alkoxychlorosulfuranes can be generated[80], e.g.:

$$\text{C}_5\text{H}_{10}\text{S} + Me_3COCl \xrightarrow[CH_2Cl_2]{-78^\circ C} \text{C}_5\text{H}_{10}\text{S(Cl)(OCMe}_3) \qquad \text{(bh)}$$

In sulfur-nitrogen chemistry, no reactions of hypohalites is known with OF_2 and R_fNSF_2 (besides oxygenation), and the same results are obtained[81,82] as with F_2:

$$R_fNSF_2 + OF_2 \xrightarrow{h\nu} R_fNSOF_2 + R_fNSF_2NR_f + SF_4 \qquad \text{(bi)}$$

where $R_f = CF_3$, C_2F_5, i-C_3F_7, $ClCF_2CF_2$, $BrCF_2CF_2$[81], COF[82]. Sulfur fluorides themselves may act as oxidizing agents, e.g.[44]:

$$SCF_2 + SF_4 \xrightarrow{200^\circ C} CF_3SF_3 \qquad \text{(bj)}$$

$$(CF_3)_2S_x + SF_4 \xrightarrow{200^\circ C} CF_3SF_3 \qquad \text{(bk)}$$

(with x = 1, 2, 3). The disproportionation and decomposition reactions of lower sulfur fluorides can also be placed in this category, e.g.:

$$16\,S_2F_2 \rightarrow 8\,SF_4 + 3\,S_8 \qquad \text{(bl)}^{83}$$

$$FSSF \xrightarrow{\text{glow discharge}} F_3SSF + SF_4 + S_8 \qquad \text{(bm)}^{84,85}$$

$$SF_2 + SSF_2 \xrightarrow{\text{cocondensation}} F_3SSSF \qquad \text{(bn)}^{86}$$

$$2\,CF_3SF \rightarrow CF_3SF_2SCF_3 \xrightarrow{CF_3SF} CF_3SF_3 + CF_3SSCF_3 \qquad \text{(bo)}^{85,87}$$

$$3\,H_3CSF \rightarrow H_3CSF_3 + CH_3SSCH_3 \qquad \text{(bp)}^{88}$$

$$3\,SF_4 \xrightarrow{500^\circ\text{–}800^\circ C} 2\,SF_6 + \tfrac{1}{8}\,S_8 \qquad \text{(bq)}^{89}$$

High-frequency[90] or microwave discharges[91] decompose SF_4 in the same way.

Pyrolysis of S_2F_{10} is catalyzed by NO or CHCl=CHCl, and SF_6 is formed irreversibly[92,93]. Two steps are involved:

$$S_2F_{10} \rightarrow 2\,[SF_5]^{\cdot} \rightarrow SF_4 + SF_6 \qquad \text{(br)}$$

The second step is rate determining.

Sulfuryl chloride is used as a mild chlorinating agent with S—N heterocycles, e.g.:

$$3\ S_4N_4 + 6\ SO_2Cl_2 \rightarrow 4\ S_3N_3Cl_3 + 6\ SO_2 \qquad (bs)^{94}$$

$$(RCN)_n(SN)_{3-n} + SO_2Cl_2 \xrightarrow[-SO_2]{} (RCN)_n(NSCl)_{3-n} \qquad (bt)$$

where R = Ph, n = 2[95]; R = CF_3, n = 1[12].

$$Ph_2P(NSN)_2(S{-}S)PPh_2 + SO_2Cl_2 \xrightarrow[23°C]{CH_2Cl_2} Ph_2P(NS(Cl)N)(NSN)PPh_2 + SO_2 \qquad (bu)^{96}$$

$$Me_2NC(NSN)_2(S{-}S)CNMe_2 + SO_2Cl_2 \rightarrow Me_2NC(NS(Cl)N)_2CNMe_2 \qquad (bv)^{97}$$

$$Ph_2P(N)PPh_2(NSNSN) + SO_2Cl_2 \xrightarrow{MeCN} Ph_2P(N)PPh_2(NS(Cl)N) + SO_2 \qquad (bw)^{96}$$

Simultaneous S—S bond cleavage and oxidation is found with disulfides[98]:

$$RSSR + 3\ SO_2Cl_2 \xrightarrow{AcOH} \underset{(98\text{–}100\%)}{RS(O)Cl} + AcCl + SO_2 + HCl \qquad (bx)$$

where R = Me, i-Pr, $PhCH_2$, Ph, 4-MeC_6H_4, $AcOCH_2CH_2$

$$RR'C{=}NSCl + SCl_2 \xrightarrow[40°C,\ 2\ days]{activated\ charcoal} RR'CClN{=}SCl_2 \qquad (by)^{99}$$

The first step in Eq. (bq) involves decomposition of SCl_2 into Cl_2 and S_2Cl_2.

Except for NF_3, nitrogen fluorides in their oxidizing behavior are similar to hypofluorites. Whereas NF_3 reacts with sulfides only at high T[100,101], SF_4 is oxidized by ONF_3[102], NO_3F[103] or N_2F_2[104,105] to give SF_6, and S_2F_{10} is also observed[104]. Under irradiation OSF_2 and N_2F_2[105] or N_2F_4[106] give OSF_4, whereas with SF_4 and N_2F_4, S—N and S—F bonds are simultaneously formed:

$$4\ SF_4 + 3\ N_2F_4 \xrightarrow{h\nu,\ 80\ h} 4\ SF_5NF_2 + \tfrac{3}{2}\ N_2 + SF_6/NF_3 \qquad (bz)^{107,108}$$

$$SF_4 + N_2F_2 + N_2F_4 \xrightarrow[100\ \%]{100°C} SF_5NF_2 + N_2 + NF_3 \qquad (ca)^{105}$$

$$CF_3SF_3 + N_2F_4 \xrightarrow{h\nu} CF_3SF_4NF_2 + SF_5NF_2, SF_4, CF_4 \qquad (cb)^{107}$$

$$SF_4 + (CF_3)_2NCl \xrightarrow[4\ days]{h\nu} (CF_3)_2NSF_5 \qquad (cc)^{109}$$

Irradiation of 1:2.5 $(CF_3)_2NCl$ and SF_4 forms the pentafluorosulfanylamine in 10% yield. Nitryl chloride oxidizes the $[SF_5]^-$ anion to SF_5Cl[110], and N-dichloroamines[111] and dibromoamines[112] are used for oxidative iminations, e.g.:

$$\underset{\mathbf{I}}{Ph(F_3C)C{=}NCl} + SCl_2 \xrightarrow[60°C/5\,h]{CCl_4} \left[ClC(Ph)(CF_3){-}N(Cl)SCl \right] \rightarrow ClC(Ph)(CF_3)N{=}SCl_2 \quad (75\%) \qquad (cd)^{113}$$

$$\mathbf{I} + PhSCl \xrightarrow[60°C,\ 25\,h]{CCl_4} ClC(Ph)(CF_3)N{=}S(Ph)Cl \quad (80\%) \qquad (ce)^{113}$$

$$R(R'O)C{=}NCl + R''SCl \xrightarrow[1\,h]{Et_2O} R(R'O)C{=}NSCl_2R'' \qquad (cf)^{114}$$

$$PhC({=}NSO_2Ar)NHCl \xrightarrow[benzene,\ 5°{-}20°C,\ 1\,h]{SCl_2-py} PhC({=}NSO_2Ar)N{=}SCl_2 \quad (90{-}92\%) \qquad (cg)^{115}$$

$$RNCl_2 + S_2Cl_2 \rightarrow RN{=}SCl_2 \qquad (ch)^{111}$$

where (R = CF_3[9,112] $PhC{=}NSO_2Ar$[113]). N-Chloro- and N-bromosuccinimide among other reagents halogenate organic sulfides to form chloro-[116] and bromosulfoniumions, $[RR'SX]^+$ are formed as primary products[117]. Organic sulfides are oxidized by PCl's to give chlorosulfonium salts, and organophosphorus halides can be useful in S—N heterocyclic chemistry:

$$PCl_{3-n}Ph_n + S_4N_4 \xrightarrow[reflux,\ 22\,h]{MeCN} \text{cyclo-}[N{=}P(Ph_nCl_{2-n}){-}N{=}S(Cl){-}N{=}P(Ph_nCl_{2-n}){-}] \qquad (ci)^{118}$$

(R. MEWS)

1. A. G. Streng, A. D. Kirshenbaum, L. V. Streng, A. V. Grosse, in *Noble Gas Compounds*, H. H. Hyman, ed., Univ. Chicago Press, 1963, p. 79.
2. N. Bartlett, F. Sladky, *J. Chem. Soc., Chem. Commun.*, 1046 (1968).
3. R. K. Marat, A. F. Janzen, *Can. J. Chem.*, *55*, 3031 (1977).
4. Yu. L. Yagupol'skii, T. I. Savina, *J. Org. Chem. USSR (Engl. Transl.)*, *15*, 386 (1979).
5. A. M. Forster, A. J. Downs, *J. Chem. Soc., Dalton Trans.*, 2827 (1984).
6. J. Varwig, R. Mews, *J. Chem. Res., (M)*, *1*, 2744 (1977); *(S)*, *1*, 245 (1977).
7. H. Steinbeißer, R. Mews, *J. Fluorine Chem.*, *17*, 505 (1981).
8. C. Lensch, O. Glemser, *Z. Naturforsch., Teil B*, *37*, 401 (1982).
9. W. Leidinger, W. Sundermeyer, *Chem. Ber.*, *115*, 2892 (1982).
10. R. Mews, P. Kricke, I. Stahl, *Z. Naturforsch., Teil B*, *36*, 1993 (1981).

11. W. F. V. Brooks, G. K. Mac Lean, J. Passmore, P. S. White, C. M. Wong, *J. Chem. Soc., Dalton Trans.*, 1961 (1983).
12. R. Maggiulli, R. Mews, to be published (1987).
13. R. C. Kumar, S. A. Kinkead, J. M. Shreeve, *Inorg. Chem.*, *23*, 3112 (1984).
14. H. Puchelt, B. R. Sabels, T. C. Hoering, *Geochim. Cosmochim. Acta*, *35*, 675 (1971).
15. C. Lau, J. Passmore, *J. Fluorine Chem.*, *6*, 77 (1975).
16. H. L. Roberts, Ger. Pat. 1,105,857 (1960); *Chem. Abstr.*, *56*, 8295 (1962).
17. F. Nyman, H. L. Roberts, *J. Chem. Soc.*, 3180 (1962).
18. F. Nyman, H. L. Roberts, T. Seaton, *Inorg. Synth.*, *8*, 160 (1966).
19. J. J. Pitts, A. W. Jache, (Olin Mathieson Chem. Corp.) *US Pat.* 3,373,000 (1968); *Chem. Abstr.*, *68*, 88,678 (1968).
20. J. S. Pitts, A. W. Jache, *Inorg. Chem.*, *7*, 1661 (1968).
21. C. J. Schack, R. D. Wilson, M. G. Warner, *J. Chem. Soc., Chem. Commun.*, 1110 (1969).
22. C. J. Schack, R. D. Wilson, *US Pat.* 3,649,222 (1972); *Chem. Abstr.*, *77*, 22,415 (1972).
23. C. Naulin, R. Bougon, *J. Chem. Phys.*, *72*, 2155 (1980).
24. F. A. Cotton, J. W. George, *J. Inorg. Nucl. Chem.*, *7*, 397 (1958).
25. C. I. Merrill, M. Lustig, G. H. Cady, *AD* 285 205 (1962); *Chem. Abstr.*, *60*, 6463 (1964).
26. C. I. Merrill, *US Pat.* 3,338,685 (1967); *Chem. Abstr.*, *67*, 92,440 (1967).
27. T. A. Kovacina, A. D. Berry, W. B. Fox, *J. Fluorine Chem.*, *7*, 430 (1976).
28. K. O. Christe, E. C. Curtis, C. J. Schack, A. Roland, *Spectrochim. Acta, Part A*, *33*, 69 (1977).
29. C. J. Schack, R. D. Wilson, *Inorg. Chem.*, *9*, 311 (1970).
30. K. Seppelt, *Z. Anorg. Allg. Chem.*, *386*, 229 (1971).
31. R. Minkwitz, M. Nass, J. Sawatzki, *J. Fluorine Chem.*, *31*, 175 (1986).
32. R. Minkwitz, M. Nass, A. Radünt, H. Preut, *Z. Naturforsch. Teil B* *40*, 1123 (1985).
33. J. Passmore, C. Lau, *J. Chem. Soc., Chem. Commun.*, 950 (1971).
34. C. Lau, J. Passmore, *J. Chem. Soc., Dalton Trans.* 2528 (1973).
35. A. Haas, in *Gmelins Handbook of Inorganic Chemistry, Perfluorohalogenoorgano Compounds of the Main Group Elements*, Springer Verlag, Berlin Part. 1, Suppl. B9, p. 28.
36. G. Dahms, G. Didderich, A. Haas, M. Yazdanbakhsch, *Chem.-Ztg.*, *98*, 109 (1974).
37. T. Abe, J. M. Shreeve, *J. Fluorine Chem.*, *3*, 17 (1973/74).
38. T. Abe, J. M. Shreeve, *Inorg. Nucl. Chem. Lett.*, *9*, 465 (1973).
39. D. T. Sauer, J. M. Shreeve, *J. Fluorine Chem.*, *1*, 1 (1971/72); *J. Chem. Soc., Chem. Commun.*, 1679 (1970).
40. D. T. Sauer, J. M. Shreeve, *Inorg. Synth.*, *14*, 42 (1973).
41. T. Kitazume, J. M. Shreeve, *J. Chem. Soc., Chem. Commun.*, 184 (1978).
42. G. Haran, D. W. A. Sharp, *J. Fluorine Chem.*, *3*, 423 (1973).
43. G. A. Sprenger, A. H. Cowley, *J. Fluorine Chem.*, *7*, 333 (1976).
44. W. J. Middleton, E. G. Howard, W. H. Sharkey, *J. Org. Chem.*, *30*, 1375 (1965).
45. T. Kitazume, J. M. Shreeve, *J. Chem. Soc., Chem. Commun.*, 154 (1978).
46. S. L. Yu, D. T. Sauer, J. M. Shreeve, *Inorg. Chem.*, *13*, 484 (1974).
47. S. L. Yu, J. M. Shreeve, *J. Fluorine Chem.*, *6*, 259 (1975).
48. D. T. Sauer, J. M. Shreeve, *Z. Anorg. Allg. Chem.*, *385*, 113 (1971).
49. T. Abe, J. M. Shreeve, *J. Fluorine Chem.*, *3*, 187 (1973).
50. A. Waterfeld, R. Mews, *J. Fluorine Chem.*, *23*, 325 (1983).
51. R. S. Michalak, J. C. Martin, *J. Am. Chem. Soc.*, *102*, 5921 (1980).
52. R. S. Michalak, J. C. Martin, *J. Am. Chem. Soc.*, *103*, 214 (1981).
53. R. A. De Marco, J. M. Shreeve, *J. Chem. Soc., Chem. Commun.*, 788 (1971).
54. R. A. De Marco, J. M. Shreeve, *J. Fluorine Chem.*, *1*, 269 (1971/72).
55. H. W. Roesky, *Angew. Chem., Int. Ed. Engl.*, *10*, 265 (1971).
56. A. J. Banister, R. G. Hey, J. Passmore, M. N.I S. Rao, *J. Fluorine Chem.*, *21*, 429 (1982).
57. B. Cohen, T. R. Hooper, D. Hugill, R. D. Peacock, *Nature (London)*, *207*, 748 (1965).
58. L. Zborilova, P. Gebauer, *Z. Anorg. Allg. Chem.*, *448*, 5 (1979).
59. A. Engelbrecht, A. Nachbaur, C. Pupp, *Monatsh. Chem.*, *95*, 219 (1964).
60. J. R. Morton, K. F. Preston, *Chem. Phys. Lett.*, *18*, 98 (1973).
61. H. Oberhammer, K. Seppelt, *Angew. Chem., Int. Ed. Engl.*, *17*, 69 (1978).
62. A. G. Streng, *J. Am. Chem. Soc.*, *85*, 1380 (1963).
63. G. Pass, H. L. Roberts, *Inorg. Chem.*, *2*, 1016 (1963).
64. F. A. Hohorst, D. D. Des Marteau, L. R. Anderson, D. E. Gould, W. B. Fox, *J. Am. Chem. Soc.*, *95*, 3866 (1973).

65. S. M. Williamson, G. H. Cady, *Inorg. Chem.*, *1*, 673 (1962).
66. J. E. Smith, G. H. Cady, *Inorg. Chem.*, *9*, 1442 (1976).
67. W. P. Gilbreath, G. H. Cady, *Inorg. Chem.*, *2*, 496 (1963).
68. B. A. O'Brien, D. D. Des Marteau, *Inorg. Chem.*, *23*, 644 (1984).
69. K. K. Johri, Y. Katsuhara, D. D. Des Marteau, *J. Fluorine Chem. 19*, 227 (1982).
70. L. C. Duncan, G. H. Cady, *Inorg. Chem.*, *3*, 1045 (1964).
71. R. L. Cauble, G. H. Cady, *J. Am. Chem. Soc.*, *89*, 5161 (1967).
72. F. A. Hohorst, J. M. Shreeve, *Inorg. Chem.*, *7*, 624 (1968).
73. R. A. de Marco, W. B. Fox, *Inorg. Nucl. Chem. Lett.*, *10*, 965 (1976).
74. J. I. Solomon, A. J. Kacmarek, J. M. MacDonough, *J. Chem. Eng. Data*, *13*, 529 (1968).
75. J. R. Morton, K. F. Preston, *J. Chem. Phys.*, *58*, 2657 (1973).
76. D. B. Denney, D. Z. Denney, Y. F. Hsu, *J. Am. Chem. Soc.*, *95*, 4064 (1973).
77. D. B. Denney, D. Z. Denney, Y. F. Hsu, *J. Am. Chem. Soc.*, *95*, 8191 (1973).
78. J. M. Shreeve, *Isr. J. Chem.*, *17*, 1 (1978).
79. Q. C. Mir, K. A. Laurence, R. W. Shreeve, D. P. Babb, J. M. Shreeve, *J. Am. Chem. Soc.*, *101*, 5949 (1979).
80. D. Swern, I. Ikeda, G. F. Witfield, *Tetrahedron Lett.*, 2635 (1972).
81. I. Stahl, R. Mews, O. Glemser, *J. Fluorine Chem.*, *7*, 55 (1976).
82. I. Stahl, R. Mews, O. Glemser, *Chem. Ber.*, *110*, 2398 (1977).
83. F. Seel, *Adv. Inorg. Chem. Radiochem.*, *16*, 297 (1974).
84. F. Seel, R. Stein, *J. Fluorine Chem.*, *14*, 339 (1979).
85. W. Gombler, A. Haas, H. Willner, *Z. Anorg. Allg. Chem.*, *469*, 135 (1980).
86. H. Willner, *Z. Anorg. Allg. Chem.*, *514*, 171 (1984).
87. F. Seel, W. Gombler, *Angew. Chem.*, *Int. Ed. Engl*, *8*, 773 (1969).
88. W. Gombler, R. Budenz, *J. Fluorine Chem.*, *7*, 115 (1976).
89. E. L. Muetterties, US Pat. 2,883,267 (1959); *Chem. Abstr.*, *53*, 15,505 (1959).
90. H. J. Emeléus, B. Tittle, *J. Chem. Soc.*, 1644 (1963).
91. W. C. Smith, V. A. Engelhardt, *J. Am. Chem. Soc.*, *82*, 3838 (1960).
92. W. R. Trost, R. L. McIntosh, *Can. J. Chem.*, *29*, 508 (1951).
93. S. W. Benson, J. Bott, *Int. J. Chem. Kinet.*, *1*, 451 (1969).
94. G. G. Alange, A. J. Banister, B. Bell, *J. Chem. Soc.*, *Dalton Trans.*, 2399 (1972).
95. R. T. Boeré, A. W. Lordes, P. J. Hayes, R. T. Oakley, R. W. Reed, W. T. Pennington, *Inorg. Chem.*, *25*, 2445 (1986).
96. N. Burford, T. Chivers, M. N. S. Rao, J. F. Richardson, *Inorg. Chem.*, *23*, 1946 (1984).
97. T. Chivers, J. F. Richardson, N. R. M. Smith, *Inorg. Chem.*, *25*, 47 (1986).
98. J.-H. Youu, R. Herrmann, *Tetrahedron Lett.*, *27*, 1493 (1986).
99. T. M. Höfs, R. Mews, M. Noltemeyer, G. M. Sheldrick, M. Schmidt, G. Henkel, B. Krebs, *Z. Naturforsch.*, *Teil B*, *38*, 454 (1983).
100. O. Glemser, J. Wegener, R. Mews, *Chem. Ber.*, *100*, 2474 (1967).
101. A. Tasaka, O. Glemser, *Doshisha Daigaku Rikogaku Kenkyu Hokoku*, *14*, 175 (1973); *Chem. Abstr.*, *83*, 52,660 (1975).
102. W. B. Fox, C. A. Wamser, R. Eibeck, D. K. Huggins, J. S. MacKenzie, R, Juurik, *Inorg. Chem.*, *8*, 1247 (1969).
103. B. Tittle, G. H. Cady, *Inorg. Chem.*, *4*, 259 (1965).
104. H. W. Roesky, O. Glemser, D. Bormann, *Chem. Ber.*, *99*, 1589 (1966).
105. M. Lustig, *Inorg. Chem.*, *4*, 104 (1965).
106. M. Lustig, C. L. Bumgardner, J. K. Ruff, *Inorg. Chem.*, *3*, 917 (1964).
107. A. L. Logothetis, G. N. Sausen, R. J. Shozda, *Inorg. Chem.*, *2*, 173 (1963).
108. G. W. Fraser, J. M. Shreeve, M. Lustig, C. L. Bumgardner, *Inorg. Synth.*, *12*, 299 (1970).
109. R. C. Dobbie, *J. Chem. Soc.*, *A*, 1555 (1966).
110. M. D. Vorob'ev, A. S. Filatov, M. A. Englin, *J. Gen. Chem. USSR (Engl. Transl.)*, *43*, 2371 (1973).
111. A. V. Kirsanov, E. S. Levchenko, L. N. Markovskii, in *Organic Sulfur Chemistry*, R. Kl. Freidlina, A. E. Skorova, eds., Pergamon Press, Oxford, 1981.
112. M. Geisel, Ph.D. Diss., Univ. Göttingen, 1984.
113. Yu. G. Shermolovich, V. S. Talamov, V. V. Pirozhenko, L. N. Markovskii, *J. Org. Chem. USSR (Engl. Transl.)*, *18*, 2240 (1982).
114. E. S. Lerchenko, T. N. Dubinina, L. V. Budnik, E. A. Romanenko, *J. Org. Chem. USSR (Engl. Trans)*, *21*, 1166 (1985).

115. G. S. Borovikova, E. S. Levchenko, E. I. Borovik, E. A. Darmokhval, *J. Org. Chem. USSR (Engl. Transl.)*, *20*, 170 (1984).
116. cf., e.g., W. R. Hardstaff, R. F. Langler, in *Sulfur in Organic and Inorganic Chemistry*, A. Senning, ed., Vol. 4, Marcel Dekker, New York, 1982, pp. 193ff.
117. cf., e.g., P. S. Magee, in *Sulfur in Organic and Inorganic Chemistry*, A. Senning, ed., Vol. 4, Marcel Dekker, New York, 1982, pp. 283ff.
118. T. Chivers, M. N. S. Rao, *Inorg. Chem.*, *23*, 3605 (1984).

2.3.11.3.2. to Give the Se–Halogen Bond.

Oxidative additions employ XeF_2, chlorine and bromine fluorides, cationic iodine species, hypochlorites, sulfur halides and chloroamines. Organoselenium derivatives are oxidized by XeF_2 to give Se—F bonds in quantitative yield; e.g.:

$$p\text{-}RC_6H_4SeCF_3 + XeF_2 \xrightarrow{HF} p\text{-}RC_6H_4SeF_2CF_3 + Xe \qquad (a)^1$$

where R = H, Br, Me, CF_3. Halogen fluorides attack low-valent inorganic and organic derivatives; e.g., SeF_5Cl forms in 30% yield from SeF_4 and ClF at 350°C [2], and slow reaction is observed even at RT [3]:

$$SeF_4 + ClF \xrightarrow[8\ days]{20°C} SeF_5Cl + SeF_6 \qquad (b)^3$$
$$(95\%)$$

$$SeF_4 + ClF_3 \rightarrow SeF_5Cl + SeF_6 \qquad (c)^3$$
$$(10\%)$$

$$Cs[SeF_5] + ClOSO_2F \rightarrow SeF_5Cl + CsOSO_2F \qquad (d)^3$$
$$(92\%)$$

The best method for SeF_5Cl is by Eq. (d); $Cs[SeF_5]$ and ClF only give SeF_6. Perfluoroalkylselenium trifluorides add ClF either directly or in the presence of CsF:

$$R_fSeF_3 + ClF \rightarrow trans\text{-}R_fSeF_4Cl \qquad (e)$$

where $R_f = CF_3$ [4], C_2F_5 [5];

$$Cs[C_2F_5SeF_4] + ClF \rightarrow trans\text{-}C_2F_5SeF_4Cl + CsF \qquad (f)^5$$

$$(R_f)_2Se + ClF \rightarrow (R_f)_2SeF_2 \qquad (g)^6$$

where $R_f = CF_3, C_2F_5$;

$$(C_2F_5)_2SeF_2 + ClF \rightarrow trans\text{-}C_2F_5SeF_4Cl + C_2F_5Cl \qquad (h)^6$$

The trifluorides R_fSeF_3 are prepared from $(R_fSe)_2$ and ClF [5] or BrF_3 [7]. The expected selenenyl fluorides are not detected.

Whereas Se(IV) fluoro cations can be prepared by fluoride abstraction from stable, neutral derivatives, oxidative addition of $[I_2]^+$ or $[I_3]^+$ gives the iodo species:

$$(C_2F_5)_2Se + [I_2][Sb_2F_{11}] + xs\ SbF_5 \rightarrow [(C_2F_5)_2SeI][Sb_2F_{11}] \qquad (i)^8$$

$$(C_2F_5Se)_2 + [I_2][Sb_2F_{11}]\ (or\ [I_3][AsF_6]) \rightarrow [C_2F_5SeI_2][Sb_2F_{11}]\ (or\ [AsF_6]^-) \qquad (j)^8$$

Oxidative addition of hypofluorites to low-valent Se species should be possible, e.g., t-butylhypochlorite and selenides give the expected products in 70–99% yield[9,10]:

$$RSeR' + (CH_3)_3COCl \rightarrow [(CH_3)_3CO]SeRR'Cl \quad (k)^{9}$$

where R = R' = Ph, p-tolyl, $PhCH_2$; or R = CF_3, R' = Ph, p-tolyl; or R = n-C_3F_7, R' = Ph or R = $PhCH_2$, R' = p-Br-C_6H_4;

$$R_2Se_2 + SO_2Cl_2 \xrightarrow[-SO_2]{} 2\,[RSeCl] \xrightarrow[-SO_2]{SO_2Cl_2} 2\,RSeCl_3 \quad (l)$$

where R = O_2N–(2-nitrophenyl with NO_2)–[11], $ClCH_2$–[12];

$$[R_4N][SeCN] + SO_2Cl_2 \xrightarrow[-SO_2]{} [R_4N][SeCl_2CN] \xrightarrow[-SO_2]{SO_2Cl_2} [R_4N][SeCl_3] + ClCN \quad (m)^{13}$$

$$SePR(OR')_2 + SO_2Cl_2 \xrightarrow[-SO_2]{} [RP(SeCl)(OR')_2]Cl \quad (n)^{14}$$

(NC)(CN)C–C ring: ClSN=C–Se–C=NSCl + SCl_2 → same ring with $SeCl_2$ (o)[15]

$$Se_2Cl_2 + 2\,S_2Cl_2 \rightarrow SeCl_4 + \tfrac{1}{2}\,S_8 \quad (p)^{16,17}$$

Decomposition[18] of Se_2F_2 forms SeF_4; Se_2Cl_2 disproportionates in $H_2S_2O_7$[19–21] or HSO_3Cl solution[22] to give $[SeCl_3]^+$ (and Se_4^{2+}) ions and decomposes to give $SeCl_2$ in the gas phase[23]. Similarly, $[SeBr_3]^+$ is formed in solution[19,20] from Se_2Br_2, and $SeBr_2$ is present in the gas phase over Se_2Br_2[24,25]. Addition of Se_2Br_2 to CSe_2 produces a heterocycle[26,27]:

$$Se_2Br_2 + CSe_2 \xrightarrow[CH_2Cl_2,\ 3\ days,\ 20°C]{NH_4Br} \text{Se–Se–Se–C(=SeBr}_2\text{) ring } (96\%) \quad (q)^{26}$$

N-Chloroamines undergo oxidative chlorination or chloroimination, e.g.:

$$2\,Se_2Cl_2 + 3\,RNCl_2 \rightarrow 3\,RN{=}SeCl_2 + SeCl_4 \quad (r)$$

where R = SF_5[28], CF_3, C_2F_5[29];

$$PhCONHCl + ArSeCl \rightarrow PhCONHSeCl_2Ar \quad (s)^{30,31}$$

$$2\,ArC(O)NCl_2 + 2\,(Ar'Se)_2 \rightarrow ArC(O)N{=}SeClAr' + Cl_2 \quad (t)^{32}$$

(R. MEWS)

1. Yu. L. Yagupol'skii, T. I. Savina, *Russ. J. Org. Chem. (Engl. Transl.)*, *15*, 386 (1979).
2. S. Colton, J. L. Margrave, P. W. Wilson, *Synth. React. Inorg. Metal-Org. Chem.*, *1*, 149 (1971).

3. C. J. Schack, R. D. Wilson, J. F. Hon, *Inorg. Chem.*, *11*, 208 (1972).
4. A. Haas, H.-U. Weiler, *Chem. Ber.*, *118*, 943 (1985).
5. C. D. Desjardins, C. Lau, J. Passmore, *Inorg. Nucl. Chem. Lett.*, *9*, 1037 (1973).
6. C. Lau, J. Passmore, *J. Fluorine Chem.*, *7*, 261 (1976).
7. E. Lehmann, *J. Chem. Res.*, *(S)*, *1*, 42 (1978).
8. J. Passmore, P. Taylor, *J. Chem. Soc.*, *Dalton Trans.*, 804 (1976).
9. N. Ya. Derkach, N. P. Tishchenko, V. G. Voloshchuk, *Russ. J. Org. Chem. (Engl. Transl.)*, *14*, 896 (1978).
10. M. R. Detty, *J. Org. Chem.*, *45*, 274 (1980).
11. D. D. Lawson, N. Kharash, *J. Org. Chem.*, *24*, 857 (1959).
12. R. Paetzold, D. Knaust, *Z. Chem.*, *10*, 269 (1970).
13. K. J. Wynne, J. Golden, *Inorg. Chem.*, *13*, 185 (1974).
14. E. V. Bayandina, R. Kh. Giniyatullin, I. A. Nuretdivov, *Acad. Sci. USSR Bull. Chem. Sci. (Engl. Transl.)*, 131 (1977).
15. F. Wudl, E. T. Zellers, *J. Am. Chem. Soc.*, *102*, 5430 (1980).
16. N. S. Fortunatov, N. I. Timoshchenko, Z. A. Fokina, *Ukr. Khim. Zh.*, *37*, 6 (1971); *Chem. Abstr.*, *74*, 146,840 (1971).
17. N. I. Timoshchenko, N. Fortunatov, Z. A. Fokina, *Nauchn. Tr. Gos. Nauchno-Issled. Proektn. Inst. Redkomet. Prom.-sti.*, *58*, 119 (1974); *Chem. Abstr.*, *83*, 90,184 (1975).
18. A. Haas, H. Willner, *Z. Anorg. Allg. Chem.*, *454*, 17 (1979)
19. A. Bali, K. C. Malhotra, *J. Inorg. Nucl. Chem.*, *39*, 957 (1977.
20. R. C. Paul, V. P. Kapila, J. K. Puri, K. C. Malhotra, *J. Chem. Soc.*, *A*, 2132 (1971).
21. R. C. Paul, R. D. Sharma, R. K. Verma, K. C. Malhotra, *Indian J. Chem.*, *10*, 737 (1972).
22. R. C. Paul, D. S. Dhillon, D. Konwer, J. K. Puri, *Indian J. Chem.*, *Sect. A*, *19*, 473 (1980).
23. M. Lundkvist, M. Lellep, *Acta Chem. Scand.*, *22*, 291 (1968).
24. K. Högberg, M. Lundkvist, *Acta Chem. Scand.*, *24*, 255 (1970).
25. G. V. Gombkova, E. S. Petrov, A. N. Kanev, *Izv. Sibirsk. Otd. Akad. Nauk SSR*, *Ser. Khim. Nauk*, 51 (1976); *Chem. Abstr.*, *84*, 173,074 (1976).
26. S. Larsen, L. Henriksen, *Acta Chem. Scand.*, *Ser. A*, *38*, 289 (1984).
27. K. Tominatsu, T. Kataoka, H. Shimizu, M. Hori, *Heterocycles*, *21*, 473 (1984).
28. J. S. Thrasher, K. Seppelt, *Z. Anorg. Allg. Chem.*, *507*, 7 (1983).
29. J. S. Thrasher, C. W. Bauknight Jr., D. D. DesMarteau, *Inorg. Chem.*, *24*, 1598 (1985).
30. N. Ya. Derkach, T. V. Lyapina, E. S. Levchenko, *Russ. J. Org. Chem.* (*Engl. Transl.*), *10*, 145 (1974).
31. N. Ya. Derkach, T. V. Lyapina, E. S. Levchenko, *Russ. J. Org. Chem.* (*Engl. Transl.*), *14*, 256 (1978).
32. N. Ya. Derkach, T. V. Lyapina, E. S. Levchenko, *Russ. J. Org. Chem.* (*Engl. Transl.*), *16*, 31 (1980).

2.3.11.3.3. the Te– and Po–Halogen Bonds.

Similar to the lighter homologues, XeF_2, interhalogens, hypochlorites and sulfur halides are used for oxidative additions; XeF_2 is especially useful in inorganic and organic Te chemistry:

$$Te(OTeF_5)_4 + XeF_2 \xrightarrow[-Xe]{} \text{cis- and trans-}F_2Te(OTeF_5)_4 + FTe(OTeF_5)_5 \quad (a)^{1,2}$$

$$(R_f)_2Te + XeF_2 \xrightarrow[-Xe]{CFCl_3,\ MeCN} (R_f)_2TeF_2 \quad (b)^{3}$$

where $R_f = CF_3$ [3,4], C_6F_5 [5]

$$Ph_{4-n}TeF_n + XeF_2 \xrightarrow[-Xe]{} Ph_{4-n}TeF_{n+2} \quad (c)^{6,7}$$

with $n = 0$–3;

$$Ph_3TeCl + XeF_2 \xrightarrow[-Xe]{} Ph_3TeF_2Cl + Ph_3TeF_3 \quad (d)^{6}$$

With halogen fluorides the (4+) and (6+) oxidation state is reached, depending on conditions:

$$TeF_4 + \text{xs ClF} \xrightarrow{\text{RT}} \underset{(72\%)}{TeF_5Cl} \qquad \text{(e)}^{8}$$

$$TeCl_4 + \text{xs ClF} \xrightarrow{\text{RT}} \underset{(84\%)}{TeF_5Cl} \qquad \text{(f)}^{8}$$

$$(CF_3)_2Te + ClF(BrF_3) \rightarrow (CF_3)_2TeF_2 \qquad \text{(g)}^{3}$$

$$(C_2F_5)_2Te + 2\ ClF \rightarrow (C_2F_5)_2TeF_2 + Cl_2 \qquad \text{(h)}^{9}$$

$$(C_2F_5)_2Te + \text{xs ClF} \rightarrow \text{trans-}C_2F_5TeF_4Cl + \text{trans-}(C_2F_5)_2TeF_4 + TeF_5Cl \qquad \text{(i)}^{9}$$

$$C_2F_5TeF_3 + ClF \rightarrow \text{trans-}C_2F_5TeF_4Cl + TeF_5Cl \qquad \text{(j)}^{9}$$

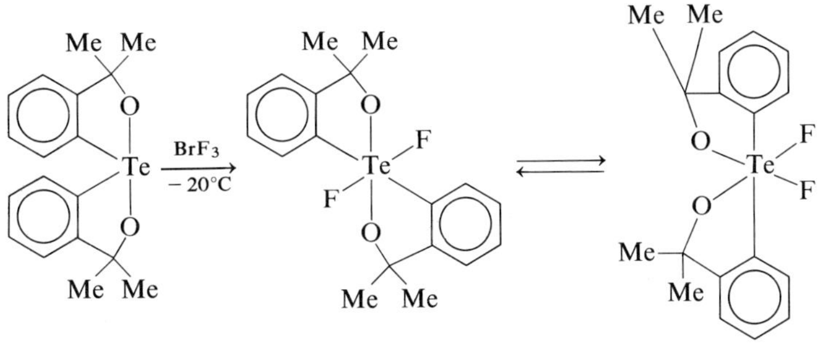

$\xrightarrow[-20°C]{BrF_3}$ (k)[10]

$\xrightarrow[-20°C]{BrF_3}$ (l)[10]

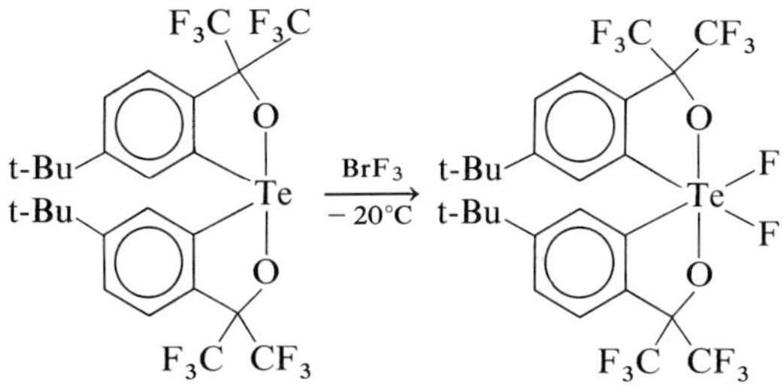

$$Ar_2Te + ICl \xrightarrow{5°C} Ar_2TeICl \qquad \text{(m)}^{11}$$

With other interhalogens only the 4+ oxidation state is obtained.

Hypochlorites give 1:1 adducts if stoichiometric amounts are used. With xs ROCl, Cl_2 is evolved and $R_2Te(OR')_2$ forms, e.g.:

$$Ar_2Te + (CH_3)_3COCl \rightarrow Ar_2Te(Cl)OC(CH_3)_3 \qquad \text{(n)}^{12}$$

$$(CF_3)_2Te + ClONO_2 \rightarrow (CF_3)_2Te(Cl)ONO_2 \qquad \text{(o)}^{13}$$

Sulfur halides, such as O_2SCl_2 or $OSCl_2$, oxidize R_2Te_2 and R_2Te (or R_2Po) to give $RTeCl_3$[14] and R_2TeCl_2[14] (or R_2PoCl_2[15,16]), respectively. With SF_4, difluorides form in high yield (72–87%), e.g.:

$$2\ (RC_6H_4)_2Te_n + SF_4 \rightarrow (RC_6H_4)_2TeF_2 + S_8 \qquad (p)^{17}$$

where R = H, 4-MeO; n = 1, 2;

$$Ar_2Te_n + TeX_4 \rightarrow Ar_2TeX_2 + n\ Te \qquad (q)^{18,19}$$

where X = Cl, Br, I. Tellurium tetrahalides also act as oxidizing agents toward aryl-Te derivatives.

The Cu-catalyzed reaction of diazonium halides with tellurides gives diaryl dihalides:

$$R_2Te + R'N_2X \rightarrow \underset{(27\text{–}95\%)}{RR'TeX_2} \qquad (r)^{20}$$

(R. MEWS)

1. D. Lentz, H. Pritzkow, K. Seppelt, *Angew. Chem., Int. Ed. Engl., 16*, 729 (1977).
2. D. Lentz, H. Pritzkow, K. Seppelt, *Inorg. Chem., 17*, 1926 (1978).
3. D. Naumann, S. Herberg, *J. Fluorine Chem., 19*, 205 (1982).
4. P. L. Baxter, A. J. Downs, A. M. Forster, M. J. Goode, D. W. A. Rankin, H. E. Robertson, *J. Chem. Soc. Dalton Trans*, 941 (1985).
5. G. Klein, D. Naumann, *J. Fluorine Chem., 30*, 259 (1985).
6. K. Alam, A. F. Janzen, *J. Fluorine Chem., 27*, 467 (1985).
7. A. S. Secco. K. Alam, B. J. Blackburn, A. F. Janzen, *Inorg. Chem., 25*, 2607 (1986).
8. C. Lau, J. Passmore, *Inorg. Chem., 13*, 2278 (1974).
9. C. D. Desjardins, C. Lau, J. Passmore, *Inorg. Nucl. Chem. Lett., 10*, 151 (1974).
10. R. S. Michalak, S. R. Wilson, J. C. Martin, *J. Am. Chem. Soc., 106*, 7529 (1984).
11. T. N. Srivastava, R. C. Srivastava, M. Singh, *J. Organomet. Chem., 160*, 449 (1978).
12. V. I. Naddaka, V. P. Garkin, I. D. Sadekov, V. P. Krasnov, V. I. Minkin, *Russ. J. Org. Chem. (Engl. Transl.), 15*, 804 (1979).
13. S. Herberg, D. Naumann, *Z. Anorg. Allg. Chem., 494*, 151 (1982).
14. K. J. Irgolic, in *The Organic Chemistry of Tellurium*, New York, 1974, pp. 68ff, 104ff.
15. V. D. Nefedov, V. E. Zhuravlev, M. A. Toropova, *J. Gen. Chem. USSR (Engl. Transl.), 34*, 3769 (1964).
16. V. D. Nefedov, M. A. Toropova, V. E. Zhuravlev, A. V. Levchenko, *Radiokhimiya, 7*, 203 (1965).
17. I. D. Sadekov, A. Ya. Bushkov, L. N. Markovskii, V. I. Minikin, *J. Gen. Chem. USSR (Engl. Transl.), 46*, 1677 (1976).
18. N. Petragnani, M. de Moura Campos, *Chem. Ber., 94*, 1759 (1961).
19. K. J. Irgolic, R. Zingaro, *Organomet. React., 2*, 117, 145 (1971).
20. I. D. Sadekov, A. A. Maksimenko, B. B. Rivkin, *Russ. J. Org. Chem. (Engl. Transl.), 19*, 541 (1983).

2.3.11.4. with Metal Halides

2.3.11.4.1. to Give the Sulfur–Halogen Bond.

Only higher valent metal fluorides undergo oxidative addition to sulfur compounds, e.g., CoF_3, AgF_2, and in some reactions also HgF_2 and a few other transition-metal fluorides:

$$FSSF + AgF_2(CoF_3) \rightarrow SF_4, SF_6 \qquad (a)^1$$

$$SSF_2 + AgF_2(CoF_3) \xrightarrow{\Delta} SF_4, SF_6 \qquad (b)^2$$

$$SSF_2 + VF_5 \xrightarrow{<25^\circ C} SF_4, SF_6 \qquad (c)^{3,4}$$

$$SF_4 + AgF_2 \xrightarrow{30^\circ C} SF_6 \qquad (d)^{5}$$

Sulfur tetrafluoride is oxidized by PuF_6 at RT, and by UF_6 at 500°C where UF_6 starts to dissociate[6]. The SF_6 is prepared from S_nCl_2 and $VClF_3$[7], or UF_6[8], or from CS_2(COS) and CrF_5[9], UF_6[10] or CoF_3[11]. These reactions are unimportant for syntheses, but in organosulfur and in sulfur–nitrogen chemistry AgF_2 is useful:

$$CS_2 + AgF_2 \xrightarrow[40\text{–}50\,\%]{} CF_3SF_3 \qquad (e)^{12}$$

$$RSSR + 6\,AgF_2 \rightarrow 2\,RSF_3 + 6\,AgF \qquad (f)$$

where R = CF_3[13], C_6H_5[12], C_6F_5[14], $H_3CC_6H_4$[12], m-$NO_2C_6H_4$[12], p-$NO_2C_6H_4$[12], n-C_4H_8F[12];

$$(RC_6H_4S)_2 + 2\,AgF_2 \xrightarrow[50^\circ C]{CFCl_2CF_2Cl} 2\,RC_6H_4SF \xrightarrow{AgF_2} 2\,RC_6H_4SF_3 \qquad (g)^{15}$$

where R = H, F, CF_3;

$$CF_2ClSCl + xs\,AgF_2 \rightarrow CF_2ClSF_3 + CF_3SF_3 + CF_3SF_5 \qquad (h)^{16}$$

$$(CH_3)_2S + 2\,AgF_2 \xrightarrow[-23^\circ C]{CDCl_3} (CH_3)_2SF_2 + 2\,AgF \qquad (i)^{17}$$

$$RSSR + 10\,AgF_2 \rightarrow 2\,RSF_5 + 10\,AgF \qquad (j)^{18}$$

$$\left(\text{–S–}C_6H_3(NO_2)\text{–S–}\right)_n + 10n\,AgF_2 \rightarrow n\ F_5S\text{–}C_6H_3(NO_2)\text{–}SF_5 + 10n\,AgF \qquad (k)^{18}$$

$$C_6H_4(S)(O)C{=}O + 2\,AgF_2 \rightarrow C_6H_4(SF_2)(O)C{=}O + 2\,AgF \qquad (l)^{19}$$

From bis-aryldisulfides the primarily expected sulfenyl fluorides are obtained with stoichiometric[15] AgF_2; xs AgF_2 gives the trifluorides, RSF_3, and pentafluorides, RSF_5, in reasonable yields, depending on conditions; $(CH_3)_2SF_2$ decomposes at <RT[17]. Reactions with AgF_2 can be carried out neat (e.g., for CF_3SF_3 or CF_3SF_5) or in a solvent (e.g., $CFCl_2CF_2Cl$ for $ArSF_3$ and $ArSF_5$). Similar results can be obtained with CoF_3; e.g., from CH_3SH, CS_2 and $(CF_3S)_2$, CF_3SF_5 is isolated at 150°–200°C[20–23].

In sulfur–nitrogen chemistry AgF_2 and HgF_2 are used for oxidative fluorination:

$$2\,(CF_3)_2C{=}NSF + 2\,HgF_2 \rightarrow (CF_3)_2CFNSF_2 + 2\,Hg_2F_2 \qquad (m)^{24}$$

$$S_4N_4 + 4\,AgF_2 \xrightarrow{CCl_4} S_4N_4F_4 + 4\,AgF \qquad (n)^{25}$$

$$S_4N_4 + 4\,HgF_2 \xrightarrow[reflux]{CCl_4} 4\,NSF + 2\,Hg_2F_2 \qquad (o)^{26}$$

In Eq. (o) AgF_2[27,28] or CoF_3[27] give poorer yields. Thiazyl trifluoride is prepared from S_4N_4 directly, or by fluorination of NSF, $OCFNSF_2$ or $SCFNSF_2$ with AgF_2:

$$S_4N_4 + 12\ AgF_2 \xrightarrow{CCl_4,\ reflux} 4\ NSF_3 + 12\ AgF \qquad (p)^{26,27}$$

$$NSF + 2\ AgF_2 \xrightarrow{20^\circ-100^\circ C} NSF_3 + 2\ AgF \qquad (q)^{28,29}$$

$$X{=}CFNSF_2 + 2\ AgF_2 \rightarrow NSF_3 + X{=}CF_2 + 2\ AgF \qquad (r)$$

where X = O[30], S[31]. The last reaction is the most suitable for the preparation of pure NSF_3.

(R. MEWS)

1. O. Glemser, U. Biermann, J. Knaak, A. Haas, *Chem. Ber., 98*, 446 (1965).
2. R. D. Brown, G. P. Pez, *Aust. J. Chem., 20*, 2305 (1967).
3. J. H. Canterford, T. A. O'Donnell, *Inorg. Chem., 5*, 1442 (1965).
4. J. H. Canterford, T. A. O'Donnell, *Inorg. Chem., 6*, 541 (1966).
5. W. Gombler, F. Seel, *J. Fluorine Chem., 4*, 333 (1974).
6. C. E. Johnson, J. Fisher, M. Steindler, *J. Am. Chem. Soc., 83*, 1620 (1961).
7. J. H. Canterford, T. A. O'Donnell, *Aust. J. Chem., 21*, 1421 (1968).
8. L. B. Asprey, E. M. Foltyn, *J. Fluorine Chem., 20*, 277 (1982).
9. T. A. O'Donnell, D. F. Stewart, *Inorg. Chem., 5*, 1434 (1966).
10. L. E. Trevorrow, J. Fischer, W. H. Gunther, *Inorg. Chem., 2*, 1281 (1963).
11. G. A. Silvey, G. H. Cady, *J. Am. Chem. Soc., 74*, 5792 (1952).
12. W. A. Sheppard, *J. Am. Chem. Soc., 84*, 3058 (1962).
13. E. W. Lawless, L. D. Harman, *Inorg. Chem., 7*, 391 (1968).
14. W. A. Sheppard, S. A. Foster, *J. Fluorine Chem., 2*, 53 (1972/73).
15. G. G. Furin, T. V. Terent'eva, G. G. Yakobson, *Izv. Sibirsk. Otd. Akad. Nauk SSR Ser. Khim Nauk, 78* (1972); *Chem. Abstr., 78*, 83,964 (1973).
16. W. Gombler, *J. Fluorine Chem., 9*, 233 (1977).
17. A. M. Forster, A. J. Downs, *J. Chem. Soc., Dalton Trans.*, 2827 (1984).
18. W. A. Sheppard, *J. Am. Chem. Soc., 84*, 3064 (1962).
19. L. D. Martin, E. F. Perozzi, J. C. Martin, *J. Am. Chem. Soc., 101*, 3595 (1979).
20. G. A. Silvey, G. H. Cady, *J. Am. Chem. Soc., 72*, 3624 (1950).
21. G. A. Silvey, G. H. Cady, US Pat. 2,697,726 (1954); *Chem. Abstr., 49*, 5794 (1955).
22. G. A. Brandt, H. J. Emeléus, R. N. Haszeldine, *J. Chem. Soc.*, 2198 (1952).
23. E. A. Tyczkowski, *US NTIS PB Rep.*, No. 233.149/4 GA (1974), *Chem. Abstr., 82*, 124,632 (1975).
24. J. Varwig, R. Mews, *J. Chem. Res., (M), 1*, 2744 (1977); *(S), 1*, 245 (1977).
25. O. Glemser, H. Schröder, H. Haeseler, *Naturwissenschaften, 42*, 44 (1955).
26. O. Glemser, H. Meyer, A. Haas, *Chem. Ber.,97*, 1704 (1964).
27. O. Glemser, H. Richert, *Z. Anorg. Allg. Chem., 307*, 313 (1961).
28. O. Glemser, H. Richert, F. Rogowski, *Naturwissenschaften, 47*, 94 (1960).
29. O. Glemser, H. Schröder, *Z. Anorg. Allg. Chem., 284*, 97 (1956).
30. A. F. Clifford, J. W. Thompson, *Inorg. Chem., 5*, 1424 (1966).
31. Yeda Research and Development, *Isr. IL* 53,536 (1981), *Chem. Abstr., 97*, 127,026 (1982).

2.3.11.4.2. to Give the Se–, Te– and Po–Halogen Bond.

As in sulfur chemistry, only metal fluorides, e.g., AgF_2, are important for syntheses:

$$R_2Se + 2\ AgF_2 \xrightarrow{CFCl_2CF_2Cl} R_2SeF_2 + 2\ AgF \qquad (a)$$

where R = CH_3[1,2], C_2H_5[2], n-C_3H_7[2], i-C_3H_7[2], C_6H_5[2], C_6F_5[3]; p-$CF_3C_6F_4$[3]; R_2 = $(CH_2)_4$[2];

$$(XC_6H_4Se)_2 + 6\ AgF_2 \rightarrow 2\ XC_6H_4SeF_3 + 6\ AgF \qquad (b)^4$$

where X = p-H, —CH_3, —F, —OC_2H_5, —NO_2.

$$CF_3SeF_3 + 2\ AgF_2 \rightarrow CF_3SeF_5 + 2\ AgF \qquad (c)^5$$

Higher Os and Ir fluorides oxidize SeF_4 to give SeF_6[6]. The adducts $PtF_4 \cdot 2\ SeF_4$ and $SeF_4 \cdot AuF_3$ decompose at 350°C to give SeF_6 and Pt, AuF_3 or Au[7]; SeF_6 is formed[8] from $PdF_3 \cdot SeF_4$.

In organotellurium and -polonium chemistry, $(CH_3C_6H_4)_3BiF_2$ is a fluorinating agent:

$$(H_3CC_6H_4)_3BiF_2 + Ar_2Te\,(Po) \rightarrow Ar_2Te(Po)F_2 + (H_3CC_6H_4)_3Bi \qquad (d)^{9-11}$$

where (Ar = p-$CH_3C_6H_4$[9], α-$C_{10}H_7$[10], 4-$CH_3OC_6H_4$[11]. The reaction between Ar_2Te and $FeCl_3$, AgCl, $CuCl_2$, $HgCl_2$ in glacial acetic acid forms diaryldichlorides:

$$(p\text{-}CH_3C_6H_4)_2Te + 2\ MCl_n \rightarrow (p\text{-}CH_3C_6H_4)_2TeCl_2 + 2\ MCl_{n-1} \qquad (e)^{12}$$

(R. MEWS)

1. K. J. Wynne, J. Puckett, *J. Chem. Soc., Chem. Commun.*, 1932 (1968).
2. K. J. Wynne, *Inorg. Chem.*, *9*, 299 (1970).
3. G. G. Furin, O. I. Andreevskaya, G. G. Yakobson, *Izv. Sib. Otd. Akad. Nauk SSR, Ser. Khim. Nauk*, 141 (1981); *Chem. Abstr.*, *95*, 132,412 (1981).
4. W. M. Maxwell, K. J. Wynne, *Inorg. Chem.*, *20*, 1707 (1981).
5. A. Haas, H. M. Weiler, *Chem. Ber.*, *118*, 943 (1985).
6. M. A. Hepworth, P. L. Robinson, G. Westland, *Chem. Ind.* (*London*), 1516 (1955).
7. N. Bartlett, P. L. Robinson, *J. Chem. Soc.*, 3417 (1961).
8. N. Bartlett, M. A. Hepworth, *Chem. Ind.* (*London*), 1425 (1956).
9. V. D. Nefedov, V. E. Zhuravlev, M. A. Toporova, S. A. Grachev, A. V. Levchenko, *J. Gen. Chem, USSR* (*Engl. Transl.*), *35*, 1440 (1965).
10. V. D. Nefedov, M. A. Toporova, V. E. Zhuravlev, A. V. Levchenko, *Radiokhimiya*, *7*, 203 (1965).
11. V. D. Nefedov, V. E. Zhuravlev, M. A. Toporova, L. N. Gracheva, A. V. Levchenko, *Radiokhimiya*, *7*, 245 (1965).
12. M. de Moura Campos, E. L. Suranyi, J. de Andrade Jr., N. Petragnani, *Tetrahedron*, *20*, 2797 (1964).

2.3.11.5. with Organic Halides.

Low oxidation number group-VIB compounds are unreactive toward organic halides. Oxidative addition is observed only in special reactions.

$$(BrCH_2)_2 + Ph_2Te \rightarrow [(Ph_2TeCH_2)_2]Br_2 \xrightarrow{+(BrCH_2)_2}$$

$$Ph_2TeBr_2 + 2\ CH_2{=}CH_2 \qquad (a)^1$$

$$R'X + R_2E \rightarrow [R'R_2E]X \qquad (b)$$

where (X = halide; R, R′ = organic substituent; E = S, Se, Te). Tris(organo)sulfonium[2], -selenonium[3,4], and -telluronium[5,6] compounds are prepared by addition of alkylhalides R′X (X expecially Br, I) to disulfides, diselenides and ditellurides. Because these compounds do not have a covalent group VIB–halogen bond, they are not discussed further.

(R. MEWS)

1. M. de Moura Campos, N. Petragnani, C. Thome, *Tetrahedron Lett.* 5 (1960).
2. J. Goerdeler, in *Houben-Weyl Methoden der Organischen Chemie*, Vol. IX, *S, Se, Te Verbindungen*, E. Müller, O. Bayer, H. Meerwein, K. Ziegler, eds., Georg Thieme Verlag. Stuttgart, 1955, p. 175ff.
3. H. Rheinboldt, in ref. 2, p. 1034ff.
4. K. J. Irgolic, M. V. Kudchadker, in *Selenium*, R. A. Zingaro, W. C. Cooper, eds., Van Nostrand-Reinhold, New York, 1974.
5. A. Rheinboldt, in ref. 2, pp. 1075ff.
6. K. J. Irgolic, in *The Organic Chemistry of Tellurium*, Gordon and Breach, New York, 1974, p. 177.

2.3.12. from Metathetical Reactions (Halogen–Halogen Exchange)

This chapter is divided into §2.3.12.1., halogenation by metal halides, and §2.3.12.2., by nonmetal halides (metathesis with simultaneous oxidative addition; cf. also §2.3.11.1. and §2.3.11.3.). Another differentiation between these two subsections is in terms of ionic and covalent halides; e.g., similar reactions are observed for the chloro- electron-pair acceptor acids BCl_3, $AlCl_3$, $SiCl_4$, $SnCl_4$, $SbCl_5$, or for the ionic fluorides CsF, tris(dialkylamino)sulfonium fluoride, $[(R_2N)_3S][Me_3SiF_2]$[1].

(R. MEWS)

1. W. J. Middleton, US Pat. 3,940,402 (1976); *Chem. Abstr.*, *85*, 6388 (1976); *Org. Synth.*, *64*, 221 (1985).

2.3.12.1. by Halogenation with Metal Halides

2.3.12.1.1. to Give the Oxygen–Halogen Bond.

No reaction of this type is known, although metathesis of hypofluorites with metal halides should be possible according to bond polarities and lattice energies:

$$\mathrm{R}\overset{\delta+}{\mathrm{O}}\!-\!\overset{\delta-}{\mathrm{F}} + \mathrm{MCl} \rightleftarrows \mathrm{MF} + \mathrm{RO{-}Cl} \tag{a}$$

Following the same criteria, reactions of other hypohalites are expected to be more complex; e.g., $ClOSO_2F$ should react with CsBr:

$$\begin{array}{ccc} \mathrm{CsBr} + \mathrm{ClOSO_2F} \rightleftarrows & \mathrm{Cs[OSO_2F]} & + \mathrm{BrCl} \\ & \updownarrow & \\ & \mathrm{CsCl} + \mathrm{BrOSO_2F} & \end{array} \tag{b}$$

The product distribution is controlled by the equilibria shown. Oxidative additions to the halide of lower electronegativity are also possible.

(R. MEWS)

2.3.12.1.2. to Give the Sulfur–Halogen Bond.

Because the chlorides are the most easily prepared sulfur halides, they serve as starting materials for metatheses, e.g., Cl–F exchange is possible by metal fluorides, e.g., alkali-metal fluorides, HgF_2, AgF, AgF_2, ZnF_2 and PbF_2, in addition to more covalent fluorides, such as SbF_3 or SbF_5. The potassium salts are mostly applied for Cl–Br or Cl–I metatheses.

The preparation of sulfur fluorides from chlorides is achieved in neat with active KF[1,2] from thermal decomposition of KSO_2F[1,2], Ag and Hg fluorides. The nucleophilic exchange of Cl by NaF in polar aprotic solvents [e.g., CH_3CN or $(CH_2)_4SO_2$][3] or the system crown-ether–alkali-metal fluoride may be used likewise:

$$ClSSCl + MF \xrightarrow[-MCl]{140°\text{–}150°C,\ <10^3\ Pa} ClSSF \xrightarrow[-MCl]{+MF} FSSF(+ SSF_2) \qquad (a)^{4,5}$$

$$ClSSCl/N_2 + 2\ KF\ (KSO_2F) \xrightarrow{145°C} SSF_2 + 2\ KCl \qquad (b)^{2,4,6}$$

$$ClS_nCl + 2\,KF \rightarrow FSSF + \tfrac{(n-2)}{8} S_8 \qquad (c)^{7}$$

$$S_2Br_2 + HgF_2 \rightarrow S_2F_2 + HgBr_2 \qquad (d)^{8}$$

$$SCl_2 + 2\ MF(MF_2) \xrightarrow[low\ P]{} SF_2\ (+ S_2ClF_3, S_2F_4, SF_4, SSF_2, FSSCl, FSSF) \qquad (e)^{4,6,9,10}$$

$$3\ SCl_2 + 4\ MF \xrightarrow[CH_3CN]{40\text{–}80°C} SF_4 + S_2Cl_2 + 4\ MCl \qquad (f)^{11-16}$$

$$SCl_2 + Cl_2 + 4\ NaF \xrightarrow[CH_3CN]{40°\text{–}80°C} SF_4 + 4\ NaCl \qquad (g)^{14}$$

$$SCl_2 + 4\ UF_6 \xrightarrow{CH_3CN} SF_4 + Cl_2 + UF_5 \qquad (h)^{15}$$

$$S_2Cl_2 + VF_5 \rightarrow SF_4\ (+ VClF_3, Cl_2, VF_4) \qquad (i)^{17,18}$$

A mixture of FSSF and SSF_2 is obtained from gaseous S_2Cl_2 and AgF, Hg_2F_2, HgF_2 at 140°–150°C and $<10^3$ Pa. The primary exchange product, ClSSF, is also observed. At higher T mainly isomerization takes place, and SSF_2 is formed. In higher chlorosulfanes the S—S bond is cleaved by KF as fluorinating agent, and only FSSF and S_8, but no F_2S_3, is formed[7]. Similarly, S_2Br_2 reacts with HgF_2 to form[8] S_2F_2. From SCl_2 vapor and KF at 170°C, KF or HgF_2 at 150°C, SF_2 results, besides a large number of byproducts. At RT AgF forms SF_2 from SCl_2 mixed with Ar or N_2 (1:200–1:1000). Side reactions giving S_2F_4, SSF_2 and SF_4 depend on conditions. For matrix investigations SCl_2 is fluorinated at $<10^{-2}$ Pa and short contact time with the fluorinating agents AgF and HgF_2 to give[10] pure SF_2.

When metatheses of sulfur chlorides with metal halides are carried out in polar solvents, SF_4 is produced in good yield via SF_2 and SF_3SF. Best results are obtained with NaF, whereas KF, CsF, CuF_2 and ZnF_2 are less useful. From S, Cl_2 and NaF even SF_6 is produced[19] at high T. Similarly, sulfur oxifluorides are prepared[3,20,21]:

$$OSCl_2 + xs\ NaF \xrightarrow[\text{or } (CH_2)_4SO_2]{CH_3CN} OSF_2 + 2\ NaCl \quad (j)^{3}$$

$$O_2SCl_2 + NaF \xrightarrow{CH_3CN} O_2SClF + NaCl \quad (k)^{3}$$

$$2\ O_2SCl_2 + PbF_2 \xrightarrow[\text{reflux}]{CH_3CN} \underset{(99.5\%\ \text{purity})}{2\ O_2SFCl} + PbCl_2 \quad (l)^{22}$$

$$O_2SClF + xs\ NaF \xrightarrow[\text{autoclave}]{CH_3CN} O_2SF_2 \quad (m)^{3}$$

$$OSCl_2 + SbF_3 \rightarrow OSClF + OSF_2 + SbCl_3 \quad (n)^{23}$$

$$OSCl_2 + 2\ UF_6 \rightarrow OSF_2 + 2\ UF_5 + Cl_2 \quad (o)^{24}$$

In molten salts (45% KF, 45% LiF, 10% NaF; mp 454°C) OSF_2 is obtained in good yields[25,26]. Formation of O_2SF_2 from O_2SCl_2 needs more drastic conditions:

$$OSCl_2 + 2\ MF \xrightarrow[\text{melt}]{600^\circ C} \underset{(80\%)}{OSF_2} \quad (p)$$

$$O_2SCl_2 + SbF_3\text{-}SbCl_5 \xrightarrow{260^\circ C} \underset{(25\%)}{O_2SF_2} \quad (q)^{27}$$

Organosubstituted sulfur chlorides and oxichlorides give results similar to their inorganic congeners:

$$CF_3SSCl + KF \xrightarrow{120^\circ C} CF_3SSF \quad (r)^{28}$$

$$CF_nCl_{3-n}SCl + KF\ (\text{active}) \rightarrow CF_nCl_{3-n}SF \quad (s)^{29}$$

where n = 0, 1-3;

$$F_3CSCl + \tfrac{1}{2}\ HgF_2(AgF) \xrightarrow{100^\circ\text{–}130^\circ C} CF_3SF + CF_3SF_2SCF_3 \quad (t)^{30-32}$$

$$i\text{-}C_3F_7SCl + AgF \xrightarrow[\text{autoclave}]{125^\circ\text{–}160^\circ C} \underset{(\text{low yield})}{i\text{-}C_3F_7SF} + AgCl \quad (u)^{33}$$

$$F_2NCCl_2SCl + AgF_2 \xrightarrow{25^\circ C} F_2NCCl_2SF + F_2NCCl_2SF_3 \quad (v)^{34}$$

Perfluoroalkylsulfenylfluorides are unstable and disproportionate to give disulfides and S(IV)-derivatives[4], CH_3SF[35] is even more unstable. Bulky groups (e.g., i-C_3F_7) stabilize the S—F group[36], and F_2NCCl_2SF is also stable[34]. Organic derivatives of OSF_2 and O_2SF_2 are isolated pure:

$$CF_3S(O)X + NaF \xrightarrow[24\ h]{RT} F_3CS(O)F + NaX \qquad (w)^{37,38}$$

where X = Cl, Br;

$$ROSCl + AgF \rightarrow [RO{-}SF] \rightarrow decomp \qquad (x)^{39,40}$$

$$R_fSO_2Cl + NaF \rightarrow R_fSO_2F + NaCl \qquad (y)^{41,42}$$

$$CF_2ClCFClSO_2Cl + CsF \rightarrow CF_2ClCFCl{-}SO_2F + CsCl \qquad (z)^{43}$$

$$RSO_2Cl + KF \xrightarrow{H_2O} RSO_2F + KCl \qquad (aa)^{44-48}$$

Alkyl and perfluoroalkylsulfonylfluorides are prepared from the corresponding chlorides and KF in H_2O.

In sulfur–nitrogen–halogen chemistry, fluorides are formed from chlorides mainly via AgF_2:

$$S_3N_3Cl_3 + AgF_2 \rightarrow S_3N_3F_3 + Ag(F, Cl) \qquad (ab)^{49}$$

$$S_5N_5Cl + AgF_2 \rightarrow S_5N_5F + Ag(F, Cl) \qquad (ac)^{50}$$

$$(NSCl)_2NSOCl + AgF_2 \rightarrow (NSF)_2NSOCl \qquad (ad)^{51}$$

$$(NSOCl)_n(NPX_2)_{3-n} + AgF_2 \rightarrow (NSOF)_n(NPCl_2)_{3-n} \qquad (ae)^{52,54}$$

$$R_2NSCl + AgF \xrightarrow[low\ P]{120^\circ C} R_2N{-}SF \qquad (af)^{40}$$

where R = CH_3, C_2H_5;

$$R_2NSCl + HgF_2 \rightarrow RNSF_2 + RF \qquad (ag)^{40}$$

$$(CF_3)_2C{=}NSCl + AgF_2(CsF) \rightarrow (CF_3)_2CFNSF_2 \qquad (ah)^{55}$$

$$RN{=}SCl_2 + 2\ AgF_2 \rightarrow RNSF_2 + 2\ AgF + Cl_2 \qquad (ai)^{56}$$

Alkali-metal fluorides are used only in special cases:

$$RN{=}SCl_2 + 2\ CsF \rightarrow R{-}N{=}SF_2 + 2\ CsCl \qquad (aj)^{57}$$

$$NSCl + CsF \xrightarrow{110^\circ C} NSF + CsCl \qquad (ak)^{58}$$

$$R_fN{=}SOFCl + CsF \rightarrow R_fN{=}SOF_2 + CsCl \qquad (al)^{59,60}$$

where R_f = CF_3, C_2F_5, i-C_3F_7

$$S_4N_4Cl_2 + 2\ NaF \rightarrow S_4N_4F_2 + 2\ NaCl \qquad (am)^{61}$$

$$(NSOCl)_3 + 3\ MF \rightarrow (NSOF)_3 + 3\ MCl \qquad (an)^{62-64}$$

where M = Na, K

$$ClSO_2NH_2 + KF \xrightarrow{CH_3CN} FSO_2NH_2 + KCl \qquad (ao)^{65}$$

$$ClSO_2NCCl_2 + 2\ AgF_2 \rightarrow FSO_2N(Cl)CF_3 + 2\ AgCl \qquad (ap)^{66}$$

A useful reagent for the introduction of fluorine is freshly sublimed SbF_3 or a SbF_3–$SbCl_5$ mixture:

$$ClSO_2NCCl_2 + SbF_3Cl_2 \rightarrow FSO_2NCCl_2 + Sb(F, Cl)_5 \quad (aq)^{67}$$

$$(NSOCl)_3 + SbF_3 \xrightarrow[-SbCl_3]{} (NSOF)_n(NSOCl)_{3-n} \quad (ar)^{68}$$

where n = 1 – 3;

$$(NSCl)_n(NSOCl)_{3-n} + SbF_3 \xrightarrow[-SbCl_3]{} (NSF)_n(NSOF)_{3-n} \quad (as)^{69}$$

$$(NSOCl)_n(NPCl_2)_{3-n} + SbF_3 \xrightarrow[-SbCl_3]{} (NSOF)_n(NPCl_2)_{3-n} \quad (at)^{52}$$

In these reactions separation of $SbCl_3$ from the products sometimes is difficult. Addition of $SbCl_5$ favors cationic intermediates. Addition of halides to such ionic intermediates, is a straight forward route[70]:

$$S_4N_5Cl + AgAsF_6 \rightarrow [S_4N_5][AsF_6] + AgCl \quad (au)^{70}$$

$$[S_4N_5][AsF_6] + CsF \rightarrow S_4N_5F + CsAsF_6 \quad (av)^{70}$$

Exchange of F—Cl in S(IV) derivatives is possible with covalent chlorides of metals that form strong fluorine bonds:

$$3\ R_fNSCl_2 + 2\ AlCl_3 \xrightarrow{CH_3NO_2} 3\ R_fNSCl_2 + 2\ AlCl_3 \quad (aw)^{71}$$

where $R_f = CF_3, C_2F_5$;

$$2\ SF_5NSF_2 + MCl_4 \rightarrow 2\ SF_5NSCl_2 + MF_4 \quad (ax)^{72}$$

$$SF_5NSF_2 + SbCl_5 \xrightarrow{100°C} SF_5NSCl_2\ (+\ SF_5NSClF) + Sb(F, Cl)_5 \quad (ay)^{72}$$

Yields in the last reaction are poor, but here also the chlorofluoride is observed.

Fluorosulfuric acid is a good fluorinating agent. Metal chlorides form metal fluorides and HSO_3Cl[73]:

$$MCl_x + x\ HSO_3F \rightarrow MF_x + x\ HSO_3Cl \quad (az)$$

The reverse formation of metal fluorosulfates from metal fluorides and HSO_3Cl is described in a patent[74]. Sulfur(VI) derivatives do not undergo metatheses without decomposition, e.g., SF_6 and $AlCl_3$ will give S_2Cl_2 and Cl_2 at 200°C[75]:

$$2\ SF_6 + 4\ AlCl_3 \rightarrow S_2Cl_2 + 4\ AlF_3 + 5\ Cl_2 \quad (ba)$$

Chlorine is exchanged for Br by KBr[76,77], NaBr[78] or $AlBr_3$[79]:

$$OSCl_2 + KBr \xrightarrow[-KCl]{SO_2,\ -40°\ to\ -20°C} OSBrCl \quad (bb)$$

$$2\ OSBrCl \rightarrow OSBr_2 + OSCl_2 \quad (bc)$$

$$3\ OSCl_2 + 2\ AlBr_3 \rightarrow 3\ OSBr_2 + 2\ AlCl_3 \quad (bd)$$

In liq SO_2, from $OSCl_2$ and slight xs KBr, $OSBr_2$ is formed in 50% yield. There is mass spectrometric and IR evidence for the existence of OSBrCl[80] as the primary product, but this compound is present only in the equilibrium mixture. The reaction with $AlBr_3$ is vigorous and the yield of $OSBr_2$ poor. Sulfur iodides are unstable, and only S_2I_2 exists. It can be prepared from neat S_2Cl_2 and KI, or in pentane with NaI. Iodine sulfanes S_xI_2 (x = 3–6) are formed from the corresponding chlorosulfanes when a solution of the sulfanes is quickly pressed through KI powder. The preparation of SICl and SI_2 from SCl_2 and KI is claimed[81].

$$S_2Cl_2 + 2\ KI \rightarrow S_2I_2 + 2\ KCl \qquad \text{(be)}^{82,83}$$

$$S_2Cl_2 + 2\ NaI \xrightarrow{\text{pentane}} S_2I_2 + NaCl \qquad \text{(bf)}^{84}$$

$$S_nCl_2 + 2\ KI \rightarrow S_nI_2 + 2\ KCl \qquad \text{(bg)}^{85}$$

In cyclic systems the S—I bond is more stabilized:

$$\text{cyclo-}[Ph_2P{=}N{-}S(Cl){=}N{-}PPh_2{=}N] \xrightarrow[\text{MeCN, 23°C, 4 h}]{\text{KI}} \text{cyclo-}[Ph_2P{=}N{-}S(I){=}N{-}PPh_2{=}N] \qquad \text{(bh)}^{86,87}$$

(R. MEWS)

1. W. Kwasnik, in *Handbook of Preparative Inorganic Chemistry*, G. Brauer, ed., Academic Press, New York, 1975, p. 340.
2. F. Seel, D. Gölitz, *Z. Anorg. Allg. Chem.*, *327*, 32 (1964).
3. C. W. Tullock, D. D. Coffman, *J. Org. Chem.*, *25*, 2016 (1960).
4. F. Seel, *Adv. Inorg. Chem. Radiochem.*, *16*, 297 (1974).
5. A. Haas, H. Willner, *Spectrochim. Acta, Part A*, *35*, 953 (1979).
6. F. Seel, R. Budenz, W. Gombler, *Chem. Ber.*, *103*, 1701 (1970).
7. F. Seel, R. Budenz, W. Gombler, H. Seitter, *Z. Anorg. Allg. Chem.*, *380*, 262 (1971).
8. W. Hückel, *Nachr. Akad. Math. Phys. Wiss. Göttingen*, *IIa*, 36 (1946); *Chem. Abstr.*, *43*, 6793 (1949).
9. F. Seel, E. Heinrich, W. Gombler, R. Budenz, *Chimia*, *23*, 73 (1969).
10. A. Haas, H. Willner, *Spectrochim. Acta, Part A 34*, 541 (1978).
11. L. R. Belohlav, E. T. McBee, *Ind. Eng. Chem.*, *51*, 1102 (1959).
12. C. W. Tullock, F. S. Fawcett, W. C. Smith, D. D. Coffman, *J. Am. Chem. Soc.*, *82*, 539 (1960); cf. *Inorg. Synth.*, *7*, 119 (1963).
13. W. C. Smith, C. W. Tullock, E. L. Muetterties, W. R. Hasek, F. S. Fawcett, V. A. Engelhardt, D. D. Coffman, *J. Am. Chem. Soc.*, *81*, 3165 (1959).
14. E. I. duPont de Nemours, *Br. Pat.* 824,142 (1959), *Chem. Abstr.* *54*, 21,683 (1960).
15. L. B. Asprey, E. M. Foltyn, *J. Fluorine Chem.*, *20*, 277 (1982).
16. C. W. Tullock, US Pat. 2,992,073 (1961); *Chem. Abstr.* *55*, 27,813 (1961).
17. J. H. Canterford, T. A. O'Donnell, *Inorg. Chem.*, *5*, 1442 (1966).
18. J. H. Canterford, T. A. O'Donnell, *Austr. J. Chem.*, *21*, 1421 (1968).
19. E. L. Muetterties Br., Pat. 805,860 (1958); *Chem. Abstr.*, *53*, 10,685 (1959).
20. C. W. Tullock, US Pat. 2,928,720 (1960); *Chem. Abstr.*, *54*, 25,641 (1960).
21. R. Dlaske, G. Furcht, H. Hannig, East Ger. Pat. 139,935 (1980); *Chem. Abstr.*, *93*, 97840 (1980).
22. D. K. Padma, V. S. Bhat, A. R. Vasuchevamurthy, *J. Fluorine Chem.*, *11*, 187 (1978).
23. H. Booth, F. C. Mericola, *J. Am. Chem. Soc.*, *62*, 640 (1940).
24. B. Moncelon, T. Kikinday, *C.R. Hebd. Seances Acad. Sci., Ser. C*, *267*, 1485 (1968).
25. Farbenfabriken Bayer AG, Br. Pat. 990,649 (1965); *Chem. Abstr.*, *63*, 2660 (1965).
26. W. Sundermeyer, *Z. Anorg. Allg. Chem.*, *314*, 100 (1962).
27. H. J. Emeléus, J. F. Wood, *J. Chem. Soc.*, 2183 (1948).
28. W. Gombler, *Z. Anorg. Allg. Chem.*, *416*, 235 (1975).
29. F. Seel, W. Gombler, R. Budenz, *Angew. Chem., Int. Ed. Engl.*, *6*, 706 (1967).

30. F. Seel, W. Gombler, *Angew. Chem., Int. Ed. Engl., 8*, 773 (169).
31. W. Gombler, *Z. Anorg. Allg. Chem., 493*, 193 (1978).
32. W. Gombler, A. Haas, H. Willner, *Z. Anorg. Allg. Chem., 469*, 135 (1980).
33. E. Kober, *J. Am. Chem. Soc., 81*, 4810 (1959).
34. L. M. Zaborowski, J. M. Shreeve, *Inorg. Chim. Acta, 5*, 311 (971).
35. W. Gombler, R. Budenz, *J. Fluorine Chem., 7*, 115 (1976).
36. R. M. Rosenberg, E. L. Muetterties, *Inorg. Chem., 1*, 756 (1962).
37. C. T. Ratcliffe, J. M. Shreeve, *J. Am. Chem. Soc., 90*, 5403 (1968).
38. C. A. Burton, J. M. Shreeve, *Inorg. Chem., 16*, 1039 (1977).
39. F. Seel, R. Budenz, W. Gombler, *Z. Naturforsch., Teil B, 25*, 885 (1970).
40. F. Seel, W. Gombler, R. Budenz, *Z. Naturforsch., Teil B, 27*, 78 (1972).
41. W. A. Sheppard, S. S. Foster, *J. Fluorine Chem., 2*, 53 (1972/73).
42. P. R. Resnick, Ger. Off. 1,959,143 (1970); *Chem. Abstr., 73*, 76,659 (1970).
43. R. E. Banks, R. N. Haszeldine, A. Peppin, *J. Chem. Soc., C*, 1171 (1966).
44. W. Davies, J. H. Dick, *J. Chem. Soc.*, 483 (1931).
45. T. Gramstadt, R. N. Haszeldine, *J. Chem. Soc.*, 173 (1956).
46. W. E. Truce, F. D. Hoerger, *J. Am. Chem. Soc. 76*, 3230 (1954).
47. J. E. Millington, G. M. Brown, F. L. M. Pattison, *J. Am. Chem. Soc. 78*, 3846 (1956).
48. S. Benefice-Malouet, H. Blancon, R. Teissedre, A. Commeyras, *J. Fluorine Chem., 31*, 319 (1986).
49. H. Schröder, O. Glemser, *Z. Anorg. Allg. Chem., 298*, 78 (1950).
50. T. Chivers, L. Fielding, W. G. Laidlaw, M. Trsic, *Inorg. Chem., 18*, 3379 (1979).
51. D. Schläfer, M. Becke-Goehring, *Z. Anorg. Allg. Chem., 362*, 1 (1968).
52. U. Klingebiel, T. P. Lin, B. Buss, O. Glemser, *Chem. Ber., 106*, 2969 (1973).
53. H. H. Baalmann, J. C. van de Grampel, *Recl. Trav. Chim. Pays-Bas, 92*, 716 (1973).
54. J. B. Van den Berg, E. Klei, B. de Ruiter, J. C. van de Grampel, C. Kruk, *Recl. Trav. Chim. Pays-Bas, 95*, 206 (1976).
55. S. Metcalf, J. M. Shreeve, *Inorg. Chem., 11*, 1631 (1972).
56. A. F. Clifford, R. G. Goel, *Inorg. Chem., 8*, 2004 (1969).
57. S. P. v. Halasz, O. Glemser, *Chem. Ber., 103*, 553 (1979).
58. A. J. Banister, R. G. Hey, J. Passmore, M. N. S. Rao, *J. Fluorine, Chem., 21*, 429 (1982).
59. R. Mews, P. Kricke, I. Stahl, *Z. Naturforsch., Teil B, 36*, 1093 (1981).
60. T. Abe, J. M. Shreeve, *Inorg. Chem., 19*, 3063 (1980).
61. L. Zborilova, P. Gebauer, J. Struad, *Z. Chem., 19*, 255 (1979).
62. T. Moeller, A. Ouchi, *J. Inorg. Nucl. Chem., 28*, 2147 (1966).
63. A. J. Banister, B. Bell, *J. Chem. Soc. A.*, 1659 (1970).
64. F. Seel, J. Simon, *Z. Naturforsch. Teil B, 19*, 354 (1964).
65. R. Appel, W. Senkpiel, *Angew. Chem., 70*, 572 (1958).
66. H. W. Roesky, *Angew. Chem., Int. Ed. Engl., 7*, 218 (1968).
67. H. W. Roesky, U. Biermann, *Angew. Chem., Int. Ed. Engl., 6*, 882 (1967).
68. T. P. Lin, U. Klingebiel, O. Glemser, *Angew. Chem., Int. Ed. Engl., 11*, 1095 (1972).
69. J. Weiss, R. Mews, O. Glemser, *J. Inorg. Nucl. Chem., Suppl.*, 213 (1976).
70. W. Isenberg, R. Mews, G. M. Sheldrick, R. Bartetzko, R. Gleiter, *Z. Naturforsch., Teil B, 38*, 1563 (1983).
71. M. Lustig, *Inorg. Chem., 5*, 1317 (1982).
72. J. S. Thrasher, N. S. Hosmane, D. E. Maurer, A. F. Clifford, *Inorg. Chem., 21*, 2506 (1982).
73. E. Hayek, J. Puschmann, A. Czaloun, *Monatsh. Chem., 85*, 359 (1954).
74. R. K. Iler, US Pat. 2,312,413 (1943); *Chem. Abstr., 37*, 4864 (1973)
75. J. R. Case, F. Nyman, *Nature (London), 193*, 473 (1962).
76. F. Feher, in *Handbook of Preparative Inorganic Chemistry*, G. Brauer, ed., Academic Press, New York, 1975, p. 340.
77. M. J. Fraser, W. Gerrard, *Chem. Ind. (London)*, 280 (1954).
78. P. J. Hartog, W. E. Sims, *Chem. News, 67*, 82 (1893).
79. A. Besson, *C.R. Hebd. Seances Acad. Sci., 123*, 884 (1896).
80. R. Steudel, D. Lautenbach, *Z. Naturforsch., Teil B, 24*, 350 (1969).
81. S. N. Nabi, M. S. Amin, *J. Chem. Soc., A*, 1018 (1966).
82. M. R. A. Rao, *Proc. Indian Acad. Sci., Sect. A 11*, 162 (1940).
83. H. J. Mäusle, R. Steudel, *Z. Anorg. Allg. Chem., 463*, 27 (1980).
84. K. Manzel, R. Minkwitz, *Z. Anorg. Allg. Chem., 441*, 165 (1978).
85. F. Feher, H. Münzner, *Chem. Ber., 96*, 1150 (1963).

86. T. Chivers, M. N. S. Rao, J. F. Richardson, *J. Chem. Soc., Chem. Commun.*, 700 (1983).
87. T. Chivers, M. N. S. Rao, *Inorg. Chem.*, *23*, 3605 (1984).

2.3.12.1.3. to Give the Se–Halogen Bond.

Lower fluorides of Se are not obtained by metathesis from Se_2Cl_2 and KF or AgF; instead, only SeF_4 is isolated[1]:

$$2\ Se_2Cl_2 + 4\ MF \rightarrow SeF_4 + 3\ Se + 4\ MCl \qquad (a)$$

where M = K, Ag;

$$SeCl_4 + 4\ AgF \rightarrow SeF_4 + 4\ AgCl \qquad (b)$$

$$OSeCl_2 + 2\ AgF \rightarrow OSeF_2 + 2\ AgCl \qquad (c)$$

A better approach[2] to the tetrafluoride is Cl—F exchange in $SeCl_4$ by AgF. **On heating $OSeCl_2$ with AgF to 140°C a violent reaction takes place.** The $OSeF_2$ formed is purified by distillation[3]. Fluorination of $OSeCl_2$ with KF and KSO_2F[4] also yields $OSeF_2$ through intermediate salts $K[OSeCl_nF_{3-n}]$[4]:

$$\underset{(1:3)}{OSeCl_2 + KF} \rightarrow [KOSeCl_nF_{3-n}] \rightarrow OSeF_2 + KCl \qquad (d)^4$$

When $OSeCl_2$ and NaF are heated in TMSO to 60°–100°C (0.8 h) and 100–115°C (0.5 h) on a molar scale[5], $OSeF_2$ forms:

$$OSeCl_2 + xs\ NaF \xrightarrow{TMSO} \underset{(28\%)}{OSeF_2} + 2\ NaCl \qquad (e)^5$$

By these methods organo and alkoxy derivatives of SeF_4 can also be obtained:

$$Ph_3SeCl + AgF \rightarrow Ph_3SeF + AgCl \qquad (f)^6$$

$$(H_3CO)_3SeCl + AgF \rightarrow (CH_3O)_3SeF + AgCl\ [(H_3CO)_2SeO + CH_3F] \qquad (g)^7$$

$$C_2H_5OSe(O)Cl + KF \rightarrow (C_2H_5O)Se(O)F + KCl \qquad (h)^{8,9}$$

The colorless liq $(H_3CO)_3SeF$ is unstable at RT and decomposes on distillation (bp 78°C, 1.8×10^3 Pa)[7]; however, Ph_3SeF is stable to ≤145°C[6]. Excess SeF_4 fluorinates potassium halides[10,11]:

$$SeF_4 + 4\ KX \rightarrow SeX_4 + 4\ KF \qquad (i)$$

where X = Cl, Br;

$$SeF_4 + 4\ KI \rightarrow Se + 4\ KF + 2\ I_2 \qquad (j)$$

However, potassium fluoroselenate(IV), $KSeF_5$, is formed, and mixed halides cannot be ruled out[12]. The driving force is the high lattice energy of the alkali-metal fluoride.

Similarly[11], with $BiBr_3$:

$$3\ SeF_4 + 4\ BiBr_3 \xrightarrow{Et_2O} 3\ SeBr_4 + 4\ BiF_3 \qquad (k)$$

With SeF_6 drastic conditions are necessary:

$$SeF_6 + 6\ KCl \xrightarrow{500°C} 6\ KF + SeCl_4 + Cl_2 \qquad (l)^{13}$$

$$SeF_6 + 12\ KBr \xrightarrow{300°C} 12\ KF + Se_2Br_2 + 5\ Br_2 \qquad (m)^{13}$$

$$SeF_6 + 6\ KI \xrightarrow{RT} 6\ KF + Se + 3\ I_2 \qquad (n)^{13,14}$$

Owing to the high solvation energy of the chloride ion, Cl is exchanged for Br in diorganoselenium dichlorides[15,16]:

$$R_2SeCl_2 + 2\ KBr \rightarrow R_2SeBr_2 + 2\ KCl \qquad (o)$$

$$R_2SeCl_2 + 2\ KI \rightarrow R_2SeI_2 + 2\ KCl \qquad (p)$$

$$R_2SeBr_2 + 2\ KI \rightarrow R_2SeI_2 + 2\ KBr \qquad (q)$$

The reaction proceeds quantitatively in H_2O or EtOH. Diiodides can be prepared similarly[15,16].

Addition of $AlBr_3$ to $OSeCl_2$ gives deep-brown $OSeBr_2$[17]:

$$3\ OSeCl_2 + 2\ AlBr_3 \rightarrow 2\ AlCl_3 + 3\ OSeBr_2 \qquad (r)$$

(R. MEWS)

1. A. Haas, H. Willner, *Z. Anorg. Allg. Chem., 454*, 17 (1979).
2. E. B. R. Prideaux, C. B. Cox, *J. Chem. Soc.*, 28 (1927); 1063 (1928).
3. E. B. R. Prideaux, C. B. Cox, *J. Chem. Soc.*, 739 (1928).
4. R. Paetzold, K. Aurich, *Z. Anorg. Allg. Chem., 315*, 72 (1962).
5. C. W. Tullock, D. D. Coffman, *J. Org. Chem., 25*, 2016 (1960).
6. H. J. Emeléus, H. G. Heal, *J. Chem. Soc.*, 1126 (1946).
7. M. Reichenbacher, R. Paetzold, *Z. Anorg. Allg. Chem., 400*, 176 (1973).
8. A. Paetzold, K. Aurich, *Z. Chem., 2*, 60 (1962).
9. R. Paetzold, K. Aurich, *Z. Anorg. Allg. Chem., 317*, 151 (1962).
10. E. E. Aynsley, R. D. Peacock, P. L. Robinson, *J. Chem. Soc.*, 1231 (1952).
11. C. Dagron, *C. R. Hebd. Seances Acad. Sci. Ser. C, 244*, 1779 (1957).
12. R. D. Peacock in B. Cohen and R. D. Peacock, *Adv. Fluorine Chem., 6*, 343, (1970).
13. C. Dagron, *C. R. Hebd. Seances Acad. Sci. Ser. C, 242*, 1027 (1956).
14. A. A. Banks, A. J. Rudge, *Nature (London), 171*, 390 (1953).
15. W. R. Gaythwaite, J. Kenyon, H. Philips, *J. Chem. Soc.*, 2292 (1928).
16. W. Kunckell, *Chem. Ber., 23*, 610 (1895).
17. J. Klikorka, J. Pavlik, V. Krejcik, *Sb. Ved. Praci Vys. Sc. Chem. Technol. Pardubice*, 41 (1960); *Chem. Abstr., 56*, 10,172 (1961).

2.3.12.1.4. to Give the Te–Halogen Bond.

In organotellurium chemistry, metathesis with metal halides proceeds similarly to their Se analogues. From chlorides, bromides and iodides the corresponding fluorides are formed:

$$(CH_3)_2TeI_2 + 2\ AgF \rightarrow (CH_3)_2TeF_2 + 2\ AgI \qquad (a)^1$$

$$(CF_3)_2TeX_2 + 2\ NaF \rightarrow (CF_3)_2TeF_2 + 2\ NaX \qquad (b)^2$$

$$Ar_2TeCl_2 + 2\ AgF \rightarrow Ar_2TeF_2 + 2\ AgCl \qquad (c)^3$$

where Ar = p-$MeOC_6H_4$, p-$EtOC_6H_4$;

$$CH_3TeI_3 + 3\ AgF \rightarrow CH_3TeF_3 + 3\ AgI \qquad (d)^1$$

$$ArTeCl_3 + 3\ AgF \rightarrow ArTeF_3 + 3\ AgCl \qquad (e)^3$$

Aryltellurium triiodes are obtained from the trichlorides in >90% yields[4]:

$$ArTeCl_3 + 3\ KI \rightarrow ArTeI_3 + 3\ KCl \quad (f)$$

(R. MEWS)

1. H. J. Emeléus, J. G. Heal, *J. Chem. Soc.*, 1126 (1964).
2. S. Herberg, D. Naumann, *Z. Anorg. Allg. Chem.*, *494*, 159 (1982).
3. F. J. Berry, E. H. Kustan, M. Roshani, B. C. Smith, *J. Organomet. Chem.*, *99*, 115 (1975).
4. N. Petragnani, *Tetrahedron Lett.*, *11*, 15 (1960).

2.3.12.2. by Halogenation with Nonmetal Halides

2.3.12.2.1. to Give the Oxygen–Halogen Bond.

Hypohalites, $\overset{\delta-}{RO}-\overset{\delta+}{X}$ ($X \neq F$), are reactive because of the positively polarized halogen, e.g., with covalent halides, R′—X′, elimination of X—X′ and formation of ROR′ is observed. With hypofluorites a possible pathway is oxidative addition with subsequent elimination of R′X:

$$ROX + R'X \rightarrow ROR' + X{-}X' \quad (a)$$

when X and X′ are not F;

$$ROF + R'X' \rightarrow [RO\overset{F}{X'}R'] \rightarrow ROX' + R'F \quad (b)$$

$$n\text{-}C_3F_7I + 2\ ClOSO_2F \xrightarrow[-Cl_2]{} n\text{-}C_3F_7I(OSO_2F)_2 \rightarrow n\text{-}C_3F_7OSO_2F + IOSO_2F \quad (c)^1$$

The formation of $IOSO_2F$ from $ClOSO_2F$ and $n\text{-}C_3F_7I$ is related to this mechanism.

From F-t-butylhypochlorite and perfluoroiodoalkanes stable I(III) derivatives are obtained[2]:

$$R_fI + 2\ (CF_3)_3COCl \xrightarrow[50\text{–}80\ h]{0°\text{–}10°C} R_fI[OC(CF_3)_3]_2 + Cl_2 \quad (d)$$

where $R_f = CF_3$, $F_5SCF_2CF_2$, C_6F_5

$$ICF_2CF_2I + 4\ (CF_3)_3COCl \xrightarrow[4\ d]{10°C} [(CF_3)_3CO]_2ICF_2CF_2I[OC(CF_3)_3]_2 + 2\ Cl_2 \quad (e)$$

Metatheses are performed with halogens or interhalogens, e.g.:

$$2\ CF_3OCl + F_2 \xrightarrow{65°C} 2\ CF_3OF + Cl_2 \quad (f)^{13}$$

$$Br_2 + 3\ FSO_2OF \xrightarrow{25°C} Br_2\cdot 3\ FSO_2OF[BrF_{3-n}(OSO_2F)_n] \quad (g)^4$$

$$2\ Br_2 + 3\ SeF_5OF \rightarrow BrF_3 + 3\ SeF_5OBr \quad (h)^5$$

$$I_2 + 6\ FSO_2OF \xrightarrow{25°C} 2\ IF_3(OSO_2F)_2 + S_2O_6F_2 \quad (i)^6$$

Whereas fluorine may exchange with chlorine to give hypofluorites, bromine and iodine do insert into the O—F bond, and Br(III) or I(V) derivatives are formed. These oxidation products undergo exchange and decomposition, leading to the direct metathesis products [e.g., Eq. (h)]. Hypochlorites give hypobromites and iodooxygen derivatives:

$$2\ CF_3SO_2OCl + Br_2 \rightarrow 2\ CF_3SO_2OBr + Cl_2 \qquad \text{(j)}^7$$

$$TeF_5OCl + Br_2 \rightarrow TeF_5OBr + BrCl \qquad \text{(k)}^8$$

$$2\ FSO_2OCl + I_2 \rightarrow 2\ FSO_2OI + Cl_2 \qquad \text{(l)}^{1,9}$$

$$SeF_5OCl + ICl \rightarrow SeF_5OI + Cl_2 \qquad \text{(m)}^5$$

$$3\ SeF_5OCl + ICl_3 \rightarrow (SeF_5O)_3I + 3\ Cl_2 \qquad \text{(n)}^5$$

$$TeF_5OCl + ICl \rightarrow TeF_5OI + Cl_2 \qquad \text{(o)}^8$$

$$3\ TeF_5OCl + ICl_3 \rightarrow (TeF_5O)_3I + 3\ Cl_2 \qquad \text{(p)}^8$$

Cleavage of the O—O bond in $S_2O_6F_2$ gives FSO_2OX derivatives (cf. §2.3.8.1.1), whereas SeF_5O and TeF_5O derivatives are accessible by metathesis. The Te compounds are more stable than their Se analogues.

(R. MEWS)

1. C. J. Schack, K. O. Christe, *J. Fluorine Chem.*, *16*, 63 (1980).
2. J. A. M. Canich, M. E. Lechev, G. L. Gard, J. M. Shreeve, *Inorg. Chem.*, *25*, 3030 (1986).
3. C. J. Schack, W. Maya, *J. Am. Chem. Soc.*, *91*, 2902 (1969).
4. W. P. Gilbreath, G. H. Cady, *Inorg. Chem.*, *2*, 496 (1963).
5. K. Seppelt, *Chem. Ber.*, *106*, 157 (1972).
6. J. E. Robert, G. H. Cady, *J. Am. Chem. Soc.*, *82*, 354 (1960).
7. Y. Katsuhara, R. M. Hammaker, D. D. Des Marteau, *Inorg. Chem.*, *19*, 607 (1980).
8. K. Seppelt, *Chem. Ber.*, *106*, 1920 (1973).
9. C. J. Schack, K. O. Christe, *J. Fluorine Chem.*, *20*, 283 (1982).

2.3.12.2.2. to Give the Sulfur–Halogen Bond.

Sulfur fluorides form from readily obtainable chlorides using nonmetal fluorides in metatheses. Halogen–fluorine exchange is possible with F_2 itself, with oxidizing fluorides, such as halogen fluorides; with nonoxidizing fluorides, such as AsF_3, HF, etc.; with ionic fluorides, such as NOF, $NOF\cdot(HF)_x$, $NR_3\cdot(HF)_x$, $py\cdot(HF)_x$; or with stabilized fluoride ions, such as $[FSO_2]^-$ and $[Me_3SiF_2]^-$.

When sulfur halides with sulfur in oxidation state $<6+$ react with F_2 or halogen fluorides, halogen–fluorine exchange and simultaneous oxidative addition result (cf. §2.3.9.). Under controlled conditions, simple metatheses are possible only for S(IV) derivatives, whereas lower valent sulfur chlorides and bromides are oxidized to the (4+) and (6+) oxidation state, e.g.:

$$2\ OSBr_2 + F_2 \rightarrow 2\ OSBrF + Br_2 \qquad \text{(a)}^1$$

$$OSCl_2 + F_2 \rightarrow OSF_2 + Cl_2 \qquad \text{(b)}^2$$

$$OSCl_2 + IF_5 \rightarrow OSClF + OSF_2 + ICl_x + Cl_2 \qquad \text{(c)}^3$$

$$SCl_4 + 4\ ClF \rightarrow SF_4 + 4\ Cl_2 \qquad \text{(d)}^4$$

With xs sulfur chlorides, F_2 and halogen fluorides, SF_4 is produced. The reaction is catalyzed by PCl_3 or $SbCl_3$ and gives SF_4 in $\leq 99\%$ purity[5,6]:

$$S_2Cl_2 + F_2 \xrightarrow[90°C]{SbCl_3} SF_4 \quad (e)^{5,6}$$

$$SCl_2 + F_2\text{-}N_2 \rightarrow SF_5Cl, SF_6 \quad (f)^{7,8}$$

$$SF_5X + F_2 \rightarrow SF_6 + XF_n \quad (g)^{9}$$

where X = Cl, Br.

With a higher $F_2:S_nCl_2$ ratio, SF_6 and small amounts of SF_5Cl are formed. Depending on the conditions, the electrochemical fluorination leads to SF_4 or SF_6, e.g.:

$$S_nCl_2 + F_2 \rightarrow SF_4 + SF_6 \quad (h)^{10-12}$$

$$SCl_2 + ClF \rightarrow SF_4 + Cl_2 \quad (i)^{13}$$

$$S_nCl_2 + ClF_3 \rightarrow SF_4 + Cl_2 \quad (j)^{14}$$

$$S_2Br_2 + IF_5 \xrightarrow{200°C} SF_4 + Br_2 + I_2 \quad (k)^{15}$$

$$S_2O_5Cl_2 + IF_5 \rightarrow SO_2FCl + SO_2F_2 \quad (l)^{15}$$

Similar results are obtained with halogen fluorides; e.g., oxidative fluorination of S_2Br_2 with IF_5 at 200°C gives SF_4 in $\leq 80\%$ yield, in $S_2O_5Cl_2$ besides metathesis, S—O bond cleavage is also observed [Eq. (l)]. Organic derivatives behave similarly, the S—C and S—O bonds are preserved, e.g.:

$$CF_3SCl + ClF_3 \rightarrow CF_3SF_3 + \cdots \quad (m)^{16}$$

$$CF_2ClSCl + BrF_3 \xrightarrow[-BrCl]{} CF_3SF_3 + ClF_2CSF_3 \quad (n)^{17}$$

$$R_fOS(O)Cl + ClF \xrightarrow[-Cl_2]{} R_fOS(O)F \quad (o)^{18}$$

The $(CF_2NCl)_3$ trimer behaves like tamed ClF, e.g.:

$$3\ R_fS(O)Cl + (CF_2NCl)_3 \rightarrow 3\ R_fS(O)F + 3\ Cl_2 + (FCN)_3 \quad (p)^{19}$$

$$RSO_2Cl \xrightarrow[\text{fluorination}]{\text{electrochemical}} R_fSO_2F \quad (q)^{20-24}$$

In the electrochemical fluorination of alkylsulfonyl chlorides, perfluoroalkylsulfonyl fluorides are formed. Yields decrease with increasing chain length. In the oxidative addition of ClF to organosulfur(II) and -(IV) compounds (cf. §2.3.11), chlorofluorides are formed in the first step. These undergo Cl—F exchange to give the observed perfluoro derivatives, e.g.:

$$R_fR'_fSO + ClF \rightarrow [R_fR'_fS(O)ClF] \quad (r)^{25,26}$$

$$R_fR'_fSO + ClF \xrightarrow[-Cl_2]{ClF} R_fR'_fS(O)F_2 \quad (s)$$

Metatheses with halogen fluorides are possible in S—N chemistry, but because of the polar nature of the X—F and S—N bonds, S—N bond cleavage and formation of SF_4 or SF_6 are mainly observed:

$$(CF_3)_2C{=}NSCl + 3\ ClF \rightarrow (CF_3)_2CFNSF_2 + 2\ Cl_2 \qquad (t)^{27}$$

$$(NSCl)_3 + 12\ ClF \rightarrow 3\ SF_4 + \tfrac{3}{2}\ N_2 + \tfrac{15}{2}\ Cl_2 \qquad (u)^{28}$$

$$NSCl + ClF \rightarrow NSF + Cl_2 + \cdots \qquad (v)^{28}$$

Similar to the halogen fluorides, perfluoroalkylhypofluorites, especially the easily accessible and nonexplosive CF_3OF, is used:

$$2\ CF_3SCl + 3\ CF_3OF \xrightarrow{-78^\circ C} 2\ CF_3SF_3 + Cl_2 + 3\ OCF_2 \qquad (w)^{29}$$

Potassium fluorosulfinate, KSO_2F (prepared from KF and SO_2[30]), is a sulfur-containing metathesis reagent that acts as activated potassium fluoride. The Cl—F exchange in S_2Cl_2, SCl_2, $OSCl_2$, sulfinic acid chlorides or sulfonyl chlorides occurs, e.g.:

$$S_2Cl_2 + 2\ KSO_2F \rightarrow SSF_2 + 2\ SO_2 + 2\ KCl \qquad (x)^{30}$$

$$SCl_2 + 2\ KSO_2F \rightarrow SSF_2 + \cdots \qquad (y)^{30}$$

$$OSCl_2 + 2\ KSO_2F \rightarrow OSF_2 + 2\ SO_2 + 2\ KCl \qquad (z)^{31,32}$$

$$ROS(O)Cl + KSO_2F \xrightarrow{100^\circ C} ROS(O)F + SO_2 + KF \qquad (aa)^{33,34}$$

$$C_6H_5SO_2Cl + KSO_2F \rightarrow C_6H_5SO_2F + SO_2 + KCl \qquad (ab)^{31}$$

Similarly, the isoelectronic $[NSF_2]^-$ ion, besides transfering the sulfur difluoride imide group also undergoes Cl—F exchange in the reaction of sulfur chlorides with $Hg(NSF_2)_2$, e.g.:

$$OSCl_2 + Hg(NSF_2)_2 \rightarrow FS(O)NSF_2 + NSF + HgCl_2 \qquad (ac)^{35}$$

$$F_5SNSCl_2 + Hg(NSF_2)_2 \rightarrow FS(NSF_5)NSF_2 + NSF + HgCl_2 \qquad (ad)^{35}$$

A useful and easy to handle group VB fluoride is AsF_3, whereas NF_3 is unreactive. The oxidative power of ONF_3 is similar to hypofluorites:

$$OSCl_2 + NF_3 \xrightarrow{360^\circ C} OSF_2 + N_2 + Cl_2 \qquad (ae)^{36}$$

$$O_2SCl_2 + NF_3 \xrightarrow{265\text{–}360^\circ C} SO_2FCl + SO_2F_2 + N_2 + Cl_2 \qquad (af)^{36}$$

$$(CF_3)_2CNSCl + ONF_3 \rightarrow (CF_3)_2CFSF_2 + NO + \tfrac{1}{2}\ Cl_2 \qquad (ag)^{27}$$

$$3\ OSCl_2 + 2\ AsF_3 \rightarrow 3\ OSF_2 + 2\ AsCl_3 \qquad (ah)^{37-39}$$

$$3\ (ClSO_2)_2NH + 2\ AsF_3 \rightarrow 3\ (FSO_2)_2NH + 2\ AsCl_3 \qquad (ai)^{40,41}$$

$$p\text{-}CH_3C_6H_4SO_2Br + AgAsF_6 \rightarrow p\text{-}CH_3C_6H_4SO_2F + AgBr + AsF_5 \qquad (aj)^{42}$$

In the last reaction the intended stabilization of a sulfonium cation was not possible. Exchanges by NOF and ammonium fluorides are discussed with the chemistry of HF.

Formation of OSF_2 from $OSCl_2$ and an acyl fluoride is not synthetically useful:

$$(CF_3)_2CHCOF{\cdot}NR_3 + OSCl_2 \rightarrow OSF_2 + \cdots \qquad \text{(ak)}^{43}$$

$$SO_2Cl_2(SO_2FCl) + [(Me_2N)_3S][Me_3SiF_2] \xrightarrow{RT} SO_2F_2 + [(Me_2N)_2S]Cl + Me_3SiF \qquad \text{(al)}^{44,45}$$

A powerful fluorinating agent is tris(dimethylamino)sulfonium fluoride[44].

In industrial chemistry, HF is the main reagent for metatheses:

$$O_nSCl_2 + 2\ HF \rightarrow O_nSF_2 + 2\ HCl \qquad \text{(am)}^{46-50}$$

where n = 1[46–49], 2[50];

$$SCl_4 + 4\ HF \xrightarrow{CCl_4} SF_4 + 4\ HCl \qquad \text{(an)}^{51}$$

$$SO_2Cl_2 + 2\ HF \rightarrow SO_2F_2 + 2\ HCl \qquad \text{(ao)}^{47}$$

$$SO_2 + Cl_2 + \text{xs HF} \xrightarrow[\text{activated C}]{\text{MF on}} O_2SF_2(O_2SFCl) + 2\ HCl \qquad \text{(ap)}^{50,52,53}$$

$$SCl_2 + Cl_2 + \text{xs HF} \xrightarrow[-HCl]{} SF_4 + \underset{\text{(trace)}}{SF_6} \qquad \text{(aq)}^{54}$$

Sulfur fluorides are prepared from the corresponding chlorides and HF. The preparation of SO_2F_2 or SO_2ClF from SO_2, Cl_2 and HF (alkaline-earth fluorides on active carbon, or active carbon alone as catalysts) at 125°–400°C is patented[52,53]. From SCl_4 in the presence of CCl_4 in organic solvents at −60° to 0°C, SF_4 is formed in ≤92% yields. After the reaction is finished, tertiary amines (e.g., Et_3N, py) are added to bind the resulting HCl or xs HF[51]. Also, under drastic conditions, from SCl_2, Cl_2 and HF, only SF_4 if formed, and SF_6 is detected only spectroscopically[54]. Hydrogen fluoride engages in Cl—F exchange in sulfonyl chlorides:

$$ClSO_2N{=}PCl_3 + \text{xs HF} \xrightarrow[-HCl]{BF_3} FSO_2NH_2 + P(Cl,F)_5 \qquad \text{(ar)}^{55}$$

$$ClSO_2N{=}CCl_2 + \text{xs HF} \xrightarrow[-HCl]{} FSO_2NHCF_3 \qquad \text{(as)}^{56}$$

$$ClSO_2NRR' + \text{xs HF} \xrightarrow[-HCl]{} FSO_2NRR' \qquad \text{(at)}^{57,58}$$

$$RSO_2Cl + KHF_2 \xrightarrow{H_2O} RSO_2F \qquad \text{(au)}^{59}$$

The reactivity of HF is modified by the presence of nitrogen derivatives, e.g., in the presence of amines or NOF, glass is not attacked. The action of $NH_3{\cdot}(HF)_x$, tertiary amine hydrofluorides or tris(hydrofluorides) and pyridinium poly(hydrogenfluoride) with sulfur chlorides is a simple and convenient preparation of S(IV) and S(VI) compounds, e.g.:

$$SO_2Cl_2 + NH_4F \xrightarrow{65°–70°C} SO_2FCl + NH_4Cl \qquad (av)^{60}$$

$$SO_2Cl_2 + NH_4HF_2 \xrightarrow{90°–150°C} O_2SF_2 + NH_4Cl + HCl \qquad (aw)^{61}$$

$$3\ SCl_2 + py(HF)_x \rightarrow SF_4 + S_2Cl_2 \qquad (ax)^{62}$$

$$SO_2Cl_2 + py(HF)_x \rightarrow SO_2ClF \qquad (ay)^{63}$$

$$ClSO_2CF_2C(O)F + [Et_3NH]F \rightarrow FSO_2CF_2C(O)F \qquad (az)^{64}$$

Pyridium poly(hydrogenfluoride), prepared from py and anhyd HF at −78°C in polyethylene, is a good medium for these reactions, which proceed smoothly under mild conditions (45°C, atm). The $py \cdot (HF)_x$ is polar and serves as a reservoir of fluoride ions[63]. Similar are tertiary amine tris(hydrofluorides), which are stable complexes, distillable in vacuo. They do not attack borosilicate glass[64].

Similar behavior is found for NOF·3 HF[54,65,66] and NOF itself[54,66,67], e.g.:

$$2\ OSCl_2 + NOF \cdot 3\ HF \rightarrow 2\ OSF_2 + NOCl + 3\ HCl \qquad (ba)^{65}$$

$$SCl_2 + NOF(NOF \cdot 3\ HF) \rightarrow SF_4, SF_5Cl, SF_6 \qquad (bb)^{54,66}$$

$$SCl_2 + xs\ NOF \xrightarrow{-23°\text{ to }50°C} OSF_2 + NOCl + S_nCl_m \qquad (bc)^{67}$$

Although under special conditions HF exchanges Cl for F in sulfur halides, HX transforms the fluorides into chlorides or bromides because of the higher H—F bond energy. The same rules govern the exchange of Cl by Br, or Cl (Br) by I, e.g.:

$$SSF_2 + 2\ HCl \rightarrow ClSSCl + 2\ HF \qquad (bd)^{30,68}$$

No trace of the isomeric $SSCl_2$ is found. The S_2Cl_2 formed reacts with xs SSF_2 to give an FSSCl and FSSF mixture[69].

Similar exchanges of HCl with perfluoroorganosulfur derivatives and SF_4 are possible, although thermodynamic data favor the formation of SF_4 from SCl_4 and HF[54], e.g.:

$$SF_4 + 4\ HCl \xrightarrow[-4\ HF]{-78°C} [SCl_4] \rightarrow SCl_2 + Cl_2 \qquad (be)^{70,71}$$

$$(CF_3)_2SF_2 + 2\ HCl \rightarrow [(CF_3)_2SCl_2] \xrightarrow[-2\ HCl]{H_2O} (CF_3)_2SO \qquad (bf)^{72,73}$$

$$CF_3S(O)F + HCl \xrightarrow[-HF]{} CF_3SOCl \qquad (bg)^{29}$$

In S—N—F chemistry, exchange of fluorine bonded to sulfur by hydrogen halides is not possible because the S—N bond is attacked. Chlorides of nonmetals, which form strong bonds to fluorine (e.g., P, Si, B), are used in metatheses:

$$R_fNSF_2 + PCl_5(PCl_3) \rightarrow [R_fNSClF] \qquad (bh)^{74-79}$$

$$R_fNSF_2 + PCl_5(PCl_3) \rightarrow R_fNSCl_2 + P(Cl, F)_{3,5} \qquad (bi)$$

where R_f = perfluoroalkyl[74,75,79], SF_5[76–78];

$$R_fN{=}S(R_f)F + PCl_5 \rightarrow R_fNS(R_f)Cl + P(Cl, F)_5 \qquad \text{(bj)}^{80}$$

$$2\ R_fNSF_2 + SiCl_4 \rightarrow 2\ R_fNSCl_2SiF_4 \qquad \text{(bk)}^{81}$$

$$4\ [Re(CO)_5NSF]^+ + SiX_4 \rightarrow 4\ [Re(CO)_5NSX]^+ + SiF_4 \qquad \text{(bl)}^{82}$$

where X = Cl, Br;

$$NSF_3 + BCl_3 \xrightarrow[-BF_3]{} [NSCl_3] \rightarrow [N(SCl)_2][BCl_4] \qquad \text{(bm)}^{83}$$

$$3\ SF_4 + 4\ BCl_3 \rightarrow 3\ SCl_2 + 4\ BF_3 + 3\ Cl_2 \qquad \text{(bn)}^{84,85}$$

$$CF_3NSFCF_3 + \tfrac{1}{3}\ BCl_3 \rightarrow CF_3NSClCF_3 + \tfrac{1}{3}\ BF_3 \qquad \text{(bo)}^{86}$$

$$F_5TeNSF_2 + \tfrac{2}{3}\ BCl_3 \rightarrow F_5TeNSCl_2 + \tfrac{2}{3}\ BF_3 \qquad \text{(bp)}^{87}$$

S(IV) derivatives undergo this exchange, OSF_2, SF_4 react similarly, and $OPCl_3$ works[75]: S(VI) derivatives are either not attacked at ≤180°C[75], or the products undergo decomposition [Eq. (bm)]. Metatheses with NOCl is possible by chloride ion addition to cationic intermediates, e.g.:

$$COFNSF_2 + AsF_5 \rightarrow [SF_2NCO][AsF_6] \xrightarrow[-NOAsF_6]{NOCl} COFNSFCl \qquad \text{(bq)}^{88}$$

From S_2Br_2 and ICl, S_2Cl_2 is formed:

$$S_2Br_2 + 2\ ICl \rightarrow S_2Cl_2 + 2\ IBr \qquad \text{(br)}^{89}$$

Mixed halides are detected in the thionyl series:

$$OSCl_2 + OSBr_2 \rightleftharpoons 2\ OSBrCl \qquad \text{(bs)}^{90}$$

$$ClSSCl + X_2 \xrightarrow[matrix]{h\nu} SX_2 + SCl_2 \qquad \text{(bt)}^{91}$$

where X = Br, I. Gaseous S_2Cl_2 passed through a microwave discharge with Br_2 or I_2 gives SBr_2 and SI_2. The halogens add to the intermediate $[SSCl]^{\bullet}$ radical[91].

From sulfur chlorides and HBr, S_2Br_2 is formed; in reaction (bu) SBr_2 may be the first product:

$$\underset{(48\%)}{SCl_2 + 2\ aq\ HBr} \xrightarrow[-2\ HCl]{} [SBr_2] \rightarrow \tfrac{1}{2}\ S_2Br_2 + \tfrac{1}{2}\ Br_2 \qquad \text{(bu)}^{92}$$

$$S_nCl_2 + 2\ HBr \xrightarrow[-2\ HCl]{} S_nBr_2 \qquad \text{(bv)}^{93-95}$$

The latter exothermic reaction gives S_2Br_2 (or S_nBr_2) quantitatively, the mixed halide ClSSBr being formed initially[96]. This compound is also present in S_2Cl_2—S_2Br_2

mixtures according to vibrational spectroscopy[97]. Sulfenylchlorides, thionylchloride, RS(O)Cl and $BrSO_2Cl$ react similarly:

$$RSCl + HBr \xrightarrow[-HCl]{} RSBr \quad (bw)^{98-102}$$

$$RSSCl + HBr \xrightarrow[-HCl]{} RSSBr \quad (bx)^{101}$$

$$OSCl_2 + 2\ HBr \xrightarrow[-2\ HCl]{} OSBr_2 \quad (by)^{103-108}$$

$$R_fS(O)Cl + HBr \xrightarrow[-HCl]{} R_fS(O)Br \quad (bz)^{29}$$

$$ArSO_2Cl + HBr \xrightarrow[-HCl]{} ArSO_2Br \quad (ca)^{109}$$

$$S_xN_yCl_z + HBr \rightarrow [S_4N_3]Br \quad (cb)^{110}$$

Chlorinated thiazyl heterocycles will be decomposed by HBr, giving $[S_4N_3]Br$ as stable final products.

Borontribromide is an effective reagent for metatheses:

$$3\ OSX_2 + 2\ BBr_3 \rightarrow 3\ OSBR_2 + 2\ BX_3 \quad (cc)^{111,112}$$

where X = F, Cl.

$$3\ CF_3NSF_2 + 2\ BBr_3 \rightarrow 3\ CF_3NSBr_2 + 2\ BF_3 \quad (cd)^{79}$$

This method has been successfully applied even for the preparation of sulfurbromideimides[79].

Rather unstable sulfur iodides SI_2 [71,91] and S_2I_2 [113–117] are formed from SF_4 [71] or S_2Cl_2 [113–117] and HI in $CFCl_3$ at −78 °C.

$$SF_4 + 4\ HI \rightarrow [SI_2] + I_2 + 4\ HF \quad (ce)$$

$$S_2Cl_2 + 2\ HI \rightarrow S_2I_2 + 2\ HCl \quad (cf)$$

$$OSCl_2 + 2\ HI \rightarrow OSI_2 + 2\ HCl \quad (cg)$$

The initial product in Eq. (cf) is a mixture of S_2I_2 and $S_2I_2 \cdot I_2$ [115]. OSI_2 has not been isolated, but there is spectroscopic evidence for this compound[116]. The existence of O_2SI_2 could not be confirmed.

(R. MEWS)

1. I. Ruppert, *J. Fluorine Chem., 20*, 75 (1982).
2. H. Moissan, P. Lebeau, *Ann. Phys. Chim., 26*, 145 (1902).
3. H. Jonas, *Z. Anorg. Allg. Chem., 265*, 273 (1951).
4. C. Lau, J. Passmore, *J. Fluorine Chem., 6*, 77 (1975).
5. W. Becher, J. Massonne, W. Pohlmeyer, Ger. Offen. 2,217,971 (1973); *Chem. Abstr., 80*, 38,921 (1974).
6. W. Becher, R. Massonne, *Chem.-Ztg., 98*, 117 (1974).
7. H. L. Roberts, N. H. Ray, *J. Chem. Soc.*, 665 (1960).
8. H. L. Roberts, Ger. Offen. 1,087,576 (1959); *Chem. Abstr., 56*, 1145 (1962).
9. C. I. Merrill, M. Lustig, G. H. Cady, US Dept. Comm., Office Tech. Serv., AD 285-205 (1962); *Chem. Abstr., 60*, 6463 (1964).

10. S. Nagase, H. Baba, K. Kodaira, T. Abe, *Bull. Chem. Soc. Jpn.*, *46*, 3435 (1973).
11. S. Nagase, T. Abe, H. Baba, *Bull. Chem. Soc., Jpn.*, *42*, 2062 (1969).
12. E. L. Muetterties, US Pat. 2,937,123 (1960); *Chem. Abstr.*, *54*, 17,123 (1960).
13. T. Abe, J. M. Shreeve, *J. Fluorine Chem.*, *3*, 187 (1973/74).
14. F. Nyman, H. L. Roberts, *J. Chem. Soc.*, 3180 (1962).
15. W. Schmidt, *Monatsh. Chem.*, *85*, 452 (1954).
16. G. H. Sprenger, A. H. Cowley, *J. Fluorine Chem.*, *7*, 333 (1976).
17. W. Gombler, *J. Fluorine Chem.* *9*, 233 (1977).
18. R. A. De Marco, T. A. Kovacina, W. B. Fox, *J. Fluorine Chem.* *6*, 93 (1975).
19. R. L. Kirchmeier, G. H. Sprenger, J. M. Shreeve, *Inorg. Nucl. Chem. Lett.*, *11*, 699 (1975).
20. A. Haas, in *Gmelins Handbook of Inorganic Chem.*, *Perfluorohalogenoorgano Compounds of the Main Group Elements*, Vol. 12/2, Verlag Chemie, Weinheim, 1973, pp. 137ff.
21. T. Gramstadt, R. N. Haszeldine, *J. Chem. Soc.*, 173 (1956).
22. T. Gramstadt, R. N. Haszeldine, *J. Chem. Soc.*, 2640 (1957).
23. J. Burdon, I. Farazmand, M. Stacey, J. C. Tatlow, *J. Chem. Soc*, 2574 (1957).
24. T. J. Brice, P. W. Trott, US Pat. 2,732,398 (1956); *Chem. Abstr.*, *50*, 13,982 (1956).
25. D. T. Sauer, J. M. Shreeve, *Z. Anorg. Allg. Chem.*, *385*, 113 (1971).
26. T. Abe, J. M. Shreeve, *J. Fluorine Chem.*, *3*, 17 (1973/74).
27. S. G. Metcalf, J. M. Shreeve, *Inorg. Chem.*, *11*, 1631 (1972).
28. A. J. Banister, R. J. Hey, J. Passmore, M. N. S. Rao, *J. Fluorine Chem.*, *21*, 429 (1982).
29. C. T. Ratcliffe, J. M. Shreeve, *J. Am. Chem. Soc.*, *90*, 5403 (1968).
30. F. Seel, D. Gölitz, *Z. Anorg. Allg. Chem.*, *327*, 32 (1964).
31. F. Seel, H. Jonas, L. Riehl, W. Langer, *Angew. Chem.*, *67*, 32 (1955).
32. F. Seel, L. Riehl, *Z. Anorg. Allg. Chem.*, *282*, 293 (1955).
33. A. Zappel, *Chem. Ber.*, *94*, 873 (1961).
34. F. Seel, J. Boudier, W. Gombler, *Chem. Ber.*, *102*, 443 (1969).
35. I. Stahl, R. Mews, O. Glemser, *Chem. Ber.*, *113*, 2430 (1980).
36. O. Glemser, U. Biermann, *Chem. Ber.*, *100*, 2484 (1967).
37. H. Moissan, P. Lebeau, *C. R. Hebd. Seances Acad. Sci.*, *130*, 1436 (1900).
38. W. Steinkopf, J. Herold, *J. Prakt. Chem.*, *101*, 79 (1921).
39. W. C. Smith, E. L. Muetterties, *Inorg. Synth.*, *6*, 162 (1960).
40. J. K. Ruff, *Inorg. Chem.*, *6*, 2108 (1967).
41. J. K. Ruff, M. Lustig, *Inorg. Synth.*, *11*, 138 (1968).
42. E. Lindner, H. Weber, *Z. Naturforsch.*, *Teil B*, *22*, 1243 (1967).
43. Yu. A. Cheburkov, N. Mukhamadaliev, I. L. Knunyants, *Acad. Sci. USSR Bull. Chem. Sci.*, 2049 (1966).
44. W. J. Middleton, US Pat. 3,940,402 (1976); *Chem. Abstr.*, *85*, 6388 (1976).
45. W. Heilemann, R. Mews, unpublished results; W. Heilemann, Staatsexamensarbeit, Göttingen, 1985.
46. J. Söll, *FIAT Rev. Ger. Sci.*, *23*, 192 (1949).
47. K. Wiechert, *Z. Anorg. Allg. Chem.*, *261*, 310 (1950).
48. U. Wannagat, G. Mennicken, *Z. Anorg. Allg. Chem.*, *278*, 310 (1955).
49. G. Urban, Ger. Offen. 2,831,413 (1978); *Chem. Abstr.*, *92*, 183,091 (1980).
50. L. J. Belf, Br. Pat. 727,062 (1955); *Chem. Abstr.* *49*, 11,970 (1955).
51. R. Appel, Ger. Offen. 2,363,679 (1975); *Chem. Abstr.*, *83*, 181,599 (1975).
52. E. S. Jones, M. A. Robinson, R. E. Eibeck, US Pat. 4,003,984 (1977); *Chem. Abstr.*, *86*, 108,788 (1977); Ger. Offen. 2,643,521 (1977); *Chem. Abstr.*, *87*, 55,235 (1977).
53. D. M. Cook, D. C. Gustafson, US Pat. 4,102,987 (1978); *Chem. Abstr.*, *90*, 124,053 (1979).
54. J. L. Russell, A. W. Jache, *J. Fluorine Chem.*, *7*, 205 (1976).
55. L. K. Huber, H. C. Mandell, Jr., US Pat. 3,431,088 (1969), *Chem. Abstr.*, *70*, 89,299 (1969).
56. H. W. Roesky, *Angew. Chem., Int. Ed. Engl.*, *7*, 63 (1968).
57. K. Grohn, Ger. Offen. 1,943,233 (1971); *Chem Abstr.*, *74*, 111,546 (1971).
58. H. W. Roesky, S. Tutkunkardes, *Z. Anorg. Allg. Chem.*, *379*, 147 (1970).
56. T. Gramstadt, R. N. Haszeldine, *J. Chem. Soc.*, 173 (1956).
60. J. Cueilleron, Y. Monteil, *Bull. Soc. Chim. Fr.*, 2172 (1965).
61. C. Woole, R. O. Michael, US Pat. 3,687,626; *Chem. Abstr.*, *78*, 45,815 (1973).
62. D. C. England, M. A. Dietrich, R. V. Lindsey, *J. Am. Chem. Soc.*, *82*, 6181 (1960).
63. G. A. Olah, M. R. Bruce, J. Welch, *Inorg. Chem.*, *16*, 2637 (1977).
64. R. Franz, *J. Fluorine Chem.*, *15*, 423 (1980).

65. F. Seel, W. Birnkraut, D. Werner, *Angew. Chem.*, *73*, 806 (1961).
66. R. E. Eibeck, R. E. Booth, US Pat. 4,082,839 (1978); *Chem. Abstr.*, *89*, 26,943 (1978); Ger. Offen. 2,629,264 (1977); *Chem Abstr.*, *86*, 173,741 (1977).
67. L. G. Anello, C. Woolf, US Pat. 3,074,781 (1963); *Chem. Abstr.*, *58*, 12,212 (1963).
68. F. Seel, *Adv. Inorg. Chem. Radiochem.*, *16*, 297 (1974).
69. F. Seel, *Chimia*, *22*, 79 (1968).
70. R. Appel, J. R. Lundehn, E. Lassmann, *Chem. Ber.*, *109*, 2442 (1976).
71. D. K. Padma, *Phosphorus Sulfur*, *3*, 19 (1977).
72. D. T. Sauer, J. M. Shreeve, *J. Fluorine Chem.*, *1*, 1 (1971/72).
73. T. Abe, J. M. Shreeve, *J. Fluorine Chem.*, *3*, 423 (1973).
74. H. W. Roesky, R. Mews, *Angew. Chem.*, *Int. Ed. Engl.*, *7*, 217 (1968).
75. M. D. Vorob'ev, A. S. Filatov, M. A. Englin, *J. Gen. Chem. USSR* (*Engl. Transl.*), *42*, 1935 (1972).
76. R. Höfer, O. Glemser, *Z. Anorg. Allg. Chem.*, *416*, 263 (1975).
77. A. F. Clifford, A. Shanzer, *J. Fluorine Chem.*, *7*, 65 (1976).
78. I. Stahl, R. Mews, O. Glemser, cited in J. M. Shreeve, in *Sulfur in Organic and Inorganic Chemistry*, A. Senning, ed., Vol. 4, Marcel Dekker, New York, 1982, pp. 131ff.
79. W. Leidinger, W. Sundermeyer, *Chem. Ber.*, *115*, 2892 (1982).
80. C. Lentsch, O. Glemser, *Z. Naturforsch.*, *Teil B*, *37*, 306 (1982).
81. O. Glemser, S. P. v. Halasz, U. Biermann, *Inorg. Nucl. Chem. Lett.*, *4*, 591 (1968).
82. G. Hartmann, R. Mews, *Z. Naturforsch*, *Teil B*, *40*, 343 (1985).
83. O. Glemser, B. Krebs, J. Wegener, E. Kindler, *Angew. Chem. Int. Ed. Engl.*, *8*, 598 (1969).
84. F. A. Cotton, J. W. George, *J. Inorg. Nucl. Chem.*, *7*, 397 (1958).
85. N. Bartlett, P. L. Robinson, *J. Chem. Soc.*, 3417 (1961).
86. W. Leidinger, W. Sundermeyer, *Z. Naturforsch.*, *Teil B*, *37*, 781 (1982).
87. H. Hartl, P. Huppmann, D. Lentz, K. Seppelt, *Inorg. Chem.*, *22*, 2183 (1983).
88. R. Mews, *J. Fluorine Chem.*, *4*, 445 (1974).
89. J. B. Hannay, *J. Chem. Soc.*, 823 (1873).
90 R. Steudel, D. Lautenbach, *Z. Naturforsch.*, *Teil B*, *24*, 350 (1969).
91. M. Feuerhahn, G. Vahl, *Inorg. Nucl. Chem. Lett.*, *16*, 5 (1982).
92. J. N. Ospenson, US Pat. 2,979,383 (1961); *Chem. Abstr.*, *55*, 22,736 (1961).
93. F. Fehér, G. Rempe, *Z. Anorg. Allg. Chem.*, *281*, 161 (1955).
94. F. Fehér, S. Ristic, *Z. Anorg. Allg. Chem.*, *293*, 311 (1958).
95. F. Fehér, in *Handbook of Preparative Inorganic Chemistry*, G. Brauer, ed., 2nd ed., Academic Press, New York, 1975, p. 386.
96. B. Solouki, Ph.D. Diss., Univ. Frankfurt-am-Main (1974).
97. R. Forneris, C. E. Hennies, *J. Mol. Struct.*, *5*, 449 (1970).
98. E. Kühle, E. Klauke, F. Grewe, *Angew. Chem.*, *76*, 807 (1964).
99. J. N. Ospenson, US Pat. 2,821,554 (1958); *Chem. Abstr.*, *52*, 10,145 (1958).
100. J. N. Ospenson, US Pat. 2,824,136 (1958); *Chem. Abstr.*, *52*, 11,890 (1958)..
101. P. S. Magee, in *Sulfur in Organic and Inorganic Chemistry*, A. Senning, ed., Vol. 4, Marcel Dekker, New York, 1982, pp. 284ff.
102. A. Haas, W. Klug, *Chem. Ber.*, *101*, 2609 (1968).
103. A. Besson, *C. R. Hebd. Seances Acad. Sci.*, *122*, 320 (1896).
104. H. A. Mayes, J. R. Partington, *J. Chem. Soc.*, 2594 (1926).
105. F. Govaert, M. Hansens, *Naturw. Tijdschr.* (*Ghent*), *20*, 77 (1938).
106. H. Hibbert, J. C. Pullmann, *Inorg. Synth.*, *1*, 113 (1939).
107. R. C. Elderfield et al., *J. Am. Chem. Soc.*, *68*, 1579 (1946).
108. F. Fehér, in *Handbook of Preparative Inorganic Chemistry* G. Brauer, ed., 2nd ed., Academic Press, New York, 1975, p. 390.
109. O. Lukashevich, *Dokl. Akad. Nauk SSR*, *103*, 627 (1955).
110. H. Vincent, M. P. Berthet, *Z. Anorg. Allg. Chem.*, *471*, 233 (1980).
111. P. M. Druce, M. F. Lappert, P. N. K. Riley, *J. Chem. Soc.*, *Chem. Commun.*, 486 (1967).
112. P. M. Druce, M. F. Lappert, *J. Chem. Soc.*, *A*, 3595 (1971).
113. A. R. Vasudeva Murthy, *Proc. Indian Acad. Sc.*, *A*, *37*, 17 (1952).
114. D. K. Padma, *Indian J. Chem.*, *12*, 417 (1978).
115. G. Vahl, R. Minkwitz, *Z. Anorg. Allg. Chem.*, *443*, 217 (1978).
116. K. Manzel, R. Minkwitz, *Z. Anorg. Allg. Chem.*, *441*, 165 (1978).
117. G. Krummel, K. Minkwitz, *Inorg. Nucl. Chem. Lett.*, *13*, 213 (1978).

2.3.12.2.3. to Give the Se–Halogen Bond.

Exchanges comparable to those with sulfur occur. The Se—X bonds (X = Cl, Br, I) are more favored than in the sulfur system, but metatheses with F_2 or halogen fluorides also lead to a complete exchange and formation of SeF_4:

$$Se_2Cl_2 + 4\ F_2 \rightarrow 2\ SeF_4 + Cl_2 \qquad \text{(a)}^{1}$$

$$SeCl_4 + 3\ ClF_3 \rightarrow SeF_4 + \underset{\text{(small amounts)}}{SeF_5Cl} + 3\ Cl_2 \qquad \text{(b)}^{2}$$

$$OSeCl_2 + F_2 \rightarrow F_5SeOOSeF_5 + F_5SeOF \qquad \text{(c)}^{3}$$

From $OSeCl_2$, the peroxide and small amounts of SeF_5OF are isolated. Nonoxidative Cl—F exchanges occur with the fluorosulfinate ion and HF:

$$6\ NOSO_2F + SeCl_4 \rightarrow (NO)_2SeF_6 + 4\ NOCl \qquad \text{(d)}^{4}$$

$$2\ KSO_2F\ (\text{in xs}) + OSeCl_2 \rightarrow OSeF_2 + 2\ KCl + 2\ SO_2 \qquad \text{(e)}^{5}$$

$$KHF_2 + OSeCl_2 \xrightarrow{150°C} SeOF_2 + KCl + HCl \qquad \text{(f)}^{5}$$

$$SeOCl_2 + \text{xs}\ HF \rightarrow OSeFCl + OSeF_2 + HCl \qquad \text{(g)}^{6}$$

Yields in reactions (e) and (f) are poor. The solid residue of Eq. (g) contains the mixed halide; OSeFCl is also obtained by redistribution of $OSeF_2$ and $OSeCl_2$:

$$OSeCl_2 + OSeF_2 \rightleftharpoons 2\ OSeFCl \qquad \text{(h)}^{7,8}$$

$$2\ Hg(OSeF_5)_2 + SeCl_4 \xrightarrow[-2\ HgCl_2]{CFCl_3,\ 10°C} Se(OSeF_5)_4$$

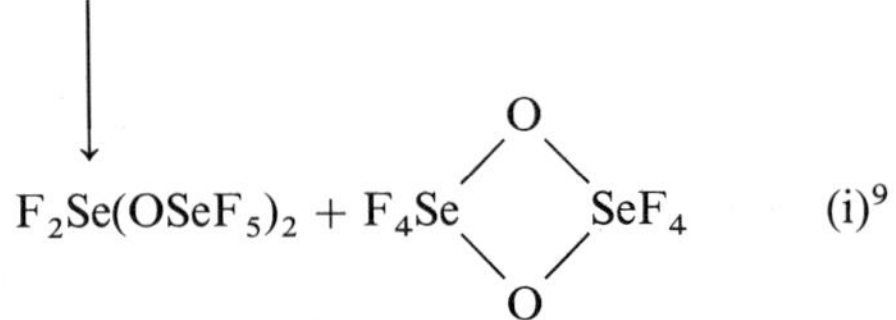

In Eq. (h) the $OSeF_5$ group acts as exchange reagent. Similar to SF_4, SeF_4 can be applied in organic chemistry in metatheses, e.g.:

$$SeF_4 + 4\ PhC(O)Cl \rightarrow SeCl_4 + 4\ PhC(O)F \qquad \text{(j)}^{5}$$

Boron halides transform SeF into other halogen derivatives:

$$3\ SeF_4 + 4\ BCl_3 \rightarrow 3\ SeCl_4 + 4\ BF_3 \qquad \text{(k)}^{10,11}$$

Metathesis of F–Br is possible via HBr:

$$C_5H_5N\cdot SeF_4\cdot OEt_2 + 6\ HBr + C_5H_5N \xrightarrow[-4\ HF]{} [C_5H_5NH]_2[SeBr_6] \qquad \text{(l)}^{12}$$

A Cl–Br exchange is observed in the $OSeBr_2$–$OSeCl_2$ equilibrium by ^{77}Se NMR:

$$OSeCl_2 + OSeBr_2 \rightleftharpoons 2\ OSeBrCl \qquad \text{(m)}^{7,8}$$

The reaction of $OSeCl_2$ with I_2 gives OSeClI and ICl:

$$OSeCl_2 + I_2 \rightarrow OSeClI + ICl \qquad \text{(n)}^{13}$$

Hydrogen halides are used for metathesis, e.g.:

$$R_2SeCl_2 \underset{\text{conc HCl}}{\overset{\text{conc HBr}}{\rightleftarrows}} R_2SeBr_2 \qquad (o)^{14}$$

(R. MEWS)

1. P. L. Goggin, *J. Inorg. Nucl. Chem., 28*, 661 (1966).
2. C. Lau, J. Passmore, *J. Fluorine Chem., 6*, 77 (1975).
3. G. Mitra, G. H. Cady, *J. Am. Chem. Soc., 81*, 2646 (1959).
4. F. Seel, H. Massat, *Z. Anorg. Allg. Chem., 280*, 186 (1955).
5. R. Paetzold, K. Aurich, *Z. Anorg. Allg. Chem., 315*, 72 (1962).
6. K. Wiechert, *Z. Anorg. Allg. Chem., 261*, 310 (1950).
7. T. Birchall, R. J. Gillespie, S. L. Vekris, *Can. J. Chem., 43*, 1672 (1965).
8. J. Milne, *Spectrochim. Acta Part A 38*, 569 (1982).
9. R. Damerius, R. Huppmann, D. Lentz, K. Seppelt, *J. Chem. Soc., Dalton Trans.*, 2821 (1984).
10. R. D. Peacock, *J. Chem. Soc.*, 3617 (1953).
11. N. Bartlett, P. L. Robinson, *J. Chem. Soc.*, 3417 (1961).
12. E. E. Aynsley, G. Hetherington, *J. Chem. Soc.*, 4695 (1954).
13. K. Niendorf, from: K. Niendorf, R. Paetzold, *J. Mol. Struct., 19*, 693 (1973).
14. H. Funk, W. Papenroth, *J. Prakt. Chem., 8*, 256 (1959).

2.3.12.2.4. to Give the Te–Halogen Bond.

In the reaction of $TeCl_2$ and $TeBr_2$ with F_2 diluted by N_2(1:9 or 1:10), the tetrafluoride is an intermediate[1,2] and the final product is TeF_6:

$$TeX_2 + 2\,F_2\text{–}N_2 \rightarrow (TeF_4) + X_2 \xrightarrow{F_2} TeF_6 \qquad (a)^{1,2}$$

where X = Cl, Br;

$$TeX_4 + F_2\text{–}N_2 \xrightarrow{25°C} TeF_6 + TeF_5X + X_2 \qquad (b)^{3}$$

where X = Cl, Br

$$TeI_4 + F_2\text{–}N_2 \xrightarrow{25°C} TeF_6 + IF_5 \qquad (c)^{3}$$

Large amounts of TeF_6 are formed in the preparation of TeF_5Cl and TeF_5Br from the tetrahalides and F_2 at RT, the Te—I bond is cleaved. The chlorofluoride is prepared in high yield from $TeCl_4$ and ClF when ClF is condensed onto $TeCl_4$ in small portions and the mixture warmed to ambient T:

$$TeCl_4 + 5\,ClF \rightarrow F_5TeCl + 2\,Cl_2 \qquad (d)^{4}$$

Xenon difluoride reacts with Ph_3SeCl to exchange halogen and give a simultaneous oxidation:

$$2\,Ph_3TeCl + 3\,XeF_2 \rightarrow 2\,Ph_3TeF_3 + 3\,Xe + Cl_2 \qquad (e)^{5}$$

Nitrosylfluorosulfinate undergoes metathesis with $TeCl_4$ in liq SO_2:

$$TeCl_4 + 6\,[NO][SO_2F] \rightarrow (NO)_2TeF_6 + 4\,NOCl + 6\,SO_2 \qquad (f)^{6}$$

whereas hydrogen halides transform the Te—F into the Te—X bond:

$$2\,[C_5H_5NH][TeF_5] + 10\,HX \rightarrow [C_5H_5NH]_2[TeX_6] + [TeX_4] + 10\,HF \qquad (g)^{7}$$

where X = Cl, Br, I. Tellurium–Cl derivatives are formed from the corresponding bromides and Cl_2:

$$2\ TeF_5Br + Cl_2 \xrightarrow{h\nu} 2\ TeF_5Cl + Br_2 \qquad (h)^3$$

$$\text{“}TeBr_2\text{”} + Cl_2 \xrightarrow[-Br_2]{} [TeCl_2] \rightarrow TeCl_4 \qquad (i)^2$$

$$3\ TeCl_4 + 4\ BBr_3 \rightarrow 3\ TeBr_4 + 4\ BCl_3 \qquad (j)^8$$

For $TeBr_4$, the quantitative conversion of $TeCl_4$ by BBr_3 in benzene can be used[8].

Equilibration is observed when a mixture of $TeCl_4$ and $TeBr_4$ is heated to 300°C for a few minutes in a closed system[9]:

$$TeCl_4 + TeBr_4 \rightleftharpoons 2\ TeCl_2Br_2 \qquad (k)^{9,10}$$

For dialkyltellurium dibromides, HBr is used with the corresponding hydroxy chlorides:

$$R_2Te(OH)Cl + 2\ HBr \rightarrow R_2TeBr_2 + HCl + H_2O \qquad (l)^{11}$$

From $TeBr_4$ and organic iodides, TeI_4 is formed:

$$TeBr_4 + 4\ C_2H_5I \rightarrow TeI_4 + 4\ C_2H_5Br \qquad (m)^{12}$$

$$TeBr_4 + 4\ ICN \rightarrow TeI_4 + 4\ BrCN \qquad (n)^{13}$$

(R. MEWS)

1. E. E. Aynsley, *J. Chem. Soc.*, 3016 (1953).
2. E. E. Aynsley, R. H. Watson, *J. Chem. Soc.*, 2603 (1955).
3. G. W. Fraser, R. D. Peacock, P. M. Walker, *J. Chem. Soc., Chem. Commun.*, 1257 (1968).
4. C. Lau, J. Passmore, *Inorg. Chem.*, *13*, 2278 (1974).
5. K. Alam, A. F. Janzen, *J. Fluorine Chem.*, *27*, 467 (1985).
6. F. Seel, H. Massat, *Z. Anorg. Allg. Chem.*, *280*, 186 (1955).
7. E. E. Aynsley, G. Hetherington, *J. Chem. Soc.*, 2802 (1953).
8. M. T. Chen, J. W. George, *J. Inorg. Nucl. Chem.*, *34*, 3261 (1972).
9. R. Beattie, O. Bizri, H. E. Blayden, S. O. Brumbach, A. Bukorzky, T. R. Gilson, R. Moss, B. A. Phillips, *J. Chem. Soc., Dalton Trans.*, 1147 (1974).
10. G. A. Ozin, A. van der Voet, *J. Chem. Soc., Chem. Commun.*, 1489, (1970); *Can. J. Chem. 49*, 704 (1971).
11. Cf., e.g., R. A. Zingaro, K. Irgolic, in *Tellurium*, W. C. Cooper, ed., van Nostrand, Reinhold, New York, 1971, p. 241.
12. E. Montignie, *Ann. Pharm. Fr.*, *5*, 239 (1947).
13. E. Montignie, *Z. Anorg. Allg. Chem.*, *315*, 102 (1962).

2.3.13. Preparation of Astatine–Group-VIB Element Bonds

2.3.13.1. Preparation of Astatine–Oxygen Bonds.

The $[AtO_3]^-$ ion is formed from lower oxidation states of At by treatment with $[OCl]^-$ in base or with H_5IO_6, $[S_2O_8]^{2-}$ or Ce(IV) in acid[1,2]. Heating to 100°C may be required to effect complete oxidation. Astatine(VII) in H_5AtO_6 or $[AtO_4]^-$ is formed by oxidation of At with XeF_2 in base[3] or by electrochemical oxidation[4]. Oxidation with $[S_2O_8]^{2-}$ or $[OCl]^-$ may also produce At(VII)[3]. Oxidation of aq At with Br_2 or

$[Cr_2O_7]^{2-}$, or with Cl_2 in Cl^- produces intermediate oxidation states of At, some of which may have At—O bonds[1,2], but these species are not confirmed. Astatine in these intermediate states forms neutral and anionic complexes with oxyanions such as $[NO_3]^-$, $[SO_4]^{2-}$ and $[Cr_2O_7]$[1-5]. Intermediate positive states of At coprecipitate with insoluble oxysalts of monovalent cations, such as the dichromates, iodates and phosphotungstates of Ag^+, Tl^+ and Cs^+, presumably to form salts between cationic At and the oxyanion[6].

The $[AtO_2]^-$ ion has been claimed to result from alkaline hydrolysis of the halocomplexes formed by treating persulfate-oxidized At with HX (see §2.2.6.3)[7].

Mass spectral lines attributed to $[AtO]^+$, $[AtOH]^+$ and $[AtOH_2]^+$ have been found after passage through a plasma ion source[8]. This indicates the formation of these cationic moieties in the source presumably by reaction with impurities containing O and H.

The compound $C_6H_5AtO_2$ is prepared by treating $C_6H_5AtCl_2$ with alkaline $[OCl]^-$[9]:

$$C_6H_5AtCl_2 + [OCl]^- + 2\ OH^- \rightarrow C_6H_5AtO_2 + 3\ Cl^- + H_2O \qquad \text{(a)}$$

The $C_6H_5AtCl_2$ is formed by chlorination of C_6H_5At:

$$C_6H_5At + Cl_2 \rightarrow C_6H_5AtCl_2 \qquad \text{(b)}$$

(W. D. LEE)

1. G. L. Johnson, R. F. Leininger, E. Segrè, *J. Chem. Phys.*, *17*, 1 (1949).
2. E. H. Appelman, *J. Am. Chem. Soc.*, *83*, 805 (1961).
3. V. A. Khalkin, Yu. V. Norseev, V. D. Nefedov, M. A. Toropova, W. I. Kuzin, *Dokl. Akad. Nauk SSSR, Ser. Khim.*, *195*, 855 (1970).
4. G. A. Nagy, P. Groz, V. A. Khalkin, D. K. Tyung, Yu. V. Norseev, *Reports Central Research Inst. Phys. (Budapest)*, *18*, 173 (1970); *Chem. Abstr.*, *75*, 14,412 (1971).
5. D. K. Tyung, I. V. Dudova, V. A. Khalkin, *Radiokhimiya*, *15*, 552 (1973).
6. Yu. V. Norseev, V. A. Khalkin, *Chem. Zvesti*, *21*, 602 (1967); *Chem. Abstr.*, *68*, 88,482 (1968).
7. R. Dreyer, I. Dreyer, V. A. Khalkin, M. Milanov, *Radiochem. Radioanal. Lett.*, *40*, 145 (1979).
8. N. A. Golovkov, I. I. Gromova, M. Janicki, Yu. V. Norseev, V. G. Sandukovskii, L. Vasáros, *Radiochem. Radioanal. Lett.*, *44*, 67 (1980).
9. V. D. Nefedov, Yu. V. Norseev, Kh. Savlevich, E. N. Sinotova, M. A. Toropova, V. A. Khalkin, *Dokl. Akad. Nauk SSSR, Ser. Khim.*, *144*, 806 (1962).

2.3.13.2. Other Astatine–Group-VIB Element Bonds.

When $[S_2O_3]^{2-}$ decomposes in acid solution containing At^- obtained by reduction with alkaline $[SO_3]^{2-}$, the At is bound to the precipitated sulfur. Formation of the $[S_7At]^+$ species is suggested[1].

Reaction of At^- with proteins containing S–H groups in the presence of H_2O_2 results in incorporation of the At in high yield[2]. The stability of the astatoprotein product is enhanced for proteins containing S–H groups. Formation of S–At bonds is indicated.

Addition of an astatine solution to an electrolyte solution of $HClO_4$ and $NH_4[SCN]$ results in the formation of an anionic At-compound formulated as $[At(SCN)_2]^-$[3].

Cationic astatine compounds with thiourea(THS) and thioacetamide(TAA) of stoichiometry $[At(THS)_2]^+$ and $[At(TAA)_2]^+$ form[4].

A series of astatine-selenocarbamide coordination compounds form from the reaction of astatine or AtX_2 ($X = Cl^-$, Br^-) with a neutral selenocarbamide ligand in $HClO_4$ at pH ranges of 1–5[5].

When At^- ion coprecipitates with Te formed in situ by reduction with SO_2 in 3 M HCl[6], an At–Te bond may be formed. Acidification of astatine with 2 N nitric acid, mixture with finely ground elemental Te and washing with distilled water produces an At–Te colloid[7].

(W. D. LEE)

1. G. W. M. Visser, E. L. Diemer, *Radiochimica Acta, 33* 145 (1983).
2. G. W. M. Visser, E. L. Diemer, F. M. Kaspersen, *Int. J. Appl. Radiat. Isot., 32*, 905 (1981).
3. R. Dreyer, I. Dreyer, M. Pfeiffer, F. Roesch, *Radiochem. Radioanal. Lett., 55*, 207 (1983).
4. R. Dreyer, I. Dreyer, S. Fischer, H. Hartmann, F. Roesch, *J. Radioanal. Nucl. Chem., 96*, 333 (1985).
5. S. Fischer, R. Dreyer, H. Hussein, M. Weber, H. Hartmann, *J. Radioanal. Nucl. Chem., 119*, 181 (1987).
6. D. R. Corson, K. R. MacKenzie, E. Segrè, *Phys. Rev., 58*, 672 (1940).
7. W. D. Bloomer, R. D. Neirinckx, S. J. Adelstein, P. R. Gordon, T. J. Ruth, A. P. Wolf, *Science, 212*, 340 (1981).

2.4. Formation of Bonds between Halogens and Group VB (N, P, As, Sb, Bi) Elements

2.4.1. Introduction

Formation of the halides of the group-VB elements can utilize the elementary forms in direct combination with the halogens themselves or with hydrogen or other inorganic and organic halides. The process can also proceed by cleavage of a group-VB hydride, or other compounds containing bonds to an element drawn from the main groups headed by carbon, oxygen, nitrogen, etc. Further oxidative additions to low oxidation state group-VB compounds takes place with halogens, hydrogen or other inorganic or organic halides. Metatheses exchange one halide for another.

(J. J. ZUCKERMAN)

2.4.2. from the Elements

2.4.2.1. to Give Group-VB Fluorides.

The direct combination of N_2 and F_2 is only observed in glow discharges. At low pressure and $-196°C$ NF_3 and N_2F_4 are formed[1–3].

Both white and red phosphorus ignite spontaneously in F_2 at RT to yield[4] PF_5.

Arsenic pentafluoride is obtained from the reaction of F_2 with As[5]. Incomplete fluorination of Sb forms[6,7] SbF_3 and $Sb_{11}F_{43}$.

When powdered Bi is fluorinated by F_2, BiF_5 is found[8]. Molten Bi gives[9] BiF_5 that contains some BiF_3.

(G.-V. RÖSCHENTHALER)

1. W. Maya, *Inorg. Chem., 3*, 1063 (1964).
2. I. V. Nikitin, V. Ya. Rosolovskii, *Izv. Akad. Nauk SSSR, Ser. Khim.*, 165 (1969).
3. I. V. Nikitin, V. Ya. Rosolovskii, *Izv. Akad. Nauk SSSR, Ser. Khim.*, 1464 (1970).
4. P. Gross, C. Hayman, D. L. Levi, M. C. Stuart, *US Dept. Comm., Office Tech. Serv., P. B. Rep.*, 153, 445, 1(1960); *Chem. Abstr., 58*, 7435 (1963).
5. F. Seel, O. Detmer, *Z. Anorg. Allg. Chem., 301*, 113 (1959).
6. A. J. Hewitt, J. H. Holloway, B. Frlec, *J. Fluorine Chem., 5*, 169 (1975).
7. A. J. Edwards, D. R. Slim, *J. Chem. Soc., Chem. Commun.*, 178 (1974).
8. O. Glemser, *Angew. Chem., Int. Ed. Engl., 4*, 446 (1965).
9. J. Fischer, E. Rudzitis, *J. Am. Chem. Soc., 81*, 6375 (1959).

2.4.2.2. to Give Group-VB Chlorides.

Phosphorus trichloride is synthesized on a large scale by the chlorination of xs white phosphorus dissolved in a refluxing solution. In the laboratory preparation, red phosphorus is used[1,2].

At elevated T, $AsCl_3$ is obtained[3] from As and Cl_2. Stoichiometric Sb reacts with Cl_2 to give[4] $SbCl_3$. Molten Bi and Cl_2 give[5] $BiCl_3$.

(G.-V. RÖSCHENTHALER)

1. M. C. Forbes, C. A. Rosewell, R. N. Maxson, *Inorg. Synth.*, *2*, 145 (1946).
2. V. Ya. Chernykh, N. D. Talanov, I. N. Smirnova, *Khim. Khim. Tekhnol.*, 220 (1963); *Chem. Abstr.*, *61*, 6619 (1964).
3. R. C. Smith, *Ind. Eng. Chem.*, *11*, 109 (1919).
4. I. Lindquist, A. Niggli, *J. Inorg. Nucl. Chem.*, *2*, 345 (1956).
5. H. A. Skinner, L. E. Sutton, *Trans. Faraday Soc.*, *36*, 681 (1940).

2.4.2.3. to Give Group-VB Bromides.

Phosphorus tribromide is synthesized[1] by bromination of xs white phosphorus in PBr_3 as the solvent or in benzene or CCl_4.

Arsenic and Br_2 react to give[2] $AsBr_3$. Antimony powder and Br_2 in MeC(O)OEt yield[3] $SbBr_3$. **(This reaction avoids the hazards of using CS_2 as a solvent.)** High-purity $SbBr_3$ is obtained in vacuum at 150°C[4].

Bromine and a suspension of Bi in MeOH yield[5] $BiBr_3$.

(G.-V. RÖSCHENTHALER)

1. M. C. Forbes, C. A. Rosewell, R. N. Maxson, *Inorg. Synth.*, *2*, 145 (1946).
2. E. Jory, *J. Pharm. Chim.*, *12*, 312 (1900).
3. N. K. Jha, A. Kumari, *Educ. Chem.*, *15*, 164 (1964).
4. G. G. Gospodinov, D. B. Gospodinova, *Russ. J. Inorg. Chem. (Engl. Trans.)*, *22*, 1839 (1977).
5. H. M. Haendler, F. A. Johnson, D. S. Crocket, *J. Am. Chem. Soc.*, *80*, 2662 (1958).

2.4.2.4. to Give Group-VB Iodides.

Phosphorus triiodide is prepared by direct action of I_2 on white phosphorus in CS_2 or on red phosphorus[1]. Tetraiododiphosphine is obtained by heating a mixture of I_2 with red phosphorus at 180°–190°C, or from I_2 with white phosphorus in CS_2[2].

Arsenic triiodide is synthesized from I_2 and As in CS_2 or Et_2O[3] or under vacuum[4]. When As and I_2 are heated in a sealed tube at 260°C in octahydrophenanthrene, As_2I_4 is obtained[5,6].

In toluene, or without a solvent, Sb and I_2 form[4,7] SbI_3. Iodine and Bi react to give[4,8] BiI_3 at 150°–180°C.

(G.-V. RÖSCHENTHALER)

1. F. E. E. Germann, R. N. Traxler, *J. Am. Chem. Soc.*, *49*, 307 (1927).
2. M. Baudler, *Z. Naturforsch., Teil B*, *13b*, 266 (1958).
3. E. Bamberger, J. Philipp, *Ber. Dtsch. Chem. Ges.*, *14*, 2643 (1881).
4. D. P. Belotskii, Ya. A. Lyuter, T. N. Sushkevich, *Ukr. Khim. Zh.*, *40*, 989 (1974); *Chem. Abstr.*, *81*, 162,741 (1974).
5. M. Baudler, H. J. Stassen, *Z. Anorg. Allg. Chem.*, *343*, 244 (1966).
6. M. Baudler, H. J. Stassen, *Z. Anorg. Allg. Chem.*, *345*, 182 (1966).
7. J. C. Bailar, P. F. Cundy, *Inorg. Synth.*, *1*, 104 (1939).
8. G. W. Watt, W. W. Hakki, G. R. Choppin, *Inorg. Synth.*, *4*, 114 (1953).

2.4.3. from Halogenation of Group-VB Elements

2.4.3.1. by Hydrogen Fluoride To Give Group-VB Fluorides.

With HF in a sealed system at 200°C, red phosphorus forms[1] PF_3. Conversion to PF_3 can be increased by traces of soluble fluorides. Arsenic, Sb and Bi yield[2] AsF_3, SbF_3 and BiF_3.

(G.-V. RÖSCHENTHALER)

1. S. Kongpricha, A. W. Jache, *J. Fluorine Chem.*, *1*, 79 (1971).
2. E. L. Muetterties, J. E. Castle, *J. Inorg. Nucl. Chem.*, *18*, 148 (1961).

2.4.3.2. by Other Inorganic Halides to Give Group-VB Halides

2.4.3.2.1. from Group-IIIB Halides.

Boron triiodide reacts with As to give[1] AsI_3:

$$BI_3 + \tfrac{1}{2} As_4 \rightarrow AsI_3 + BAs \qquad \text{(a)}$$

(G.-V. RÖSCHENTHALER)

1. J. Bonix, R. Hillel, *J. Less-Common Met.*, *47*, 67 (1976).

2.4.3.2.2. from Group-IVB Halides.

When CF_3I and white or red phosphorus are heated in a steel cylinder at 170°–220°C, CF_3PI_2 and $(CF_3)_2PI$[1,2] are formed. With powdered As or Sb, CF_3AsI_2, $(CF_3)_2AsI$[3,4], CF_3SbI_2 and $(CF_3)_2SbI$[5] are obtained, respectively.

The reaction of white phosphorus exposed to γ-rays in CCl_4 gives[6] CCl_3PCl_2. Similar results are obtained using visible light[6]. White phosphorus and CBr_3H yield CBr_3PBr_2 and $(CBr_3)_2PBr$ at 190°C[7], whereas the diphosphines $CCl_3(Br)PPBr(CCl_3)$ and $CCl_3(Br)PPBr_2$ are synthesized from CCl_3Br and P_4 in 40% yield[8].

The phosphines $(CHF_2)_2PI$ and CHF_2PI_2 are made in high yield from P_4 and CHF_2I at 190°C[9].

(G.-V. RÖSCHENTHALER)

1. F. W. Bennett, G. R. A. Brandt, H. J. Emeléus, R. N. Haszeldine, *Nature (London)*, *166*, 225 (1950).
2. A. B. Burg, W. Mahler, A. J. Bilbo, C. P. Haber, D. L. Herring, *J. Am. Chem. Soc.*, *79*, 247 (1957).
3. F. W. Bennett, H. J. Emeléus, R. N. Haszeldine, *J. Chem. Soc.*, 1565 (1953).
4. E. G. Walaschewski, *Chem. Ber.*, *86*, 272 (1953).
5. J. W. Dale, H. J. Emeléus, R. N. Haszeldine, J. H. Moss, *J. Chem. Soc.*, 3708 (1957).
6. K. D. Asmus, A. Henglein, G. Meissner, D. Perner, *Z. Naturforsch., Teil B*, *19*, 549 (1964).
7. P. L. Airey, H. Drawe, A. Henglein, *Z. Naturforsch., Teil B*, *23*, 916 (1968).
8. P. L. Airey, *Z. Naturforsch., Teil B*, *24*, 1393 (1969).
9. A. B. Burg, *Inorg. Chem.*, *24*, 3342 (1985).

2.4.3.2.3. from Group-VB Halides.

Nitrogen trifluoride oxidizes Bi to give[1] BiF_3:

$$2\,Bi + 6\,NF_3 \rightarrow 2\,N_2F_4 + 2\,BiF_3 \quad (a)$$

Nitrylchloride reacts to form $BiCl_3$[2].

$$2\,Bi + 3\,NO_2Cl \rightarrow BiCl_3 + Bi(NO_2)_3 \quad (b)$$

Tetrachlordiphosphine is obtained[3] by passing an electric discharge through the gas phase above a solution of white phosphorus in PCl_3. The formation of Sb_2I_4 is observed in the Sb–SbI_3 system[4].

(G.-V. RÖSCHENTHALER)

1. C. B. Colburn, A. Kennedy, *J. Am. Chem. Soc., 80*, 5004 (1958).
2. H. H. Baley, H. H. Sisler, *J. Am. Chem. Soc., 74*, 3408 (1952).
3. A. A. Sandoval, H. C. Moser, *Inorg. Chem., 2*, 27 (1963).
4. B. L. Bruner, J. D. Corbett, *J. Inorg. Nucl. Chem., 20*, 62 (1961).

2.4.3.2.4. from Group-VIB Halides.

Fluorosulfuric acid produces[1] AsF_3 with As. Phosphorus pentafluoride is obtained[2] from the element and SF_5Cl. Bismuth trichloride is formed from Bi and $SOCl_2$[3] or S_2Cl_2[4].

(G.-V. RÖSCHENTHALER)

1. A. Engelbrecht, A. Aignesberger, E. Hayek, *Monatsh. Chem., 86*, 470 (1955).
2. H. J. Emeléus, *Angew. Chem., Int. Ed. Engl., 1*, 189 (1962).
3. R. A. Hubbard, W. F. Luder, *J. Am. Chem. Soc., 73*, 1327 (1951).
4. H. Funk, K.-H. Berndt, G. Henze, *Wiss. Z. Univ. Halle-Wittenberg, Math.-Naturwiss. Reihe, 6*, 815 (1956/57); *Chem. Abstr., 54*, 12,860 (1960).

2.4.3.2.5. from Group-VIIB Halides.

Phosphorus pentafluoride is formed[1] by ClF_3 with red phosphorus in liq HF. Bismuth trichloride and tribromide are obtained from Bi and ICl[2], or IBr[3], respectively.

(G.-V. RÖSCHENTHALER)

1. A. F. Clifford, A. G. Morris, *J. Inorg. Nucl. Chem., 5*, 71 (1957).
2. V. Gutmann, *Z. Anorg. Allg. Chem., 264*, 169 (1951).
3. V. Gutmann, *Monatsh. Chem., 82*, 280 (1951).

2.4.3.3. by Organic Halides to Give Group-VB Halides

2.4.3.3.1. from Alkyl Halides.

Methyl chloride passed through a mixture of red phosphorus and Cu at 260°–400°C yields[1,2] $MePCl_2$ and Me_2PCl. Addition of activated carbon offers[3] a direct route to Me_2PCl. Both are prepared similarly, $MeAsCl_2$ and Me_2AsCl[4]. The bromides $MePBr_2$[1], Me_2PBr[1], $MeAsBr_2$[4], Me_2AsBr[4], $MeSbBr_2$[5], Me_2SbBr[5] and $SbBr_3$[5] are synthesized from MeBr, P, As or Sb and a Cu catalyst at 350°C. A better yield of $MePCl_2$ and

Me_2PCl is obtained with white phosphorus and H_2 in the gas phase over an activated carbon catalyst. The separations of $MePCl_2$ and Me_2PCl is achieved[5] by saturating the mixture with HCl. Ethyl chloride and red phosphorus give $EtPCl_2$ and Et_2PCl under similar conditions[1]. n-Octylbromide, white phosphorus, PBr_3 and I_2 as a catalyst yield[7] $n\text{-}C_8H_{17}PBr_2$ and $(n\text{-}C_8H_{17})_2PBr$.

In a steel cylinder, $PhCH_2Cl$ reacts with white phosphorus and PCl_3 in the presence of I_2 at 300°C to form $PhCH_2PCl_2$ [2,7] and $(PhCH_2)_2PCl$[7]. Under similar conditions, $PhCH_2PBr_2$ and $(PhCH_2)_2PBr$ are formed with $PhCH_2Br$ [8,9], p-$(Cl_2PCH_2)_2C_6H_4$ with p-$(ClCH_2)_2C_6H_4$ [10], $Cl_2PCH_2CH_2C_6H_5$ with $ClCH_2CH_2C_6H_5$ [10] and o-, p-isomers of $MeC_6H_4CH_2PCl_2$ with $MeC_6H_4CH_2Cl$ [10].

Passing $ClCH_2CH_2OMe$ through a tube containing red phosphorus and Cu_2Cl_2 at 360°C gives[7] $MePCl_2$ and $Me_2P(O)Cl$ in 12 h.

The interaction of $Me_2Si(R)CH_2Cl$ and Sb in the presence of a quaternary ammonium or phosphonium iodide at 150°–200°C yields $[Me_2Si(R)CH_2]_2SbCl$ (R = Me, Cl)[12].

Red phosphorus and C_2F_4 react in the presence of I_2 at 200°C to yield 4% octafluoro-1-iodophospholane and 21% octafluoro-1,4-diiodo-1,4-diphosphorinane[11]:

$$P + C_2F_4 \xrightarrow[200^\circ C]{I_2} \quad (a)$$

F2 F2 / F2 P F2 / I + I / P / F2 F2 / F2 P F2 / I

When C_2F_5I and red phosphorus are heated to 219°C in a steel cylinder, $C_2F_5PI_2$ and $(C_2F_5)_2PF$ are obtained[14]. Under similar conditions $C_2F_5AsI_2$ and $(C_2F_5)_2AsI$ are synthesized[15]. From C_3F_7I and red phosphorus, $C_3F_7PI_2$ (30%) and $(C_3F_7)_2PI$ (70%) are formed[16] at 230°C. The perfluoroalkyl iodides, $n\text{-}C_4F_9I$, $C_6F_{13}I$, $C_8F_{17}I$ and $C_{10}F_{21}I$, react with red phosphorus in an autoclave at 200°–280°C to form $C_nF_{2n+1}PI_2$ and $(C_nF_{2n+1})_2PI$ in 90% yield (n = 4, 6, 8, 10)[17].

Red phosphorus gives $EtSP(O)Cl_2$ and $(EtS)_2P(O)Cl$ or $n\text{-}BuSP(O)Cl_2$ and $(BuS)_2P(O)Cl$ with EtSCl or BuSCl in liq SO_2 at −50 to −30°C [18].

(G.-V. RÖSCHENTHALER)

1. L. Maier, *Helv. Chim. Acta, 46*, 2026 (1963).
2. B. M. Gladshtein, L. N. Shitov, B. G. Kovalev, L. Z. Soborovskii, *J. Gen. Chem. USSR (Engl. Transl.), 35*, 1574 (1965).
3. F. W. Parrett, M. S. Sun, *Synth. Inorg. Metal.-Org. Chem., 6*, 115 (1976).
4. L. Maier, *Inorg. Synth., 7*, 82 (1963).
5. L. Maier, E. G. Rochow, W. C. Fernelius, *J. Inorg. Nucl. Chem., 16*, 213 (1961).
6. A. G. Knappsack, Fr. Pat. 1,547,575 (1968); *Chem. Abstr., 71*, 61,546 (1969); 1,561,018 (1969); *72*, 79,226 (1970).
7. K. A. Petrov, V. V. Smirnov, A. K. Tsareva, V. I. Emelyanov, *J. Gen. Chem. USSR (Engl. Transl.), 31*, 2823 (1961).
8. A. I. Titov, P. O. Gitel, *Dokl. Akad. Nauk SSSR, 158*, 1380 (1964).
9. Yu. I. Baranov, S. V. Gorelenko, *J. Gen. Chem. USSR (Engl. Transl.), 39*, 799 (1969).
10. Yu. I. Baranov, O. F. Filippov, S. L. Varshavskii, M. I. Kabachnik, *Dokl. Akad. Nauk SSSR, 182*, 337 (1968).
11. B. M. Gladshtein, L. N. Shitov, *J. Gen. Chem. USSR (Engl. Transl.), 37*, 2461 (1967).
12. V. P. Kochergin, V. I. Shiryaev, V. F. Mironov, *J. Gen. Chem. USSR (Engl. Transl.), 48*, 1312 (1978).
13. C. G. Krespan, C. M. Langkammerer, *J. Org. Chem., 27*, 3584 (1962).

14. A. H. Cowley, T. A. Furtsch, D. S. Diersdorf, *J. Chem. Soc., Chem. Commun.*, 523 (1970).
15. P. B. Ayscough, H. J. Emeléus, *J. Chem. Soc.*, 3381 (1954).
16. H. J. Emeléus, J. D. Smith, *J. Chem. Soc.*, 375 (1959).
17. H. Brecht, Ger. Pat. 2,110,769 (1972); *Chem. Abstr.*, *78*, 18,654 (1973).
18. K. A. Petrov, A. A. Neimysheva, G. V. Dotsev, A. G. Varich, *J. Gen. Chem. USSR (Engl. Transl.)*, *31*, 1265 (1961).

2.4.3.3.2. from Aryl Halides.

By reacting PhBr or m-$CH_3C_6H_4Br$ with white phosphorus PBr_3, $PhPBr_2$ and Ph_2PBr, m-$CH_3C_6H_4PBr_2$ and (m-$CH_3C_6H_4)_2PBr$ are synthesized[1]. p-Dichlorobenzene with white phosphorus, PCl_3 and I_2 as a catalyst form p-$(Cl_2P)_2C_6H_4$ and $ClC_6H_4PCl_2$ on heating to 340°C in a steel autoclave[2].

(G.-V. RÖSCHENTHALER)

1. K. A. Petrov, V. V. Smirnov, A. K. Tsareva, V. I. Emelyanov, *J. Gen. Chem. USSR (Engl. Transl.)*, *31*, 2833 (1961).
2. Yu. I. Baranov, O. F. Filippov, S. L. Varshovskii, M. I. Kabachnik, *Dokl. Akad. Nauk SSSR*, *182*, 337 (1968).

2.4.4. from Cleavage of the Group-VB–Hydrogen Bond

2.4.4.1. by Halogens

2.4.4.1.1. to Give Group-VB Fluorides.

In xs N_2F_2 reacts with NH_3 in a packed Cu reactor to give NF_3 and NH_4F. With xs NH_3, N_2F_4, N_2F_2 and NHF_2 are also obtained[1,2]. Ammonium fluoride and F_2 yield NF_3 (96%) and HF at 20°–70°C; in the presence of NaF or KF, N_2F_4 (25%) is formed. In Eqs. (a)–(d) $NFCl_2$ or NF_2Cl are obtained[3]:

$$NH_4F + 2\ F_2 + Cl_2 \rightarrow NFCl_2 + 4\ HF \quad (a)$$

$$2\ NH_4F + 5\ F_2 + Cl_2 \rightarrow 2\ NF_2Cl + 8\ HF \quad (b)$$

$$NH_4F + 2\ NaCl + 3\ F_2 \rightarrow NFCl_2 + 2\ NaF + 4\ HF \quad (c)$$

$$NH_4F + NaCl + 3\ F_2 \rightarrow NF_2Cl + NaF + 4\ HF \quad (d)$$

Argon-diluted F_2 reacts with a slight xs of HN_3 gas to yield[4] FN_3.

The amine $C_6H_{11}NH_2$ is fluorinated in H_2O to $C_6H_{11}NF_2$ in 66% yield; the diamine $H_2N(CH_2)_6NH_2$ gives[5] $F_2N(CH_2)_6NF_2$. Fluorine, diluted by N_2, reacts with $C_6H_2(NO_2)_3NH_2$ to form[6] $C_6H_2(NO_2)_3NF_2$ at −5 to 0°C.

Fluorination of H_2NCN in a phosphate-buffered H_2O produces F_2NCN in 20% yield[7]:

$$H_2NCN + 2\ F_2 \xrightarrow{F_2-H_2O} \underset{(20\%)}{F_2NCN} + 2\ HF \quad (e)$$

Acyclic and cyclic amides, $HC(O)NFCH_3$, $\overline{CH_2(CH_2)_2C(O)NH}$ and $\overline{CH_2(CH_2)_3C(O)NH}$ are fluorinated at −30°C to give $HC(O)NFCH_3$, $\overline{CH_2(CH_2)_2C(O)NF}$ and $\overline{CH_2(CH_2)_3C(O)NF}$ in 17–31% yields[8].

Urethanes, $ROC(O)NH_2$, are converted by aq fluorination at 0°–5°C into monofluoro derivatives[8], $ROC(O)NHF$ (R = Me, Et, i-Pr). In CH_3CN i-$PrOC(O)NF_2$ is obtained from i-$PrOC(O)NH_2$ at −10° to −15°C[8].

Fluorination of $H_2NC(O)NH_2$ in CH_3CN gives $H_2NC(O)NHF$ in 53% yield[9] at 5°C and, in H_2O, $H_2NC(O)NF_2$ in 74% yield[9] at 0°–5°C. A monofluoride, $CH_3NFC(O)NHCH_3$, is obtained in 24% yield when $(CH_3NH)_2CO$ is reacted with F_2. The reaction of $EtNHC(O)NH_2$ and $(EtNH)_2CO$ with F_2 in H_2O yields $EtNHC(O)NF_2$[9] and EtNHCONFEt[10] at 0°–5°C.

At 0°–5°C, EtOC(O)NHMe is fluorinated to give EtOC(O)NFMe in 53% yield[11].

The direct fluorination of $(CF_3)_2C{=}NH$ in the presence of CsF produces $(CF_3)_2CFNF_2$ and $(CF_3)_2C{=}NF$ in >30% yield, whereas KF gives rise only to $(CF_3)_2C{=}NF$ (70%)[12]; without CsF and xs F_2, $(CF_3)_2CFNF_2$ is obtained in 97% yield[12].

Aminoiminomethanesulfonic acid is fluorinated in the presence of NaF at 0°C to $CF_2(NF_2)_2$[13]:

$$HN(H_2N)CSO_2H + 6\,F_2 \rightarrow (F_2N)_2CF_2 + SO_2F_2 + 4\,HF \tag{f}$$

Guanidine hydrofluoride mixed with NaF is converted at 0°C to $F_2NC({=}NF)NF_2$ (in 25% yield) which is **an extremely sensitive explosive**[14].

Sulfonamide is fluorinated in high yield to form[15] $H_2NSO_2NF_2$. The synthesis of $FN(SO_2F)_2$ from $HN(SO_2F)_2$ is quantitative[16].

Fluorine diluted by N_2 (30% F_2) reacts with perfluorocarboxylic acid imino esters $RC({=}NH)OCH_2CF_2CF_2H$ (R = CF_3, $CF(NO_2)_2$) to form N-fluoroiminoesters $RC({=}NF)OCH_2CF_2CF_2H$ in 45–55% yield at −5°C[17].

Fluorination of cytosin yields N,N-Difluoroaminocytosin and N,N-Difluoroamino-5-fluorocytosin within 0.5-1 h[18].

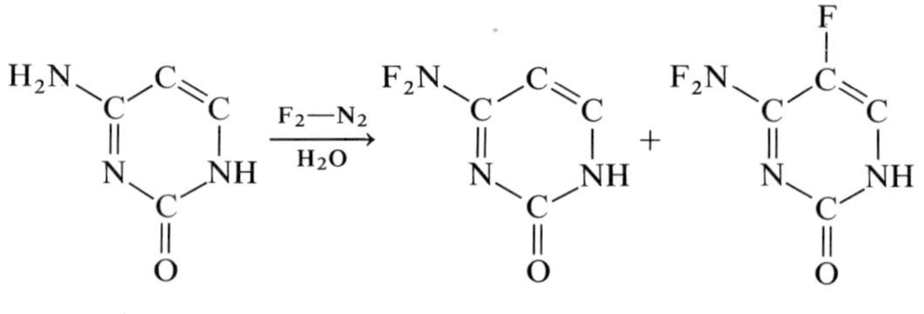

$$\text{cytosine } (H_2N\text{–}C_4H_2N_2(NH)O) \xrightarrow[H_2O]{F_2\text{—}N_2} F_2N\text{-cytosine} + F_2N\text{-5-fluorocytosine}$$

(G.-V. RÖSCHENTHALER)

1. S. I. Morrow, D. D. Perry, M. S. Cohen, *J. Am. Chem. Soc.*, *81*, 6338 (1959).
2. S. I. Morrow, D. D. Perry, M. S. Cohen, C. Schoenfelder, *J. Am Chem. Soc.*, *82*, 5301 (1960).
3. A. V. Pankratov, O. M. Sokolov, D. S. Miroshnichenko, *Russ. J. Inorg. Chem. (Engl. Transl.)*, 1618 (1968).
4. D. E. Milligan, M. E. Jacox, *J. Chem. Phys.*, *40*, 2461 (1964).
5. C. Sharts, *J. Org. Chem.*, *33*, 1008 (1968).
6. C. L. Coon, M. E. Hill, D. L. Ross, *J. Org. Chem.*, *33*, 1387 (1968).
7. M. D. Meyers, S. Frank, *Inorg. Chem.*, *5*, 1455 (1966).
8. V. Grakauskas, K. Baum, *J. Org. Chem.*, *35*, 1545 (1970).
9. V. Grakauskas, K. Baum, *J. Am. Chem. Soc.*, *92*, 2096 (1970).
10. R. E. Banks, R. N. Haszeldine, J. P. Lahn, *J. Chem. Soc.*, *C*, 1514 (1966).
11. V. Grakauskas, K. Baum, *J. Org. Chem.*, *34*, 2840 (1969).
12. J. K. Ruff, *J. Org. Chem.*, *32*, 1675 (1967).
13. R. S. Koshar, D. R. Husted, R. A. Meiklejohn, *J. Org. Chem.*, *31*, 4232 (1966).

14. R. A. Davis, T. L. Kroon, D. A. Rausch, *J. Org. Chem.*, *32*, 1662 (1967).
15. R. A. Wiesboeck, J. K. Ruff, *Inorg. Chem.*, *4*, 123 (1965).
16. M. Lustig, C. L. Bumgardner, F. A. Johnson, J. K. Ruff, *Inorg. Chem.*, *3*, 1165 (1964).
17. A. V. Fokin, Yu. N. Studnev, V. P. Stolyarov, N. N. Baranov, *Izv. Akad. Nauk SSSR, Ser. Khim.*, 937 (1982).
18. H. Meinert, U. Gross, S. Rüdiger, *J. Fluorine Chem.*, *24*, 355 (1984).

2.4.4.1.2. to Give Group-VB Chlorides.

(i) Formation of the N—Cl Bond. The gas-phase reaction of NH_3 and Cl_2 requires a large xs of NH_3 for good (75–90%), yields[1,2]:

$$2\ NH_3 + Cl_2 \rightarrow NH_2Cl + NH_4Cl \qquad \text{(a)}$$

The disproportionation of Cl_2 in liq NH_3 gives NH_2Cl in 50–60% yield[3]. Nitrogen trichloride is obtained safely from $[NH_4]^+$ salts and Cl_2 in the presence of an organic solvent that is inmiscible with H_2O, such as CCl_4, for NCl_3 passes into the organic phase[4]. **Nitrogen trichloride is highly unstable pure**[5].

Chlorine and NHF_2 in the presence of alkali metal fluorides react to give[6] $ClNF_2$. The N—H bond in alkylamines, RNH_2, is cleaved by Cl_2 in aq $NaHCO_3$ at 8°–12°C to give $RNCl_2$ (R = i-Pr, n-Bu, s-Bu, c-C_6H_{11}, n-C_8H_{17}) quantitatively. From $H_2NC_2H_4NH_2$, the tetrachloride, $Cl_2NC_2H_4NCl_2$, is formed under the same conditions[7].

In the presence of NaF, Cl_2 reacts with $(CF_3)_2NH$ to yield $(CF_3)_2NCl$ in 64% yield at 100°–325°C[8].

Cesium fluoride catalyzes the conversion of N—H to N—Cl in $(CF_3)_2C{=}NH$. The yield of $(CF_3)_2C{=}NCl$ is 97%[9]. The analogous $Ph_2C{=}NCl$ is obtained[10] in aq $NaHCO_3$.

Tetramethylguanidine gives $(Me_2N)_2C{=}NCl$ with Cl_2 in CCl_4 at 0°–10°C[11].

Chlorine in aq acetate buffer reacts with $EtOC(O)NH_2$ and n-$BuOC(O)NH_2$ to form $EtOC(O)NCl_2$[12,13] and n-$BuOC(O)NCl_2$[13], respectively.

In triphenylphosphineimine, Cl_2 cleaves the N—H bond and $Ph_3P{=}NCl$ is obtained in benzene[14]. Dialkylphosphoramidates, e.g., $(RO)_2P(O)NH_2$ and (R′O)(MeO)P(O)NHBu-t are chlorinated in an acetate buffer at 10°–15°C to the corresponding N-chloro derivatives, $(RO)_2P(O)NCl_2$[15] (R = Me, Et, i-Pr) and (R′O)(MeO)P(O)NClBu-t[16] (R′ = n-Pr, n-Bu) in good yield. Under the same conditions, from Cl_2 and $(EtO)_2P(O)NHOR$, $(EtO)_2P(O)NClR$ (R = Me, Et, n-Bu, CH_2Ph) are obtained (90%)[17].

In a mixture of an acetate buffer and $CHCl_3$ $Me_2NSO_2NCl_2$ is synthesized[18] in 96% yield from $Me_2NSO_2NH_2$. The imide $Me_2S(NH)_2$ and Cl_2 give $Me_2S(NCl)_2$ in 19% yield.

Chlorinated ammine complexes of Pt(IV) are obtained with Cl_2 in H_2O at RT[19–23], e.g.:

$$[PtCl_3(NH_3)_3]Cl + 2\ Cl_2 \rightarrow PtCl_3(NH_3)_2NCl_2 + 3\ H^+ + 3\ Cl^- \qquad \text{(a)}$$

(ii) Formation of the P—Cl Bond. Chlorine and the phosphanes $CHF_2CH_2PH_2$, $CHF_2CF_2PH_2$, $(CHF_2CH_2)_2PH$, $(CHF_2)_2PH$ and $H_2PCF_2CF_2PH_2$ are chlorinated in an autoclave at low T to give[24] $CHF_2CH_2PCl_2$, $CHF_2CF_2PCl_2$, $(CHF_2CH_2)_2PCl$, $(CHF_2CF_2)_2PCl$ and $Cl_2PCF_2CF_2PCl_2$.

In chlorobenzene, $Me_2P(O)H$ is converted to $Me_2P(O)Cl$ at −20°C; better yields are found[25] with the $Me_2P(O)H \cdot HCl$ adduct.

The P—H cleavage of n-$Bu_2P(S)H$, $(RO)_2P(O)H$ and $(EtO)_2P(S)H$ in CCl_4 furnishes n-$Bu_2P(S)Cl$[26], $(RO)_2P(O)Cl$[27] (R = Me, Et, n-Pr, i-Pr, n-Bu, i-Bu, s-Bu, n-C_5H_{11}) and $(EtO)_2P(S)Cl$[28].

Bis(1-adamantyl)phosphite reacts with Cl_2 to yield (74%) $(1\text{-}AdO)_2P(O)Cl$[29]. Using Cl_2-CCl_4 at −10°C $(Me_2N)_3P{=}NCl$ is formed from $(Me_2N)_3P{=}NH$[30].

(G.-V. RÖSCHENTHALER)

1. H. H. Sisler, G. M. Omietanski, *Inorg. Synth.*, *5*, 92 (1957).
2. R. G. Laughlin, *Chem.-Ztg.*, *92*, 383 (1968).
3. J. Jander, *Z. Anorg. Allg. Chem.*, *280*, 264 (1955).
4. W. A. Noyes, *Inorg. Synth.*, *1*, 65 (1939).
5. P. Kovacic, M. K. Lowerby, K. W. Field, *Chem. Rev.*, *70*, 639 (1970).
6. W. C. Firth, *Inorg. Chem.*, *4*, 254 (1965).
7. L. K. Jackson, G. N. R. Smart, G. F. Wright, *J. Am. Chem. Soc.*, *69*, 1539 (1947).
8. C. W. Tullock, US Pat. 3,052,723 (1962); *Chem. Abstr.*, *58*, 10,090 (1963).
9. J. K. Ruff, *J. Org. Chem.*, *32*, 1675 (1967).
10. L. Spialter, D. H. O'Brian, *J. Org. Chem.*, *32*, 223 (1967).
11. A. J. Papa, *J. Org. Chem.*, *31*, 1426 (1966).
12. T. Foglia, D. Swern, *J. Org. Chem.*, *31*, 3625 (1966).
13. F. Boberg, G.-J. Wentrup, M. Koepke, *Synthesis*, 502 (1975).
14. R. Appel, A. Hauss, G. Büchler, *Z. Naturforsch.*, *Teil B*, *16*, 405 (1961).
15. A. Zwierzak, A. Koziara, *Tetrahedron*, *26*, 3521 (1970).
16. M. Okahara, K. Ozawa, T. Yaginuma, M. Miki, I. Ikeda, *J. Org. Chem.*, *42*, 617 (1977).
17. A. Zwierzak, J. Brylikowska, *Synthesis*, 712 (1975).
18. D. Graetbanks, T. P. Seden, R. W. Turner, *Tetrahedron Lett.*, 4863 (1968).
19. R. Appel, D. Hänssgen, *Angew. Chem.*, *79*, 96 (1967).
20. Y. N. Kukushkin, *J. Gen. Chem. USSR (Engl. Transl.)*, *5*, 947 (1960).
21. Y. N. Kukushkin, *J. Gen. Chem. USSR (Engl. Transl.)*, *6*, 899 (1961).
22. Y. N. Kukushkin, *J. Gen. Chem. USSR (Engl. Transl.)*, *7*, 397 (1962).
23. Y. N. Kukushkin, *J. Gen. Chem. USSR (Engl. Transl.)*, *10*, 325 (1965).
24. G. M. Burch, H. Goldwhite, R. N. Haszeldine, *J. Chem. Soc.*, 1083 (1963).
25. H. J. Kleiner, *Justus Liebigs Ann. Chem.*, 751 (1974).
26. G. Peters, *J. Org. Chem.*, *27*, 2198 (1962).
27. A. M. De Roos, H. J. Toet, *Recl. Trav. Chim.*, *Pays-Bas*, *77*, 946 (1958).
28. J. Michalski, C. Krawiecki, *Rocz. Chem.*, *31*, 715 (1957); *Chem. Abstr.*, *48*, 3243 (1954).
29. R. I. Yurchenko, T. I. Klepa, O. V. Lozovitskaya, V. P. Tikhonov, *J. Gen. Chem. USSR (Engl. Transl.)*, *53*, 2206 (1983).
30. A. P. Marchenko, G. N. Koidan, A. M. Pinchuk, *J. Gen. Chem. USSR (Engl. Transl.)*, *53*, 583 (1983).

2.4.4.1.3. to Give Group-VB Bromides.

(i) Formation of the N—Br Bond. Bromamines in Et_2O are obtained when NH_3 is passed into cooled Br_2 in ether. Ammonium bromide precipitates, and the ether contains mainly NH_2Br with xs NH_3[1]. An equilibrium between NH_2Br, NBr_2H and NH_3 exists[2]. When the yellow solution is poured into xs pentane, black-violet NH_2Br precipitates at −120°C. Solutions of NH_2Br in liq NH_3 can be prepared by addition of Br_2 at −78°C[3]; NBr_3 is obtained[4,5] at low T and high concentration of Br_2.

Difluoramine bubbled through Br_2 in H_2O yields NF_2Br when HgO is present[6]. Using xs Br_2 the N—H bond is cleaved in $(CF_3)_2C{=}NH$ in the presence of KF to give[7] 68% $(CF_3)_2C{=}NBr$. In aq $NaHCO_3$ $Ph_2C{=}NBr$ is synthesized from its N—H precursor[8], whereas $(Me_2N)_2C{=}NBr$ forms in 59% yield from $(Me_2N)_2C{=}NH$ with Br_2 at 0°–10°C in CCl_4[9]:

$$2\,(Me_2N)_2C{=}NH + Br_2 \xrightarrow{CCl_4} (Me_2N)_2C{=}NBr + (Me_2N)_2C{=}NH{\cdot}HBr \qquad \text{(a)}$$

In the presence of Ag_2O, Br_2 reacts with $R_fC(O)NH_2$ in $CF_3C(O)OH$ at 20°C to form $R_fC(O)NHBr$ ($R_f = CF_3$, C_2F_5, n-C_3F_7, n-C_4F_9)[10,11]. Under similar conditions $(CF_2CO)_2NH$ is converted[12] to $(CF_2CO)_2NBr$.

N-Bromocarbamates RNBrC(O)OR′ (R = H, R′ = Me, Et; R = R′ = Me) in yields >81% are formed from RNHC(O)OR′ with $Br_2/NaOH/CHCl_3/H_2O$ at 25°C within 1 h.

Triphenylphosphinimine undergoes N—H bond cleavage in benzene to yield[14,15] $Ph_3P{=}NBr$.

In H_2O, $(RO)_2P(O)NBr_2$ (R = Et, n-Pr, i-Pr, n-Bu, t-Bu) are obtained from the amido precursor[16], whereas the reaction of Br_2 and $(EtO)_2P(O)NHR'$ is carried out in CH_2Cl_2 to form $(EtO)_2P(O)NBrR'$ (R′ = Me, Et, n-Bu, s-Bu) in excellent yield[17].

The halogenation of $Me_2S(NH)_2$ and $Me_2S(O)(NH)$ in buffered H_2O gives $Me_2S(NH)NBr$[18], $Me_2S(NBr)_2$[19] and $Me_2S(O)NBr$[18].

Ammine complexes of Pt(IV) are brominated in H_2O to form corresponding complexes containing NBr_2 ligands[20], e.g.:

$$Pt(py)(NH_3)(NH_2)Cl_3 + 2\,Br_2 \rightarrow Pt(py)(NH_3)(NBr_2)Cl_3 + 2\,HBr \qquad \text{(b)}$$

(ii) Formation of the P—Br Bond. Bromine and $CHFClCF_2PH_2$ yield[21] $CHFClCF_2PBr_2$. Phenylphosphane is converted by Br_2 in $CHCl_3$ at 20°C to $PhPBr_2$ in 22% yield. The pentaerithritol derivative,

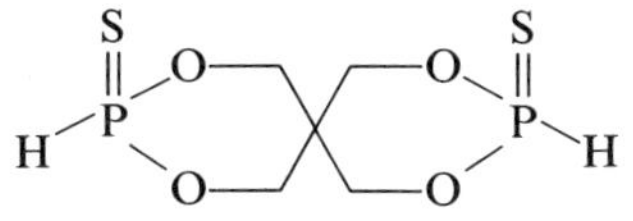

reacts to give the P—Br compound[22].

(G.-V. RÖSCHENTHALER)

1. W. Moldenhauer, M. Burger, *Ber. Dtsch. Chem. Ges.*, *62*, 1615 (1929).
2. J. Jander, C. Lafrenz, *Z. Anorg. Allg. Chem.*, *349*, 57 (1967).
3. J. Jander, E. Kurzbach, *Z. Anorg. Allg. Chem.*, *296*, 117 (1958).
4. J. K. Johannesson, *J. Chem. Soc.*, 2998 (1959).
5. J. Jander, *Adv. Inorg. Chem. Radiochem.*, *19*, 48 (1976).
6. E. W. Lawless, I. C. Smith, in *Inorganic High-Energy Oxidizers*, Edward Arnold, London, 1968, p. 86.
7. J. K. Ruff, *J. Org. Chem.*, *32*, 1675 (1967).
8. L. Spialter, D. H. O'Brian, *J. Org. Chem.*, *32*, 223 (1967).
9. A. J. Papa, *J. Org. Chem.*, *31*, 1426 (1966).
10. J. D. Park, H. J. Gerjovich, W. R. Lycan, J. R. Lacher, *J. Am. Chem. Soc.*, *74*, 2189 (1952).
11. J. D. Park, W. R. Lycan, J. R. Lacher, *J. Am. Chem. Soc.*, *76*, 1388 (1954).
12. A. L. Henne, W. F. Zimmer, *J. Am. Chem. Soc.*, *73*, 1103 (1951).
13. V. B. Mochalin, T. N. Maksimova, N. I. Filenko, O. V. Bakova, T. V. Khenkina, *Zh. Grg. Khim*, *18*, 1202 (1982).
14. R. Appel, A. Hauss, G. Büchler, *Z. Naturforsch., Teil B*, *16*, 405 (1961).
15. R. Appel, A. Hauss, *Z. Anorg. Allg. Chem.*, *311*, 290 (1961).
16. S. Zawadzki, A. Zwierzak, *Tetrahedron*, *29*, 315 (1973).
17. K. Osowski, A. Zwierzak, *Synthesis*, *1979*, 577.

18. R. Appel, H. W. Fehlhaber, D. Hänssgen, R. Schöllhorn, *Chem. Ber., 99*, 3108 (1966).
19. R. Appel, D. Hänssgen, *Angew. Chem., Int. Ed. Engl., 6*, 91 (1967).
20. Y. N. Kukushkin, *Russ. J. Inorg. Chem. (Engl. Transl.), 8*, 420 (1963).
21. G. M. Burch, H. Goldwhite, R. N. Haszeldine, *J. Chem. Soc.*, 1083 (1963).
22. H. J. Delventhal, W. Kuchen, *Z. Naturforsch., Teil B, 26*, 190 (1971).

2.4.4.1.4. to Give Group-VB Iodides.

Liquid NH_3 reacts with I_2 below −75°C to form $NH_2I{\cdot}NH_3$ and $NI_3{\cdot}3\ NH_3$; at −70°C to −30°C, only $NI_3{\cdot}3\ NH_3$, which is stable to −25°C, is obtained[1].

At 0°C, $MeNH_2$ or Me_2NH and I_2 react in H_2O to give $MeNI_2$ and Me_2NI; at −55° to −45°C the neat amines form the corresponding iodides[2]:

$$3\ MeNH_2 + 2\ I_2 \rightarrow MeNI_2 + 2\ [MeNH_3]I \qquad \text{(a)}$$

N-Iodotriphenylphosphine imine is formed when $Ph_3P{=}NH$ and I_2 are reacted[3]. The iodination of $Me_2S(NH)_2$ gives $Me_2S(NI)_2$ in 97% yield in carbonate-buffered H_2O[4].

The P—H bond in CF_3PH_2 is cleaved by I_2 to form $CF_3P(I)H$, which undergoes ligand exchange[5]:

$$2\ CF_3PH_2 + I_2 \xrightarrow[-HI]{} 2\ CF_3P(I)H \rightleftharpoons CF_3PH_2 + CF_3PI_2 \qquad \text{(b)}$$

The iodine-containing compound, $Me_2C(CH_2O)_2P(S)I$, is obtained[6] from $Me_2C(CH_2O)_2P(S)H$.

Methylarsine is converted[7] by I_2 to MeAsHI and $MeAsI_2$.

Iodine reacts with $RSbH_2$ to form $RSbI_2$ and the polymeric compounds $(RSbI_{0.4})_x$ (R = Me, Et, n-Bu)[8]; with $PhSbH_2$ in ether $PhSbI_2$ is found[9].

Succinimide is iodinated by I_2 in the presence of $AgOC(O)CH_3$ to form N-iodosuccinimide[10] in 94% yield.

(G.-V. RÖSCHENTHALER)

1. J. Jander, U. Engelhardt, in *Developments in Inorganic Nitrogen Chemistry* C. W. Colburn, ed., Vol. 2, Elsevier, Amsterdam, 1973, p. 70.
2. J. Jander, K. Knuth, W. Renz, *Z. Anorg. Allg. Chem., 392*, 143 (1972).
3. R. Appel, A. Hauss, G. Büchler, *Z. Naturforsch., Teil B, 16*, 405 (1961).
4. R. Appel, D. Hänssgen, *Angew. Chem., Int. Ed. Engl., 6*, 97 (1967).
5. R. C. Dobbie, P. D. Gosling, B. P. Straughan, *J. Chem. Soc., Dalton Trans.*, 2368 (1975).
6. R. S. Edmundson, *Chem. Ind. (London), 1965*, 1220.
7. A. L. Rheingold, J. M. Bellama, *J. Chem. Soc., Chem. Cummun.*, 1058 (1969).
8. P. Choudhury, M. F. El-Shazley, C. Spreng, A. L. Rheingold, *Inorg. Chem., 18*, 543 (1979).
9. E. Wiberg, H. Nöth, *Z. Naturforsch., Teil B, 12*, 128 (1957).
10. T. R. Beebe, J. W. Wolfe, *J. Org. Chem., 35*, 2056 (1970).

2.4.4.2. by Other Halides

2.4.4.2.1. to Give Group-VB Halides from Hypohalites.

(i) by Inorganic Hypohalites. Dilute aq NH_2Cl forms from aq NH_3 with cold aq NaOCl. The more dilute, the higher the yields[1,2]. Difluoramine is converted into NF_2Cl (80–100% yield)[3].

The amines RNH_2 or ammonium salts, $[RNH_3]Cl$, are converted to give $RNCl_2$ (R = Me[4], Et[5], i-Pr[4], t-Bu[6]) in good yield. With xs amine, RNHCl (R = Me[4], Et[7], t-Bu[4]) is synthesized. In cold aq NaOCl the secondary amines R_2NH give R_2NCl (R = Me[4], Et[7], n-Pr[7], n-Bu[7]; $R_2 = CH_2CH_2$[8]). More than 90% yield of 1-$AdNCl_2$[9] and 1-$AdCH_2NCl_2$[10] is observed from the corresponding amines in cold aq $Ca(OCl)_2$ which contains CH_2Cl_2. Only 50% t-$BuCH_2NCl_2$ is isolated under similar conditions[10].

From aq $RNHCH_2CN$ and $Ca(OCl)_2$ $RNClCH_2CN$ (R = Me, Et, i-Pr, t-Bu) are obtained[11].

Cold aq NaOCl reacts with carboxamides, $RC(O)NH_2$, and carbamates, $R'OC(O)NH_2$, to form R'C(O)NHCl[12] (R = H, Me, Et, $ClCH_2CH_2$, $BrCH_2$, $ClCH_2$, FCH_2, Cl_2CH, Cl_3C, F_3C) and $R'OC(O)NH_2$[12] (R' = Et, n-Pr, $MeOCH_2CH_2$, $ClCH_2CH_2$, Cl_3CCH_2, $PhCH_2$) in good yields.

The phosphorus derivatives i-$Pr_2P(O)NH_2$, t-$Bu_2P(O)NH_2$, t-$Bu_2P(O)NHCl$ and the phosphetanes $Me_2\overline{CCH(Me)CMe_2P}(O)NH_2$ and $Me_2\overline{CCH(Me)CMe_2P}(O)NHMe$ are converted to i-$Pr_2P(O)PNCl_2$, t-$Bu_2P(O)NCl_2$, $Me_2\overline{CCH(Me)CMe_2P}(O)NCl_2$ and $Me_2\overline{CCH(Me)CMe_2P}(O)NClMe$ in >60% yield[13].

Sodium hypochlorite reacts with $HN(SO_3K)_2$ to give[14] $ClN(SO_3K)_2$.

From aq NaOBr and $MeNH_2$, Me_2NH, $\overline{CH_2CH_2N}H$, 1-$AdNH_2$, $RC(O)NH_2$ and $R'OC(O)NH_2$, $MeNBr_2$[4], Me_2NBr[4], $\overline{CH_2CH_2N}Br$[8], 1-AdNHBr[15], 1-$AdNBr_2$[16], RC(O)NHBr[12] (R' = Et, n-Bu, $ClCH_2CH_2$, $BrCH_2$, $ClCH_2$, FCH_2, Cl_2CH, Cl_3C, F_3C) and R'OC(O)NHBr[12] (R' = Et, Cl_3CCH_2, $PhCH_2$) are obtained.

The N—H bond in p-$RC_6H_4S(O)Me(NH)$ is cleaved by aq NaOBr to form[17] p-$RC_6H_4S(O)Me({=}NBr)$ (R, Me).

Aqueous NaOBr brominates $(MeSO_2)_2NH$ to $(MeSO_2)_2NBr$ at 0°C[18]

(ii) by Organic Hypohalites. Chlorodifluoramine is synthesized[19] quantitatively by reacting NHF_2 with t-BuOCl.

The N—H bonds in t-$BuNH_2$, $C_6H_{11}NH_2$ and $(PhCH_2)_2NH$ are cleaved to give t-$BuNCl_2$[20], $C_6H_{11}NCl_2$[21] and $(PhCH_2)_2NCl$[22].

Fluorinated anilines, $C_6HF_4NH_2$, $C_6F_5NH_2$, C_6F_5NHR and the biphenyl derivative $H_2NC_6F_4C_6F_4NH_2$ are converted in CCl_4 to $C_6HF_4NCl_2$[23], $C_6F_5NCl_2$[24], C_6F_5NClR[25] (R = Me, MeCO, CF_3CO) and $Cl_2NC_6F_4C_6F_4NCl_2$[25] in high yield.

The compounds RNHCN are converted[26] to RNClCN (R = Me, i-Pr, t-Bu) at 0°C.

Treatment of $[MeC(O)]_2NH$, $[EtC(O)O]_2NH$ and maleimide gives[22] $[MeC(O)]_2NCl$, $[EtC(O)O]_2NCl$ and N-chloromaleimide. N-chlorosuccinimide and N-chlorophthalimide are obtained from the N—H precursors[20]. t-Butylhypochlorite reacts with parabanic acid to give of the N,N'-dichloroderivative in 96% yield[21]:

2 t-BuOCl + (O, O, H–N, N–H, O) → (O, O, Cl–N, N–Cl, O) + 2 t-BuOH (a)

t-Butylhypochlorite in Et_2O reacts with Ph_3SiNH_2 and t-$Bu(Me_3Si)NH$ to form[27] $Ph_3SiNHCl$, Ph_3SiNCl_2 and t-$Bu(Me_3Si)NCl$.

The N—H bond cleavage in $(R'NH)_2P(O)R$ by t-BuOCl produces (R'NCl)(R'NH)P(O)R (R = Me, R' = t-Bu; R = Me, R' = 1-Ad; R = t-Bu, R' = Me; R = t-Bu, R' = t-Bu; R = OPh, R' = t-Bu; R = NHBu-t, R' = t-Bu)[28].

The chlorination of $i\text{-}Pr_2P(O)NH_2$, $t\text{-}Bu_2P(O)NH_2$, $t\text{-}Bu_2P(O)NHCl$ and the phosphetanes $Me_2CCH(Me)CMe_2P(O)NH_2$ and $Me_2CCH(Me)CMe_2P(O)NHMe$ in CH_2Cl gives better yields for NaOCl[13].

The reaction of RSO_2NHR' and t-BuOCl in CCl_4 furnishes[22,26,29] RSO_2NClR' (R = R' = Me[29]; R = Me, R' = t-Bu[29]; R = n-Bu; R' = t-Bu[29]; R = $t\text{-}BuCH_2$, R' = t-Bu[26]; R = $p\text{-}MeC_6H_4$, R' = CO_2Et[22]) in good yields.

Solutions of $(Me_3Si)_2NH$ and $[(MeO)_3Si]_2NH$ in ether react to form $(Me_3Si)_2NCl$ (40–50%) and $[(MeO)_3Si]_2NCl$ (98%)[27].

The tosylate $[Me_3NNH_2][OTOS]$ is transferred by t-BuOCl in MeOH (82% yield) to form $[Me_3NN(Cl)H][OTOS]$[30].

Triamidophosphozochlorides $(R_2N)_3P{=}NCl$ (R = Et, Pr, Bu) are obtained in 90% yield using aq NaOCl[31].

Sodium cyanamide, NaHNCN and t-BuOCl react to give NaN(Cl)CN[32].

Mesylhyohorylamines $MeSO_2NHOR$(R = Me, SO_2Me) react with t-BuOCl/acetone to furnish $MeSO_2N(Cl)OR$ (yield >83%)[33].

3-Fluoro- and 3-chlorocyclophosphamide are formed in 20% and 87% yield, respectively, from the hydrogen precursor.

H, N, O, P, O, $N(CH_2CH_2Cl)_2$ —(CF_3OF, CH_2Cl_2; −78 till 20°C)→ F, N, O, P, O, $N(CH_2CH_2Cl)_2$

H, N, O, P, O, $N(CH_2CH_2Cl)_2$ —(NaOCl, H_2O, $CHCl_3$; 22h, 20°C)→ Cl, N, O, P, O, $N(CH_2CH_2Cl)_2$

The atiridines RR'C—NH are chlorinated by t-BuOCl to yield RR'C—NCl (R = R = H, Me, Ph)[34].

Bis(alkysulfonyl)amines $(RSO_2)_2NH$ and t-BuOCl in aq t-BuOH form $(RSO_2)_2NCl$ (R = Me, Et) at −25°C[18].

t-Butylhypobromite with NH_3 or $t\text{-}BuNHSO_2Bu\text{-}n$ gives NBr_3[35] and $t\text{-}BuNBrSO_2Bu\text{-}n$[26].

Acetylperbromite, $CH_3C(O)OBr$, brominates in almost quantitative yield $CF_3C(O)NH_2$, $p\text{-}O_2NC_6H_4C(O)NHR$ (R = H, Me), succinimide and phthalimide to give $CF_3C(O)NHBr$, $p\text{-}O_2NC_6H_4C(O)NBrR$ (R = H, Me, Br), N-bromosuccinimide and N-bromophthalimide, respectively[36].

(G.-V. RÖSCHENTHALER)

1. L. F. Audrieth, R. A. Rowe, *J. Am. Chem. Soc.*, *77*, 4726 (1955).
2. L. F. Audrieth, R. A. Rowe, *J. Am. Chem. Soc.*, *78*, 563 (1956).
3. E. W. Lawless, I. C. Smith, *Inorganic High-Energy Oxidisers*, Edward Arnold, London, 1968.
4. V. L. Heasley, P. Kovacic, R. M. Lange, *J. Org. Chem.*, *31*, 3050 (1966).

5. W. S. Metcalf, *J. Chem. Soc.*, 148 (1942).
6. H. Bock, K.-L. Kompa, *Z. Anorg. Allg. Chem.*, *332*, 238 (1964).
7. G. H. Coleman, *J. Am. Chem. Soc.*, *55*, 3001 (1933).
8. A. F. Graefe, R. E. Meyer, *J. Am. Chem. Soc.*, *80*, 3939 (1958).
9. P. Kovacic, P. D. Roskos, *J. Am. Chem. Soc.*, *91*, 6457 (1969).
10. J. T. Roberts, B. R. Rittberg, P. Kovacic, F. V. Scalzi, *J. Org. Chem.*, *45*, 5239 (1980).
11. J. H. Boyer, J. Kovi, *J. Am. Chem. Soc.*, *98*, 1099 (1976).
12. C. Bachand, H. Driguez, J. P. Paton, D. Touchard, J. Lessard, *J. Org. Chem.*, *39*, 3136 (1974).
13. M. J. Hager, M. A. Stephen, *J. Chem. Soc.*, *Perkin Trans. 1*, *1980*, 705.
14. F. Raschig, *Z. Anorg. Allg. Chem.*, *147*, 1 (1925).
15. H. Stetter, E. Smulders, *Chem. Ber.*, *104*, 917 (1971).
16. S. J. Padeginas, P. Kovacic, *J. Org. Chem.*, *37*, 2672 (1972).
17. G. Dauphin, A. Kergomard, A. Scarset, *Bull. Soc. Chim. Fr.*, 862 (1976).
18. D. Koch, A. Blaschette, *Z. Anorg. Allg. Chem.*, *454*, 5 (1979).
19. K. O. Christe, *Inorg. Chem.*, *8*, 1539 (1969).
20. H. Zimmer, L. F. Audrieth, *J. Am. Chem. Soc.*, *76*, 3856 (1954).
21. G. H. Alt, W. S. Knowles, *Org. Synth.*, *Collect. Vol.*, *5*, 208 (1973).
22. S. C. Cottrell, C. Abrams, D. Swern, *Org. Prep. Proced. Int.*, *8*, 25 (1976); *Chem. Abstr.*, *85*, 32,390 (1976).
23. R. E. Banks, M. G. Barlow, J. C. Hornby, T. J. Noakes, *J. Fluorine Chem.*, *13*, 179 (1979).
24. R. E. Banks, T. J. Noakes, *J. Chem. Soc.*, *Perkin Trans. 1*, *1976*, 143.
25. R. E. Banks, M. G. Barlow, T. J. Noakes, M. M. Saleh, *J. Chem. Soc.*, *Perkin Trans. 1*, *1977*, 1746.
26. R. S. Neale, N. L. Marcus, *J. Org. Chem.*, *34*, 1808 (1969).
27. N. Wiberg, F. Raschig, *J. Organomet. Chem.*, *10*, 15 (1967).
28. H. Quast, M. Heuschmann, M. O. Abdel-Rahman, *Justus Liebigs Ann. Chem.*, 943 (1981).
29. H. Quast, F. Kees, *Chem. Ber.*, *114*, 774 (1981).
30. G. V. Shustov, N. B. Tavakalyan, R. G. Kostyanovsky, *Tetrahedron*, *41*, 575 (1985).
31. A. P. Marchunko, G. N. Koidan, A. M. Pinchuk, *Zh. Obshch. Khim.*, *53*, 670 (1983).
32. M. G. K. Hutchins, D. Swern, *J. Org. Chem.*, *47*, 4847 (1982).
33. K. Brink, R. Mattes, *Chem. Ztg.*, *109*, 10 (1985).
34. M. Bucciarelli, A. Forin, I. Moretti, G. Torre, *J. Org. Chem.*, *48*, 2640 (1983).
35. C. Walling, A. Padna, *J. Org. Chem.*, *27*, 2976 (1962).
36. T. R. Beebe, J. W. Wolfe, *J. Org. Chem.*, *35*, 2056 (1970).

2.4.4.2.2. to Give Group-VB Halides from Hydrogen Halides.

Difluoramine reacts with HCl to form[1] $ClNF_2$.

The stibines Me_2SbH and Ph_2SbH undergo an Sb—H bond cleavage by HCl to give the corresponding chloro derivatives[2,3]:

$$R_2SbH + HCl \rightarrow R_2SbCl + H_2 \tag{a}$$

where (R = Me, Ph).

(G.-V. RÖSCHENTHALER)

1. E. A. Lawton, J. Q. Weber, *J. Am. Chem. Soc.*, *85*, 3595 (1963).
2. A. Burg, L. R. Grant, *J. Am. Chem. Soc.*, *81*, 1 (1959).
3. A. N. Nesmeyanov, A. E. Borisov, N. V. Novikova, *Izv. Akad. Nauk SSSR, Ser. Khim.*, 815 (1967).

2.4.4.2.3. to Give Group-VB Halides from Group-IIIB Halides.

Difluoramine and BCl_3 yield a complex, $HNF_2 \cdot BCl_3$, at −130°C, which decomposes to form $ClNF_2$, HCl, Cl_2 and BF_3 on warming to RT[1].

The gas-phase reaction of Me_2SbH and B_2H_5Br at −78°C produces[2] B_2H_6 and Me_2SbBr.

(G.-V. RÖSCHENTHALER)

1. R. C. Petry, *J. Am. Chem. Soc.*, *82*, 1400 (1960).
2. A. B. Burg, L. R. Grant, *J. Am. Chem. Soc.*, *81*, 1 (1959).

2.4.4.2.4. to Give Group-VB Halides from Group-IVB Halides.

The N—H bond in NHF_2 is cleaved by $COCl_2$ to give[1] $NClF_2$.

i-Butylphosphine and n-Bu_2PH are converted by $COCl_2$ into i-$BuPCl_2$ and n-Bu_2PCl in $CHCl_3$ at −50°C or in CH_2Cl_2 at −30°C, respectively[2], e.g.:

$$\text{i-BuPH}_2 + 2\ \text{COCl}_2 \rightarrow \text{i-BuPCl}_2 + 2\ \text{CO} + 2\ \text{HCl} \qquad \text{(a)}$$

By the same chlorinating agent, $Me(PhCH_2)PH$, Et(Ph)PH, $Ph(PhCH_2)PH$, $PhPH(CH_2)_6P(H)Ph$, p-$[MeP(H)CH_2]_2C_6H_4$ and p-$[PhP(H)CH_2]_2C_6H_4$ are converted[3] at −30°C in 1,2-dichlorethane to the corresponding chloro derivatives $Me(PhCH_2)PCl$, Et(Ph)PCl, $Ph(PhCH_2)PCl$, p-$[MeP(H)CH_2]_2C_6H_4$, p-$[PhP(Cl)CH_2]_2C_6H_4$ and $PhP(Cl)(CH_2)_6P(Cl)Ph$. Phenylphosphine and Ph_2PH undergo P—H bond cleavage with phosgene[4,5] to form $PhPCl_2$ and Ph_2PCl in >75% yield.

Treatment of the cyclic phosphine I:

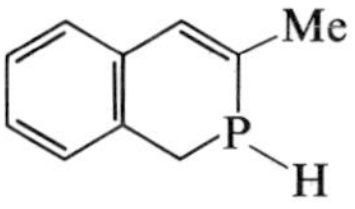

I 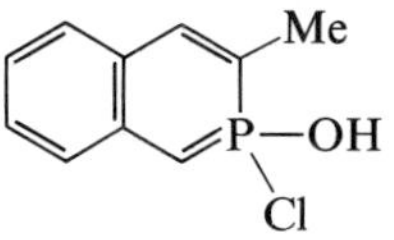 II 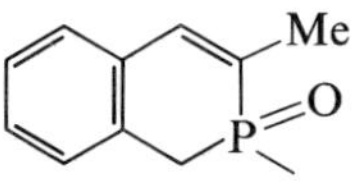 III

with 1 equiv of phosgene gives the secondary chlorophosphine. Excess $COCl_2$ and Et_3N at −70°C in toluene leads to the phosphorine, II, which is also formed from the phosphinoxide, III, with $COCl_2$ under the same conditions[6].

In good yields, $(EtO)_2P(O)Cl$ and $(BuO)_2P(O)Cl$ are obtained from $COCl_2$ and the corresponding P—H precursors[7].

The P—H bond in $Et_2P(S)H$ and $(MeO)_2P(S)H$ is cleaved in the presence of Et_3N to CCl_4 to form[8] $Et_2P(S)Cl$ and $(MeO)_2P(S)Cl$:

$$\text{Et}_2\text{P(S)H} + \text{CCl}_4 \rightarrow \text{Et}_2\text{P(S)Cl} + \text{HCCl}_3 \qquad \text{(b)}$$

The P—H bond of the phosphazene, IV, reacts with CCl_4, which replaces hydrogen by Cl[9]:

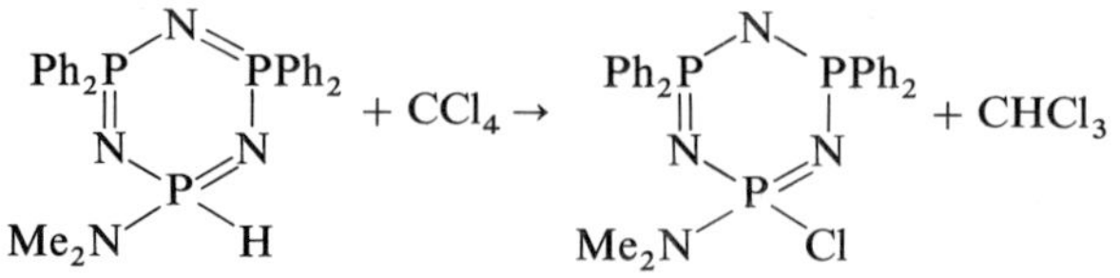

IV + CCl_4 → (P–Cl analogue) + $CHCl_3$ (c)

The stibines $RSbH_2$ are converted by CCl_4 into $RSbCl_2$ (R = Me, Et, n-Bu), by CI_4 into $RSbI_2$ and $(RSbI_{0,4})_x$ (R = Me, Et, n-Bu)[10].

Dimethylarsine abstracts a chlorine from halo substituted cycloolefins $ClC{=}CH(CF_2)_nCF_2$ (n = 1, 2) and $ClC = C(OMe)CF_2CF_2$ to form Me_2AsCl at 150°C in >45% yield[11] and $HC{=}CH(CF_2)_nCF_2$ (n = 1, 2) and $HC{=}C(OMe)CF_2CF_2$. Methylphenylarsine is also chlorinated by $ClC{=}CHCF_2CF_2$, but in only 15% yield[11].

Dimethylamine reacts with COF_2 in the presence of NEt_3-CH_2Cl_2 with formation of Me_2NF (45%); $(RO)_2P(O)H$(R = Et, Bu) forms $(RO)_2P(O)F$ (760%).

(G.-V. RÖSCHENTHALER)

1. E. A. Lawton, J. Q. Weber, *J. Am. Chem. Soc.*, *85*, 3595 (1963).
2. W. A. Henderson, S. A. Buckler, N. E. Day, M. Grayson, *J. Org. Chem.*, *26*, 4770 (1961).
3. E. Steininger, *Chem. Ber.*, *96*, 3184 (1963).
4. A. Michaelis, F. Dittler, *Ber. Dtsch. Chem. Ges.*, *12*, 338 (1879).
5. E. Hoffmann, Br. Pat. 904,086 (1962); *Chem. Abstr.*, *57*, 16,661 (1962).
6. T. Klebach, C. Jongsma, F. Bickelhaupt, *Recl. Trav. Chim. Pays-Bas*, *98*, 14 (1979).
7. A. F. Kolomiets, P. S. Khokhlov, S. G. Zhemchuzhin, *J. Gen. Chem. USSR (Engl. Transl.)*, *37*, 1279 (1967).
8. W. Lorenz, G. Schrader, Ger. Pat. 1,067,017 (1958); *Chem. Abstr.*, *56*, 1482 (1962).
9. J. Ebeling, M. A. Leva, H. Stary, A. Schmidpeter, *Z. Naturforsch., Teil B*, *26*, 650 (1971).
10. P. Choudhury, M. F. El-Shazley, C. Spreng, A. L. Rheingold, *Inorg. Chem.*, *18*, 543 (1979).
11. W. R. Cullen, P. S. Dhaliwal, *Can. J. Chem.*, *45*, 719 (1967).
12. O. D. Gupta, J. M. Shreeve, *J. Chem. Soc. Chem. Commun.*, 416 (1984).

2.4.4.2.5. to Give Group-VB Halides from Group-VB Halides.

(i) Fluorides. Diphenyltrifluorophosphorane, Ph_2PF_3, is obtained[1] when Ph_2PH is treated with N_2F_4.

(ii) Chlorides. Coordinated $PhPH_2$ in h^5-$C_5H_5Mn(CO)_2PPhH_2$ is chlorinated by $C_6H_{11}NCl_2$ to give[2] h^5-$C_5H_5Mn(CO)_2PPhHCl$ and h^5-$C_5H_5Mn(CO)_2PPhCl_2$.

From 2,2-dimethylaziridine and 2,2,3,3-tetramethylaziridine with N-chlorosuccinimide (NCS), the N-chloro derivatives Me_2CCH_2NCl and Me_2CCMe_2NCl are synthesized, respectively, in ether at RT[3].

The N—H bond in t-Bu_2NH is cleaved by N-chlorosuccinimide[4]. The yield of t-Bu_2NCl is 90%.

N-chlorosuccinimide reacts with $Ph_2P(O)H$ at 0°C in $CHCl_3$, with Me(i-PrO)P(O)H and Me(i-PrO)P(S)H in CCl_4 and with $(PhCH_2O)_2P(O)H$ to give $Ph_2P(O)Cl$[5], Me(i-PrO)P(O)Cl[6,7], Me(i-PrO)P(S)Cl[6] and $(PhCH_2O)_2P(O)Cl$[8], respectively.

N-Trichloroisocyanuric acid, with succinimide, phthalimide and $(MeNH)_2CO$ at RT in acetonitrile, lead to N-chlorosuccinimide, N-chlorophthalimide and $(MeNCl)_2CO$, respectively[9], e.g.:

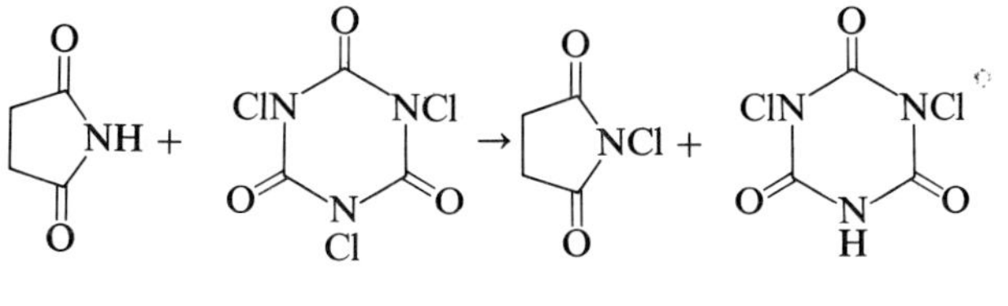

(a)

Only with $(MeNH)_2CO$ are all three chlorines transferred[9]:

$$3\,(MeNH)_2CO + [ClNC(O)]_3 \rightarrow 3\,MeNHC(O)NClMe + [HNC(O)]_3 \qquad \text{(b)}$$

From PCl_3 and $(n\text{-}C_8H_{17})_2P(O)H$, $i\text{-}C_4H_9PH(O)ONa$, $C_6H_{11}PH(O)ONa$, $PhPH(O)OH$, $p\text{-}MeC_6H_4PH(O)OH$, $p\text{-}ClC_6H_4PH(O)OH$ and $Ph_2P(O)H$, the corresponding P—Cl compounds $(n\text{-}C_8H_{17})_2PCl$ [10], $i\text{-}C_4H_9PCl$ [11], $C_6H_{11}PCl_2$ [11], $PhPCl_2$ [12], $p\text{-}CH_3C_6H_4PCl_2$ [13], $p\text{-}ClC_6H_4PCl_2$ [14] and Ph_2PCl [10] are prepared.

Ferrocenyldichlorophosphine $(h^5\text{-}C_5H_4PCl_2)(h^5\text{-}C_5H_5)Fe$ is synthesized[15] from $[h^5\text{-}C_5H_4PH(O)OH](h^5\text{-}C_5H_5)Fe$ and PCl_3 in benzene.

Phosphorus pentachloride chlorinates $Bu_2P(O)H$ [5,16]:

$$n\text{-}Bu_2P(O)H + PCl_5 \rightarrow n\text{-}Bu_2P(O)Cl + HCl + P(O)Cl_3 \qquad (c)$$

Ethylstibine and $MeAsCl_2$ undergo rapid H-Cl exchange[17]:

$$EtSbH_2 + 2\ MeAsCl_2 \rightarrow EtSbCl_2 + 2\ MeAsHCl \qquad (d)$$

From $MeSbH_2$ and Me_2AsI, the polymeric product $(MeSbI_{0.4})_x$ is obtained[178].

(iii) Bromides. Nitrogen tribromide with NH_3 at $-87°C$ in CH_2Cl_2 gives[18] NH_2Br:

$$NBr_3 + 2\ NH_3 \rightarrow 3\ NH_2Br \qquad (e)$$

t-Butyldibromamine converts[19] $t\text{-}BuNH_2$ into t-BuNHBr in iso-pentane.

N-Bromosuccinimide (NBS) cleaves the N—H bonds in $t\text{-}BuNH_2$, Me_2CCH_2NH, Me_2CCMe_2NH, $t\text{-}Bu(Me_3Si)NH$ and $(Me_3Si)_2NH$ to form $t\text{-}BuNBr_2$ [19], Me_2CCH_2NBr [3], Me_2CCMe_2NBr [3], $t\text{-}Bu(Me_3Si)NBr$ [20] and $(Me_3Si)_2NBr$ [20], respectively.

Dibromoisocyanuric acid (DBI) and $MeNH_2$, $t\text{-}BuNH_2$ or $H_2NCH_2CH_2NH_2$ in a mixed CH_2Cl_2 and aq NaOH at 0°C give[21] $MeNBr_2$, $t\text{-}BuNBr_2$ and $Br_2NCH_2CH_2NBr_2$, respectively, e.g.:

$$\text{(DBI: O; Br–N, N–Br; O, N–H, O)} + MeNH_2 \rightarrow \text{(O; HN, NH; O, N–H, O)} + MeNBr_2 \qquad (f)$$

With the amides $RC(O)NH_2$ in CH_2Cl_2 or $ClCH_2CH_2Cl$ gives the N-bromo and N,N-dibromo derivatives, $RC(O)NHBr$ [22] (R = Me, Et, $ClCH_2$, Cl_2CH, Cl_3C, F_3C, t-Bu, Ph) and $RC(O)NBr_2$ [21] (R = Me, $ClCH_2$, Cl_2CH, H, Ph), respectively, in over 80% yield. N,N,N′,N′-Tetrabromosuccindiamide is prepared similarly[21].

The cyclic $(H_2NCN)_3$ is converted by dibromoisocyanuric acid in conc H_2SO_4 to $(Br_2NCN)_3$ in 67% yield[23].

(iv) Iodides. The iodides NI_3NH_3 and $NI_3 \cdot 3\ NH_3$ react with liq $MeNH_2$ and Me_2NH to form[24,25] $MeNI_2 \cdot MeNH_2$ and Me_2NI at $-15°C$, e.g.:

$$3\ Me_2NH + NI_3 \cdot NH_3 \rightarrow 3\ MeNI + 2\ NH_3 \qquad (g)$$

The diamines $H_2NCH_2CH_2CH_2NH_2$, $Me_2NCH_2CH_2NH_2$, $MeNHCH_2CH_2NHMe$ and piperazine and $NI_3 \cdot NH_3$ form $I_2NCH_2CH_2NI_2$, $Me_2NCH_2CH_2I_2$, $MeNICH_2CH_2NIMe$ and N,N′-diodopiperazine[25], respectively.

N-Iodosuccinimide and $MeNH_2$ or Me_2NH give[26] $MeNI_2$ and Me_2NI, respectively.

The synthesis of low-boiling primary N-chloramines HRNCl (R = Me, Et, n-Pr, n-Bu, $CH_2CH{=}CH_2$, $CH_2C{\equiv}CH$), secondary N-chloramines RR'NCl (R = R' = Me, R = R' = Et, RR' = $-(CH_2)-$, n = 2 − 5) and N,N-dichloramines $RNCl_2$ (R' = Me, Et, n-Pr, n-Br, $CH_2CH{=}CH_2$, $CH_2C{\equiv}CH$) is carried out by a vacuum gas/solid N-chlorination of the amine over solid N-chlorosuccinimide at RT[27]. Primary mono and dihalophosphines $X(H)PC(CF_3)_2OSiMe_3$ and $X_2PC(CF_3)_2OSiMe_3$ (X = Cl, Br) are obtained in >60% yield from the reaction of $H_2PC(CF_3)_2$, $OSiMe_3$ and the appropriate amount of N-chloro or N-bromo-succinimid within 1 h without a solvent[28]. The secondary holophosphines $XP[Cl(CF_3)_2OSiMe_3]_2$ (X = Cl, Br) are synthesized using the same hologenating agents in 90% yield[29].

(G.-V. RÖSCHENTHALER)

1. W. C. Firth, S. Frank, M. Gerber, V. P. Wystrach, *Inorg. Chem.*, *4*, 765 (1965).
2. G. Huttner, H.-D. Müller, *Angew. Chem., Int. Ed. Engl.*, *14*, 571 (1975).
3. S. J. Brois, *J. Am. Chem. Soc.*, *90*, 506 (1968).
4. T. G. Back, D. H. R. Barton, *J. Chem. Soc., Perkin Trans. 1*, 924 (1977).
5. B. B. Hunt, B. C. Saunders, *J. Chem. Soc.*, 1957, 2413.
6. L. J. Szafraniec, L. P. Rieff, H. S. Aaron, *J. Am. Chem. Soc.*, *92*, 6391 (1970).
7. L. P. Reiff, H. S. Aaron, *J. Am. Chem. Soc.*, *92*, 5275 (1970).
8. O. Nagase, *Chem. Pharm. Bull.*, *15*, 648 (1967).
9. H. Leimeister, K. Dehnicke, *Z. Anorg. Allg. Chem.*, *415*, 115 (1975).
10. L. D. Quin, H. G. Anderson, *J. Org. Chem.*, *31*, 1206 (1966).
11. E. E. Nifant'ev, M. P. Koreteev, *J. Gen. Chem. USSR (Engl. Transl.)*, *37*, 1293 (1967).
12. A. W. Frank, *J. Org. Chem.*, *26* 850 (1961).
13. R. E. Montgomery, L. D. Quin, *J. Org. Chem.*, *30*, 2393 (1965).
14. A. I. Bokanov, V. A. Plakhov, *J. Gen. Chem. USSR (Engl. Transl.) 35*, 350 (1965).
15. G. P. Sollott, B. Howard, *J. Org. Chem.*, *29*, 2451 (1964).
16. E. V. Kuznetsov, R. K. Valetdinov, T. Ya. Roitburd, *J. Gen. Chem. USSR (Engl. Transl.)*, *33*, 143 (1963).
17. P. Choudhury, M. F. El-Shazley, C. Spring, A. L. Rheingold, *Inorg. Chem.*, *18*, 543 (1979).
18. J. Jander, C. Lafrenz, *Z. Anorg. Allg. Chem.*, 349, 57 (1967).
19. W. Gottardi, *Monatsh. Chem.*, *104*, 1681 (1973).
20. N. Wiberg, F. Raschig, *J. Organomet. Chem.*, *10*, 15 (1967).
21. W. Gottardi, *Monatsh. Chem.*, *104*, 421 (1973).
22. W. Gottardi, *Monatsh. Chem.*, *106*, 611 (1975).
23. W. Gottardi, *Monatsh. Chem.*, *103*, 878 (1972).
24. J. Jander, U. Engelhardt, G. Weber, *Angew. Chem., Int. Ed. Engl. 1*, 46 (1962).
25. J. Jander, K. Knuth, K.-U. Trommsdorff, *Z. Anorg. Allg. Chem.*, *394*, 225 (1972).
26. J. Jander, K. Knuth, W. Renz, *Z. Anorg. Allg. Chem.*, *392*, 143 (1972).
27. J. C. Guillemin, J. M. Denis, *Synthesis*, 1131 (1985).
28. H. Kischkel, G.-V. Röschenthaler, *Chem. Ber.*, *118*, 4842 (1985).
29. H. Kischkel, R. Francke, G.-V. Röschenthaler, *Rev. Chim. Minerale*, *23*, 690 (1986).

2.4.4.2.6. to Give Group-VB Halides from Group-VIB Halides.

The N-fluoro derivative is obtained when 2,2,6,6-tetramethylpiperidin-4-on reacts[1] with $FOClO_3$.

In the presence of Et_3N, SO_2ClF[2] or SO_2F_2[3] will fluorinate P—H bonds:

$$(EtO)_2P(O)H + SO_2FX \xrightarrow{CH_2Cl_2} (EtO)_2P(O)F \qquad (a)$$

At 2:1 $MeAsH_2{:}S_2Cl_2$, $MeAsCl_2$, H_2S and $(MeAs)_x$ form; however, at 1:2 only $MeAsCl_2$, HCl and S_8 can be isolated[2].

Trifluoromethylsulfenylchloride and Me_2AsH react to give[3] Me_2AsCl. Trichloromethylsulfenylchloride converts[4] $(RO)_2P(O)H$ into $(RO)_2P(O)Cl$ (R = Me, Et, n-Pr, i-Pr, n-Bu, i-Bu) at $-5°C$[4]:

$$(RO)_2P(O)H + ClSCCl_3 \rightarrow (RO)_2P(O)Cl + CSCl_2 + HCl \qquad \text{(b)}$$

Thionyl chloride and $PhPH_2$ form $PhP(S)Cl_2$ in benzene[5]; $Ph_2P(O)H$ reacts to give $Ph_2P(O)Cl$ in 32% yield[6].

Sulfurylchloride in CCl_4 or benzene cleaves the P—H bonds in PhPH(O)OBu-n, (n-BuO)(PhO)P(O)H, $(EtO)_2P(Se)H$ and $(PhO)_2P(O)H$ to give PhPCl(O)OBu-n[7], (n-BuO)(PhO)P(O)Cl[8], $(EtO)_2P(Se)Cl$[9] and $(PhO)_2P(O)Cl$[10] in good yields.

(G.-V. RÖSCHENTHALER)

1. D. M. Gardner, R. Helitzer, D. H. Rosenblatt, *J. Org. Chem., 32*, 1115 (1967).
2. A. Lopusiuski, J. Michalski, *Angew. Chem., 94*, 302 (1982).
3. T. Mahmood, J. M. Shreeve, *Inorg. Chem., 24*, 1395 (1985).
4. P. Choudhury, A. L. Rheingold, *Inorg. Chim. Acta, 28*, L127 (1978).
5. W. R. Cullen, P. S. Dhaliwal, *Can. J. Chem., 45*, 379 (1967).
6. V. Ettel, M. Zbirovsky, *Chem. Listy, 50*, 1265 (1956).
7. L. Anschütz, H. Wirth, *Naturwissenschaften., 43*, 16 (1956).
8. B. B. Hunt, B. C. Saunders, *J. Chem. Soc.* 2413 (1957).
9. E. A. Markevich, A. I. Razumov, A. D. Reshetnikova, *J. Gen. Chem. USSR (Engl. Transl.), 27*, 2455 (1957).
10. F. L. Maklyaev, M. I. Druzin, I. V. Palagina, R. Ya. Aleksandrova, V. K. Prokhodtseva, R. A. Kamidulina, *J. Gen. Chem. USSR (Engl. Transl.), 32*, 3357 (1962).
11. C. Krawiecki, J. Michalski, R. A. Y. Jones, A. R. Katritzky, *Rocz. Chem., 43*, 869 (1969); *Chem. Abstr., 71*, 61,484 (1969).
12. E. N. Walsh, *J. Am. Chem. Soc., 81*, 3023 (1959).

2.4.4.2.7. to Give Group-VB Halides from Group-VIIB Halides.

Chlorine trifluoride with NHF_2 forms[1] $NClF_2$; ClF and ClF_5 give $NClF_2$, too[1]:

$$ClF_3 + 3\ NHF_2 \rightarrow ClNF_2 + N_2F_4 + 3\ HF \qquad \text{(a)}$$

Iodine chloride exchanges H for I in $H_2NCH_2CH_2NH_2$, $M_2NCH_2CH_2NH_2$, $MeHNCH_2CH_2NHMe$ and piperazine[2]. From $Ph_3P{=}NH$ and ICl, $Ph_3P{=}NI$ is obtained[3]; from $RSbH_2$ and ICl, the polymer $(RSbI_{0.4})_x$ and $RSbI_2$ (R = Me, Et, n-Bu)[4].

The iodochloride I gives on heating the N-chloro derivative II[5]:

$$\text{I} \rightarrow \text{II} \qquad \text{(b)}$$

I: $C_6H_4(ICl_2)C(O)NH_2$ — II: cyclic I(Cl)–N(Cl)–C(=O)

(G.-V. RÖSCHENTHALER)

1. D. Pilipovich, C. J. Schack, *Inorg. Chem., 7*, 386 (1968).
2. J. Jander, K. Knuth, K.-U. Trommsdorff, *Z. Anorg. Allg. Chem., 394*, 225 (1972).
3. R. Appel, A. Hauss, G. Büchler, *Z. Naturforsch., Teil B, 16*, 405 (1961).
4. P. Choudhury, M. F. El-Shazley, C. Spreng, A. L. Rheingold, *Inorg. Chem., 40*, 2129 (1975).

2.4.4.2.8. to Give Group-VB Halides from Transition-Metal Halides.

Silver difluoride cleaves the N—H bond in $CF_3SF_4NHR_f$ at 70°C to form $CF_3SF_4NFR_f$ ($R_f = CF_3$, C_2F_5)[1].

Mercury(II) chloride initiates the formation of a betaine from $(PhO)_2P(O)H$ and pyridine[2]:

$$HgCl_2 + (PhO)_2P(O)H + 2\ C_5H_5N \rightarrow [C_5H_5N][PCl(O)(OPh)_2] + Hg + [C_5H_5N]Cl \quad (a)$$

(G.-V. RÖSCHENTHALER)

1. S.-L. Yu, J. M. Shreeve, *J. Fluorine Chem.*, *7*, 85 (1976).
2. N. Yamazaki, F. Higashi, *Tetrahedron Lett.*, 415 (1972).

2.4.5. from Cleavage of the Group VB–Carbon Bond

2.4.5.1. by Halogens

2.4.5.1.1 to Give Group-VB Fluorides.

Methyldifluoramine is formed from F_2 diluted by N_2 with aq $(MeNH)_2CO$, MeNHC(O)Me and $MeNHCO_2Et$ at 0°–5°C[1] in 5–50% yield. Fluorine also cleaves the C—N and N—H bonds in $EtNHC(O)NH_2$ to form $EtNF_2$ in 49% yield[2] and in MeOC(O)NHCH(Me)Et to form $EtCH(Me)NF_2$ in CH_3CN in 46% yield at −20°C[3]. The cyclic amide, $\overline{CH_2(CH_2)_2C(O)NH}$, gives $F_2N(CH_2)_3CO_2H$ at −30°C in 11% yield[4].

The fluorination of $HC(O)NMe_2$ furnishes $(CF_3)_2NF$ besides other products[5]. A C—N bond cleavage takes place when guanylurea sulfate,

$$H_2NC(NH)NHC(O)NH_2 \cdot H_2SO_4,$$

is treated with F_2. The fluoramines $CF(NF_2)_3$ and $(F_2N)_2C{=}NF$ are obtained.[6]

At −60° to −28°C F_2 and $(CF_3)_3As$ give[7] AsF_3 and CF_4.

(G.-V. RÖSCHENTHALER)

1. R. E. Banks, R. N. Haszeldine, J. P. Lalu, *J. Chem. Soc., C*, 1514 (1966).
2. V. Grakauskas, K. Baum, *J. Am. Chem. Soc.*, *92*, 2096 (1970).
3. V. Grakauskas, K. Baum, *J. Org. Chem.*, *34*, 2840 (1969).
4. V. Grakauskas, K. Baum, *J. Org. Chem.*, *35*, 1545 (1970).
5. J. A. Attaway, R. H. Groth, L. A. Bigelow, *J. Am. Chem. Soc.*, *81*, 3599 (1959).
6. R. J. Koshar, D. R. Husted, C. D. Wright, *J. Org. Chem.*, *32*, 3859 (1967).
7. H. J. Emeléus, R. N. Haszeldine, E. G. Walaschewski, *J. Chem. Soc.*, 1552 (1953).

2.4.5.1.2. to Give Group-VB Chlorides.

The chlorination of $(CF_3)_3As$ gas at 20°C gives a mixture of CF_3Cl, $AsCl_3$ and $(CF_3)_2AsCl$; at 125°C CF_3Cl and $AsCl_3$ are obtained quantitatively[1].

The reaction of 1:2 $(CF_3)_2NAs(CF_3)_2$:Cl_2 at $-50°C$ gives $(CF_3)_2NAsCl_2$ and CF_3Cl. With xs Cl_2, $AsCl_3$, CF_3Cl and $(CF_3)_2NCl$ are formed[2]:

$$(CF_3)_2NAs(CF_3)_2 + 2\ Cl_2 \rightarrow (CF_3)_2NAsCl_2 + 2\ CF_3Cl \quad \text{(a)}$$

Treatment of Ph_2SbCCl_3 with Cl_2 gives[3] Ph_2SbCl_3 and CCl_4.

Chlorine cleaves the Bi—C bond in Me_3Bi to form[4] Me_2BiCl and MeCl[4].

(G.-V. RÖSCHENTHALER)

1. H. J. Emeléus, R. N. Haszeldine, E. G. Walaschewski, *J. Chem. Soc.*, 1552 (1953).
2. H. G. Ang, *J. Inorg. Nucl. Chem.*, *31*, 3311 (1969).
3. R. Müller, S. Reichel, C. Dathe, *J. Prakt. Chem.*, *311*, 930 (1969); *Chem. Abstr.*, *72*, 67,025 (1970).
4. A. M. Marquardt, *Ber. Dtsch. Chem. Ges.*, *20*, 1516 (1887).

2.4.5.1.3. to Give Group-VB Bromides.

Bromine with $(CF_3)_3As$ yields[1] $AsBr_3$, $(CF_3)_2AsBr$ and CF_3AsBr_2.

The As—C bonds in $Me(Ph)AsCH_2Ph$ and $MeAs(CH_2Ph)_2$ are cleaved by Br_2 at $-20°C$ to give Me(Ph)AsBr and $Me(PhCH_2)AsBr$, respectively[2]:

$$Me(Ph)AsCH_2Ph + Br_2 \rightarrow Me(Ph)AsBr + BrCH_2Ph \quad \text{(a)}$$

Mixing $(CF_3)_3Sb$ and Br_2 produces $(CF_3)_2SbBr$, CF_3SbBr_2, $SbBr_3$ and CF_3Br at RT[3].

From Me_3Bi, Et_3Bi, $(i\text{-}Bu)_3Bi$ and $(CH_2{=}CMe)_3Bi$ and Br_2 in pet. ether or $CHCl_3$ the bromobismuthines Me_2BiBr[4], Et_2BiBr[4], $(i\text{-}Bu)_2BiBr$[5] and $CH_2{=}CMeBiBr_2$[6], respectively, are obtained. Dibromomethylbismuthine is formed also from Me_3Bi and Br_2 in 15–20% yield but cannot be separated from the other products[7].

(G.-V. RÖSCHENTHALER)

1. H. J. Emeléus, R. N. Haszeldine, E. G. Walaschewski, *J. Chem. Soc.*, 1552 (1953).
2. A. Schulze, S. Samaan, L. Horner, *Phosphorus*, *5*, 265 (1975).
3. J. W. Dale, H. J. Emeléus, R. N. Haszeldine, J. H. Moss, *J. Chem. Soc.*, 3708 (1957).
4. A. Marquardt, *Ber. Dtsch. Chem. Ges.*, *20*, 1516 (1887).
5. A. Marquardt, *Ber. Dtsch. Chem. Ges.*, *21*, 2038 (1888).
6. A. E. Borisov, M. A. Osipova, A. N. Nesmeyanov, *Izv. Akad. Nauk SSSR, Ser. Khim.*, 1507 (1963).
7. O. Scherer, P. Hornig, M. Schmidt, *J. Organomet. Chem.*, *6*, 259 (1966).

2.4.5.1.4. to Give Group-VB Iodides.

The phosphine $(CF_3)_3P$ undergoes P—C bond clevage by I_2 at 180°C. Typically, the products are PI_3 (30%), $(CF_3)_2PI$ (16%) and CF_3PI_2 (4%)[1]. Similarly, for $(CF_3)_3As$, $(CF_3)_2AsI$ (5%), CF_3AsI_2 (5%) and AsI_3 (8%) are obtained at 100°C[2]. If $(CF_3)_3Sb$ is treated with I_2 at 20°C, $(CF_3)_2SbI$ (54%) and CF_3SbI_2 (14%) are found[2]. Iodine and Ph_3Bi(1:1.35) in Et_2O give $PhBiI_2$ and PhI in 50% yield[3,4].

(G.-V. RÖSCHENTHALER)

1. F. W. Bennett, H. J. Emeléus, R. N. Haszeldine, *J. Chem. Soc.*, 1565 (1953).
2. G. R. A. Brandt, H. J. Emeléus, R. N. Haszeldine, J. H. Moss, *J. Chem. Soc.*, 3708 (1957).
3. J. F. Wilkinson, F. Challenger, *J. Chem. Soc.*, *125*, 854 (1924).
4. G. Wittig, D. Hellwinkel, *Chem. Ber.*, *97*, 789 (1964).

2.4.5.2. by Hydrogen Halides

2.4.5.2.1. to Give Group-VB Fluorides.

The interaction of Me_5Sb with HF or KHF_2 yields[1] Me_4SbF:

$$Me_5Sb + HF \rightarrow Me_4SbF + MeH \quad (a)$$

Pentaphenylantimony is cleaved by HF to give triphenylantimony difluoride[2]:

$$Ph_5Sb + 2\ HF \rightarrow Ph_3SbF_2 + 2\ PhH \quad (b)$$

(G.-V. RÖSCHENTHALER)

1. H. Schmidbaur, K. H. Mitschke, *Chem. Ber.*, *106*, 1226 (1973).
2. G. A. Olah, P. Schilling, I. M. Gross, *J. Am. Chem. Soc.*, *96*, 876 (1974).

2.4.5.2.2. to Give Group-VB Chlorides.

Treating $ClRPCH_2PRCl$ with HCl gives 30–88% of RPClMe and $RPCl_2$ (R = Me, i-Pr, i-Bu, c-C_6H_{11}, Ph)[1]. Under the same conditions, $Ph_2PCH_2PCl_2$ and HCl form[1] Ph_2PMe and PCl_3.

Hydrogen chloride and $(CF_2{=}CF)_3As$ form $CF_2{=}CFAsCl_2$ in 93% yield[2]. Arsenic trichloride and Ph_2NH are obtained from[3]:

The C—N bond in $Ph_2Sb_2CH_2$ is cleaved by HCl in $CHCl_3$ at 0°C to give $(Cl_2Sb)_2CH_2$ in 100% yield[4]. Refluxing Ph_3Sb in CH_3OH that is saturated with HCl gives[5] Ph_2SbCl.

Concentrated HCl and Me_3Bi yield[6] $BiCl_3$ and CH_4.

Liquid HCl cleaves all three phenyl groups of Ph_3Bi to form $BiCl_3$ and benzene[7]. Treatment of $Ph_4BiOSiPh_3$ with ethanolic HCl results in the reaction[8]:

$$Ph_4BiOSiPh_3 + 2\ HCl \rightarrow Ph_3BiCl_2 + Ph_3SiOH + PhH \quad (a)$$

(G.-V. RÖSCHENTHALER)

1. A. Prishchenko, Z. S. Novikova, I. F. Lutsenko, *J. Gen. Chem. USSR (Engl. Transl.). 50*, 787 (1980).
2. R. N. Sterlin, L. N. Pinkina, R. D. Yatsenko, I. L. Knunyants, *Khim. Nauk Prom.*, *4*, 800 (1959); *Chem. Abstr.*, *54*, 14,103 (1960).
3. G. A. Razuvaev, M. M. Koton, *Zh. Obshch. Khim.*, *2*, 529 (1932); *Chem. Abstr.*, *27*, 984 (1933).
4. Y. Matsumura, R. Okawara, *Inorg. Nucl. Chem. Lett.*, *7*, 113 (1971).
5. K. Issleib, B. Hamann, *Z. Anorg. Allg. Chem.*, 343, 196 (1966).
6. A. Marquardt, *Ber. Dtsch. Chem. Ges.*, *20*, 1516 (1887).
7. M. E. Peach, T. C. Waddington, *J. Chem. Soc.*, 1238 (1961).
8. G. A. Razuvaev, N. A. Osanova, V. V. Sharutin, *Dokl. Akad. Nauk SSSR*, *225*, 581 (1975).

2.4.5.2.3. to Give Group-VB Bromides.

The diarsine **I** reacts with HBr[1]:

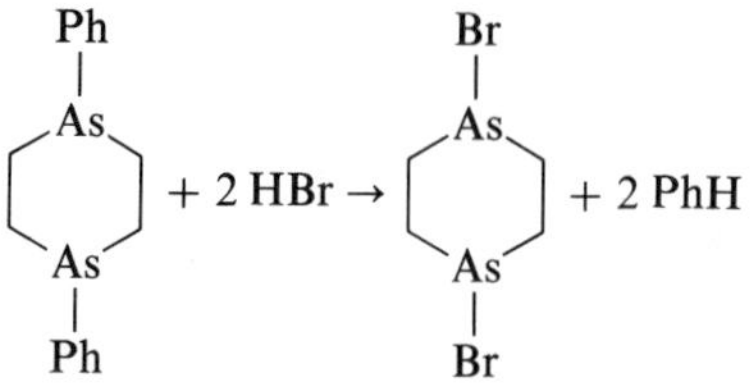

$$\text{I (1,4-diphenyl-1,4-diarsacyclohexane)} + 2\,HBr \rightarrow \text{1,4-dibromo-1,4-diarsacyclohexane} + 2\,PhH \qquad (a)$$

Liquid HBr reacts with Ph_3Sb and Ph_3Bi to form Ph_2SbBr, $PhSbBr_2$, $SbBr_3$ and $BiBr_3$, respectively[2]:

$$Ph_3Bi + 3\,HBr \rightarrow BiBr_3 + 3\,PhH \qquad (b)$$

(G.-V. RÖSCHENTHALER)

1. E. R. H. Jones, F. G. Mann, *J. Chem. Soc.*, 401 (1955).
2. M. E. Peach, *J. Inorg. Nucl. Chem.*, *39*, 565 (1977).

2.4.5.2.4. to Give Group-VB Iodides.

The phenylalkylarsines Me_2AsPh, $MeAsPh_2$, $EtAsPh_2$, $(CH_2)_n(AsPh_2)_2$ (n = 1, 2) and $C(CH_2AsPh_2)_4$ react with liq or gaseous HI in nonaq solvents, e.g., CH_2Cl_2 or dimethylsulfoxide (DMSO) by selective cleavage of the As–phenyl bonds, yielding[1] Me_2AsI, $MeAsI_2$, $EtAsI_2$, $(CH_2)_n(AsI_2)_2$ (n = 1,2) and $C(CH_2AsI_2)_4$. The phenyl—As bond is cleaved[2–4] in **I**, **II** and **III**[2–4]:

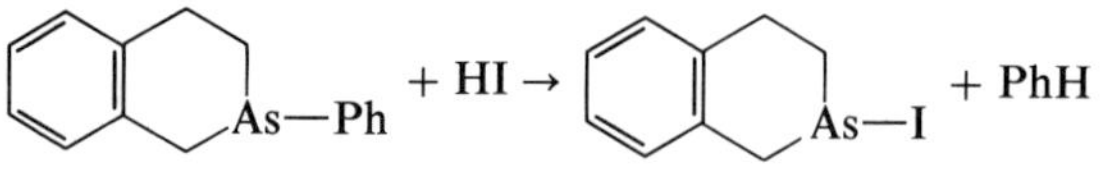

$$\text{I (2-phenyl-1,2,3,4-tetrahydro-2-benzarsinine, As—Ph)} + HI \rightarrow \text{As—I analogue} + PhH \qquad (a)$$

$$\text{II (1,4-diphenyl-1,4-diarsacyclohexane)} + 2\,HI \rightarrow \text{1,4-diiodo-1,4-diarsacyclohexane} + 2\,PhH \qquad (b)$$

$$\text{III (5-phenyl-10,11-dihydrodibenzarsepine)} + 3\,HI \rightarrow PhCH_2CH_2Ph + AsI_3 + PhH \qquad (c)$$

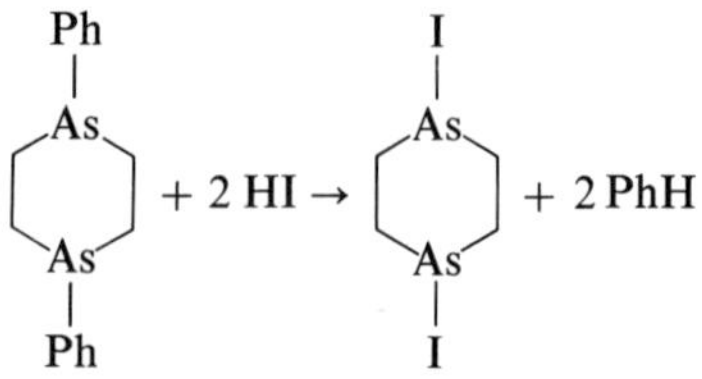

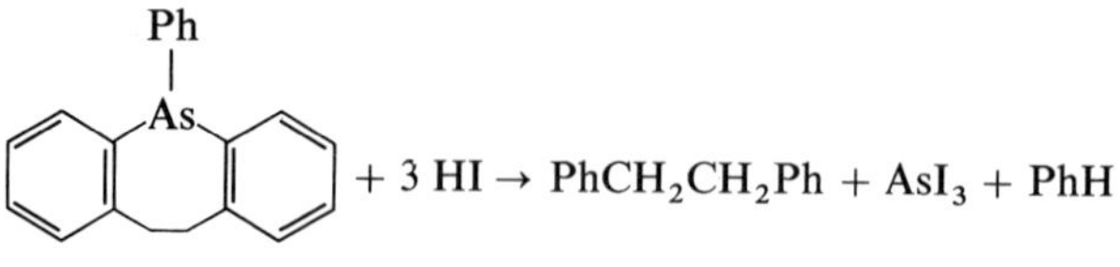

(G.-V. RÖSCHENTHALER)

1. J. Ellermann, H. Schlössner, A. Haag, H. Schödel, *J. Organomet. Chem.*, *65*, 33 (1974).
2. M. H. Beeby, G. H. Cookson, F. G. Mann, *J. Chem. Soc.*, 1917 (1950).
3. E. R. H. Jones, F. G. Mann, *J. Chem. Soc.*, 401 (1955).
4. F. G. Mann, I. T. Millar, B. B. Smith, *J. Chem. Soc.*, 1130 (1953).

2.4.5.3. by Other Inorganic Halides

2.4.5.3.1. of Group IIIB.

Trifluoromethyl groups are lost from $(CF_3)_3As$ on heating with BI_3 or AlI_3. From BI_3 are found BF_3, CF_3I, $(CF_3)_2AsI$ and AsI_3, whereas AlI_3 produces CF_3I, $(CF_3)_2AsI$ at 115°C; and CF_4, C_2F_6, AsF_3 and As at 150°C[1].

Thallium(III) chloride cleaves the C—Bi bond in Ph_3Bi to give[2] Ph_2BiCl, $BiCl_3$ and Ph_2TlCl.

(G.-V. RÖSCHENTHALER)

1. W. R. Cullen, *Can. J. Chem.*, *41*, 317 (1963).
2. F. Kh. Solomakhina, *Tr. Tashk. Farm. Inst.*, *1*, 321 (1957); *Chem. Abstr.*, *55*, 15,389 (1961).

2.4.5.3.2. of Group IVB.

The main products from n-Bu_5Sb and CCl_4 at 100°C are[1] $CHCl_3$ and n-Bu_3SbCl_2.

Triphenylbismuthine is cleaved by $SiCl_4$[2] or $SnCl_4$[3] to give $BiCl_3$ and Ph_2BiCl. Tetramethylantimony fluoride results from[4] Me_5Sb and Me_3SnF:

$$Me_5Sb + Me_3SnF \rightarrow Me_4SbF + Me_4Sn \tag{a}$$

(G.-V. RÖSCHENTHALER)

1. A. N. Nesmeyanov, A. E. Borisov, N. G. Kizim, *Izv. Akad. Nauk SSSR, Ser. Khim.*, 1672 (1974).
2. F. Kh. Solomakhina, *Tr. Tashk. Farm. Inst.*, *2*, 317 (1960); *Chem. Abstr.*, *57*, 11,230 (1962).
3. F. Kh. Solomakhina, *Tr. Tashk. Farm. Inst.*, *1*, 321 (1957); *Chem. Abstr.*, *55*, 15,389 (1961).
4. H. Schmidbaur, J. Weidlein, K.-H. Mitschke, *Chem. Ber.*, *102*, 4136 (1969).

2.4.5.3.3. of Group VB.

(i) Nitrogen Halides. The ylids $Ph_3P{=}CHCO_2R$ (R = Me, Et) are converted[1] into Ph_3PF_2 with N_2F_4 at RT:

$$Ph_3P{=}CHCO_2R + N_2F_4 \rightarrow N_2 + Ph_3PF_2 + CHF_2CO_2R \tag{a}$$

(ii) Phosphorus Halides. Phosphorus trichloride cleaves the C—P bonds in $Ph_2P(CH_2)_nPPh_2$ to give $Cl_2P(CH_2)_nPCl_2$ (n = 1–4) and $PhPCl_2$ at 280°C[2]. Triphenylbismuthine and PCl_3 (1:1) give Ph_2BiCl in 98% yield from pet. ether[3].

When heated to 95°C with PCl_5 chloromethylphosphonic dichloride gives rise to CCl_4, $P(O)Cl_3$ and PCl_3, quantitatively[4]. With $(ClCH_2)_2P(O)Cl$ or $(ClCH_2)_3PO$, PCl_3 and $CCl_3P(O)Cl_2$ or $Cl_3P(CH_2Cl)_2$ are formed[4].

(iii) Arsenic Halides. With h^1-$C_5H_5AsR_2$ (R = Me, t-Bu), and h^1-$C_5H_5SbMe_2$ arsenic trifluoride forms h^1-$C_5H_5AsF_2$, R_2AsF and h^1-$C_5H_5SbF_2$[5].

Arsenic trichloride cleaves the C—As bonds in $(CH_2{=}CH)_3As$, $(CF_3)_3As$, Ph_2AsCH_2Cl, $Ph_2AsCH_2CH_2AsPh_2$ and in Ph_3As to form $CH_2{=}CHAsCl_2$[6]

and $(CH_2{=}CH)_2AsCl$[6], CF_3AsCl_2[7] and $(CF_3)_2AsCl$[7], $ClCH_2AsCl_2$[8], $Cl_2AsCH_2CH_2AsCl_2$[9] and Ph_2AsCl[10], respectively by redistribution, e.g.:

$$2\ Ph_3As + AsCl_3 \rightarrow 3\ Ph_2AsCl \qquad \text{(b)}$$

A mixture of $PhAsCl_2$, Ph_2AsCl and Ph_2BiCl is found[11] when $AsCl_3$ reacts with Ph_3Bi.

The arsenic halide $AsCl_3$ converts[5] h^1-$C_5H_5AsMe_2$ and h^1-$C_5H_5SbMe_2$ to h^1-$C_5H_5AsCl_2$ and Me_2AsCl and h^1-$C_5H_5SbCl_2$.

When $(CH_2{=}CH)_3As$ reacts with $AsBr_3$ the respective dibromo and monobromoarsines, $CH_2{=}CHAsBr_2$ and $(CH_2{=}CH)_2AsBr$, are obtained[5].

From $AsBr_3$ and h^1-$C_5H_5AsMe_2$, or h^1-$C_5H_5SbMe_2$, the bromides h^1-$C_5H_5EBr_2$(E = As, Sb) are obtained[5]. In correct proportions, 2:1 $AsBr_3$ and Et_3As react to give[6] $EtAsBr_2$.

The cleavage of the C—As bond in $(CF_3)_3As$, h^1-$C_5H_5AsMe_2$ and h^1-$C_5H_5SbMe_2$ is caused by AsI_3. The reaction products are $(CF_3)_2AsI$ and CF_3AsI_2 at 230–240°C[12] and h^1-$C_5H_5EI_2$ (E = As, Sb)[5].

(iv) Antimony Halides. Excess $SbCl_3$ reacts with Me_3Sb in N,N-dimethylformamide (DMF) at 100°C in a sealed tube to give[13] mainly $MeSbCl_2$. Redistribution of $SbCl_3$ and $(CH_2{=}CH)_3Sb$ gives[6] $CH_2{=}CHSbCl_2$ and $(CH_2{=}CH)_2SbCl$. Other redistributions are carried out with $SbCl_3$ and $(EtO_2CCH_2)_3Sb$ in benzene at RT[14], with $SbCl_3$ and $(p\text{-}MeC_6H_4)_3Sb$ in xylene at 245°C[15] and with $SbCl_3$ and Ph_3SbCl in refluxing CH_2Cl_2[16] to give $EtO_2CH_2SbCl_2$ in 93% yield, p-$CH_3C_6H_4SbCl_2$ and, almost quantitatively, $PhSbCl_2$.

From h^1-$C_5H_5AsMe_2$, h^1-$C_5H_5SbMe_2$ and h^1-C_5H_5SbBu-t_2 with $SbCl_3$, h^1-$C_5H_5ECl_2$ (E = As, Sb) and Me_2SbCl, or, for the last, t-Bu_2SbCl, are found[5].

Antimony trichloride cleaves the C—Bi bond in Ph_3Bi to give[11] Ph_2BiCl and Ph_3SbCl_2.

A butyl group is transferred from Bu_3Sb onto $PhSbCl_2$, which gives[17] n-Bu_2SbCl and Ph(Bu)SbCl.

The redistribution of Et_3Sb and $SbBr_3$, or $(CH_2{=}CH)_3Sb$ and $SbBr_3$, gives $EtSbBr_2$ as a major product in $CH_3C(O)NEt_2$ at 100°C[18] or $(CH_2{=}CH)_2SbBr$[6], respectively.

Antimony triiodide and h^1-$C_5H_5AsMe_2$, h^1-$C_5H_5SbMe_2$ or h^1-C_5H_5SbBu-t_2 produce[5] h^1-$C_5H_5EI_2$ (E = As, Sb) and Me_2SbI or t-Bu_2SbI. In a sealed tube, $(CF_3)_3Sb$ and SbI_3 form[19] $(CF_3)_2SbI$.

(v) Bismuth Halides. From 2:1 $BiCl_3$ and R_3Bi in Et_2O, C_6H_6, $CHCl_3$, CH_3CO_2H or CH_3COOH_3, dichlorobismuthines are synthesized[20] (R = Me[21,22], Et[21], Ph[23]):

$$R_3Bi + 2\ BiCl_3 \rightarrow 3\ RBiCl_2 \qquad \text{(c)}$$

With 1:2 R_3Bi:$BiCl_3$, monochlorobismuthines are obtained (R = Ph[24], o-MeC_6H_4[25], p-$C_6H_5C_6H_4$[26], p-ClC_6H_4[25], p-BrC_6H_4[27]):

$$2\ R_3Bi + BiCl_3 \rightarrow 3\ R_2BiCl \qquad \text{(d)}$$

Dichlorobismuthines are obtained by the disproportionation of chlorobismuthines in $CHCl_3$ or tetrahydrofuran (THF) at RT or in liq NH_3[27], e.g., p-$BrC_6H_4BiCl_2$, p-$MeOC_6H_4BiCl_2$ and p-$Me_2NC_6H_4BiCl_2$.

Dibromobismuthines are formed like dichlorobismuthines [see Eq. (c)] (R = Me[21,22]; R = i-Bu[28]; R = Ph[24], o-, p-MeC_6H_4[25], p-FC_6H_4[29]). Bromobismuthines are prepared according to Eq. (d) (R = i-Bu[28], i-C_5H_{11}[28], Ph[23], p-FC_6H_4[29], p-ClC_6H_4[30]).

The redistribution reaction of $(HC{\equiv}C)_3Bi$ and BiI_3 yields[31] $(HC{\equiv}C)_2BI$.

(G.-V. RÖSCHENTHALER)

1. A. V. Fokin, Yu. N. Studnev, L. D. Kuznetsova, A. F. Kolomiets, *J. Gen. Chem. USSR (Engl. Transl.)*, *39*, 2303 (1969).
2. K. Sommer, *Z. Anorg. Allg. Chem.*, *376*, 37 (1970).
3. E. O. Fischer, S. Schreiner, *Chem. Ber.*, *93*, 1417 (1960).
4. A. W. Frank, *Can. J. Chem.*, *46*, 3573 (1968).
5. P. Jutzi, M. Kuhn, *J. Organomet. Chem.*, *173*, 221 (1979).
6. L. Maier, D. Seyferth, F. G. A. Stone, E. G. Rochow, *J. Am Chem. Soc.*, *79*, 5884 (1957).
7. W. R. Cullen, *Can. J. Chem.*, *41*, 317 (1963).
8. K. Sommer, *Z. Anorg. Allg. Chem.*, *377*, 128 (1970).
9. K. Sommer, *Z. Anorg. Allg. Chem.*, *376*, 150 (1970).
10. H. Weingarten, J. R. van Wazer, *J. Am. Chem. Soc.*, *88*, 2700 (1966).
11. F. Kh. Solomakhina, *Tr. Tashk. Farm. Inst.*, *1*, 321 (1957). *Chem. Abstr.*, *55*, 15,389 (1961).
12. H. J. Emeléus, R. N. Haszeldine, E. G. Walaschewski, *J. Chem. Soc.*, 1552 (1953).
13. H. Weingarten, J. R. Van Wazer, *J. Am. Chem. Soc.*, *88*, 2700 (1966).
14. E. A. Besolova, V. L. Foss, I. F. Lutsenko, *J. Gen. Chem. USSR (Engl. Transl.)*, *38*, 1523 (1968).
15. J. Hasenbäumer, *Ber. Dtsch. Chem. Ges.*, *31*, 2910 (1898).
16. M. and T. Chemicals, Neth. Pat. 65-05216 (1965); *Chem. Abstr.*, *64*, 9766 (1965).
17. J. C. Summers, H. H. Sislers, *Inorg. Chem.*, *9*, 862 (1970).
18. H. I. Weingarten, W. A. White, U.S. Pat. 3,366,655 (1968); *Chem. Abstr.*, *68*, 95,979 (1968).
19. J. W. Dale, H. J. Emeléus, R. N. Haszeldine, J. H. Moss, *J. Chem. Soc.*, 3708 (1957).
20. L. D. Freedman, G. O. Doak, *Chem. Rev.*, *82*, 15 (1982).
21. A. Marquardt, *Ber. Dtsch. Chem. Ges.*, *20*, 1516 (1887).
22. E. Amberger, *Chem. Ber.*, *94*, 1447 (1961).
23. S. Faleschini, P. Zanella, L. Doretti, G. Faraglia, *J. Organomet. Chem.*, *44*, 317 (1972).
24. R. Okawara, K. Yasuda, M. Inoue, *Bull. Chem. Soc. Jpn.*, *39* 1823 (1966).
25. H. Gilman, H. L. Yablunky, *J. Am. Chem. Soc.*, *63*, 207 (1941).
26. D. E. Worrall, *J. Am. Chem. Soc.*, *58*, 1820 (1936).
27. H. Hartmann, G. Habenicht, W. Reiss, *Z. Anorg. Allg. Chem.*, *317*, 54 (1962).
28. A. Marquardt, *Ber. Dtsch. Chem. Ges.*, *21*, 2038 (1888).
29. S. I. Pombrik, D. N. Kravtsov, B. A. Kvasov, E. I. Fedin, *J. Organomet. Chem.*, *136*, 185 (1977).
30. F. Challenger, L. R. Ridgway, *J. Chem. Soc.*, *121*, 104 (1922).
31. K. Moedritzer, W. Groves, J. R. Van Wazer, H. Weingarten, U.S. Pat. 3,504,005; *Chem. Abstr.*, *72*, 121,707 (1970).

2.4.5.3.4. of Group VIB.

Thionyl and sulfuryl chlorides cleave one phenyl group from Ph_3Bi to give[1,2] Ph_2BiCl.

(G.-V. RÖSCHENTHALER)

1. F. Challenger, *J. Chem. Soc.*, *109*, 250 (1916).
2. F. Challenger, L. R. Ridgway, *J. Chem. Soc.*, *121*, 104 (1922).

2.4.5.3.5. of Group VIIB.

Iodine monochloride cleaves the C—Bi bond in Ph_3Bi to give[1] Ph_2BiCl and PhI. From IBr, Ph_2BiBr and PhI are obtained[2].

(G.-V. RÖSCHENTHALER)

1. F. Challenger, C. F. Allpress, *J. Chem. Soc.*, *107*, 16 (1915).
2. A. D. Beveridge, G. S. Harris, F. Inglis, *J. Chem. Soc.*, *A*, 520 (1966).

2.4.5.3.6. of Transition-Metal Halides.

When $(o\text{-}MeC_6H_4)_3Sb$ is heated with $HgCl_2$ in tetrahydrofuran, one of the o-tolyl groups migrates to the Hg atom to form[1] $(o\text{-}MeC_6H_4)_2SbCl$ and $o\text{-}MeC_6H_4HgCl$.

From $TiCl_4$ and Ph_3Bi in refluxing $CHCl_3$, Ph_2BiCl and $BiCl_3$ are formed[2].

When $(CF_3)_3As$ vapor is in contact with CoF_3 at 100°C, $(CF_3)_2AsF$, AsF_3 and CF_4 are obtained[3].

Copper(II) chloride and $ZnCl_2$ cleave a phenyl group from Ph_3Bi to give[4] Ph_2BiCl.

At 0°C in Et_2O, one of the aryl group of Ph_3Bi, $(o\text{-}MeC_6H_4)_3Bi$ or $(p\text{-}MeC_6H_4)_3Bi$ is transferred to the Hg atom[1], e.g.:

$$Ph_3Bi + HgCl_2 \rightarrow Ph_2BiCl + PhHgCl \tag{a}$$

(G.-V. RÖSCHENTHALER)

1. G. Deganello, G. Dolcetti, M. Ginstiniani, U. Belluco, *J. Chem. Soc.*, *A*, 2138 (1969).
2. F. Kh. Somomakhina, *Tr. Tashk. Farm. Inst.*, *2*, 317 (1960); *Chem. Abstr.*, *57*, 11,230 (1962).
3. H. J. Emeléus, R. N. Haszeldine, E. G. Walaschewski, *J. Chem. Soc.*, 1552 (1953).
4. F. Kh. Solomakhina, *Tr. Tashk. Farm. Inst.*, *1*, 321 (1957); *Chem. Abstr.*, *55*, 15,389 (1961).

2.4.5.4. by Organic Halides.

Methyl iodide reacts with Me_3Bi in a sealed tube at 200°C to form[1] $MeBiI_2$ and C_2H_6.

Acetylchloride cleaves the C—As bonds in $p\text{-}O_2NC_6H_4AsCl_2$ and R_2PhAsS to give $p\text{-}O_2NC_6H_4COCH_3$ and $AsCl_3$ [2], and R(Ph)AsCl and CH_3COSR [3] (R = n-Pr, C_6H_{13}).

The bismuthine $(p\text{-}MeC_6H_4)_3Bi$ and MeCOCl produce[4] $(p\text{-}MeC_6H_4)_2BiCl$ and $p\text{-}MeC_6H_4C(O)Me$.

(G.-V. RÖSCHENTHALER)

1. A. Marquardt, *Chem. Ber.*, *20*, 1516 (1887).
2. M. S. Malinovskii, *Zh. Obshch. Khim.*, *5*, 1355 (1935); *Chem. Abstr.*, *30*, 1037 (1936).
3. G. M. Usacheva, G. Kh. Kamai, *Izv. Akad. Nauk SSSR, Ser. Khim.*, 413 (1968).
4. F. Challenger, L. R. Ridgway, *J. Chem. Soc.*, *121*, 104 (1922).

2.4.6. from Cleavage of the Group VB–Other Group IVB Element Bond

2.4.6.1. by Halogens

2.4.6.1.1. to Give Nitrogen Halides.

Primary and secondary organic N-haloamines[1] can be prepared by cleavage of the Si—N bond with halogens in nonaq media, e.g., $CHCl_3$, CH_2Cl_2, Et_2O, or in the absence of solvent:

$$Me_3SiNR_2 + X_2 \xrightarrow{-20 \text{ to } -70^\circ C} R_2NX + Me_3SiX \tag{a}$$

where R = Me, Et, n-Pr, i-Pr, X = Cl, and R = Et, n-Pr; X = Br, I. Yields are, e.g., 90% for Et_2NCl and 60% for Me_2NCl; the latter is preferably prepared from $Me_2Si(NMe_2)_2$ and Cl_2. The dichloroamine, $EtNCl_2$, and monochloroamine, EtNHCl, are obtained similarly[1].

The cleavage of the Sn—N bond in Et_3SnNMe_2 with Cl_2 gives N-chlorodimethylamine in 88% yield[2].

Reaction (a) is also applied to the synthesis of N-iododimethylamine[3] employing ICl as the halogenating agent:

$$Me_3SiNMe_2 + ICl \xrightarrow[CHCl_3]{-60°C} Me_2NI + Me_3SiCl \tag{b}$$

Nitrogen tribromide is synthesized by[4]:

$$(Me_3Si)_2NBr + 2\ BrCl \xrightarrow[pentane]{-87°C} NBr_3 + 2\ Me_3SiCl \tag{c}$$

(thermally unstable; a suspension in mineral oil explodes at −100°C by mechanical shock):

Similarly, N-trimethylsilylthionylimide, Me_3SiNSO, undergoes cleavage with halogens at −70°C[5]:

$$Me_3SiNSO + X_2 \rightarrow XNSO + Me_3SiX \tag{d}$$

where X = F, Cl, Br. Yields are high for the preparation of ClNSO, and low for the corresponding fluorine compound. With BrNSO decomposition is at RT and is observed during distillation. Reaction (d) is also applied to a sulfur(VI) derivative, $Me_3SiNS(O)F_2$, which is cleaved by Cl_2 to give[6,7] $ClNS(O)F_2$.

The reaction of $Me_3SiNSNSiMe_3$ with Cl_2 at −70°C leads to $Me_3SiNSNCl$ in 80% yield, no ClNSNCl being observed. With Br_2 or I_2 only decomposition products are obtained **(Caution: these products may be explosive)**[8].

(M. FILD)

1. K. Seppelt, W. Sundermeyer, *Z. Naturforsch., Teil B, 24*, 774 (1969).
2. T. A. George, M. F. Lappert, *J. Chem. Soc., A*, 992 (1969).
3. H. Hartl, H. Pritzkow, J. Jander, *Chem. Ber., 103*, 652 (1970).
4. J. Jander, J. Knackmuss, K. U. Thiedemann, *Z. Naturforsch., Teil B, 30*, 464 (1975).
5. W. Verbeek, W. Sundermeyer, *Angew. Chem., Int. Ed. Engl., 8*, 376 (1969).
6. K. Seppelt, W. Sundermeyer, *Angew. Chem., Int. Ed. Engl., 9*, 905 (1970).
7. J. K. Ruff, *Inorg. Chem., 5*, 1787 (1966).
8. W. Lidy, W. Sundermeyer, W. Verbeek, *Z. Anorg. Allg. Chem., 406*, 228 (1974).

2.4.6.1.2. to Give Phosphorus Halides.

The Si—P bond in Ph_2PSiMe_3 is cleaved by Cl_2 or Br_2 at −80°C to give diphenylphosphinous halides in >95% yields[1]:

$$Ph_2PSiMe_3 + X_2 \rightarrow Ph_2PX + Me_3SiX \tag{a}$$

where X = Cl, Br. Similarly iodine monochloride and monobromide all give diphenylphosphinous iodide, Ph_2PI, and the corresponding halogenosilane[1].

Bromine reacts with Ph_2PGeMe_3 to yield[2] Ph_2PBr and Me_3GeBr, or with a heterocyclic to give a Ge-substituted phosphinous bromide[3]:

$$\text{Me}_2\text{Ge}\overline{(\text{CH}_2)_3}\text{PPh} + Br_2 \rightarrow Me_2Ge(Br)(CH_2)_3P(Br)Ph \quad \text{(b)}$$

(M. FILD)

1. E. W. Abel, R. A. N. McLean, I. H. Sabherwal, *J. Chem. Soc., A*, 2371 (1968).
2. E. H. Brooks, F. Glockling, K. A. Hooton, *J. Chem. Soc.*, 4283 (1965).
3. C. Couret, J. Escudie, J. Satge, G. Redoules, *Synth. React. Inorg. Met.-Org. Chem.*, *7*, 99 (1977).

2.4.6.1.3. to Give As and Bi Halides.

Halogens split Si— and Sn—As bonds to yield arsinous halides[1]:

$$Me_3MAsMe_2 + X_2 \rightarrow Me_2AsX + Me_3MX \quad \text{(a)}$$

where M = Si, Sn; X = Br, I. In accord with the polarity of the Si—As bond, ICl with $Me_3SiAsMe_2$ leads[1] to a mixture of Me_2AsI and Me_3SiCl.

Treating Ge derivatives of Bi with Br_2 gives $BiBr_3$ in nearly quantitative yields[2], cleaving both the Ge— and C—Bi bonds (see §2.4.5), e.g.:

$$[(C_6F_5)_3Ge]_nBiEt_{3-n} \xrightarrow{3\ Br_2} n(C_6F_5)_3GeBr + BiBr_3 + (3-n)EtBr \quad \text{(b)}$$

where n = 1, 2;

$$[(C_6F_5)_2GeBiEt]_2 \xrightarrow{6\ Br_2} 2\,(C_6F_5)_2GeBr_2 + 2\,BiBr_3 + 2\,EtBr \quad \text{(c)}$$

(M. FILD)

1. E. W. Abel, S. M. Illingworth, *J. Chem. Soc., A*, 1094 (1969).
2. M. N. Bochkarev, N. I. Gurev, G. A. Razuvaev, *J. Organomet. Chem.*, *162*, 289 (1978).

2.4.6.2. by Hydrogen Halides

2.4.6.2.1. to Give Bismuth Chlorides.

The Ge—Bi bond is cleaved with HCl to give Bi chlorides in high yields[1]:

$$[(C_6F_5)_3Ge]_nBiEt_{3-n} \xrightarrow{n\ HCl} Cl_nBiEt_{3-n} + n\,(C_6F_5)_3GeH \quad \text{(a)}$$

(n = 1, 2) e.g., $EtBiCl_2$ (64%) and Et_2BiCl (96%). With $[(C_6F_5)_2GeBiEt]_2$ the cleavage with HCl only results in the formation of elemental Bi.

(M. FILD)

1. M. N. Bochkarev, N. I. Gurev, G. A. Razuvaev, *J. Organomet. Chem.*, *162*, 289 (1978).

2.4.6.3. by Other Halides

2.4.6.3.1. to Give Nitrogen Halides.

Silylated N-haloimines react with thionyl chloride[1]:

$$(Me_3Si)_2NX + SOCl_2 \xrightarrow{-10°C} XN{=}S{=}O + 2\ Me_3SiCl \quad (a)$$

Reaction (a) is performed at −10°C in $CFCl_3$ or other inert solvents to give good yields of the thionyl imides, IN=S=O and BrN=S=O; the corresponding chloride, ClN=S=O, is obtained in 25% yield but is difficult to separate from Me_3SiCl.

(M. FILD)

1. K. Seppelt, W. Sundermeyer, *Naturwissenschaften, 56*, 281 (1969).

2.4.6.3.2. to Give Phosphorus Halides.

Phosphinous chlorides are synthesized in high yields (73–93%) from silylated phosphines with hexachloroethane[1]:

$$RR'PSiMe_3 + C_2Cl_6 \rightarrow RR'PCl + C_2Cl_4 + Me_3SiCl \quad (a)$$

where R′ = Ph; R′ = Me, Et, n-Pr, n-Bu, Ph, c-C_6H_{11}. Pentachloroethane or Ph_3PCl_2 may also be employed, although yields are lower. This is a cleavage of the P—P bond, for it can be shown, by varying mol ratios, that diphosphines are formed that are cleaved with hexachloroethane[1].

Phosphonous dichlorides, e.g. t-BuPCl_2[2], are formed from t-BuP$(SiMe_3)_2$:

$$t\text{-}BuP(SiMe_3)_2 \xrightarrow[0°C]{C_2Cl_6} t\text{-}BuP(Cl)SiMe_3 \xrightarrow{C_2Cl_6} t\text{-}BuPCl_2 \quad (b)$$

The first product observed, in nearly quantitative yields, is the unstable intermediate, t-BuP(Cl)$SiMe_3$. Hexachloroethane is also used for the synthesis of P—C-unsaturated compounds, e.g. t-BuP=C($OSiMe_3$)P(t-Bu)Cl[13] and $Me_3SiO(Me_3C)C{=}PCl$[14].

Silylated phosphines may be converted to phosphonous dichlorides as the final stable products using phosgene as the halogenating agent[3], e.g.:

$$PhP(SiMe_3)_2 + 2\ COCl_2 \rightarrow PhPCl_2 + 2\ Me_3SiCl + 2\ CO \quad (c)$$

Intermediates are isolated if the reaction is performed stepwise[3]:

$$PhP(SiMe_3)_2 \xrightarrow{COCl_2} PhP{=}C(OSiMe_3)P(Ph)SiMe_3 \xrightarrow{COCl_2} [(PhP)_2(COSiMe_3)_2]_2 \xrightarrow{COCl_2} (PhP)_n \quad (d)$$

The P—P bonds in the polyphosphine, $(PhP)_n$, are cleaved by $COCl_2$ to yield $PhPCl_2$ as given in the overall reaction (c) (see §2.4.10.). Similar experiments utilize hexachloroethane[4].

With Ge or Sn tetrachlorides the P—Si bond in $t\text{-}Bu_2PSiMe_3$ reacts to yield bis(t-butyl)phosphinous chloride[5,6]:

$$t\text{-}Bu_2PSiMe_2 \xrightarrow[-Me_3SiCl]{+MCl_4} t\text{-}Bu_2PMCl_3 \xrightarrow{\Delta} t\text{-}Bu_2PCl + MCl_2 \quad \text{(e)}$$

where M = Sn, Ge. The intermediate $t\text{-}Bu_2PSnCl_3$ is unstable at 20°C, whereas the Ge derivative decomposes under reduced pressure (<1 torr) at 100°C. Two products are formed in the reaction with $GeCl_4$ (or $GeBr_4$)[7]:

$$t\text{-}Bu_2PSiMe_3 + GeX_4 \longrightarrow \begin{cases} t\text{-}Bu_2PGeX_3 \\ t\text{-}Bu_2P{<}^{GeX_2}_{X} \end{cases} \quad \text{(f)}$$

where X = Cl, Br.

By mixing the Ge derivatives $t\text{-}Bu_2PGeX_3$ (X = Cl, Br) with a tertiary phosphine in toluene at RT, the GeX_3 group is replaced[7]:

$$t\text{-}Bu_2PGeX_3 + R_3P \rightarrow t\text{-}Bu_2PX + R_3PGeX_2 \quad \text{(g)}$$

Phosphinimines containing a P—Si or P—Ge bond react with CCl_4 to form chlorine-substituted acyclic phosphazenes in high yields[8,9]:

$$t\text{-}Bu_2P({=}NSiMe_3)(MMe_3) + CCl_4 \rightarrow t\text{-}Bu_2P({=}NSiMe_3)(Cl) + Me_3MCCl_3 \quad \text{(h)}$$

where M = Si, Ge.

Using nitrosyl chloride as a halogenating agent, a cleavage and oxidation process is observed[10], viz.:

$$Me_2PSiMe_3 \xrightarrow{NOCl} Me_2P(O)Cl + Me_3SiCl + (Me_3Si)_2O \text{ (+ other products)} \quad \text{(i)}$$

Similarly, sulfuryl chloride leads to the same phosphorus-containing product[11]:

$$Me_2PSiMe_3 \xrightarrow{SO_2Cl_2} Me_2P(O)Cl + Me_3SiCl \text{ (+ other products)} \quad \text{(j)}$$

The P—F bond forms[12] using PF_5:

$$(CF_3)_2PSiH_3 \xrightarrow[<25^\circ C]{PF_5} (CF_3)_2PF \text{ (+ several other products)} \quad \text{(k)}$$

which has no preparative value. A cleavage and oxidation reaction is observed using SF_4[15]:

$$2,4,6\text{-}t\text{-}Bu_3C_6H_2P(SiMe_3)_2 \xrightarrow{SF_4} 2,4,6\text{-}t\text{-}Bu_3C_6H_2PF_4 \quad \text{(l)}$$

(M. FILD)

1. R. Appel, K. Geisler, H. Schöler, *Chem. Ber.*, *110*, 376 (1977).
2. R. Appel, W. Paulen, *Angew. Chem., Int. Ed. Engl.*, *20*, 869 (1981).
3. R. Appel, V. Barth, *Angew. Chem., Int. Ed. Engl.*, *18*, 469 (1979).
4. R. Appel, V. Barth, M. Halstenberg, G. Huttner, J. von Seyerl, *Angew. Chem., Int. Ed. Engl.*, *18*, 872 (1979).
5. W. W. DuMont, H. Schumann, *Angew. Chem., Int. Ed. Engl.*, *14*, 368 (1975).
6. H. Schumann, W. W. DuMont, *Chem. Ber.*, *108*, 2261 (1975).
7. W. W. DuMont, H. Schumann, *J. Organomet. Chem.*, *128*, 99 (1977).
8. O. J. Scherer, G. Schieder, *Chem. Ber.*, *101*, 4184 (1968).
9. O. J. Scherer, G. Schieder, *Angew. Chem., Int. Ed. Engl.*, *7*, 75 (1968).
10. J. R. Byrne, C. R. Russ, *J. Organomet. Chem.*, *22*, 357 (1970).
11. J. R. Byrne, C. R. Russ, *J. Organomet. Chem.*, *38*, 319 (1972).
12. L. Maya, A. B. Burg, *Inorg. Chem.*, *14*, 698 (1975).
13. R. Appel, W. Paulsen, *Chem. Ber.*, *116*, 109 (1983).
14. R. Appel, V. Barth, F. Knoch, *Chem. Ber.*, *116*, 938 (1983).
15. R. Appel, L. Krieger, *J. Fluorine Chem.*, *26*, 445 (1984).

2.4.6.3.3. to Give As Halides.

With $Me_2AsSiMe_3$ and nitrosyl chloride, dimethylarsinous chloride, Me_2AsCl, is formed[1], besides $(Me_3Si)_2O$ and N_2O. Nearly quantitative yields of Me_2AsCl are obtained if sulfuryl chloride is employed[2]:

$$Me_2AsSiMe_3 + SO_2Cl_2 \rightarrow Me_2AsCl + Me_3SiCl + SO_2 \quad \text{(a)}$$

(M. FILD)

1. J. E. Byrne, C. R. Russ, *J. Organomet. Chem.*, *22*, 357 (1970).
2. J. E. Byrne, C. R. Russ, *J. Organomet. Chem.*, *38*, 319 (1972).

2.4.7. from Cleavage of the Group VB–Oxygen Bond

2.4.7.1. by Halogens.

Nitrous oxide can be fluorinated at 400–700°C to give[1] NF_3:

$$N_2O + 2\,F_2 \rightarrow NF_3 + NOF \quad \text{(a)}$$

The handling of F_2 at elevated T is a potentially hazardous operation and should be carried out only by an experienced person. Special apparatus is required.

Reaction of $(F_3C)_2POP(CF_3)_2$ with Cl_2 gives[2] $(F_3C)_2P(O)Cl$ and $(F_3C)_2PCl$. Cleavage of the P—O bond in $[OCH(CH_3)CH(CH_3)O]_2POCHCH_2$ with Br_2 affords[3] $[OCH(CH_3)CH(CH_3)O]_2PBr$.

The action of Br_2 on As_2O_3 in the presence of elemental sulfur gives[2] $AsBr_3$. The mixture is heated for 7 h until the vapors lose the color of Br_2[2]:

$$2\,As_2O_3 + 3\,S + 6\,Br_2 \rightarrow 4\,AsBr_3 + 3\,SO_2 \quad \text{(b)}$$

Similarly As_2O_3, sulfur and I_2 give[4] AsI_3.

Bromoethoxyphenylarsine is formed from Br_2 with diethoxyphenylarsine in 45–50% yield[5]:

$$2\,PhAs(OEt)_2 + Br_2 \xrightarrow{-EtBr} PhAs(Br)OEt + PhAs(O)(OEt)_2 \quad \text{(c)}$$

Cleavage of both As—O and As—S bonds is observed with Br_2 and 2-phenyl-1,3,2-oxathiaarsolane in CCl_4 to give[6] $PhAsBr_2$:

$$\text{(2-phenyl-1,3,2-oxathiaarsolane)} + Br_2 \rightarrow PhAsBr_2 + \text{products} \quad \text{(d)}$$

(H.J. BREUNIG)

1. K. Jones, in *Comprehensive Inorganic Chemistry*, Vol. 2, A. F. Trotman-Dickenson, ed., Pergamon Press, Oxford, 1973, p. 295.
2. J. E. Griffiths, A. B. Burg, *J. Am. Chem. Soc.*, *84*, 3442 (1962).
3. T. N. Kudratvtseva, M. B. Karlstedt, M. V. Proskurmina, V. A. Frolovskii, J. F. Lutsenko, *J. Gen. Chem. USSR* (*Engl. Transl.*), *54*, 486 (1984).
4. P. W. Schenk, in *Handbook of Preparative Inorganic Chemistry* G. Brauer, ed., Academic Press, London, 1963.
5. A. Schultze, S. Samaan, L. Horner, *Phosphorus*, *5*, 265 (1975).
6. N. A. Chadaeva, G. Kamai, K. A. Mamakov, *J. Gen. Chem. USSR* (*Engl. Transl.*), *43*, 821 (1973).

2.4.7.2. by Hydrogen Halides

2.4.7.2.1. to Give the Nitrogen–Halogen Bond.

Liquid HF converts NO_2 at 25°C to a ternary mixture of HF, HNO_3 and nitrosylfluoride NOF[1]:

$$2\, NO_2 + (x + 1)\, HF \rightarrow NOF{\cdot}(HF)_x + HNO_3 \quad \text{(a)}$$

The conc aq mixture of HCl and HNO_3 known as aqua regia evolves nitrosyl chloride, NOCl, and Cl_2:

$$HNO_3 + 3\, HCl \rightarrow NOCl + Cl_2 + H_2O \quad \text{(b)}$$

It is, however, difficult to isolate NOCl[2]. Nitrosyl chloride is formed in 56% yield by the action of HCl on $NaNO_2$[3]:

$$NaNO_2 + 2\, HCl \rightarrow NOCl + NaCl + H_2O \quad \text{(c)}$$

Nitrosylsulfuric acid, which is obtained from SO_2 and HNO_3, reacts with HCl at −45°C to form[4] NOCl and H_2SO_4:

$$SO_2 + HNO_3 \rightarrow HOSO_2ONO \xrightarrow{HCl} NOCl + H_2SO_4 \quad \text{(d)}$$

Nitryl chloride, NO_2Cl, and H_2SO_4 are formed by adding $ClSO_3H$ to HNO_3 at 0°C. The yield of NO_2Cl is 80–90%[5].

Handling HF, NOCl and NO_2Cl is a potentially hazardous operation. Special equipment and experienced personnel are required.

(H.J. BREUNIG)

1. F. Seel, G. Fuchs, D. Werner, *Chem. Ber.*, *96*, 179 (1963).
2. J. D. Richards, in *Mellor's Comprehensive Treatise on Inorganic and Theoretical Chemistry*, Vol. VIII, Suppl. II, Part II, A. A. Eldridge, ed., Longmans, Green, London, 1967, p. 420.

3. J. R. Morton, H. W. Wilcox, *Inorg. Synth.*, *4*, 48 (1953).
4. G. H. Coleman, G. A. Lillis, G. E. Goheen, *Inorg. Synth.*, *1*, 55 (1939).
5. R. Kaplan, H. Shechter, *Inorg. Synth. 4*, 52 (1953).

2.4.7.2.2. to Give Phosphorus–Halogen Bond.

Hydrogen fluoride reacts with P_4O_{10} in a sequence of steps, and product distribution is dependent on reactant concentration; e.g., slow addition of HF to P_4O_{10} (6:1) converts[1,2] the oxide to a mixture of HPO_2F_2 and H_2PO_3F. Anhydrous difluorphosphoric acid may be recovered from the mixture by a vacuum distillation. Hexafluorphosphoric acid, HPF_6, is the major product[1] from 24:1 HF:P_4O_{10}. Commercial monofluorophosphoric acid is prepared from 69% aq HF and P_4O_{10}. It contains[3,4] 15–20% each of H_3PO_4 and HPO_2F_2. The relationship between the fluorophosphoric acids and H_3PO_4 is:

$$H_3PO_4 \xrightarrow[-H_2O]{+HF} H_2PO_3F \xrightarrow[-H_2O]{HF} HPO_2F_2 \xrightarrow[-2\ H_2O]{4\ HF} HPF_6 \qquad \text{(a)}$$

The handling of HF is a potentially hazardous operation. Special equipment and apparatus are required.

Reaction of $(PhO)_2POCH_2CMe_2$ with HCl gives[5] a 78% yield of $PhOP(Cl)OCH_2CMe_3$ and PhOCl.

(H.J. BREUNIG)

1. E. L. Mutterties, C. W. Tullock, *Prep. Inorg. React.*, *2*, 284 (1965).
2. L. C. Mosier, W. E. White, *Ind. Eng. Chem.*, *43*, 246 (1951).
3. W. E. White, C. Pupp, in *Kirk-Othmer Encyclopedia of Chemical Technology*, R. E. Kirk, D. F. Othmer, eds., 2nd ed., Vol. IX, Wiley-Interscience, New York, 1966, p. 826.
4. W. Lange, *Inorg. Synth.*, *2*, 155 (1946).
5. H. R. Hudson, A. Kow, J. C. Roberts, *Phosphorus, Sulfur*, *19*, 375 (1984).

2.4.7.2.3. to Give the Arsenic–Halogen Bond.

Arsenic trifluoride is prepared from As_4O_6 and HF, which is obtained from CaF_2 and conc H_2SO_4 in 85% yield[1,2]:

$$As_4O_6 + 12\ HF \rightarrow 4\ AsF_3 + 6\ H_2O \qquad \text{(a)}$$

The H_2O is taken up by the H_2SO_4. When anhyd HF reacts with As_2O_3 at 140°C in an Fe distillation apparatus, AsF_3 is obtained in 80% yield at −18°C[3].

With Ph_3AsO, HF gives[4] Ph_3AsF_2. Distillation of As_4O_6 with conc HCl in a stream of HCl gives[5] $AsCl_3$:

$$As_4O_6 + 12\ HCl \rightarrow 4\ AsCl_3 + 6\ H_2O \qquad \text{(b)}$$

Presumably $AsCl_3$ is an intermediate in the formation of AsI_3 by reaction of As_4O_6 with HCl and KI[6].

Saturation of an arsonic acid, $RAsO_3H_2$, in HCl with SO_2 gives organoarsenic dichlorides:

$$RAsO_3H_2 \xrightarrow[HCl]{SO_2} RAsCl_2 \qquad \text{(c)}$$

where R = CH_3[7,8], C_2H_5[7], C_2F_5[9], C_3H_7[7], C_6H_5[10]. The reactions are catalyzed by I^- ions. The alkylarsenic dichloride syntheses are at RT. The yield is 50–70% for CH_3AsCl_2, 60–65% for $C_2H_5AsCl_2$ and 47% for $C_3H_7AsCl_3$. The reduction of $PhAsO_3H_2$ is carried out at elevated T to give $PhAsCl_2$ in 95% yield.

Analogous reduction and halogenation of arsinic acids, R_2AsO_2H, with SO_2 in HCl gives diorganoarsenic chlorides[7,11,12]:

$$R_2AsO_2H \xrightarrow[HCl]{SO_2} R_2AsCl \tag{d}$$

where R = C_2H_5 (60%)[7], C_6H_5[11,12].

Dimethylarsenic chloride is obtained by reduction of $(CH_3)_2AsO_2H$ with H_3PO_2 and halogenation with HCl[15].

Organoarsenic dibromides are formed by reduction of arsonic acids with SO_2, followed by treatment with HBr_3:

$$RAsO_3H_2 \xrightarrow[HBr]{SO_2} RAsBr_2 \tag{e}$$

where R = CH_3 (80%)[8], C_2H_5[7], 4-BrC_6H_4 (86%)[14].

Reaction of SO_2 and HBr with arsinic acids gives diorganoarsenic bromides:

$$RR'AsO_2H \xrightarrow[HBr]{SO_2} RR'AsBr \tag{f}$$

where if R = C_6H_5, R' = C_4H_9 (75% yield)[15]; if R = CH_3, R' = 3-$NO_2C_6H_4$ (54% yield)[16].

Organoarsenic diiodides, and diorganoarsenic iodides are obtained from reduction of the corresponding arsonic or arsinic acids with SO_2 in HCl in the presence of 1:1 KI. Halide–halide exchange is likely in these reactions, and, therefore, they belong in §2.4.11.

Organoarsenic diiodides, however, are obtained from conc HI with arsonic acids. Here, HI is both the reducing and the halogenating agent; e.g., 2'-diiodoarsino-4-methoxy-benzophenon forms, from the corresponding arsonic acid[17]. Iodination of $CH_3AsO_3H_2$ by conc HI, however, gives[18] CH_3AsI_4.

Arsenic halides, HF and other chemicals used are strong poisons. Special equipment must be used. Contact of As halides with the skin must be avoided.

(H.J. BREUNIG)

1. A. A. Woolf, N. N. Greenwood, *J. Chem. Soc.*, 2200 (1950).
2. C. J. Hoffman, *Inorg. Synth.*, *4*, 150 (1953).
3. W. Kwasnik, in *Handbook of Preparative Inorganic Chemistry*, G. Brauer, ed., Academic Press, London, 1963.
4. G. S. Harris, I. M. Mack, J. S. McKechnie, *J. Fluorine Chem.*, *11*, 481 (1978).
5. P. W. Schenk, in *Handbook of Preparative Inorganic Chemistry*, G. Brauer, ed., Academic Press, London 1963, pp. 596, 608, 621.
6. P. W. Schenk, in *Handbook of Preparative Inorganic Chemistry*, G. Brauer, ed., Academic Press, London 1963, p. 534.
7. C. K. Banks, J. F. Morgan, R. L. Clark, E. B. Hatlelid, H. W. Paxton, E. J. Cragoe, R. D. Andres, B. Elpern, R. F. Coles, J. Cawhead, C. S. Hamilton, *J. Am. Chem. Soc.*, *69*, 927 (1947).
8. G. P. Kelen, *Bull. Soc. Chim. Belg.*, *65*, 343 (1956).
9. A. B. Bruker, T. G. Spiridonova, L. Z. Soborovski, *J. Gen. Chem. USSR*, *28*, 350 (1958).
10. R. L. Barker, E. Booth, W. E. Jones, A. F. Millidge, F. N. Woodward, *J. Soc. Chem. Ind. (London)*, *68*, 289 (1949); *Chem. Abstr.*, *44*, 3451 (1950).

11. H. Bart, *Justus Liebigs Ann. Chem.*, *429*, 55 (1922).
12. M. P. Osipova, G. K. Kamai, N. A. Chadaeva, *J. Gen. Chem. USSR* (*Engl. Transl.*), *37*, 1660 (1967).
13. F. Kober, W. J. Rühl, *Z. Anorg. Allg. Chem.*, *406*, 52 (1974).
14. F. F. Blicke, S. R. Saphir, *J. Am. Chem. Soc.*, *63*, 575 (1941).
15. G. Kamai, O. N. Belerossova, *Izv. Akad. Nauk USSR*, 191 (1947); *Chem. Abstr.*, *42*, 4133 (1948).
16. E. J. Cragoe, R. J. Andres, R. F. Coles, B. Elpern, J. F. Morgan, C. S. Hamilton, *J. Am. Chem. Soc.*, *69*, 925 (1947).
17. W. L. Lewis, H. C. Cheetham, *J. Am. Chem. Soc.*, *45*, 510 (1923).
18. H. Klinger, A. Kreutz, *Justus Liebigs Ann. Chem.*, *249*, 147 (1888).

2.4.7.2.4. to Give the Sb– and Bi–Halogen Bond.

Antimony trifluoride is obtained[1] from gaseous HF and Sb_4O_6 at elevated T. Bismuth trifluoride is prepared from aq HF and $Bi(OH)_3$[1,2] or Bi_2O_3[3], and $BiCl_3$ results from Bi_2O_3 and HCl[1].

The action of HCl in the presence of catalytic KI is useful for the formation of Sb—Cl bonds by cleavage of Sb—O bonds in organoantimony compounds. If a reducing agent, such as SO_2 or $SnCl_2$, is used, Sb(V) compounds are converted to Sb(III).

Phenylantimony dichloride is obtained from $PhSbO_3H_2$ by reduction with SO_2 in conc HCl at 0°C[4,5]:

$$PhSbO_3H_2 + SO_2 + HCl \rightarrow PhSbCl_2 + H_2SO_4 + 2\ H_2O \qquad \text{(a)}$$

Examples are given in Table 1.

When the reducing agent is omitted there is no change in Sb oxidation number, and the number of Sb—Cl bonds is determined by the number of organo groups on Sb.

Diphenylantimony trichloride is the product of the conversion of Ph_2SbO_2H with conc HCl[10,11]

$$Ph_2SbO_2H + 3\ HCl \rightarrow Ph_2SbCl_3 + 2\ H_2O \qquad \text{(b)}$$

Examples of Sb—Cl bonds in organoantimony compounds formed without redox are given in Table 2.

Similarly synthesized is Ph_3BiCl_2 from $Ph_3Bi(ONO_2)_2$ with HCl in glacial acetic acid[14].

Both Bi—C and Bi—O bond cleavage is observed in $Ph_4BiOSiPh_3$ with HCl in C_2H_5OH yielding Ph_4BiCl and Ph_3SiOH and C_6H_6 as byproducts[15].

TABLE 1. REACTIONS OF THE Sb—O BOND WITH HCl UNDER REDUCTIVE CONDITIONS

Oxostibine[a]	Reducing agent	Product[a]	T (°C)	Yield (%)	Ref.
$2\text{-}BrC_6H_4SbO_3H_2$	SO_2	$2\text{-}BrC_6H_4SbO_3H_2$		75	6
$1\text{-}C_{10}H_7SbO_3H_2$	SO_2	$1\text{-}C_{10}H_7SbO_3H_2$			7
$PhSbO_3H_2$	SO_2	$PhSbCl_2$		58	4, 5
$PhSbO_3H_2$	$SnCl_2$	$PhSbCl_2$		35	4
Ph_2SbO_2H	SO_2	Ph_2SbCl	40–70	90	8
$RR'SbO_2H$	$SnCl_2$	$RR'SbCl$		87	9

[a] R = 4-methylphenyl, R′ = biphenyl.

TABLE 2. REACTIONS OF THE Sb—O BOND WITH HCl

Oxostibine	Product	Yield (%)	Ref.
$PhSbO_3H_2$	$PhSbCl_4$		10, 11
$2\text{-}ClC_6H_4SbO_3H_2$	$2\text{-}ClC_6H_4SbCl_4$	95	10, 11
Ph_2SbO_2H	Ph_2SbCl_3		10, 11
$(4\text{-}F\text{-}C_6H_4)_2SbO_2H$	$(2\text{-}FC_6H_4)_2SbCl_3$		12
$Ph_3Sb[OC(O)CH_3]_2$	Ph_3SbCl_2		13

TABLE 3. REACTIONS OF THE Sb—O BOND WITH HI

Oxostibine	Product	Ref.
$(4\text{-}BrC_6H_4)_2SbOC(O)CH_3$	$(4\text{-}BrC_6H_4)_2SbI$	20
$(4\text{-}CH_3C_6H_4)_2SbOC(O)CH_3$	$(4\text{-}CH_3C_6H_4)_2SbI$	20
$[(2\text{-}C_6H_5CH_2C_6H_4)_2Sb]_2O$	$(2\text{-}C_6H_5CH_2C_6H_4)_2SbI$	21
$[(1\text{-}C_{10}H_7)_2Sb]_2O$	$(1\text{-}C_{10}H_7)_2SbI$	22

Using HBr instead of HCl results in the formation of Sb— or Bi—Br bonds; e.g., $SbBr_3$ is synthesized from Sb_4O_6 and HBr[1,16], and $BiBr_3$ from Bi_2O_3 and HBr[17].

A facile synthesis of Sb—I bonds is the reaction of Sb_4O_6 or organoantimony oxides with HI yielding SbI_3[16,18] or organoantimony iodides, respectively.

Methylantimony diiodide is prepared[19] from CH_3SbO and HI:

$$1/x(CH_3SbO)_x + 2\ HI \rightarrow CH_3SbI_2 + 2\ HI \tag{c}$$

Reaction of HI and $[(1\text{-}C_{10}H_7)_2Sb]_2O$ gives $(1\text{-}C_{10}H_7)_2Sb$. Examples are given in Table 3.

Organoantimony halides and HF are strong poisons. Special skill and precautions are necessary to avoid danger.

(H.J. BREUNIG)

1. G. Brauer, ed., *Handbook of Preparative Inorganic Chemistry*, Academic Press, New York, 1963.
2. H. v. Wartenberg, *Z. Anorg. Anal. Chem., 244*, 344 (1940).
3. B. Aurivillius, *Acta Chem. Scand., 9*, 1206 (1955).
4. G. O. Doak, H. H. Jaffé, *J. Am. Chem. Soc., 72*, 3025 (1950).
5. H. H. Jaffé, G. O. Doak, *J. Am. Chem. Soc., 71*, 602 (1949).
6. B. R. Cook, C. A. McAuliffe, D. W. Meek, *Inorg. Chem., 10*, 2676 (1971).
7. P. Pfeiffer, P. Schmidt, *J. Prakt. Chem., 152*, 27 (1939); *Chem. Abstr 33*, 3347 (1939).
8. A. B. Bruker, *J. Gen. Chem. USSR* (*Engl. Transl.*), *27*, 2223 (1957).
9. I. G. M. Campbell, *J. Chem. Soc.*, 3109 (1952).
10. H. Schmidt, *Justus Liebigs Ann. Chem., 421*, 174 (1920).
11. H. Schmidt, *Justus Liebigs Ann. Chem., 429*, 123 (1922).
12. G. O. Doak, J. M. Summy, *J. Organomet. Chem., 55*, 143 (1973).
13. A. N. Nesmeyanov, O. A. Reutov, O. A. Ptitsyna, P. A. Tsurkan, *Izv. Akad. Nauk USSR*, 1435 (1958).
14. J. F. Wilkinson, F. Challenger, *J. Chem. Soc., 125*, 854 (1924).
15. G. A. Razuvaev, N. A. Osanova, V. V. Sharutin, *Dokl. Akad. Nauk SSSR, 225*, 581 (1975).
16. P. M. Druce, M. F. Lappert, *J. Chem. Soc., A*, 3595 (1971).
17. D. Cubicciotti, *Inorg. Chem., 7*, 208 (1968).
18. J. C. Bailar, *Inorg. Synth. 1*, 103 (1939).
19. G. T. Morgan, G. R. Davies, *Proc. R. Soc. London, Ser. A, 110*, 523 (1926).

20. F. F. Blicke, U. O. Oakdale, *J. Am. Chem. Soc.*, *55*, 1198 (1933).
21. G. T. Morgan, G. R. Davies, *Proc. R. Soc. London Ser. A*, *143*, 38 (1933).
22. G. T. Morgan, G. R. Davies, *Proc. R. Soc. London, Ser. A.*, *127*, 1 (1930).

2.4.7.3. by Other Halides.

Nitric acid reacts with $ClSO_3H$ to give[1] NO_2Cl in 80–90% yield and H_2SO_4:

$$HNO_3 + ClSO_3H \rightarrow H_2SO_4 + NO_2Cl \quad \text{(a)}$$

Conversion of P— and As—O bonds into P—F and As—F, respectively, can be accomplished by treatment with SF_4 at elevated T. Phenylphosphonic acid reacts with SF_4 to give[2] $PhPF_4$:

$$PhPO_3H_2 + 3\ SF_4 \rightarrow PhPF_4 + 3\ SOF_2 + 2\ HF \quad \text{(b)}$$

Examples are given in Table 1.

A laboratory substitute for HF is NH_4F, for it can be more easily handled as a solid.

Heating P_4O_{10} and NH_4F in a Ni or Cu crucible until the reaction starts results in the formation of $NH_4PO_2F_2$ and $(NH_4)_2PO_3F$. After workup, the yield is 70%[4–6]:

$$P_4O_{10} + 6\ NH_4F \rightarrow 2\ NH_4PO_2F_2 + 2\ (NH_4)_2PO_3F \quad \text{(c)}$$

Oxygen replacement in P_4O_{10} with FSO_3H gives F_3PO in 80% yield[7].

Direct conversion of Ph(t-Bu)P(O)OH into Ph(t-Bu)P(O)F occurs on treatment with sulfonyl chloride fluoride in the presence of Et_3N in CH_2Cl_2 at −40°C[8].

$$Ph(t\text{-}Bu)P(O)OH + ClSO_2F + Et_3N \rightarrow Ph(t\text{-}Bu)PF + [Et_3NH][SO_2Cl] \quad \text{(d)}$$

Reaction of $(EtO)_3P$ with BF_3 gives[9] $EtOPF_2$ in 79% yield:

$$(EtO)_3P + 2\ BF_3 \rightarrow EtOPF_2 + \tfrac{2}{3}\ (EtOBF_2)_3 \quad \text{(e)}$$

The same procedure affords[9] 67% $(PhO)_2PF$ from $(PhO)_3P$. Action of PF_5 on $Me_3SiOP(O)[OC(CF_3)_2C(CF_3)_2O]$ gives[10] $F_3P[OC(CF_3)_2C(CF_3)_2O]$.

Partial fluorination occurs as well in the reaction of $Mg_2P_2O_7$ with MgF_2 at >750°C giving[11] OPF_3. Reaction of solid $K_2[SiF_6]$ and $Mg_2[P_2O_7]$ gives[12] OPF_3 and SiF_4.

Bis(trifluoromethyl)arsenic fluoride is obtained[13] from methoxybis(trifluoromethyl)arsine with BF_3 at 25°C:

$$(CF_3)_2AsOCH_3 + BF_3 \rightarrow (CF_3)_2AsF + BF_2OCH_3 \quad \text{(f)}$$

Reaction of P_4O_{10} with IF_5 gives[14] OPF_3.

TABLE 1. CLEAVAGE REACTIONS OF THE GROUP VB—O BOND WITH SF_4

Reactant	Fluoride	T (°C)	Yield (%)	Ref.
$PhPO_3H_2$	$PhPF_4$	150	58	2
Ph_3PO	Ph_3PF_2	150	67	2
$PhAsO_3H_2$	$PhAsF_4$	70	45	2
$[F_2P(O)]_2CH_2$	$(F_4P)_2CH_2$	−20	76	3

TABLE 2. CLEAVAGE OF THE P—O BOND WITH PCl_5

Reactant	Product	Yield (%)	Ref.
POF_2OH	POF_2Cl	79	15
$P_2O_3F_4$	POF_2Cl	95	15
$POF(OH)_2$	$POFCl_2$	60	16
$[Me_2P(O)]_2O$	$Me_2P(O)Cl$	87	17

Conversion of a P—O into a P—Cl bond occurs with PCl_5. Difluorophosphoric acid with PCl_5 gives POF_2Cl in 79% yield[15]:

$$F_2POH + PCl_5 \rightarrow POF_2Cl + POCl_3 \tag{g}$$

Examples are given in Table 2.

Mixtures of P_4O_{10} with $CaCl_2$ or NaCl at 500°C give $POCl_3$ as the major volatile product. With CaF_2 and NaCl together, mixed phosphoryl halides (e.g., $POCl_3$, $POCl_2F$, $POClF_2$, POF_3) are formed[18]. Cleavage of $(n\text{-}Bu)_2POPh$ with PhC(O)Cl gives[19] $n\text{-}Bu_2PCl$ and PhC(O)OPh. Reaction of PCl_3 with $(i\text{-}PrO)_3P$ at 20°C affords[20] $i\text{-}PrOPCl_2$ and $(i\text{-}PrO)_2PCl$. Conversion of $(i\text{-}PrO)_2PCl$ to $i\text{-}PrOPCl_2$ by PCl_3 is accomplished[20] after 8 hrs. at 100°C. Action of PCl_3 on $(n\text{-}BuO)_2PF$ at 20°–40°C gives[21] n-BuOPFCl in 34% yield.

Phenylarsonic acid dichloride is converted to $PhAsCl_4$ with $SOCl_2$ at 60°C[22]:

$$PhAsOCl_2 + SOCl_2 \rightarrow PhAsCl_5 + SO_2 \tag{h}$$

Triphenylarsine oxide reacts with $SOCl_2$ or $COCl_2$ to form[23] Ph_3AsCl_2.

Action of $SOCl_2$ on phenylarsonic acid or its diethyl ester at 80°C gives $PhAsOCl_2$ in 60% yield[22], e.g.:

$$PhAsO(OC_2H_5)_2 + 2\ SOCl_2 \rightarrow PhAsOCl_2 + SO(OC_2H_5)_2 \tag{i}$$

Examples for the conversion of a P—O into a P—Cl bond by $SOCl_2$ are given in Table 3. Another effective reagent for the cleavage of P—O bonds is $COCl_2$, e.g. $[Me_2P(O)]_2O$ reacts with phosgeneto form[28] $Me_2P(O)Cl$ and CO_2:

$$Me_2P(O)OP(O)Me_2 + COCl_2 \rightarrow 2\ Me_2P(O)Cl + CO_2 \tag{j}$$

Examples are given in Table 4.

Oxalylchloride is used[31] for the conversion of $MeP(O)(OMe)_2$ to MeP(O)(OMe)Cl.

TABLE 3. CLEAVAGE OF THE P—O BOND WITH $SOCl_2$

Reactant	Product	Yield (%)	Ref.
Me_3PO	Me_3PCl_2	95	24
$(C_6H_{11})_2P(O)OH$	$(C_6H_{11})_2P(O)Cl$	82	25
MePO(OEt)OH	MePO(OEt)Cl	80	26
$Me_2P(O)CH_2C(CF_3)_2OH$	$Me_2P(Cl)[OC(CF_3)_2CH_2]$		27

TABLE 4. CLEAVAGE OF THE P—O BOND WITH $COCl_2$

Reactant	Product	Ref.
$Et_2P(O)OH$	$Et_2P(O)Cl$	29
$Me_2P(O)OMe$	$Me_2P(O)Cl$	30
$Ph_2P(O)OMe$	$Ph_2P(O)Cl$	30
MeEtP(O)OMe	MeEtP(O)Cl	30

Bis(dichloroarsino)methane is obtained[32] in a complex reaction from As_4O_6 and CH_3COCl in the presence of $AlCl_3$:

$$As_4O_6 + 10\ CH_3COCl \xrightarrow{AlCl_3} 2\ H_2C(AsCl_2)_2 + 4\ (CH_3CO)_2O + 2\ HCl + CO_2\,(k)$$

Conversion of P=O to P—Br is accomplished by PBr_5, e.g. Me_3PO gives[33] Me_3PBr_2 with PBr_5.

When P_4O_{10} is heated with PBr_5, $OPBr_3$ is obtained in 80% yield[34]

$$P_4O_{10} + 6\ PBr_5 \xrightarrow{100°C} 10\ OPBr_3 \tag{1}$$

Action of PBr_5 on POF_2OH gives OPF_2Br, $OPBr_3$ and HBr in 60% yield[35].

Addition of Me_3SiI to $(C_2H_5)_3PO$ in CH_2Cl_2 gives $Me_3SiOP(C_2H_5)_3I$ in 88% yield[36].

Group-VB halides and halogenating agents, especially fluorinating agents, are poisonous. Their handling requires special experience and precautions.

(H.J. BREUNIG)

1. R. Kaplan, H. Shechter, *Inorg. Synth., 4* (1953).
2. W. C. Smith, *J. Am. Chem. Soc., 82,* 6176 (1960).
3. W. Althoff, M. Fild, R. Schmutzler, *Chem. Ber., 114,* 1082 (1981).
4. W. Kwasnik, in *Handbook of Preparative Inorganic Chemistry,* G. Brauer, ed., Academic Press, London, 1963.
5. W. Lange, *Ber. Dtsch. Chem. Ges., 62,* 790 (1929).
6. H. S. Booth, G. G. Seegmiller, in *Inorg. Synth., 2,* 155, 157 (1946).
7. E. Hayek, A. Aignesberger, A. Engelbrecht, *Monatsh. Chem., 86,* 735 (1955).
8. A. Lopusinski, J. Michalski, *Angew. Chem., Int. Ed. Engl., 21,* 294 (1982).
9. H. Binder, R. Fischer, *Z. Naturforsch., Teil B, 27,* 753 (1972).
10. R. Bohlen, R. Franke, J. Heine, R. Schmutzler, G.-V. Röschenthaler, *Z. Anorg. Allg. Chem., 533,* 18 (1986).
11. J. Berak, I. Tomczak, *Rocz. Chem., 39,* 1761 (1965); *Chem. Abstr., 64,* 18487 (1966).
12. D.-H. Menz, L. Kolditz, *Z. Chem., 26,* 185 (1986).
13. A. B. Burg, J. Singh, *J. Am. Chem. Soc., 88,* 718 (1966).
14. E. E. Aynsley, R. Nichols, P. L. Robinson, *J. Chem. Soc.,* 623 (1953).
15. H. W. Roesky, *Inorg. Synth., 15,* 195 (1974).
16. H. W. Roesky, *Inorg. Synth., 15,* 196 (1974).
17. K. Moedritzer, *J. Am. Chem. Soc., 83,* 4381 (1961).
18. G. Tarbutton, E. P. Egan, S. G. Frary, *J. Am. Chem. Soc., 63,* 1782 (1941).
19. T. K. Gazizov, R. U. Belyalov, V. A. Kharlamov, A. N. Pudovik, *J. Gen. Chem., USSR (Engl. Transl.), 50,* 232 (1980).

20. R. U. Belyalov, A. M. Kibardin, T. Kh. Gazizov, A. N. Pudovik, *J. Gen. Chem. USSR* (*Engl. Transl.*), *51*, 19 (1981).
21. H. Binder, R. Fischer, *Z. Naturforsch., Teil B*, *34*, 794 (1979).
22. A. F. Kolomiets, G. S. Levskaya, *J. Gen. Chem. USSR* (*Engl. Transl.*), *36*, 2024 (1966).
23. R. Appel, D. Rebhahn, *Chem. Ber.*, *102*, 3955 (1969).
24. J. Goubeau, R. Baumgärtner, *Z. Elektrochem.*, *64*, 598 (1960).
25. K. Issleib, A. Brack, *Z. Anorg. Allg. Chem.*, *277*, 258 (1954).
26. Z. Delchowicz, *J. Chem. Soc.*, 238 (1961).
27. H. Kischkel, G. V. Röschenthaler, *Phosphorus, Sulfur*, *27*, 371 (1986).
28. K. Weissermel, H. J. Kleiner, M. Finke, U. H. Felcht, *Angew. Chem. Int.* (*Ed. Engl.*), *93*, 223 (1981).
29. A. A. Neimyshev, I. L. Knunyants, *J. Gen. Chem. USSR* (*Engl. Transl.*) *36*, 1105 (1966).
30. U.-H. Felcht, in Houben-Weyl, Methoden der organischen Chemie, Vol. E2, Phosphorverbindungen II, M. Regitz Ed. p. 164, G. Thieme, Stuttgart, New York 1982, 4th ed.
31. Z. Pelchowicz, *J. Chem. Soc.*, 238 (1966).
32. H. Gutbier, H.-G. Plust, *Ber. Chem.*, *88*, 1777 (1955).
33. J. Goubeau, R. Baumgärtner, *Z. Elektrochem.*, *64*, 598 (1960).
34. H. S. Booth, C. G. Seegmiller, *Inorg. Synth.*, *2* (1946).
35. H. W. Roesky, *Inorg. Synth.*, *15*, 197 (1974).
36. V. D. Romanenko, V. I. Tovstenko, L. N. Markovskii, *J. Gen. Chem. USSR* (*Engl. Transl.*), *49*, 1680 (1979).

2.4.8. from Cleavage of the Group VB–Other Group VIB Element Bond

2.4.8.1. by Halogens.

Cleavage of the As—S bond in $Ph_2AsSC_2H_5$ with Br_2 gives[1] Ph_2AsBr:

$$Ph_2AsSC_2H_5 + Br_2 \rightarrow Ph_2AsBr + C_2H_5SBr \qquad (a)$$

Diphenylbromoarsine is also obtained[1] by action of Br_2 on Ph_2AsSPh.

Triorganoarsenic sulfides react with I_2 to form the corresponding diiodides[2]. Iodination of Et_3AsS gives[2] $[Et_3AsI]I_3$:

$$Et_3AsS + I_2 \rightarrow [Et_3AsI]I_3 + \text{products} \qquad (b)$$

Me_3AsI_4 and Ph_3AsI_4 are prepared similarly[2].

Addition of Cl_2 to $Ph_2P(S)SH$ in CCl_4 at 20–30°C during 2–3 h gives Ph_2PCl_3 in 98% yield[3]:

$$Ph_2\overset{\overset{\displaystyle S}{\|}}{P}SH + 3\,Cl_2 \rightarrow Ph_2PCl_3 + S_2Cl_2 + HCl \qquad (c)$$

Chlorination of $Ph_2P(O)SH$ in C_6H_6 at 28–72°C gives[4,5] $Ph_2P(O)Cl$:

$$Ph_2\overset{\overset{\displaystyle O}{\|}}{P}SH + 2\,Cl_2 \rightarrow Ph_2\overset{\overset{\displaystyle O}{\|}}{P}Cl + HCl + SCl_2 \qquad (d)$$

Selective cleavage of the P—S bond in $MeP(S)(OEt)SH$ by Cl_2 gives[6] $MeP(S)(OEt)Cl$. At 22°C solid $(C_6F_5)_3PS$ reacts[7] with gaseous Cl_2 to yield $(C_6F_5)_3PCl_2$

and SCl_2. The P═S bond is converted to the P—Cl bond in the chlorination of $Me_2P(S)NHP(S)Me_2$ to give[8] $[Cl(Me_2)P{=}NPMe_2Cl]Cl$:

$$Me_2\overset{S}{P}NH\overset{S}{P}Me_2 + 2\ Cl_2 \rightarrow [Me_2\overset{Cl}{P}{=}N\overset{Cl}{P}Me_2] + HCl + 2\ S \qquad (e)$$

Action of Br_2 on $(EtS)_3P$ affords[9] $(EtS)_2PBr$ in 45% yield. Bromination of $Me_2(H_2N)P{=}S$ gives[10] 78% of $[BrMe_2P{=}NPMe_2Br]Br_3$:

$$2\ Me_2\overset{S}{P}NH_2 + 4\ Br_2 \rightarrow [Me_2\overset{Br}{P}{=}N\overset{Br}{P}Me_2]Br_3 \qquad (f)$$

(H.J. BREUNIG)

1. N. A. Chadaeva, G. Kamai, K. A. Mamakov, *Izv. Akad. Nauk USSR*, 1612 (1972).
2. R. A. Zingaro, E. A. Meyers, *Inorg. Chem.*, *1*, 771 (1962).
3. W. A. Higgins, P. W. Vogel, W. G. Craig, *J. Am. Chem. Soc.*, *77*, 1864 (1955).
4. K. Sasse, in *Houben-Weyl, Methoden der Organischen Chemie*, E. Mueller, ed., Vol. 12, Part 1, *Organische Phosphorverbindungen*, G. Thieme Verlag, Stuttgart, 1963, p. 244.
5. W. G. Craig, U.S. Pat. 2,724,726 (1954); *Chem. Abstr.*, *50*, 10,129 (1956).
6. G. Stähler, D. E. 2920172 (1979/1980); *Chem. Abstr.*, *94*, 192457 (1981).
7. H. J. Emeleus, J. M. Miller, *J. Inorg. Nucl. Chem.*, *28*, 662 (1966).
8. A. Schmidpeter, J. Ebeling, *Chem. Ber.*, *101*, 815 (1968).
9. O. G. Sinyashin, Sh. Karimullin, D. A. Pudovik, E. S. Batyeva, A. N. Pudovik, *J. Gen. Chem. USSR* (*Engl. Transl.*), *54*, 2195 (1984).
10. A. Schmidpeter, N. Schindler, *Chem. Ber.*, *102*, 2160 (1969).

2.4.8.2. by Hydrogen Halides.

Cleavage of the P—S bond in $[Me_2P(S)S]_2$ by HF at −20°C affords[1] 50% of MeP(S)(SH)F:

$$Me\overset{S}{P}\langle S_2 \rangle\overset{S}{P}Me + 2\ HF \rightarrow 2\ Me\overset{S}{P}(SH)F \qquad (a)$$

Cleavage of the As—S bonds in phenylthioarsonic acid anhydride with HCl gives[2] $PhAsCl_2$:

$$PhAsS + 2\ HCl \rightarrow PhAsCl_2 + H_2S \qquad (b)$$

The reaction of Sb_2S_3 with conc HCl synthesize[3] $SbCl_3$. The action of HCl on $PhAs(SBu)_2$ gives[4] $PhAsCl_2$ and BuSH.

The P—S bond is converted to the P—Cl bond by HCl, e.g., addition of dry HCl to molten $Ph_2P(S)SH$ at 200°C results in $Ph_2P(S)Cl$[5].

(H.J. BREUNIG)

1. H. W. Roesky, *Chem. Ber.*, *101*, 3679 (1968).
2. L. Anschütz, H. Wirth, *Chem. Ber.*, *89*, 1530 (1956).
3. P. W. Schenk, in *Handbook of Preparative Inorganic Chemistry*, G. Brauer, ed., Academic Press, New York, 1963.
4. N. A. Chadaeva, G. Kamai, K. A. Mamakov, *Izv. Akad. Nauk. SSSR*, 1612 (1972).
5. K. Sasse, in *Houben-Weyl, Methoden der Organischen Chemie*, Vol. 12, Part 1, G. Thieme Verlag, Stuttgart, 1963, p. 274.

2.4.8.3. by Other Halides

2.4.8.3.1. to Give Phosphorus Fluorides from P—S and P—Se Bonds.

Fluorophosphoranes result from tertiary phosphine sulfides with SbF_3. Tributylphosphine sulfide reacts with SbF_3 to eliminate sulfur as Sb_2S_3 and form[1,2] Bu_3PF_2:

$$3\ n\text{-}Bu_3PS + 2\ SbF_3 \rightarrow 3\ n\text{-}Bu_3PF_2 + Sb_2S_3 \qquad (a)$$

Tetramethylbiphosphine disulfide is cleaved by SbF_3 to eliminate sulfur as Sb_2S_3 and form dimethyltrifluorophosphorane[1]:

$$2\ Me_2\overset{S}{P}\text{—}\overset{S}{P}Me_2 + 6\ SbF_3 \rightarrow 6\ Me_2PF_3 + 2\ Sb + 2\ Sb_2S_3 \qquad (b)$$

Examples are given in Table 1.

TABLE 1. CLEAVAGE OF THE P—S BOND WITH SbF_3

Phosphine sulfide	T (°C)	Time (h)	Yield (%)	Ref.
Et_3PS	25–120	2, 5	73	1
$n\text{-}Bu_3PS$	25–130	2	61	1, 2
$[Me_2P(S)]_2$	25–220	2	82	1
$[Et_2P(S)]_2$	25–140		84	1
$[Bu_2P(S)]_2$	25–150	2	43	1

Carbonyl fluoride and P_2S_5 at 300°C give SPF_3 in high yields[3].

Triphenyldifluorophosphorane results from Ph_3PS or Ph_3PSe using sulfuryl chloride fluoride[4]:

$$Ph_3PS + 2\ FSO_2Cl \rightarrow Ph_3PF_2 + 2\ SO_2 + SCl_2 \qquad (c)$$

Reaction of $[Et_4N]F$ and $[Et_4N]SH$ with P_4S_{10} gives[5] 57% of $[Et_4N]_2PS_3F$.

(H.J. BREUNIG)

1. R. Schmutzler, *Inorg. Chem., 3*, 421 (1964).
2. R. Schmutzler, *Inorg. Synth. 9*, (1967).
3. E. L. Mutterties, C. W. Tullock, *Prep. Inorg. React., 2*, 286 (1965).
4. A. Lopusinski, J. Michalski, *J. Am. Chem. Soc., 104*, 290 (1982).
5. L. Kolditz, U. Calov, A.-R. Grimmer, S. I. Trojanow, *Z. Chem., 20*, 309 (1980).

2.4.8.3.2. to Give Group-VB–Chlorine Bonds.

Replacement of sulfur by Cl and vice versa occurs with P_2S_5 and PCl_5 [1]:

$$P_2S_5 + 3\ PCl_5 \xrightarrow{150°C} \underset{(70\%)}{5\ PSCl_3} \qquad (a)$$

Cleavage of RMeO(EtO)P(O)SMe with SO_2Cl_2 in C_6H_6 at 9°C gives[2] 51% of RMeO(EtO)P(O)Cl. Action of PCl_5 on $(n\text{-}PrS)_3P$ affords[3] $(n\text{-}Pr)_2PCl$. Phosgene reacts with [MeP(O)(OPr-i)S]Na to give[4] the P—Cl bond in MeP(O)(OPr-i)Cl. Another reagent for the cleavage of the P—S bond is $SOCl_2$, e.g. $(4\text{-}MeOC_6H_4)_2P(S)OH$ reacts with $SOCl_2$ to form[5] $(4\text{-}MeOC_6H_4)_2P(O)Cl$. Conversion of a P—Se into a P—Cl bond occurs[6] in the reaction of $(EtO)_2P(Se)OH$ with SO_2Cl_2 to give[6] $(EtO)_2P(O)Cl$.

Action of MeCl on P_4S_3 at 325°C for 8 h affords[7] $Me_2P(S)Cl$ and PCl_3.

Conversion of the As—S to the As—Cl bond is accomplished with PCl_5 or SO_2Cl_2. Sulfuryl chloride or PCl_5 react with $PhAs(SBu\text{-}n)_2$ to form[2] $PhAsCl_2$, e.g.:

$$PhAs(SBu\text{-}n)_2 + SO_2Cl_2 \rightarrow PhAsCl_2 + (n\text{-}BuS)_2 + SO_2 \qquad \text{(b)}$$

Reaction of Ph(Et)AsSEt with Ph(Et)PCl gives[8] Ph(Et)AsCl:

$$Ph(Et)AsSEt + Ph(Et)PCl \rightarrow Ph(Et)AsCl + Ph(Et)PSEt \qquad \text{(c)}$$

Bis[1,2-thiaarsolanyl(2)]sulfide reacts with $AsCl_3$ to form 2-chloro-1,2-thiarsolan[9]:

$$\text{[S–As ring]}As—S—As\text{[As–S ring]} + 4AsCl_3 \rightarrow 10\ \text{[S–As ring]}As—Cl + As_4S_5Cl_2 \qquad \text{(d)}$$

Optical active (+)Ph(4-HOOC—C_6H_4)AsBr is obtained from the corresponding propyl thio ester with C_3H_7Br in a time-consuming (50 h) preparation at 20°C:

$$(-)Ph(HOOCC_6H_4)AsSC_3H_7 + C_3H_7Br \xrightarrow[50\ h]{20°C}$$

$$(+)Ph(HOOCC_6H_4)AsBr + (C_3H_7)_2S \qquad \text{(e)}$$

Acetyl chloride converts the As—S and Sb—S bonds, e.g., ethylarsenic dichloride is obtained[2] from $CH_3C(O)Cl$ and $EtAs(SBu\text{-}n)_2$. Trimethylstibine sulfide reacts with $CH_3C(O)Cl$ with quantitative formation[10] of $Me_3Sb(SCOCH_3)Cl$:

$$Me_3SbS + CH_3C(O)Cl \xrightarrow{CHCl_3} Me_3Sb(Cl)SCOCH_3 \qquad \text{(f)}$$

Chlorination of the As—S bond is also accomplished by $HgCl_2$, e.g., pentafluorophenylarsenic dichloride is obtained from $(C_6F_5AsS)_4$ with $HgCl_2$ in boiling Et_2O in 89% yield[11]:

$$(C_6F_5AsS)_4 + 4\ HgCl_2 \rightarrow 4\ C_6F_5AsCl_2 + 4\ HgS \qquad \text{(g)}$$

Refluxing $HgCl_2$ and $[(C_6F_5)_2As]_2S$ in Et_2O for 0.5 h gives $(C_6F_5)_2AsCl$ and HgS in 96% yield[5].

(H.J. BREUNIG)

1. D. R. Martin, W. M. Duval, *Inorg. Synth.*, *4*, 73 (1953).
2. C. R. Hall, T. D. Inch, *J. Chem. Soc.* (*Perkin I*), 1104 (1979).
3. O. G. Sinyashin, Sh. Karimullin, D. A. Pudovik, E. S. Batyeva, A. N. Pudovik, *J. Gen. Chem. (USSR) (Engl. Transl.)*, *54*, 2195 (1984).
4. H. S. Aaron, R. T. Ugeda, H. F. Frack, J. I. Miller, *J. Am. Chem. Soc.*, *84*, 617 (1962).
5. M. I. Kabachnik, T. A. Mastryakova, T. A. Melenteva, *J. Gen. Chem. USSR (Engl. Transl.)*, *33*, 375 (1963).
6. A. Markowska, *Bull. Acad. Polon. Sci., Ser. Sci. Chim.*, *15*, 153 (1967), *Chem. Abstr.*, *67*, 90891 (1967).
7. A. D. F. Toy, E. H. Uhing, US 4076746 (1976); *Chem. Abstr.*, *89*, 43771 (1978).
8. N. A. Chadaeva, G. Kamai, K. A. Mamakov, *Izv. Akad. Nauk SSSR*, 1612 (1972).

9. K. Sommer, *Z. Anorg. Allg. Chem.*, *375*, 55 (1970).
10. J. Otera, R. Okawara, *J. Organomet. Chem.*, *17*, 353 (1969).
11. M. Green, D. Kirkpatrick, *J. Chem. Soc.*, *A*, 483 (1968).

2.4.8.3.3. to Give the Group-VB—Br and —I Bond.

Conversion of the P—S bond to the P—Br bond is accomplished with PBr_5, e.g., action of PBr_5 on P_2S_5 gives[1] of $PSBr_3$ in 80% yield:

$$P_2S_5 + 3\ PBr_5 \xrightarrow{100°C} 5\ PSBr_3 \tag{a}$$

Ethylbromide reacts with P_2S_5 and $AlBr_3$ to form[2] $(EtS)_2P(S)Br$.

Bromination of a As—S bond in 2-phenyl-1,3,2-oxathiaarsolane with CH_3COBr gives[3] $PhAsBr_2$:

$$\text{(2-phenyl-1,3,2-oxathiaarsolane: S–As(Ph)–O ring)} + 2H_3CCOBr \xrightarrow[4\ h]{80°C} PhAsBr_2 + H_3COOCOCH_2CH_2SCOOCH_3 \tag{b}$$

Reaction of $[Me(CF_3)P]_2S$ with MeI gives[4] $Me(CF_3)PI$ and $Me_2(CF_2)PS$. Methyl iodide and trimethylantimony sulfide form[5] Me_3SbI_2:

$$2\ Me_3SbS + MeI \rightarrow Me_3SbI_2 + Me_3Sb + (MeS)_2 \tag{c}$$

In addition to MeI, EtI and $PhCH_2I$ are also used[5].

Treatment of $PhAs(SEt)_2$ with MeBr or MeI gives[6] $PhAsBr_2$ or $PhAsI_2$:

$$PhAs(SEt)_2 + 2\ MeI \rightarrow PhAsI_2 + 2\ MeSEt \tag{d}$$

(H.J. BREUNIG)

1. H. S. Booth, C. A. Seabright, *Inorg. Synth. 2*, (1946).
2. I. V. Muravev, I. S. Federovich, *J. Gen. Chem., USSR* (*Engl. Transl.*), *46*, 1241 (1976).
3. N. A. Chadaeva, K. A. Mamakov, G. K. Kamai, *J. Gen. Chem. USSR* (*Engl. Transl.*), *43*, 821 (1973).
4. A. B. Burg, D. K. Kang, *J. Am. Chem. Soc.*, *92*, 1901 (1970).
5. J. Otera, R. Okawara, *J. Organomet. Chem.*, *16*, 335 (1968).
6. N. H. Chadaeva, G. K. Kamai, K. A. Mamakov, M. P. Osipova, *J. Gen. Chem. USSR* (*Engl. Transl.*), *42*, 125 (1972).

2.4.9. from Cleavage of the Group VB–Nitrogen Bond

2.4.9.1. by Halogens.

The formation of group VB—halogen bonds by cleavage of group VB—nitrogen bonds with halogens is not a general method.

The reaction of F_2 with HN_3 in a stream of N_2 gives[1] NH_4F and FN_3:

$$2\ F_2 + 4\ HN_3 \rightarrow FN_3 + N_2 + NH_4F \tag{a}$$

Handling F_2 requires special apparatus and experienced personnel; HN_3 and its derivatives are potentially explosive.

Chlorination of $PhP(NCS)_2$ gives[2] $PhPCl_4$, S and ClCN:

$$PhP(NCS)_2 + 3\ Cl_2 \rightarrow PhPCl_4 + 2\ S + ClCN \qquad (b)$$

(H.J. BREUNIG)

1. K. Dehnicke, *Angew. Chem., Int. Ed. Engl., 6,* 240 (1967).
2. A. Michaelis, *Justus Liebigs Ann. Chem., 293,* 193 (1896).

2.4.9.2. by Hydrogen Halides.

Conversion of the P—N into the P-halogen bond is accomplished by hydrogen halides; e.g., F_2PNMe_2 and HCl form[1,2] F_2PCl:

$$F_2PNMe_2 + 2\ HCl \rightarrow F_2PCl + Me_2NH_2Cl \qquad (a)$$

Examples are given in Table 1.

Arsenic—nitrogen bonds are converted by hydrogen halides to As—halogen bonds. The action of HCl on Et_2AsNEt_2 gives Et_2AsCl in 64% yield:

$$Et_2AsNEt_2 + 2\ HCl \rightarrow Et_2AsCl + [Et_2NH_2]Cl \qquad (b)$$

Condensation of HCl on $MeAs(NMe_2)_2$ gives $MeAsCl_2$ in 99% yield[5]:

$$MeAs(NMe_2)_2 + 2\ HCl \rightarrow MeAsCl_2 + 2\ [Me_2NH_2]Cl \qquad (c)$$

The ease of the cleavage of the As—N bond by hydrogen halides allows the synthesis of arsonic acid dihalides with different halogen atoms, e.g., conversion of the Me_2N group in $MeAs(I)NMe_2$ with HBr gives[15] MeAsIBr:

$$MeAs(I)NMe_2 + 2\ HBr \rightarrow MeAs(I)Br + [Me_2NH_2]Br \qquad (d)$$

Examples are given in Table 2.

TABLE 1. CLEAVAGE OF THE P—N BOND WITH HYDROGEN HALIDES

Hydrogen halide	Compound with the P—N Bond	T (°C)	Product	Yield	Ref.
HCl	F_2PNMe_2	25/−15	F_2PCl	99	1, 2, 9
HBr	F_2PNMe_2	25	F_2PBr	99	1
HI	F_2PNMe_2	25	F_2PI	94	1, 3
HCl	$PhP(NMe_2)_2$	−10	$PhPCl_2$		4
HCl	Bu_2PNEt_2	0	Bu_2PCl	85	5
HF	$MeP(O)(OPr\text{-}i)NMe_2$	23	$MeP(O)(OPr\text{-}i)F$	81	6
HF	$CF_3P(NEt_2)_2$		CF_3PF_2		7
HCl	$(cyclo\text{-}Pr)_2PNEt_2$	20	$(cyclo\text{-}Pr)_2PCl$	67	8
HCl	$H_2C{=}CHP(NEt_2)_2$	−90/20	$H_2C{=}CHPCl_2$	30	10
HBr	$H_2C{=}CHP(NEt_2)_2$	−90/20	$H_2C{=}CHPBr_2$	30	10
HF	$(F_2P{=}N)_3$	25	$H_3N \cdot PF_5$	41	11
HBr	$[CH_2CH_2P(NEt_2)_2]_2$	−90	$[CH_2CH_2PBr_2]_2$	81	12
HCl	$p\text{-}[(Me_2N)_2P]_2C_6H_4$	0	$p\text{-}(Cl_2P)_2C_6H_4$	75	13, 14

TABLE 2. CLEAVAGE OF THE As—N BOND WITH HYDROGEN HALIDES

Hydrogen halide	Compound with the As—N Bond	T (°C)	Product	Yield (%)	Ref.
HCl	Et_2AsNEt_2	0	Et_2AsCl	64	6
HCl	Pr_2AsNEt_2	0	Pr_2AsCl	58	6
HCl	Bu_2AsNEt_2	0	Bu_2AsCl	83	6
HCl	$MeAs(NMe_2)_2$	25	$MeAsCl_2$	99	7
HCl	$MeAs(Br)NMe_2$	25	MeAsClBr	80	8
HCl	$MeAs(I)NMe_2$	25	MeAsClI	65	8
HBr	$MeAs(I)NMe_2$	25	MeAsBrI	70	8

Organo phosphorus and organoarsenic halides are poisons. All operations should be carried out in a well-vented hood. Gloves should be worn to avoid contact with the skin.

(H.J. BREUNIG)

1. J. G. Morse, K. Cohn, R. W. Rudolph, R. W. Parry, *Inorg. Synth.*, *10*, (1967).
2. R. G. Cavell, *J. Chem. Soc.*, 1992 (1964).
3. R. W. Rudolph, J. G. Moorse, R. W. Parry, *Inorg. Chem.*, *5*, 1464 (1966).
4. A. B. Burg, R. I. Wagner, U.S. Pat. 2,934,564 (1960); *Chem. Abstr.*, *54*, 18,437 (1960).
5. K. Issleib, W. Seidel. *Chem. Ber.*, *92*, 2681 (1959).
6. R. Greenhalgh, J. R. Blanchfield, *Can. J. Chem.*, *44*, 501 (1966).
7. W. Vollbach, I. Ruppert, *Tetrahedron Lett.*, *24*, 5509 (1983).
8. H. Schmidbaur, A. Schier, *Synthesis*, 372 (1983).
9. W. Albers, W. Krüger, W. Storzer, R. Schmutzler, *Synth. React. Inorg. Met.-org. Chem.*, *15*, 187 (1985).
10. K. Diemert, B. Kottwitz, W. Kuchen, *Phosphorus and Sulfur*, *26*, 307 (1986).
11. W. Storzer, D. Schomburg, G.-V. Röschenthaler, R. Schmutzler, *Chem. Ber.*, *116*, 367 (1983).
12. K. Diemert, W. Kuchen, J. Kutter, *Chem. Ber.*, *115*, 1947 (1982).
13. K. Drewelies, H. P. Latscha, *Angew. Chem., Int. Ed. Engl.*, *21*, 638 (1982).
14. H. J. Wörz, H. P. Latscha, *Chem. Ztg.*, *108*, 329 (1984).
15. A. Tzschach, W. Lange, *Z. Anorg. Allg. Chem.*, *326*, 280 (1964).
16. F. Kober, *Z. Anorg. Allg. Chem.*, *397*, 97 (1973).
17. J. Kaufmann, F. Kober, *J. Organomet. Chem.*, *96*, 243 (1975).

2.4.9.3. by Other Halides.

Cleavage of a P—N bond and conversion to a phosphorus halogen bond is accomplished by group-VB halides (Table 1) e.g., addition of PCl_3 to $(Me_2N)_3P$ gives $(Me_2N)_2PCl$ in 92% yield[1].

Intramolecular fluorination occurs when $(F_3C)_2NP(CF_3)_2Cl_2$ decomposes on standing to form[5] $(CF_3)_2PCl_2F$:

$$(F_3C)(F_2C{-}F)N{-}P(CF_3)_2Cl_2 \rightarrow (CF_3)_2PCl_2F + CF_3N{=}CF_2 \quad \text{(a)}$$

Reaction of $(Me_2N)_3PO$ with NH_4F at 180–220°C gives[6] $(Me_2N)_2P(O)F$. Sulphurtetrafluoride cleaves the P—N bonds in $MeP(O)(NMe_2)_2$ to give[7] $MeP(O)F_2$. Boron

TABLE 1. CLEAVAGE OF THE P—N BOND WITH GROUP-VB HALIDES

Compound with the P—N Bond	Group-V Halide	Product	Ref.
$MeP(S)(NCO)_2$	SbF_3	$MeP(S)(NCO)F$	2
$CF_3P(NEt_2)_2$	PCl_3	$CF_3P(NEt_2)Cl$	3
$CF_3P(NEt_2)_2$	PBr_3	$CF_3P(NEt_2)Br$	3
$CF_3P(NEt_2)_2$	PI_3	$CF_3P(NEt_2)I$	3
$CF_3P(NEt_2)Cl$	PCl_3	CF_3PCl_2	3
Et_2NPF_2	PCl_3	$ClPF_2$	4
$(Et_2N)_2PF$	PCl_3	Cl_2PF	4
Et_2NPF_2	PBr_3	$BrPF_2$	4

trifluoride and $(Me_2N)_3P$ form[8] 74% of Me_2NPF_2. Action of Me_3SiBr on $(Me_2N)_3P$ gives[9] $(Me_2N)_2PBr$ and Me_2NSiMe_3 in equilibrium with the starting materials. The analogous reaction of Me_3SiI with $(Me_2N)_3P$ affords[9] $(Me_2N)_2PI$. Reaction of $Ph_2P(O)NEt_2$ with $HSiCl_3$ in presence of py gives[10] $Ph_2P(O)Cl$. The P=N bond is transformed[11] to P—Cl in:

```
            Ph                                   Ph
           /                                    /
       N=C                                  N=C
      /   \                                /   \
Me—N       N                         Me—N       N
     \P//          + PCl5 →              \ /   \
     /  \                          Cl—P         PCl3 + HCl      (b)
Me—N      NH                             / \   /
    \    /                           Me—N       N
     N=C                                 \     /
        \                                 N=C
         Ph                                  \
                                              Ph
```

Conversion of the As—N to the As—I bond is accomplished with CH_3I; e.g., $MeAs(NMe_2)_2$ with MeI in benzene gives[12] $MeAsI_2$:

$$MeAs(NMe_2)_2 + 4\ MeI \rightarrow MeAsI_2 + 2\ Me_3N \tag{c}$$

Diethylarsenic iodide is obtained from Et_2AsNEt_2 and MeI in 62% yield[4]:

$$Et_2AsNEt_2 + 2\ MeI \rightarrow Et_2AsI + [Me_2NEt_2]I \tag{d}$$

Examples are given in Table 2.

TABLE 2. CLEAVAGE OF THE As—N BOND WITH METHYL IODIDE

Compound with the As—N Bond	Product	Ref.
$MeAs(NMe_2)_2$	$MeAsI_2$	12
	$PhAsI_2$	13
n-Bu_2AsN⬡	Bu_2AsI	14
Et_2AsNEt_2	Et_2AsI	14

Organophosphorus and arsenic halides are poisons. All operations should be carried out in a well-vented hood by experienced persons.

(H.J. BREUNIG)

1. H. Nöth, H. J. Vetter, *Chem. Ber. 94*, 1505 (1961).
2. Z. M. Ivanova, S. K. Mikhailik, G. I. Derkach, *J. Gen. Chem., USSR* (*Engl. Transl.*) *40*, 1459 (1970).
3. W. Volbach, I. Ruppert, *Tetrahedron Lett. 24*, 5509 (1983).
4. W. Albers, W. Krüger, W. Storzer, R. Schmutzler, *Synth. React. Inorg. Met. org. Chem., 15*, 187 (1985).
5. H. G. Aug, *J. Inorg. Allg. Chem., 31*, 3311 (1969).
6. H. Liberda, DE 2418984 (1974/1975); *Chem. Abstr., 84*, 30482 (1976).
7. D. H. Brown, K. D. Crosbie, J. I. Darragh, D. S. Ross, D. W. A. Sharp, *J. Chem. Soc., (A)* 914 (1970).
8. H. Nöth, H. D. Veller, *Chem. Ber., 96*, 1109 (1963).
9. M. Cypruk, J. Choinowski, J. Michalski, *Tetrahedron, 41*, 2471 (1985).
10. L. D. Quin, J. Szewzyk, *Phosphorus and Sulfur, 21*, 161 (1984).
11. A. Schmidpeter, M. Nayibi, P. Mayer, H. Tautz, *Chem. Ber., 116*, 1468 (1983).
12. F. Kober, *Z. Anorg. Allg. Chem., 397*, 97 (1973).
13. E. Fluck, G. Jakobson, *Z. Anorg. Allg. Chem., 369*, 178 (1969).
14. A. Tzschach, W. Lange, *Z. Anorg. Allg. Chem., 326*, 280 (1964).

2.4.10. from Cleavage of the Group VB–Other Group VB Element Bond

2.4.10.1. by Halogens

2.4.10.1.1. to Give Phosphorus Halides.

The P—P bond in the diphosphines, $R_2P—PR_2$, and cyclopolyphosphines, $(RP)_n$, is cleaved with halogens; cf. §2.4.2, where the P—P bond in elemental phosphorus is split.

Diphosphines react 1 : 1 with halogens in inert solvents, such as CCl_4, CH_2Cl_2, Et_2O or benzene, to form the corresponding phosphinous halides in almost quantitative yields:

$$R_2P—PR_2 + X_2 \rightarrow 2\ R_2PX \qquad (a)$$

where R = organic group; X = Cl, Br, I. This is a versatile method for phosphinous halides[1–9,25].

Excess halogen must be avoided to prevent formation of phosphoranes (see §2.4.13.). Cleavage of the P—P bond in

$$\begin{array}{c} CF_3—C{=}C—CF_3 \\ \quad | \quad\; | \\ CF_3—P—P—CF_3 \end{array}$$

with I_2 does not occur[10], whereas bromination of $(CF_3)_2PP(CF_3)_2$ at 80°C[1] and of $CCl_3(Br)PP(Br)CCl_3$ at RT[11] simultaneously cleaves the P—P and P—C bonds, respectively.

Similarly, cyclopolyphosphines undergo ring cleavage with halogens to yield alkyl- or arylphosphonous dihalides[9]:

$$(RP)_n + nX_2 \rightarrow nRPX_2 \qquad (b)^{2-5,10,12-14}$$

The reaction is stepwise; e.g., it is possible to obtain iodinated P—P derivatives[2]:

$$(PhP)_5 + I_2 \rightarrow Ph(I)P{-}P(I)Ph \qquad (c)$$

The triphosphine $(t\text{-}BuP)_3$ reacts with halogens similarly:

$$(t\text{-}BuP)_3 + X_2 \xrightarrow[\text{pentane}]{-60°C} t\text{-}Bu(X)P{-}(t\text{-}Bu)P{-}P(X)Bu\text{-}t \qquad (d)$$

where X = Cl, Br, I. Whereas the iodine derivative may be isolated in low yields, the reaction proceeds to give t-Bu(X)PP(X)Bu-t and finally $t\text{-}BuPX_2$ (X = Cl, Br, I)[15,26].

Diphosphorus compounds with phosphorus atoms in different oxidation states may be brominated to yield phosphinous bromide and thiophosphinic bromide[7]:

$$R_2P{-}P(S)R_2 + Br_2 \rightarrow R_2PBr + R_2P(S)Br \qquad (e)$$

Similarly, nitrogen-substituted diphosphorus derivatives are cleaved[16]:

$$Ph_2P{-}P(Ph)_2{=}N{-}P(S)Ph_2 + Br_2 \rightarrow Ph_2PBr + Ph_2P(S)N{=}P(Br)Ph_2 \qquad (f)$$

An important synthetic route to thiophosphinic halides, $R_2P(S)X$, is based on the reaction of diphosphine disulfides with halogens[7,17–21,27] in inert solvents to give excellent yields[22]:

$$R_2P(S)P(S)R_2 + X_2 \rightarrow 2\ R_2P(S)X \qquad (g)$$

where R = organic group; X = Cl, Br, I.

Nitrogen-substituted derivatives are also brominated[16], e.g.:

$$Ph_2P(S){-}P(Ph)_2{=}N{-}P(S)Ph_2 + Br_2 \rightarrow Ph_2P(S)Br + Ph_2P(S)N{=}P(Br)Ph_2 \qquad (h)$$

Excess halogen must be avoided, otherwise phosphoranes are formed by the subsequent cleavage of the P—S bond (see §2.4.8). Under more vigorous conditions xs halogens may cause C—H bond halogenation and oxidation to phosphoranes[23], e.g.:

$$Me_2P(S)P(S)Me_2 + 17\ Cl_2 \rightarrow 2\ (CCl_3)_2PCl_3 + 2\ SCl_2 + 12\ HCl \qquad (i)$$

The As—P bond is split by Br_2 to give phosphinous bromide[24], e.g.:

$$(CF_3)_2PAsH_2 + Br_2 \xrightarrow{-196°C\ \text{to RT}} (CF_3)_2PBr + [AsH_2Br] \qquad (j)$$

The cleavage of halogenated diphosphines leads to scrambling reactions (see §2.4.12).

The cyclotetraphosphine $(i\text{-}Pr_2NP)_4$ is cleaved with X_2 to give $i\text{-}Pr_2NPX_2$ (where X = Cl, Br)[28]. The phosphene RP=PR (with R = $(Me_3Si)_3C$, 2,4,6-t-BuC_6H_2) reacts with X_2 to RPX_2 (where X = Cl, Br, I) followed by oxidation to RPX_4[29,30].

(M. FILD)

1. A. B. Burg, J. E. Griffiths, *J. Am. Chem. Soc.*, *82*, 3514 (1960).
2. H. Hoffmann, R. Grünewald, *Chem. Ber.*, *94*, 186 (1961).
3. K. Issleib, B. Mitscherling, *Z. Naturforsch., Teil B*, *15*, 267 (1960).
4. K. Issleib, W. Seidel, *Chem. Ber.*, *92*, 2681 (1959).

5. K. Issleib, W. Seidel, *Z. Anorg. Allg. Chem.*, *303*, 155 (1960).
6. W. Kuchen, H. Buchwald, *Chem. Ber.*, *91*, 2871 (1958).
7. L. Maier, *J. Inorg. Nucl. Chem.*, *24*, 275 (1962).
8. H. Vetter, H. Nöth, *Chem. Ber.*, *96*, 1816 (1963).
9. M. Fild, R. Schmutzler, in *Organic Phosphorus Compounds*, G. M. Kosolapoff, L. Maier, eds., Vol. 4, Wiley, New York, 1972, p. 75.
10. W. Mahler, *J. Am. Chem. Soc.*, *86*, 2306 (1964).
11. P. L. Airey, *Z. Naturforsch.*, *Teil B*, *24*, 1393 (1969).
12. W. Kuchen, H. Buchwald, *Chem. Ber.*, *91*, 2296 (1958).
13. W. Kuchen, W. Grünewald, *Angew. Chem.*, *75*, 576 (1963).
14. W. Mahler, A. B. Burg, *J. Am. Chem. Soc.*, *79*, 251 (1957).
15. M. Baudler, J. Hellmann, *Z. Anorg. Allg. Chem.*, *480*, 129 (1981).
16. L. Meinel, H. Nöth, *Z. Anorg. Allg. Chem.*, *373*, 36 (1970).
17. H. J. Harwood, K. A. Pollart, *J. Org. Chem.*, *28*, 3430 (1963).
18. W. Kuchen, H. Buchwald, *Angew. Chem.*, *71*, 162 (1959).
19. L. Maier, *Chem. Ber.*, *94*, 3043 (1961).
20. L. Maier, *Chem. Ber.*, *94*, 3051 (1961).
21. K. A. Pollart, H. J. Harwood, *J. Org. Chem.*, *27*, 4444 (1962).
22. M. Fild, R. Schmutzler, in *Organic Phosphorus Compounds*, G. M. Kosolapoff, L. Maier, eds., Vol. 4, Wiley, New York, 1972, p. 155.
23. H. Reinhardt, D. Bianchi, O. Mölle, *Chem. Ber.*, *90*, 1656 (1957).
24. A. Breckner, J. Grobe, *Z. Anorg. Allg. Chem.*, *414*, 269 (1975).
25. A. B. Burg, *Inorg. Chem.*, *20*, 3731 (1981).
26. M. Baudler, J. Hellmann, *Z. Anorg. Allg. Chem.*, *509*, 53 (1984).
27. N. G. Feshchenko, N. D. Gomelya, A. G. Matyusha, *J. Gen. Chem. USSR* (*Engl. Transl.*), *52*, 202 (1982).
28. R. B. King, N. D. Sadanani, *J. Org. Chem.*, *50*, 1719 (1985).
29. J. Escudie, C. Couret, H. Ranaivonjatova, J. Satge, J. Jaud, *Phosphorus Sulfur*, *17*, 221 (1983).
30. M. Yoshifuji, I. Shima, K. Shibayama, N. Inamoto, *Tetra. Lett.*, *25*, 411 (1984).

2.4.10.1.2. to Give As, Sb and Bi Halides.

Tetramethyldiarsine is cleaved by I_2 to give[1] Me_2AsI:

$$Me_2AsAsMe_2 + I_2 \xrightarrow{Et_2O} 2\ Me_2AsI \tag{a}$$

The reaction of I_2 with $(PhAs)_6$ in benzene yields[2] PhAs(I)—As(I)Ph, which on further addition of I_2 is cleaved[3] to $PhAsI_2$.

The distibine $Me_2SbSbMe_2$ is converted[4,5] to Me_2SbX by X_2:

$$Me_2SbSbMe_2 + X_2 \rightarrow 2\ Me_2SbX \tag{b}$$

where X = Cl, Br, I.

Iodine cleaves $Ph_2BiBiPh_2$ to diphenyliodobismuthine[6], and phenylbromobismuthine is obtained from the polymeric $(PhBi)_n$ and Br_2 vapor[7]:

$$(PhBi)_n + n\,Br_2 \rightarrow n\,PhBiBr_2 \tag{c}$$

(M. FILD)

1. C. T. Mortimer, H. A. Skinner, *J. Chem. Soc.*, 4331 (1952).
2. F. F. Blicke, *J. Am. Chem. Soc.*, *54*, 3353 (1932); *J. Am. Chem. Soc.*, *55*, 1161 (1933).
3. F. F. Blicke, L. D. Powers, *J. Am. Chem. Soc.*, *55*, 315 (1933).
4. F. Paneth, H. Loleit, *J. Chem. Soc.*, 366 (1935).
5. H. J. Breuning, H. Jawad, *J. Organomet. Chem.*, *243*, 417 (1983).
6. F. Calderazzo, R. Poli, G. Pelizzi, *J. Chem. Soc. Dalton*, 2365 (1984).
7. E. Wiberg, K. Moedritzer, *Z. Naturforsch.*, *Teil B*, *12*, 132 (1957).

2.4.10.2. by Hydrogen Halides

2.4.10.2.1. to Give Phosphorus Halides.

Diphosphines are treated with HCl to give phosphinous chlorides[1–3,11]:

$$R_2PPR_2 + HCl \rightarrow R_2PCl + R_2PH \quad \text{(a)}$$

where R = Me, CF_3. Similarly, unsymmetrical diphosphines are cleaved[4,12]:

$$Me_2PP(CF_3)_2 + HCl \rightarrow Me_2PCl + (CF_3)_2PH \quad \text{(b)}^4$$

$$2\ Ph_2PP(NMe)_2 + 6\ HCl \xrightarrow{-70°C} 2\ Ph_2PCl + \frac{1}{n}[MeNP(Cl)]_n + 3\ [Me_2NH_2]Cl \quad \text{(c)}^5$$

Anhydrous HCl also reacts with the cyclopolyphosphine, $(PhP)_6$, to yield[6] $PhPCl_2$ and $PhPH_2$. Reactions are carried out in the gas or liq phase or in inert solvents, such as Et_2O or benzene.

Diphosphorus compounds with phosphorus atoms in different oxidation states also react:

$$Me_2PP(Me)_2{=}NSiMe_3 + 4\ HCl \rightarrow 2\ Me_2PCl + Me_3SiCl + NH_4Cl \quad \text{(d)}^7$$

$$Et_2PP(Et)_2{=}CRR' + HCl \rightarrow Et_2PCl + Et_2PCHRR' \quad \text{(e)}^8$$

where R = R′ = COOMe, COOEt; or R = H, R′ = SO_2CF_3).

A mixture of $Me_2P(S)Cl$ and Me_2PCl is obtained from the cleavage of the P—P bond with HCl, followed by a subsequent reaction at the P—N bond[5]:

$$Me_2P(S)P(Me)_2{=}NSiMe_3 + 4\ HCl \xrightarrow{Et_2O}$$

$$Me_2P(S)Cl + Me_2PCl + Me_3SiCl + NH_4Cl \quad \text{(f)}$$

By treating the bis(iminophosphorane) with HF in $CFCl_3$ a fluorophosphetidine, as the dimer of an intermediate $Me_2PF{=}NSiMe_3$, is formed[9]:

$$Me_3SiN{=}P(Me)_2{-}P(Me)_2{=}NSiMe_3 + HF \xrightarrow{CFCl_3}$$

$$\tfrac{1}{2}\ [Me_2PF{-}NSiMe_3]_2 + Me_2P(CFCl_2){=}NSiMe_3 + HCl \quad \text{(g)}$$

In Et_2O the P—P bond is retained and the phosphorane $Me_2(F)_2PP(F)_2Me_2$ forms, which on further addition of HF breaks down[9] to Me_2PHF_2 and Me_2PF_3 (§2.4.13.).

The diphosphine P_2H_4 also reacts with HF to give unstable PH_2F, which undergoes oxidative addition and disproportionation in the presence of HF[10] (§2.4.13.).

Treatment of the cyclotetraphosphine $(i\text{-}Pr_2NP)_4$ with HCl at 0°C gives $i\text{-}Pr_2NPCl_2$ and $(i\text{-}Pr_2N)_2P_2Cl_2$ as a mixture[13]. The phosphene RP=PR (where R = $(Me_3Si)_3C$, 2,4,6-t-BuC_6H_2) reacts with HCl to give RP(H)Cl[14–16].

(M. FILD)

1. A. B. Burg, W. Mahler, *J. Am. Chem. Soc.*, *79*, 4242 (1957).
2. H. Nöth, *Z. Naturforsch., Teil B.*, *15*, 327 (1960).
3. F. Seel, K. D. Vellemann, *Chem. Ber.*, *104*, 2967 (1971).
4. L. R. Grant, A. B. Burg, *J. Am. Chem. Soc.*, *84*, 1834 (1962).
5. H. J. Vetter, H. Nöth, *Chem. Ber.*, *96*, 1816 (1963).

6. L. Maier, *Helv. Chim. Acta, 49*, 1119 (1966).
7. R. Appel, R. Milker, *Chem. Ber., 107*, 2658 (1974).
8. O. Kolodyazhnyi, *J. Gen. Chem. USSR (Engl. Transl.), 49*, 88 (1979).
9. R. Appel, R. Milker, I. Ruppert, *Z. Anorg. Allg. Chem., 429*, 69 (1977).
10. F. Seel, K. Vellemann, *Z. Anorg. Allg. Chem., 385*, 123 (1971).
11. A. B. Burg, *Inorg. Chem., 24*, 3342 (1985).
12. A. B. Burg, *Inorg. Chem., 20*, 3734 (1981).
13. R. B. King, N. D. Sadanani, *J. Org. Chem., 50*, 1719 (1985).
14. J. Escudie, C. Couret, H. Ranaivonjatova, J. Satge, J. Jaud, *Phosphorus Sulfur, 17*, 221 (1983).
15. A. H. Cowley, J. E. Kilduff, N. C. Norman, M. Pakulski, J. L. Atwood, W. E. Hunter, *J. Am. Chem. Soc., 105*, 4845 (1983).
16. A. H. Cowley, J. E. Kilduff, N. C. Norman, M. Pakulski, *J. Chem. Soc. Dalton*, 1801 (1986).

2.4.10.2.2. to Give As and Sb Halides.

Tetramethyldiarsine with hydrogen halides gives arsinous halides in an equilibrium[1,2]:

$$Me_2AsAsMe_2 + HX \rightleftharpoons Me_2AsX + Me_2AsH \tag{a}$$

where X = Cl[1,2], Br[2], I[2].

Arsinous chloride is formed[3] in high yields from cleavage of the As—P bond by HCl:

$$(Me_2As)_2PCF_3 + 2\ HCl \xrightarrow{-196^\circ C\ to\ RT} 2\ Me_2AsCl + CF_3PH_2 \tag{b}$$

Tetramethyldistibine reacts with HCl at RT to give[4] Me_2SbCl and H_2.

(M. FILD)

1. W. R. Cullen, *Can. J. Chem., 41*, 322 (1963).
2. A. L. Rheingold, E. J. Plean, W. T. Ferrat, *Inorg. Chim. Acta, 22*, 215 (1977).
3. A. H. Cowley, D. S. Dierdorf, *J. Am. Chem. Soc., 91*, 6609 (1969).
4. A. B. Burg, L. R. Grant, *J. Am. Chem. Soc., 81*, 1 (1959).

2.4.10.3. by Other Halides

2.4.10.3.1. to Give Phosphorus Halides.

When tetraalkyldiphosphines are treated with alkyl halides, monophosphonium salts form[1,2]. Heating with xs alkyl halide to 200°C[1] ruptures P—P. Cleavage rather than quaternization is observed if trifluoromethyl-[3] or phenyl-substituted[4] diphosphines are employed, e.g.:

$$Ph_2PPPh_2 + MeI \rightarrow Ph_2PI + Ph_2PMe \tag{a}$$

$$(CF_3)_2PP(CF_3)_2 + MeI \rightarrow (CF_3)_2PI + (CF_3)_2PMe \tag{b}$$

Replacing the alkyl group by bulkier t-butyl—[5] c-C_6H_{11}—[2] or the CF_3— groups[3,6] in the alkyl halide has a similar effect.

Cyclopolyphosphines are also cleaved with alkyl halides, preferably iodides[7], or with CF_3I[8] and C_3F_7I[9] to give phosphonous iodides in good yields.

Diphosphines are cleaved[10] with CCl_4:

$$R_2PPR_2 + CCl_4 \xrightarrow{120°C} R_2PCl + R_2PCCl_3 \qquad (c)$$

where R = Ph, c-C_6H_{11}, n-Bu. Complete fission of the P—P bond with CCl_4 is observed[11] for $(PhP)_4$:

$$(PhP)_4 + 4\ CCl_4 \rightarrow PhP(CCl_3)Cl \qquad (d)$$

whereas (c-$C_6H_{11}P)_4$ gives a diphosphine:

$$(c\text{-}C_6H_{11}P)_4 + 2\ CCl_4 \rightarrow c\text{-}C_6H_{11}(Cl)PP(CCl_3)C_6H_{11}\text{-}c \qquad (e)$$

The cyclopolyphosphines $(MeP)_5$ and $(EtP)_{4,5}$ produce phosphonium salts by addition of CCl_4, but (t-BuP)$_4$ is inert[11]. The reaction of (t-BuP)$_3$ with PBr_3 yields acyclic [t-Bu(Br)P]$_3$P[34].

A controlled reaction of the cyclopolyphosphine (t-BuP)$_3$ with phosphorus pentahalides[12] cleaves one P—P bond, e.g.:

$$(t\text{-}BuP)_3 + PX_5 \rightarrow t\text{-}Bu(X)P\text{—}P(t\text{-}Bu)\text{—}P(X)Bu\text{-}t \qquad (f)$$

where X = Cl, Br. The chloro derivative, which is stable at $\leq 50°C$, decomposes to t-Bu(Cl)PP(Cl)Bu-t, t-BuPCl_2, and (t-BuP)$_4$.

Silyl phosphines react stepwise with C_2Cl_6 or $COCl_2$ to form phosphinous or phosphonous chlorides (see §2.4.6.3.2.) with diphosphines or cyclopolyphosphines as intermediates, which are cleaved in the final step at the P—P bond[13–15].

The thermally initiated decomposition of phosphonium salts containing a P—P bond is related to the above reaction of diphosphines with alkyl halides, e.g.:

$$[R_3P{=}NPPh_2PPh_2]Cl \xrightarrow{\Delta} [(R_3P{=}N)_2PPh_2]Cl + Ph_2PCl + Ph_2PPPh_2 \qquad (g)^{16}$$

where R = Et, Ph;

$$[Ph_2PPPh_2{=}NPPh_2PPh_2]Cl \xrightarrow{\Delta} Ph_2PCl + Ph_2PN{=}PPh_2PPh_2 \qquad (h)^{17}$$

1,1-Dialkoxy-2,2-dialkyldiphosphines react with alkyl halides forming alkylphosphorohalodites and phosphonium salts[18]:

$$i\text{-}Pr_2P\text{—}P(OR)_2 + 2\ R'X \rightarrow (RO)_2PX + [i\text{-}Pr_2PR'_2]X \qquad (i)$$

where (R = n-Bu, Ph; R′ = Me, Et, i-Pr, PhCH$_2$; X = Cl, Br, I. The exothermic reaction gives 70–95% yields, the rate being dependent on the nature of the halogen and organic groups RR′.

This reaction is also performed with (PhO)$_2$PCl as halogen source[18], e.g.:

$$i\text{-}Pr_2PP(OBu\text{-}n)_2 + (PhO)_2PCl \rightarrow i\text{-}Pr_2PP(OPh)_2 + (n\text{-}BuO)_2PCl \qquad (j)$$

Similarly, cleavage occurs with i-Pr$_2$PCl or Ph$_2$PCl to give phosphorochloridites[18].

Phosphorochloridites are also obtained[19] from diphosphites with SO_2Cl_2 or PCl_3:

$$(RO)_2PP(OR)_2 + SO_2Cl_2 \rightarrow 2\ (RO)_2PCl + SO_2 \qquad (k)$$

Acid chlorides react in a similar way, e.g.:

$$(RO)_2PP(OR)_2 + MeCOCl \rightarrow (RO)_2PCl + (RO)_2PCOMe \qquad (l)$$

where R = Et, n-Pr, i-Bu, n-Bu, to give phosphorochloridites in ≤90% yields. Other acid chlorides employed[20,21] are Me_3CCOCl and Et_3SiCH_2COCl. This reaction is also applied to heterocycles[22]:

$$\text{(cyclic O}_2\text{P)–P(cyclic O}_2\text{)} + RCOCl \rightarrow \text{(cyclic O}_2\text{)PCl} + \text{(cyclic O}_2\text{)PCOR} \qquad \text{(m)}$$

The P—P bond may also be cleaved with transition-metal carbonyl halides, e.g., the pentacarbonyl halides $Re(CO)_5X$ and $Mn(CO)_5X$ (X = Cl, Br, I) react[23] with $(CF_3)_2PP(CF_3)_2$:

$$(CF_3)_2PP(CF_3)_2 + 2\ M(CO)_5X \xrightarrow[Et_2O]{60\text{–}90^\circ C} (CF_3)_2PX + Mn_2(CO)_8P(CF_3)_2X + 2\ CO \qquad \text{(n)}$$

where M = Mn, Re; X = Cl, Br, I.

Diphosphorus compounds with phosphorus atoms in different oxidation states are split at the P—P bond with, e.g., Ph_2PCl[24]:

$$Ph_2PP(O)(OMe)Me + Ph_2PCl \rightarrow MeP(O)(OMe)Cl + Ph_2PPPh_2 \qquad \text{(o)}$$

or $PhICl_2$[25]:

$$t\text{-}Bu_2PP(O)Ph_2 + PhICl_2 \rightarrow t\text{-}Bu_2PCl + Ph_2P(O)Cl + PhI \qquad \text{(p)}$$

Diphosphine disulfides are transformed into thiophosphinic chlorides in high yields using[26,27] 1:1 SO_2Cl_2

$$R_2P(S)P(S)R_2 + SO_2Cl_2 \rightarrow 2\ R_2P(S)Cl + SO_2 \qquad \text{(q)}$$

Thionyl chloride[28], SCl_2[26], PCl_5[32] and $HgCl_2$[26] also serve as halogen source. With xs SO_2Cl_2 or $SOCl_2$ and at elevated T, sulfur is replaced by oxygen to give phosphinic chlorides[27–29]:

$$R_2P(S)P(S)R_2 + 2\ SOCl_2 \rightarrow 2\ R_2P(O)Cl + 2\ S + S_2Cl_2 \qquad \text{(r)}$$

The P—P bond is cleaved by phosphorus halides such as PCl_3, PBr_3 and $POCl_3$[30]; e.g., dimethylthiophosphinic chloride is obtained in 93% yield:

$$Me_2P(S)P(S)Me_2 + PCl_3 \rightarrow 2\ Me_2P(S)Cl + [(PCl)_x] \qquad \text{(s)}$$

Substituted phosphonous or phosphinous chlorides, e.g., Ph_2PCl[26] or $PhPCl_2$[31], cleave the P—P bond in $Me_2P(S)P(S)Me_2$ with subsequent removal of sulfur to give Me_2PCl, but phosphonic chloride converts diphosphine disulfides to a mixture of phosphonous and thiophosphinic chlorides[30]:

$$2\ n\ RP(O)Cl_2 + n\ R_2P(S)P(S)R_2 \rightarrow n\ RPCl_2 + 2\ n\ R_2P(S)Cl + [RP(O)O]_n \qquad \text{(t)}$$

Alkyl halides and alkylene dihalides at 120–200°C are also used[30]:

$$R_2P(S)P(S)R_2 + R'X \rightarrow R_2P(S)X + R_2P(S)R' \qquad \text{(u)}$$

where X = Cl, Br, I;

$$2\ R_2P(S)P(S)R_2 + X\text{—}(CH_2)_n\text{—}X \rightarrow 2\ R_2P(S)X + R_2P(S)(CH_2)_nP(S)R_2 \qquad \text{(v)}$$

where X = Cl, Br, I.

Chlorinated carbosilanes cleave P—P bonds[35,36]:

$$(Cl_3Si)_2CCl_2 + Me_2PPMe_2 \rightarrow (Cl_3Si)_2C{=}P(Cl)Me_2 + Me_2PCl \quad (w)$$

$$(Me_3Si)_2CCl_2 + (Me_3Si)_2C{=}P(Me)_2PMe_2 \rightarrow 2\ (Me_3Si)_2C{=}P(Cl)Me_2 \quad (x)$$

(M. FILD)

1. H. Nöth, *Z. Naturforsch., Teil B, 15*, 327 (1960).
2. K. Issleib, W. Seidel, *Chem. Ber., 92*, 2681 (1959).
3. W. R. Cullen, *Can. J. Chem., 38*, 439 (1960).
4. H. Hoffmann, R. Grünewald, L. Horner, *Chem. Ber., 93*, 861 (1960).
5. K. Issleib, M. Hoffmann, *Chem. Ber., 99*, 1320 (1966).
6. R. Demuth, J. Apel, J. Grobe, *Spectrochim. Acta, Sect. A, 34*, 357 (1978).
7. H. Hoffmann, R. Grünewald, *Chem. Ber., 94*, 186 (1961).
8. M. A. A. Beg, H. C. Clark, *Can. J. Chem., 39*, 564 (1961).
9. I. G. Maslennikov, A. N. Lavrentev, V. I. Shibaev, E. G. Sochilin, *J. Gen. Chem. USSR (Engl. Transl.), 46*, 1841 (1976).
10. R. Appel, R. Milker, *Chem. Ber., 108*, 1783 (1975).
11. R. Appel, R. Milker, *Z. Anorg. Allg. Chem., 417*, 161 (1975).
12. M. Baudler, J. Hellmann, *Z. Anorg. Allg. Chem., 480*, 129 (1981).
13. R. Appel, K. Geisler, H. Schöler, *Chem. Ber., 110*, 376 (1977).
14. R. Appel, V. Barth, *Angew. Chem., Int. Ed. Engl., 18*, 469 (1979).
15. R. Appel, V. Barth, M. Halstenberg, G. Huttner, J. von Seyerl, *Angew. Chem., Int. Ed. Engl., 18*, 872 (1979).
16. W. Wolfsberger, *J. Organomet. Chem., 86*, C3 (1975).
17. H. Nöth, L. Meinel, *Z. Anorg. Allg. Chem., 349*, 225 (1967).
18. V. L. Foss, Y. A. Veits, I. F. Lutsenko, *J. Gen. Chem. USSR (Engl. Transl.), 48*, 1558 (1978).
19. I. F. Lutsenko, *Phosphorus Sulfur, 1*, 99 (1976).
20. I. F. Lutsenko, M. V. Proskurnina, A. L. Chekhun, *Phosphorus, 4*, 57 (1974).
21. M. V. Proskurnina, A. L. Chekhun, I. F. Lutsenko, *J. Gen. Chem. USSR (Engl. Transl.), 44*, 1216 (1974).
22. E. E. Nifantev, I. V. Komlev, I. P. Konyaeva, A. I. Zavalishina, V. M. Tulchinskii, *J. Gen. Chem. USSR (Engl. Transl.), 43*, 2353 (1973).
23. J. Grobe, W. Mohr, *J. Fluorine Chem., 8*, 341 (1976).
24. K. M. Abraham, J. R. Van Wazer, *Inorg. Chem., 14*, 1099 (1975).
25. V. L. Foss, V. A. Solodenko, I. F. Lutsenko, *J. Gen. Chem. USSR (Engl. Transl.), 49*, 2134 (1979).
26. L. Maier, *Chem. Ber., 94*, 3051 (1961).
27. R. Coelln, G. Schrader, Ger. Pat. 1,056,606 (1959); *Chem. Abstr., 56*, 11,621 (1962).
28. K. A. Pollart, H. J. Harwood, *J. Org. Chem., 27*, 4444 (1962).
29. L. Maier, *Chem. Ber., 94*, 3056 (1961).
30. E. N. Tsvetkov, T. A. Chepaikina, M. I. Kabachnik, *Bull. Acad. Sci. USSR*, 394 (1979).
31. G. W. Parshall, *J. Inorg. Nucl. Chem., 12*, 372 (1960).
32. W. Kuchen, H. Buchwald, K. Strolenberg, J. Metten, *Justus Liebigs Ann. Chem., 652*, 28 (1962).
33. M. Baudler, J. Hellmann, *Z. Anorg. Allg. Chem., 509*, 53 (1984).
34. M. Baudler, J. Hellmann, *Z. Anorg. Allg. Chem., 490*, 11 (1982).
35. G. Fritz, U. Braun, W. Schick, W. Hönle, H. G. v. Schnering, *Z. Anorg. Allg. Chem., 472*, 45 (1981).
36. G. Fritz, W. Schick, *Z. Anorg. Allg. Chem., 511*, 108 (1984).

2.4.10.3.2. to Give As and Sb Halides.

Cyclopolyarsines react with alkyl halides to form arsonium salts (§4.3.8), but the As—As bond in the cyclic $(t\text{-BuAs})_4$ is cleaved[1] with xs MeI:

$$(t\text{-BuAs})_4 + 6\ MeI \rightarrow 2\ t\text{-BuAsI}_2 + 2\ [Me_3(t\text{-Bu})As]I \quad (a)$$

Similarly, the cyclic diarsines react[2] with MeI:

$$(CH_2)_n\langle As(Ph)-As(Ph)\rangle + 2MeI \rightarrow [PhAs(I)(CH_2)_nAs(Me)_2Ph]I \quad \text{(b)}$$

with n = 3, 4.

Diarsines, R_2AsAsR_2, are converted by alkyl halides to arsonium salts[3], although the formation of arsonium salts together with arsinous halides, R_2AsX is observed[4].

A different course of reaction is observed if CF_3I is the halogenating agent. Arsenobenzene, $(PhAs)_6$, reacts at 115°C[5]:

$$(PhAs)_6 \xrightarrow{CF_3I} PhAsI_2 + Ph(CF_3)AsI + Ph(CF_3)_2As \quad \text{(c)}$$

The reaction T may be lowered if UV irradiation is used.

No quaternary salt of As is observed.

Arsinous iodides[6] are formed in reactions of CF_3I with diarsines:

$$Me_2AsAsMe_2 + CF_3I \xrightarrow{RT} Me_2AsI + Me_2(CF_3)As \quad \text{(d)}$$

$$(CF_3)_2AsAs(CF_3)_2 + CF_3I \xrightarrow{75°C} (CF_3)_2AsI + (CF_3)_3As \quad \text{(e)}$$

With xs CF_3I, reaction (d) is nearly complete, whereas reaction (e) gives $(CF_3)_2AsI$ in 60% yield. The latter asseninous iodide is also obtained in 10% yield from $(CF_3)_2AsAs(CF_3)_2$ and MeI:

$$(CF_3)_2AsAs(CF_3)_2 + MeI \xrightarrow[14\,h]{105°C} \underset{(10\%)}{(CF_3)_2AsI} + (CF_3)_2AsMe \quad \text{(f)}$$

The As—As bond also cleaves with SO_2Cl_2[7], e.g.:

$$(PhAs)_6 + 6\,SO_2Cl_2 \rightarrow 6\,PhAsCl_2 + 6\,SO_2 \quad \text{(g)}$$

Similarly, arsinous chlorides are obtained from diarsines and halogenating agents, such as CF_3SCl[8], $SOCl_2$[9] and $COCl_2$[9]; e.g., with CF_3COCl[10]:

$$Me_2AsAsMe_2 + CF_3COCl \rightarrow Me_2AsCl + Me_2AsCOCF_3 \quad \text{(h)}$$

Halocarbons may also serve as a halide source for the fission of the As—As bond[5]:

$$Me_2AsAsMe_2 + CF_2{=}CFBr \rightarrow Me_2AsBr + Me_2AsCF{=}CF_2 \quad \text{(i)}^{5}$$

$$Me_2AsAsMe_2 + X\overline{C{=}CXCF_2C}F_2 \rightarrow Me_2AsX + Me_2As\overline{C{=}CXCF_2C}F_2 \quad \text{(j)}^{5,11}$$

where X = F, Cl.

Transition-metal carbonyl halides react with the As—As bond[12]:

$$(CF_3)_2AsAs(CF_3)_2 + 2\ M(CO)_5X \xrightarrow[\text{ether}]{60\text{–}90°C} (CF_3)_2AsX + M_2(CO)_8As(CF_3)_2X + 2\ CO \quad \text{(k)}$$

where X = Cl, Br, I; M = Mn, Re. The substituted Re octacarbonyl compounds may react further with diarsines[12]:

$$Re_2(CO)_8M(CF_3)_2X + (CF_3)_2AsAs(CF_3)_2 \rightarrow (CF_3)_2AsX + Re_2(CO)_8M(CF_3)_2As(CF_3)_2 \quad \text{(l)}$$

where if M = As, X = Br; if M = P, X = Cl, I.

The cleavage of the As—P bond can also yield As halides:

$$(CF_3)_2AsPH_2 + X_2PI \rightarrow (CF_3)_2AsI + X_2PPH_2 \quad \text{(m)}$$

where X = F[13], CF_3[13];

$$RAs[P(O)(OMe)_2]_2 + 2\ Ph_2PCl \rightarrow RAsCl_2 + 2\ Ph_2PP(O)(OMe)_2 \quad \text{(n)}$$

where R = Me[14], Ph[14].

The Sb—Sb bond in distibines is cleaved with MeI[15,16,18]:

$$R_2SbSbR_2 + MeI \rightarrow R_2SbI + R_2(Me)Sb \quad \text{(o)}$$

where R = alkyl, aryl. This reaction has no preparative value because the products are difficult to separate.

Tetraalkyldistibines are cleaved with SO_2Cl_2 in CH_2Cl_2 to give high yields of chlorides[17]:

$$R_2SbSbR_2 + SO_2Cl_2 \xrightarrow{-78\text{ to }20°C} 2\ R_2SbCl + SO_2 \quad \text{(p)}$$

where R = Me, Et, Pr. Excess SO_2Cl_2 must be avoided because oxidation to trichlorostiboranes occurs (see §2.4.13).

(M. FILD)

1. A. Tzschach, V. Kiesel, *J. Prakt. Chem.*, *313*, 259 (1973).
2. A. Tzschach, G. Pacholke, *Z. Anorg. Allg. Chem.*, *336*, 270 (1965).
3. W. R. Cullen, *Can. J. Chem.*, *41*, 322 (1963).
4. P. Borgstrom, M. M. Dewar, *J. Am. Chem. Soc.*, *44*, 2915 (1922).
5. W. R. Cullen, N. K. Hota, *Can. J. Chem.*, *42*, 1123 (1964).
6. W. R. Cullen, *Can. J. Chem.*, *38*, 439 (1960).
7. L. Anschütz, H. Wirth, *Naturwissenschaften*, *43*, 59 (1956).
8. W. R. Cullen, P. S. Dhaliwal, *Can. J. Chem.*, *45*, 379 (1967).
9. G. T. Morgan, D. C. Vining, *J. Chem. Soc.*, *117*, 777 (1920).
10. W. R. Cullen, G. E. Styan, *Can. J. Chem.*, *44*, 1225 (1966).
11. W. R. Cullen, D. S. Dawson, P. S. Dhaliwal, G. E. Styan, *Chem. Ind.* (*London*), 502 (1964).
12. J. Grobe, W. Mohr, *J. Fluorine Chem.*, *8*, 341 (1976).
13. R. Demuth, J. Grobe, *J. Fluorine Chem.*, *2*, 299 (1972/73).
14. K. M. Abraham, J. R. Van Wazer, *Inorg. Chem.*, *15*, 857 (1976).
15. K. Issleib, B. Hamann, *Z. Anorg. Allg. Chem.*, *339*, 289 (1965).
16. K. Issleib, B. Hamann, *Z. Anorg. Allg. Chem.*, *343*, 196 (1966).
17. H. A. Meinema, H. F. Martens, J. G. Noltes, *J. Organomet. Chem.*, *51*, 223 (1973).
18. H. J. Breuning, H. Jawad, *J. Organomet. Chem.*, *243*, 417 (1983).

2.4.11. by Halide–Halide Exchange (Metathesis) (with or without Concomitant Redox)

2.4.11.1. by Metal Halides

2.4.11.1.1. of Group IA.

Metathesis reactions using metal fluorides are discussed in §2.4.11.3.

Conversion of an Sb—Cl to a Sb—Br bond is accomplished by NaBr; e.g., $Me_3Sb[SC(O)Me]Cl$ with NaBr in refluxing MeOH gives[1] $Me_3Sb[SC(O)Me]Br$:

$$Me_3Sb[S(CO)CH_3]Cl + NaBr \rightarrow Me_3Sb[SC(O)Me]Br + NaCl \tag{a}$$

Analogously, $Me_3Sb[SC(O)Ph]Br$ is obtained[1].

Group VB–I bonds are synthesized by the exchange with alkali iodides; e.g., LiI converts $MePCl_2$ to $MePI_2$ in 40–60% yield[2]. t-Butylphosphorus diiodide is obtained in 93% yield from $t\text{-}BuCl_2$ and LiI in benzene at 20°C[3]:

$$t\text{-}BuPCl_2 + 2\ LiI \rightarrow t\text{-}BuPI_2 + 2\ LiCl \tag{b}$$

Phosphorus pentachloride with LiI forms PI_5 in 90% yield[4].

Sodium iodide converts the As—Cl to the As—I bond; e.g., phenylarsenic dichloride with NaI forms[5,6] $PhAsI_2$:

$$PhAsCl_2 + 2\ NaI \rightarrow PhAsI_2 + 2\ NaCl \tag{c}$$

Arsenic and Sb iodides are obtained from the corresponding oxides by reduction with SO_2 in conc HCl and metathesis with NaI; e.g., methylarsenic diiodide is prepared from $Na_2[MeAsO_3]$, HCl, SO_2 and NaI. Methylarsenic dichloride is formed in situ and converted to $MeAsI_2$ by stoichiometric NaI[11]:

$$Na_2[MeAsO_3] + SO_2 + 2\ HCl \rightarrow MeAsCl_2 + Na_2SO_4 + H_2O \tag{d}$$

$$MeAsCl_2 + 2\ NaI \rightarrow MeAsI_2 + 2\ NaCl \tag{e}$$

Instead of NaI KI is often used; e.g., phenylantimony dichloride with KI with forms KCl and $PhSbI_2$ which, being less soluble than $PhSbCl_2$, precipitates from solution[12].

$$PhSbCl_2 + 2\ KI \rightarrow PhSbI_2 + 2\ KCl \tag{f}$$

TABLE 1. CLEAVAGE OF GROUP VB—Cl BONDS WITH NaI

Group VB chloride	T (°C)	Solvent	Time (h)	Product	Yield (%)	Ref.
$PhAsCl_2$	25	$(CH_3)_2CO$	4	$PhAsI_2$	61	5, 6
$EtAsCl_2$	25	$(CH_3)_2CO$		$EtAsI_2$		6, 7
Me_2AsCl	25	$(CH_3)_2CO$	4	Me_2AsI	58	6
Ph_2AsCl	25	$(CH_3)_2CO$	4	Ph_2AsI	85	6
Ph_2SbCl	25	$(CH_3)_2CO$	12	Ph_2SbI		8, 9
Ph_2BiCl	25	C_2H_5OH	12	Ph_2BiI	65	8, 10
$(Me_2N)_2P(O)Cl$	25	$(CH_3)_2CO$		$(Me_2N)_2P(O)I$	17	

TABLE 2. CLEAVAGE OF GROUP VB CHLORIDES (FROM OXIDES WITH SO_2 OR $SnCl_2$) WITH NaI

Oxide	Reducing Agent	Chloride	Product	Yield (%)	Ref.
$Na_2[MeAsO_3]$	SO_2	$MeAsCl_2$	$MeAsI_2$	78	11
$Na[Me_2AsO_2]$	SO_2	Me_2AsCl	Me_2AsI	50	11
$PhSbO_3H_2$	$SnCl_2$	$PhSbCl_2$	$PhSbI_2$		12

TABLE 3. CLEAVAGE OF GROUP VB CHLORIDES FROM OXIDES WITH $SnCl_2$ OR SO_2 AND HCl WITH KI

Oxide	Reducing agent	Chloride	Product	Ref.
$PhSbO_3H_2$	$SnCl_2$	$PhSbCl_2$	$PhSbI_2$	12
$3\text{-}CNC_6H_4SbO_3H_2$	$SnCl_2$	$3\text{-}CNC_6H_4SbCl_2$	$3\text{-}CNC_6H_4SbI_2$	13
$(3\text{-}ClC_6H_4SbO)_x$		$3\text{-}ClC_6H_4SbCl_2$	$3\text{-}ClC_6H_4SbI_2$	14
$(4\text{-}ClC_6H_4SbO)_x$		$4\text{-}ClC_6H_4SbCl_2$	$4\text{-}ClC_6H_4SbI_2$	15

Phenylantimony dichloride is prepared[12] in situ by reduction of $PhSbO_3H_2$ with $SnCl_2$ in ag HCl. Examples of group-VB—chlorine bonds with KI are given in Table 3.

Direct metathesis of $AsCl_3$ with KI in conc HCl gives AsI_3 in 90% yield[16]:

$$AsCl_3 + 3\ KI \rightarrow AsI_3 + 3\ KCl \qquad (g)$$

Shaking Ph_2BiCl with KI in EtOH for 12 h gives[10] Ph_2BiI.

Organo-group-VB halides are poisons. All operations should be carried out in well-vented hoods by experienced persons.

(H.J. BREUNIG)

1. J. Otera, R. Okawara, *J. Organomet. Chem.*, *17*, 353 (1969).
2. E. A. Mel'nichuk, N. G. Feshchenko, *J. Gen. Chem. USSR* (*Engl. Transl.*), *49*, 1457 (1979).
3. N. G. Feshchenko, E. A. Mel'nichuk, *J. Gen. Chem. USSR* (*Engl. Transl.*), *48*, 329 (1978).
4. N. G. Feshchenko, V. G. Kostina, A. V. Kirsanov, *J. Gen. Chem. USSR* (*Engl. Transl.*), *48*, 196 (1978).
5. G. J. Burrows, E. E. Turner, *J. Chem. Soc.*, *117*, 1373 (1920).
6. W. Steinkopf, G. Schwenn, *Ber. Deutsch. Chem. Ges.*, *54*, 1437 (1921).
7. W. R. Cullen, *Can. J. Chem.*, *39*, 2486 (1961).
8. F. F. Blicke, U. O. Oakdale, F. D. Smith, *J. Am. Chem. Soc.*, *53*, 1025 (1931).
9. F. F. Blicke, U. O. Oakdale, *J. Am. Chem. Soc.*, *55*, 1198 (1933).
10. H. Gilman, H. L. Yablunky, *J. Am. Chem. Soc.*, *63*, 207 (1941).
11. I. T. Millar, H. Heaney, D. M. Heinekey, W. C. Fernelius, *Inorg. Synth.*, *6*, 113, 116, (1960).
12. H. Schmidt, *Liebigs Ann. Chem.*, *421*, 174 (1920).
13. I. G. M. Campbell, *J. Chem. Soc.*, 4 (1947).
14. G. O. Doak, H. H. Jaffé, *J. Am. Chem. Soc.*, *72*, 3025 (1950).
15. A. N. Nesmeyanov, O. A. Reutov, O. A. Ptitsyna, P. A. Tsurkan, *Izv. Akad. Nauk SSR*, 1435 (1958).
16. J. C. Bailar, *Inorg. Synth.*, *1*, 103 (1939).
17. H. J. Vetter, *Z. Naturforsch. Teil B*, *19*, 72 (1964).

2.4.11.1.2. of Other than Group IA.

Exchange of Cl in n-BuOPCl$_2$ with MgBr$_2$ in Et$_2$O gives[1] n-BuOPBr$_2$ in 82% yield. Conversion of the P—Cl to the P—I bond by MgI$_2$ is performed at 20°C in Et$_2$O; e.g., (EtO)$_2$PI is obtained[2] in 62% yield from the corresponding chloride and MgI$_2$:

$$(EtO)_2PCl + MgI_2 \rightarrow (EtO)_2PI + MgICl \tag{a}$$

Chloro- and bromophosphines metathesize with Me$_3$SiI to convert the P—Cl or P—Br to the P—I bond; e.g., Me$_3$SiI with PCl$_3$ in C$_6$H$_6$ at 20°C gives PI$_3$ after 0.5 h in 92% yield[3]:

$$PCl_3 + 3\ Me_3SiI \xrightarrow[0.5\ h]{20°C} PI_3 + 3\ Me_3SiCl \tag{b}$$

Conversion of the group VB—I to the group VB—Cl bond is accomplished by AgCl; e.g., (CF$_3$)$_2$AsCl is obtained quantitatively by shaking (CF$_3$)$_2$AsI with AgCl for 2 days[7]:

$$(CF_3)_2AsI + AgCl \rightarrow (CF_3)_2AsCl + AgI \tag{c}$$

Reaction of (CF$_3$)$_2$SbI with AgCl in a sealed tube for 24 h at RT gives crude (CF$_3$)$_2$SbCl in 80% yield[8].

Phosphorus and As halides are poisonous. All operations should be carried out by experienced personnel using well-ventilated fume hoods.

(H.J. BREUNIG)

TABLE 1. CLEAVAGE[a] OF THE P—Cl BOND WITH MI$_2$

Chloride	Product	Yield (%)	Ref.
(EtO)$_2$PCl	(EtO)$_2$PI	62	2
(n-BuO)$_2$PCl	(n-BuO)$_2$PI	86	2
(i-Pr$_2$)PCl	i-Pr$_2$PI	85	2
n-BuOPCl$_2$	n-BuOPI$_2$	85	2

[a] In Et$_2$O at 20°C.

TABLE 2. CLEAVAGE OF THE P—Cl OR P—Br BOND WITH Me$_3$SiI[3]

Chloride or bromide	T (°C)	Solvent	Time (h)	Product	Yield (%)	Ref.
PCl$_3$	20	C$_6$H$_6$	0.5	PI$_3$	92	3
(Et$_2$N)PCl$_2$	−50	Hexane	0.5	Et$_2$NPI$_2$	94	3
EtPCl$_2$	20	Et$_2$O	1	EtPI$_2$	74	3, 4
t-BuPCl$_2$	−20	Et$_2$O	0.5	t-BuPI$_2$	97	3
Ph$_2$PCl	20	C$_6$H$_6$	1	Ph$_2$PI	77	3, 5
PCl$_5$	20	C$_6$H$_6$	1	PI$_5$	98	3
Ph$_3$PBr$_2$	0	CH$_2$Cl$_2$	1.5	Ph$_3$PI$_2$	86	3, 6

1. Z. S. Novikova, M. M. Kabachnik, A. A. Prishchenko, I. F. Lutsenko, *J. Gen. Chem. USSR* (*Engl. Transl.*), *44*, 1825 (1974).
2. M. M. Kabachnik, A. A. Prishchenko, Z. S. Novikova, I. F. Lutsenko, *J. Gen. Chem. USSR* (*Engl. Transl.*), *49*, 1264 (1979).
3. W. D. Romanenko, V. I. Tovstenko, L. N. Markovskii, *Synthesis*, 823 (1980).
4. K. Issleib, B. Mitschevling, *Z. Naturforsch., Teil B*, *15*, 267 (1960).
5. K. Issleib, W. Seidel, *Chem. Ber.*, *92*, 2681 (1959).
6. A. D. Beveridge, G. S. Harris, D. S. Payne, *J. Chem. Soc. A.*, 726 (1966).
7. W. R. Cullen, N. K. Hota, *Can. J. Chem.*, *42*, 1123 (1964).
8. J. W. Dale, H. J. Emeléus, R. N. Haszeldine, J. H. Moss, *J. Chem. Soc.*, 3708 (1957).

2.4.11.2. by Nonmetal Halides.

Metathesis with nonmetal fluorides are considered in §2.4.11.3. Reactions with group VB halides are discussed in §2.4.11.1.2.

The P—Cl bonds in $PhPCl_2$ are converted[1] to P—Br bonds in 80% yield by HBr in refluxing PBr_3:

$$PhPCl_2 + 2\ HBr \rightarrow PhPBr_2 + 2\ HCl \tag{a}$$

Conversion of P—Cl to P—Br is also achieved by PBr_3 and BBr_3 (Table 1).

From $BiCl_3$ in conc HCl BiI_3 is obtained[7] with conc HI:

$$BiCl_3 + 3\ HI \rightarrow BiI_3 + 3\ HCl \tag{b}$$

The handling of haloid acids should be done in a well-ventilated fume hood. Phenylphosphorus halides are poisonous.

(H.J. BREUNIG)

1. P. Nannelli, G. R. Feistel, T. Moeller, *Inorg. Synth.*, *9*, 73 (1967).
2. W. Kuchen, W. Grünwald, *Chem. Ber.*, *98*, 480 (1965).
3. A. Hinke, W. Kuchen, *Phosphorus, Sulfur*, *15*, 93 (1983).
4. K. Diemert, W. Kuchen, J. Kutter, *Chem. Ber.*, *115*, 1947 (1982).
5. B. A. Arbuzov, V. V. Krupnova, A. O. Vizel, *Izv. Akad. Nauk SSSR, Ser. Khim.*, (*Engl. Transl.*) 1147 (1972).
6. P. M. Druce, M. F. Lappert, *J. Chem. Soc.*, 3595 (1971).
7. P. W. Schenk, in *Handbook of Preparative Inorganic Reactions*, G. Brauer, ed., Academic Press, New York, 1963.

TABLE 1. PHOSPHORUS CHLORIDES WITH BROMIDES

Reactant	Bromide	Yield (%)	T. (°C)	Product	Ref.
$PhPCl_2$	HBr	80	190	$PhPBr_2$	1
$PhPCl_2$	PBr_3	86	190	$PhPBr_2$	2, 3
Ph_2PCl	PBr_3	87	190	Ph_2PBr	3
$EtPCl_2$	PBr_3	70	190	$EtPBr_2$	2
$CH_2(PCl_2)_2$	HBr	98	−85	$CH_2(PBr_2)_2$	4
$PhP(S)Cl_2$	PBr_3	89		$PhP(S)Br_2$	5
$MeP(O)Cl_2$	PBr_3	90		$MeP(O)Br_2$	5
$PhPCl_2$	BBr_3	100	20	$PhPBr_2$	6

2.4.11.3. by Fluorine Derivatives

2.4.11.3.1. of Hydrogen, Ammonium and Potassium.

Hydrogen fluoride converts group VB-halogen bonds into the group VB–F bond. **The handling of HF gas, however, is difficult and hazardous.** Laboratory substitutes for HF are NH_4F and KHF_2, which are solids. The only limitation in their use is the reactivity of the product fluoride toward F^-. They cannot be used to prepare group VB fluorides that are electron-pair acceptors, because the product will be the complex salt[1] instead.

Gaseous HF is used for the conversion of PCl_3 to PF_3 at 50–65°C in an Fe or SiO_2 vessel. The PF_3 is condensed at −183°C in 90% yield[2]:

$$PCl_3 + 3\ HF \rightarrow PF_3 + 3\ HCl \qquad (a)$$

A 90% yield of POF_3 is obtained from HF gas with $POCl_3$ at 65°C with 5 wt % $SbCl_5$ as catalyst[2]. Antimony pentafluoride is prepared by exchange between $SbCl_5$ and HF at 150°C. The yield of SbF_5 is ≤90%[1,3]:

$$SbCl_5 + 5\ HF \rightarrow SbF_5 + 5\ HCl \qquad (b)$$

A mixture of solid NH_4F and PCl_5 heated in a glass tube forms[2] NH_4PF_6:

$$PCl_5 + 6\ NH_4F \rightarrow [NH_4][PF_6] + 5\ NH_4Cl \qquad (c)$$

This operation should be carried out in a well-vented fume hood using safety glasses.

Action of HF on $[MePCl_3][AlCl_4]$ in an autoclave gives[4] $MePF_4$. Hydrogen fluoride and $EtPCl_2$ react in $FCCl_3$ at −30°C to form $EtPF_2$. Subsequent addition of HF gives[5] 85% $EtP(H)F_3$.

Action of CCl_4 and adducts of HF on KF, NH_3, Et_3N or PhNCO yield trialkyldifluoro phosphoranes; e.g. Me_3P reacts with CCl_4 and HF·PhNCO to give[6] Me_3PF_2:

$$Me_3P + CCl_4 + 2\ HF\cdot PhNCO \rightarrow Me_3PF_2 + HCCl_3 + HCl\cdot PhNCO + PhNCO \qquad (d)$$

Exchange of Br in $[Ph_4P]Br$ with an anion exchange resin gives $[Ph_4P][HF_2]$, that forms[7] $[Ph_4P]F$ on heating at 300°C.

Methylarsenic dichloride is converted to $MeAsF_2$ by NH_4F at 80–90°C. The yield of $MeAsF_2$ is 90%[8]:

$$MeAsCl_2 + 2\ [NH_4]F \rightarrow MeAsF_2 + 2\ [NH_4]Cl \qquad (e)$$

Equation (d) is also used[8] to synthesize $EtAsF_2$, ClCH=CHAsF_2 and $PhAsF_2$.

Reaction of Me_2AsCl with NH_4F in the presence of SbF_3 at 85°C gives Me_2AsF in 40% yield[9].

The Bi—Br bonds in $BiBr_3$ are converted to Bi—F bonds by the action of NH_4F; NH_4BiF_4 is formed in quantitative yield[10]:

$$BiBr_3 + 4\ [NH_4]F \xrightarrow{25°C} [NH_4][BiF_4] + 3\ [NH_4]Br \qquad (f)$$

Antimony pentachloride and $K[HF_2]$ react to form[9] $K[SbF_6]$:

$$3\ K[HF_2] + SbCl_5 \rightarrow K[SbF_6] + 2\ KCl + 3\ HCl \qquad (g)$$

Owing to this reaction, $K[HF_2]$ is not suitable for converting $SbCl_5$ to SbF_5[7].

The use of HF as fluorinating agent requires experience and apparatus of special materials. These operations should be carried out by experienced persons.

(H.J. BREUNIG)

1. E. L. Mutterties, C. W. Tullock, *Prep. Inorg. React.*, *2*, 272 (1965).
2. W. Kwasnik, in *Handbook of Preparative Inorganic Chemistry*, G. Brauer, ed., Academic Press, New York, 1963.
3. L. Kolditz, H. Daunicht, *Z. Anorg. Allg. Chem.*, *302*, 230 (1959).
4. H. Coates, P. R. Carter, Brit. 734187 (1955), *Chem. Abstr.*, *50*, 7123 (1956).
5. R. Appel, A. Gilak, *Chem. Ber.*, *108*, 2693 (1975).
6. R. Appel, A. Gilak, *Chem. Ber.*, *107*, 2169 (1974).
7. S. I. Brown, J. H. Clark, *J. Chem. Soc. Chem. Commun.*, 1256 (1983).
8. L. H. Long, H. J. Emeléus, H. V. A. Briscoe, *J. Chem. Soc.*, 1123 (1946).
9. E. G. Claeys, *J. Organomet. Chem.*, *5*, 446 (1966).
10. H. M. Haendler, F. A. Johnson, D. S. Crockett, *J. Am. Chem. Soc.*, *80*, 2662 (1958).

2.4.11.3.2. of Alkali Metals.

Because of their high lattice energies the alkali-metal fluorides are inactive at moderate T in most halogen exchange reactions. At elevated T they participate in exchange, but their utility in the synthesis of binary fluorides is restricted because the alkali-metal fluorides react further to give ternaries[1]. Sodium fluoride with PCl_5 gives[2] $NaPF_6$:

$$6\ NaF + PCl_5 \rightarrow NaPF_6 + 5\ NaCl \quad (a)$$

The activity of alkali-metal fluorides in halogen exchange may be increased by slurrying the fluorides in a polar organic liquid that has an appreciable dielectric constant. Sodium fluoride is the most reactive of the alkali-metal fluorides in such exchanges. Polar solvents, such as nitriles, amides, alcohols, sulfones or nitrobenzene, are used.

Triethylbromophosphonium bromide is converted[3] to Et_3PF_2 by NaF in MeCN:

$$[Et_3PBr]Br + 2\ NaF \rightarrow Et_3PF_2 + 2\ NaBr \quad (b)$$

Reaction of NaF with $PhOP(O)Cl_2$ in acetone at 40–45°C in presence of benz-15-crown-5 gives[10] $PhOP(O)F_2$.

TABLE 1. THE PHOSPHORUS–HALOGEN BOND WITH NaF

Phosphorus halide	T (°C)	Solvent	Yield (%)	Product	Ref.
$[Et_3PBr]Br$		MeCN	62	Et_3PF_2	3
$[n\text{-}Bu_3PBr]Br$		MeCN	85	$n\text{-}Bu_3PF_2$	3
$POCl_3$	215		43	POF_3	4
$PSCl_3$	170		53	PSF_3	4
$PhP(O)Cl_2$	120		65	$PhP(O)F_2$	4
$t\text{-}BuPCl_2$	80–180	Sulfolane[a]	90	$t\text{-}BuPF_2$	5
$t\text{-}Bu_2PCl$	150–180	Sulfolane[a]		$t\text{-}Bu_2PF$	5
Me_2NPCl_2	50–60	Sulfolane	53	Me_2NPF_2	6, 7
$PhPCl_2$		MeCN		$PhPF_2$	8, 9

[a] Sulfolane = tetramethylene sulfone.

Triphenylbismuth difluoride is obtained[11] from Ph_3BiCl_2 with KF in wet EtOH over 3 h.

(H.J. BREUNIG)

1. E. L. Mutterties, C. W. Tullock, *Prep. Inorg. React.*, *2*, 273 (1965).
2. W. Lange, G. v. Krueger, *Chem. Ber.*, *65*, 1253 (1932).
3. R. Bartsch, O. Stelzer, R. Schmutzler, *J. Fluorine Chem.*, *20*, 85 (1982).
4. C. W. Tullock, D. D. Coffman, *J. Org. Chem.*, *25*, 2016 (1960).
5. O. Stelzer, R. Schmutzler, *Inorg. Synth.*, *18*, 173 (1978).
6. J. G. Morse, K. Cohn, R. W. Rudolph, R. W. Parry, *Inorg. Synth.*, *10*, 142 (1967).
7. R. Schmutzler, *Inorg. Chem.*, *3*, 415 (1964).
8. M. Fild, R. Schmutzler, *J. Chem. Soc.*, *A*, 2359 (1970).
9. M. Fild, R. Schmutzler, *J. Chem. Soc.*, *A*, 840 (1969).
10. F. Effenberger, G. König, H. Klenk, *Synthesis*, 70 (1981).
11. F. Challenger, J. F. Wilkinson, *J. Chem. Soc.*, *121*, 91 (1922).

2.4.11.3.3. of Antimony.

Antimony(III) fluoride alone or in combination with an antimony(V) halide replaces Cl by F in group-VB compounds. These systems are versatile, but some exchanges are sluggish, and mixtures of fluorides and halofluorides may be produced.

Reaction of PCl_3 and SbF_3 with an $SbCl_5$ catalyst results in complete or partial replacement of Cl by F and PF_3, PF_2Cl and/or $PFCl_2$ are formed[1–3]. The threshold T of fluorination decreases with each fluorine introduced into the PCl_3 molecule. Under controlled conditions of T and P, such that the unfluorinated liq halide remains at reflux, the fluorinated product distills as it is formed from the reaction zone. The reactions are stepwise, and major yields of mixed halides can be obtained[2–6], e.g.:

$$PCl_3 + SbF_3 \rightarrow PF_3 + SbCl_3 \quad \text{(a)}$$

$$PCl_3 + SbF_3 \rightarrow PCl_2F + SbClF_2 \quad \text{(b)}$$

Examples of exchange using SbF_3 are given in Table 1.

Antimony trifluoride reacts with $MeOPCl_2$ to form MeOPClF (8%) and $MeOPF_2$ (40%)[8]. Action of SbF_3 on $H(Br)PC(CF_3)_2OSiMe_3$ gives[9] $H(F)PC(CF_3)_2OSiMe_3$.

TABLE 1. PHOSPHORUS HALIDES WITH ANTIMONY TRIFLUORIDE

Reactant	T (°C)	Pressure (torr)	Catalyst	Yield (%)	Product	Ref.
PCl_3	25	760	$SbCl_5$	94	PF_3	1, 2, 7
PCl_3	30	280	$SbCl_5$	40	PF_2Cl	3
PCl_3	42	125	$SbCl_5/PCl_5$	40	$PFCl_2$	3, 2, 6
$POCl_3$	75	190	$SbCl_5$	55	POF_3	4
$POCl_3$	75	190	$SbCl_5$	5	POF_2Cl	4
$POCl_3$	75	190	$SbCl_5$	40	$POFCl_2$	4
$POBr_3$	100	25	$SbCl_5$	60	POF_3	5
$POBr_3$	100	25	$SbCl_5$	30	$POFBr_2$	5
$POBr_3$	100	25	$SbCl_5$	10	POF_2Br	5

In many cases SbF_3 reacts not only in halogen exchange, but also as an oxidizing agent. The reaction of $PhPCl_2$ with SbF_3 gives[10] $PhPF_4$ and Sb together with $SbCl_3$:

$$3\ PhPCl_2 + 4\ SbF_3 \rightarrow 3\ PhPF_4 + 2\ Sb + 2\ SbCl_3 \qquad (c)$$

Reaction of t-Bu_2PCl, with SbF_3 gives[11] t-Bu_2PF_3.

Redox is not observed when t-$BuPCl_2$ reacts with SbF_3. Synthesis of t-$BuPF_4$ is carried[13] out by halide exchange of t-$BuPCl_4$ with SbF_3.

Controlled low-T fluorination of PCl_2F_3 with SbF_3 gives[13] $PClF_4$.

Oxidative fluorination of a phosphorus(III) halide can be accomplished by SbF_5. A 79% yield of $ClCH_2PF_4$ is obtained[10] from $ClCH_2PCl_2$ with SbF_5:

$$3\ ClCH_2PCl_2 + 3\ SbF_5 \rightarrow 3\ ClCH_2PF_4 + 2\ SbCl_3 + SbF_3 \qquad (d)$$

Phosphorus fluorides are poisonous. Manipulations must be done in a well-vented hood using special equipment.

(H.J. BREUNIG)

1. G. Tarbutton, E. P. Egan, S. G. Frary, *J. Am. Chem. Soc., 63,* 1982 (1941).
2. H. S. Booth, A. R. Bozarth, *J. Am. Chem. Soc., 61,* 2927 (1939).
3. R. R. Holmes, W. P. Gallagher, *Inorg. Chem., 2,* 433 (1963).
4. H. S. Booth, F. B. Dutton, *J. Am. Chem. Soc., 61,* 2937 (1939).
5. H. S. Booth, C. G. Seegmiller, *J. Am. Chem. Soc., 61,* 3120 (1939).
6. W. Kwasnik, in *Handbook of Preparative Inorganic Chemistry,* G. Brauer, ed., Academic Press, New York, 1963.
7. F. Swarts, *Bull. Soc. Chim. Fr., 35,* 1556 (1924).
8. D. R. Martin, P. J. Pizzolato, *Inorg. Synth., 4,* (1953).
9. H. Kischkel, G.-V. Röschenthaler, *Chem. Ber., 118,* 4842 (1985).
10. R. Schmutzler, *Inorg. Synth., 9,* 64 (1967).
11. R. Schmutzler, O. Stelzer, in *MTP International Review of Science,* Vol 2, *Main-Group Elements: Groups V and VI,* C. C. Addison, D. B. Sowerby eds., *Inorganic Chemistry,* Series 1, H. J. Emeléus, ed., Butterworths, London, 1972.
12. M. Fild, R. Schmutzler, *J. Chem. Soc., A,* 2359 (1970).
13. R. P. Carter, R. R. Holmes, *Inorg. Chem., 4,* 738 (1965).

2.4.11.3.4. of Arsenic(III).

Arsenic(III) fluoride is less active than SbF_3 but is easier to handle being a mobile liquid, bp 62.8°C:

Addition of AsF_3 to PCl_5 at 25°C gives[1] PF_5:

$$3\ PCl_5 + 5\ AsF_3 \rightarrow 3\ PF_5 + 5\ AsCl_3 \qquad (a)$$

Addition of AsF_3 to PCl_5 in $AsCl_3$ as a solvent gives $[PCl_4][PF_6]$ quantitatively[1,2]. Its formation reflects the ionic nature of PCl_5:

$$[PCl_4][PCl_6] + 2\ AsF_3 \rightarrow [PCl_4][PF_6] + 2\ AsCl_3 \qquad (b)$$

Reaction of $[PCl_4][PF_6]$ with AsF_3 at RT is a source of PF_5[2]:

$$3\ [PCl_4][PF_6] + 4\ AsF_3 \rightarrow 6\ PF_5 + 4\ AsCl_3 \qquad (c)$$

Interaction of $SbCl_5$ and AsF_3 gives[5] crystalline $SbCl_4F$.

TABLE 1. METATHESIS OF PHOSPHORUS HALIDES BY AsF_3

Halide	T (°C)	Solvent	Product	Ref.
PCl_3	25		PF_3	3
PCl_5	25		PF_5	1, 2
PCl_5	10	$AsCl_3$	$[PCl_4][PF_6]$	1, 2
PBr_5	30–50	CCl_4	$[PBr_4][PF_6]$	4
$[CH_2P(O)Cl_2]_2$	70	CCl_4	$[CH_2P(O)F_2]_2$	5

Arsenic and phosphorus halides are poisonous. Operations should be carried out in a well-vented fume hood.

(H.J. BREUNIG)

1. W. Kwasnik, in *Handbook of Preparative Inorganic Chemistry*, G. Brauer, ed., Academic Press, New York, 1963.
2. L. Kolditz, *Z. Anorg. Allg. Chem., 284*, 144 (1956).
3. C. J. Hoffman, *Inorg. Synth., 4*, 149 (1953).
4. L. Kolditz, A. Feltz, *Z. Anorg. Allg. Chem., 293*, 155 (1957).
5. W. Althoff, M. Fild, R. Schmutzler, *Chem. Ber., 114*, 1082 (1981).
6. L. Kolditz, *Z. Anorg. Allg. Chem., 302*, 230 (1959).

2.4.11.3.5. of Other Metals.

Calcium fluoride is a poor fluorinating agent because of its low kinetic and thermodynamic reactivity; however, CaF_2 undergoes halogen exchange with PCl_5 at 400°C, yielding PF_5 quantitatively[1]:

$$2\ PCl_5 + 5\ CaF_2 \rightarrow 2\ PF_5 + 5\ CaCl_2 \tag{a}$$

Zinc fluoride is a mild fluorinating agent, used to prepare PF_3[2], $Ph(O)PF_2$ (92%) and $PhPF_4$ (40% yield) spontaneously from the corresponding chlorides[3]. The ZnF_2 synthesis of PF_3 from PCl_3 uses inexpensive, available reagents to give a controlled supply of pure product[2]:

$$2\ PCl_3 + 3\ ZnF_2 \rightarrow 2\ PF_3 + 3\ ZnCl_2 \tag{b}$$

Lead difluoride converts[4] PCl_3 to PF_3:

$$2\ PCl_3 + 3\ PbF_2 \rightarrow 2\ PF_3 + 3\ PbCl_2 \tag{c}$$

Silver(I) fluoride is reactive but does not offer unusual or unique fluorinating action in halogen exchanges. It is used to convert the As—Cl into the As—F bond. Arsenic pentafluoride is obtained[5] from $[AsCl_4][AsF_6]$ and AgF:

$$[AsCl_4][AsF_6] + 4\ AgF \rightarrow 2\ AsF_5 + 4\ AgCl \tag{d}$$

Reaction of Me_3SbBr_2 with AgF in hot H_2O gives AgBr and Me_3SbF_2 in 42% yield[9]:

$$Me_3SbBr_2 + 2\ AgF \rightarrow Me_3SbF_2 + 2\ AgBr \tag{e}$$

Triphenylbismuth difluoride is obtained from Ph_3BiCl_2 and AgF in benzene[10]:

$$Ph_3BiCl_2 + 2\ AgF \rightarrow Ph_3BiF_2 + 2\ AgCl \tag{f}$$

TABLE 1. THE As—Cl BOND WITH AgF

Chloride	T (°C)	Product	Yield (%)	Ref.
$[AsCl_4][AsF_6]$		AsF_5		5
$(CF_3)_3AsCl_2$	20	$(CF_3)_3AsF_2$	52	6
Me_3AsCl_2	60	Me_3AsF_2	45	7
$(PhCH_2)_3AsCl_2$	80	$(PhCH_2)_3AsF_2$	40	8

Phosphorus and As fluorides are poisons. Operations should be carried out in a well-ventilated fume hood.

(H.J. BREUNIG)

1. E. L. Mutterties, T. A. Bither, M. W. Farlow, D. D. Coffmann, *J. Inorg. Nucl. Chem., 16*, 52 (1960).
2. A. A. Williams, *Inorg. Synth., 5*, 95 (1957).
3. E. L. Mutterties, C. W. Tullock, *Prep. Inorg. React., 2*, 277 (1965).
4. A. N. Güntz, *C. R. Hebd. Seances Acad. Sci., 103*, 58 (1886).
5. L. Kolditz, D. Weisz, U. Calov, *Z. Anorg. Allg. Chem., 316*, 261 (1962).
6. H. J. Emeléus, R. N. Haszeldine, E. G. Walaschewski, *J. Chem. Soc.*, 1552 (1953).
7. M. H. O'Brien, G. O. Doak, G. G. Long, *Inorg. Chem. Acta., 1*, 34 (1967).
8. C. G. Moreland, M. H. O'Brien, C. E. Donthit, G. G. Long, *Inorg. Chem., 7*, 834 (1968).
9. G. G. Long, G. O. Doak, L. D. Freedman, *J. Am. Chem. Soc., 86*, 209 (1964).
10. P. G. Goel, H. S. Prasad, *Can. J. Chem., 48*, 2488 (1970).

2.4.11.3.6. of Potassium Sulfinate.

Fluorosulfinates, prepared from KF and xs SO_2, undergoes halogen exchange. The sulfinate is reacted directly with the halide to be fluorinated or is suspended in an inert organic medium, such as chloro- or nitrobenzene. The upper operating limit is set by the thermal decomposition T of $>150°C$, and KSO_2F cannot be used with strong acceptors, such as SbF_5, which give[1] $KSbF_5$:

$$SbF_5 + KSO_2F \rightarrow KSbF_6 + SO_2 \quad (a)$$

Clean halogen exchange occurs between $(Cl_2PN)_4$ and KSO_2F; $(F_2PN)_4$ is obtained in 75% yield[2]:

$$(Cl_2PN)_4 + 8\ KSO_2F \rightarrow (F_2PN)_4 + 8\ KCl + SO_2 \quad (b)$$

TABLE 1. THE P—Cl BOND WITH KSO_2F

Chloride reactant	T (°C)	Product	Yield (%)	Ref.
$(Cl_2PN)_3$	100	$(F_2PN)_3$	65	2
$(Cl_2PN)_4$	80	$(F_2PN)_3$	75	2
Cl_3PS	125	F_3PS	55	3
$PhP(S)Cl_2$	100	$PhP(S)F_2$	50	3

(H.J. BREUNIG)

1. F. Seel, L. Piehl, *Z. Anorg. Allg. Chem.*, *282*, 293 (1955).
2. F. Seel, J. Langer, *Z. Anorg. Allg. Chem.*, *295*, 316 (1958).
3. F. Seel, K. Ballreich, R. Schmutzler, *Chem. Ber.*, *95*, 199 (1962).

2.4.11.3.7. of Other Nonmetals.

Phosphorus trifluoride is prepared by fluorination of PCl_3 with benzoyl fluoride[1]. Addition of $AgBF_4$ to Ph_3SbCl_2 in boiling EtOH gives Ph_3SbF_2 in 71% yield[2].

$$Ph_3SbCl_2 + 2\ AgBF_4 \rightarrow Ph_3SbF_2 + 2\ AgCl + 2\ BF_3 \qquad (a)$$

(H.-J. BREUNIG)

1. F. Seel, K. Ballreich, W. Peters, *Chem. Ber.*, *92*,2117 (1959).
2. G. O. Doak, G. G. Long, C. D. Freedman, *J. Organomet. Chem.*, *4*, 82 (1965).

2.4.12. from Cleavage of Other Element–Group VB Bonds (Including Scrambling)

2.4.12.1. by Halogens.

Addition of Br_2 to arylphosphorus diiodides results in the cleavage of the P—I bond and formation of arylphosphorus dibromides and arylphosphorus tetraiodides; e.g., p-$ClC_6H_4PI_2$ reacts with Br_2 to form p-$ClC_6H_4PBr_2$, in 60% and p-ClC_6H_4PI, in 80% yield[1]:

$$2\ p\text{-}ClC_6H_4PI_2 + Br_2 \rightarrow p\text{-}ClC_6H_4PBr_2 + p\text{-}ClC_6H_4PI_4 \qquad (a)$$

The cleavage of Me_3GePPh_2 with Br_2 gives Ph_2PBr (of little synthetic use)[2].

$$Me_3GePPh_2 + Br_2 \rightarrow Me_3GeBr + Ph_2PBr \qquad (b)$$

(H.J. BREUNIG)

1. T. V. Kovaleva, N. G. Feshchenko, *J. Gen. Chem. USSR* (*Engl. Trasl.*), *49*, 476 (1979).
2. E. H. Brooks, F. Glockling, K. A. Hooton, *J. Chem. Soc.*, *A*, 4283 (1965).

2.4.12.2. by Other Halides from Scrambling.

Phosphorus trimixed halides are formed by redistribution of the pure trihalides. Both $PClBr_2$ and PCl_2Br are formed from PBr_3 and PCl_3. The system reaches equilibrium in 1–1.5 h[1,2]. The products are too unstable to isolate:

$$PBr_3 + PCl_3 \rightleftharpoons PClBr_2 + PCl_2Br \qquad (a)$$

but the fluorine-containing halides are sufficiently stable to isolate[2]:

$$PF_3 + PCl_3 \rightarrow PF_2Cl + PFCl_2 \qquad (b)$$

The reaction of $PhOPCl_2$ with PBr_3 at 170–180°C gives[3] PCl_3 and $PhOPBr_2$.

Halogen exchange equilibria exist between different triorganoantimony dihalides in $CHCl_3$. The position of the equilibrium is T dependent[4,5]. The action of Me_3SbF_2 on Me_3SbCl_2 gives[4] Me_3SbFCl:

$$Me_3SbF_2 + Me_3SbCl_2 \rightleftharpoons 2\ Me_3SbClF \qquad (c)$$

Similarly, $Me_3SbClBr$, Me_3SbClI and Me_3SbBrI are formed[4]. Reaction of Me_3AsF_2 and Me_3AsCl_2 forms[4] Me_3AsFCl.

(H.J. BREUNIG)

1. M. L. Delwaulle, M. Bridoux, *C. R. Hebd. Seances Acad. Sci. 248*, 1342 (1959).
2. D. S. Payne, *Top. Phosphorus Chem.*, *4*, 85 (1967).
3. B. A. Arbusov, V. K. Krupnov, A. O. Vizel, *Izv. Akad. Nauk SSSR, Ser. Khim.* (*Engl. Transl.*); 1147 (1972).
4. G. G. Long, C. G. Moreland, G. O. Doak, M. Miller, *Inorg. Chem.*, *5*, 1358 (1966).
5. C. G. Moreland, M. H. O'Brien, C. D. Douthit, G. G. Long, *Inorg. Chem.*, *7*, 834 (1968).

2.4.13. by Oxidative Halogenation of Low-Valent Group-VB Compounds

2.4.13.1. by Halogens

2.4.13.1.1. to Give the Nitrogen–Halogen Bond.

The nitrosyl halides FNO, ClNO and BrNO are prepared by direct halogenation of NO; e.g., F_2 with NO forms[1] FNO:

$$2\ NO + F_2 \rightarrow 2\ FNO \qquad (a)$$

Addition of Cl_2 to NO givesClNO, and BrNO is obtained[1] from Br_2 and NO.

An exothermic reaction of F_2 and NO_2 in a tube made from Ni gives[2,3] NO_2F:

$$2\ NO_2 + F_2 \rightarrow 2\ FNO_2 \qquad (b)$$

Chlorination of NaN_3 gives[4,5] ClN_3:

$$NaN_3 + Cl_2 \rightarrow ClN_3 + NaCl \qquad (c)$$

When aq NaN_3 is mixed with aq NaOCl underlaid with CCl_4 and acetic acid is added, ClN_3 can be extracted into CCl_4.

Operations should be carried out behind a shield because ClN_3 is explosive.

(H.J. BREUNIG)

1. P. W. Schenk, in *Handbook of Preparative Inorganic Chemistry*, G. Brauer, ed., 2nd. ed., Vol. 1, Academic Press, New York, 1963, p. 457.
2. W. Kwasnick, in *Handbook of Preparative Inorganic Chemistry*, G. Brauer, ed., 2nd. ed., Vol. 1, Academic Press, New York, 1963.
3. O. Ruff, W. Menzel, W. Neumann, *Z. Anorg. Allg. Chem.*, *208*, 298 (1932).
4. P. W. Schenk, in *Handbook of Preparative Inorganic Chemisty*, G. Brauer, ed., 2nd. ed., Vol. 1, Academic Press, New York, 1963, p. 428.
5. W. J. Frierson, J. Konrad, A. W. Browne, *J. Am. Chem. Soc.*, *65*, 1696 (1943).

2.4.13.1.2. to Give the Group VB–Fluorine Bond.

Direct oxidative addition of F_2 to $(PhO)_2PMe$ gives $(PhO)_2PF_2Me$[1]:

$$(PhO)_2PMe + F_2 \rightarrow (PhO)_2PF_2Me \tag{a}$$

TABLE 1. OXIDATIVE ADDITION OF F_2 TO PHOSPHORUS(III) ESTERS

Reactant	Product	Ref.
$(PhO)_2PMe$	$(PhO)_2PF_2Me$	1
$(PhO)_2PF$	$(PhO)_2PF_3$	1
$(MeO)_3P$	$(MeO)_3PF_2$	1
$MeOPPh_2$	$MeOPF_2Ph_2$	1

Addition of F_2 to SbF_3 gives[2,3] SbF_5.

$$SbF_3 + F_2 \rightarrow SbF_5 \tag{b}$$

The synthesis is carried out in an SiO_2 apparatus. Gaseous F_2 is added through a Ni tube to boiling SbF_3, which reacts partially with the formation of flames, and SbF_5 distills[3].

Direct fluorination of BiF_3 with F_2 gives[2,4,5] BiF_5 at 550°C:

$$BiF_3 + F_2 \rightarrow BiF_5 \tag{c}$$

To prepare BiF_5, a fire clay tube is used. The BiF_3 is placed in CaF_2 boats in the tube and F_2 added through a Cu capillary[4,5].

Handling F_2 is hazardous. Special equipment is necessary. Operations with F_2 should be carried out only by experienced persons.

(H.J. BREUNIG)

1. I. Ruppert, *Z. Anorg. Allg. Chem., 477*, 59 (1981).
2. L. Kolditz, *Adv. Inorg. Chem. Radiochem.*, 7, 2 (1965).
3. W. Kwasnik, in *Handbook of Preparative Inorganic Chemistry*, G. Brauer ed., 2nd ed., Vol. 1, Academic Press, New York, 1963, p. 192.
4. W. Kwasnik, in *Handbook of Preparative Inorganic Chemistry*, G. Brauer, ed., 2nd ed., Vol. 1, Academic Press, New York, p. 193.
5. H. v. Wartenberg, *Z. Anorg. Allg. Chem., 244*, 344 (1940).

2.4.13.1.3. to Give the Phosphorus–Chlorine Bond.

Phosphorus(V) chlorofluorides are synthesized in a glass apparatus by condensing xs Cl_2 on the appropriate phosphorus(III) chlorofluoride cooled in liq N_2 and subse-

TABLE 1. PHOSPHORUS(III) CHLOROFLUORIDES WITH Cl_2

Phosphorus(III) reagent	T (°C)	Time (h)	Product	Yield (%)	Ref.
PCl_2F	−78	0.5	PCl_4F		1
$PClF_2$	−78	0.5	PCl_3F_2		1
PF_3	−127	0.5	PCl_2F_3		1, 2, 3
PCl_3	0		PCl_5	100%	4, 5

TABLE 2. ADDITION OF Cl_2 TO ORGANO(HALO)PHOSPHINES

Phosphine reagent	Product	Ref.
$MeP(CF_3)_2$	$MePCl_2(CF_3)_2$	6
$(CF_3)_2NP(CF_3)_2$	$(CF_3)_2NP(CF_3)_2Cl_2$	7
CF_3PF_2	$CF_3PF_2Cl_2$	8
$(CF_3)_2PF$	$(CF_3)_2PFCl_2$	8
Ph_2PCl	Ph_2PCl_3	4
Ph_3P	Ph_3PCl_2	9
$Me_3SiOP[OC(CF_3)_2C(CF_3)_2O]$	$Cl_3P[OC(CF_3)_2C(CF_3)_2O]$	10
$(Me_2N)_3P$	$(Me_2N)_3PCl_2$	11

quent fractionation in vacuo after holding 30 min at a somewhat higher T. Addition of Cl_2 to $PFCl_2$ and reaction at $-78°C$, e.g., gives[1] $PFCl_4$:

$$PFCl_2 + Cl_2 \rightarrow PFCl_4 \qquad (a)$$

Chlorination of organo(halo)phosphines with Cl_2 is a common method of synthesizing chlorophosphoranes, e.g., addition of Cl_2 to $MeP(CF_3)_2$ gives[6] $MePCl_2(CF_3)_2$:

$$MeP(CF_3)_2 + Cl_2 \rightarrow MePCl_2(CF_3)_2 \qquad (b)$$

Phosphorus chlorides are poisons. Operations should be carried out in a well-ventilated fume hood.

(H.J. BREUNIG)

1. R. R. Holmes, W. P. Gallagher, *Inorg. Chem.*, *2*, 433 (1963).
2. W. Kwasnik, in *Handbook of Preparative Inorganic Chemistry*, G. Brauer ed., 2nd ed., Vol. 1, Academic Press, New York, 1963, p. 185.
3. V. Schomaker, J. B. Hatscher, *J. Am. Chem. Soc.*, *60*, 1937 (1938).
4. M. Bermann, J. R. Van Wazer, *Inorg. Synth.*, *15*, 199 (1974).
5. R. N. Maxson, *Inorg. Synth.*, *1*, 99 (1939).
6. K. I. The, R. G. Cavell, *Inorg. Chem.*, *16*, 1463 (1977).
7. H. G. Ang, *J. Inorg. Nucl. Chem.*, *31*, 3311 (1969).
8. J. F. Nixon, *J. Inorg. Nucl. Chem.*, *31*, 1615 (1969).
9. M. F. Ali, G. S. Harris, *J. Chem. Soc., Dalton Trans.*, 1545 (1980).
10. R. Bohlen, G. V. Röschenthaler, *Z. Anorg. Allg. Chem.*, *513*, 199 (1984).
11. H. Nöth, H. J. Vetter, *Chem. Ber.*, *98*, 1981 (1965).

2.4.13.1.4. to Give the As—Cl Bond.

Addition of Cl_2 to As(III) compounds synthesizes As(V) chloro derivatives at low T in inert solvents, such as saturated hydrocarbons, CCl_4, CH_2Cl_2, Et_2O or benzene.

Triphenylarsine reacts with Cl_2 in CCl_4 to form[1,2] Ph_3AsCl_2:

$$Ph_3As + Cl_2 \rightarrow Ph_3AsCl_2 \qquad (a)$$

Addition of Cl_2 to $(CF_3)_3As$ results in oxidation and cleavage of an As—C bond, giving $(CF_3)_2AsCl_3$ in 30% yield after 30 days[8]:

$$(F_3C)_3As + 2\ Cl_2 \rightarrow (F_3C)_2AsCl_3 + F_3CCl \qquad (b)$$

TABLE 1. REACTION OF Cl_2 WITH TERTIARY ARSINES

Arsine	Product	Yield Ref.
Ph_3As	Ph_3AsCl_2	1, 2
Me_3As	Me_3AsCl_2	3
$(PhCH_2)_3As$	$(PhCH_2)_3AsCl_2$	4
Me_2PhAs	$Me_2PhAsCl_2$	5
$(CF_3)_2AsN(SiMe_3)_2$	$[(CF_3)_2(Cl)As{=}NSiMe_3]_2$	6[a]
$(CF_3)_2AsN(SiMe_3)_2$	$(CF_3)_2(Me_3Si)AsCl_2$	7

[a] xs Cl_2.

Diorganoarsenic chlorides react with Cl_2 to form the corresponding trichlorides. The method gives organoarsenic dichlorides from the thermal elimination of the alkyl chloride from the arsenic trichloride. Chlorination of $(C_6H_{11})_2AsCl$ gives $(C_6H_{11})_2AsCl_3$ which is not isolated but converted to $(C_6H_{11})AsCl_2$ by elimination of $C_6H_{11}Cl$[9]:

$$(C_6H_{11})_2AsCl + Cl_2 \xrightarrow{+Cl_2} (C_6H_{11})_2AsCl_3 \xrightarrow{-C_6H_{11}Cl} (C_6H_{11}AsCl_2 \quad \text{(c)}$$

The overall yield of $C_6H_{11}AsCl_2$ is 70%. The reaction is carried out in cold pet. ether.

Reactions of inorganic arsenic halides with Cl_2 are carried out in $AsCl_3$ or Cl_2 as a solvent.

Arsenic pentachloride is formed by addition of Cl_2 to $AsCl_3$ in Cl_2 at $-100°C$. The reaction mixture is irradiated in order to activate Cl_2[10].

$$AsCl_3 + Cl_2 \xrightarrow{h\nu,\ -100°C} AsCl_5 \quad \text{(d)}$$

Arsenic trifluoride and Cl_2 do not react in the usual way. With traces of water or $AsCl_3$ as catalysts $[AsCl_4][AsF_6]$ is formed. The catalytic activity of $AsCl_3$ is associated with the chorinating action of an $AsCl_3$–Cl_2 mixture[11,12].

$$2\ AsF_3 + 2\ Cl_2 \rightarrow [AsCl_4][AsF_6] \quad \text{(e)}$$

Arsenic halides and organoarsenic halides are extremely poisonous. All the handling should be done in a well-vented hood.

(H.J. BREUNIG)

1. A. D. Beveridge, G. S. Harris, *J. Chem. Soc.*, 6076 (1964).
2. A. Michaelis, *Justus Liebigs Ann. Chem.*, *321*, 141 (1902).
3. M. H. O'Brien, G. O. Doak, G. Long, *Inorg. Chim. Acta*, *1*, 34 (1967).
4. C. G. Moreland, M. H. O'Brien, C. E. Douthit, G. G. Long, *Inorg. Chem.*, *7*, 834 (1968).
5. W. Steinkopf, W. Mieg, *Ber. Deutsch. Chem. Ges.*, *53*, 1016 (1920).
6. H. W. Roesky, R. Bohra, W. S. Sheldrick, *J. Fluorine Chemistry*, *22*, 199 (1983).
7. R. Bohra, H. W. Roesky, *J. Fluorine Chemistry*, *25*, 145 (1984).
8. H. J. Emeléus, R. N. Haszeldine, E. G. Walaschewski, *J. Chem. Soc.*, 1552 (1953).
9. W. Steinkopf, H. Dudeck, S. Schmidt, *Ber. Deutsch. Chem. Ges.*, *61*, 1911 (1928).
10. K. Seppelt, *Z. Anorg. Allg. Chem.*, *434*, 5 (1977).
11. H. M. Dessy, R. W. Parry, G. L. Vidale, *J. Am. Chem. Soc.*, *78*, 5730 (1956).
12. L. Kolditz, *Z. Anorg. Allg. Chem.*, *289*, 128 (1957).

2.4.13.1.5. to Give the Sb– and Bi–Chlorine Bond.

Both inorganic and organic Sb(V) and Bi(V) chlorides are obtained by direct chlorination of Sb(III) or Bi(III) compounds with Cl_2. These oxidative additions of Cl_2 are performed in inert organic solvents, e.g., pet. ether, or in melts of the reagent. Although good yields are obtained, the method has the disadvantage of handling Cl_2.

Reaction of a $SbCl_3$ melt with Cl_2 gives $SbCl_5$ in 85% yield. The pentachloride is isolated by distillation at 14 torr[1].

$$SbCl_3 + Cl_2 \rightarrow SbCl_5 \tag{a}$$

A quantitative yield of $SbCl_2F_3$ is obtained when Cl_2 is added to SbF_3 in a steel cylinder. Chlorine is quickly absorbed by SbF_3[2]:

$$SbF_3 + Cl_2 \rightarrow SbF_3Cl_2 \tag{b}$$

Oxidation of tertiary stibines with Cl_2 gives good yields of triorganoantimony dichlorides; e.g., Me_3SbCl_2 is obtained from Me_3Sb and Cl_2 in Et_2O at $-20°C$[3]:

$$Me_3Sb + Cl_2 \rightarrow Me_3SbCl_2 \tag{c}$$

TABLE 1. OXIDATIVE ADDITION OF Cl_2 TO TERTIARY STIBINES

Sb reagent	Product	Ref.
Me_3Sb	Me_3SbCl_2	3-5
$(CF_3)_3Sb$	$(CF_3)_3SbCl_2$	6
Pr_3Sb	Pr_3SbCl_2	7
Ph_3Sb	Ph_3SbCl_2	8, 9

Examples of the chlorination of tertiary stibines are given in Table 1.

Diorganoantimony chlorides react with Cl_2 in inert solvents at low T to form the corresponding trichlorides. Dimethylantimony trichloride is obtained[4,10] in 68% yield by addition of Cl_2 to Me_2SbCl in CS_2:

$$Me_2SbCl + Cl_2 \rightarrow Me_2SbCl_3 \tag{d}$$

Addition of Cl_2 to Ph_2SbCl in Et_2O gives[11,12] Ph_2SbCl_3.

At 0°C, Cl_2 cleave the Bi—C bonds of trialkylbismutines. The stable triarylbismuth-dichlorides are obtained from triarylbismutines and Cl_2. Oxidative chlorination of tertiary arylbismuthines is preparative method; e.g., triphenylbismuth dichloride is obtained from Cl_2 and $BiPh_3$ in pet ether[13]:

$$Ph_3Bi + Cl_2 \rightarrow Ph_3BiCl_2 \tag{e}$$

Examples of oxidative chlorination of triarylbismuthines are given in Table 2.

Antimony and Bi halides are poisons. Operations must be carried out in a well-ventilated hood.

TABLE 2. OXIDATIVE CHLORINATION OF TRIARYLBISMUTHINES

Bi reagent	Solvent	Product	Ref.
Ph_3Bi	Pet. ether	Ph_3BiCl_2	13–15
$(o\text{-}CH_3C_6H_4)_3Bi$	$CHCl_3$	$(o\text{-}CH_3C_6H_4)_3BiCl_2$	14
$(1\text{-}C_{10}H_7)Ph_2Bi$	$CHCl_3$	$(1\text{-}C_{10}H_7)Ph_2BiCl_2$	16
$(p\text{-}ClC_6H_4)_3Bi$	Pet. ether/$CHCl_3$	$(p\text{-}ClC_6H_4)_3BiCl_2$	17

(H.J. BREUNIG)

1. P. W. Schenk, in *Handbook of Preparative Inorganic Chemistry*, G. Brauer ed., 2nd. ed., Vol. 1, Academic Press, New York, 1960, p. 544.
2. W. Kwasnik in *Handbook of Preparative Inorganic Chemistry*, G. Brauer ed., 2nd ed., Vol. 1, Academic Press, New York, 1963, p. 192.
3. G. O. Doak, G. G. Long, M. E. Key, *Inorg. Synth., 9*, 92 (1967).
4. G. T. Morgan, G. R. Davies, *Proc. R. Soc. London, Ser. A, 110*, 523 (1926).
5. G. G. Long, G. O. Doak, L. D. Freedman, *J. Am. Chem. Soc., 86*, 209 (1964).
6. H. J. Emeléus, J. H. Moss, *Z. Anorg. Allg. Chem., 282*, 24 (1955).
7. W. J. C. Dyke, W. J. Jones, *J. Chem. Soc.*, 1921 (1930).
8. A. Michaelis, A. Reese, *Justus Liebigs Ann. Chem., 233*, 49 (1886).
9. H. H. Willard, L. R. Perkins, F. F. Blicke, *J. Am. Chem. Soc., 70*, 737 (1948).
10. O. J. Scherer, P. Hornig, M. Schmidt, *J. Organomet. Chem., 6*, 259 (1966).
11. A. Michaelis, A. Günther, *Ber. Deutsch. Chem. Ges., 44*, 2319 (1911).
12. G. T. Morgan, F. M. Micklethwait, *J. Chem. Soc., 99*, 2292 (1911).
13. A. Michaelis, A. Polis, *Ber. Deutsch. Chem. Ges., 20*, 55 (1887).
14. A. Michaelis, A. Marquardt, *Justus Liebigs Ann. Chem., 251*, 323 (1889).
15. K. A. Jensen, *Z. Anorg. Allg. Chem., 250*, 257 (1943).
16. F. Challenger, J. F. Wilkinson, *J. Chem. Soc., 121*, 91 (1922).
17. F. Challenger, L. R. Ridgway, *J. Chem. Soc., 121*, 104 (1922).

2.4.13.1.6. to Give the Phosphorus–Bromine Bond.

Addition of Br_2 to phosphorus(III) compounds forms the corresponding bromides. If the product is thermally unstable, low T is required; otherwise reactions are carried out at RT in inert solvents. Bromine offers many advantages, and, therefore, no substitute is required.

Bromination of PBr_2 with liq Br_2, gives[1] 80% of PBr_5. The mixture is cooled with H_2O:

$$PBr_3 + Br_2 \rightarrow PBr_5 \qquad (a)$$

Action of Br_2 on PF_3 gives PBr_2F_3, which decomposes at 15°C[2,3]. Passing Br_2 vapor in a N_2 gas stream through liq PBr_2F at −75°C gives PBr_4F which decomposes at RT[3]:

$$PBr_2F + Br_2 \rightarrow PBr_4F \qquad (b)$$

Reaction of Br_2 with $FP(CF_3)_2$ gives[4] $FBr_2P(CF_3)_2$.

Addition of Br_2 to tertiary phosphines synthesizes tertiary phosphine dihalides; e.g., Br_2 in benzene reacts with Et_3P in $Et_2O\text{–}C_6H_6$ to form Et_3PBr_2 in 87% yield[5].

Action of Br_2 on $(EtO)_3P$ at $-100°C$ gives[6] $(EtO)_2P(O)Br$ and EtBr. The analogous bromination of $(Me_3SiO)_3P$ at $-110°C$ yields[6] $(Me_3SiO)_2P(O)Br$ and Me_3SiBr. Addition of Br_2 to Na(18)crown-6- $P(CN)_2$ gives[7] Na(18)crown-6-$P(CN)_2Br_2$.

Oxidative conversion of P—P to P—Br bonds is accomplished from diphosphines or polyphosphines with Br_2, e.g., selective cleavage of one P—P bond in $(t\text{-}BuP)_3$ gives[8] t-BuP[P(t-Bu)Br]$_2$:

$$(t\text{-}BuP)_3 + Br_2 \rightarrow BrP(t\text{-}Bu)P(t\text{-}Bu)P(t\text{-}Bu)Br \qquad (c)$$

Addition of Br_2 to $Me_2P(S)P(S)Me_2$ in CCl_4 at $<30°C$ gives $Me_2P(S)Br$ in 85–92% yield[8]:

$$Me_2\overset{S}{P}—\overset{S}{P}Me_2 + Br_2 \rightarrow 2\ Me_2\overset{S}{P}Br \qquad (d)$$

Phosphorus bromides are poisons. Experimental work must be carried out in a well-ventilated fume hood.

(H.J. BREUNIG)

1. R. N. Maxson, *Inorg. Synth.*, *1*, 99 (1939).
2. H. Moissan, *C. R. Hebd. Seances Acad. Sci.*, *100*, 1348 (1885).
3. L. Kolditz, K. Bauer, *Z. Anorg. Allg. Chem.*, *302*, 241 (1959).
4. K. I. The, R. G. Cavell, *Inorg. Chem.*, *16*, 1463 (1977).
5. K. Issleib, W. Seidel, *Z. Anorg. Allg. Chem.*, *288*, 201 (1956).
6. J. Michalski, M. Pakulski, A. Skowronska, *J. Chem. Soc.* (*Perkin I*), 833 (1980).
7. A. Schmidpeter, F. Zwaschka, *Angew. Chem.*, *Int. Ed. Engl. 19*, 411 (1979).
8. M. Baudler, J. Hellmann, *Z. Anorg. Allg. Chem.*, *480*, 129 (1981).
9. R. Schmutzler, *Inorg. Synth.*, *12*, 287 (1970).

2.4.13.1.7. to Give the As—Br Bond.

Bromine adds to tertiary arsines in CH_3CN, CCl_4 or other inert solvents to form the corresponding triorganoarsenic dibromides. Exactly stoichiometric Br_2 must be used; otherwise triorganoarsenic tetrahalides form. Excess Br_2 with Me_3As in Et_2O yields[1] a precipitate of red Me_3AsBr_4:

$$Me_3As + 2\ Br_2 \rightarrow Me_3AsBr_2 \cdot Br_2 \qquad (a)$$

Equimolar Br_2 and Ph_3As in CCl_4 or CH_3CN give[2,3] Ph_3AsBr_2:

$$Ph_3As + Br_2 \rightarrow Ph_3AsBr_2 \qquad (b)$$

Examples are given in Table 1.

TABLE 1. ADDITION OF Br_2 TO TERTIARY ARSINES

Arsine	T (°C)	Solvent	Product	Ref.
Ph_3As	10	CCl_4–MeCN	Ph_3AsBr_2	2, 3
Pr_3As		CCl_4	Pr_3AsBr_2	4
$(Me_3SiCH_2)_3As$		Pet. ether	$(Me_3SiCH_2)_3AsBr_2$	5
Ph_2MeAs			$Ph_2MeAsBr_2$	6

Diorganoarsenic tribromides are synthesized most easily by adding Br_2 to diorganoarsenic bromides in nonpolar solvents, such as Et_2O, CCl_4 or C_6H_6; e.g., Br_2 with Ph_2AsBr gives Ph_2AsBr_3, which adds Br_2 to form[7] the perbromide $Ph_2AsBr_3 \cdot Br_2$.

$$Ph_2AsBr \xrightarrow{Br_2} Ph_2AsBr_3 \xrightarrow{Br_2} Ph_2AsBr_3 \cdot Br_2 \quad (c)$$

Arsenic bromides are toxic. Operations should be carried out in a well-ventilated hood.

(H.J. BREUNIG)

1. A. Hantzsch, H. Hibbert, *Ber. Deutsch. Chem. Ges., 40*, 1508 (1907).
2. A. Michaelis, *Justus Liebigs Ann. Chem., 321*, 141 (1902).
Æ. A. D. Beveridge, G. S. Harris, *J. Chem. Soc.*, 6076 (1964).
4. W. J. C. Dyke, G. Davies, W. J. Jones, *J. Chem. Soc.*, 185 (1931).
5. D. Seyferth, *J. Am. Chem. Soc., 80*, 1336 (1958).
6. A. N. Nesmayanov, A. E. Borisov, A. I. Borisova, *Izv. Akad. Nauk USSR*, 1199 (1962).
7. C. P. A. Kappelmeier, *Recl. Trav. Chim. Pays-Bas, 49*, 57 (1930).

2.4.13.1.8. to Give the Sb– or Bi–Br Bond.

Addition of Br_2 to $SbBr_3$ fails to give $SbBr_5$, $[NH_4]_2SbBr_6$ being formed instead quantitatively in the presence of NH_4Br in conc HBr as solvent[1]:

$$SbBr_3 + \tfrac{1}{2} Br_2 + 2\ NH_4Br \rightarrow [NH_4]_2SbBr_6 \quad (a)$$

Oxidative bromination of tertiary stibines with Br_2 synthesizes triorgano dibromides. Excess Br_2 must be avoided; otherwise perbromides are formed as in the bromination of tertiary arsines (see §2.4.13.7). The reactions are carried out in Et_2O, pet. ether or CH_2Cl_2. When CCl_4 is used as solvent, Sb chlorides may be formed as byproducts of the bromides owing to halogen exchange with the solvent. Adding Br_2 to Me_3Sb in Et_2O is the best method[2] for Me_3SbBr_2.

$$Me_3Sb + Br_2 \rightarrow Me_3SbBr_2 \quad (b)$$

Examples of tertiary stibines with Br_2 are given in Table 1.

Reaction of Br_2 with Ph_2SbBr in CCl_4 at RT or in CH_2Cl_2 at $-97°C$ gives[6,7] Ph_2SbBr_3:

$$Ph_2SbBr + Br_2 \rightarrow Ph_2SbBr_3 \quad (c)$$

Addition of Br_2 to Ph_2SbCl at $-196°C$ in CH_2Cl_2 gives Ph_2SbBr_2Cl in 85% yield[6].

TABLE 1. TERTIARY STIBINES WITH Br_2

Stibine	T (°C)	Solvent	Product	Yield (%)	Ref.
Me_3Sb	25	Et_2O	Me_3SbBr_2	80	2
Et_3Sb	25	Et_2O	Et_3SbBr_2	90	3
$i\text{-}Pr_3Sb$	−30	C_5H_{12}	$i\text{-}Pr_3SbBr_2$	99	4
Ph_3Sb	25		Ph_3SbBr_2	quant.	5

Oxidative cleavage of the Sb—Sb bond in distibines with Br_2 forms pure diorganoantimony bromides; e.g., Me_4Sb_2 reacts with Br_2 in benzene to give Me_2SbBr quantitatively[8,9].

$$Me_2Sb—SbMe_2 + Br_2 \rightarrow 2\ Me_2SbBr \tag{d}$$

Oxidative addition of Br_2 to trialkylbismuthines fails to form stable Bi(V) derivatives. Trimethylbismuth dibromide is an intermediate in Me_3Bi with Br_2, which gives Me_2BiBr and MeBr as products[10]. Bromination of triarylbismuthines, however, forms stable Bi(V) derivatives. Triphenylbismuth dibromide is obtained quantitatively by reaction of Ph_3Bi with Br_2 in pet. ether[11,12]:

$$Ph_3Bi + Br_2 \rightarrow Ph_3BiBr_2 \tag{e}$$

TABLE 2. TERTIARY BISMUTHINES WITH Br_2

Bismuthine	Product	Ref.
Ph_3Bi	Ph_3BiBr_2	10
$(o\text{-}CH_3C_6H_4)_3Bi$	$(o\text{-}CH_3C_6H_4)_3BiBr_2$	11
$(m\text{-}CH_3C_6H_4)_3Bi$	$(m\text{-}CH_3C_6H_4)_3BiBr_2$	12
$(p\text{-}ClC_6H_4)_3Bi$	$(p\text{-}ClC_6H_4)_3BiBr_2$	13
$(p\text{-}CH_3OC_6H_4)_3Bi$	$(p\text{-}CH_3OC_6H_4)_3BiBr_2$	11

Organoantimony and Bi bromides are toxic. Operations should be carried out in an inert atmosphere in dry solvents in an effective hood.

(H.J. BREUNIG)

1. P. W. Schenk, in *Handbook of Preparative Inorganic Chemistry*, G. Brauer, ed., 2nd ed., Vol. 1, Academic Press, New York, 1963, p. 547.
2. G. O. Doak, G. G. Long, M. E. Key, *Inorg. Synth.*, *9*, 92 (1967).
3. K. Issleib, B. Hamann, *Z. Anorg. Allg. Chem.*, *339*, 289 (1965).
4. H. J. Breunig, W. Kanig, *Phosphorus Sulfur*, *12*, 149 (1982).
5. A. Michaelis, A. Reese, *Justus Liebigs Ann. Chem.*, *233*, 39 (1886).
6. S. P. Bone, D. B. Sowerby, *J. Chem. Soc., Dalton Trans.*, 715 (1979).
7. T. Severengiz, H. J. Breunig, *Chem.-Zeit.*, *104*, 202 (1980).
8. F. A. Paneth, H. Loleit, *J. Chem. Soc.*, 366 (1935).
9. H. J. Breunig, H. Jawad, *J. Organomet. Chem.*, *243*, 417 (1983).
10. A. Marquardt, *Ber. Deutsch. Chem. Ges.*, *20*, 1516 (1887).
11. A. Gillmeister, *Ber. Deutsch. Chem. Ges.*, *30*, 2843 (1897).
12. F. Challenger, F. Pritchard, J. R. A. Jinks, *J. Chem. Soc.*, *125*, 864 (1924).
13. F. Challenger, L. R. Ridgway, *J. Chem. Soc.*, *121*, 104 (1922).

2.4.13.1.9. to Give the Group VB–Iodine Bond.

Elemental I_2 forms group VB-iodine bonds by oxidative addition to trivalent group-VB compounds or by cleavage of element–element bonds. Iodinations are carried out in Et_2O or benzene at RT.

Treatment of Me_3P with I_2 in C_6H_6 gives[3] Me_3PI_2. Iodination of Ph_3P gives[4] Ph_3PI_2. Action of I_2 on t-Bu_3P gives[5] t-Bu_3PI_2, that is formulated as [t-Bu_3PI]I or t-Bu_3P-I-I.

Addition of I_2 in benzene to t-BuPI$_2$ at 20°C gives of t-BuPI$_4$ in 93% yield[1]:

$$\text{t-BuPI}_2 + I_2 \rightarrow \text{t-BuPI}_4 \quad (a)$$

Cleavage of a P—P bond in (t-BuP)$_3$ by I_2 gives[2] t-BuP[P(t-Bu)I]$_2$:

$$(\text{t-BuP})_3 + I_2 \rightarrow \text{I(t-BuP)}_3\text{I} \quad (b)$$

Addition of I_2 to tertiary arsines and stibines gives the corresponding triorgano-element dihalides. Iodine and Ph_3As in dry pet. ether form[6] Ph_3AsI_4, which precipitates first, and Ph_3AsI_2[7]:

$$Ph_3As \xrightarrow{I_2} Ph_3AsI_2 \xrightarrow{I_2} Ph_3AsI_2 \cdot I_2 \quad (c)$$

TABLE 1. ADDITION OF IODINE TO TERTIARY ARSINES AND STIBINES

Reagent	Solvent	T (°C)	Product	Ref.
Ph_3As	pet. ether	25	$Ph_3AsI_2 \cdot I_2$	6
Ph_3As	pet. ether	25	Ph_3AsI_2	7
Me_3Sb	Et_2O	25	Me_3SbI_2	8

(H.J. BREUNIG)

1. N. G. Feshchenko, E. A. Mel'nichuk, *J. Gen. Chem. USSR (Engl. Transl.)*, *48*, 329 (1978).
2. M. Baudler, J. Hellman, *Z. Anorg. Allg. Chem.*, *480*, 129 (1981).
3. J. Goubeau, R. Baumgärtner, *Z. Elektrochem.*, *64*, 598 (1960).
4. A. B. Beveridge, G. S. Harris, F. Inglis, *J. Chem. Soc., A.*, 520 (1966).
5. W.-W. du Mont, M. Bätcher, S. Pohl, W. Saak, *Angew. Chem., Int. Ed. Engl.*, *26*, 912 (1987).
6. A. Michaelis, *Justus Liebigs Ann. Chem.*, *321*, 141 (1902).
7. A. D. Beveridge, G. S. Harris, *J. Chem. Soc.*, 6076 (1964).
8. G. O. Doak, G. G. Long, M. E. Key, *Inorg. Synth.*, *9*, 92 (1967).

2.4.13.2. by Hydrogen Halides.

Chlorination of $(CH_3)_3Sb$ in a sealed tube with HCl gives[1,2] $(CH_3)_3SbCl_2$.

$$(CH_3)_3Sb + 2\,HCl \rightarrow (CH_3)_3SbCl_2 + H_2 \quad (a)$$

Conversion of $MePCl_2$ with HF and H_2O at -5°C in Et_2O gives[3] 47% of MeP(O)(H)F. Cleavage of the P—P bond in diphosphanes with HCl gives[4] P—Cl; e.g. $Ph_2PP(NMe_2)_2$ with HCl gives Ph_2PCl.

(H.J. BREUNIG)

1. A. B. Burg, L. R. Grant, *J. Am. Chem. Soc.*, *81*, 1 (1959).
2. H. Landolt, *J. Prakt. Chem.*, *84*, 328 (1861).
3. U. Ahrens, H. Falius, *Chem. Ber.*, *105*, 3317 (1972).
4. H. J. Vetter, H. Nöth, *Chem. Ber.*, *96*, 1816 (1963).

2.4.13.3. by Other Halides

2.4.13.3.1. to Give the Fluorine–Group-VB Bond.

Fluorination of NO with AgF_2 gives[1] FNO:

$$NO + AgF_2 \rightarrow FNO + AgF \tag{a}$$

Fluorination of Ph_3P with NF_3 or perfluoropiperidine gives[2] Ph_3PF_2. Difluordiazirine reacts with trivalent organophosphorus derivatives to form N-cyano phosphorus imides and difluorophosphoranes; e.g. Ph_3P with F_2CN_2 gives[3] Ph_3PF_2 and $Ph_3P{=}NC{\equiv}N$. Reaction of Carbonyldifluoride with Ph_3P in CH_2Cl_2 gives[4] 80% of Ph_3PF_2 and CO. Addition of perfluoropropene, $F_2C{=}CFCF_3$ to Me_3P gives[5] 83% $Me_3P(F)CF{=}CFCF_3$. Hexafluoromolybdenum with n-Bu_3P forms[6] n-Bu_3PF_2. An exothermic reaction results from AsF_3 with Ph_2PCl to give Ph_2PF_3 in 72% yield. The mixture should be cooled with ice water to keep T $\leq 50°C$[7]:

$$3\ Ph_2PCl + 3\ AsF_3 \rightarrow 3\ Ph_2PF_3 + 2\ As + AsCl_3 \tag{b}$$

Fluorination of $Me_2P(S)P(S)Me_2$ with SbF_3 at 80–250°C gives Me_2PF_3 in 76% yield. The SbF_3 is partially reduced to Sb and partially transformed to Sb_2S_3[8]:

$$3\ Me_2\overset{S}{P}{-}\overset{S}{P}Me_2 + 6\ SbF_3 \rightarrow 6\ Me_2PF_3 + 2\ Sb + 2\ Sb_2S_3 \tag{c}$$

The system must be protected from moisture[8].

The fluorophosphoranes $CF_2{=}CFPF_4$ and $(CF_2{=}CF)_2PF_3$ are prepared[9] by oxidative fluorination of the chlorophosphines $CF_2{=}CFPCl_2$ and $(CF_2{=}CF)_2PCl$, respectively, with SbF_5.

From Bu_3P with $CF_3(F)C{=}CF_2$ $CF_3(F)C{=}C(F)PBu_3F$ is formed[10]:

$$(F_3C)(F)C{=}C(F)(F) + n\text{-}Bu_3P \rightarrow (F_3C)(F)C{=}C(F)(FP(Bu\text{-}n)_3) \tag{d}$$

Phosphorus(III) fluoride and cis-N_2F_2 react in a stainless steel reactor at RT to form PF_5 and N_2. Most of the PF_3 is converted[11] to PF_5:

$$PF_3 + N_2F_2 \xrightarrow{17\ h} PF_5 + N_2 \tag{e}$$

Fluorination of Ph_3As with SF_4 in benzene at 60–130°C in a steel reactor with high-pressure equipment gives[12] Ph_3AsF_2 in 46% yield:

$$2\ Ph_3As + SF_4 \rightarrow 2\ Ph_3AsF_2 + S \tag{f}$$

Oxidative fluorination of Ph_3Sb with XeF_2 in CH_2Cl_2 at 25°C gives Ph_3SbF_2 in 95% yield[13]:

$$Ph_3Sb + XeF_2 \rightarrow Ph_3SbF_2 + Xe \tag{g}$$

Group-VB fluorides are toxic.

(H.J. BREUNIG)

1. K. Jones, in *Comprehensive Inorganic Chemistry*, Vol. 2, Trotman-Dickenson, ed., Pergamon Press, Oxford, 1973, p. 300.
2. R. E. Banks, R. N. Haszeldine, R. Hatton, *Tetrahedron Lett.*, 3993 (1967).
3. R. A. Mitsch, *J. Amer. Chem. Soc.*, *89*, 6297 (1967).
4. O. D. Gupta, J. M. Shreeve, *J. Chem. Soc., Chem. Commun.*, 416 (1984).
5. U. v. Allwörden, G.-V. Röschenthaler, *Chemiker-Zeitung*, *109*, 81 (1985).
6. F. Mathey, J. Bensoam, *C. R. Acad. Sci. 274 C*, 1095 (1972); *Chem. Abstr.*, *76*, 127096 (1972).
7. R. Schmutzler, *Inorg. Synth.*, *9*, 64 (1967).
8. R. Schmutzler, *Inorg. Synth.*, *9*, 67 (1967).
9. K. Ramaswamy, B. Krishna Rao, *Z. Phys. Chem.*, *242*, 18 (1969).
10. R. G. Cavell, J. A. Gibson, K. I. The, *J. Am. Chem. Soc.*, *99*, 7841 (1977).
11. M. Lustig, *Inorg. Chem.*, *4*, 104 (1965).
12. W. C. Smith, *J. Am. Chem. Soc.*, *82*, 6176 (1960).
13. L. M. Yagupol'skii, V. I. Popov, N. V. Kontratenko, B. L. Korsunskii, N. N. Aleinikov, *Russ. J. Inorg. Chem.* (*Engl. Transl.*), *11*, 459 (1975).

2.4.13.3.2. to Give the Chlorine–Group-VB Bond.

Hexachlorethane with tertiary phosphines, such as Me_3P or Ph_3P, forms in refluxing CH_3CN Me_3PCl_2 or Ph_3PCl_2 quantitatively. Compared with Cl_2, hexachloroethane is a mild and easy to handle chlorinating agent. The byproduct, tetrachloroethylene, is easily separated from the products and the solvent[1]:

$$Me_3P + C_2Cl_6 \rightarrow Me_3PCl_2 + Cl_2C{=}CCl_2 \tag{a}$$

Action of CCl_4 on Me_3P forms[2] Me_3PCl_2 and $[Me_3P{-}C(Cl){=}PMe_3]Cl$.

$$3\ Me_3P + CCl_4 \rightarrow Me_3PCl_2 + [Me_3P{-}\overset{\overset{\displaystyle Cl}{|}}{C}{=}PMe_3]Cl \tag{b}$$

Action of CCl_4 on $(Me_3Si)_2NP(i\text{-}Pr)_2$ at 25°C gives[3] $Me_3SiN{=}P(Cl)(i\text{-}Pr)CMe_2SiMe_3$ and $CHCl_3$.

Ring cleavage of $(t\text{-}BuP)_3$ with PCl_5 gives[4] $ClP(t\text{-}Bu)P(t\text{-}Bu)P(t\text{-}Bu)Cl$.

Another substitute for Cl_2 is SO_2Cl_2, which is easy to store and handle. The byproduct of chlorinations with SO_2Cl_2 is SO_2, which evaporates from the mixture. Action of SO_2Cl_2 on $(Me_2PS)_2$ at 20°C gives[5] 86% of Me_2PSCl and SO_2.

$$Me_2(S)PP(S)Me_2 + SO_2Cl_2 \rightarrow 2\ Me_2P(S)Cl + SO_2 \tag{c}$$

TABLE 1. OXIDATIVE CHLORINATION BY SULFURYL CHLORIDE

Reagent	T (°C)	Product	Yield (%)	Ref.
$(Me_2PS)_2$	20	$Me_2P(S)Cl$	86	5
$(Me_2Sb)_2$	−78	Me_2SbCl	60	6
Ph_3Sb	0–20	Ph_3SbCl_2	90	7
Et_2SbCl	−78	Et_2SbCl_3	100	6

Conversion of P—P to P—Cl is also achieved by $SOCl_2$; $(Me_2PS)_2$ and $SOCl_2$ give[9] $Me_2P(S)Cl$.

Group VB chlorides chlorinate tertiary stibines in the laboratory; e.g., Et_3Sb with $SbCl_3$ in hexane at 70°C forms Sb and Et_3SbCl_2:

$$3\ Et_3Sb + 2\ SbCl_3 \rightarrow 3\ Et_3SbCl_2 + 2\ Sb \qquad (d)$$

TABLE 2. OXIDATIVE CHLORINATION BY GROUP-VB CHLORIDES

Chlorinating agent	Substrate	T (°C)	Product	Yield (%)	Ref.
PCl_5	$(t\text{-}BuP)_3$		$t\text{-}BuP(P(t\text{-}Bu)Cl)_2$		4
$SbCl_3$	Et_3Sb	70	Et_3SbCl_2	80	8
PCl_3	Me_3P	25	Me_3PCl_2		10
$PhAsCl_2$	Bu_3Sb	25	Bu_3SbCl_2	60	10
PCl_5	$[n\text{-}Pr_2P(S)]_2$	77	$n\text{-}Pr_2P(S)Cl$	81	11
$PhPCl_5$	$[Me_2P(S)]_2$	200	$Me_2P(S)Cl$		12
$SbCl_5$	$ClP[OC(CF_3)_2C(CF_3)_2O]$		$Cl_3P[OC(CF_3)_2C(CF_3)_2O]$		13

Reaction of ICl_3 with Ph_3P affords[14] Ph_3PCl_2, $CuCl_2$ and Me_2NPF_2 give[15] 84% of $Me_2NPF_2Cl_2$. Action of $GeCl_4$ or $SnCl_4$ on $t\text{-}Bu_3P$ forms[16] $[t\text{-}Bu_3PCl][GeCl_3]$ or $[t\text{-}Bu_3PCl][SnCl_3]$.

Group-VB chlorides are poisons.

(H.J. BREUNIG)

1. R. Appel, H. Schöler, *Chem. Ber.*, *110*, 2382 (1977).
2. R. Appel, R. Milker, I. Ruppert, *Chem. Ber.*, *110*, 2385 (1977).
3. R. R. Ford, M. A. Goodman, R. H. Neilson, A. K. Roy, U. G. Wettermark, P. Wilsan-Neilson, *Inorg. Chem.*, *23*, 2063 (1984).
4. M. Baudler, J. Hellmann, *Z. Anorg. Allg. Chem.*, *480*, 129 (1981).
5. G. W. Parshall, *Inorg. Synth.*, *15*, 192 (1974).
6. H. A. Meinema, H. F. Martens, J. G. Noltes, *J. Organomet. Chem.*, *51*, 223 (1973).
7. A. J. Banister, L. F. Moor, *J. Chem. Soc.*, *A*, 1137 (1968).
8. Y. Takashi, I. Aishima, *J. Organomet. Chem.*, *8*, 209 (1967).
9. K. A. Pollart, H. J. Harwood, *J. Org. Chem.*, *27*, 4444 (1962).
10. J. C. Summers, H. H. Sisler, *Inorg. Chem.*, *9*, 862 (1970).
11. W. Kuchen, H. Buchwald, K. Strolenberg, J. Metten, *Justus Liebigs Ann. Chem.*, *652*, 28 (1962).
12. S. O. Grün, J. D. Mitchell, *Phosphorus*, *6*, 89 (1976).
13. R. Bohlen, G.-V. Röschenthaler, *Z. Anorg. Allg. Chem.*, *513*, 199 (1984).
14. M. F. Ali, G. S. Harris, *J. Chem. Soc. Dalton Trans.*, 1545 (1980).
15. K. Kohn, R. W. Parry, *Inorg. Chem.*, *7*, 46 (1968).
16. W.-W. duMont, *Z. Anorg. Allg. Chem.*, *458*, 85 (1979).

2.4.13.3.3. to Give Br and I Group-VB Bonds.

Reaction of $(t\text{-}BuP)_3$ with PBr_5 gives[1] $t\text{-}BuP[(PBu\text{-}t)Br]_2$. A 2:3 PBr_3 and Me_3P mixture in CH_3CN forms[2] phosphorus and Me_3PBr_2:

$$3\ Me_3P + 2\ PBr_3 \rightarrow 3\ Me_3PBr_2 + 2/x\ P_x \qquad (a)$$

This reaction is also useful for Sb–halogen bonds; e.g., oxidation of Me_3Sb with AsI_3 in toluene affords[2] Me_2SbI_2 and As, and Bu_3SbI_2 is obtained[2] from Ph_2AsI or $PhAsI_2$:

$$Bu_3Sb + PhAsI_2 \rightarrow Bu_3SbI_2 + \tfrac{1}{6}(PhAs)_6 \quad \text{(b)}$$

(H.J. BREUNIG)

1. M. Baudler, J. Hellmann, *Z. Anorg. Allg. Chem.*, *480*, 129 (1981).
2. J. C. Summers, H. H, Sisler, *Inorg. Chem.*, *9*, 862 (1970).

2.4.14. Preparation of Astatine—Group-VB Element Bonds

The salts $[At(py)_2][ClO_4]$ and $[At(py)_2][NO_3]$ are coprecipitated with the homologous iodine compounds[1]. The perchlorate is obtained by reaction of $[Ag(py)_2]ClO_4$ with an AtI solution in $CHCl_3$ containing excess I_2 and py. After removal of AgI, the $[At(py)_2]ClO_4$ and $[I(py)_2]ClO_4$ are coprecipitated by addition of ether. The nitrates are prepared similarly, except that a mixture of $AgNO_3$ and py is used instead of $[Ag(py)_2]NO_3$, which cannot be isolated.

(W. D. LEE)

1. J. J. C. Schats, A. H. W. Aten, Jr., *J. Inorg. Nucl. Chem.*, *15*, 197 (1960).

2.5. The Formation of the Halogen–Group-IVB (C, Si, Ge, Sn, Pb) Element Bond

2.5.1. Introduction

Group-IVB halides are a common starting material for many group-IVB compounds. These bonds can be formed from the elements or from other bonds.

(A.P. HAGEN)

2.5.2. from the Elements.

Free halogens react with various allotropic forms of carbon.

Fluorine is the only halogen to react with powdered amorphous carbon to give CF_4 gas (bp −128°C):

$$C_{amorph} + 2\ F_2 \rightarrow CF_4\ [-160\ kcal(-670\ kJ)] \quad (a)$$

Higher fluoroalkanes are also formed as side products, from C_2F_6 to C_7F_{16}; CF_4 is usually obtained by action of Ag fluoride on CCl_4 at 300°C. Slightly soluble in water, it is stable toward heat and is chemically inert.

Whereas diamond does not react, graphite absorbs F_2 and gives two compounds at 400–650°C, CF and C_4F, which are also chemically inert. The carbon–carbon bond length is 1.54 Å, as in saturated compounds. These polymeric graphite intercalation compounds, graphite fluorides, decompose when heated to give small molecules CF_4, C_2F_6 and pulverized carbon. Other halides, except I_2, also form[1,2] $(C_8X)_n$ insertion compounds at low T, which are stable only under the saturation pressure of the halogens. Mixed-halogen carbides formed, e.g., by ICl, are known.

At 400°C powdered carbon reacts with Cl_2 to form CCl_4; however, CCl_4 (bp 76.8°C) is prepared by chlorination of CS_2 by Cl_2 or SCl_2 in the presence of Sb chloride or Fe sulfide as catalysts. Carbon tetrabromide (mp 93.7°C) and CI_4 (mp 171°C) are prepared indirectly from the hydrides or from decomposition of CCl_4 at 100°C by $AlBr_3$, AlI_3 or BiI_3; C_2I_4 is formed by elimination of I_2 when CI_4 is heated or exposed to sunlight.

Teflon, C_2F_4 and Freons are not prepared from the elements. Freon, CCl_2F_2, is manufactured by the action of HF on CCl_4 in the presence of $SbCl_3$. Mixed halides[3,4] are known, e.g., $CClF_3$, CCl_2F_2.

The simple Si halides, SiX_4, can be obtained by the direct union of the constituent elements. The compounds are stable but are hydrolyzed immediately, even at RT, in accordance with the great affinity of Si for oxygen.

Stability increases from the iodide to the fluoride. Silicon inflammes at RT in F_2 to give SiF_4 (bp −86°C), which is prepared by heating Ba fluorosilicate to its decomposition T:

$$Ba[SiF_6] \rightarrow BaF_2 + SiF_4 \quad (b)$$

The fluoride is able to add additional F^- ions, forming complex fluorosilicate ions, in particular the hexafluorosilicate anion:

$$SiF_4 + 2\ F^- \rightarrow [SiF_6]^{2-} \qquad (c)$$

which is stable in H_2O. The other Si halides are not capable of adding further halogen ions.

Heating Si in a stream of Cl_2 gives $SiCl_4$ (bp 57.5°C). A mixture of SiO_2 and carbon is also used. At 450°C higher halides are formed such as Si_2Cl_6, $Si_3Cl_8, \ldots, Si_6Cl_{14}$:

$$3\ SiCl_4 + Si \rightarrow 2\ Si_2Cl_6 \qquad (d)$$

Si_2Cl_6 (mp −1°C) is formed when the products are suddenly chilled. Silicon chlorides with long chains are obtained by the hot- and cold-tube method, from the pyrolysis of $SiCl_4$ vapor in N_2 at 1250°C; e.g., $Si_{25}Cl_{52}$ is a resinous, plastic substance. Silicon powder with 4% Cu dust as a catalyst, produces $SiBr_4$ (mp 5.4°C) at 500°C.

Passing I_2 vapor in a stream of CO_2 above 600°C over red hot Si forms SiI_4 (mp 120.5°C). If the tetraiodide is heated with finely divided Ag metal to 290–300°C, Si_2I_6 is obtained[5] (dec 250°C). Bromine with Si_2I_6 in CS_2 to displace I_2:

$$Si_2I_6 + 3\ Br_2 \rightarrow Si_2Br_6 + 3\ I_2 \qquad (e)$$

The hexachloride, Si_2Cl_6, is obtained in the same way. Silicon hexahalides are unstable at RT, but the rate of the reaction is small, and decomposition occurs only on moderate heating.

By thermal decomposition of higher halides such as $Si_{10}Cl_{22}$, silicon monochloride, $(SiCl)_n$, is formed, as is Si monoiodide, $(SiI)_n$, from the similar iodides.

As with Si, the simple Ge halides, GeX_4, can be obtained from the elements. A better way to GeF_4 (subl −37°C) is the action of conc HF on GeO_2 from which hydrated $GeF_4 \cdot 3\ H_2O$ separates. The anhydr fluoride is formed on heating. Fluorogermanates like $K_2[GeF_6]$ separate from a solution of GeF_4 to which KF is added; $GeCl_4$ (bp 83.1°C) is made[6] by burning elemental Ge in a stream of Cl_2 at 500–600°C, or better prepared pure by warming GeO_2 with fuming HCl. In conc HCl, $GeCl_4$ is present as chlorogermanic acid, $H_2(GeCl_6)$. By heating $GeCl_4$ in a hot and cold tube the Ge compound $(GeCl)_n$ corresponding to Si monochloride is prepared. The dichloride, $GeCl_2$, and the unstable hexachloride Ge_2Cl_6, are formed at the same time. Germanium tetrachloride undergoes partial reduction to the dichloride when it is heated with elemental Ge.

Germanium tetrabromide $GeBr_4$ (mp 26°C) and GeI_4 (mp 144°C) can be prepared by methods similar to those for the chloride; GeI_4 begins to decompose into GeI_2 and I_2 above its mp and, like $GeBr_4$, is vigorously decomposed by H_2O.

Tin combines with the free halogens forming the tetrahalides, even at RT, for Cl_2 and Br_2. It does not react with F_2 except above 100°C, when it does so violently, catching fire. The coating of SnF_4 (subl 705°C) on Sn prevents further attack; SnF_4 is also prepared by introducing SnF_4 into anhydr HF and heating gently. As the reaction is an equilibrium, xs HF must be present.

Fluorostannates can be obtained and correspond to $M_2^{(I)}[SnF_6]$. With NH_4F a salt of composition $[NH_4]_4[SnF_8]$ is also formed. Tin tetrachloride. (bp 114.2°C) dissolves in H_2O and forms hydrates. Aqueous $SnCl_4$ is hydrolyzed:

$$SnCl_4 + H_2O \rightleftharpoons SnO_2 + 4\ HCl \qquad (f)$$

The HCl formed combines with undecomposed $SnCl_4$ to produce hexachlorostannic acid, $H_2[SnCl_6]$. Chlorostannates, $M_2^{(I)}[SnCl_6]$, are known as the $[NH_4]$ so-called pink salt.

Tin tetrabromide, (mp 31°C) crystallizes from H_2O as the tetrahydrate, $SnBr_4 \cdot 4\ H_2O$. Hexabromostannic acid, $H_2[SnBr_6] \cdot 8\ H_2O$, forms and salts $M_2^{(I)}[SnBr_6]$ are known. Tin tetraiodide (mp 144.5°C) is obtained either by direct union of Sn with I_2, preferably dissolved in CS_2 and gently warmed, or by precipitation from a conc $SnCl_4$ by KI.

Higher halides, e.g., Sn_2Cl_6; are unknown for Sn.

Lead combines directly when it is heated with the halogens but forms predominantly bivalent products. However, like the other group-IVB elements, Pb can also be quadrivalent, although it is less stable in this state than they are. Lead tetraiodide and $PbBr_4$ are unknown, although bromoplumbates, $M_2^{(I)}[PbBr_6]$ and iodoplumbates, $M_2^{(I)}[PbI_6]$ are obtained by the reaction of chloroplumbates with KBr or KI. Cotunnite, $PbCl_2$, and matlockite, PbFCl, are minerals.

Lead(II) fluoride[7] (mp 855°C) is obtained from the elements, and as for Sn a protective coating forms at RT. Lead(IV) fluoride[8] (mp 600°C) is prepared by passing F_2 over PbF_2 above 250°C. Lead(II) chloride (mp 498°C) may be also formed by the direct combination of Pb and Cl at 300°C, or better by the action of HCl on the oxide PbO. Lead(IV) chloride (mp −15°C) is formed by passing Cl_2 into a suspension of $PbCl_2$ in conc HCl at 10–15°C. It decomposes at RT and loses half its Cl. Chloroplumbates, $M_2^{(I)}[PbCl_6]$, derived from $PbCl_4$ are more stable than $PbCl_4$. The complex salts $M_4^{(I)}[PbF_8]$ are derived from PbF_4.

(M. A. DELMAS, J. C. MAIRE)

1. W. Rüdorff, *Z. Anorg. Allg. Chem.*, *245*, 383 (1941).
2. G. R. Henning, *J. Chem. Phys.*, *20*, 1443 (1952).
3. O. Ruff, *Z. Anorg. Allg. Chem.*, *201*, 245 (1931).
4. M. Hauptschein, *J. Am. Chem. Soc.*, *74*, 1347 (1952).
5. C. Friedel, A. Ladenburg, *C.R. Hebd. Séances Acad. Sci.*, *68*, 920 (1869).
6. L. S. Foster, J. W. Drenan, A. F. Williston, *Inorg. Synth.*, *2*, 109 (1946).
7. H. Moissan, *Ann. Chim. Phys.*, (*6*), 24 (1891).
8. Wartenberg, *Z. Anorg. Allg. Chem.*, *244*, 337 (1940).

2.5.3. from Halogenation of the Group-IVB Elements

2.5.3.1. by Hydrogen Halides.

Halogenation by hydrogen halides of the group-IVB elements begins with Si and gives subvalent oxidation state (II) products.

The reaction of HCl with Si is related to the direct synthesis of organochlorosilanes and gives trichlorosilane, $HSiCl_3$ (bp 33°C, 785 torr), an important precursor for semiconductors. In the gas phase and with Cu as a catalyst, HCl reacts with elemental Si at 200–400°C:

$$Si + 3\ HCl \rightarrow HSiCl_3 + H_2 \qquad (a)$$

At increasing T, a side reaction produces $SiCl_4$. In addition, small amounts of dichlorosilane and chlorosilane are also formed[1]. Hydrogen bromide and HI give similar reactions[2]; $HSiBr_3$ (bp 109°C) and $HSiI_3$ (mp 8°C) decomposes more readily; $HSiF_3$ (bp −131°C) is prepared by the action of SbF_3 on $HSiCl_3$. Germanium forms with HF at 225°C a mixture of gases, principally GeF_4 with some Ge difluoride, GeF_2 (dec > 350°C), produced by dismutation

Trichlorogermane, $HGeCl_3$ (mp −71°C), is obtained by passing HCl over gently heated Ge powder

$$Ge + 3\ HCl \rightarrow HGeCl_3 + H_2 \qquad (b)$$

Passing $GeCl_4$ over heated Ge and treating the product consisting chiefly of $GeCl_2$ with HCl is better:

$$GeCl_2 + HCl \rightarrow HGeCl_3 \qquad (c)$$

By heating $GeCl_2$, $GeCl_4$ and elementary Ge are immediately formed. In $HGeCl_3$ Ge must be present in the bivalent state since I_2 has an oxidizing action that leads to $GeICl_3 \cdot HGeCl_3$, which hydrolyzes, forming Ge(II) oxide hydrate. Germanium(II) bromide (mp 122°C) is obtained by the action of HBr on elemental Ge at 400°C; $HGeBr_3$ (mp −24°C) is formed simultaneously. It may be reduced with Zn metal.

Germanium(II) bromide combines with HBr to form $HGeBr_3$, which decomposes at high T.

Anhydrous tin(II) chloride (mp 246°C) can be prepared by heating Sn metal in a current of HCl. It crystallizes from H_2O as $SnCl_2 \cdot 2\ H_2O$ (mp 40.5°C). This is the so-called tin salt manufactured by dissolving Sn turnings in hydrochloric acid. In dilute acid Sn is only slowly attacked in accordance with its normal potential and that of hydrogen.

The hydrate $SnCl_2 \cdot 2\ H_2O$ is dehydrated by heating to redness in a current of HCl. Tin(II) dichloride is a reducing agent which precipitates Ag and Au from their solutions as the metals. Oxidation of aq solutions is prevented by the addition of Sn metal. Crystalline chlorostannates $M^{(I)}[SnCl_3]$ and $M^{(II)}[SnCl_4]$ are obtained from solutions of $SnCl_2$ to which alkali-metal chlorides are added.

Tin(II) bromide, $SnBr_2$ (mp 232°C), is prepared by dissolving Sn powder in conc HBr and concentrating the solution. Pure $SnBr_2$ is formed on further heating.

Tin(II) iodide (mp 320°C) and tin(II) fluoride are similar to the chloride in their behavior. With the iodide, only the complexes $M^{(I)}[SnI_3]$ are known.

Lead(II) fluoride is formed[3] by the action of HF on Pb powder at 160°C. Lead comes immediately above hydrogen in the electrochemical potential series; and, although this implies that it is more basic than hydrogen, Pb does not dissolve in dilute acids. Furthermore, lead may be protected from dissolution by the formation of an insoluble coating on its surface, as with HF. Lead(II) bromide (mp 373°C) and lead(II) iodide (mp 412°C) are formed only when Br^- and I^- anions are added to Pb salt solutions.

Lead is not attacked by hydrochloric acid <100°C. On dissolving granulated Pb in dil HNO_3 $PbCl_2$ precipitates on adding hydrochloric acid:

$$3\ Pb + HNO_3 \rightarrow 3\ Pb(NO_3)_2 + 2\ NO + 4\ H_2O \qquad (d)$$

$$Pb(NO_3)_2 + 2\ HCl \rightarrow PbCl_2 + 2\ HNO_3 \qquad (e)$$

The $PbCl_2$ (bp 954°C) can be distilled without decomposition in a current of CO_2 and can combine with additional Cl^- ions, forming chloroplumbates(II), $M^{(I)}[PbCl_3]$, $M_2^{(II)}[PbCl_4]$ and $M^{(I)}[Pb_2Cl_5]$ in the solid state.

In the presence of atmospheric O_2 Pb metal is attacked by acids and H_2O to form lead(II) oxide. Finely divided Pb is pyrophoric, although compact Pb is only superficially attacked by atmospheric O_2.

(M. A. DELMAS, J. C. MAIRE)

1. A. Besson, L. Fournier, *C.R. Hebd. Séances Acad. Sci.*, *144*, 555 (1909).
2. W. C. Schumb, R. C. Young, *J. Am. Chem. Soc.*, *52*, 1464 (1930).
3. E. L. Muetterties, *Inorg. Chem.*, *1*, 342 (1962).

2.5.3.2. by Various Halides.

Group-IVB elements react more readily with various halides as their metallic character increases from C to Pb. Carbon halides are obtained not from metal halides and carbon, but from CCl_4 and Ag(I) fluoride, Al(III) bromide, Al(III) or Bi(III) iodide; CCl_4 is converted to CF_4, CBr_4 or CI_4, respectively. When amorph Si is heated to redness with AgF, ZnF_2 or PbF_2, SiF_4 is formed[1]. Similarly, CuI or HgI_2 forms SiI_4. Calcium fluoride in the presence of stoichiometric Fe(II) sulfide gives also SiF_4 by means of an electric furnace[2]:

$$Si + 2\ CaF_2 + 2\ FeS \rightarrow SiF_4 + 2\ CaS + 2\ Fe \tag{a}$$

Heating Si with anhydr metal chlorides, such as CuCl, $CuCl_2$ and $PbCl_2$, gives[3] $SiCl_4$. Copper(I) bromide gives the corresponding tetrabromosilane.

Few reactions with halides are known for Ge. Elemental Ge or GeS_2 heated with Hg(II) chloride gives[4] $GeCl_4$, and $HgBr_2$ forms $GeBr_4$. Conversion to $GeCl_4$ is either from[5] GeF_4 by means of $AlCl_3$, $MgCl_2$ and $FeCl_3$ or from[6] GeI_4 heated with $AlCl_3$, $BiCl_3$ or, e.g. $SbCl_5$:

$$GeI_4 + 2\ SbCl_5 \rightarrow GeCl_4 + 2\ SbCl_3 + 2\ I_2 \tag{b}$$

Heating phenylmercury(II) chloride with Sn powder gives triphenyltin chloride in 47% yield:

$$4\ Sn + 6\ C_6H_5HgCl \rightarrow 2\ (C_6H_5)_3SnCl + SnCl_2 + 6\ Hg \tag{c}$$

Other Hg(II) compounds give analogous reactions.

Lead metal reduces tin(II) bromide at 360°C to form $PbBr_2$ and Sn metal[7].

Various other metal halides react with Pb at different T to give $PbCl_2$, such as $AsCl_3$ at 15.2°C[8], AgCl at 420°C[9], CuCl at 418–444°C[10]; AgCl, CuCl and Pb melted to gether[11], $BiCl_3$[12], and $FeCl_3$[13] are used in the same way.

Anodic polarization[14] of lead in ag KI precipitates PbI_2, PbI_2 and $PbBr_2$ are obtained by electrolysis[15] of aq KI, KBr or HBr[16] using a Pb anode.

(M. A. DELMAS, J. C. MAIRE)

1. E. Vigouroux, *C.R. Hebd. Séances Acad. Sci.*, *120*, 367 (1895).
2. F. T. Sisco, *Chem. Met. Eng.*, *26*, 17 (1922).
3. W. C. Schumb, *Chem. Rev.*, *31*, 587 (1942).
4. A. Tchakirian, H. Volkringer, *C.R. Hebd. Séances Acad. Sci.*, *200*, 1758 (1935).
5. W. C. Schumb, D. W. Break, *J. Am. Chem. Soc.*, *74*, 1754 (1952).
6. T. Karantassio, *C.R. Hebd. Séances Acad. Sci.*, *196*, 1894 (1933).
7. Yu. K. Delimarskii, *Zh. Obshch. Khim.*, *11*, 1081 (1941).

8. E. Montignie, *Bull. Soc. Chim. Fr.*, (*5*) *3*, 190 (1936).
9. A. P. Palkin, Yu. P. Afinogenov, *Russ. J. Inorg. Chem.* (*Engl. Transl.*) *8*, 196 (1963).
10. A. P. Palkin, Yu. P. Afinogenov, *Russ. J. Inorg. Chem.* (*Engl. Transl.*) *5*, 110 (1960).
11. A. P. Palkin, Yu. P. Afinogenov, E. S. Mushenko, *Russ. J. Inorg. Chem.* (*Engl. Transl.*) *8*, 1352 (1963).
12. K. A. Seleznev, *Tr. Khim. Khim. Tekhnol.*, *2*(3), 513 (1959); *Chem. Abstr.*, *57*, 5562 (1962).
13. S. G. Rublev, *Zh. Khim. Prom. 9*, 31 (1932); *Chem. Abstr.*, *27*, 2588 (1933).
14. G. W. D. Briggs, W. F. K. Wynne-Jones, *J. Chem. Soc.*, 2966 (1956).
15. J. A. Wilkinson, *J. Phys. Chem.*, *13*, 691 (1908/09).
16. K. Elbs, R. Nübling, *Z. Elektrochem.*, *9*, 776 (1903).

2.5.3.3. by Organic Halides

2.5.3.3.1. with Elemental Si.

The so-called direct synthesis concerns alkyl, aryl, unsaturated and numerous miscellanous organohalosilanes. While a Si—halogen bond is formed, a C—Si bond is also formed. This method is important industrially, and with the simple and cheap reaction of CH_3Cl gas with elemental Si, with or without a catalyst[1,2] such as metallic Cu, dimethyldichlorosilane forms, a starting material for silicone rubbers, oils and resins.

$$2\ CH_3Cl + Si \xrightarrow[280–340°C]{Cu} (CH_3)_2SiCl_2 + \text{ca. 40 other compounds} \qquad (a)$$

When the reaction starts T is easily maintained, eventually with cooling, but heat is required to initiate the attack. Good yields are obtained (80%), but products from $Si(CH_3)_4$ to $SiCl_4$; or even with Si—H bonds such as $HSiCl_3$ are encountered.

Separation is difficult as bps are close, e.g., $(CH_3)_2SiCl_2$ (bp 70.0°C), CH_3SiCl_3 (bp 66.0°C), $(CH_3)_3SiCl$ (bp 57.7°C), etc., and requires chemical and physical procedures. When working on a large scale separation is by rectification on efficient columns. To increase the yield of monohalide[3], the incorporation of such metals as Zn or Al, as a catalyst to the contact mass of Si is necessary.

Use of other gases mixed with the halide, e.g., H_2, HCl, Cl_2 or N_2, may also change amounts of particular products such as the halohydrides. When H_2 is mixed with CH_3Cl at 390°C, the yield of CH_3SiHCl_2 reaches 17%[4]. Free-radical mechanisms[5,6] remain a good model for this reaction[7,8]. An unstable organo intermediate, RCuCl, is formed by the alkyl halide and Cu metal catalyst. After its decomposition, an alkyl free radical is formed and the Cu chloride produces surface chlorination of the elemental Si. The free radicals react with this surface and give alkylchlorosilanes.

The rate halogenation[9] for EtCl is slower than with MeBr, but the required T is too high, and decomposition prevents the formation of iodosilanes from MeI. Methyl fluoride is untested. For alkyl or aryl halides, the product composition depends upon the purity and granularity of the elemental Si, the nature and proportion of the catalyst, the contact time, the treatment of the contact mass, the presence of inert gases and the T. Derivatives of iodine, bromine or fluorine have only limited laboratory uses. The nature of the organic groups is more important than that of the halogen, but the halogen manifest itself in the activation of the elemental Si. Higher alkyl groups show a lower stability under the conditions of the direct synthesis and give decomposition products such as olefins. Yields are acceptable with Pr- or BuCl[10]. With chlorobenzene >450°C is required, and decomposition also takes place. The proportion of diphenyldichlorosilane obtained is much lower that that of dimethyldichlorosilane with methyl chloride. Silver is the catalyst. No halohydride is formed, and triphenylchlorosilane is obtained in small amounts.

(M. A. DELMAS, J. C. MAIRE)

1. E. G. Rochow, *J. Am. Chem. Soc., 67*, 963 (1945).
2. W. F. Gilliam, R. M. Meals, R. O. Sauer, *J. Am. Chem. Soc. 68*, 1161 (1946).
3. D. T. Hurd, U.S. Pat. 2,427,605 (1947); *Chem. Abstr. 42*, 202 (1948).
4. M. M. Spring, W. F. Gilliam, U.S. Pat. 2,380,999 (1945); *Chem. Abstr., 42*, 202 (1945).
5. J. Joklik, V. Bàzant, *Coll. Czech, Chem. Commun., 38*, 3176 (1973).
6. S. A. Golubtsov, K. A. Andrianov, N. T. Ivanova, R. A. Turetskaya, I. M. Podgornyi, N. S. Feldstein, *J. Gen. Chem. USSR (Engl. Transl.), 43*, 1985 (1973).
7. A. L. Klebanskii, V. S. Fikhtengol'ts, *J. Gen. Chem. USSR (Engl. Transl.), 26*, 2502 (1956).
8. P. Trambouze, *Bull. Soc. Chim. Fr.*, 1756 (1956).
9. J. Joklik, M. Kraus, V. Bàzant, *Coll. Czech. Chem. Commun., 26*, 427 (1961).
10. A. D. Petrov, N. P. Smetankina, G. I. Nikishim, *J. Gen. Chem. USSR (Engl. Transl.), 25*, 2332 (1955).

2.5.3.3.2. with Elemental Ge.

The features of the reaction of Si are paralleled by Ge. Organohalogermanes are prepared by direct synthesis from alkyl halide vapors and elemental Ge at 300–350°C.

Methylchloride reacts at 340°C in the presence of Cu as a catalyst to give dimethyldichlorogermane in 56% yield together with mono- and trimethylgermanium chloride and traces of hydrocarbons such as CH_4. In contrast to Si, the equivalant halohydride, $(CH_3)_2GeHCl$, is not formed. The reaction proceeds[1] also without Cu metal, but needs higher T, near 460°C[1]. The reaction can be controlled to give the mono- or the trimethylgermanium chloride. The yield of CH_3GeCl_3 is increased by operating[2] with a greater proportion of Cu metal and higher T. At 510°C Ge powder dispersed over glass wool yields 70% of methyltrichlorogermane[3] without catalysis. Adding Ga metal to the Cu catalyst, and at 400°C, a mixture with reverse proportions is formed with yields of $(CH_3)_3GeCl$ (85%), $(CH_3)_2GeCl_2$ (10%) and CH_3GeCl_3 (5%)[4].

A Germanium heterocycle is produced when elemental Ge metal reacts with methylene chloride[5] in the presence of Cu:

$$CH_2Cl_2 + Ge \xrightarrow[380^{\circ}C]{Cu} (Cl_2GeCH_2)_3 + CH_2(GeCl_3)_2 + CH_3GeCl_3 \qquad (a)$$

(19%) (23%) (27%)

The direct synthesis works with various alkyl or aryl halides. With EtCl[6] and MeBr, reactions are Cu catalyzed. As for Si, chlorobenzene requires a higher T and the presence of an Ag metal catalyst. Diphenyldichlorogermane is the principal product.

Related to the direct synthesis is the quantitative reaction between GeI_2 and the lower alkyl iodides[7]:

$$RI + GeI_2 \rightarrow RGeI_3 \qquad (b)$$

in a sealed tube with CH_3I or C_2H_5I at 110°C. At ≥140°C, ethyliodide gives GeI_4. Instead of Ge dihalides; Ge tetrahalides are also used.

(M. A. DELMAS, J. C. MAIRE)

1. E. G. Rochow, *J. Am. Chem. Soc., 69*, 1729 (1947).
2. E. G. Rochow, R. Didtchenko, R. C. West, *J. Am. Chem. Soc., 73*, 5486 (1951).
3. M. Wieber, C. D. Frohning, M. Schmidt, *J. Organometl. Chem., 6*, 427 (1966).
4. G. Ya Zueva, N. V. Lukyankina, V. A. Ponomarenko, *Izv. Akad. Nauk. SSSR, Ser. Khim.*, 2777 (1971).
5. V. F. Mironov, T. K. Gar, *Izv. Akad. Nauk. SSSR, Ser. Khim.*, 1970 (1964).
6. G. Ya Zueva, N. V. Luk'yankina, A. G. Kechina, V. A. Ponomarenko, *Izv. Akad. Nauk SSSR*, 1780 (1966).
7. M. Lesbre', P. Mazerolles, G. Manuel, *C.R. Hebd. Séances Acad. Sci., 257*, 2303 (1963).

2.5.3.3.3. with Elemental Tin.

The most economical method for synthesizing organotin halides on an industrial scale is also the direct synthesis. Alkyl halides and Sn metal can be heated in a sealed tube at 130–220°C during 1–2 days[1]:

$$2\ RX + Sn \rightarrow R_2SnX_2 \tag{a}$$

Yields are poor, and small amounts of the monohalide R_3SnX are also formed. Under these conditions reactivity is Cl > Br > I. The introduction of higher alkyl groups is feasible. Tin powder in suspension in toluene or H_2O and benzylchloride without catalyst at 110°C gives di- or tribenzyltin chlorides[2].

Use of dihalides such as methylene halides, CH_2X_2, leads to methyltrihalostannanes[3].

If the sealed glass tube is irradiated with γ-rays from a ^{60}Co source[4] n-Bu_2SnBr_2, e.g., is obtained from n-bromobutane and Sn powder in 80% yield. During this radiation-induced synthesis[5], side products are formed such as n-Bu_3SnBr, n-$BuSnBr_3$ and $SnBr_4$. Redistribution of n-Bu_4Sn and $SnBr_4$ is the simplest way to pure n-Bu_2SnBr_2. Allowing alkyl halide gases to bubble through molten Sn at 350–450°C without a catalyst gives alkyltin halides with MeCl in 10% yield. With methyl chloride and methyl bromide, the dihalide remains the principal product, whereas with MeI, CH_3SnI_3 is obtained. Adding Cu or Zn catalyst enhances the yield[6]. With Li metal or LiBr as a catalyst, n-BuI and Sn metal in refluxing n-butanol give n-Bu_2SnI_2 in 90% yield[7].

Magnesium or Zn in the presence of alcohols or THF can be incorporated under high pressure. In an autoclave at 100°C, MeBr gives Me_2SnCl_2 in high yield[8].

Some alkyltin halides are prepared by the electrolysis[9] of alkyl halides in a solvent, using an Mg cathode and an Sn anode. Related to the direct synthesis are the use of Na-Sn, Na-Sn-Zn[10] and Ms-Sn[11] alloys, which lead to a mixture of trialkyltin halides and tetraalkylstannanes, and the reaction between organic chlorides and tin(II) chloride in the presence of a trialkylantimony catalyst[12]. Organotin trichlorides results from addition:

$$RCl + SnCl_2 \xrightarrow{SbCl_3} RSnCl_3 \tag{b}$$

At 160°C, MeI and tin(II) iodide give[13] CH_3SnI_3. The direct synthesis, which does not lead to tetrasubstituted derivatives, is used in the preparation of $(C_6F_4)_4Sn$ by heating C_6F_5I with Sn metal at 240°C.

(M. A. DELMAS, J. C. MAIRE)

1. G. Grüttner, E. Krause, M. Wiernick, *Chem. Ber.*, *50*, 1549 (1917).
2. K. Sisido, Y. Takeda, Z. Kinugawa, *J. Am. Chem. Soc.*, *83*, 538 (1961).
3. K. A. Kochevshkov, *Chem. Ber.*, *61*, 1659 (1928).
4. L. V. Abramova, N. J. Sheverdina, K. A. Kocheshkov, *Industrial Use of Large Radiation Source*, Vol. 1, IAEA, Vienna, 1961, p. 83.
5. A. F. Fentiman, R. E. Wyant, J. L. McFarling, J. F. Kircher, *J. Organomet. Chem.*, *6*, 645 (1966).
6. A. C. Smith, E. G. Rochow, *J. Am. Chem. Soc.*, *75*, 4105 (1953).
7. V. Oakes, R. E. Hutton, *J. Organomet. Chem.* *9*, 133 (1966).
8. R. Irmscher, W. Knöpke, H. Kunze, *Ge Pat.*, 1,050,336 (1959); *Chem. Abstr.*, *55*, 2, 485 (1961).
9. L. V. Armenskaya, K. N. Korotaevskii, É. N. Lysenko, L. M. Monastyrskii, USSR Pat. *184*, 853 (1966); *Chem. Abstr.*, *66*, 71949 (1967).
10. T. Harada, *Sci. Pap. Inst. Phys. Chem. Res. Tokyo*, *35*, 290 (1939).
11. G. J. M. Van der Kerk, J. G. A. Luijten, *J. Appl. Chem.*, *4*, 307 (1954).
12. E. J. Bulten, *J. Organomet. Chem.* *97*, 167 (1975).
13. H. Gilman, S. D. Rosenberg, *J. Am. Chem. Soc.*, *74*, 531 (1952).

2.5.3.3.4. with Lead Metal.

Unlike the organometal halides of other group-IVB metals, the organolead halides are not prepared by direct synthesis. Direct synthesis gives organometal halides with elementary Si, Ge and Sn and produces tetrasubstituted derivatives of Pb:

$$4\ RX + 3\ Pb \rightarrow PbR_4 + 2\ PbX_2 \tag{a}$$

The addition of $Na[Et_4Al]$ converts all the Pb metal to organic products.

Lead-sodium alloy[1] reacts more efficiently with alkyl halides, but gives also tetraalkyllead. Ethyl iodide in alcoholic alkalis is electrolyzed with a Pb cathode and produces Et_4Pb. The reactions applied to tin(II) chloride and iodide[2] are related to the direct synthesis. In the presence of trialkylantimony as a catalyst, monoalkyllead triodide $RPbI_3$, is formed from a organic halide and lead(II) iodide:

$$RI + PbI_2 \rightarrow RPbI_3 \tag{b}$$

(M. A. DELMAS, J. C. MAIRE)

1. B. C. Saunders, G. J. Stacey, *J. Chem. Soc.*, 919 (1949).
2. G. Ghobert, M. Devaud, *J. Organomet. Chem.*, *153*, C23 (1978).

2.5.4. by Other Halogenating Agents.

Interhalogens such as ClF_3 may be used with a fluidized carbon bed to form a mixture of fluorochlorocarbons; ClF_3 diluted with N_2 gives[1] CF_4, $CClF_3$, CCl_2F_2 and C_2F_6 at 412°C. Mixed fluorides of Ge are obtained[2] by the fluorination of $GeCl_4$ with SbF_3 in the presence of $SbCl_3$; PCl_3 and $SbCl_3$ are used[3] partially to chlorinate $SiBr_4$.

Passing a mixture of Cl_2 and Br_2 over elemental Si at 700°C yields[4] trichlorobromosilane, $SiCl_3Br$ (bp 0°C), together with $SiCl_2Br_2$ and $SiClBr_3$. Silicon gives halocompounds that are reactive and short-lived divalent radicals, $[SiX_2]^{\cdot}$, corresponding to[5] CX_2, SiX_2, GeX_2 with X = F, Cl, Br, I and $[SnI_2]^{\cdot}$. The reducing agent for SiF_4, e.g., is the element itself[6]:

$$Si(s) + SiF_4(g) \xrightarrow{1000^\circ C} 2\ SiF_2(g) \tag{a}$$

at low P and high T as there is an increase in entropy. At 1400°C, 85% conversion is obtained; $SiCl_2$ and $SiBr_2$ are also formed, but SiI_2 immediately decomposes.

Iodine reacts with SiF_2 to give F_2SiI_2; $SiCl_2$ reacts at low P and >1300°C with ECl_3 compounds such as PCl_3, BCl_3[7] to give Cl_3ECl_2 compounds and with CCl_4 to give[8] Cl_3SiCCl_3. Silicon warmed with IF_5 gives SiF_4[9]; IF_7 is also used[10].

Perfluoroalkyl halides react with an Si–Cu contact mass at 500–1000°C; CF_3Cl, C_2F_5Cl and C_4F_7Cl afford fluorochlorosilanes[11] such as F_3SiCl or F_2SiCl_2. With CF_3Cl, CF_3Br, CF_3I, such compounds as $(CF_3)_2SiX_2$ are also obtained[12]. Mixed halides, e.g., CF_3SiBrF_2 occur. Higher fluorinated compounds are formed but decompose immediately at ≈ 250°C. Silicon heated with CCl_4 chloride at 80–300°C in the presence of metal catalysts, e.g., Cu, Ni, Sn, Sb, Mn, Ag or Ti[13,14] gives $SiCl_4$, C_2Cl_4 and Si_2Cl_6. With a Cu-metal catalyst (10%) and Fe and Al, the products are[15]:

$$\underset{}{Si + CCl_4} \xrightarrow[10\ \%\ \text{Cu catal}]{210\text{–}410^\circ C} \underset{(15\%)}{Si_2Cl_6} + \underset{(14\%)}{Cl_3SiCCl{=}CClSiCl_3} + \underset{(11\%)}{Cl_3SiC{\equiv}CSiCl_3} + \underset{(38\%)}{Cl_2C{=}CCl_2} \ \ldots \qquad \text{(b)}$$

A reaction related to the direct synthesis involves haloorganohalosilanes such as α-chloroalkylsilanes. At 300°–400°C, α-chloromethyltrichlorosilane reacts with Si, in the presence of a Cu catalyst to form[16] $Cl_3SiCH_2SiCl_3$ (30%), among other products.

Germanium(II) fluoride is made[17] by passing GeF_4 gas over Ge powder at 100°C; GeF_2 is best prepared at 150°C under reduced P. Germanium(II) chloride is formed[19] by passing $GeCl_4$ vapor over heated Ge; $GeCl_2$ is a solid sensitive to moisture and atmospheric O_2:

$$Ge + GeCl_4 \xrightarrow{350\text{–}430^\circ C} 2\ GeCl_2 \qquad \text{(c)}$$

The colorless product turns orange when warmed. Decomposition occurs on stronger heating to give Ge and $GeCl_4$ again. A poor yield of Ge(II) iodide is obtained by passing GeI_4 vapor over heated Ge, since it disproportionates rapidly.

Powdered Ge and IBr, mixed together for 30 min and heated to reflux, produce[20] $GeCl_4$ (50%); $GeCl_4$ is also produced from $SOCl_2$ and GeO_2 at 300°C in a sealed tube, but not from elemental Ge.

Elemental Ge reacts with NO_2F_2 prepared from $NaNO_2$ and F_2, to give[21] a white solid $[NO_2]_2GeF_6$, at RT. Decomposition of unsymmetrical diaryliodonium chlorides in the presence of Sn powder and tin(II) chloride[22] yields organotin halides:

$$2\ RR'ICl{\cdot}SnCl_2 + Sn \rightarrow R_2SnCl_2 + 2\ R'I + 2\ SnCl_2 \qquad \text{(d)}$$

The resulting diorganotin dichloride is that containing the more electronegative aryl group.

The reaction of SeF_6 on Pb metal forms[23] PbF_2. If powdered Pb reacts with S_2Cl_2[24], $PbCl_2$ is obtained:

$$3\ Pb + S_2Cl_2 \xrightarrow{400^\circ C} \underset{(65.5\%)}{PbCl_2} + \underset{(34.5\%)}{2\ PbS} \qquad \text{(e)}$$

(M. A. DELMAS, J. C. MAIRE)

1. R. M. Mantell, H. J. Passino, W. O. Teeters, US Pat. 2,686,987 (1954); *Chem. Abstr.*, *49*, 10,354 (1955).
2. H. S. Booth, W. C. Morris, *J. Am. Chem. Soc.*, *58*, 90 (1936).

3. W. C. Schumb, H. H. Anderson, *J. Am. Chem. Soc., 58*, 994 (1936).
4. A. G. MacDiarmid, *Prep. Inorg. React., 1*, 165 (1964).
5. P. L. Timms, *Prep. Inorg. React., 4*, 59 (1968).
6. D. C. Pease, US Pat., 2,840,588, 3,026, 173 (1958); *Chem. Abstr. 52*, 19,245 (1958).
7. A. G. Massey, D. S. Urch, *Proc. Chem. Soc. (London)*, 284 (1964).
8. P. A. DiGiorgio, L. H. Sommer, F. C. Whitmore, *J. Am. Chem. Soc., 70*, 3512 (1958).
9. H. Moissan, *Bull. Soc. Chim. Fr.*, (3)29, 6 (1903).
10. O. Ruff, R. Kheim, *Z. Anorg. Allg. Chem., 193*, 176 (1930).
11. E. F. Izard, S. L. Kwoleck, *J. Am. Chem. Soc., 73*, 1156 (1951).
12. H. J. Passino, L. C. Rubin, US Pat. 2,686, 194 (1954); *Chem Abstr., 49*, 1368 (1955).
13. British Thomson-Houston Co. Ltd., *Br. Pat.* 575,669, 575,672 (1946); *Chem Abstr., 41*, 6894 (1947).
14. W. I. Patnode, R. W. Schiessler, *Fr. Pat.* 949,100 (1947/49).
15. R. Muller, H. Beyer, *Chem. Ber., 92*, 1957 (1959).
16. S. I. Sadykh-Zhade, E. A. Chernyshev, V. F. Mironov, *Dokl. Akad. Nauk SSSR, 105*, 496 (1955)
17. L. M. Dennis, A. W. Laubengayer, *Z. Phys. Chem., 130*, 520 (1927).
18. H. C. Clark, C. J. Willis, *J. Am. Chem. Soc., 84*, 898 (1962).
19. L. M. Dennis, L. H. Hunter, *J. Am. Chem. Soc., 51*, 1151 (1929).
20. V. Gutmann, *Monatsh. Chem., 82*, 280 (1951).
21. E. E. Aynsley, G. Hetherington, P. L. Robinson, *J. Chem. Soc.*, 1119 (1954).
22. H. C. Clark, R. J. Puddephatt, in *The Bond to Halogens and halogenoids*, A. G. MacDiarmid ed., Marcel Dekker, New York, 1972, p.74.
23. C. Dragon, *C.R. Hebd. Séances Acad. Sci., 241*, 418 (1955).
24. H. Fink, K. H. Berndt, G. Henze, *Wiss. Z. Martin-Luther Univ. Halle-Wittenberg, 6*, 815 (1956/57); *Chem. Abstr.*, 54, 12,860 (1960).

2.5.5. from Cleavage of the Group-IVB–Hydrogen Bond

2.5.5.1. by Halogens.

Fluorine gas reacts violently with CH_4 even at low T. Chlorine and CH_4 at RT explode under UV light with decomposition of the hydrocarbon:

$$CH_4 + 2\ Cl_2 \rightarrow C + 4\ HCl \qquad \text{(a)}$$

but it is possible to realize partially substituted chloromethanes by dilution of the gas with CO_2. Bromine has the same reactivity, but I_2 none.

Cyanogen halides are prepared by substitution of aq cyanohydric acid. With Cl_2, cyanogen chloride is formed with some cyanogen. The cleavage of C—H bonds is the provinence of organic chemistry.

Free-radical photochemical halogenation of saturated hydrocarbons is used industrially. Light produces halogen atoms which easily abstract hydrogen atoms from saturated carbon atoms[1], e.g., the steps of the chain reactions, for CH_4:

$$Cl_2 \rightarrow 2\ Cl^{\bullet} \qquad \text{initiation} \qquad \text{(b)}$$

$$\left.\begin{array}{r} Cl^{\bullet} + CH_4 \rightarrow HCl + [CH_3]^{\bullet} \quad \text{(c)} \\ CH_3^{\bullet} + Cl_2 \rightarrow CH_3Cl + Cl^{\bullet} \quad \text{(d)} \end{array}\right\} \text{propagation}$$

$$\left.\begin{array}{r} 2\ Cl^{\bullet} \rightarrow Cl_2 \quad \text{(e)} \\ 2\ [CH_3]^{\bullet} \rightarrow C_2H_6 \quad \text{(f)} \\ [CH_3]^{\bullet} + Cl^{\bullet} \rightarrow CH_3Cl \quad \text{(g)} \end{array}\right\} \text{termination}$$

Petroleum hydrocarbons give mixtures of alkyl halides by photochlorination. The Br_2 reactions are slow because the chains are short.

Tertiary C—H bonds are more susceptible to attack by halogens than secondary and primary positions. Aromatic side chains are activated in the α position. For saturated carbons, the α positions of carbonyl and nitro compounds are susceptible to halogenation in the presence of catalysts by electrophilic substitution. Enols or enolate ions are intermediates in the process. Two ways are possible. Acid-catalyzed bromination, e.g., of dimethylketone, gives α-halogenation. But base-catalyzed halogenation of methylene and methylketones cannot be stopped at the monohaloketone step. Haloform cleavage occurs after complete halogenation.

Polyhaloketones are cleaved under basic conditions forming carboxylic acids and haloforms.

Unsaturated and aromatic hydrocarbons undergo addition of chlorine. By photochlorination benzene gives isomers of hexachlorocyclohexane. However, with alkylbenzenes, such as toluene, acid catalyst allows a more favorable competition between an electrophilic substitution mechanism which, gives nuclear attack, and the free-radical halogenation mechanism, which results in side-chain attack. Substituted-aromatic systems are halogenated readily. Reactive, simple substrates like aniline or phenol give 2,3,5-tribromo compounds quantitatively with Br_2, and this method is the most useful for introducing other substituents.

Free halogens are used to convert Si—H to Si halides. The ease of halogenating Si—H bonds decreases Cl > Br > I. Whereas for I_2 it may be necessary to accelerate the reaction by catalysts like $AlCl_3$, bromination and especially chlorination are nonselective. The reaction[2] is first order in Cl_2, second order in Br_2 and third order in I_2. Organic groups attached to Si can be simultaneously halogenated, but the Si—H bond is more reactive than the C—H bond, so only organic derivatives are halogenated, and only minute amounts of side products are formed. The rate of halogenation of Si hydrides is not altered by bulky organic substituents.

Triorganochlorosilanes are rarely produced from corresponding organosilicon hydrides, since reduction of these compounds is the source of the hydrides.

Conducted in the vapor or liquid and without solvent, **the reaction of halogens with Si hydrides proceeds violently and explosions can occur**. Fluorination is, of course, impossible, and organofluorosilanes are not prepared by this method.

The solvent employed in the halogenation affects the rate. Cleavages are carried out in inert, nonpolar solvents such as CCl_4 at low T, where, e.g., $HSiCl_3$ with Cl_2 gives[3] $SiCl_4$. Halogenated Si hydrides are obtained by partial halogenation of Si hydrides containing Si—H bonds[4].

Under mild conditions in a suitable solvent, Si—H bonds can be halogenated selectively in the presence of reactive Si—O, Si—S and Si—Si bonds[5]. For polysilanes, it is possible to halogenate Si—H bonds keeping Si—Si bonds intact. If the organosilane contains Si-phenyl bonds which are cleaved by halogens in preference to Si—H bonds, it is not possible to obtain phenylchlorosilanes in this way. The reactions of Si hydrides with Cl_2 and Br_2 proceed with retention of configuration[6].

Free halogens also convert Ge—H bonds. Trialkylgermanes give substitution quantitatively with either I_2, Br_2 or Cl_2. The resulting trialkylmonohalogermanes are pure. However, organogermanium halohydrides are of greatest interest.

With I_2 under mild conditions it is possible to control the halogenation so as to replace only one H atom in a di- or trihydride[7]. In the absence of catalyst at RT, only one

Ge—H of n-$BuGeH_3$ is substituted to give n-$BuGeIH_2$. Refluxing with Cu metal, or with xs halogen, both di- and trihydrides are fully halogenated. Bromine is also able to substitute di- and trihydrides partially. As with Si, Ph—Ge bonds are cleaved.

Starting from phenylgermanium hydrides, the final product of the attack by I_2 is GeI_4.

Chloride substitutes in optically active germanes[8] with retention of configuration.

Halogenation of the Sn—H bond is not known, since tin hydrides are prepared from the corresponding halogenostannanes. Moreover, considering also the great reactivity of the Sn—C bond toward halogens, only tetrahalogenotins would form.

Lead chemistry has not shared in this development because of the instability of the Pb—H bond. A few monohydrides are known, e.g., Me_3PbH prepared from the chloride by LiH_4. Their thermal instability is greater than their chemical reactivity. These compounds decompose far below 0°C to form H_2, Pb metal, tetraalkyllead and hexaalkyldilead.

(M. A. DELMAS)

1. G. Sosnovsky, *Free Radical Reactions in Preparative Organic Chemistry*, Macmillan, New York, 1964.
2. L. H. Sommer, *Stereochemistry Mechanism and Silicon*, McGraw-Hill, New York, 1965.
3. C. Friedel, A. Laudenbourg, *Ann. Chim. Phys.*, *23*, (4), 430 (1871).
4. N. Viswanathan, C. H. Van Dyke, *J. Chem. Soc.*, *A*, 487 (1968).
5. N. N. Sokolov, S. M. Akimova, *J. Gen. Chem. USSR* (*Engl. Transl.*), *26*, 2216 (1956).
6. L. H. Sommer, C. L. Frye, M. C. Musolf, G. A. Parker, P. G. Rodewald, K. W. Michael, Y. Okaya, R. Pepinsky, *J. Am. Chem. Soc.*, *83*, 2210 (1961).
7. H. H. Anderson, *J. Am. Chem. Soc.*, *82*, 3016 (1960).
8. G. Glocking, *The Chemistry of Germanium*, Academic Press, New York, 1969, p. 131.

2.5.5.2. by Hydrogen Halides.

These reactions are less vigorous than the halogenation of group-IVB—hydrogen bonds. Hydrocarbons are not sensitive to hydrochloric acid. Hydrogen halides give selective and partial halogenation of Si hydrides. Halogenation is less exothermic. Often, elevated T and use of $AlCl_3$ or related catalysts are required.

The Si—H bond of triorganosilanes is converted to Si—F or Si—Cl bonds by HF or HCl in the presence of a Pd catalyst.

However, when organohydrosiloxanes are treated with HF Si—O—Si bonds and Si phenyl groups are cleaved. If only phenyl groups are linked to Si, it is possible to obtain mono- and dihalosilanes, since phenyl–Si bonds are easily cleaved in preference to Si—H bonds:

$$(C_6H_5)_2SiH_2 + 2\ HX \rightarrow SiH_2X_2 + 2\ C_6H_6 \quad \text{(a)}$$

With HCl, unlike Cl_2, there is no chlorination of organic groups attached to Si, which is an advantage.

Hydrogen halides cleave Ge—H bonds more easily; e.g., HCl with an $AlCl_3$ catalyst reacts with GeH_4 to give a mixture of mono- and dichlorogermane. With CH_3GeH_3, mono-, di- and trichloromethylgermane and H_2 are obtained. Partial substitution is possible for $(CH_3)_2GeH_2$ and the halohydride is formed, but at 100°C the disubstituted product appears. Hydrogen bromide under the same conditions also leads to $(CH_3)_2GeHBr$. Reactivity with the Ge—H bonds increases HI > HBr > HCl.

Stannane is unstable, but progressive substitution with alkyl or aryl groups increases stability. Hydrogen halides halogenate tin hydrides, but redistribution is often used. Tin–carbon cleavage may occur instead of Sn—H, as with phenylstannanes. With Si and Ge it is possible to form the R_2SnHX; e.g., aq HCl in dioxane at RT reacts with n-Bu_2SnH_2 to give n-Bu_2SnClH.

Monobromation of tin trihydride occurs at −78°C[1]. With strong HCl reactions are rapid and quantitative so that thermochemical measurements to derive heats of formation of the hydrides are possible.

The Pb hydride, PbH_4, is so unstable thermally that no reactions are known. Trimethyllead hydride decomposes at −37°C. With HCl, the chloride derivative and H_2 are formed quantitatively.

(M. A. DELMAS)

1. G. Fritz, H. Scheer, *Z. Anorg. Allg. Chem.*, *1*, 338 (1965).

2.5.5.3. by Other Halides.

Several reactions with halides involve the cleavage of the group-IVB—hydrogen bonds so that classification is difficult; e.g., bromination with N-bromosuccinimide substitutes C—H bonds adjacent to olefinic, aromatic or carbonyl groups. These radical-chain reactions are initiated by light or by decomposition of catalysts such as labile azocompounds and peroxides, e.g.:

$$\text{cyclohexene} + \text{Br—N}\begin{matrix}\text{C(=O)—CH}_2\\ |\\ \text{C(=O)—CH}_2\end{matrix} \xrightarrow[\text{azo catalyst}]{CCl_4,\ 60°C} \text{3-bromocyclohexene}\ (70\%) + \text{H—N}\begin{matrix}\text{C(=O)—CH}_2\\ |\\ \text{C(=O)—CH}_2\end{matrix} \quad \text{(a)}$$

Other N-bromoamides, including N-bromohydantoins and caprolactam, are also used[1] to produce this allylic halogenation.

Although the most common halogenation agents for cleaving Si—H bonds are free halogens and hydrogen halides, inorganic and organic halides are also used.

Halogenation of SiH_4, Si_2H_6, CH_3SiH_3, $(CH_3)_2SiH_2$, $(CH_3)_3SiH$ can be carried out[2] with Ag halides at 260–280°C:

$$-\overset{|}{\underset{|}{Si}}-H + 2\,AgX \rightarrow -\overset{|}{\underset{|}{Si}}-X + HX + 2\,Ag \qquad \text{(b)}$$

Partial halogenation is possible. Refluxing trialkylsilanes with halides of the transition elements results in good yields of the corresponding alkylhalosilanes. The starting metal halide is reduced to a lower oxidation state or sometimes to the free metal[3], as e.g., with $CuCl_2$, VCl_4, $PtCl_4$, $RuCl_2$, $PdCl_2$, etc.

Cleavage by fluorides like PF_5 or AgF transforms the Si—H bond to an Si—F bond[4]:

$$(CH_3)_3SiH + PF_5 \xrightarrow{\text{liq}} (CH_3)_3SiF + HF + PF_3 \qquad \text{(c)}$$

Chlorination by inorganic halides is used with PCl_5 and partial chlorination of mono and diphenylsilane is achieved by this agent[5]:

$$C_6H_5SiH_3 \xrightarrow{PCl_5} \underset{88\%}{C_6H_5SiH_2Cl} \tag{d}$$

Boron trichloride and PCl_5 are also sometimes used[6]:

$$Si_2H_6 \xrightarrow{BCl_3} Si_2H_5Cl \tag{e}$$

Boron trichloride does not react[7] with $HSiCl_3$ even at 150°C. With SO_2Cl_2, the reaction rate decreases as substitution of Cl on Si increases $R_3SiH > R_2SiHCl \gg HSiCl_3$. Group-IVB tetrachlorides are used as halogenating agents, the Pb derivative excepted. By means of CCl_4 or $HCCl_3$, the polysilicon hydrides Si_2H_6, Si_3H_8 and Si_4H_{10} are halogenated[8] with $AlCl_3$ as a catalyst at ca. 50°C. A Pd catalyst is superior and also works with CH_2Cl_2. Benzoyl peroxide allows monochlorination of trialkylsilanes and especially monophenylsilane[9] to give $C_6H_5SiH_2Cl$, while $GeCl_4$ and $SnCl_4$ lead to partial halogenation and, e.g., $SnCl_4$ reacts[10] in near 70% yield, with $ClCH_2SiH_3$ to form $ClCH_2SiH_2Cl$. Bromination is less used than chlorination. N-Bromosuccinimide brominates SiH_4, but more highly brominated derivatives than SiH_3Br are formed. At 0°C BBr_3 reacts with SiH_4 and gives mainly monobromination[11]. The polysilane Si_4H_{10} reacts with HCI_3 or PI_3 without a catalyst to form polyiodotetrasilanes[12]. The reaction in which the Si—H bond is halogenated by alkenyl or aryl halides is usually run to introduce organic groups into organohalosilicon hydrides or $HSiCl_3$. The gas-phase thermal condensation that increases organic substitution competes with a side reaction which involves exchange between Si—H bonds and alkyl, aryl and acyl halides.

Under mild conditions in Et_2O substitution is possible for trialkylsilanes without a catalyst, either with allyl halides[13]:

$$(C_2H_5)_3SiH + BrCH_2CH{=}CH_2 \rightarrow (C_2H_5)_3SiBr + CH_3CH{=}CH_2 \tag{f}$$

or with acylhalides[14]:

$$(C_6H_5CH_2)_3SIH + Cl{-}\overset{\displaystyle O}{\overset{\|}{C}}{-}C_6H_5 \rightarrow (C_6H_5CH_2)_3SiCl + C_6H_5{-}\overset{\displaystyle O}{\overset{\|}{C}}{-}H \tag{g}$$

Palladium metal is also an effective catalyst for halogenation using aryl halides[15].

Cleavage of the Ge—H bonds in organosubstituted germanes uses the same reagents as for Si—H cleavage, and many organometallic and inorganic halides convert Ge—H to Ge—X bonds[16], e.g., $HgCl_2$, $AlCl_3$, SO_2Cl_2, CCl_4, $GeCl_4$; yields are 75–90%. With $GeCl_4$ redistribution occurs which lead to $HGeCl_3$ and chlorination of the R_3GeH.

Dialkyldihalogermanes give a similar reaction with an AlX_3 catalyst:

$$R_3GeH + R_2GeX_2 \xrightarrow{AlX_3} R_2GeHX + R_3GeX \tag{h}$$

At RT, Hg(II) halides quantitatively monohalogenate organogermanes:

$$C_2H_5GeH_3 \xrightarrow{HgCl_2} C_2H_5GeH_2Cl \tag{i}$$

Using chloromethylmethyl ether chlorination may be taken one stage further within $AlCl_3$ as a catalyst:

$$C_2H_5GeH_2Cl \xrightarrow[AlCl_3]{ClCH_2OCH_3} C_2H_5GeHCl_2 + CH_3OCH_3 \qquad (j)$$

In some cases $ClCH_2OCH_3$ combines without a catalyst[17].

Organic reagents, such as chloroacetone, trichloroacetic acid and acid chlorides, halogenate R_3GeH. Highly selective free-radical reactions occur when N-chloro-, N-bromo- or N-iodosuccinimide are used with phenyl- and diphenylgermane[18]. The N-halosuccinimides give also monosubstitution for alkyltrihydrides like ethylgermane[19]. Organic halides, including acyl halides, react with alkylgermanium hydrides and monohalogenation occurs[20]:

$$(C_2H_5)_2GeH_2 \xrightarrow{C_2H_5I} C_2H_5GeHI + C_2H_6 \qquad (k)$$

Alkyl iodides give the best yields. Reactivity increases Cl to I. Substitution by organic groups or halogens reduces Ge bond reactivity. In the more reactive cases the reaction is exothermic, but mono- and dihydrides require high T to initiate.

Tin hydrides are used to produce halogenostannanes by cleavage of the Sn—H bond with inorganic and organic halides; e.g., the easily handled triphenyltin hydride gives triphenylchlorostannane in high yields with PCl_3, $HgCl_2$ and $SbCl_3$[21] and Ph_2PCl[22]; CCl_4 and $CHCl_3$ react exothermically with trialkylstannanes also to give good yields of trialkyltinchlorides[21,23]. N-bromosuccinimide converts triphenylstannane to triphenylbromostannane in 86% yield[24]. Other group-IVB halides react, e.g., $(CH_3CO_2CH_2)_3GeBr$ with triethylstannane[25] to give $(CH_3CO_2CH_2)_3GeH$ and triethylbromostannane.

Redistribution between tin hydrides and tin halides affords mixed substituted stannanes. When n-Bu_2SnX_2 react[26,27] with n-Bu_2SnH_2 to give n-Bu_2SnHX, these products may lose H_2 to give as decomposition products the unstable distannanes, n-$Bu_2SnXSnXBu_2$-n. The Sn—H bond is substituted when organic hydrides reduce alkyl halides[28] in exothermic reactions:

$$R_3SnH + RX \rightarrow R_3SnX + RH \qquad (l)$$

Even with allyl bromide, reduction occurs in preference to addition across the double bond, and triphenylstannane gives Ph_3SnBr and propene quantitatively.

The scope of these reactions is known[29,30], and kinetic results are consistent with a free-radical mechanism in which organotin radicals attack the alkyl halide. The catalysis by azobis-isobutyronitrile supports this mechanism. The order of reactivity is $RI > RBr > RCl > RF$, and for the tin hydrides $(C_6H_5)_2SnH_2 \approx$ n-$BuSnH_3 >$ $(C_6H_5)_3SnH_5 \approx$ n-$Bu_2SnH_2 >$ n-Bu_3SnH. Electronic and solvent effects on aromatic halides show similar results[31]. However, aryl halides require high T to be reduced by organotin hydrides.

Organotin halides are formed in good yields from organostannanes with acid chlorides or bromides[32,33]. The organic products obtained, esters or aldehydes, depend upon the structure of the organic part of the acid halide used.

The thermal stability of the group-IVB hydrides decrease from carbon to Pb. Thus preparation of Pb halides is realized by halogenation of Pb—C bonds, which are adequately reactive, instead of Pb—H bonds. The reactivity of the Pb—H bond is used to realize quantitative estimation of organolead hydrides. Ethyl iodide reacts with trialkyllead hydrides to give ethane and trialkyllead iodide[34].

(M. A. DELMAS)

1. B. Taub, J. B. Hino, *J. Org. Chem.*, *25*, 263 (1960).
2. R. P. Hollandworth, W. M. Ingle, M. A. Ring, *Inorg. Chem.*, *6*, 844 (1969).
3. H. H. Anderson, *J. Am. Chem. Soc.*, *80*, 5083 (1958).
4. S. K. Gondal, A. G. MacDiarmid, F. E. Saalfeld, M. V. Dowell, *Inorg. Nucl. Chem. Lett.*, *5*, 351 (1969).
5. S. J. Mawaziny, *J. Chem. Soc., A.*, 1641 (1970).
6. J. E. Drake, N. Goddard, *J. Chem. Soc., A.*, 2587 (1970).
7. O. Ruff, C. Albert, *Chem. Ber.*, *38*, 2222 (1905).
8. A. Stock, P. Stiebeler, *Chem. Ber.*, *56*, 1087 (1923).
9. Y. Nagai, T. Yoshihara, S. Nakaido, *Bull. Chem. Soc. Jpn.*, *40*, 2214 (1967).
10. N. S. Nametkin, T. I. Chernysheva, O. V. Kuz'min, K. I. Kobrakov, *Dokl. Chem.* (*Engl. Transl.*), *178*, 1332 (1968).
11. J. E. Drake, J. Simpson, *Inorg. Nucl. Chem. Lett.*, *2*, 219 (1966).
12. H. J. Emeléus, A. G. Maddock, *J. Chem. Soc.*, 1131 (1946).
13. H. Westermark, *Acta Chem. Scand.*, *8*, 1089 (1954).
14. J. W. Jenkins, H. W. Post, *J. Org. Chem.*, *15*, 559 (1950).
15. J. D. Citron, J. E. Lyons, L. H. Sommer, *J. Org. Chem.*, *34*, 638 (1969).
16. M. Massol, J. Satgé, *C.R. Hebd. Séances Acad. Sci.*, *261*, 170 (1965).
17. P. Rivière, J. Satgé, *Bull. Soc. Chim. Fr.*, 4039 (1967).
18. P. Rivière, J. Satgé, *Bull. Soc. Chim. Fr.*, 1773 (1966).
19. M. Massol, J. Satgé, *Bull. Soc. Chim. Fr.*, 2737 (1966).
20. J. Satgé, *Ann. Chim.* (*Paris*), *6*, 519 (1961).
21. A. E. Borisov, A. N. Abramova, *Izv. Akad. Nauk SSSR, Ser. Khim.* 844 (1964).
22. D. Seyferth, Y. Sata, M. Takamizawa, *J. Organomet. Chem.* *2*, 367 (1964).
23. H. Kriegsmann, K. Ullricht, *Z. Chem.* *3*, 67 (1963).
24. E. J. Kupchik, T. Lanigan, *J. Org. Chem.*, *27*, 3661 (1962).
25. Yu. I. Bankov, G. S. Burlachenkow, Yu. I. Belavin, I. F. Lutsenko, *J. Gen. Chem. USSR* (*Engl. Transl.*) *38*, 1846 (1968).
26. A. K. Sawyer, H. G. Kuivila, *Chem. Ind.* (*London*), 260 (1961).
27. A. K. Sawyer, J. E. Brown, H. G. Kuivila, *J. Organomet. Chem.* *3*, 464 (1965).
28. J. G. Noltes, G. J. M. Van der Kerk, *Chem. Ind.* (*London*), 294 (1959).
29. H. G. Kuivila, L. W. Menapace, *J. Org. Chem.*, *28*, 2165 (1963).
30. H. G. Kuivila, L. W. Menapace, C. R. Warner, *J. Am. Chem. Soc.* *84*, 3584 (1962).
31. D. H. Lorentz, P. Shapiro, A. Stern, E. I. Becker, *J. Org. Chem.*, *28*, 2332 (1963).
32. E. J. Wash, H. G. Kuivila, *J. Am. Chem. Soc.*, *86*, 3047 (1964).
33. E. J. Kupchik, R. J. Kiesel, *J. Org. Chem.*, *31*, 456 (1966).
34. W. P. Neumann, K. Kühlen, *Angew. Chem., Int. Ed. Engl.*, *4*, 784 (1965).

2.5.6. from Cleavage of the Group IVB–Carbon Bond

2.5.6.1. the Si–Carbon Bond

2.5.6.1.1. by Halogens.

Halogenation of the alkyl group by Cl_2 and Br_2 occurs without cleavage, an atomic mechanism being involved[1] (X = Cl, Br):

$$R_4Si + X_2 \rightarrow R_3SiX + RX \qquad (a)$$

Higher alkyl groups (Et, Pr, etc.) react readily with Br_2 but less readily with Cl_2. Under UV irradiation Cl_2 reacts with methyl compounds of Si easily, whereas Br_2 requires the presence of traces of Cl_2 for reaction to occur. More highly substituted compounds react more easily; disubstituted derivatives are almost always formed as well.

Iodine in the presence of catalytic AlI_3 cleaves the C—Si bond in tetraalkylsilanes with the formation of alkyliodosilanes and reacts one step further with the ethyl group to form Et_2SiI_2:

$$Et_4Si + I_2 \xrightarrow{AlI_3} Et_3SiI + EtI \tag{b}$$

$$Et_3SiI + I_2 \xrightarrow{AlI_3} Et_2SiI_2 + EtI \tag{c}$$

The ease of displacement of alkyl groups is Ph $\gg$ Me > Et > n-Pr > i-Pr [2]. Owing to steric effects, $(i\text{-}Pr)_3SiMe$ and $(i\text{-}Pr)_2SiMe_2$ react:

$$Me_2Si(Pr\text{-}i)_2 + I_2 \xrightarrow{AlI_3} Me(Pr\text{-}i)_2SiI + MeI \tag{d}$$

Anomalous results can be obtained owing to the relative ease of cleavage of, e.g., the C_3F_7 group[3]:

$$Me_2Si(C_3F_7)_2 + 2\,I_2 \xrightarrow[C_6H_5CH_3]{reflux} Me_2SiI_2 + C_3F_7I \tag{e}$$

Cleavage of tetraethylsilane by I_2 in the presence of AlI_3 may be due to the predominant electrophilic attachment of I^+ on carbon with AlI_3 assisting in the ionization of the halogen[4].

Like I_2, Br_2 cleaves allyltrimethylsilane, although some electrophilic reagents add to the double bond without cleavage[5,6]:

$$Me_3SiCH_2CH{=}CH_2 + Br_2 \rightarrow Me_3SiBr + BrCH_2CH{=}CH_2 \tag{f}$$

The first step in the electrophilic cleavage is the addition of a positive fragment to form a β-carbonium ion, this being followed by attack of a negative ion, either at the β-carbon (resulting in addition) or at Si (resulting in cleavage)[7]:

$$Me_3SiCH_2CH{=}CH_2 + Br_2 \rightarrow [Me_3SiCH_2CHCH_2Br]^+ + Br^- \tag{g}$$

$$Br^- + [Me_3SiCH_2CHCH_2Br]^+ \rightarrow Me_3SiBr + CH_2{=}CHCH_2Br \tag{h}$$

Fission of the Si—Ph bond in $PhSiMe_3$ by I_2 takes place[8] on refluxing for 12 h, yielding Me_3SiI:

$$Me_3SiPh + I_2 \rightarrow Me_3SiI + PhI \tag{i}$$

Iodination of $PhSiMe_3$ in dry acetic acid at 25°C by ICl yields PhI and hexamethyldisiloxane in 90% yield:

$$Me_3SiPh + ICl \rightarrow PhI + Me_3SiCl \tag{j}$$

The effectiveness of ICl is decreased by the formation of the complex ion:

$$ICl + Cl^- \rightleftharpoons [ICl_2]^- \tag{k}$$

Iododesilylation also occurs with an interaction between Si and Cl_2 in a four-center transition state[9].

(J. M. BELLAMA)

1. C. Eaborn, *Organosilicon Compounds*, Butterworths, London, 1960, p. 379.
2. C. Eaborn, *J. Chem. Soc.*, 2755 (1949).
3. H. C. Clark, J. T. Kwon, D. Whyman, *Can. J. Chem.*, *41*, 2628 (1963).
4. C. Eaborn, R. W. Bott, in *Organometallic Compounds of the Group IV Elements*, A. G. MacDiarmid, ed., Vol. 1 part 1, Marcel Dekker, New York, 1968.
5. C. L. Agre, W. Hilling, *J. Am. Chem. Soc.*, *74*, 3895 (1952).
6. D. Grafstein, *J. Am. Chem. Soc.*, *77*, 6650 (1955).
7. L. H. Sommer, L. J. Tyler, F. C. Whitmore, *J. Am. Chem. Soc.*, *70*, 2872 (1948).
8. B. O. Pray, L. H. Sommer, G. M. Goldberg, G. T. Kerr, P. A. Di Giorgio, F. C. Whitmore, *J. Am. Chem. Soc.*, *70*, 433 (1948).
9. W. E. Evison, F. S. Kipping, *J. Chem. Soc.*, 2774 (1931).

2.5.6.1.2. by Hydrogen and Alkyl Halides.

In the presence of an alkylation catalyst, such as $AlCl_3$, one methyl group is split from Me_4Si by HCl[1]:

$$Me_4Si + HCl \xrightarrow{AlCl_3} Me_3SiCl + CH_4 \qquad (a)$$

The relative rates of splitting of alkyl groups are[2] $CH_3 \gg C_2H_5 > n\text{-}C_4H_9, n\text{-}C_3H_7 > i\text{-}C_3H_7$.

At 60°C a mixture of MeBr and $AlBr_3$ react with Me_4Si to form ethane:

$$Me_4Si + MeBr \xrightarrow{AlBr_3} Me_3SiBr + C_2H_6 \qquad (b)$$

The mechanism involves the slight polarization of the Si–methyl or of the Si–halogen bond produced to yield an intermediate with sufficient energy to attack another molecule, thereby producing redistribution products[3]:

$$\underset{\textbf{I}}{(CH_3)_3\overset{\delta+}{Si} \text{---} R\text{---}\overset{\delta-}{Al}Br_3} \quad \text{or} \quad \underset{\textbf{II}}{\overset{\delta+}{C}H_3 \text{---} Br\text{---}\overset{\delta-}{Al}Br_3}$$

Cleavage of one Si—C bond occurs[4] by refluxing for 5 h a freshly prepared solution of $n\text{-}Bu_4Si$ and i-amylchloride in the presence of 4% $AlCl_3$:

$$(n\text{-}C_4H_9)_4Si + i\text{-}C_5H_{11}Cl \xrightarrow[\text{reflux}]{AlCl_3} (n\text{-}C_4H_9)_3SiCl \qquad (c)$$

Tetraethylsilane undergoes cleavage with the evolution of heat when refluxed for several hours with i-PrCl in the presence of catalytic $AlCl_3$ to give[5] Et_3SiCl (92.2%) and a mixture of alkanes (C_2H_6, C_3H_8, $i\text{-}C_5H_{12}$) and C_2H_4:

$$Et_4Si + i\text{-}C_3H_7Cl \xrightarrow[\text{reflux}]{AlCl_3} \underset{(92.2\%)}{Et_3SiCl} + \text{alkanes} \qquad (d)$$

Anhydr HI with p-$ClC_6H_4SiH_3$ in a sealed tube for 12 h at RT forms the cleavage product H_3SiI in 60% yield[6]:

$$\text{p-ClC}_6\text{H}_4\text{SiH}_3 + \text{HI} \rightarrow \text{C}_6\text{H}_5\text{Cl} + \text{H}_3\text{SiI} \quad \text{(e)}$$

Liquid HCl cleaves a phenyl group from $PhSiH_3$ at −78°C:

$$\text{PhSiH}_3 + \text{HCl} \xrightarrow{-78°\text{C}} \text{PhH} + \text{H}_3\text{SiCl} \quad \text{(f)}$$

The reaction is slow (8% in 48 h at −78°C; no reaction in 15 h at 130°C). An Al metal catalyst redistributes the product. The ease of splitting of the Si—Ph bond by halogen acids decreases[7] with increasing negative character of the substituents of Si.

The cleavage of $PhSiH_2I$ into benzene and H_2SiI_2 is brought about[8] by liq HI at −30°C:

$$\text{PhSiH}_2\text{I} + \text{HI} \rightarrow \text{PhH} + \text{H}_2\text{SiI}_2 \quad \text{(g)}$$

This is a simple method for preparing H_2SiI_2 but does not proceed as readily as the corresponding reactions[9] of $C_6H_5SiH_3$.

Halosilanes are obtained by the cleavage[10] of phenylsilanes with HCl; e.g., diphenylsilane and xs dry HCl are reacted for 42 h at −78°C:

$$\text{Ph}_2\text{SiH}_2 + \text{HCl} \rightarrow \text{PhH} + \underset{(50\%)}{\text{PhSiH}_2\text{Cl}} \quad \text{(h)}$$

Cleavage of the Si—Ph bond in phenylsilane is brought about[7,11] quantitatively by liq HI at −40°C.

$$\text{PhSiH}_3 + \text{HI} \xrightarrow{-40°\text{C}} \text{PhH} + \text{H}_3\text{SiI} \quad \text{(i)}$$

Hydriodic acid cleaves bis-1,1-trimethylsilylethene at −80°C:

$$(\text{Me}_3\text{Si})_2\text{C}{=}\text{CH}_2 + \text{HI} \xrightarrow{-80°\text{C}} \text{Me}_3\text{SiI} + \text{Me}_3\text{SiCH}{=}\text{CH}_2 \quad \text{(j)}$$

The cleavage, as contrasted to addition of HI to $R_3SiCH{=}CH_2$, is attributed to the effect of the vinyl group on the two silyl groups and explained by the interaction of electrons of the double bond with the 3d-orbitals of the adjacent Si atoms[12].

(J. M. BELLAMA)

1. P. D. George, U.S. Pat. 2,802,852 (1957); *Chem. Abstr., 51*, 17,982 (1957).
2. G. A. Russell, K. L. Hagpal, *Tetrahedron Lett.*, 421 (1961).
3. G. A. Russell, *J. Am. Chem. Soc., 81*, 4831 (1959).
4. Dao-Huy-Giao, *C. R. Hebd. Séances Acad. Sci., 260*, 6937 (1965).
5. N. S. Vyazankin, G. A. Razuvaev, O. S. Dyachkovskaya, *J. Gen. Chem. USSR (Engl. Transl.), 33*, 613 (1963).
6. B. J. Aylett, I. A. Ellis, *J. Chem. Soc.*, 3415 (1960).
7. G. Fritz, D. Kumar, *Z. Anorg. Allg. Chem., 308*, 105 (1961).
8. G. Fritz, D. Kumar, *Z. Anorg. Allg. Chem., 306*, 191 (1960).
9. G. Fritz, D. Kumar, *Z. Anorg. Allg. Chem., 310*, 327 (1961).
10. G. Fritz, D. Kumar, *Chem. Ber., 94*, 1143 (1961).

11. G. Fritz, D. Kumar, *Z. Anorg. Allg. Chem., 304*, 322 (1960).
12. G. Fritz, J. Grobe, *Z. Anorg. Allg. Chem., 309*, 98 (1961).

2.5.6.1.3. by Other Halides.

The cleavage of an alkyl group from Me_4Si is effected by molten $GaCl_3$ and refluxing at 70–90°C for 10 min under N_2. Trimethylchlorosilane and $MeGaCl_2$ are produced:

$$(CH_3)_4Si + GaCl_3 \rightarrow \left[(CH_3)_3Si\overset{CH_3}{\underset{Cl}{\rightleftarrows}}GaCl_2\right] \rightarrow \underset{(90\%)}{(CH_3)_3SiCl + CH_3GaCl_2} \qquad \text{(a)}$$

Dimethylpolysiloxanes are also cleaved by anhydr $GaCl_3$ by warming slightly above RT under N_2. Simultaneous transalkylation also occurs:

$$[(CH_3)_2SiO]_x + GaCl_3 \rightarrow \underset{(85\%)}{CH_3GaCl_2} + [CH_3SiClO]_x \qquad \text{(b)}$$

This also provides a simple method for the monoalkylation of Ga trihalides[1].

Cleavage of an alkyl group from Si occurs with $HgCl_2$ in a reaction analogous to redistribution of tetraalkylsilanes. Tetraethylsilane and $HgCl_2$ heated for 2 h at 146–150°C yields 15% EtMgCl, although the Si derivative is not isolated[2]:

$$Et_4Si + HgCl_2 \rightarrow \underset{(15\%)}{EtHgCl} + Et_3SiCl \qquad \text{(c)}$$

Some mono- and bis(trifluorosilyl)methane and -chloromethane may be prepared from the corresponding chloro compounds by fluorination with SbF_3.

Tris(trifluorosilyl)chloromethane is obtained[3] by adding $(Cl_3Si)_3CCl$ in dry n-Bu_2O into SbF_3 in the same solvent at −70°C, and warming to 5–10°C:

$$(Cl_3Si)_3CCl + SbF_3 \rightarrow (F_3Si)_3CCl + SiF_4 \qquad \text{(d)}$$

Adding $(Cl_3Si)CH$ in dry xylene with stirring and cooling into SbF_3 in the same solvent and warming to RT forms[4] $(F_3Si)_3CH$ after 1–2 h:

$$(Cl_3Si)_3CH + SbF_3 \rightarrow SiF_4 + (F_3Si)_3CH \qquad \text{(e)}$$

Siloxane bis(β-diketone), when treated with anhyd $BeCl_2$ in a nonpolar solvent such as benzene, xylene, or decalin, cleaves, yielding a polymer with a chelated Be main chain alternating regularly with trimethylene and dimethylsiloxane groups[5]:

$$2\,Me_3SiCH(C(=O)Me)(C(=O)OEt) + BeCl_2 \rightarrow Be\left[O{=}C(Me){-}CH{=}C(OEt){-}O\right]_2 + 2\,Me_3SiCl \qquad \text{(f)}$$

Cleavage of an alkyl group from Si in (n-chloropropyl)methylsilacyclopentane is effected[6] by heating the cyclopentane with 1–5% $AlCl_3$ under Ar at atm P.

$$\text{(cyclo-}C_4H_8\text{)Si(Me)}CH_2CH_2CH_2Cl + AlCl_3 \rightarrow \text{(cyclo-}C_4H_8\text{)Si(Me)Cl} + \text{cyclo-}(CH_2)_3 \tag{g}$$

Slow addition of $AlCl_3$ to a solution of β-trimethylsilylpropionyl chloride at RT yields trimethylchlorosilane, ethylene, and CO:

$$Me_3SiCH_2CH_2COCl + AlCl_3 \rightarrow \underset{(87\%)}{Me_3SiCl} + \underset{(30\%)}{CH_2{=}CH_2} + CO \tag{h}$$

Both acylation and elimination are involved and proceed through the same intermediate; $Me_3SiCH_2CH_2COCl—AlCl_3$, the fate of which depends upon the availability of an acylable substrate[7].

Cleavage of Ph_2SiCl_2 by $AlCl_3$ occurs at RT on addition of EtBr in five installments over a period of 2 days. The mixture is warmed gently, driving off halogen acids to form[8] $PhSiCl_3$:

$$SiPh_2Cl_2 + AlCl_3 \rightarrow PhAlCl_2 + SiPhCl_3 \tag{i}$$

On heating at 150°–200°C for 7 h, diphenyldichlorosilane with $SbCl_5$ produces[9] $PhSiCl_3$ and $SbCl_3$:

$$Ph_2SiCl_2 + SbCl_5 \rightarrow PhSiCl_3 + PhSbCl_4 \rightarrow PhSiCl_3 + SbCl_3 + PhCl \tag{j}$$

The phenyl is recovered as PhCl in a disproportionation with the intermediate formation of unstable $PhSbCl_4$, which cleaves to $SbCl_3$ and PhCl.

Diphenyldichlorosilane with dearyllation reagents, such as $FeCl_3$, loses[10] a Ph radical to form $PhSiCl_3$ and $[PhFeCl_2]$. The cyclohexane solution obtained by mixing Ph_2SiCl_2 and $FeCl_3$ at 0°C is heated to 35–40°C for 7 h:

$$Ph_2SiCl_2 + FeCl_3 \rightarrow PhSiCl_3 + [PhFeCl_2]. \tag{k}$$

(J. M. BELLAMA)

1. H. Schmidbaur, W. Findeiss, *Angew. Chem. Int. Ed. Engl.*, *3*, 696 (1964).
2. Z. M. Manulkin, *Zh. Obshch. Khim.*, *16*, 235 1946); *Chem. Abstr.*, *41*, 90 (1947).
3. R. Müller, S. Reichel, R. Dathe, *Chem. Ber.*, *97*, 1673 (1964).
4. R. Müller, W. Müller, *Chem. Ber.*, *97*, 1111 (1964).
5. A. Hofer, H. Kuckertz, M. Sander, *Makromol. Chem.*, *90*, 38 (1966).
6. V. M. Vdovin, N. S. Nametkin, V. I. Zav'yalov, K. S. Pushchevaya, *J. Prakt. Chem.*, *23*, 281 (1964).
7. E. V. Wilkus, W. H. Rauscher, *J. Org. Chem.*, *30*, 2889 (1965).
8. L. M. Stock, A. R. Spector, *J. Org. Chem.*, *28*, 3272 (1963).
9. A. Ya. Yakubovich, G. V. Motsarev, *J. Gen. Chem. USSR* (*Engl. Transl.*), *23*, 1414 (1953).
10. A. Ya. Yakubovich, G. V. Motsarev, *J. Gen. Chem. USSR* (*Engl. Transl.*), *23*, ,1059 (1953).

2.5.6.2. the Ge–Carbon Bond

The Ge—C bond in tetraalkylgermanes can be cleaved by halogens, halogen acids and alkyl halides.

(J. M. BELLAMA)

2.5.6.2.1. by Halogens.

Tetraethylgermane can be brominated to give Et_3GeBr by Br_2 at 40°C for 6 days[1]. The Br_2 can be added in anhyd EtBr to Et_4Ge at 0°C and then the mixture maintained at 40–50°C for 24 h[2]:

$$(C_2H_5)_4Ge + Br_2 \xrightarrow{C_2H_5Br} (C_2H_5)_3GeBr + C_2H_5Br \qquad (a)$$
$$(82\%)$$

Similarly, Me_3GeBr is produced from Me_4Ge and Br_2 (ca. 0.2 mol xs) in a sealed tube for a week and subsequent fractional distillation[3]:

$$(CH_3)_4Ge + Br_2 \xrightarrow[RT]{1\ week} (CH_3)_3GeBr + CH_3Br \qquad (b)$$

Bromotrimethylgermane is produced in good yield when Br_2 is added in separate portions to Me_4Ge in n-PrBr. The reactants are boiled for 20 h during which time the T rises from 47° to 78°C. Fractionation gives Me_3GeBr in 98% yield[4]:

$$(CH_3)_4Ge + Br_2 \xrightarrow[n\text{-}C_3H_7Br]{reflux\ 20\,h} (CH_3)_3GeBr + CH_3Br \qquad (c)$$
$$(98\%)$$

With catalytic $AlBr_3$, Me_2GeBr_2 can be prepared in the same way by boiling Me_4Ge, and i-PrBr for 1 h until the T rises to 110°C[4]:

$$(CH_3)_3GeBr + Br_2 \xrightarrow[AlBr_3]{reflux\ 1\ h} (CH_3)_2GeBr_2 + 2\ CH_3Br \qquad (d)$$
$$(79\%)$$

With an electron-pair acceptor acid, cleavage by I_2 occurs and also allows access to the halogenated derivatives. Specifically, I_2 cleaves the Ge—C bond in presence of AlI_3 in only a few minutes[5]:

$$R_4Ge + I_2 \xrightarrow{AlI_3} R_3GeI + RI \qquad (e)$$

where R = CH_3, C_2H_5. However, with heavier tetraalkylgermanes, the reaction of I_2 is 2:1:

$$2\ R_4Ge + I_2 \xrightarrow{AlI_3} 2\ R_3GeI + R_2 \qquad (f)$$

where R = n-C_4H_9, i-C_4H_9, n-C_5H_{11}.

Homologous compounds larger than n-Pr_4Ge give reaction (f). Tetra-n-propylgermane follows both pathways and gives a mixture of n-propyliodide and alkanes[5]:

$$(n\text{-}C_3H_7)_4Ge + I_2 \rightarrow (n\text{-}C_3H_7)_3GeI + n\text{-}C_3H_7I + n\text{-}C_6H_{14} \qquad (g)$$

Iodine can cleave two or even three Ge—C(alkyl) bonds with a catalyst but does so alone only with difficulty. The action of the halogen varies as the number of substituents increases, e.g.:

$$(C_2H_5)_4Ge + I_2 \xrightarrow{AlI_3} (C_2H_5)_3GeI + C_2H_5I \qquad (h)$$

If the reaction is allowed to proceed further, I_2 will simultaneously follow both the pathways noted above:

$$2\ (C_2H_5)_3GeI + I_2 \rightarrow 2\ (C_2H_5)_2GeI_2 + n\text{-}C_4H_{10} \quad \text{(i)}$$

$$(C_2H_5)_3GeI + I_2 \rightarrow (C_2H_5)_2GeI_2 + C_2H_5I \quad \text{(j)}$$

The reaction proceeds further[5] in the presence of xs AlI_3:

$$2\ (C_2H_5)_2GeI_2 \rightarrow 2\ C_2H_5GeI_3 + n\text{-}C_4H_{10}$$

At low T olefinic derivatives add a Br_2 molecule, but cleavage of the olefinic chain occurs under the same conditions for derivatives containing an allyl, methallyl, or butenyl group. Cyclopentadienyl groups are cleaved[6] by Br_2:

$$(C_2H_2)_3Ge\text{—}C_5H_5 + Br_2 \rightarrow (C_2H_5)_3GeBr + C_5H_5\text{—}Br \quad \text{(k)}$$

Owing to ring strain and polarizability of the intracyclic Ge—C bonds, germacyclobutanes possess higher chemical reactivity than do cyclobutanes. Bromine cleaves one intracyclic Ge—C bond at 0°C even in the absence of catalyst[7]:

$$(C_2H_5)_2Ge\langle\text{cyclo-}(CH_2)_3\rangle + Br_2 \xrightarrow{0°C} (C_2H_5)_2Ge(Br)(CH_2)_3Br \quad \text{(l)}$$

Since there is no strain in the saturated germacyclopentane ring, it resists reaction. Bromine in EtBr has no action on diethylgermacyclopentane, but the ring is cleaved with an $AlBr_3$ catalyst:

$$(C_2H_5)_2Ge\langle\text{cyclo-}(CH_2)_4\rangle + Br_2 \xrightarrow{AlBr_3} (C_2H_5)_2GeBr_2 + C_4H_8 \quad \text{(m)}$$

Such reactions illustrate the effect of the ring size and group type on chemical reactivity. Thus, Br_2 will cleave a Ge—C_6H_5 bond in preference to a five-membered ring, and a four-membered ring is even more reactive, e.g., in the action of Br_2 on 4-germaspiro[3.4] octane[8]:

$$\text{4-germaspiro[3.4]octane} + Br_2 \xrightarrow[0°C]{EtBr} \text{cyclo-}(CH_2)_4Ge(Br)(CH_2)_3Br \quad \text{(n)}$$

One or two phenyl groups are cleaved from Ph_4Ge by Br_2. Because of angular tension, Ph—Ge is cleaved preferentially by Br_2 over alkyl or cyclic groups, e.g.:

$$\text{cyclo-}(CH_2)_4GePh_2 + 2\ Br_2 \rightarrow \text{cyclo-}(CH_2)_4GeBr_2 \quad \text{(o)}$$

$$\text{cyclo-}(CH_2)_4GePh_2 + Br_2 \xrightarrow[0°]{EtBr} \text{cyclo-}(CH_2)_4Ge(Ph)Br \quad \text{(p)}$$

Phenylbromogermacyclopentane is produced by the reaction of Br_2 and diphenylgermacyclopentane in EtBr at 0°C. Dibromogermacyclopentane is obtained in 80% yield by the slow addition of Br_2 in EtBr for 10 h in the absence of light:

$$\text{(CH}_2)_4\text{Ge(C}_6\text{H}_5)_2 + Br_2 \xrightarrow[C_2H_5Br]{0°C} \text{(CH}_2)_4\text{Ge(C}_6\text{H}_5)\text{Br} \xrightarrow[C_2H_5Br]{Br_2} \text{(CH}_2)_4\text{GeBr}_2 \qquad (q)$$

The reactivity of the Ge—C(phenyl) bond may be used in producing ethylbromogermacyclopentane from Br_2 with ethylphenylgermacyclopentane in EtBr, like the production of bromophenylgermacyclopentane cited above:

$$\text{(CH}_2)_4\text{Ge(C}_2\text{H}_5)(\text{C}_6\text{H}_5) + Br_2 \rightarrow \underset{(85\%)}{\text{(CH}_2)_4\text{Ge(C}_2\text{H}_5)\text{Br}} + C_6H_5Br \qquad (r)$$

Similarly, phenylbromogermacyclohexane results from the action of Br_2 on diphenylgermanecyclohexane[9]:

$$\text{(CH}_2)_5\text{Ge(C}_6\text{H}_5)_2 \xrightarrow{Br_2} \text{(CH}_2)_5\text{Ge(C}_6\text{H}_5)\text{Br} \qquad (s)$$

Unsaturated compounds like $R_3GeCH_2CH{=}CH_2$ are cleaved by Br_2 under mild conditions[10]:

$$(n\text{-}C_4H_9)_3GeCH{=}CH_2 + Br_2 \xrightarrow{-80°C} (n\text{-}C_4H_9)_3GeBr + CH_2{=}CHCH_2Br \qquad (t)$$

Vinylic compounds such as styrylgermane react similarly[11] with Br_2:

$$(n\text{-}C_4H_9)_3GeCH{=}CHC_6H_5 \xrightarrow{Br_2} (n\text{-}C_4H_9)_3GeBr + C_6H_5CH{=}CHBr \qquad (u)$$

Bromine addition to the double bond can occur instead of cleavage[10]:

$$(n\text{-}C_4H_9)_3GeCH{=}CH_2 + Br_2 \rightarrow (n\text{-}C_4H_9)_3GeCHBrCH_2Br \qquad (v)$$

Organogermanes with substituents in the β position are cleaved when Br_2 in $CHCl_3$ is heated[12]:

$$(n\text{-}C_4H_9)_3GeCH_2CO_2CH_3 + Br_2 \rightarrow \underset{(88.5\%)}{(n\text{-}C_4H_9)_3GeBr} + \underset{(71\%)}{BrCH_2CO_2CH_3} \qquad (w)$$

(J. M. BELLAMA)

1. C. A. Kraus, E. A. Flood, *J. Am. Chem. Soc.*, *54*, 1635 (1932).
2. C. Eaborn, K. C. Pande, *J. Chem. Soc.*, 3200 (1960).
3. M. P. Brown, G. W. A. Fowler, *J. Chem. Soc.*, 2811 (1958).
4. V. F. Mironov, A. L. Kranchenko, *Bull. Acad. Sci. USSR, Ser. Chem.*, 988 (1965).
5. M. Lesbré, P. Mazerolles, *C. R. Hebd. Séances Acad. Sci.*, *246*, 1708 (1958).
6. M. Lesbré, P. Mazerolles, G. Manuel, *C. R. Hebd. Séances Acad. Sci.*, *255*, 544 (1962).
7. P. Mazerolles, *Bull. Soc. Chim. Fr.*, 1907 (1962).
8. P. Mazerolles, J. Dubac, M. Lesbré, *J. Organomet. Chem.*, *5*, 35 (1966).

9. P. Mazerolles, G. Manuel, *Bull. Soc. Chim., Fr.*, 327 (1966).
10. P. Mazerolles, M. Lesbré, *C. R. Hebd. Séances Acad. Sci., 248*, 2018 (1959).
11. M. Lesbré, J. Satgé, *C. R. Hebd. Séances Acad. Sci., 250*, 2220 (1960).
12. I. F. Lutsenko, Yu. I. Baukov, B. N. Xasapov, *J. Gen. Chem. USSR* (*Engl. Transl.*), *33*, 2724 (1963).

2.5.6.2.2. by Hydrogen and Alkyl Halides.

Iodine monochloride cleaves the cyclobutane ring of dibutylgermacyclobutane. Adding a solution of ICl in EtI to germacyclobutane, followed by distillation, yields[1]:

$$(n\text{-}C_4H_9)_2Ge\diamond + ICl \rightarrow (n\text{-}C_4H_9)_2Ge(Cl)(CH_2)_3I \qquad \text{(a)}$$

Gaseous HCl or HBr cleaves Me_4Ge, although the yield is poor. Only one methyl group is removed. The yield is more satisfactory with an $AlCl_3$ or $AlBr_3$ catalyst[2,3]:

$$R_4Ge + HX \xrightarrow{AlX_3} R_3GeX + RH \qquad \text{(b)}$$

where X = Cl, Br.

Trimethylchlorogermane is prepared by cleavage of a Ge—C bond by boiling a mixture of Me_4Ge, i-PrCl and $AlCl_3$ catalyst for ca. 90 min. The bp of the mixture rises to 95°C and is followed by distillation[4]:

$$(CH_3)_4Ge + i\text{-}C_3H_7Cl \xrightarrow[90\ min]{AlCl_3} (CH_3)_3GeCl \qquad \text{(c)}$$

Trimethyliodogermane is prepared from Me_3GePh by preferential cleavage of the phenyl group by n-BuI with catalytic $AlCl_3$. The cleavage does not occur with Me_4Ge. Dropwise addition of n-BuI to Me_3GePh and $AlCl_3$ followed by boiling for 1 h produces Me_3GeI in 79% yield[4]:

$$(CH_3)_3GeC_6H_5 + n\text{-}C_4H_9I \xrightarrow{AlCl_3} (CH_3)_3GeI \qquad \text{(d)}$$

Cleavage of n-Bu_4Ge by refluxing i-AmBr or i-AmCl for 5 h in the presence of 4% $AlBr_3$ or $AlCl_3$ yields the corresponding bromide or chloride[5]:

$$(n\text{-}C_4H_9)_4Ge + i\text{-}C_5H_{11}X \xrightarrow{AlX_3} (n\text{-}C_4H_9)_3GeX \qquad \text{(e)}$$

where X = Cl, Br. Tetramethylgermane is cleaved by aq HF at −10°C in a Cu vessel. The solution is warmed to 25°C and then finally to 55°C for 2–3 h. Excess HF is removed by KF; fractional distillation gives Me_3GeF in 71% yield. Triethylfluorogermane is obtained similarly[6]:

$$R_4Ge + HF \rightarrow R_3GeF + RH \qquad \text{(f)}$$

where R = CH_3, C_2H_5. The rate of cleavage of tetraarylgermanes by HBr in $CHCl_3$ at RT increases $C_6H_5CH_2— < C_6H_5— < m\text{-}CH_3C_6H_4— < p\text{-}CH_3C_6H_4—$.

Chloroform solutions of the Ge compounds are treated with HBr gas for 24 h at RT, the $CHCl_3$ is distilled, and Ph_3GeBr is obtained from $(C_5H_5)_3GeC_6H_4CH_3$-p or -m by fractional distillation[7]:

$$(C_6H_5)_3Ge(m\text{- or }p\text{-}CH_3C_6H_4) \xrightarrow{HBr} (C_6H_5)_3GeBr + C_6H_5CH_3 \qquad \text{(g)}$$

Similarly, $(m\text{-}CH_3C_6H_4)_3GeC_6H_4CH_3\text{-}p$ in $CHCl_3$ reacts with HBr gas to yield $(m\text{-}CH_3C_6H_4)_3GeBr$, purified by vacuum distillation[7]:

$$(m\text{-}CH_3C_6H_4)_3Ge(p\text{-}CH_3C_6H_4) \xrightarrow[CHCl_3]{HBr} (m\text{-}CH_3C_6H_4)_3GeBr + C_6H_5CH_3 \quad \text{(h)}$$

The reaction of unsaturated organogermanes with HBr depends on conditions. While allyl groups are cleaved by HBr through the formation of an addition intermediate:

$$(n\text{-}C_4H_9)_3GeCH_2CH{=}CH_2 + HBr \rightarrow (n\text{-}C_4H_9)_3GeBr + CH_3CH{=}CH_2 \quad \text{(i)}$$

vinyl groups add HBr[8]:

$$(n\text{-}C_4H_9)_3GeCH{=}CH_2 + HBr \rightarrow (n\text{-}C_4H_9)_3GeCH_2CH_2Br \quad \text{(j)}$$

By halogenation of R_4Ge with HBr, the R_3GeX compounds can be prepared. Gaseous HBr and Me_4Ge react with catalytic anhyd $AlBr_3$ in a glass bulb for 2 days. Fractional distillation at $-100°C$ produces[3] pure Me_3GeBr, mp $-25°C$:

$$(CH_3)_4Ge + HBr \xrightarrow{AlBr_3} (CH_3)_3GeBr + CH_4 \quad \text{(k)}$$

Similarly, Me_3GeCl is prepared[2] from Me_4Ge HCl gas with catalytic $AlCl_3$. Fractional distillation at $-80°C$ gives the product, bp 102°C:

$$(CH_3)_4Ge + HCl \xrightarrow[1\ h]{AlCl_3} \underset{(33\%)}{(CH_3)_3GeCl} + CH_4 \quad \text{(l)}$$

Addition of i-PrX (X = Br, I) to equimol Et_4Ge with 2% $AlCl_3$ and refluxing for 4 h results in Et_3GeX (X = Be, I) in 60–93% yield[2]:

$$(C_2H_5)_4Ge + i\text{-}C_3H_7X \xrightarrow{AlCl_3} (C_2H_5)_3GeX \quad \text{(m)}$$

where X = Cl, Br; Me_4Ge behaves similarly. Refluxing a mixture of Me_4Ge and $i\text{-}C_3H_7X$ in the presence of AlX_3 for 90 min until T rises to 95°C gives Me_3GeX. When X = Cl, the yield is 95% (bp 97.8°C); when Br, 91.5% (bp 113°); when I, the reaction fails[9]:

$$(CH_3)_4Ge + i\text{-}C_3H_7X \xrightarrow{AlX_3} (CH_3)_3GeX \quad \text{(n)}$$

where X = Cl, Br. Trimethylchlorogermane may also be obtained from Me_3GePh. The Ge–phenyl group is cleaved more readily than the alkyl group; $AlCl_3$ catalyzes the cleavage by alkyl halides, e.g., refluxing for 1 h the solution obtained by gradual addition of i-BuCl to $(CH_3)_3GeC_6H_5$ in the presence of $AlCl_3$ gives $(CH_3)_3GeCl$ in 85% yield[10]:

$$(CH_3)_3GeC_6H_5 + i\text{-}C_4H_9Cl \xrightarrow{AlCl_3} (CH_3)_3GeCl \quad \text{(o)}$$

Iso-amyl halides cleave the Ge—C bond of Bu_4Ge. Refluxing a freshly prepared solution of Bu_3GeX with 4% $AlCl_3$ yields the corresponding Bu_3GeX (X = Cl, Br)[10]:

$$(C_4H_9)_4Ge + i\text{-}C_5H_{11}X \xrightarrow{AlX_3} (C_4H_9)_3GeX \quad \text{(p)}$$

(J. M. BELLAMA)

1. P. Mazerolles, J. Dubac, M. Lesbré, *J. Organomet. Chem.*, *5*, 35 (1966).
2. J. E. Griffiths, M. Onyszchuk, *Can. J. Chem.*, *39*, 339 (1961).
3. L. M. Dennis, W. I. Patnode, *J. Am. Chem. Soc.*, *52*, 2779 (1930).
4. V. F. Mironov, A. L. Kravchenko, *Bull. Acad. Sci. USSR, Ser. Chem.*, 988 (1965).
5. Dao-Huy-Giao, *C. R. Hebd. Séances Acad. Sci.*, *260*, 6937 (1965).
6. B. M. Gladstein, V. V. Rode, L. Z. Soborovskii, *J. Gen. Chem. USSR (Engl. Transl.)*, *29*, 2155 (1959).
7. J. K. Simons, *J. Am. Chem. Soc.*, *57*, 1299 (1935).
8. P. Mazerolles, M. Lesbré, *C. R. Hebd. Séances Acad. Sci.*, *248*, 2018 (1959).
9. G. A. Razuvaev, N. S. Vyazankin, O. S. D'yachkovskaya, I. G. Kiseleva, Yu. I. Dergunov, *J. Gen. Chem. USSR (Engl. Transl.)*, *31*, 4056 (1961).
10. V. F. Mironov, A. L. Kravchenko, *Bull. Acad. Sci. USSR, Ser. Chem.*, 988 (1965).

2.5.6.2.3. by Other Halides.

One Bu group is exchanged for a Cl atom when Bu_4Ge and $SnCl_4$ are heated at 200°C without a catalyst[1]:

$$Bu_4Ge + SnCl_4 \xrightarrow{200^\circ C} Bu_3GeCl + BuSnCl_3 \quad \text{(a)}$$

The reaction proceeds slowly even at 250°C, where decomposition occurs. No catalyst is known for this reaction. Tin(IV) chloride is more reactive than $GeCl_4$[1].

Phenyltrichlorogermanium is obtained[2] in 75% yield by redistribution of Ph_4Ge and $GeCl_4$ by heating at 350°C for 36 h:

$$(C_6H_5)_4Ge + 3\ GeCl_4 \rightarrow 4\ G_6H_5GeCl_3 \quad \text{(b)}$$

Anhydrous $GaCl_3$ acts as an electron-pair acceptor acid and cleaves one Ge—C bond of Me_4Ge to form Me_3GeCl and $MeGaCl_2$ in 92% yield[3]:

$$Me_4Ge + GaCl_3 \rightarrow Me_3GeCl + MeGaCl_2 \quad \text{(c)}$$

The equilibrium exchange of Me and Cl groups on Ge at 300°C using 0.1% catalytic $AlCl_3$ takes 4 days with an equimol mixture but requires 30 days with xs $GeCl_4$:

$$Me_4Ge + GeCl_4 \rightarrow Me_3GeCl + MeGeCl_3 \quad \text{(d)}$$

Since Me_2GeCl_2 is the major equilibrium species when equimol Me_4Ge and $GeCl_4$ are employed, Eq. (e) is 100 times faster than Eq. (d)[4]:

$$Me_3GeCl + MeGeCl_3 \rightarrow 2\ Me_2GeCl_2 \quad \text{(e)}$$

Other reactions may occur, e.g.[4]:

$$Me_4Ge + CH_3GeCl_3 \rightarrow Me_3GeCl + Me_2GeCl_2 \quad \text{(f)}$$

but are usually slower than Eq. (e).

Allylic compounds, e.g., $Bu_3GeCH_2CH{=}CH_2$, are cleaved by $HgCl_2$ in alcohol at 20°C:

$$(C_4H_9)_3GeCH_2CH{=}CH_2 + HgCl_2 \rightarrow CH_2{=}CHCH_2HgCl + (C_4H_9)_3GeCl \quad \text{(g)}$$

Mercury(II) chloride, an electron-pair acceptor acid, catalyzes the reaction. An open (S_E2) rather than a cyclic (S_F) transition state is suggested, whereas cleavage in CH_3CN acetonitrile is a simple bimolecular electrophilic substitution of Hg for Ge. The reaction is >100 times faster than that of the corresponding silanes[5].

(J. M. BELLAMA)

1. J. G. A. Luijten, F. Rijkens, *Recl. Trav. Chim. Pays. Bas.*, *83*, 857 (1964).
2. R. Schwarz, M. Schmeisser, *Chem. Ber.*, *69*, 579 (1936).
3. H. Schmidbaur, W. Findeiss, *Chem. Ber.*, *99*, 2187 (1966).
4. G. M. Burch, J. R. Van Wazer, *J. Chem. Soc.*, *A*, 586 (1966).
5. P. Mazerolles, M. Lesbré, *C. R. Hebd. Séances Acad. Sci.*, *248*, 2018 (1959).

2.5.6.3. Cleavage of the Tin–Carbon Bond

2.5.6.3.1. by Halogens.

Iodine cleaves the Sn—C bond with replacement of one phenyl group when Ph_4Sn and I_2 in $CHCl_3$ are refluxed. The product compound is purified by extraction with ether[1].

$$SnPh_4 + I_2 \xrightarrow[CHCl_3]{reflux} SnPh_3I + PhI \qquad (a)$$

One Me group (in preference to the CF_3 group) in Me_3SnCF_3 is cleaved by Cl_2:

$$(CH_3)_3SnCF_3 + Cl_2 \rightarrow CF_3(CH_3)_2SnCl + CH_3Cl \qquad (b)$$

Owing to the high electronegativity of the CF_3 group, the F_3C—Sn bond is highly polarized and, therefore, differs from bonds formed when other organic groups are attached to Sn. The effective electronegativity of the CF_3 group lies between those of F and Cl. Thus, the CF_3 group possesses halogenlike character[2].

Tin tetraalkyls with Br_2 at $-40°$ to $-30°C$ form[3] R_3SnBr:

$$SnR_4 + Br_2 \xrightarrow[-40° \text{ to } -30°C]{CCl_4} SnR_3Br + RBr \qquad (c)$$

The cyclics containing five carbon atoms and one tin cleave at the Sn—C bond and are opened by Br_2 at 0°C in ethyl acetate[4]:

$$\text{(cyclo-}(CH_2)_5\text{)}SnMe_2 + Br_2 \xrightarrow[EtAc]{0°C} BrCH_2CH_2CH_2CH_2CH_2Sn(Me)_2Br \qquad (d)$$

Iodometalation of tetraalkylstannanes follows a S_E2 mechanism and involves cleavage of a Sn—C bond by I_2:

$$R_4Sn + I_2 \xrightarrow{CH_3OH} SnR_3I + RI \qquad (e)$$

The rate of iodometalation decreases owing to steric hinderance with larger organic groups, methyl > ethyl > n-propyl. Iodide ion has no effect on the rate[5]. The interaction of $R_3SnCH_2CH{=}CH_2$ with I_2 also involves a S_E2 mechanism:

$$R_3SnCH_2CH{=}CH_2 + I_2 \rightarrow [R_3Sn]^+ + CH_2{=}CH{-}CH_2I + I^- \qquad (f)$$

Again, I^- concentration has no effect on the rate[5].

(J. M. BELLAMA)

1. R. F. Chambers, P. C. Scherer, *J. Am. Chem. Soc.*, *48*, 1054 (1926).
2. R. D. Chambers, H. C. Clark, C. J. Willis, *Can J. Chem.*, *39*, 131 (1961).

3. G. Grüttner, E. Krause, *Chem. Ber.*, *50*, 1802 (1917).
4. G. Grüttner, E. Krause, M. Wiernik, *Chem. Ber.*, *50*, 1549 (1917).
5. M. Gielen, J. Nasielski, *Bull. Soc. Chim. Belg. 71*, 32 (1962).

2.5.6.3.2. by Hydrogen Halides.

By saturating of tetraalkyltins in ether with dry HCl at RT, the Sn—C bond is cleaved and the corresponding R_3SnCl compounds are obtained[1]:

$$R_4Sn + HCl \xrightarrow[RT]{ether} R_3SnCl + RH \tag{a}$$

where R = n-dodecyl, n-tetradecyl, n-hexadecyl, n-octadecyl.

Triphenyltin chloride is obtained[2] by passing HCl gas through Ph_4Sn in $CHCl_3$ at RT:

$$Ph_4Sn + HCl \xrightarrow{CHCl_3} Ph_3SnCl + PhH \tag{b}$$

The tin–allyl group in R_3SnCH_2-$CH{=}CH_2$ is cleaved by HCl in MeOH in preference to vinyl or phenyl groups:

$$R_3SnCH_2CH{=}CH_2 + HCl \xrightarrow{CH_3OH} [R_3Sn]^+ + CH_2{=}CHCH_2 + Cl^- \tag{c}$$

where R = phenyl or methyl. An S_E2 mechanism is involved[3].

Gaseous HBr with alkylstannanes at −78°C gives[4] white crystalline $RSnH_2Br$:

$$RSnH_3 + HBr \xrightarrow{-78°C} RSnH_2Br + H_2 \tag{d}$$

At temperatures > −65°C, a polymeric, yellow compound is formed[4]:

$$RSnH_2Br \xrightarrow{> -65°C} H_2 + (RSnBr)_n \tag{e}$$

(J. M. BELLAMA)

1. R. N. Meals, *J. Org. Chem.*, *9*, 211 (1944).
2. G. Bahr, *Z. Anorg. Allg. Chem.*, *256*, 107 (1948).
3. H. G. Kuivila, J. A. Verdone, *Tetrahedron Lett.*, *2*, 119 (1964).
4. G. Fritz, H. Scheer, *Z. Naturforsch.*, *Teil B*, *19*, 537 (1964).

2.5.6.3.3. by Organic Halides.

The cleavage of Sn—C bonds in R_4Sn can be accomplished by halogen-transfer agents such as aryl or sulfuryl chlorides. The products are either R_3SnX or R_2SnX_2 or a mixture of both.

Boiling Ph_4Sn and acetyl chloride (at 100°C for 15 h), benzoyl chloride (at 200°C for 10 h), or benzene sulfonyl chloride (at 200°C for 18 h) under reduced pressure and elevated T in a sealed tube produces appreciable amounts of product ketones (acetophenone, benzophenone, diphenylsulfone, respectively). The yield is, however, poor for acetyl chloride[1]:

$$SnR_4 + acyl\text{-}Cl \rightarrow acyl\text{-}R + R_3SnCl \tag{a}$$

Tetraphenyltin and benzoyl chloride treated with powdered anhyd $AlCl_3$ and heated at 80°C for ca. 30 min gives Ph_2CO in 70% yield with fission of the Sn—C bond:

$$Ph_4Sn + PhC(O)Cl \xrightarrow{AlCl_3} \underset{(70\%)}{Ph_2CO} + \text{oxime} \qquad \text{(b)}$$

The use of CCl_4 as solvent gives better yields. Similarly, an acyl chloride and Ph_4Sn with an $AlCl_3$ catalyst give BzMe and a semicarbazone[2].

The result with Ph_4Sn and sulfuryl chloride is not conclusive. Reflux in benzene forms only small amounts of Ph_3SnCl, leaving most of the Ph_4Sn unreacted[3].

Heating Ph_4Sn and alkyl halides such as t-BuBr, i-AmBr, gives Ph_3SnBr in poor yield. The alkyl halides tend to split off halogen acids[3].

$$Ph_4Sn + RX \xrightarrow{\text{heat}} R_3SnX + \text{unsaturated hydrocarbons} + \text{halogen acids} \qquad \text{(c)}$$

Alkyltin compounds, e.g., Et_4Sn, Et_6Sn_2, or Et_3SnCl, with benzene spontaneously exchange with cleavage of the Sn—C bond e.g.[4]:

$$Et_4Sn + \text{i-PrCl} \rightarrow Et_3SnCl \qquad \text{(d)}$$

$$Et_3SnCl + \text{i-PrCl} \rightarrow \underset{(35\%)}{Et_2SnCl_2} \qquad \text{(e)}$$

$$Et_6Sn_2 + \text{i-PrCl} \rightarrow \underset{(55\%)}{Et_3SnCl} + \underset{(22\%)}{Et_4Sn} \qquad \text{(f)}$$

Tetraethyltin reacts photochemically with PrBr, which proceeds via a free-radical mechanism with fission of the Sn—C bond:

$$R_4Sn \xrightarrow[N_2]{h\nu} [R_3Sn]^{\cdot} + R\cdot \qquad \text{(g)}$$

$$[R_3Sn]^{\cdot} + PrBr \rightarrow R_3SnBr + Pr\cdot \qquad \text{(h)}$$

where R = ethyl. Free-radical initiation with benzoyl peroxide also results in the same reaction type[5].

Attempts to prepare sulfonic acid derivatives of alkyltin compounds are unsuccessful[6].

(J. M. BELLAMA)

1. R. W. Bost, P. Borgstrom, *J. Am. Chem. Soc., 51*, 1922 (1929).
2. A. P. Skoldinov, K. A. Kocheshkov, *Zh. Obshch. Khim., 12*, 398 (1942); *Chem. Abstr., 37*, 3064 (1942).
3. R. W. Bost, P. Borgstrom, *J. Am. Chem. Soc., 51*, 1922 (1929).
4. G. A. Razuvaev, N. S. Vyazankin, Yu. I. Dergunov, O. S. D'yachkovskava, *Dokl. Akad. Nauk SSSR, 132*, 364 (1960).
5. G. A. Razuvaev, N. S. Vyazankin, Yu. I. Dergunov, E. N. Gladyshev, *Bull. Acad. Sci. USSR, Ser. Chem., 2*, 794 (1964).
6. T. A. Smith, F. S. Kipping, *J. Chem. Soc., 101*, 2553 (1912).

2.5.6.3.4. by Other Halides.

Alkyltins react with halogenated electron-pair acceptor acids to exchange an acidic halide for an alkyl group. The stoichiometric interaction of Me_4Sn and BF_3 at 100°C for

2 h in a sealed tube produces $[(CH_3)_3Sn][BF_4]$, which melts at 89°C in the sealed tube or at 100°C in vacuo to form the white sublimate $(CH_3)_3SnF$ with loss of BF_3:

$$(CH_3)_4Sn + 2\ BF_3 \xrightarrow{100^\circ C} [(CH_3)_3Sn][BF_4] + CH_3BF_2 \qquad \text{(a)}$$

Despite its low mp, trimethyltinfluoroborate was reported to have an ionic structure[1], but bridging $\gt Sn{-}F{-}B\lt$ would seem to be more likely.

Warming 1:1 BF_3 and Me_3SnCF_3 in CCl_4 produces a 1 :1 adduct, trimethyltin trifluoromethylfluoroborate:

$$(CH_3)_3SnCF_3 + BF_3 \rightarrow [(CH_3)_3Sn][CF_3BF_3] \qquad \text{(b)}$$

With KF the adduct yields Me_3SnF, trimethyltinfluoride:

$$[(CH_3)_3Sn][CF_3BF_3] + KF \rightarrow (CH_3)_3SnF + K[CF_3BF_3] \qquad \text{(c)}$$

The $[CF_3BF_3]^-$ ion is stable, which is attributed to the saturation of the acceptor tendencies of the boron atom and the delocalization of its charge[2].

Tetraphenyltin and BF_3 converts Ph_4Sn into Ph_3SnF; Me_4Sn and BF_3 at 140°C for 40 h produce Me_3SnBF_4:

$$R_4Sn + 2\ BF_3 \rightarrow R_3SnBF_4 + RBF_2 \qquad \text{(d)}$$

which on heating[3] gives Ph_3SnF:

$$R_3SnBF_4 \xrightarrow{\text{heat}} R_3SnF + BF_3 \qquad \text{(e)}$$

Tetramethyltin reacts with $GaCl_3$ exothermically at 0°C; this reaction produces $(CH_3)_3SnCl$ and CH_3GaCl_2 as a coordination compound[4]:

$$(CH_3)_4Sn + GaCl_3 \xrightarrow{0^\circ C} (CH_3)_3SnCl \cdot CH_3GaCl_2 \qquad \text{(f)}$$

Tetraphenyltin undergoes rearrangement with $AlCl_3$:

$$R_3SnR' + Al_2Cl_6 \rightarrow R_3SnCl + R'Al_2Cl_5 \qquad \text{(g)}$$

The reaction is dependent on T and the catalyst concentration, requiring >2 mol % catalyst for completion. At 120°C several hours are needed; at 170°C it takes 2 min, and at >170°C decomposition occurs[5,6].

Redistribution of Me_4Sn and $SnCl_4$ in CCl_4 at 0–50°C provides 1:1 $(CH_3)_3SnCl$ and $(CH_3)SnCl_3$[7]:

$$(CH_3)_4Sn + SnCl_4 \rightarrow (CH_3)_3SnCl + CH_3SnCl_3 \qquad \text{(h)}$$

For $Z > 2$ ($Z = CH_3/Sn = 2$) some $(CH_3)_2SnCl_2$ is also formed. After several weeks at 120°C, equilibrium is reached in scrambling of the M groups and Cl atoms.

(J. M. BELLAMA)

1. A. B. Burg, J. R. Spielman, *J. Am. Chem. Soc.*, *83*, 2667 (1961).
2. R. D. Chambers, H. C. Clark, C. J. Willis, *J. Am. Chem. Soc.*, *82*, 5298 (1960).
3. D. W. A. Sharp, J. M. Winfield, *J. Chem. Soc.*, 2278 (1965).

4. H. Schmidbaur, W. Findeiss, *Angew. Chem., Int. Ed. Engl., 3,* 696 (1964).
5. F. H. Pollard, G. Nickless, P. C. Undev, *J. Chromatogr., 19,* 28 (1965).
6. F. H. Pollard, G. Nickless, D. J. Cooke, *J. Chromatogr., 17,* 472 (1965).
7. D. Grant, J. R. Van Wazer, *J. Organomet. Chem., 4,* 229 (1965).

2.5.6.4. of the Lead–Carbon Bond

2.5.6.4.1. by Halogens.

Iodine is taken up by Ph_2Pb in dry Et_2O at $-20°C$. Refluxing for 1 h cleaves the Pb—C bond and yields[1] PbI_2, which is crystallized from EtOH:

$$(C_6H_5)_2Pb + 2\ I_2 \rightarrow PbI_2 + 2\ C_6H_5I \qquad (a)$$

Reaction of Br_2 and Ph_4Pb in pyridine at $-15°C$ gives Ph_2PbBr_2 and Ph_3PbBr in $>50\%$ yield:

$$(C_6H_5)_4Pb + Br_2 \xrightarrow{-15°C} (C_6H_5)_2PbBr_2 + (C_6H_5)_3PbBr \qquad (b)$$

Unsaturated organoleads, such as $Ph_3PbC{\equiv}CPbPh_3$, are cleaved by I_2 to form[2] R_3PbI:

$$R_3PbC{\equiv}CPbR_3 + 2\ I_2 \rightarrow 2\ R_3PbI + IC{\equiv}CI \qquad (c)$$

where R = o-tolyl, cyclohexyl and phenethyl.

Triphenyllead bromide is obtained in almost quantitative yield by mixing a suspension of Ph_4Pb in py and Br_2 in pyridine above $-40°C$. Triphenyllead bromide separates from alcohol in silky needles and decomposes[3]:

$$(C_6H_5)_4Pb + Br_2 \xrightarrow[py]{-40°C} (C_6H_5)_3PbBr + C_6H_5Br \qquad (d)$$

(J. M. BELLAMA)

1. L. C. Willemsens, G. J. M. Van der Kerk, *Investigations in the Field of Organolead Chemistry,* International Lead-Zinc Research Organization, New York, 1965.
2. H. Hartmann, W. Eschenbach, *Naturwissenschaften., 46,* 321 (1959).
3. G. Grüttner, *Chem, Ber., 51,* 1298 (1918).

2.5.6.4.2. by Hydrogen and Alkyl Halides.

Hexaorganodileads react with halogen acids to yield organolead halides. The halogenation of hexalkyldileads to trialkyllead halides proceeds cleanly, but the reaction of hydrogen halides with $Ph_3PbPbPh_3$ is complex. Hydrogen chloride with R_3PbPbR_3 in 95% EtOH gives[1–3]:

$$Ph_6Pb_2 + 3\ HCl \rightarrow Ph_3PbCl + PbCl_2 + 3\ PhH \qquad (a)$$

in which the Pb—C bond cleaves instead of the Pb—Pb bond.

The reaction with dry HCl in $CHCl_3$/EtOH occurs by two pathways[4]:

$$Ph_6Pb + 2\ HCl \rightarrow Ph_4Pb_2Cl_2 \rightarrow Ph_4Pb\,(+PbCl_2) \xrightarrow{HCl} Ph_3PbCl \qquad (b)$$

$$Ph_6Pb_2 + 3\ HCl \rightarrow Ph_3Pb_2Cl_3 \rightarrow Ph_3PbCl + PbCl_2 \qquad (c)$$

Pathway (b) is preferred because of the role of Ph_3Pb intermediates in other electrophilic reactions[4].

However, 1:1 Ph_6Pb_2 and HCl at RT prefers pathway (c) and the following[5]:

$$Ph_6Ph_2 + HCl \rightarrow Ph_3PbCl + Ph_2Pb \xrightarrow{HCl} Ph_3PbCl + PbCl_2 \tag{d}$$

Tetraorganoleads undergo reaction with anhyd HX to yield R_3PbX, R_2PbX_2 or PbX_2, depending upon solvent and T. Thus, Ph_3PbCl can be prepared[6] by the addition of HCl to Ph_3PbCl in warm $CHCl_3$ until the first appearance of Ph_2PbCl_2. The small amount of Ph_2PbCl_2 is filtered, the filtrate evaporated to dryness, and the residue extracted with EtOH to leave unreacted Ph_4Pb:

$$Ph_4Pb + HCl \xrightarrow[50-60°C]{CHCl_3} \underset{(75\%)}{Ph_3PbCl} \tag{e}$$

Analogously, Ph_3PbBr can be prepared[5] from HBr gas and Ph_4Pb. Yields are higher (70–80%) than with Ph_3PbCl (55–75%) because of the better solubility of the bromide in EtOH:

$$Ph_4Pb + HBr_{(g)} \xrightarrow{EtOH} Ph_3PbBr \tag{f}$$

Triethyllead chloride can be obtained[7] by bubbling HCl into Et_4Pb in Et_2O at RT:

$$Et_4Pb + HCl_{(g)} \xrightarrow{Et_2O} Et_3PbCl \tag{g}$$

However, hexane is a better solvent for this reaction; it gives high yields of purer product because of the lower solubility of Et_3PbCl. With both solvents, Et_3PbCl precipitates almost immediately upon addition of HCl. The reaction[7] is monitored by periodic filtration of the Et_3PbCl formed, followed by further treatment of the filtrate with additional HCl.

Organolead salts can also be prepared from R_4Pb with aq HX but this way is slower because of the lower solubility of R_4Pb in the aq acid. Addition of Et_4Pb to HCl at 30–33°C forms[8] Et_3PbCl:

$$Et_4Pb + HCl \rightarrow \underset{86\%}{Et_3PbCl} \tag{h}$$

Silica gel is a catalyst.

(J. M. BELLAMA)

1. P. R. Austin, *J. Am. Chem. Soc.*, *53*, 3514 (1931).
2. H. Gilman, J. C. Bailie, *J. Am. Chem. Soc.*, *61*, 731 (1939).
3. H. Gilman, E. B. Towne, *J. Am. Chem. Soc.*, *61*, 739 (1939).
4. H. J. Emeléus, P. R. Evans, *J. Chem. Soc.*, 511 (1964).
5. L. C. Willemsens, G. J. M. van der Kerk, *Investigations in the Field of Organometallic Chemistry*, International Lead-Zinc Research Organization, New York, 1965, p. 26.
6. H. Gilman, J. Robinson, *J. Am. Chem. Soc.*, *51*, 3112 (1929).
7. H. Gilman, J. Robinson, *J. Am. Chem. Soc.*, *52*, 1975 (1930).
8. O. H. Browne, E. E. Reid, *J. Am. Chem. Soc.*, *49*, 830 (1927).

2.5.6.4.3. by Other Halides.

Tetraorganoleads redistribute with $AlCl_3$ at ca. 80°C in N_2. The reaction is affected by the R group and by catalysts, catalyst concentration, solvent, T, and time. Aluminum chloride is effective at 0.50 mol %. For an aryl compound, no catalyst is necessary. The effect of T on the rate is marked, but the effect of solvent is not pronounced[1]:

$$R_4Pb + AlCl_3 \rightleftharpoons R_3PbCl + RAlCl_2 \quad \text{(a)}$$

Metatheses of R_4Pb with metal salts cleave one or more Pb–carbon bonds to form organolead salts:

$$R_4Pb + MX_n \rightarrow R_{4-x}PbX_x + RMX_{n-x} \quad \text{(b)}$$

where M = As, Sb, Bi. Tetraalkyls and tetraaryls react with halides of Bi, Sb, As under mild conditions. With Et_4Pb at 25°C, two Et groups react to yield Et_2PbCl_2; at 100°C three Et groups are used and $PbCl_2$ is formed:

$$Et_4Pb + 3\ MCl_3 \xrightarrow{100°C} 3\ EtMCl_2 + EtCl + PbCl_2 \quad \text{(c)}$$

where M = As or Sb. Tetraphenyllead with $AsCl_3$ and $BiCl_3$ gives Ph_2AsCl and Ph_2BiCl, respectively. Chlorides of sulfur and Te react[2] with R_4Pb:

$$TeCl_4 + Ph_4Pb \rightarrow Ph_2PbCl_2 + Ph_2TeCl_2 \quad \text{(d)}$$

$$S_2Cl_2 + 2\ Et_4Pb \rightarrow 2\ Et_3PbCl + EtSSEt \quad \text{(e)}$$

$$SOCl_2 + Et_4Pb \rightarrow Et_3PbCl + EtSOCl \quad \text{(f)}$$

The Pb–carbon bond is cleaved to form Et_3PbCl when S_2Cl_2 is added with agitation to Et_4Pb in pet. ether, the mixture is heated to 30–40°C, and moisture is excluded. After 15 h the product is filtered and recrystallized from pet. ether and benzene[3]:

$$2\ (C_2H_5)_4Pb + S_2Cl_2 \rightarrow 2\ (C_2H_5)_3PbCl + C_2H_5SSC_2H_5 \quad \text{(g)}$$

(J. M. BELLAMA)

1. G. Calingaert, A. A. Beatty, H. Soroos, *J. Am. Chem. Soc.*, *62*, 1099 (1940).
2. L. C. Willemsens, G. J. M. van der Kirk, in *Organometallic Compounds of the Group IV Elements*, A. G. MacDiarmid, ed., Vol. 1, Part II, Marcel Dekker, New York, 1968, p. 215.
3. H. D. Lutz, *Z. Naturforsch., Teil B*, *20*, 1011 (1965).

2.5.7. from Cleavage of the Group IVB–Other Group-IVB Element Bond

2.5.7.1. the Si–Si Bond.

Compared to the analogous C—C bond, the Si—Si bond in $Ph_3SiSiPh_3$ is more stable, and cleavage by Br_2 is slow. Bromotriphenylsilane is obtained in yields 33% and 55% by refluxing $Ph_3SiSiPh_3$ and Br in CCl_4 for 42 h and 6 days, respectively[1]:

$$[(C_6H_5)_3Si]_2 + Br_2 \xrightarrow{CCl_4} 2\ (C_6H_5)_3SiBr \quad \text{(a)}$$

Hexaphenyldisilane does not react with I_2 or HCl under ordinary conditions[1]:

$$(Ph_3Si)_2 + X_2, HX \rightarrow \text{no reaction} \quad (b)$$

Trimethylhalosilanes may be prepared in high yield by treating $Me_3SiSiMe_3$ (**I**) with halogens:

$$(R_3Si)_2 + X_2 \rightarrow 2\ R_3SiX \quad (c)$$

where X = Br, I, Cl. Bromo-, iodo-, and chlorotrimethylsilane can be made (a) by adding Br_2 to (**I**) with cooling and distillation of the product, (b) by heating I_2 and (**I**) and distilling and (c) by passing dry Cl_2 into (**I**) in CCl_4[2].

Hexamethyldisilane and substituted methyldisilanes with I_2 and Br_2 form addition products by halogen attack on the Si—Si bond[3,4].

Hexaethyldisilane is also cleaved by halogens; e.g., Br_2 added to $Et_3SiSiEt_3$ in Et_2O forms[5] Et_3SiBr.

Also, Et_3SiI is obtained by stirring I_2 and $Et_3SiSiEt_3$ in CCl_4 for 3 days[5].

Iodine and Br_2 also cleave the Si—Si bond in alkylchlorodisilanes. At RT the reaction goes to completion with Br_2, whereas with I_2 only 20% in 48 h. With catalytic AlI_3, cleavage occurs quantitatively[6].

$$Cl(CH_3)_2SiSi(CH_3)_2Cl + Br_2 \rightarrow \underset{(20\%)}{2\ (CH_3)_2SiClBr} \quad (d)$$

$$Cl(CH_3)_2SiSi(CH_3)_2Cl + I_2 \xrightarrow{AlI_3} \underset{(100\%)}{2\ (CH_3)_2SiClI} \quad (e)$$

Cleavage provides an excellent method for the preparation of mixed halosilanes.

The Si—Si bond cleaves[6] at high T with 1:1 HCl passed at 0.083 mol/h into a tube at 350–700°C containing $Me_3SiSiMe_3$:

$$[(CH_3)_3Si]_2 + HCl \xrightarrow{350-700^\circ C} (CH_3)_3SiH + (CH_3)_3SiCl \quad (f)$$

With catalytic tertiary organic amines at 500°C, $ClMe_2SiSiMe_3$ undergoes Si—Si cleavage[7] as in Eq. (f):

$$Cl(CH_3)_2SiSi(CH_3)_3 + HCl \xrightarrow{500^\circ C} (CH_3)_2SiHCl + (CH_3)_3SiCl \quad (g)$$

A triethyltrimethyl-substituted disilane is cleaved by the addition of Br_2:

$$CH_3(C_2H_5)_2SiSi(CH_3)_2C_2H_5 \xrightarrow{Br_2} CH_3(C_2H_5)_2SiBr + (CH_3)_2(C_2H_5)SiBr \quad (h)$$

From kinetics, addition products form by attack of halogen on the Si—Si bond[8].

Cleavage of the Si—Si bond to form[9] an almost theoretical yield of $Me_2EtSiBr$ results from the addition of Br_2 to 1-(chloromethyl)-2-ethyltetramethyldisilane in EtBr and subsequent refluxing for 3 h:

$$2\ ClCH_2(CH_3)_2SiSi(CH_3)_2C_2H_5 + Br_2 \rightarrow 2\ BrCH_2(CH_3)_2SiBr + 2\ C_2H_5(CH_3)_2SiBr \quad (i)$$

Pentamethylbromodisilane is obtained in high yield by treating octamethyltrisilane in $CHCl_3$ with Br_2 in $HCCl_3$ at $-40°C$ under N_2:

$$(CH_3)_3SiSi(CH_3)_2Si(CH_3)_3 + Br_2 \xrightarrow[CHCl_3]{-40°C} (CH_3)_3SiSi(CH_3)_2Br + (CH_3)_3SiBr \quad (j)$$

The Si—Si bond is protected by the negative Me substituents against further attack by halogens[10].

Methylsilicon telomer reacts with Br_2 in CCl_4 at $-30°$ to $-50°C$ for 2 h and at RT for 1 h with scission of the Si—Si bond[11]:

$$CH_3[(CH_3)_2Si]_3CH_3 + Br_2 \xrightarrow{-50°C} (CH_3)_3SiBr + Br(CH_3)_2SiSi(CH_3)_2Br \quad (k)$$

Decamethyltetrasilane in $CHCl_3$ added at $-40°C$ under N_2 to Br_2 in $CHCl_3$ forms cleavage products[12]:

$$(CH_3)_3SiSi(CH_3)_2Si(CH_3)_2Si(CH_3)_3 + Br_2 \xrightarrow[CHCl_3]{-40°C}$$

$$(CH_3)_3SiSi(CH_3)_2Br + (CH_3)_3SiSi(CH_3)_2Si(CH_3)_2Br + (CH_3)_3SiBr \quad (l)$$

The protected Si—Si bond is not attacked by Br_2.

The Si—Si bond in the unsymmetrical $(Me_3Si)_3SiCH_3$ undergoes cleavage by Cl_2 in CCl_4 to give a good yield of chlorinated products[13]:

$$[(CH_3)_3Si]_3SiCH_3 + Cl_2 \xrightarrow{CCl_4} [(CH_3)_3Si]_2(CH_3)SiCl + (CH_3)_3SiCl \quad (m)$$

1-Bromo-1-methyl-1-silacyclopentane is obtained in 73% yield from $(CH_2)_4Si(CH_3)Si(CH_3)(CH_2)_4$ with Br_2 in an Si—Si bond cleavage[14]:

$$(CH_2)_4Si(CH_3)\text{—}Si(CH_3)(CH_2)_4 + Br_2 \rightarrow 2\ (CH_2)_4Si(CH_3)\text{—}Br \quad (n)$$

Tris(trimethylsilyl)chlorosilane reacts rapidly with Cl_2 at $-30°$ to $-20°C$ to give $(Me_3Si)_2SiCl_2$ in 70–76% yield[15]:

$$[(CH_3)_3Si]_3SiCl \xrightarrow[Cl_2]{1\ h} [(CH_3)_3Si]_2SiCl_2 \quad (o)$$

The disappearance of $[(CH_3)_3Si]_3SiCl$ can be followed by vapor-phase chromatography to minimize secondary cleavage owing to xs contact time with the chlorinating agent.[15]

Like $(Me_3Si)_3SiMe$, $(Me_3Si)_3SiH$ with Cl_2 in CCl_4 cleaves the Si—Si bond[13]:

$$[(CH_3)_3Si]_3SiH + Cl_2 \rightarrow [(CH_3)_3Si]_2SiCl_2 + (CH_3)_3SiCl \quad (p)$$

The Si—Si bond in mono- and diethylchlorodisilanes is cleaved[16] by $[NH_4]^+$ halides at 120°–130°C:

$$C_2H_5SiCl_2SiCl_3 \xrightarrow{NH_4Cl} C_2H_5SiCl_3 + HSiCl_3 \quad (q)$$

$$C_2H_5(Cl)SiSi(Cl)_2C_2H_5 \xrightarrow{NH_4Cl} C_2H_5SiCl_3 + C_2H_5SiHCl_2 \quad (r)$$

The Si—Si bond of cyclic alkyl or arylpolysilanes is cleaved by halogens and hydrogen halides with formation of α,ω-dihalopolysilanes[17,18]. The periphenylated compounds are more easily cleaved than their alkyl counterparts:

$$[(C_6H_5)_2Si]_4 + HX \rightarrow X[(C_6H_5)_2Si]_4H \tag{s}$$

where X = Cl, Br, I,

$$[(C_6H_5)_2Si]_5 + X_2 \rightarrow X[(C_6H_5)_2Si]_5X \tag{t}$$

where X = Br, Cl.

Organometallic disilanes reduce many metal salts with scission of the Si—Si bond. The cleavage of the Si—Si bond depends upon the strength of the electron-pair acceptor acid, e.g., Me_3SiCl is produced by refluxing $(Me_3Si)_2$ with metal salts under N_2 for several hours. The duration of refluxing required varies from metals to metal salts[19,20]; e.g., the higher oxidation state chlorides of Sn, Cu, Ni, Sb, P, Te, Ti, Se, Ge and Fe and $SnBr_4$ react with $(CH_3)_6Si_2$ at 150–200°C to yield $(CH_3)_3SiX$ and the reduced form of starting halide ($SnCl_2$, CuCl, etc.)

Dimethyltetramethoxydisilane in $(EtOCH_2CH_2)_2O$ added over 1 h to a vigorously stirred solution of $CuCl_2$ in the same solvent at 150°C, followed by refluxing for 2 h under identical conditions, reacts[20]:

$$Me(MeO)_2SiSi(MeO)_2Me + CuCl_2 \rightarrow \underset{(87\%)}{2\ Me(MeO)_2SiCl} + Cu \tag{u}$$

Similarly, dimethyltetraethyldisilane on refluxing for 18 h at 190°–200°C with $NiBr_2$ under identical conditions reacts[19]:

$$CH_3(C_2H_5)_2SiSi(C_2H_5)CH_3 + NiBr_2 \xrightarrow{190-200\ ^\circ C} \underset{(44\%)}{2\ CH_3(C_2H_5)_2SiBr} + Ni \tag{v}$$

N-Bromosuccinimide, a brominating agent, attacks the Si—Si bond in hexaorganodisilanes directly through a four-center transition state. The cleavage occurs by addition of the hexaorgano compound to N-bromosuccinimide in dry $CHCl_3$ or CCl_4 under mild conditions[19,21]:

$$R_3Si{-}SiR_3 + Br{-}N\begin{matrix}CO{-}CH_2\\ |\\ CO{-}CH_2\end{matrix} \rightarrow \begin{matrix}\diagdown\,|\quad|\,\diagup\\ {-}Si\cdots Si{-}\\ \vdots\qquad\vdots\\ Br\cdots N(COCH_2)_2\end{matrix} \rightarrow$$

$$-\overset{|}{\underset{|}{Si}}{-}Br + {-}\overset{|}{\underset{|}{Si}}{-}N\begin{matrix}CO{-}CH_2\\ |\\ CO{-}CH_2\end{matrix} \tag{w}$$

(J. M. BELLAMA)

1. T. C. Wu, H. Gilman, *J. Org. Chem.*, *23*, 913 (1958).
2. T. T. Tsai, W. L. Lehn, C. J. Marshall, Jr., *J. Org. Chem.*, *31*, 3047 (1966).
3. M. Kumada, K. Shiina, M. Yamaguchi, *J. Chem. Soc., Jpn., Ind. Chem. Sect.*, *57*, 230 (1954).
4. A. Taketa, M. Kumada, K. Tarama, *Nippon Kagaku Zasshi*, *78*, 999 (1957); *Chem. Abstr.*, *52*, 8942 (1958).
5. H. Gilman, R. I. Ingham, A. G. Smith, *J. Org. Chem.*, *18*, 1743 (1953).
6. U. G. Stolberg, *Chem. Ber.*, *96*, 2798 (1963).

7. K. Shiina, M. Kumada, *J. Chem. Soc. Jpn., Ind. Chem. Sect., 60*, 1395 (1957).
8. A. Taketa, M. Kumada, K. Tarama, *Nippon. Kagaku Zasshi, 78*, 999 (1957); *Chem. Abstr., 52*, 8942 (1958).
9. M. Kumada, J. I. Nakajima, M. Ishikawa, Y. Yamamoto, *J. Org. Chem., 23*, 292 (1958).
10. U. Stolberg, *Chem. Ber., 96*, 2798 (1963).
11. M. Kumada, M. Ishikawa, B. Murai, *J. Chem. Soc. Jpn., Ind. Chem. Sect., 66*, 637 (1963).
12. U. Stolberg, *Chem. Ber., 96*, 2798 (1963).
13. H. Gilman, C. L. Smith, *J. Organomet. Chem., 8*, 245 (1967).
14. N. Hagihara, M. Kumada, R. Okawara, *Handbook of Organometallic Compounds*, W. A. Benjamin, New York, 1968.
15. H. Gilman, R. L. Harrell, *J. Org. Chem., 5*, 199 (1966).
16. C. J. Wilkins, *J. Chem. Soc.*, 3409 (1953).
17. H. Gilman, D. R. Chapman, G. L. Schwebke, *J. Organomet. Chem., 14*, 267 (1968).
18. H. Gilman, D. R. Chapman, *J. Organomet. Chem., 8*, 451 (1967).
19. R. Calas, E. Frainnet, Y. Dentone, *C. R. Hebd. Séances Acad. Sci., 259*, 3777 (1964).
20. R. C. Paul, A. Arreja, S. P. Narula, *Inorg. Nucl. Chem. Lett., 5*, 1013 (1969).
21. H. Sakurai, A. Hosomi, J. I. Nakajima, M. Kamada, *Bull. Chem. Soc. Jpn., 39*, 2263 (1966).

2.5.7.2. the Si–Sn Bond.

Cleavage of the Si—Sn bond occurs by adding 1 equiv of Br_2 to an Et_2O solution of the compound, followed by evaporation and vacuum distillation[1]:

$$Ph_3SiSnMe_3 + Br_2 \rightarrow Ph_3SiBr + Me_3SnBr \qquad (a)$$

(J. M. BELLAMA)

1. C. A. Kraus, H. Eatough, *J. Am. Chem. Soc., 55*, 5008 (1933).

2.5.7.3. the Ge–Ge Bond.

Although the Ge—Ge bond is more stable than, e.g., the Sn—Sn bond, Br_2 in $CHCl_3$ or CCl_4 cleaves R_3GeGeR_3 quantitatively:

$$R_3GeGeR_3 + Br_2 \rightarrow R_3GeBr + RBr \qquad (a)$$

where R = aryl or alkyl. In 1,2-dibromoethane solvent, Ph_2GeBr_2 is obtained from the $R = C_6H_5$ derivative.

Bromine in $CHCl_3$ cleaves hexaalkyl compounds, e.g., hexavinyldigermane, yielding $(H_2C{=}CH)_3GeBr$. Alkyldigermanes react more readily with halogens than do aryldigermanes[1–4].

Hexaalkyldigermanes react with halogens. Refluxing $(H_2C{=}CH)_3GeGe(CH{=}CH_2)_3$ with I_2 for 1 h in Et_2O or 4 h in $CHCl_3$ cleaves the carbon-Ge bond[5]:

$$(CH_2{=}CH)_6Ge_2 + I_2 \rightarrow \underset{(45.6\%)}{2\,(CH_2{=}CH)_3GeI} \qquad (b)$$

Iodine in $CHCl_3$ cannot cleave $Ph_3GeGePh_3$ but in Na-dried xylene is refluxed for 24 h with 3 drops of quinoline[6]:

$$Ph_6Ge_2 + I_2 \rightarrow Ph_3GeI \qquad (c)$$

(J. M. BELLAMA)

1. D. Seyferth, *J. Am. Chem. Soc.*, *79*, 2738 (1957).
2. P. Mazerolles, *Bull. Soc. Chim. Fr.*, 1911 (1961).
3. O. H. Johnson, O. M. Harris, *J. Am. Chem. Soc.*, *52*, 4031 (1930).
4. F. Glockling, K. A. Hooton, *J. Chem. Soc.*, 1849 (1963).
5. D. Seyferth, *J. Am. Chem. Soc.*, *79*, 2738 (1957).
6. H. Gilman, C. W. Gerow, *J. Am. Chem. Soc.*, *77*, 5509 (1955).

2.5.7.4. the Ge–Sn Bond.

Triphenylgermyltriphenyltin, an analogue of Ph_3CCPh_3, undergoes cleavage of the Ge—Sn bond upon slow addition at RT of I_2 in $CHCl_3$. Triphenyliodogermane is the only product isolated. No tin compound is recovered[1].

$$Ph_3GeSnPh_3 + I_2 \xrightarrow[CHCl_3]{RT} Ph_3GeI \qquad \text{(a)}$$

Trimethylstannyltriphenylgermane reacts with Br_2 in $CHCl_3$ to yield a mixture of triphenylgermanium bromide and trimethyltin bromide. The cleavage products are recoverable in almost quantitative amounts[2]:

$$Ph_3GeSnMe_3 + Br_2 \rightarrow Ph_3GeBr + Me_3SnBr \qquad \text{(b)}$$

(J. M. BELLAMA)

1. H. Gilman, C. W. Gerow, *J. Org. Chem.*, *22*, 336 (1957).
2. C. A. Kraus, L. S. Foster, *J. Am. Chem. Soc.*, *49*, 457 (1927).

2.5.7.5. the Ge–Pb Bond.

A phenyl group attached to the Ge facilitates the cleavage of the Ge—Pb bond in Pb_3GePbR_3 by I_2 and provides a quantitative method[1] for titration of Ge by I_2:

$$R_3PbGePh_3 + I_2 \rightarrow R_3PbI + Ph_3GeI \qquad \text{(a)}$$

(J. M. BELLAMA)

1. W. P. Neumann, K. Kuehlein, *Tetrahedron Lett.*, *29*, 3619 (1966).

2.5.7.6. the Sn–Sn Bond

2.5.7.6.1. by Halogens.

The reactions of halogens with Sn—Sn bonds are exemplified by[1–3]:

$$R_3SnSnR_3 + X_2 \rightarrow 2\ R_3SnX \qquad \text{(a)}$$

which can be used for quantitative titration of the Sn—Sn bond with Br_2 or I_2. The cleavage mechanism is that of an acyclic, four-centered intermediate, formed after attachment of the solvent in a fast preliminary step. The rates vary for different R groups, with the aryl compound falling into the slow group, the alkyl into the fast group and the two mixed ditin compounds between. The Sn—CH_3 or Sn—Br (or both) bond energies are increased in Me_3SnBr compared to Me_4Sn and $SnBr_4$. The reaction of I_2 on

R_3SnSnR_3 (R = Me, Bu) is attributed to the formation of an unsymmetrical intermediate complex which results from electrophilic attack of I_2 on one of the Sn atoms of the ditin[3].

Iodine in benzene reacts quantitatively with $(Ph_2Sn)_6$ to form Ph_2SnI_2 as the cleavage product:

$$[(C_6H_5)_2Sn]_6 \xrightarrow[C_6H_6]{I_2} (C_6H_5)_2SnI_2 \tag{b}$$

This quantitative reaction is useful for titrimetric purposes[4].

Similarly, I_2 in benzene quantitatively cleaves[5] $(Bz_2Sn)_4$ to Bz_2SnI_2:

$$[(C_6H_5CH_2)_2Sn]_4 + I_2 \rightarrow (C_6H_5CH_2)_2SnI_2 \tag{c}$$

Halogens act on polycyclic heterocycles:

$$[R_2Sn]_p + X_2 \rightarrow I[R_2Sn]_nI \tag{d}$$

With xs X_2, the Sn—Sn bond cleaves and thus n = 1 in the above product; e.g., nonameric $(Et_2Sn)_9$ undergoes degradation with Cl_2 and Br_2 at $-60°$ to $-70°C$ to give[6] Et_2SnX_2:

$$(Et_2Sn)_9 + 9\ X_2 \rightarrow 9\ Et_2SnX_2 \tag{e}$$

where X = Br, Cl.

The di-n-butyltin product from dry $SnCl_2$ and n-BuLi in Et_2O–C_6H_6 at $-10°C$ is octameric and reacts with Br_2 to yield[7] n-Bu_2SnBr_2:

$$(n\text{-}Bu_2Sn)_8 + 8\ Br_2 \rightarrow 8\ n\text{-}Bu_2SnBr_2 \tag{f}$$

Tetrameric tins react similarly. Excess I_2 in refluxing benzene with (t-$Bu_2Sn)_8$ yields[8] t-Bu_2SnI_2 in 20 min:

$$(t\text{-}Bu_2Sn)_4 + I_2 \rightarrow t\text{-}Bu_2SnI_2 \tag{g}$$

However, when 1 :1 I_2 and oligomeric $(R_2Sn)_n$ is present, cleavage of only one Sn—Sn bond occurs to give the ring-opened α,ω-substituted polystannane product.

The cyclic tetramer reacts with equimol I_2 in benzene at 20°C to form a diiodo compound[9]:

$$(t\text{-}Bu_2Sn)_4 + I_2 \rightarrow I(t\text{-}Bu_2Sn)_4I \tag{h}$$

Similarly, I_2 reacts with $(Et_2Sn)_9$ without cleavage of the polymeric chain[10]:

$$(Et_2Sn)_9 + I_2 \rightarrow I(Et_2Sn)_9I \tag{i}$$

(J. M. BELLAMA)

1. J. B. Pedley, H. A. Skinner, C. L. Chernick, *Trans. Faraday Soc., 53*, 1612 (1957).
2. S. Bone, M. Gielen, J. Nasielski, *Bull. Soc. Chim. Belg., 72*, 864 (1964).
3. G. Tagliavini, S. Faleschini, G. Pilloni, G. Plazzogna, *J. Organomet. Chem., 5*, 136 (1966).
4. W. P. Neumann, K. Konig, *Justus Liebigs Ann. Chim.*, , 677, 1 (1964).
5. W. P. Neumann, K. Konig, *Angew. Chem. Int. Ed. Engl., 3*, 751 (1964).
6. *W. P. Neumann, J. Pedain, Justus Liebigs Ann. Chim., 672*, 36 (1964).
7. N. N. Zemlyanskii, E. M. Penov, K. A. Kocheshkov, *Dokl. Akad. Nauk SSSR, 146*, 1335 (1962).
8. W. V. Farrar, H. A. Skinner, *J. Organomet. Chem., 1*, 433 (1964).
9. W. V. Farrar, H. A. Skinner, *J. Organomet. Chem., 1*, 434 (1964).
10. W. P. Neumann, J. Pedain, *Justus Liebigs Ann. Chim., 672*, 341 (1964).

2.5.7.6.2. by Hydrogen, Alkyl and Other Halides.

Hydriodic and hydrobromic acid cleave[1] di-n-butyltin $(n\text{-}Bu_2Sn)_2$ in an inert atm or in vacuo. Boiling $(n\text{-}Bu_2Sn)_2$ and 45% HI for 5 h at 120°C yields 65%; $n\text{-}Bu_2SnI_2$ and 40% HBr at 120°C yields 32% $n\text{-}Bu_2SnBr_2$:

$$[(n\text{-}Bu)_2Sn]_2 + HX \xrightarrow{120°C} X(n\text{-}Bu)_2SnX \qquad (a)$$

where X = Br_3I.

The Sn—Sn bond in $Me_3SnSnMe_3$ cleaves with HCl in MeOH at 85°C to form Me_3SnCl:

$$[(CH_3)_3Sn]_2 + HCl \rightarrow (CH_3)_3SnCl \qquad (b)$$

but reaction with $AuCl_3$ also produces[2] Me_2SnCl_2:

$$[(CH_3)_3Sn]_2 + Au(III)Cl \rightarrow (CH_3)_3SnCl + (CH_3)_2SnCl_2 \qquad (c)$$

Treatment of dialkyl- and diaryltin with an alkyl halide also cleaves the Sn—Sn bond:

$$(R_2Sn)_n + R'X \rightarrow R_2R'SnX \qquad (d)$$

where R = n-Bu, Et, Ar.

Only R_3SnX compounds are produced when the starting material is a dialkyltin, whereas Ph_3SnX, $PhRSnX_2$, PhR_2SnX, and Ph_2SnX_2 compounds are formed with a diaryltin reactant. Also, alkyl iodides give only $R_2R'SnI$ compounds, alkyl bromides give both $R_2R'SnBr$ and R_3SnBr compounds, and alkyl chlorides give only R_3SnCl compounds.

The R'_2RSnX compounds and Ph_2RSnX formed from dialkyltin and diaryltin, respectively, are produced via an addition–cleavage of the ring, the simple R_3SnX is formed via redistribution, and Ph_3SnX, Ph_2SnX_2, etc., are formed via both addition and redistribution[3–6].

Finally, the Sn—Sn bond in R_3SnSnR_3 is cleaved by R_fI to give R_3SnR_f derivatives[7]:

$$R_6Sn_2 + R_fI \rightarrow R_3SnR_f + R_3SnI \qquad (e)$$

e.g., UV irradiation of $Me_3SnSnMe_3$ and CF_3I produces Me_3SnCF_3 and Me_3SnI. Despite the more polar nature of CF_3I, the reaction resembles the familiar halogen cleavage of R_3SnSnR_3. A four-center mechanism is proposed[8]:

$$(CH_3)_3SnSn(CH_3)_3 + CF_3I \rightleftharpoons \begin{array}{c}(CH_3)_3Sn\text{----}Sn(CH_3)_3\\ \vdots \qquad \vdots \\ CF_3\text{---}I\end{array} \rightarrow (CH_3)_3SnCF_3 + (CH_3)_3SnI \qquad (f)$$

(J. M. BELLAMA)

1. N. S. Vyazankin, V. I. Bychkov, *J. Gen. Chem. USSR* (*Engl. Transl.*), *36*, 1694 (1966).
2. G. Tagliavini, L. Belluco, G. Pillono, *Ric. Sci. Parte 2: Sez. A.*, *3*, 889, (1963); *Chem. Abstr.*, *60*, 15,900 (1964).
3. G. A. Razuvaev, Yu I. Drgunov, N. S. Vyazankin, *J. Gen. Chem. USSR* (*Engl. Transl.*), *32*, 2515 (1962).
4. V. T. Bychkov, N. S. Vyazankin, *J. Gen. Chem. USSR* (*Engl. Transl.*), *35*, 687 (1965).
5. K. Sisido, S. Kozima, T. Isibasi, *J. Organomet. Chem.*, *10*, 439 (1967).

6. K. Sisido, T. Miyanisi, K. Nabika, S. Kozima, *J. Organomet. Chem., 11*, 281 (1968).
7. H. D. Kaesz, J. R. Phillips, F. G. A. Stone, *J. Am. Chem. Soc., 82*, 6228 (1960).
8. H. C. Clark, C. J. Willis, *J. Am. Chem. Soc., 82*, 1888 (1960).

2.5.7.7. the Pb–Pb Bond

2.5.7.7.1. by Halogens.

The halogenation of hexaalkyldilead in Et_2O at $-60°C$ yields trialkyllead halides, which are unstable and are not isolated[1].

$$R_6Pb_2 + X_2 \rightarrow 2\ R_3PbX \tag{a}$$

Halogenation of $Ph_3PbPbPh_3$ is less convenient; in contrast to the halogenation of hexaalkyls such as $(n\text{-}Bu)_3PbPb(Bu\text{-}n)_3$, it is complex because of the more polar nature of the Pb—Ph bond compared to the Pb—Bu-n bond, and the greater electron-withdrawing properties of the Ph group[2].

Unlike other organoleads, $(c\text{-}C_6H_{11})_3PbPb(C_6H_{11}\text{-}c)_3$ adds I_2 in benzene even at RT. A quantitative yield of $(c\text{-}C_6H_{11})_3PbI$, mp 91.70°C, is stable for months when protected from air[3].

The reaction of hexaaryldileads with halogens is complex (see above). Bromination of hexa-p-xylyldilead in $CHCl_3$ at $-10°C$ produces[4] R_3PbX and R_2PbX_2:

$$[p\text{-}C_6H_3(CH_3)_2]_6Pb_2 \xrightarrow[CHCl_3]{Br_2} [p\text{-}C_6H_3(CH_3)_2]_3PbBr + [p\text{-}C_6H_3(CH_3)_2]_2PbBr_2 \tag{b}$$

whereas bromination of hexa-p-tolyldilead in py yields[5] $PbBr_2$ and $(p\text{-}CH_3C_6H_5)_4Pb$:

$$(p\text{-}C_6H_4CH_3)_6Pb_2 \xrightarrow[py]{Br_2} PbBr_2 + (p\text{-}C_6H_4CH_3)_4Pb \tag{c}$$

(J. M. BELLAMA)

1. L. C. Willemsens, G. J. M. Van der Kerk, *Investigations in the Field of Organolead Chemistry*, International Lead-Zinc Research Organization, New York, 1965.
2. L. C. Willemsens, G. J. M. Van der Kerk, *J. Organomet. Chem., 15*, 117 (1965).
3. E. Krause, *Chem. Ber., 54*, 2060 (1921).
4. E. Krause, N. Schmitz, *Chem. Ber., 52*, 2165 (1919).
5. E. Krause, G. G. Reissaus, *Chem. Ber., 55*, 888 (1922).

2.5.7.7.2. by Hydrogen Halides.

The reaction of $Ph_3PbPbPh_3$ with hydrohalic acids is also complex:

$$Ph_6Pb_2 + 3\ HX \rightarrow Ph_3PbX + PbX_2 + 3\ PhH \tag{a}$$

$$Ph_3PbX + HX \rightarrow Ph_2PbX_2 + PhH \tag{b}$$

Excess acid is used, but no conclusion can be drawn about intermediates or mechanism. However, dissociation into Ph_4Pb and Ph_2Pb is suggested[1–3]:

$$Ph_6Pb_2 \rightleftarrows Ph_4Pb + Ph_2Pb \tag{c}$$

Additional examples are in §2.5.6.4.2.

(J. M. BELLAMA)

1. P. R. Austin, *J. Am. Chem. Soc., 53*, 3514 (1931).
2. H. Gilman, J. C. Bailie, *J. Am. Chem. Soc., 61*, 739 (1939).
3. H. Gilman, E. B. Towne, *J. Am. Chem. Soc., 61*, 739 (1939).

2.5.7.7.3. by Other Halides.

Hexaethyldilead in toluene reacts with 1:1 S—Cl compounds e.g., SO_2Cl_2, $SOCl_2$, SCl_2, and S_2Cl_2, under mild conditions to give 97, 86, 67, and 65% yields, respectively, of Et_3PbCl. The decreasing yields are due to the increasing nucleophilic character of the sulfur atom in the series[1]:

$$Et_6Pb_2 + SO_2Cl_2 \xrightarrow{\text{toluene}} Et_3PbCl \quad (a)$$

$$Et_6Pb_2 + SOCl_2 \xrightarrow{\text{toluene}} Et_3PbCl \quad (b)$$

$$Et_6Pb_2 + SCl_2 \xrightarrow{\text{toluene}} Et_3PbCl \quad (c)$$

$$Et_6Pb_2 + S_2Cl_2 \xrightarrow{\text{toluene}} Et_3PbCl \quad (d)$$

Hexaethyldilead in MeOH reacts[2] with Cu, Hg, Au and Fe salts upon mixing for 5 h in N_2 at 25°C, e.g.:

$$Et_6Pb_2 + 2\ HgCl_2 \rightarrow Et_3PbCl + Hg_2Cl_2 \quad (e)$$

$$Et_6Pb_2 + Cu_2Cl_2 \rightarrow Et_4Pb + PbCl_2 + Cu \quad (f)$$

$$Et_6Pb_2 + 2\ CuCl_2 \rightarrow Et_4Pb + PbCl_2 + Et_3PbCl + Cu_2Cl_2 \quad (g)$$

$$2\ Et_6Pb_2 + Hg_2Cl_2 \rightarrow Et_4Pb + PbCl_2 + Hg \quad (h)$$

$$2\ Et_6Pb_2 + AuCl_3 \rightarrow Et_4Pb + PbCl_2 + Au \quad (i)$$

$$2\ Et_6Pb_2 + FeCl_3 \rightarrow Et_4Pb + PbCl_2 + FeCl_2 \quad (j)$$

Hexaphenyldilead reacts[3] with anhyd $CuCl_2$ and $CoCl_2$ in pyridine, dioxane, etc. at elevated T. With xs $CuCl_2$, an intermediate complex of Ph_6Pb_2 with $CuCl_2$ forms[3] to yield Ph_3PbCl:

$$Ph_6Pb_2 + CuCl_2 \rightarrow [Ph_6Pb_2\text{---}CuCl_2] \rightarrow Ph_3PbCl \quad (k)$$

The product undergoes further chlorination to form $PbCl_2$, i.e.:

$$Ph_3PbCl \rightarrow Ph_2PbCl_2 \rightarrow PbCl_2 + PhCl \quad (l)$$

Similarly, Ph_6Pb_2 reacts with $CoCl_2$:

$$Ph_6Pb_2 + CoCl_2 \rightarrow [Ph_6Pb_2\text{---}CoCl_2] \xrightarrow{\text{dearylation}} PbCl_2 + PhCl \quad (m)$$

These reactions proceed via a Ph_6Pb_2-MCl_2 complex in which the first step involves rupture of the Pb—Pb bond to form Ph_3PbCl.

Hexa-m-xylyldilead reacts with $TlCl_3$ to yield di-m-xylyllead dichloride and TlCl:

$$[m\text{-}C_6H_3(CH_3)_2]_6Pb_2 + TlCl_3 \rightarrow [m\text{-}C_6H_3(CH_3)_2]_2PbCl_2 + TlCl \quad (n)$$

Tri-m-xylyllead in benzene is mixed with $TlCl_3$ in Et_2O and after standing for a week crystals of the dichloride form and can be recrystallized from benzene. The product is more soluble in organic media[4] than is Ph_2PbCl_2.

Both Fe(II) and Fe(III) halides cleave the Pb—Pb bond of substituted hexaphenyldileads[5]. Thus, $(p\text{-}CH_3OC_6H_5)_3PbPb(C_6H_4OCH_3\text{-}p)_3$ and $(c\text{-}C_6H_{11})_3PbPb(C_6H_{11}\text{-}c)_3$ react with $FeCl_3$ with the formation of Ph_3PbCl almost quantitatively:

$$(p\text{-}CH_3OC_6H_4)_6Pb_2 + FeCl_3 \rightarrow (p\text{-}CH_3OC_6H_4)_3PbCl + FeCl_2 \quad \text{(o)}$$

$$(c\text{-}C_6H_{11})_6Pb_2 + FeCl_3 \rightarrow (c\text{-}C_6H_{11})_3PbCl + FeCl_3 \quad \text{(p)}$$

However, FeI_2 yields PbI_2 and $(p\text{-}CH_3OC_6H_4)_4Pb$:

$$(p\text{-}CH_3OC_6H_4)_6Pb_2 + FeI_2 \rightarrow PbI_2 + (p\text{-}CH_3OC_6H_4)_4Pb \quad \text{(q)}$$

The reaction of Pb(II) halides and organomagnesium halide reagents provides a source of R_3PbX. Hexaethyldilead with $MgBr_2$ and MgI_2 in Et_2O forms[6,7] Et_3PbX:

$$Et_6Pb_2 + MgX_2 \xrightarrow{Et_2O} Et_3PbX \quad \text{(r)}$$

where X = Br, I. The formation of R_3PbMgX as an intermediate is postulated[6,7]:

$$R_6Pb_2 + MgI_2 \rightarrow 2\ R_3PbI + Mg \quad \text{(s)}$$

$$R_3PbX + Mg \rightarrow R_3PbMgX \quad \text{(t)}$$

a suggestion that is supported by[6,7]:

$$2\ (C_2H_5)_3PbBr + 2\ Mg \xrightarrow{Et_2O} 2\ C_2H_5MgBr + 2\ (C_2H_5)_4Pb + Pb \quad \text{(u)}$$

Aluminum chloride cleaves $(o\text{-}CH_3C_6H_4)_3PbPb(C_6H_4CH_3\text{-}o)_3$ to form $PbCl_2$, tetraaryllead, and arylaluminum chlorides[8]. The reaction is carried out by heating 1:1 hexa-o-tolyllead and $AlCl_3$ on a steam bath for ca. 4 h:

$$(C_6H_4OCH_3\text{-}o)_6Pb_2 \rightarrow PbCl_2 + Pb(C_6H_4OCH_3\text{-}o)_4 + (C_6H_4OCH_3\text{-}o)AlCl_2 \quad \text{(v)}$$

On refluxing in $BrCH_2CH_2Br$ for 15 min, $Ph_3PbPbPh_3$ forms[9] Ph_4Pb (81%) and $PbBr_2$ (98%):

$$2\ Ph_6Pb_2 + C_2H_4Br_2 \rightarrow 3\ Ph_4Pb + PbBr_2 + C_2H_4 \quad \text{(w)}$$

The $Ph_3PbPbPh_3$ undergoes an initial decomposition to Ph_4Pb and elemental Pb via the unstable Ph_2Pb intermediate:

$$Ph_6Pb_2 \rightarrow Ph_4Pb + [Ph_2Pb] \quad \text{(x)}$$

$$2\ [Ph_2Pb] \rightarrow Ph_4Pb + Pb \quad \text{(y)}$$

followed by:

$$Pb + BrC_2H_4Br \rightarrow PbBr_2 + C_2H_4 \quad \text{(z)}$$

(J. M. BELLAMA)

1. U. Belluco, L. Cattalini, A. Peloso, G. Tagliavini, *Ric. Sci. Rend. Sez. A*, *3*, 1107 (1963); *Chem. Abstr.*, *61*, 677 (1964).
2. U. Belluco, G. Belluco, *Ric. Sci.*, *Parte 2*: *Sez. A*, *32*, 102 (1962); *Chem. Abstr.*, *57*, 13,786 (1962).

3. G. A. Razuvaev, M. S. Fedotov, T. B. Zavarova, N. N. Bazhenova, *Tr. Khim. Khim. Tekhnol.*, *4*, 662 (1961); *Chem. Abstr.*, *58*, 543 (1963).
4. A. E. Goddard, *J. Chem. Soc.*, *123*, 1161 (1923).
5. M. Leichtenwalter, *Iowa State Coll. J. Sci.*, *14*, 57 (1939); *Chem. Abstr.*, *34*, 6241 (1940).
6. H. Gilman, J. C. Bailey, *J. Am. Chem. Soc.*, *61*, 731 (1939).
7. W. C. Setzer, R. W. Leeper, H. Gilman, *J. Am. Chem. Soc.*, *61*, 1609 (1939).
8. H. Gilman, L. D. Appreson, *J. Org. Chem.*, *4*, 162 (1939).
9. A. W. Krebs, M. C. Henry, *J. Am. Chem. Soc*, *28*, 1911 (1966).

2.5.8. from Cleavage of the Oxygen–

2.5.8.1. Si Bond

2.5.8.1.1. by Halogens.

Cleavage of the Si—O bond in bis-monocyclohexylsilylether occurs[1] on heating with a deficiency of warm I_2. The reaction involves the gradual addition of I_2 to the refluxing siloxane to yield HI gas, cyclo-$C_6H_{11}SiH_2I$ and a high-boiling viscous liq $[(cy\text{-}C_6H_{11}SiHO)]_{3\ or\ 4}$:

$$3\ (cy\text{-}C_6H_{11}SiH_2)_2O + 3\ I_2 \rightarrow 3\ cy\text{-}C_6H_{11}SiH_2I + 3\ HI + (cy\text{-}C_6H_{11}SiHO)_3 \quad (a)$$

(J. M. BELLAMA)

1. H. H. Anderson, *J. Am. Chem. Soc.*, *81*, 4785 (1959).

2.5.8.1.2. by Halogens in the Presence of a Reducing Agent.

Iodine cleaves the Si—O bond in $(Me_3Si)_2O$ in the presence of Al metal, a reducing agent. A solution of $(R_3Si)_2O$, I_2, and Al filings is refluxed for ca. 1 h and the mixture distilled in N_2 in diffused light with occasional cooling of the exothermic reaction. The trialkyliodosilane is distilled[1] from powdered Cu:

$$(R_3Si)_2O + I_2 \xrightarrow[\text{reflux}]{\text{Al filings}} R_3SiI \quad (a)$$

where R = CH_3, C_2H_5, n-C_3H_7, n-C_4H_9 or n-C_5H_{11} [2].

Iodine reacts[3] with $(C_2H_5)_2Si(OC_6H_5)_2$ in the presence of Al filings to cleave the Si—O bond, e.g., by adding I_2 to $(C_2H_5)_2Si(OC_6H_5)_2$ and Al filings and then refluxing until the violet I_2 color disappears:

$$(C_2H_5)_2Si(OC_6H_5)_2 + I_2 \xrightarrow[\text{reflux}]{\text{Al filings}} \underset{(44\%)}{(C_2H_5)_2SiI_2} \quad (b)$$

(J. M. BELLAMA)

1. H. H. Anderson, *J. Am. Chem. Soc.*, *81*, 4785 (1959).
2. M. G. Voronkov, Yu I. Khudobin, *Izv. Akad. Nauk SSSR., Otd. Khim. Nauk*, 713 (1956).
3. B. N. Dolgov, S. N. Borisov, M. G. Voronkov, *J. Gen. Chem. USSR* (*Engl. Transl*), 2692 (1957).

2.5.8.1.3. by Hydrogen Halides.

Aqueous-alcoholic HF converts organosiloxanes into organofluorosilanes. Triethylfluorosilane is obtained from ethanolic $(Et_3Si)_2O$ with HF at 75°C for 7.5 h, pouring the product into xs H_2O and thus enabling the fluoro compound to separate[1]:

$$[(C_2H_5)_3Si]_2O + 2\ HF \rightarrow 2\ (C_2H_5)_3SiF + H_2O \tag{a}$$

This reaction, together with hydrolysis and alcoholysis of organofluorosilanes, illustrates the reversibility of the reactions[1], i.e.:

$$\equiv SiOR + HF \rightleftharpoons \equiv SiF + ROH \tag{b}$$

where R = H or Et.

The cleavage of silicones by anhyd liq HF occurs by treating hexaethylcyclotrisiloxane with anhyd $CuSO_4$ at atm pressure to form Et_2SiF_2 in almost quantitative yield[2].

$$[(C_2H_5)_2SiO]_3 + 6\ HF \rightarrow 3\ (C_2H_5)_2SiF_2 + 3\ H_2O \tag{c}$$

The above reaction is a solvolysis:

$$(Si{-}O{-}Si) + 3\ HF \rightarrow 2\ (Si{-}F) + H_2O{\cdot}HF \tag{d}$$

With no electron-donating groups in the aromatic ring, i.e., no p-Me, p-MeO, etc., the Si—Ph bond does not cleave by aq alcoholic HF in alkylphenyldisiloxanes[3]:

$$[(CH_3)_2(C_6H_5)Si]_2O + 2\ HF \rightarrow 2\ (CH_3)_2(C_6H_5)SiF + H_2O \tag{e}$$

A cleavage of the Si—O bond that involves heating sym-bis(cyclohexyl)disiloxanes with HCl gas in the presence of P_2O_5 forms the chloro product in 25% yield[4]:

$$(cyclo\text{-}C_6H_{11}SiH_2)_2O + HCl \xrightarrow{P_2O_5} cyclo\text{-}C_6H_{11}SiH_2Cl + H_2O{\cdot}P_2O_5 \tag{f}$$

Both $(RO)_4Si$ and Me_3SiOR are cleaved at the Si—O bond by passing dry HCl into the ester for 90 min at −10°C and trapping the Me_3SiCl formed at −80°C, e.g.:

$$(CH_3)_3SiOR + HCl \xrightarrow{-10\ °C} \underset{(77\text{–}90\%)}{(CH_3)_3SiCl} + ROH\ (or\ RCl) \tag{g}$$

An alcohol is formed[5] when R = n-C_4H_9, $PhCH_2CH_2$—, $PhCH_2CH_2CH_2$—, $(PhCH_2)_2CH$—, and $PhCH_2CHPh$—. Formation of an alkyl chloride occurs[5] when R = $PhCH_2$—, PhCHMe—, Ph_2CH—, and PhCH=$CHCH_2$—.

(J. M. BELLAMA)

1. C. Eaborn, *J. Chem. Soc*, 2846 (1952).
2. H. S. Booth, M. L. Freedman, *J. Am. Chem. Soc.*, *72*, 2847 (1950).
3. C. Eaborn, *J. Chem. Soc*, 2846 (1952).
4. H. H. Anderson, *J. Am. Chem. Soc.*, *81*, 4785 (1959).
5. W. Gerrard, K. D. Kilburn, *J. Chem. Soc.*, 1536 (1956).

2.5.8.1.4. by Organic Halides.

Addition of an ethereal $[(C_2H_5)_3SiO]_3P$ to an acetyl chloride in Et_2O forms:

$$[(C_2H_5)_3SiO]_3P + CH_3COCl \rightarrow (C_2H_5)_3SiCl + CH_3C(O)P(O)[OSi(C_2H_5)_3]_2 \tag{a}$$

The reaction is slightly exothermic and proceeds through an intermediate[1]

$$\begin{array}{c} R'O \quad\quad Cl^- \\ \diagdown^{+} \quad \swarrow \\ P{-}OSiR_3 \\ \diagup \quad \diagdown \\ R_3SiO \quad\quad COR \end{array} \quad \text{or} \quad \begin{array}{c} \quad\quad\quad R \\ \quad\quad\quad | \\ R'O \quad\quad C{=}O \\ \diagdown \quad\quad | \\ P \quad\quad Cl \\ \diagup | \quad\quad \searrow \\ R_3SiO \quad O{-\!\!-}SiR_3 \end{array}$$

Acetyl chloride cleaves the Si—O bond in $Me_3SiSiMe_2OEt$ at 104–110°C for 1 h in a sealed glass tube followed by fractional distillation[2]:

$$(CH_3)_3SiSi(CH_3)_2OC_2H_5 + CH_3COCl \rightarrow \underset{(80\%)}{(CH_3)_3SiSi(CH_3)_2Cl} \tag{b}$$

Similarly, 1:1 benzoyl chloride and $Me_2EtOSiSiMe_2OEt$ heated at 230–240°C with pyridine yields[2] a small amount of an unidentified, low-bp product along with $Me_2ClSiSiMe_2Cl$:

$$C_2H_5OSi(CH_3)_2Si(CH_3)_2OC_2H_5 + C_6H_5COCl \rightarrow \underset{(64\%)}{Cl(CH_3)_2SiSi(CH_3)_2Cl} \tag{c}$$

Refluxing $n\text{-}Pr_3SiOEt$ with acetyl chloride 10 min and distillation of the mixture yields[3] 85% $n\text{-}Pr_3SiCl$:

$$(n\text{-}C_3H_7)_3SiOC_2H_5 + CH_3COCl \rightarrow (n\text{-}C_3H_7)_3SiCl \tag{d}$$

Benzyl chloride cleaves[4] the Si—O bond under mild condition, e.g., upon heating $Me_2Si(H)OEt$ in a sealed vessel:

$$(CH_3)_2Si(H)OC_2H_5 + C_6H_5COCl \rightarrow (CH_3)_2SiHCl \tag{e}$$

Like the monosilanes, alkoxy derivatives of polysilanes are converted in satisfactory yields to the corresponding chloropolysilanes by treatment with acetyl chloride[5]:

$$\text{(cyclopentyl)}\overset{CH_3}{\underset{}{Si}}{-\!\!-}\overset{CH_3}{Si}(OC_2H_5)_2 + CH_3COCl \rightarrow \underset{(77\%)}{\text{(cyclopentyl)}\overset{CH_3}{Si}{-\!\!-}\overset{CH_3}{\underset{OC_2H_5}{Si}}{-}Cl} \tag{f}$$

Ethoxy to chloride conversion is achieved by refluxing ethoxynonamethylcyclopentasilane with xs acetyl chloride to form the corresponding chloro compound in 88% yield[6]:

$$\begin{array}{c} CH_3 \quad OC_2H_5 \\ \diagdown \; \diagup \\ Si \\ \diagup \quad \diagdown \\ (CH_3)_2Si \quad\quad Si(CH_3)_2 \\ \diagdown \quad \diagup \\ (CH_3)_2Si{-}Si(CH_3)_2 \end{array} + CH_3COCl \rightarrow \begin{array}{c} CH_3 \quad Cl \\ \diagdown \; \diagup \\ Si \\ \diagup \quad \diagdown \\ (CH_3)_2Si \quad\quad Si(CH_3)_2 \\ \diagdown \quad \diagup \\ (CH_3)_2Si{-}Si(CH_3)_2 \end{array} \tag{g}$$

Unlike most silanols, Bz_3SiOH is converted quantitatively into Bz_3SiCl upon warming with xs acetyl chloride[7]:

$$(C_6H_5CH_2)_3SiOH + RCOCl \rightarrow (C_6H_5CH_2)_3SiCl + RCOOH \tag{h}$$

where R = CH_3, C_6H_5.

Triphenylsilanol also forms the corresponding chloro compound upon refluxing the Ph_3SiOH with xs acetyl chloride for 2 h, followed by recrystallization from ligroin[8]:

$$(C_6H_5)_3SiOH + CH_3COCl \xrightarrow{\text{reflux}} (C_6H_5)_3SiCl + CH_3COOH \qquad \text{(i)}$$

Butoxy-Cl interconversion takes place between $Me_2Si(OBu\text{-}n)_2$ and benzoyl chloride by distilling a 1:2 mixture for ca. 10 h with quinoline phosphate to give Me_2SiCl_2 in 50% yield[9]:

$$(CH_3)_2Si(OC_4H_9)_2 + C_6H_5COCl \rightarrow (CH_3)_2SiCl_2 \qquad \text{(j)}$$

The ethoxy group of $Ph_2Si(OEt)_2$ is also chlorinated by benzoyl chloride on refluxing with a pyridine catalyst for 8 h, followed by fractional distillation under reduced pressure to yield[10] pure Ph_2SiCl_2:

$$(C_6H_5)_2Si(OC_2H_5)_2 + C_6H_5COCl \rightarrow (C_6H_5)_2SiCl_2 \qquad \text{(k)}$$

Analogously, alkoxy–fluorine interconversion takes place between hexafluoropropene epoxide and alkylalkoxysilanes, e.g., Me_3SiOMe yields Me_3SiF:

$$CF_3{-}F\overset{\overset{O}{/\ \backslash}}{C}{-}CF_2 + (CH_3)_3Si(OCH_3) \rightarrow (CH_3)_3SiF + CF_3CF_2CO_2CH_3 \qquad \text{(l)}$$

The reaction proceeds through an unstable intermediate[11]:

$$CF_3\underset{\underset{O}{\backslash\ /}}{CFCF_2} + {\equiv}SiX \rightarrow [CF_3CFXCF_2OSi] \rightarrow CF_3CFXCOF + {\equiv}SiF \qquad \text{(m)}$$

where X = OCH_3, OC_2H_5.

Warming 3:2 hexafluoropropene epoxide and $Me_2Si(OEt)_2$ at 20°C for 118 h replaces both ethoxy groups by fluorine[11]:

$$2\ CF_3F\overset{\overset{O}{/\ \backslash}}{C}{-}CF_2 + (CH_3)_2Si(OC_2H_5)_2 \rightarrow (CH_3)_2SiF + 2\ CF_3CF_2CO_2C_2H_5 \qquad \text{(n)}$$

while 2:1 hexafluoropropene epoxide and dimethyldiethoxysilane in a mol ratio at 20°C for only 40 h replaces only one ethoxy by fluorine[11]:

$$CF_3F\overset{\overset{O}{/\ \backslash}}{C}{-}CF_2 + (CH_3)_2Si(OC_2H_5)_2 \rightarrow (CH_3)_2Si(F)OC_2H_5 + CF_3CF_2CO_2C_2H_5 \qquad \text{(o)}$$

(J. M. BELLAMA)

1. N. F. Orlov, B. L. Kaufman, *J. Gen. Chem. USSR (Engl. Transl.)*, *38*, 1842 (1968).
2. M. Kumada, M. Yamaguchi, Y. Yamamoto, J. I. Nakajima, K. Shiina, *J. Org. Chem.*, *21*, 1264 (1956).
3. C. Eaborn, *J. Chem. Soc.*, 2755 (1949).
4. R. Okawara, M. Sakiyama, *J. Chem. Soc. Jpn., Ind. Chem. Sect.*, *58*, 805 (1955).
5. M. Kumada, K. Tamao, *Adv. Organomet. Chem.*, *6*, 19 (1968).
6. M. Ishikawa, M. Kumada, *J. Chem. Soc., Chem. Commun.*, 567 (1969).
7. G. Martin, F. S. Kipping, *J. Chem. Soc.*, *95*, 302 (1909).
8. U. Wannagat, H. Burger, E. Ringel, *Monatsh. Chem.*, *93*, 1363 (1962).
9. R. O. Sauer, *J. Am. Chem. Soc.*, *68*, 138 (1946).

10. E. Larsson, L. Bjellerup, *J. Am. Chem. Soc.*, *75*, 995 (1953).
11. L. Heinrich, *Z. Chem.*, *9*, 257 (1968).

2.5.8.1.5. by Other Halides.

(i) Boron Halides. With boron trihalides, $(Me_3Si)_2O$ forms an addition compound that splits into a silyl halide with scission of the Si—O bond[1,2].

$$3\ [(CH_3)_3Si]_2O + 2\ BX_3 \rightarrow 6\ (CH_3)_3SiX + B_2O_3 \qquad \text{(a)}$$

where X = F, Br, Cl. The intermediate in the reaction, a siloxylsilyl halide, degrades to the silylhalide, or, alternatively, it can react with xs siloxane. Thus, frozen 1:1 BBr_3 and $(Me_3Si)_2O$ form an addition compound at −40°C that split out $(CH_3)_3SiBr$ at RT and more $(CH_3)_3SiBr$ after 30 min heating at 80°C.

With 1,1′-dimethyl-,1,1′,2,2′-tetramethyl- or hexamethyldisiloxane, BF_3 and BCl_3 form unstable addition compounds that degrade to methylhalogenosilane and methylsiloxyboron dihalide with scission of the Si—O bond. The latter decompose to methylhalogenosilane, boron trihalide, and B_2O_3:

$$3\ [H_n(CH_3)_{3-n}Si]_2O + 2\ BX_3 \rightarrow 6\ H_n(CH_3)_{3-n}SiX + B_2O_3 \qquad \text{(b)}$$

where X = F, Cl; n = 1–3. A stepwise mechanism involving a four-center cyclic transition state is postulated.

Preparation of methylhalogenosilanes is accomplished by reacting boron trihalides and siloxanes at −78°C for 12–20 h followed by fractional distillation[3,4].

Boron trifluoride cleaves the Si—O bond in bispentamethyldisilanylether in preference to the Si—Si bond when BF_3 is passed into the refluxing ether for ca. 4 h. Distillation gives the fluoride[5] in 84% yield:

$$3\ [(CH_3)_3SiSi(CH_3)_2]_2O + 2\ BF_3 \rightarrow 6\ (CH_3)_3SiSi(CH_3)_2F + B_2O_3 \qquad \text{(c)}$$

Action of Me_2BBr on 1:1 hexamethyldisiloxane at 220°C yields Me_3SiBr and scission of the Si—O bond[6]:

$$[(CH_3)_3Si]_2O + (CH_3)_2BBr \rightarrow (CH_3)_3SiBr + (CH_3)_3SiOB(CH_3)_2 \qquad \text{(d)}$$

Unsymmetrical 1,1,1-trimethyldisiloxane cleaves its Si—O bond when it reacts with 1:1 BCl_3 at −78°C for 12 h:

$$(CH_3)_3SiOSiH_3 + BCl_3 \rightarrow (CH_3)_3SiOBCl_2 + SiH_3Cl \qquad \text{(e)}$$

A four-centered transition complex is postulated[7]:

$$\begin{array}{c} (CH_3)_3Si—O—SiH_3 \\ \downarrow \quad \uparrow \\ Cl_2B—Cl \end{array}$$

Preferential cleavage of the Si—O bond in a siloxydisilane occurs by treating BCl_3 with xs siloxane at −78°C for ca. 12 h:

$$6\ SiH_3SiH_2OSiH_3 + BCl_3 \rightarrow 6\ SiH_3SiH_2Cl + 3\ (SiH_3)_2O + B_2O_3 \qquad \text{(f)}$$

Because of the ease of cleavage of the SiH_3SiH_2—O bond, a four-center transition complex is postulated[7]:

$$\begin{array}{c} H_3Si—O—SiH_2SiH_3 \\ \downarrow \quad \uparrow \\ Cl_2B—Cl \end{array}$$

Organosiloxanes containing a 1,6-disilahexane system undergo cleavage[8] at the Si—O bond by $F_3B \cdot OEt_2$ on distilling for 3 h while T increases from 100° to 130°C with the gradual formation of Me_3SiF:

$$(CH_3)_3SiO[Si(CH_3)_2(CH_2)_4Si(CH_3)_2O]_nSi(CH_3)_3 \xrightarrow{F_3B\cdot OEt_2} 2\ (CH_3)_3SiF + n\ [F(CH_3)_2Si(CH_2)_4Si(CH_3)_2F] \quad (g)$$

where n = 1–4.

Reaction of BCl_3 and Me_3SiOBu-n replaces the n-butoxy group by Cl with fission of the Si—O bond when BCl_3 vapor is passed into refluxing Me_3SiOBu-n to give Me_3SiCl in 62% yield[9]:

$$3\ (CH_3)_3SiO(n\text{-}C_4H_9) + BCl_3 \xrightarrow{reflux} 3\ (CH_3)_3SiCl + (n\text{-}C_4H_9)_3B \quad (h)$$

Halogen–alkoxy exchange occurs between boron halides and Me_3SiOEt when BBr_3 (at 80°C) and BCl_3 (at 50°C) form the Me_3SiX in 95% yield[1]:

$$2\ (CH_3)_3SiOEt + BX_3 \rightarrow 2\ (CH_3)_3SiX + B(OC_2H_5)_2 \quad (i)$$

where X = Br, Cl.

The ethoxy group in $(C_6F_5)_3SiOEt$, which protects against total elimination during fluorination, is replaced with a reactive Cl by the action of BCl_3 with the silane[10]:

$$(C_6F_5)_3SiOC_2H_5 + BCl_3 \rightarrow (C_6F_5)_3SiCl + C_2H_5OBCl_2 \quad (j)$$

The ethoxy group of ethoxysilane is replaced by Cl with 1:1 BCl_3 at −196°C when held at −96°C for 75 min and at RT for 4 h, to form fluorosilane[11]:

$$CH_3OSiH_3 + BF_3 \rightarrow SiH_3F + CH_3OBF_2 \quad (k)$$

The reaction proceeds[12] via intermediate $(CH_3O)_2BF$.

1-Chloro-2,2,4,4,6,6-hexamethyl-5-oxa-2,4,6-trisilaheptane undergoes cleavage at the siloxane linkages by $F_3B \cdot OEt_2$ upon refluxing until a head T of 125°C is reached to form 57% 1-chloro-4-fluoro-2,2,4-trimethyl-2,4-disilapentane[13]:

$$CH_3Si(CH_3)_2OSi(CH_3)_2CH_2Si(CH_3)_2CH_2Cl \xrightarrow{F_3B\cdot EtO_2} FSi(CH_3)_2CH_2Si(CH_3)_2CH_2Cl \quad (l)$$

Hydroxyl–Br interconversion takes place exothermically when 1:3 BBr_3 and Et_3SiOH in benzene at RT form Et_3SiBr in almost quantitative yield[12]:

$$3\ (C_2H_5)_3Si(OH) + BBr_3 \rightarrow 3\ (C_2H_5)_3SiBr + B(OH)_3 \quad (m)$$

The reaction of $MeOSiF_3$ with boron halides provides trifluorosilyl halides; e.g., $MeOSiF_3$ and boron halides are condensed at low T, heated to RT and fractionated after 12 h to yield the trifluorosilyl halide in an almost quantitative yield[14]:

$$CH_3OSiF_3 + BX_3 \rightarrow SiF_3X + CH_3OBX_2 \quad (n)$$

where X = F, Cl, Br. Boron trichloride with methyl and ethyl trimers and tetramers of organocyclosiloxanes produces R_2SiCl_2 through a stepwise redistribution of the products[15]:

$$(R_2SiO)_n \xrightarrow{BCl_3} R_2SiCl_2 + (R_3SiO)_3 \quad (o)$$

where R = CH_3, C_2H_5; n = 3,4. The overall cleavage proceeds via the following sequence:

$$(R_2SiO)_n + nBCl_3 \rightarrow nR_2\underset{\displaystyle |\atop Cl}{Si}OBCl_2 \qquad (p)$$

$$3\ R_2\underset{\displaystyle |\atop Cl}{Si}{-}OBCl_2 \xrightarrow[\text{heating}]{\text{mild}} (R_2\underset{\displaystyle |\atop Cl}{Si}O)_3B + 2\ BCl_3 \qquad (q)$$

$$2\ (R_2\underset{\displaystyle |\atop Cl}{Si}O)_3B \xrightarrow{\text{short reflux}} 3\ (R_2\underset{\displaystyle |\atop Cl}{Si})_2O + B_2O_3 \qquad (r)$$

$$3\ (R_2\underset{\displaystyle |\atop Cl}{Si})O \xrightarrow[\text{distillation}]{\text{long heating}} 3\ R_2SiCl_2 + (R_2SiO)_3 \qquad (s)$$

(ii) by Al Halides. As with boron, Al halides also cleave the Si—O linkage in $(SiH_3)_2O$ at low T to give SiH_3X in 64–81% yields:

$$(SiH_3)_2O + AlX_3 \rightarrow SiH_3X + SiH_3OAlX_2 \qquad (t)$$

The chloride, bromide, and iodide of Al require different times and T. Some evidence of initial adduct formation is available[16].

Aluminum halides do not form stable adducts with $(R_3Si)_2O$ but instead cleave the Si—O bond to form R_3SiX in 76–96% yields:

$$(R_3Si)_2O + AlX_3 \rightarrow 2\ R_3SiX + AlOX_2 \qquad (u)$$

where X = Cl, Br, I; R = CH_3, C_2H_5, n-C_3H_7, by dissolving (a) $AlCl_3$ in $(Me_3Si)_2O$ and heating to 40°C overnight to give Me_3SiCl; (b) $AlBr_3$ in $(Me_3Si)_2O$ with cooling and then distilling in vacuum to give $(CH_3)_3SiBr$; and (c) AlI_3 in $(Me_3Si)_2O$ and refluxing for 30 min to obtain $(CH_3)_3SiI$. Similarly, Et and n-Pr complexes can also be obtained. An AlX_3 complex with the oxygen atom is the intermediate[17,18].

Cleavage of the Si—O bond forms halogen-containing organosilicons on heating 1:1$(R_3Si)_2O$ and AlX_3. Distillation yields 70–85% $R_3SiOAlX_2$:

$$[(CH_3)_n(C_2H_5)_{3\text{-}n}Si]_2O + AlX_3 \rightarrow (CH_3)_n(C_2H_5)_{3\text{-}n}SiX + (CH_3)_n(C_2H_5)_{3\text{-}n}SiOAlX_2 \qquad (v)$$

where X = Br, Cl. Dimeric complexes are formed as a result of cleavage of $(R_3Si)_2O$ by AlX_3:

$$(R_3Si)_2O + AlX_3 \rightarrow R_3SiX + R_3SiOAlX_2 \qquad (w)$$

where X = I, Br, Cl; R = CH_3, C_2H_5. To AlX_3 in dry cyclohexane is added with stirring 1:1 $(R_3Si)_2O$, followed by refluxing and distillation. The reactions with AlI_3 are cooled, and the resulting solution is decolorized by Cu powder[19].

Heating $(MeEt_2Si)_2O$ with $AlCl_3$ at 135°C for 1 h and finally at 210°C forms $CH_3(C_2H_5)_2SiCl$ and $CH_3(C_2H_5)_2SiOAlCl_2$. Similarly, $(CH_3)_3SiOAlCl_2$, $(CH_3)_2C_2H_5SiOAlCl_2$, and $(CH_3)_3SiOBr_2$ compounds can also be formed[20,21].

Alkoxyhalosilanes are also prepared by the cleavage of $AlCl_3$ with alkylalkoxysilanes, involving simultaneous interconversion of alkoxy groups and Cl atoms e.g., $Me_2Si(OEt)_2$ with $AlCl_3$ at 140°C yields[22] $Me_2(EtO)SiCl$:

$$(CH_3)_2Si(OC_2H_5)_2 + AlCl_3 \xrightarrow{140\ °C} (CH_3)_2Si(OC_2H_5)Cl \tag{x}$$

Analogously, Et_3SiOPr-i and $AlCl_3$ at 200–210°C, produces Et_3SiCl in good yields[22]:

$$(C_2H_5)_3SiOC_3H_7\text{-i} + AlCl_3 \rightarrow (C_2H_5)_3SiCl \tag{y}$$

Dimethylaluminum bromide reacts with $(H_3Si)_2O$ at low T to yield a volatile white solid, $Me_2Al(OSiH_3)_2$:

$$2\ (CH_3)_2AlBr + 2\ (SiH_3)_2O \rightarrow 2\ SiH_3Br + [(CH_3)_2Al(OSiH_3)]_2 \tag{z}$$

A complex $(CH_3)_{6-(x+y)}Al_2Br_x(OSiH_3)_y$ is formed, where x = 0–1; y = 2–3 depending on conditions[23].

Organopolysiloxanes containing alkylhydrogensiloxane linkages degrade to form[24] alkylhydrogenchlorosilanes from 2:1 organopolysiloxane and $AlCl_3$ at 250°C for 1 h:

$$(CH_3)_3SiO[SiH(CH_3)O]_nSi(CH_3)_3 + AlCl_3 \xrightarrow{250\ °C} \underset{(71\%)}{CH_3SiHCl_2} + (CH_3)_2SiHCl + (CH_3)_3SiCl + SiCl_4 \tag{aa}$$

(iii) by Si Halides. Trialkylsiloxy compounds can be converted to Me_3SiF; e.g., $(Me_3Si)_2O$ with ca. a threefold xs SiF_4 under 20–30 atm at 150°C for < 30 min forms[25,26]:

$$[(CH_3)_3Si]_2O + SiF_4 \rightarrow (CH_3)_3SiF + (CH_3)_3SiOSiF_3 \tag{ab}$$

Heating $(Me_3Si)_2O$ with a SiX_4 in an autoclave under elevated pressure at 250°–300°C yields Me_3SiX by rupture of the siloxane chain:

$$[(CH_3)_3Si]_2O + SiX_4 \rightarrow (CH_3)_3SiX \tag{ac}$$

where X = Cl, Br.

On the basis of $(CH_3)_3SiX$ yields, metal halides reactivities are $TiCl_4 > SiCl_4 > SnCl_4$; $SiBr_4 > SiCl_4$.

Similarly:

$$[(CH_3)_2SiO]_4 + SiCl_4 \xrightarrow{\Delta,\ 3\ h} (CH_3)_2SiCl_2 + \text{partially depolymerized product}^{27-29} \tag{ad}$$

Alkoxysilanes are converted in the gas phase to chlorosilanes by the interaction of $Me_2Si(OEt)_2$ with $SiCl_4$ with dry HCl as catalyst. A contact surface such as glass, mineral wool, or pumice stone is used[30,31] at 250°–300°C:

$$(CH_3)_2Si(OC_2H_5)_2 + SiCl_4 \xrightarrow{HCl} (CH_3)_2SiCl_2 + (C_2H_5O)_2SiCl_2 \tag{ae}$$

Trimethylalkoxysilanes react with 4:1 $SiCl_4$ at RT to replace all halogen atoms. Benzyloxytrimethylsilane reacts[32] in 29 h and the 2-phenylethylester in 2.5 h:

$$4\ (CH_3)_3SiOR + SiCl_4 \rightarrow 4\ (CH_3)_3SiCl + Si(OR)_4 \quad \text{(af)}$$

where R = $C_6H_5CH_2—$, $C_6H_5CH_2CH_2—$.

Ethoxy–Cl interconversion between $Ph_2Si(OEt)_2$ and $SiCl_4$ occurs at 400°C in a sealed tube[33]:

$$(C_6H_5)_2Si(OC_2H_5)_2 + SiCl_4 \rightarrow (C_6H_5)_2SiCl_2 \quad \text{(ag)}$$

Similarly, $MePhSi(OEt)_2$ reacts[33] with $SiCl_4$ in a sealed tube at 400°C:

$$CH_3(C_6H_5)Si(OC_2H_5)_2 + SiCl_4 \rightarrow CH_3(C_6H_5)_2SiCl_2 \quad \text{(ah)}$$

The hydroxyl group interconverts with Br by dropwise addition of $SiBr_4$ to ethereal Ph_3SiOH and filtering in N_2 to yield[34] 50% Ph_3SiBr:

$$(C_6H_5)_3SiOH \xrightarrow{SiBr_4} (C_6H_5)_3SiBr \quad \text{(ai)}$$

Halosilanes also produce an interchange between halogens and the Si—O—Si linkage in a reversible reaction by heating with a $FeCl_3$ catalyst:

$$(R_3Si)_2O + R'_3SiX \rightarrow R_3SiOSiR'_3 + R_3SiX \quad \text{(aj)}$$

where X = Cl.

$$[(CH_3)_3Si]_2O + (C_2H_5)_3SiCl \rightarrow \underset{(66\%)}{(CH_3)_3SiOSi(C_2H_5)_3} + (CH_3)_3SiCl \quad \text{(ak)}$$

The order of reactivities[35-37] is $RSiCl_3 > R_2SiCl_2 > R_3SiCl$ and $MeSiCl_3 > EtSiCl_3 > PhSiCl_3 > BuSiCl_3 > EtOSiCl_3$.

The mechanism of the cleavage of siloxanes by halosilanes with $FeCl_3$ is[36]:

$$R'_3SiX + FeX_3 \rightleftharpoons [R'_3Si]^+ + [FeX_4]^- \quad \text{(al)}$$

$$R_3SiOSiR_3 + [R'_3Si]^+ \rightleftharpoons [(R_3Si)_2—O—SiR'_3]^+ \rightleftharpoons R_3SiOSiR'_3 + [R_3Si]^+ \quad \text{(am)}$$

The catalyst becomes incapacitated during the reaction and must be replenished.

One application is forming $[(R_3Si)_2]^{\bullet}$ radicals, en route to unsymmetrical disiloxanes. Pentamethylethyldisiloxane is obtained in high yield[36] from Et_3SiCl with $(MeEt_2Si)_2O$:

$$[CH_3(C_2H_5)_2Si]_2O + (C_2H_5)_3SiCl \rightarrow CH_3(C_2H_5)_2SiOSi(C_2H_5)_3 + CH_3(C_2H_5)_2SiCl \quad \text{(an)}$$

Another application is conversion[36], of lower $(R_3Si)_2O$ into higher ones. Thus, the reaction of $(Me_3Si)_2O$ with xs Et_3SiCl gives the Et analogue:

$$[(CH_3)_3Si]_2O + (C_2H_5)_3SiCl \rightarrow \underset{(54\%)}{[(C_2H_5)_3Si]_2O} + \underset{(35\%)}{(CH_3)_3SiCl} \quad \text{(ao)}$$

Similarly:

$$[(CH_3)_3Si]_2O + 2\ (n\text{-}C_3H_7)_3SiBr \rightarrow [(n\text{-}C_3H_7)_3Si]_2O + (CH_3)_3SiBr \quad \text{(ap)}$$

The cleavage reaction affords chain lengthening of organosiloxanes[36,37]. Thus, xs $(Me_3Si)_2O$ reacts with dialkylhalosilanes under identical conditions[37]:

$$2\,[(CH_3)_3Si]O + R_2SiX_2 \rightarrow (CH_3)_3SiOSiR_2OSi(CH_3)_3 + 2\,(CH_3)_3SiX \qquad \text{(aq)}$$

where R = CH_3, C_2H_5; and C_2H_5, H, and X = Br, Cl.

Cleavage at the Si—O bond and closure of the ring takes place by refluxing a mixture of 1,2-bis(trimethylsiloxy)-1-cyclohexane and dialkyldichlorosilane for 4 h in the presence of $(NH_4)_2SO_4$ to give 1,3,6,8-tetraoxa-2,7-disila-4-9-cyclodecadienes[38]:

$$2\ \text{1,2-bis(OSi(CH}_3)_3\text{)cyclohexene} + 2\,R_2SiCl_2 + (NH_4)_2SO_4 \rightarrow \text{bis(cyclohexeno)-}[-O-SiR_2-O-]_2 \text{ ring} + 4\,(CH_3)_3SiCl \qquad \text{(ar)}$$

where R = CH_3, C_6H_5.

α,ω-Dichlorosiloxanes are prepared[39,40] by the equilibration of 4:1 Me_2SiCl_2 and $Me_2Si(OEt)_2$ at 85°C for 7 h with $FeCl_3$:

$$(CH_3)_2SiCl_2 + (CH_3)_2Si(OC_2H_5)_2 \xrightarrow{FeCl_3} ClSi(CH_3)_2[OSi(CH_3)_2]_nCl + C_2H_5Cl \qquad \text{(as)}$$

where n = 1–3.

Also, redistribution to form ethoxychlorosilanes occurs by heating alkyltriethoxysilane and alkyltrichlorosilane at 180°–200°C for 30–40 h in a sealed tube[41]:

$$RSi(OC_2H_5)_3 + RSiCl_3 \xrightarrow{180\text{–}200°C} RSi(OC_2H_5)Cl_2 + RSi(OC_2H_5)_2Cl \qquad \text{(at)}$$

where R = CH_3, 180°C, 30 h; R = C_2H_5, 200°C, 40 h.

Similarly, the interaction of $R_2Si(OEt)_2$ with R_2SiCl_2 at 180–200°C for 35–40 h in a sealed tube yields[41]:

$$R_2Si(OC_2H_5)_2 + R_2SiCl_2 \xrightarrow[35\text{–}40\ h]{180\text{–}2000°C} 2\,R_2Si(OC_2H_5)Cl \ (76\text{–}82\%) \qquad \text{(au)}$$

where when R = CH_3, T = 180°C, 35 h; when R = C_2H_5, T = 200°C, 40 h.

(iv) by Phosphorus Halides. Hexamethyldisiloxane will cleave its Si—O bond when 1:2 PBr_3 is added dropwise at RT, stirred 2 h and refluxed 45 min. Repeated fractional distillation yields[42,43] pure Me_3SiBr:

$$[(CH_3)_3Si]_2O + PBr_3 \xrightarrow{reflux} (CH_3)_3SiBr \ (80\%) \qquad \text{(av)}$$

Phosphorus trihalide also cleaves the Si—O bond of hexamethyldisiloxane[3] in the presence of a $ZnCl_2$ or $FeCl_2$ catalyst[44]:

$$3\ [R_3Si]_2O + PX_3 \xrightarrow[FeCl_2]{ZnCl_2\ or} 3\ R_3SiX + (R_3SiO)_3P \quad \text{(aw)}$$

where R = CH_3, C_2H_5; X = Cl, Br.

With a $ZnCl_2$ catalyst, $(Me_3Si)_2O$ and PBr_3 yield 91.9% Me_3SiBr and $P(OSiMe_3)_3$ in 18 h. With an $FeCl_2$ catalyst, $(Me_3Si)_2O$ and PCl_3 yield 90% Me_3SiCl and $P(OSiMe_3)_3$ in 18 h. A six-membered intermediate cyclic complex forms with the catalyst[44].

Refluxing a solution of $(n\text{-}Pr_3Si)_2O$ with PBr_3 yields $n\text{-}Pr_3SiBr$ in 56% yield[45]:

$$[(n\text{-}C_3H_7)_3Si]_2O + PBr_3 \rightarrow (n\text{-}C_3H_7)_3SiBr \quad \text{(ax)}$$

With PCl_3, $(Et_3Si)_2O$ cleaves at the Si—O bond after refluxing for 6 h to give Et_3SiBr in 88% yield[46]. Similarly, PBr_3 gives 95.6% Et_3SiBr:

$$[(C_2H_5)_3Si]_2O + PX_3 \xrightarrow{reflux} (C_2H_5)_3SiX \quad \text{(ay)}$$

where X = Cl, Br.

The action of PBr_3 on $MeEtSi(OEt)_2$ under mild conditions forms[47] $MeEtSiBr_2$

$$C_2H_5(CH_3)Si(OC_2H_5)_2 + PBr_3 \rightarrow C_2H_5(CH_3)SiBr_2 \quad \text{(az)}$$

Phosphorus trichloride with $MeOSiH_3$ at $-78°C$ for 12 h produces $ClSiH_3$, unlike the reaction with disiloxane:

$$CH_3OSiH_3 + PCl_3 \rightarrow SiH_3Cl + CH_3OPCl_2 \quad \text{(ba)}$$

A four-center transition complex forms as an intermediate, depending on the donor properties of $MeOSiH_3$ and the amphoteric character of the P(III) halide[48].

Stepwise replacement of halogens in PCl_3 is effected when a solution of 2:1 $n\text{-}BuOSiMe_3$ is added dropwise at $-10°C$ and then held at 15°C for 12 h to yield Me_3SiCl and mono-, di- and tri-butylphosphite[49]:

$$(CH_3)_3SiOC_4H_9\text{-n} + PCl_3 \rightarrow (CH_3)_3SiCl + (n\text{-}C_4H_9O)_xPCl_{3-x} \quad \text{(bb)}$$

where x = 1–3.

Dimethyldichlorosilane is obtained in high purity by cleaving a polysiloxane with PCl_3 and anhyd $FeCl_3$. Distillation at 160°C of $[Me_2SiO]_4$ and 1–5 parts of PCl_3 and 0.001–0.1 parts of anhyd $FeCl_3$ yields[50] Me_2SiCl_2:

$$[(CH_3)_2SiO]_4 + PCl_3 \xrightarrow{FeCl_3} (CH_3)_2SiCl_2 \quad \text{(bc)}$$

The action of $(Me_3Si)_2O$ on PF_5 produces cleavage at the Si—O bond in quantitative yield at RT[51]:

$$[(CH_3)_3Si]_2O + PF_5 \rightarrow 2\ (CH_3)_3SiF + POF_3 \quad \text{(bd)}$$

Diphenylphosphoric fluoride, Ph_2POF, is obtained in almost quantitative yield by refluxing Ph_2PF_3 with a slight xs of $(Me_3Si)_2O$ for 16 h:

$$[(CH_3)_3Si]_2O + R_2PF_3 \rightarrow 2\ (CH_3)_3SiF + R_2POF \quad \text{(be)}$$

where R = C_6H_5, F.

Similarly, Me_3SiF is obtained in 93% yield by refluxing $PhPF_4$ and disiloxane for 1 h at 40°C with cooling of the exothermic reaction[52,53].

By the interaction of 2:1 PCl_5 and $Me_2Si(OEt)_2$, the EtO groups are displaced[54] by Cl:

$$(CH_3)_2Si(OC_2H_5)_2 + 2\ PCl_5 \rightarrow (CH_3)_2SiCl_2 + 2\ POCl_3 + 2\ C_2H_5Cl \quad \text{(bf)}$$

Trimethylethoxysilane added to 1:1 PCl_5 and heated to 70°C yields[55] Me_3SiCl:

$$(CH_3)_3SiOC_2H_5 + PCl_5 \rightarrow (CH_3)_3SiCl + POCl_3 + C_2H_5Cl \quad \text{(bg)}$$

Phosphorus pentafluoride reacts with Me_3SiOMe at low T to form[51] Me_3SiF:

$$(CH_3)_3SiOCH_3 + PF_5 \rightarrow (CH_3)_3SiF + POF_3 + CH_3F \quad \text{(bh)}$$

Phosphorus pentachloride cleaves the Si—O bond of alkoxysilanes to form the corresponding organohalosilanes, but partial halogenation occurs in a system that contains more than one alkoxy group[56]:

$$(C_2H_5)Si(OC_2H_5)_3 + PCl_5 \rightarrow C_2H_5Si(OC_2H_5)_2Cl \quad \text{(bi)}$$

The action of alkylalkoxysilane on $POCl_3$ produces Me_3SiCl:

$$(CH_3)_3SiOR + POCl_3 \rightarrow (CH_3)_3SiCl + ROPOCl_2 \quad \text{(bj)}$$

where R = CH_3, C_2H_5, n-C_4H_9.

When R = CH_3, the reaction mixture is held at 50–60°C for 2 h; when R = C_2H_5, the reaction mixture is refluxed for 1 h at 100°C; when R = n-C_4H_9, the reaction mixture is held at 20°C for 12 h[49,57].

Equimolar Me_3SiOBu-n and $POBr_3$ at −10°C give Me_3SiBr and n-$BuOPOBr_2$ at 0°C. The latter decomposes on attempted distillation to give[49] n-BuBr:

$$3\ (CH_3)_3SiOC_4H_9\text{-n} + POBr_3 \rightarrow 3\ (CH_3)_3SiBr + (n\text{-}C_4H_9O)_3PO \quad \text{(bk)}$$

Refluxing dry $(Me_3Si)_2O$ with $POCl_3$ for 2 h yields the following compounds, separated by fractional distillation[57]:

$$[(CH_3)_3Si]_2O + POCl_3 \rightarrow (CH_3)_3SiCl + (CH_3)_3SiOPOCl_2 \quad \text{(bl)}$$

Stable pentacoordinated fluorophosphoranes containing an aryloxy group are prepared[58] by the interaction of RPF_4 with Me_3SiOPh:

$$RPF_4 + 2\ PhOSiMe_3 \rightarrow RPF_2(OPh)_2 + 2\ Me_3SiF \quad \text{(bm)}$$

where R = Me, Ph. Irrespective of the order of combination of the reactants or of their mol ratio, the disubstituted $RPF_2(OPh)_2$ derivatives are formed[58] rather than RPF_3OPh.

Similarly, Ph_2PF_3 reacts[58] with Me_3SiOPh:

$$(CH_3)_3SiOC_6H_5 + (C_6H_5)_2PF_3 \rightarrow (CH_3)_3SiF + (C_6H_5)_2PF_2OC_6H_5 \quad \text{(bn)}$$

Fluorination of Me_3SiOPr-i by $EtPF_4$ cleaves the Si—O bond when the phosphorane is added dropwise to a solution of cooled siloxane[53]:

$$(CH_3)_3SiO(i\text{-}C_3H_7) + C_2H_5PF_4 \rightarrow (CH_3)_3SiF + C_2H_5POF_2 + (CH_3)_2CHF \quad \text{(bo)}$$

(v) by Sulfur Halides, Sulfuryl Compounds, Sulfuric Acid, and in Sulfuric Acid Solution. Fluorination of organosilanes by means of SF_4 can occur by cleaving the Si—O—Si bonds and forming Si—F bonds; e.g., hexamethyldisiloxane undergoes cleavage of the Si—O bond upon heating with SF_4 in an autoclave for 2 h at 20°C or for 15 h at 13–15°C to give[59] Me_3SiF:

$$[(CH_3)_3Si]_2O + SF_4 \rightarrow 2\,(CH_3)_3SiF + SOF_2 \qquad \text{(bp)}$$

Heating $Me_2Si(OEt)_2$ with SF_4 in an autoclave at 20°C for 18 h followed by distillation at RT yields[59] 77% Me_2SiF_2:

$$(CH_3)_2Si(OC_2H_5)_2 + SF_4 \xrightarrow{20\,°C} (CH_3)_2SiF_2 + C_2H_5F + (C_2H_5O)_2SO \qquad \text{(bq)}$$

Similarly, heating Me_3SiOH with SF_4 in an autoclave for 22 h at 20°C followed by distillation at 45°C, produces Me_3SiF in 70% yield[59]:

$$(CH_3)_3Si(OH) + SF_4 \xrightarrow{20\,°C} \underset{(70\%)}{(CH_3)_3SiF} + SOF_2 + HF \qquad \text{(br)}$$

Likewise, $Ph_2Si(OH)_2$ is converted to the corresponding difluorosilanes by SF_4 in an autoclave at −50 to −20°C for 15 h and at 0°C for 1–2 h. Distillation at +10°C yields;

$$(C_6H_5)_2Si(OH)_2 + SF_4 \xrightarrow{-50\text{ to }-20\,°C} \underset{(79\%)}{(C_6H_5)_2SiF_2} + 2\,SOF_2 + 2\,HF \qquad \text{(bs)}$$

Thionyl fluoride is an effective fluorinating agent for the Si—OH group in silanols and disilanols; e.g., heating SOF_2 with silanols in an autoclave at 20°C for 50 h yields[60] 90% Me_3SiF:

$$(CH_3)_3SiOH + SOF_2 \rightarrow \underset{(90\%)}{(CH_3)_3SiF} + HF + SO_2 \qquad \text{(bt)}$$

Under the same conditions, $Ph_2Si(OH)_2$ reacts to yield[60]:

$$(C_6H_5)_2Si(OH)_2 + 2\,SOF_2 \rightarrow \underset{(80\%)}{(C_6H_5)_2SiF_2} + 2\,HF + 2\,SO_2 \qquad \text{(bu)}$$

Heating a mixture of Me_2SiOEt and $SOCl_2$ in an autoclave at 20°C for 24 h yields[60] 60–75% Me_2SiF_2:

$$(CH_3)_2Si(OC_2H_5)_2 + SOF_2 \xrightarrow{20\,°C} (CH_3)_2SiF_2 + (C_2H_5O)_2SO \qquad \text{(bv)}$$

Thionyl chloride acts on neopentyloxytriethylsilane with quinoline hydrochloride to cleave the Si—O bond on refluxing at 115°C for 23 h, forming neopentyl chloride in 60% yield. Under identical conditions, neopentylchlorosulfinate yields 47% neopentyl chloride, thus indicating that the alkyl chloride of the previous reaction results from the chlorosulfinate[61-63]:

$$(C_2H_5)_3SiOCH_2C(CH_3)_3 + SOCl_2 \xrightarrow{C_9H_7NHCl} (C_2H_5)_3SiCl + (CH_3)_3CCH_2Cl \qquad \text{(bw)}$$

Thionyl chloride also cleaves the Si—O bond of Me_3SiOR to produce Me_3SiCl and either the alkylchlorosulfinate or the dialkylsulfite, depending on the concentration of reactants. The alkylchlorosulfinates are produced by adding the ester dropwise to 1:1 $SOCl_2$ at RT and distilling after it stands overnight[61,64]:

$$(CH_3)_3SiOR + SOCl_2 \rightarrow (CH_3)_3SiCl + ROSOCl \qquad \text{(bx)}$$

where R = i-, n-, s-C_4H_9, neo-C_5H_{11}.

Similarly, dialkylsulfite is produced when 1:2 $SOCl_2$ is added to the ester at RT[61,64]:

$$2\ (CH_3)_3SiOR + SOCl_2 \rightarrow 2\ (CH_3)_3SiCl + (RO)_2SO \qquad \text{(by)}$$

where R = n-, s-C_4H_9, $C_6H_{13}CH(CH_3)$.

Equimolar Me_3SiOR and chlorosulfinates, refluxed at 80°C for 7–12 h give dialkylsulfite in 85–89% yield[61,64]:

$$(CH_3)_3SiOR + ROSOCl \rightarrow (CH_3)_3SiCl + (RO)_2SO \qquad \text{(bz)}$$

where R = n-, i-C_4H_9.

Alkoxytrimethylsilanes are also cleaved by $SOBr_2$. Equimolar ester is added at RT[64]:

$$(CH_3)_3SiOR + SOBr_2 \rightarrow (CH_3)_3SiBr + ROSOBr \qquad \text{(ca)}$$

where R = n-C_4H_9, i-C_4H_9; R ≠ s-C_4H_9.

Cleavage of organosiloxanes also occurs under UV irradiation on refluxing $(Me_3Si)_2O$ with $SOCl_2$ for 25 h. No reaction occurs without irradiation[65]:

$$[(CH_3)_3Si]_2O + SOCl_2 \xrightarrow[\text{reflux}]{\text{uv}} \underset{(94\%)}{(CH_3)_3SiCl} \qquad \text{(cb)}$$

Alkyl and cycloalkylsilicon fluorides are prepared by dropwise addition of conc FSO_3H to alkyl or cycloalkylsilicon alkoxides at low T. The interaction of n-$Bu_2Si(OEt)_2$ with FSO_3H at 3–10°C gives n-Bu_2SiF_2 good yields[66,67]:

$$(n\text{-}C_4H_9)_2Si(OC_2H_5)_2 \xrightarrow[3-10\,^{\circ}C]{FSO_3H} (n\text{-}C_4H_9)_2SiF_2 \qquad \text{(cc)}$$

Similarly, 1:3 n-$AmSi(OEt)_3$ with FSO_3H at RT produces[66,67] n-$AmSiF_3$:

$$RSi(OC_2H_5)_3 + FSO_3F \rightarrow RSiF_3 \qquad \text{(cd)}$$

where R = n-C_5H_{11}, etc.

(vi) by Hydrogen and Alkyl Halides. Hexamethyldisiloxane reacts with NH_4Cl or NH_4F in conc H_2SO_4:

$$[(CH_3)_2Si]_2O + 2\ NH_4X + H_2SO_4 \rightarrow 2\ (CH_3)_3SiX + (NH_4)_2SO_4 + H_2O \qquad \text{(ce)}$$

where X = Cl, F.

The reaction proceeds through intermediates:

$$(R_3Si)_2O + H_2SO_4 \rightarrow (R_3Si)_2SO_4 + H_2O \qquad \text{(cf)}$$

$$(R_3Si)_2SO_4 + 2\ HCl \rightarrow 2\ R_3SiCl + H_2SO_4 \qquad \text{(cg)}$$

Trimethylsilylsulfate is a crystalline solid that can be extracted continuously with pentane during the reaction[68,69].

Addition of NH_4Cl or NH_4F to cold $(Et_3Si)_2O$ in conc H_2SO_4 for 2 h with stirring, followed by stirring for 1 h more, yields triethylhalosilane[70,71]:

$$[(C_2H_5)_3Si]_2O + 2\ NH_4X + H_2SO_4 \rightarrow 2\ (C_2H_5)_3SiX + (NH_4)_2SO_4 + H_2O \quad \text{(ch)}$$

where X = Cl, F.

The interaction of $(Me_3Si)_2O$ with NaF in conc H_2SO_4 forms[72] Me_3SiF:

$$[(CH_3)_3Si]_2O + 2\ NaF + H_2SO_4 \rightarrow 2\ (CH_3)_3SiF + Na_2SO_4 \quad \text{(ci)}$$

Fluorination of $(i\text{-}Pr_2MeSi)_2O$ involves adding NH_4F slowly to disiloxane in sulfuric acid with stirring and cooling below 35°C. Crystallization of the fluoro compound occurs on pouring the solution onto ice and distilling[73]:

$$[CH_3(i\text{-}C_3H_7)_2Si]_2O + 2\ NH_4F + H_2SO_4 \rightarrow 2\ CH_3(i\text{-}C_3H_7)_2SiF + [NH_4]_2SO_4 + H_2O \quad \text{(cj)}$$

Treating a stirred divinyltetramethyldisiloxane and NH_4Cl solution with conc H_2SO_4 at 0°C forms dimethylvinylchlorosilane in 42% yield[74]:

$$[CH_2{=}CHSi(CH_3)_2]_2O + 2\ NH_4Cl + H_2SO_4 \rightarrow 2\ CH_2{=}CHSi(CH_3)_2Cl + [NH_4]_2SO_4 + H_2O \quad \text{(ck)}$$

Preparation of trialkylhalosilanes proceeds through reversible reactions:

$$(R_3Si)_2O + H_2SO_4 \rightarrow \underset{(94\%)}{(R_3Si)_2SO_4} + H_2O \quad \text{(cl)}$$

$$(R_3Si)_2SO_4 + 2\ HX \rightarrow \underset{(88\%)}{2\ R_3SiX} + H_2SO_4 \quad \text{(cm)}$$

where X = Br, Cl.

The preparation of R_3SiCl from $(R_3Si)_2O$ and HCl suggest an oxonium ion should be added to reactions (cl) and (cm) to represent fully the reaction[69].

Finally, hexamethyltrisilylene sulfate, a monomeric methylpolysilicon sulfate, is cleaved at the Si—O bond in preference to the Si—Si bond by dry HCl and hexamethyltrisilylene sulfate with NH_4Cl in benzene to yield quantitatively 1,3-dichlorohexamethyltrisilane[75]:

$$(CH_3)_2Si\begin{matrix} \diagup Si(CH_3)_2O \diagdown \\ \diagdown Si(CH_3)_2O \diagup \end{matrix}SiO_2 \xrightarrow[NH_4Cl(C_6H_6)]{HCl\ and} Cl(CH_3)_2SiSi(CH_3)_2Si(CH_3)_2Cl \quad \text{(cn)}$$

Acetyl and benzoyl chlorides are used to halogenate alkoxysilanes[76]:

$$(C_2H_5)_2Si(OC_2H_5)_2 \xrightarrow{C_6H_5COCl} (C_2H_5)_2SiCl_2 \quad \text{(co)}$$

$$(C_2H_5)_3SiOC_2H_5 \xrightarrow{CH_3COCl} (C_2H_5)_3SiCl \quad \text{(cp)}$$

These reactions are useful for separating halosilanes from organohalosilanes owing to ready conversion of halosilanes into alkoxysilanes.

(vii) Cleavage by Metal Halides. Titanium tetrachloride cleaves the siloxane bond in $(Me_3Si)_2O$ at 120°C in 1 h to yield[77,78] Me_3SiCl:

$$[(CH_3)_3Si]_2O + TiCl_4 \xrightarrow{120°C} (CH_3)_3SiCl + (CH_3)_3SiOTiCl_3 \qquad (cq)$$

Vanadium oxychloride also cleaves the Si—O bond:

$$[(CH_3)_3Si]_2O + VOCl_3 \rightarrow (CH_3)_3SiCl + (CH_3)_3SiOVOCl_2 \qquad (cr)$$

The monosubstituted vanadium compound reacts further with $[(CH_3)_3Si]_2O$ and also cleaves the Si—O bond:

$$(CH_3)_3SiOVOCl_2 + [(CH_3)_3Si]_2O \rightarrow (CH_3)_3SiCl + [(CH_3)_3SiO]_2VOCl \qquad (cs)$$

Similarly, the initial reaction also proceeds further by boiling the mixture for 1 day at 100°C with fresh $(Me_3Si)_2O$:

$$2\,[(CH_3)_3Si]_2O + VOCl_3 \rightarrow 2\,(CH_3)_3SiCl + [(CH_3)_3SiO]_2VOCl \qquad (ct)$$

Also, $[(CH_3)_3Si]_2O$ forms as a decomposition product of the unstable disubstituted intermediate[79].

Alkylalkoxy and alkylphenoxysilanes with WF_6 produce a series of methoxy and phenoxytungsten(VI) fluorides, $WF_{6-n}(OR)_n$ (R = Me, n = 1–4; R = Ph, n = 1–2).

Tungsten hexafluoride and Me_3SiOPh gives[80]:

$$(CH_3)_3SiOC_6H_5 + WF_6 \xrightarrow{20\,°C,\ 6\ h} (CH_3)_3SiF + WF_5OC_6H_5 \qquad (cu)$$

$$(CH_3)_3SiOC_6H_5 + WF_6 \xrightarrow{20\,°C,\ 12\ h} (CH_3)_3SiF + WF_4OC_6H_5 \qquad (cv)$$

$$(CH_3)_3SiOC_6H_5 + WF_6 \xrightarrow{20\,°C,\ 2\ h} (CH_3)_3SiF + WF_3OC_6H_5 + WF_4(OC_6H_5)_2 \qquad (cw)$$

Similarly, $Me_2Si(OMe)_2$ reacts[80] with WF_6:

$$\underset{(8.3\ mmol)}{(CH_3)_2Si(OCH_3)_2} + \underset{(19.5\ mmol)}{WF_6} \xrightarrow{0\,°C,\ 2\ h} (CH_3)_2SiF_2 + WF_5OCH_3 \qquad (cx)$$

$$\underset{(19.0\ mmol)}{4\,(CH_3)_2Si(OCH_3)_2} + \underset{(18.0\ mmol)}{3\,WF_6} \xrightarrow{0\,°C,\ 2\ h} 4\,(CH_3)_2SiF_2 + WF_4(OCH_3)_2 + 2\,WF_3(OCH_3)_3 \qquad (cy)$$

$$\underset{(16.0\ mmol)}{2\,(CH_3)_2Si(OCH_3)_2} + \underset{(6.1\ mmol)}{WF_6} \xrightarrow{0\,°C,\ 8\ h} 2\,(CH_3)_2SiF_2 + WF_2(OCH_3)_4 \qquad (cz)$$

Also, the interaction of $MeSi(OMe)_3$ with WF_6 cleaves the Si—O bond[80].

$$\underset{(12.5\ mmol)}{CH_3Si(OCH_3)_3} + \underset{(11.0\ mmol)}{WF_6} \xrightarrow{0\,°C,\ 2\ h} CH_3SiF_3 + WF_3(OCH_3)_3 \qquad (da)$$

$$CH_3Si(OCH_3)_3 + WF_6 \xrightarrow{0\,°C,\ 5\ h} (CH_3)SiF_3 + WF_2(OCH_3)_4 \qquad (db)$$

Quantitative cleavage of the Si—O bond and partial cleavage of the Si—H bond occur in $Me_2Si(H)OMe$ with WF_6 at 0°C for 6 h; $(CH_3)_2SiHF$, $(CH_3)_2SiF_2$, and a black solid are formed[81]:

$$(CH_3)_2SiHOCH_3 + WF_6 \xrightarrow{0\,°C,\ 6\,h} (CH_3)_2SiHF + (CH_3)_2SiF_2 \qquad \text{(dc)}$$

Tungsten tetrafluoride cleaves the siloxane chain in $(Me_3Si)_2O$ at 20°C for 24 h:

$$[(CH_3)_3Si]_2O + WF_4 \xrightarrow{20\,°C} 2\,(CH_3)_3SiF + WOF_4 \qquad \text{(dd)}$$

Tetrafluorooxotungsten is obtained as a dark brown solid, which on sublimation gives a white solid identified by its IR spectrum[80,81]

Anhydrous $GaCl_3$ also cleaves siloxanes under mild conditions:

$$n\,[(CH_3)_3Si]_2O + nGaCl_3 \rightarrow n\,(CH_3)_3SiCl + n\,CH_3GaCl_2 + [(CH_3)_2SiO]_n \qquad \text{(de)}$$

$$n\,[(CH_3)_2SiO]_n + GaCl_3 \rightarrow CH_3GaCl_2 + [CH_3SiClO]_n \qquad \text{(df)}$$

This reaction provides a method for alkylation of Ga in more that 85% yield[82].

(J. M. BELLAMA)

1. E. Wiberg, U. Krüerke, *Z. Naturforsch., Teil B, 86*, 608 (1953).
2. H. Grosse-Ruyken, *Angew. Chem., 66*, 756 (1954).
3. H. J. Emeléus, M. Onyszchuk, *J. Chem. Soc.*, 604 (1958).
4. M. Onyszchuk, *Can. J. Chem., 39*, 808 (1961).
5. J. V. Urenovitch, A. G. MacDiarmid, *J. Chem. Soc.*, 1091 (1963).
6. E. Wiberg, U. Krüerke, *Z. Naturforsch., Teil B, 86*, 610 (1953).
7. C. H. Van Dyke, A. G. MacDiarmid, *Inorg. Chem., 3*, 747 (1964).
8. L. H. Sommer, G. R. Ansul, *J. Am. Chem. Soc., 77*, 2482 (1955).
9. M. J. Frazer, W. Gerrard, J. A. Strickson, *J. Chem. Soc.*, 4701 (1960).
10. M. F. Lappert, J. Lynch, *J. Chem. Soc., Chem. Commun.*, 750 (1968).
11. B. Sternbach, A. G. MacDiarmid, *J. Am. Chem. Soc., 83*, 3384 (1961).
12. E. Wiberg, U. Krüerke, *Z. Naturforsch., Teil B, 86*, 609 (1953).
13. W. H. Knoth, R. V. Lindsey, *J. Am. Chem. Soc., 23*, 1392 (1958).
14. W. Airey, G. M. Sheldrick, *J. Inorg. Nucl. Chem., 32*, 1827 (1970).
15. P. A. McCusker, T. Ostdick, *J. Am. Chem. Soc., 80*, 1103 (1958).
16. W. A. Kriner, A. G. MacDiarmid, E. C. Evers, *J. Am. Chem. Soc., 80*, 1546 (1958).
17. A. H. Cowley, F. Fairbrother, N. Scott, *J. Chem. Soc.*, 717 (1959).
18. M. A. Voronkov, B. N. Dolgov, N. A. Dmitrieva, *Dokl. Akad. Nauk SSSR., 84*, 959 (1952).
19. H. Schmidbaur, H. Hussek, F. Schindler, *Chem. Ber., 97*, 255 (1964).
20. N. F. Orlov, *Dokl. Akad. Nauk SSSR., 114*, 1033 (1957).
21. N. F. Orlov, B. N. Dolgov, *USSR Pat. 110*, 979 (1958); *Chem. Abstr., 52*, 18,216 (1958).
22. E. Itsugi, *Jpn. Pat.* 3861 (1957); *Chem. Abstr., 52*, 8182 (1958).
23. J. F. Salmon, S. J. Evers, E. C. Evers, *J. Inorg. Nucl. Chem., 28*, 2787 (1966).
24. J. Jack, *Br. Pat.* 881,179 (1969); *Chem. Abstr., 56*, 8745 (1962); Ger. Pat. 1,111,182; (1960); *Chem. Abstr., 56*, 8746 (1962).
25. A. G. MacDiarmid, J. J. Moscony, C. R. Russ, T. Yoshioka, *Intern. Sympos. Organosilcon Chem., Sci. Commun.*, Prague, 100 (1965); *Chem. Abstr., 65*, 7214 (1966).
26. J. J. Moscony, A. G. MacDiarmid, *J. Chem. Soc., Chem. Commun.*, 307 (1965).
27. S. Maeda, E. Nojimoto, M. Nagata, Jpn. Pat. 3860 (1957); *Chem. Abstr., 52*, 5880 (1958).
28. S. Maeda, E. Nojimoto, *Kogyo Kagaku Zasshi, 62*, 522 (1959); *Chem. Abstr., 68*, 29,823 (1968).
29. V. V. Yastrebov, A. I. Chernyshev, *J. Gen. Chem. USSR (Engl. Transl.), 37*, 2140 (1967).
30. E. Larsson, B. Smith, *Sven. Kem. Tidskr., 62*, 87 (1950); *Chem. Abstr., 44*, 7224 (1950).
31. B. E. F. Smith, L. Tiganik, *Swed. Pat.*, 15,329 (1956); *Chem. Abstr., 51*, 1246 (1957).

32. W. Gerrard, R. D. Kilburn, *J. Chem. Soc.*, 1536 (1956).
33. B. Smith, Ph.D Thesis, Chalmers Technical High School, Rundquists Boktrycheri, Gothenberg, Sweden, 1951.
34. P. A. McCusker, E. L. Reilly, *J. Am. Chem. Soc.*, *75*, 1583 (1953).
35. S. Maeda. E. Nojimoto, *Kogyo Kagaku Zasshi*, *62*, 522 (1959); *Chem. Abstr.*, *57*, 8602 (1962).
36. M. G. Voronkov, L. M. Chudesova, *Izv. Akad. Nauk. SSSR, Otd. Khim. Nauk*, 1415 (1957).
37. M. G. Voronkov, L. M. Chudesova, *J. Gen. Chem. USSR* (*Engl. Transl.*), *29*, 1534 (1959).
38. K. Rühlmann, R. Volkmer, C. Michael, *Justus Liebigs Ann. Chem.*, *706*, 18 (1967).
39. R. L. Elliot, L. W. Breed, *Inorg. Chem.*, *4*, 1455 (1965).
40. L. W. Breed, M. E. Whitehead, R. L. Elliot, *Inorg. Chem.*, *6*, 1254 (1967).
41. M. Kumada, *J. Inst. Polytech. Osaka City Univ., Ser. C.*, *2*, 139 (1952).
41. M. Kumada, *J. Inst. Polytech. Osaka City Univ., Ser. C.*, *2*, 139 (1952); *Chem. Abstr.*, *48*, 11,303 (1954).
42. P. A. McCusker, E. L. Reilly, *J. Am. Chem. Soc.*, *75*, 1583 (1953).
43. W. F. Gilliam, R. N. Meals, R. O. Sauer, *J. Am. Chem. Soc.*, *68*, 1161 (1946).
44. M. G. Voronkov, Yu. I. Skorik, *J. Gen. Chem. USSR* (*Engl. Transl.*), *35*, 106 (1965).
45. N. S. Nametkin, A. V. Topchiev, L. I. Kartasheua, *Dokl. Akad. Nauk SSSR.*, *93*, 667 (1953).
46. E. A. Chernyshev, A. D. Petrov, *Dokl. Akad. Nauk SSSR*, *105*, 282 (1955).
47. R. O. Sauer, U.S. Pat. 2,647,136 (1953); *Chem. Abstr.*, *48*, 8252 (1954).
48. C. H. Van Dyke, *J. Inorg. Nucl. Chem.*, *30*, 81 (1968).
49. J. Fertig, W. Gerrard, H. Herbst, *J. Chem. Soc.*, 1488 (1957).
50. B. A. Ashby, Br. Pat. 990,657 (1965); *Chem. Abstr.*, *63*, 3076 (1965); *U.S. Pat.* 3,179,679 (1965); *Chem. Abstr.*, *63*, 9987 (1965).
51. E. W. Kifer, C. H. Van Dyke, *J. Chem. Soc., Chem. Commun.*, 1330 (1969).
52. R. Schmutzler, *Inorg. Chem.*, *3*, 410 (1964).
53. R. Schmutzler, *J. Chem. Soc.*, 4551 (1964).
54. E. Larsson, B. Smith, *Sven. Kem. Tidgkr.*, *62*, 87 (1950).
55. R. R. McGregor, E. L. Warrick, *U.S. Pat.* 2,386,488 (1945); *Chem. Abstr.*, *40*, 592 (1946).
56. A. Ladenburg, *Justus Liebigs Ann. Chim.*, *164*, 300 (1972).
57. M. Schmidt, H. Schmidbaur, A. Binger, *Chem. Ber.*, *93*, 872 (1960).
58. S. C. Peake, R. Schmutzler, *J. Chem. Soc., Chem. Commun.*, 665 (1968).
59. R. Müller, D. Mross, *Z. Anorg. Allg. Chem.*, *324*, 78 (1963).
60. R. Müller, D. Mross, *Z. Anorg. Allg. Chem.*, *324*, 86 (1963).
61. W. Gerrard, P. Tolcher, *J. Chem. Soc.*, 3460 (1954).
62. L. H. Sommer, H. D. Blankman, P. C. Miller, *J. Am. Chem. Soc.*, *76*, 803 (1954).
63. L. H. Sommer, H. D. Blankman, P. C. Miller, *J. Am. Chem. Soc.*, *73*, 3542 (1951).
64. B. R. Currell, M. J. Frazer, W. Gerrard, *J. Chem. Soc.*, 2776 (1960).
65. E. V. Kukharskaya, *J. Gen. Chem. USSR* (*Engl. Transl.*), *34*, 2092 (1964).
66. F. J. Sowa, U.S. Pat. 2,477,704 (1949); *Chem. Abstr.*, *44*, 3008 (1950).
67. H. H. Szmant, G. W. Miller, J. Makhlouf, K. C. Schreiber, *J. Org. Chem.*, *27*, 261 (1962).
68. L. H. Sommer, G. T. Kerr, F. C. Whitmore, *J. Am. Chem. Soc.*, *70*, 445 (1948).
69. L. H. Sommer, E. W. Pietrusza, G. T. Kerr, W. H. Whitmore, *J. Am. Chem. Soc.*, *68*, 156 (1946).
70. P. A. DiGiiorgio, W. A. Strong, L. H. Sommer, F. C. Whitmore, *J. Am. Chem. Soc.*, *68*, 1380 (1946).
71. E. A. Flood, *J. Am. Chem. Soc.*, *55*, 1735 (1933).
72. H. Kriegsmann, *Z. Anorg. Allg. Chem.*, *294*, 113 (1958).
73. C. Eaborn, *J. Chem. Soc.*, 2755 (1949).
74. K. A. Andrianov, A. A. Zhdanov, V. A. Odinets, *J. Gen. Chem. USSR* (*Engl. Transl.*), *32*, 1126 (1962).
75. M. Kumada, S. Maeda, *Inorg. Chim. Acta*, *1*, 105 (1967).
76. A. Ladenburg, *Chem. Ber.*, *4*, 726 (1871).
77. B. N. Dolgov, N. F. Orlov, M. G. Voronkov, *Izv. Akad. Nauk SSSR, Otd. Khim. Nauk*, 1408 (1959).
78. K. A. Andrianov, N. A. Kurasheva, *Dokl. Akad. Nauk SSSR.*, *131*, 825 (1960).
79. M. Schmidt, H. Schmidbaur, *Angew. Chem.*, *71*, 220 (1959).
80. A. M. Noble, J. M. Winfield, *J. Chem. Soc., A*, 2574 (1970).
81. N. Viswanathan, C. H. Van Dyke, *J. Organomet. Chem.*, *11*, 181 (1968).
82. H. Schmidbaur, W. Findeiss, *Angew. Chem., Int. Ed. Engl.*, *3*, 696 (1964).

2.5.8.2. Ge Bond

2.5.8.2.1. by Hydrogen Halides.

The Ge—O bond in $(PhO)_4Ge$ is cleaved by bubbling dry HCl gas into germane in benzene for 90 min and refluxing for 3 h at 100°C to form $GeCl_4$ and phenol[1]:

$$Ge(OC_6H_5)_4 + 4\ HCl \rightarrow GeCl_4 + 4\ C_6H_5OH \quad (a)$$

Anhydrous $HClO_4$ is titrated into Et_4NBr in acetic acid; the HBr produced in situ with R_3GeOR' forms bromogermane and alcohol[2].

$$R_3GeOR' + HBr \rightarrow R_3GeBr + R'OH \quad (b)$$

Trialkylalkoxygermane is cleaved at the Ge—O bond[3] by HI to yield R_3GeI:

$$R_3GeOR + 2\ HI \rightarrow R_3GeI + R'I + H_2O \quad (c)$$

Hydrogen halides cleave the Ge—O bond in $(R_3Ge)_2O$ to yield the corresponding halides. Concentrated aq hydrogen halides are required, and the cleavage products are obtained in high yields[4–10]:

$$(R_3Ge)_2O + 2\ HX \rightarrow 2\ R_3GeX + H_2O \quad (d)$$

where X = F, Cl, Br, I.

Hydrogen halides also cleave polymeric R_2GeO at the Ge—O bond, e.g., conc aq halogen acid forms the corresponding dihalides[11–13]:

$$[R_2GeO]_n + 2n\ HX \rightarrow n\ R_2GeX_2 + n\ H_2O \quad (e)$$

where R = alkyl, aryl; X = F, Cl, Br, I.

Alkyl or arylgermanoic anhydrides react with halogen acids to cleave the Ge—O bond. Alkylgermanium halides are more easily formed than aryl. n-Propylgermanoic acid anhydride reacts with conc HCl to yield n-$PrGeCl_3$. Monoarylgermanium halides result quantitatively from heating arylgermanoic acid anhydride with 1:10 by wt of conc HCl for ca. 3 h at 100°C under pressure[14–16]:

$$[RGeO]_2O + 6\ HX \rightarrow 2\ RGeX_3 + 3\ H_2O \quad (f)$$

where X = Cl, Br, I; R = alkyl or aryl.

(J. M. BELLAMA)

1. R. C. Mehrotra, G. Chandra, *J. Ind. Chem. Soc., 39*, 235 (1962).
2. J. A. Magnuson, E. W. Knaub, *Anal. Chem., 37*, 1607 (1965).
3. M. Lesbré, G. Satgé, *C.R. Hebd. Séances Acad. Sci., 254*, 1453 (1962).
4. H. H. Anderson, *J. Am. Chem. Soc., 73*, 5440 (1951).
5. H. H. Anderson, *J. Am. Chem. Soc., 79*, 326 (1957).
6. R. Fuchs, H. Gilman, *J. Org. Chem., 23*, 911 (1958).
7. C. A. Kraus, E. A. Flood, *J. Am. Chem. Soc., 54*, 1635 (1932).
8. P. Mazerolles, *Bull. Soc. Chim. Fr.*, 2737 (1966).
9. P. Mazerolles, *Bull. Soc. Chim. Fr.*, 1911 (1961).
10. W. R. Orndorff, D. L. Tabern, L. M. Dennis, *J. Am. Chem. Soc., 49*, 2512 (1927).
11. C. A. Kraus, C. L. Brown, *J. Am. Chem. Soc., 52*, 3696 (1930).
12. O. H. Johnson, D. M. Harris, *J. Am. Chem. Soc., 72*, 5564 (1950).
13. E. A. Flood, *J. Am. Chem. Soc., 54*, 1663 (1932).
14. E. A. Flood, *J. Am. Chem. Soc., 55*, 4935 (1933).

15. O. H. Johnson, L. V. Jones, *J. Org. Chem.*, *17*, 1172 (1952).
16. H. Bauer, K. Burschkies, *Chem. Ber.*, *66*, 1156 (1933).

2.5.8.2.2. by Organic Halides.

Heating R_3GeOR' with benzoyl chloride at 80°C yields R_3GeCl quantitatively[1]:

$$R_3GeOR' + C_6H_5COCl \rightarrow R_3GeCl + C_6H_5COOR' \qquad \text{(a)}$$

Trimethoxychlorogermane is prepared[2] in 65% yield by refluxing $(MeO)_4Ge$ in benzene with 1:1 acetyl chloride for ca. 1 h:

$$Ge(OMe)_4 + CH_3COCl \rightarrow \underset{(65\%)}{(MeO)_3GeCl} \qquad \text{(b)}$$

An RO–Cl exchange between $(RO)_4Ge$ and 1:1 acetyl chloride takes place upon refluxing for a few hours in $CHCl_3$ or benzene to give[3] almost quantitative yields of $(RO)_3GeCl$:

$$Ge(OR)_4 + CH_3COCl \rightarrow (RO)_3GeCl \qquad \text{(c)}$$

where R = C_2H_5, C_3H_7, n-C_4H_9.

Similarly, refluxing $(RO)_4Ge$ with 1:2 acetyl chloride in $CHCl_3$ or benzene yields $(RO)_2GeCl_2$ almost quantitatively[3]:

$$Ge(OR)_4 + 2\ CH_3COCl \rightarrow (RO)_2GeCl_2 \qquad \text{(d)}$$

RO–Br exchange occur when 1:1 acetyl bromide and $Bu_2Ge(OR)_2$ react[4]:

$$Bu_2Ge(OR)_2 + (CH_3)COBr \rightarrow Bu_2Ge\begin{matrix}\diagup Br\\ \diagdown OR\end{matrix} + CH_3C(O)R \qquad \text{(e)}$$

where R = C_2H_5, i-C_3H_7, t-C_4H_9.

Similarly, 1:2 acetyl bromide and $Bu_2Ge(OR)_2$ form Bu_2GeBr_2:

$$Bu_2Ge(OR)_2 + 2\ CH_3COBr \rightarrow Bu_2GeBr_2 + 2\ CH_3COBr \qquad \text{(f)}$$

where R = C_2H_5, i-C_3H_7, t-C_4H_9.

(J. M. BELLAMA)

1. M. Lesbré, J. Satgé, *C.R. Hebd. Séances Acad. Sci.*, *254*, 1453 (1962).
2. V. I. Adveeva, G. S. Burlachenko, Yu, I. Baukov, I. F. Lutsenko, *J. Gen. Chem. USSR (Engl. Transl.)*, *36*, 1676 (1966).
3. R. C. Mehrotra, G. Chandra, *Recl. Trav. Chim. Pays-Bas*, *82*, 683 (1963).
4. S. Mathur, G. Chandra, A. K. Rai, R. C. Mehrotra, *J. Organomet. Chem.*, *4*, 294 (1965).

2.5.8.2.3. by Other Halides.

Hexamethyldigermoxane reacts with 3:2 mol BF_3 at −80°C for 12 h to form Me_3GeF instead of a stable adduct. A four-center cyclic intermediate is postulated[1]:

$$(Me_3Ge)_2O + BF_3 \xrightarrow[12\ h]{-80°C} \underset{(97\%)}{2\ Me_3GeF} + B_2O_3 \qquad \text{(a)}$$

Like its Si analogue, $(R_3Ge)_2O$ reacts with $AlCl_3$ with scission of the Ge—O bond[2]:

$$R_3GeOGeR_3 \xrightarrow{AlCl_3} 2\ R_3GeCl \tag{b}$$

When $R = C_2H_5$, $(C_2H_5)_3GeCl$ is formed in 63% yield; when $R = CH_3$, 70% $(CH_3)_3GeCl$ and 10% $(CH_3)_4Ge$ are formed.

Exclusive cleavage of the Ge—O bond in $Me_3GeOSiMe_3$ occurs quantitatively with the exothermic action of anhyd $AlCl_3$ at 0°C in a dry box[3]:

$$(CH_3)_3SiOGe(CH_3)_3 + AlCl_3 \rightarrow (CH_3)_3GeCl + (CH_3)_3SiOAlCl_2 \tag{c}$$

The interaction of $POCl_3$ with $Me_3SiOGeMe_3$ cleaves both the Ge—O and the Si—O bonds[4]:

$$(CH_3)_3SiOGe(CH_3)_3 + POCl_3 \rightarrow \begin{matrix} (CH_3)_3GeCl + (CH_3)_3SiOPOCl_2\ (80\%) \\ (CH_3)_3SiCl + (CH_3)_3GeOPOCl_2\ (20\%) \end{matrix} \tag{d}$$

(J. M. BELLAMA)

1. J. E. Griffiths, M. Onyszchuk, *Can. J. Chem.*, *39*, 339 (1961).
2. V. F. Mironov, E. S. Sobolov, L. M. Antipin, *J. Gen. Chem. USSR* (*Engl. Transl.*), *37*, 2448 (1967).
3. H. Schmidbaur, M. Schmidt, *Chem. Ber.*, *94*, 1349 (1961).
4. M. Schmidt, H. Schmidbaur, I. Ruidisch, *Angew. Chem.*, *73*, 408 (1961).

2.5.8.3. Sn Bond

2.5.8.3.1. by Hydrogen Halides.

Lower alkyltin halides are prepared from HX with alkylstannoic acid and cleavage of the Sn—O bond[1–3]:

$$CH_3SnOOH + HX \rightarrow CH_3SnX_3 + 2\ H_2O \tag{a}$$

where X = Cl, Br, I, and where HCl reacts at 50°C, fuming HBr at 180–220°C and aq HI at 160°C. Ethyltin halides are prepared similarly. However, the compound obtained by the action of ethylstannoic acid and HBr is incorrectly reported as $C_2H_5SnBr_3$.

Triorganotin hydroxides are converted to the halides with cleavage of the Sn—O bond by shaking them in ether with aq halogen acids. Fluorides are better prepared by other methods[4]:

$$R_3SnOH + HX \rightarrow R_3SnX + H_2O \tag{b}$$

Similarly, $R_2Sn(OH)_2$ or R_2SnO may be converted to the organotin halide, e.g.:

$$R_2Sn(OH)_2 + HX \rightarrow R_2SnX_2 + 2\ H_2O \tag{c}$$

(J. M. BELLAMA)

1. J. G. F. Druce, *J. Chem. Soc.*, *119*, 758 (1921).
2. H. Kriegsmann, S. Pauly, *Z. Anorg. Allg. Chem.*, *330*, 275 (1964).
3. J. G. A. Luijten, G. J. M. Van der Kerk, *Investigations in the Field of Organotin Chemistry*, Tin Research Institute, London, 1959.
4. R. K. Ingham, S. D. Rosenberg, H. Gilman, *Chem. Rev.*, *60*, 459 (1960).

2.5.8.3.2. by Other Halides.

Dibutyltin dibutoxide and Bu_3SnOMe, tributyltin methoxide, are cleaved at the Sn—O bond by organohalosilanes, by BCl_3, by acid chlorides, and by RCl to form the corresponding organotin halides in yields of 35–90%.

Organic halides are heated under reflux for 4 h with the alkoxides:

$$Bu_2Sn(OMe)_2 + CH_2{=}CHCH_2Br \rightarrow Bu_2SnBr_2 + MeOCH_2{=}CH_2 \tag{a}$$

Such reactions are examples of nucleophilic substitution at an Sn atom by organic halides[1].

Tin alkoxides are also cleaved by organohalosilanes[2]:

$$Bu_3SnOMe + Et_3SiCl \rightarrow \underset{(90\%)}{Bu_3SnCl} + Et_3SiOMe \tag{b}$$

$$Bu_3SnOMe + Ph_3SiCl \rightarrow \underset{(70\%)}{Bu_3SnCl} + Ph_3SiMe \tag{c}$$

Tin alkoxides also react[3] exŏthermally with 1:1 aryl or benzoyl chlorides to interchange RO and Cl:

$$Bu_3SnOMe + CH_3COCl \rightarrow Bu_3SnCl + Bu_3SnOCOCH_3 \tag{d}$$

$$Bu_3SnOMe + C_6H_5COCl \rightarrow Bu_3SnCl + Bu_3SnOCOC_6H_5 \tag{e}$$

Alkoxy–Cl conversion also occurs exothermically by BCl_3 with n-Bu_2Sn(O-n-Bu)$_2$ at RT[4]:

$$\text{n-}Bu_2Sn(OBu\text{-n})_2 + 2\ BCl_3 \rightarrow \text{n-}Bu_2SnCl_2 + 2\ \text{n-}BuOBCl_2 \tag{f}$$

Aluminum chloride cleaves both the S—O and the metal–oxygen bonds in the analogous Sn (and Pb) heterosiloxanes[5,6]:

$$(CH_3)_3SiOM(CH_3)_3 + (AlCl_3)_2 \begin{cases} \rightarrow (CH_3)_3MCl + [(CH_3)_3SiOAlCl_2]_2 \\ \rightarrow 2(CH_3)_3SiCl + 2\,(CH_3)_3MCl + (AlOCl)_{2n} \end{cases} \tag{g}$$

Organotin oxides, hydroxides and alkoxides react[7] with NH_4X and NH_4NCS in an inert solvent to give the corresponding organotin halides and isothiocyanates, respectively, in high yields. The reactivity of NH_4X is X = I > Br > Cl:

$$R_2SnO + 2\ NH_4X \rightarrow R_2SnX_2 + 2\ NH_3 + H_2O \tag{h}$$

where R = n-C_4H_9, n-C_3H_7, CH_3; X = Cl, Br, I, NCS; yield = 60–92%.

Magnesium (II) bromide reacts with Et_3SnOH with fission of the Sn—O bond when $MgBr_2$ is slowly added in ether and heated for 1 h on a water bath to form[8] Et_3SnBr in 86% yield:

$$Et_3SnOH + MgBr_2 \rightarrow Et_3SnBr \tag{i}$$

Halogenation of R_2SnO is also achieved by anhyd PCl_3, PCl_5 and $SnBr_4$ by distilling under reduced pressure[9,10].

$$R_2SnO \xrightarrow[SnBr_4]{PCl_5,\,PCl_3,} R_2SnX_2 \tag{j}$$

Heating alkaline NaF with R_2SnO yields the corresponding fluorides[11].

(J. M. BELLAMA)

1. J. C. Pommier, J. Valade, *C.R. Hebd. Séances Acad. Sci., 260*, 951 (1963).
2. J. C. Pommier, M. Pereyre, J. Valade, *C.R. Hebd. Séances Acad. Sci., 260*, 6397 (1965).
3. J. Valade, M. Pereyre, *C.R. Hebd. Séances Acad. Sci., 256*, 3693 (1962).
4. W. Gerrard, R. G. Rees, *J. Chem. Soc.*, 3510 (1964).
5. H. Schmidbaur, H. Hussek, F. Schindler, *Chem. Ber., 97*, 255 (1964).
6. H. Schmidbaur, M. Schmidt, *Chem. Ber., 94*, 1349 (1961).
7. K. C. Pande, *J. Organomet. Chem., 13*, 187 (1968).
8. M. F. Shostakovskii, N. V. Komarov, V. K. Misyunas, M. I. Zainchkovskaya, *Izv. Akad. Nauk SSSR. Ser. Khim.*, 1102 (1964).
9. H. H. Anderson, *J. Org. Chem., 19*, 1766 (1954).
10. A. Cahours, *Hun. Chem., 114*, 227 (1860); 354 (1860).
11. G. Bahr, G. Zoche, *Chem. Ber., 88*, 1450 (1955).

2.5.8.4. Pb Bond.

Hydrogen halides cleave the Pb—O bond in $(Ph_3Pb)_2O$, obtained by oxidation of Ph_4Pb with $KMnO_4$. Careful neutralization with the halogen acid is required to avoid an excess. Bromo compounds are obtained in higher yield than chloro[1]:

$$(Ph_3Pb)_2O + HX \rightarrow \underset{(65\text{–}80\%)}{Ph_3PbX} \tag{a}$$

where X = Cl, Br. Exactly neutralizing alcoholic R_3PbOH with one part 33% HF in 10 parts alcohol and evaporating in vacuo over P_2O_5 yields[2] R_3PbF:

$$R_3PbOH + HF \xrightarrow{EtOH} R_3PbF \tag{b}$$

where R = CH_3, C_2H_5, C_3H_7, i-C_4H_9.

Adding conc HCl or 46% HBr to conc tetraethyldinitrosyllead dinitrate (obtained by reacting Ph_4Pb with N_2O_4) yields[3] Et_2PbCl_2 or Et_2PbBr_2[3]; e.g.,

$$R_4Pb + N_2O_4 \rightarrow [R_4Pb(NO)_3]^{2+}[NO_3]_2^- + HX \rightarrow R_2PbX_2 \tag{c}$$

where R = C_2H_5, n-C_3H_7; X = Cl, Br.

Finally, $AlCl_3$ cleaves[4] both Si—O and Pb—O bonds in $Me_3SiOPbMe_3$.

(J. M. BELLAMA)

1. L. C. Willemsens, G. J. M. van der Kerk, *Investigations in the Field of Organolead Chemistry*, International Lead-Zinc Research Organization, New York, 1965.
2. E. Krause, E. Pohland, *Chem. Ber., 55*, 1282 (1922).
3. B. Hetnarski, T. Urbanski, *Tetrahedron, 19*, 1319 (1963).
4. H. Schmidbaur, M. Schmidt, *Chem. Ber., 94*, 5349 (1961).

2.5.9. from Cleavage of the Group-IVB–Group-VIB Element Bond

2.5.9.1. by Halogens.

Elemental Br_2 cleaves the Sn—S bond, e.g., when added slowly to Me_3SnSPh in CCl_4 to yield[1] Me_3SnBr and Ph_2S_2:

$$2\ Me_3SnSR + Br_2 \rightarrow \underset{(96\%)}{Me_3SnBr} + R_2S_2 \tag{a}$$

The analogous reaction takes place quantitatively[2] with I_2.

Similarly, Cl_2, Br_2 or I_2 splits the Sn—Se bond. The halogens in benzene are added to $(Et_3Sn)_2Se$ in the same solvent and cooled with ice. Triethyltin halides and elemental Se are formed[3,4].

$$[(C_2H_5)_3Sn]_2Se + X_2 \rightarrow 2\ (C_2H_5)_3SnX + Se \qquad \text{(b)}$$

where X = Cl, Br, I.

On treatment with Br_2 in $CHCl_3$ $(p\text{-}CH_3C_6H_4S)_4Ge$ does not form an addition compound but is converted into $(p\text{-}CH_3C_6H_4S)_2GeBr$ with scission of the Ge—S bond[5]:

$$(p\text{-}CH_3C_6H_4S)_4Ge + Br_2 \rightarrow (p\text{-}CH_3C_6H_4S)_2GeBr + (p\text{-}CH_3C_6H_4S)_2 \qquad \text{(c)}$$

The cleavage of Ge—Te bond in $(Et_3Ge)_2Te$ by Br_2 forms Et_3GeBr and elemental Te. Products are obtained by adding Br_2 in benzene to ice-cold $(Et_3Ge)_2Te$ in the same solvent and subsequent crystallization[4]:

$$(Et_3Ge)_2Te + Br_2 \rightarrow \underset{(85\%)}{2\ Et_3GeBr} + \underset{(66\%)}{Te} \qquad \text{(d)}$$

Similarly, adding Br_2 in benzene to an ice-cold solution of $(Et_3Sn)_2Te$ in the same solvent results in an exothermic and rapid cleavage[5]:

$$(Et_3Sn)_2Te + Br_2 \rightarrow 2\ Et_3SnBr + Te \qquad \text{(e)}$$

(J. M. BELLAMA)

1. E. W. Abel, D. A. Armitage, *J. Chem. Soc., A*, 554 (1967).
2. M. E. Peach, *Can. J. Chem., 46*, 211 (1968).
3. M. N. Bochkarev, L. P. Sanina, N. S. Vyzankin, *J. Gen. Chem. USSR (Engl. Transl.), 39*, 135 (1969).
4. N. S. Vyazankin, L. P. Sanina, G. S. Kakinina, M. N. Bochkarev, *J. Gen. Chem. USSR (Engl. Transl.), 38*, 1800 (1968).
5. H. J. Baker, F. Stienstra, *Recl. Trav. Chim. Pays-Bas, 52*, 1033 (1933).

2.5.9.2. by Hydrogen Halides.

The Ge—Se bond in $(Et_3Ge)_2Se$ is cleaved overnight by 1:1 dry HCl in benzene to form[1] 100% H_2Se and 94% Et_3GeCl:

$$(Et_3Ge)_2Se + HCl \rightarrow Et_3GeCl + H_2Se \qquad \text{(a)}$$

(J. M. BELLAMA)

1. M. N. Bochkarev, L. P. Sanina, N. S. Vyazankin, *J. Gen. Chem. USSR (Engl. Transl.), 39*, 135 (1969).

2.5.9.3. by Other Halides.

Cleavage of the Ge—S bond by MeI occurs in Me_3GeSMe in a week at RT to give trimethylsulfonium iodide in 67% yield[1]:

$$(CH_3)_3GeSCH_3 + CH_3I \rightarrow (CH_3)_3GeI + [(CH_3)_3S]I + CH_3SCH_3 \qquad \text{(a)}$$

Germylphenyl sulfide decomposes when warmed to RT for 15 min with BF_3 to yield GeH_3F and a pale-yellow solid containing Ge—H bonds and a thioaromatic group:

$$H_3GeSPh + BF_3 \rightarrow GeH_3F + GeH_x[S(C_6H_5)]_{4-x} \qquad \text{(b)}$$

On warming to RT H_3GeSPh forms an adduct with BCl_3 which subsequently decomposes[2] to GeH_3Cl, GeH_4 and GeH_2Cl_2:

$$H_3GeSPh + BCl_3 \rightarrow GeH_2Cl_2 + GeH_3Cl + GeH_4 \qquad \text{(c)}$$

The Sn—S bonds in organotin mercaptides are cleaved by interaction with RX[3], $HgCl_2$[4], $SnCl_2$[4] and organotin halides[5] to form organotin halides and the corresponding mercaptides, e.g.:

$$[(CH_3)_2NC_6H_4S]_4Sn + CH_3I \rightarrow SnI_4 + 4\,[(CH_3)_3NC_6H_4SCH_3]I \qquad \text{(d)}$$

$$3\,(CH_3)_3SnSC_4H_9 + BCl_3 \rightarrow B(SC_4H_9)_3 + 3\,(CH_3)_3SnCl \qquad \text{(e)}$$

The interaction of Me_3SnSBu-n with nonmetal and metal halides cleaves the Sn—S bond[6]:

$$3\,(CH_3)_3SnS(C_4H_9\text{-n}) + Z \rightarrow M[S(C_4H_9\text{-n})]_3 + 3\,(CH_3)_3SnX \qquad \text{(f)}$$

where Z = BCl_3, $HgCl_2$, $CdCl_2$, X_2PtCl_4, PCl_3, $AsCl_3$, $HgBr_2$, HgI_2, $CdBr_2$.

A mercaptoacetate group attached to Ge exchanges with Cl at the bp of the solution during removal of the volatile components by distillation[7]:

$$Et_3GeSCH_2COOGeEt_3 + HgCl_2 \rightarrow \underset{(80\%)}{Et_3GeCl} + Hg(SCH_2COO) \qquad \text{(g)}$$

This is much faster and is not a redistribution. Also, the interactions of alkylthiotins such as $Me_2Sn(SR)_2$ with Ni salts form hexameric Ni mercaptides, $[Ni(SR)_2]_6$. Nickel chloride reacts with $R_2Sn(SR')_2$ at RT with fission of the Sn—S bond. The reaction proceeds[8,9] through an intermediate coordinate complex with subsequent elimination of R_2SnCl_2, e.g.:

$$6\,NiCl_2 + 6\,(CH_3)_2Sn(SC_2H_5)_2 \rightarrow [Ni(SC_2H_5)_2]_6 + 6\,(CH_3)_2SnCl_2 \qquad \text{(h)}$$

Finally, alkyl- and arylselenotin compounds react with pentacarbonyl halides of Mn and Re to produce the dimeric $[M(CO)_4SeR]_2$ species. Polynuclear, Se-bridged Mn and Re carbonyls are obtained from $Me_3SnSePh$, $Me_2Sn(SePh)_2$ and analogous organotin methyl- and ethylselenides with Mn and Re bromopentacarbonyls[10]:

$$\begin{array}{l} 2R_3SnSeR' + 2\,M(CO)_5X \rightarrow \\ R_2Sn(SeR')_2 + 2\,M(CO)_5X \rightarrow \end{array} \begin{array}{ccccc} & & R' & & \\ & CO & | & CO & \\ OC & | & Se & | & CO \\ & M & & M & \\ OC & | & Se & | & CO \\ & CO & | & CO & \\ & & R' & & \end{array} + R_2SnX_2 + 2\,CO \qquad \text{(i)}$$

where R = CH_3; R' = CH_3, C_2H_5, C_6H_5, M = Mn, Re; X = Cl, Br.

(J. M. BELLAMA)

1. K. A. Hooton, A. L. Alfred, *Inorg. Chem.*, *4*, 671 (1965).
2. C. Glidewell, D. W. H. Rankin, *J. Chem. Soc.*, *A*, 753 (1969).
3. H. Wuyls, A. Vangindertaelen, *Bull. Soc. Chim. Belg.*, *30*, 323 (1921).
4. R. C. Poller, J. A. Spillman, *J. Organomet. Chem.*, *6*, 668 (1966).
5. A. G. Davies, P. G. Harrison, *J. Chem. Soc.*, *C*, 298 (1967).
6. E. W. Abel, D. B. Brady, B. C. Crosse, *J. Organomet. Chem.*, *5*, 260 (1966).

2.5.10. from Cleavage of Group-IVB–Nitrogen Bond

The E—N bond (E = Si, Ge, Sn, Pb) is cleaved by covalent halides to produce E—X bonds (X = F, Cl, Br, I). However, in most such reactions, the species containing the E—X bond is a byproduct of little interest, and the synthetic importance is in the production of compounds containing other nitrogen bonds such as N—B or N—P. Such reactions are only given passing mention, here, particularly because the N—E bonded species are frequently formed from the halides in the first place. However, they can be used as intermediates in the transformation of an available halide such as a chloride into an iodide by cleavage of the E—N bond with HI. Examples of specific useful syntheses of this type are given below. Also, cleavage by HX to produce group-IV halides is a convenient method for characterizing or analyzing species containing the group IVB-nitrogen bond. Such a procedure becomes important when more conventional analytical techniques are inappropriate. Because the group-IVB—halide is the desired product in these cases, examples of this type of procedure are included.

(J. E. DRAKE)

2.5.10.1. by Halogens.

One of the Si—N bonds in $Me_3SiNSNSiMe_3$ is cleaved by Cl_2 to give $Me_3SiNSNCl$ and Me_3SiCl in high yield[1] even at $-70°C$ in Freon-114. Such a reaction is of no importance as a synthetic route to Me_3SiCl.

(J. E. DRAKE)

1. W. Lidy, W. Sundermeyer, W. Verbeek, *Z. Anorg. Allg. Chem.*, *406*, 228 (1974).

2.5.10.2. by Hydrogen Halides.

The E—N bond (where E = group-IVB element) is cleaved by HX to give the halogen-group-IVB bond:

$$\text{E—N}\langle + \text{HX} \rightarrow \text{E—X} + \text{HN}\langle \qquad \text{(a)}$$

This reaction can be used to synthesize organosilyl halides and is of historic interest as the method[1] first used to synthesize $HSiI_3$. Its application allows trichloro- and dichlorosilanes to be converted to the corresponding iodides:

$$RSiCl_3 + 6\ PhNH_2 \rightarrow RSi(HNPh)_3 + 3\ [PhNH_3]Cl \qquad \text{(b)}$$

$$RSi(HNPh)_3 + 3\ HI \rightarrow RSiI_3 + 3\ [PhNH_3]I \qquad \text{(c)}$$

where R = Me, Et, n-Pr, i-Pr, n-Bu, n-Amyl, $C_{12}H_{25}$ and also, for R_2SiI_2, R = Me, Et.

The Si—N species are prepared[2,3] in situ by addition of xs $C_6H_5NH_2$ to chlorosilanes in benzene, and then the Si—N bond is cleaved by anhyd HI. Each precipitation stage must be complete. Conditions range from addition of $C_6H_5NH_2$ at RT to refluxing the mixture for several hours before passage of anhyd HI for up to 4 h. The iodides are distilled from benzene solution. Similarly, REt_2SiCl is transformed[4] to the corresponding Si—Br or Si—F by conversion to REt_2SiNH_2 by treatment with liq NH_3 followed by cleavage with HBr in Et_2O or 48% HF. In principle, all hydrogen halides cleave the Si—N bond, but with HF, ionic compounds result. Thus, with $R_3Si(NHR')$ and $R_2Si(NHR')_2$, HF produces R_3SiF and R_2SiF_2, whereas with $RSi(NHR')_3$, the pentafluoroorganosilicates, $[RNH_3]_2\ [RSiF_5]$, are formed[5].

Tris-triphenylgermylsilylamine is cleaved in ET_2O by HCl to give the chloride, which can be crystallized from ethyl acetate[6]:

$$(Ph_3Ge)_3SiNH_2 + 2\ HCl \rightarrow (Ph_3Ge)_3SiCl + NH_4Cl \qquad (d)$$

The Si—N bond is cleaved preferentially in compounds that contain both Si—O and Si—N bonds. Thus, e.g., $t\text{-}BuO_2Si(NH_2)_2$ in benzene is treated with anhyd HCl for 3 h at RT or until an xs is detectable at the exit tube, NH_4Cl precipitates and $(t\text{-}BuO)_2SiCl_2$ can be distilled from the solution[7]:

$$(t\text{-}BuO)_2Si(NH_2)_2 + 4\ HCl \rightarrow (t\text{-}BuO)_2SiCl_2 + 2\ NH_4Cl \qquad (e)$$

Other mixed species such as $(EtO)_3SiNMe_2$ react similarly to yield[8] EtO_3SiCl.
1,3-dihalogenodisilazanes are obtained[9] in 40–70% yield by partial cleavage of the Si—N bonds of cyclotri- or tetra-silazanes with HX, where X = F, Cl or Br, e.g.:

$$(SiR_2NHSiR_2NH)_n + 3\ HX \rightarrow XSiR_2NHSiR_2X + NH_4X \qquad (f)$$

Under milder conditions, $R_2Si(NHSiR_2F)_2$ may be formed[10].

Partial cleavage of the Si—N bonds can also occur[11]:

$$ClSiMe_2\text{—}N\begin{matrix} SiMe_2 \\ / \quad \backslash \\ \backslash \quad / \\ SiMe_2 \end{matrix}NSiMe_2Cl \xrightarrow{+4\ HCl} (ClSiMe_2)_3N + Me_2SiCl_2 + NH_4Cl \qquad (g)$$

The cleavage of the Si—N bond with HCl or HBr produces the corresponding Si—Cl or Si—Br species; e.g., the cleavage of $(SiH_3)_2NMe$ by HCl to give SiH_3Cl quantitatively is used to characterize the amine[12] but it is not an appropriate preparation of the halide because the amine is normally prepared from it. Other examples include the action of HCl on $(SiH_3)_3N$[13] or $(SiH_3)_2NEt$[14] to produce SiH_3Cl and on $(MeSiH_2)_3N$ to produce $MeSiH_2Cl$[15], and the action of HBr on SiH_3NMe_2 to Yield SiH_3Br[16]. In addition, the unstable $(GeH_3)_3N$ cleaves with HCl to give GeH_3Cl and NH_3, although it is actually made[17] under controlled conditions by the action of GeH_3Cl on NH_3.

The Sn—N, Ge—N and Pb—N species are increasingly less available than the Si—N, so it makes even less sense to use them as intermediates in the synthesis of a halide. However, in the course of using cleavage by HX as characterization of an M—N bond, a novel M—X product may be formed. Thus, in ring expansion reactions of 1,1-dimethyl-l-stannacyclopropane, a six-membered heterocycle is formed that with M aq

HCl in Et_2O, gives the novel, functionally substituted, Sn—Cl species 1-[4-(dimethylchlorostannyl)butyl]-1,2-bis(carboethoxy)hydrazine in 34% yield[18].

$$Me_2Sn(CH_2)_4 \text{ (ring)} + EtO_2CN{=}NCCO_2Et \rightarrow Me_2Sn(CH_2)_4N(CO_2Et){-}N(CO_2Et) \text{ (ring)} \xrightarrow{HCl\ (aq)} Me_2ClSn(CH_2)_4N(CO_2Et)NH(CO_2Et) \quad (h)$$

The chloride is distilled after drying the organic phase over $MgSO_4$.

(J. E. DRAKE)

1. O. Ruff, *Chem. Ber.*, *41*, 3738 (1908).
2. H. H. Anderson, D. L. Seaton, R. P. T. Rudnicki, *J. Am. Chem. Soc.*, *73* 2144 (1951).
3. H. H. Anderson, *J. Am. Chem. Soc.*, *73* 2351 (1951).
4. D. L. Bailey, L. H. Sommer, F. C. Whitmore, *J. Am. Chem. Soc.*, *70* 435 (1948).
5. L. Tansjö, *Acta Chem. Scand.*, *15*, 1583 (1961); *18*, 456, 465 (1964).
6. J. G. Milligan, C. A. Kraus, *J. Am. Chem. Soc.*, *72*, 5297 (1950).
7. P. D. George, L. H. Sommer, F. C. Whitmore, *J. Am. Chem. Soc.*, *75*, 6308 (1953).
8. L. Rosnati, *Gazz. Chim. Ital.*, *78*, 516 (1948).
9. U. Wannagat, *Agnew, Chem., Int. Ed. Engl.*, *4*, 626 (1965).
10. U. Wannagat, E. Bogusch, F. Hofler, *J. Organomet. Chem.*, *7*, 203 (1967).
11. U. Wannagat, E. Bogusch, *Inorg. Nucl. Chem. Lett.*, *2*, 97 (1966).
12. H. J. Emeléus, N. Miller, *Nature (London)*, 142, 996 (1938).
13. A. Stock, C. Somieski, *Chem. Ber.*, *54*, 740 (1921).
14. M. J. Emeléus, N. Miller, *J. Chem. Soc.*, 819 (1939).
15. E. A. V. Ebsworth, H. J. Emeléus, *J. Chem. Soc.*, 2150 (1958).
16. S. Sujishi, S. Witz, *J. Am. Chem. Soc.*, *76*, 4631 (1954).
17. D. W. H. Rankin, *J. Chem. Soc., Chem. Commun.*, 194 (1964); *J. Chem. Soc., A*, 1926 (1969).
18. E. J. Bulten, H. A. Budding, *J. Organomet. Chem.*, *166*, 339 (1979).

2.5.10.3. by Other Halides.

Reactions:

$$M{-}N + M'{-}X \rightarrow M'{-}N + M{-}X \quad (a)$$

where M = group IVB; M′ ≠ group IVB, are legion, but the product of interest is that containing the M′—N and not the M—X bond. These reactions mostly involve Si—N or Sn—N bonded compounds. Experimental details are given in the sections on N—B, N—P, etc., bond formation.

The reaction of Me_2NSiMe_3, formed[1] from Me_3SiCl and Me_2NH, with $F_4PCH_2PF_4$ produces a diphosphorus zwitterion[2], and Me_3SiF:

$$F_4PCH_2PF_4 + 2\ Me_3SiNMe_2 \rightarrow [F_5P]^-[CH_2PF(NMe_2)_2]^+ + Me_3SiF \quad (b)$$

Metal halides also participate as illustrated by the reaction of $PhN{=}NSiMe_3$, with $BrMn(CO)_5$:

$$2\ PhN{=}NSiMe_3 + 2\ Mn(CO)_5Br \rightarrow [PhN{=}NMn(CO)_4]_2 + 2\ CO + 2\ Me_3SiBr \quad (c)$$

to give of the dimeric product in 20% yield[3].

When the Sn—N bond is cleaved by BCl_3[4] or PF_2Cl[5] to form aminoboranes and phosphines, respectively, Me_3SnCl is the byproduct e.g.:

$$RN(SnMe_3)_2 + 2\ BCl_3 \rightarrow RN(BCl_2)_2 + 2\ Me_3SnCl \tag{d}$$

$$(Me_3Sn)_3N + 3\ PF_2Cl \rightarrow N(PF_2)_3 + 3\ Me_3SnCl \tag{e}$$

The expected R_3SiX compounds are produced as the result of elimation from silylaminophosphorus or -antimony halides providing an extra means of characterization[6] rather than a useful synthetic procedure, e.g.:

$$t\text{-}BuMe_2SiN(Me)PF_4 \xrightarrow{\Delta} t\text{-}BuMe_2SiF + 1/2\ (MeNPF_3)_2 \tag{f}$$

The cleavage of the Ge—N bond in $ClGeMe_2CH_2Cl$ derivatives with Me_3SiCl produces a Ge—Cl-bonded intermediate, e.g.:

$$ClGeMe_2CH_2Cl + 2\ LiNMe_2 \rightarrow Me_2NGeMe_2CH_2NMe_2 + 2\ LiCl \tag{g}$$

$$Me_2NGeMe_2CH_2Me_2 + Me_3SiCl \rightarrow ClGeMe_2CH_2NMe_2 + Me_3SiNMe_2 \tag{h}$$

$$ClGeMe_2CH_2NMe_2 + Me_3SiPMe_2 \rightarrow Me_2PGeMe_2CH_2NMe_2 + Me_3SiCl \tag{i}$$

Again, however, the purpose is to produce the chelating ligands,[7] $Me_2MGeMe_2CH_2M'Me_2$, and not the Ge or Si halides.

(J. E. DRAKE)

1. J. Mack, C. H. Yoder, *Inorg. Chem., 8*, 278 (1969).
2. A. H. Cowley, R. C-Y. Lee, *Inorg. Chem., 18*, 60 (1979).
3. E. W. Abel, C. A. Burton, *J. Organomet. Chem., 170*, 229 (1979).
4. T. Gasparis, H. Nöth, W. Storch, *Angew Chem., Int, Ed. Engl., 18*, 326 (1979).
5. W. Krüger, R. Schmutzler, *Inorg. Chem., 18*, 871 (1979).
6. R. H. Neilson, W. A. Kusterbeck, *J. Organomet. Chem.*, 116, 309 (1979).
7. J. Grobe, J. Hendriock, *J. Organomet. Chem., 132*, 77 (1977).

2.5.11. from Cleavage of the Group-IVB–Group-VB Element Bond

The cleavage of a group-IVB—group-VB element bond is not preferred for the formation of the halogen to group-IVB bond. However, particularly with hydrogen halides, such a cleavage can be used to produce a halide as a means of characterizing products containing the group-IVB—group-VB bond.

(J. E. DRAKE)

2.5.11.1. by Halogens.

Of the few such reactions, the Si—P bonds in $(SiH_3)_3P$ are cleaved by the addition of I_2 to give only SiH_3I and P_4 as identifiable products[1]. However, this is not a suitable synthesis of SiH_3I. With GeH_3AsH_2, I_2 gives[2] GeH_3I, AsH_3 and As_2H, but again this is not a viable route to GeH_3I:

$$H_3GeAsH_2 + I_2 \rightarrow GeH_3I + \text{"}AsH_2I\text{"} \quad (a)$$

$$H_3GeAsH_2 + \text{"}AsH_2I\text{"} \rightarrow GeH_3I + H_2AsAsH_2 \quad (b)$$

$$H_2AsAsH_2 \rightarrow 6/5\ AsH_3 + 2/5\ As_2H \quad (c)$$

The cleavage of an Si—Sb bond by I_2 forms Me_3SiI:

$$2\ (Me_3Si)_3Sb + I_2 \rightarrow (Me_3Si)_4Sb_2 + 2\ Me_3SiI \quad (d)$$

but the product of interest is the distibine[3].

(J. E. DRAKE)

1. E. A. V. Ebsworth, C. Glidewell, G. M. Sheldrick, *J. Chem. Soc., A*, 352 (1969).
2. J. E. Drake, C. Riddle, *J. Chem. Soc., A*, 2452 (1968).
3. H. J. Brennig, V. Brennig-Lyriti, *Z, Naturforsch., Teil B, 34*, 926 (1979).

2.5.11.2. by Hydrogen Halides.

The M—M′ bond (group IVB—group-VB) is cleaved by hydrogen halides to give the halogen-group-IVB bond:

$$M{-}M' + HX \rightarrow M{-}X + M'{-}H \quad (a)$$

However, this is not a synthetic route to the group IVB halides since the M—M′ bond is synthesized from the halide produced in the cleavage (see, e.g., §4.3.4). The HX cleavage reactions are used to characterize products containing an M—M′ bond. This is particularly true when handling the air-sensitive, volatile hydride derivatives.

The simplest Si—P bond-containing compound is SiH_3PH_2. When 1:1 HBr (0.4 mmol) is held for 1 h at RT in a small enclosed vessel (12 ml), quantitative conversion to SiH_3Br and PH_3 is achieved[1]. This is a useful characterization but not a preparative route. A suitable preparation of SiH_3PH_2 involves the reaction of SiH_3Br with KPH_2[2] or $LiAl(PH_2)_4$[3]. Similarly, the Si—As bond in SiH_3AsH_2 can be cleaved to yield SiH_3Br quantitatively. The same result can be achieved with HCl, but the SiH_3Cl produced is not separated readily from phosphine or arsine by vacuum fractionation.

By contrast, with GeH_3PH_2 and GeH_3AsH_2, the favored reagent is HCl because GeH_3Cl can be separated. Similarly, $(SiH_3)_2PH$ reacts with HCl to give[4] SiH_3Cl and PH_3 as does[5] $(SiH_3)_3P$. With the latter, even with a deficiency of HCl, all three Si—P bonds are cleaved to give unreacted $(SiH_3)_3P$ rather than intermediate products. With HBr, $(GeH_3)_3P$ gives[6] GeH_3Br and PH_4Br, while $(GeH_3)_3Sb$ gives[7] GeH_3Br, H_2 and Sb metal, reflecting the relative instability of SbH_3. With HI, $(GeH_3)_3As$ gives the expected GeH_3I and AsH_3. Similarly, Si_2H_5Cl is produced[4] by the cleavage of $SiH_3SiH_2PH_3$.

A slight xs of HX (X = Cl, Br) quantitatively cleaves all the Si—As bonds in organosubstituted silylarsines, including SiH_3AsMe_2, $(SiH_3)_2AsMe$, SiH_3AsPh_2, $(SiH_3)_2AsPh$, $Me_3SiAs(SiH_3)_2$, $Me_3SiAs(Me)SiH_3$, $(Me_3Si)_2AsSiH_3$, $Me_3SiAsPh_2$ and $(Me_3Si)_2AsPh$, to give[8] the arsine and SiH_3X or Me_3SiX. Similarly, Me_3SnAsH_2, $(Me_3Sn)_2AsH$ or $Me_3SnAs(H)Ph$ react with HBr when mixed at RT for 1/2 h to give

Me_3SnBr and the corresponding arsine in quantitative yield[9]. Again, Me_3SnBr is a starting material used to prepare the stannylarsines in the first place.

(J. E. DRAKE)

1. J. Simpson, *PhD Thesis*, Univ. Southampton, 1967.
2. C. Glidewell, G. M. Sheldrick, *J. Chem. Soc., A*, 350 (1969).
3. A. D. Norman, *J. Chem. Soc., Chem. Commun.*, 812 (1968).
4. S. D. Gorkhale, W. L. Jolly, *Inorg. Chem.*, *4*, 596 (1965).
5. E. A. V. Ebsworth, C. Glidewell, G. M. Sheldrick, *J. Chem. Soc., A*, 352 (1969).
6. S. Craddock, E. A. V. Ebsworth, G. Davidson, L. A. Woodward, *J. Chem. Soc., A*, 1229 (1967).
7. E. A. V. Ebsworth, D. W. H. Rankin, G. M. Sheldrick, *J. Chem. Soc., A*, 2828 (1968).
8. J. W. Anderson, J. E. Drake, *J. Inorg. Nucl. Chem.*, *34*, 2455 (1972).
9. J. W. Anderson, J. E. Drake, *Can. J. Chem.*, *49*, 2524 (1971).

2.5.11.3. by Other Halides.

Although less extensively used than reactions involving group-IVB—N bonds, these reactions:

$$M{-}M' + M''{-}X \rightarrow M{-}X + M'{-}M'' \quad (a)$$

where M = group IVB; M′ and M″ ≠ group IVB, are still important for M′—M″ bonds but lack applicability to M—X bond synthesis. Experimental details are given in sections involving P—M, As—M or Sb—M bond formation.

Trimethylsilylphosphine with $BrMn(CO)_5$ produces[1] a phosphorus-bridged manganese carbonyl and Me_3SiBr:

$$(Me_3Si)_3P + 2\ (CO)_5MnBr \rightarrow Mn_2(CO)_8(\mu\text{-}Br)[\mu\text{-}P(SiMe_3)_2] + Me_3SiBr \quad (b)$$

The As—Si bond is cleaved by CF_3I to provide As—CF_3-containing compounds[2], e.g.:

$$Me_2AsSiMe_3 + CF_3I \rightarrow Me_2AsCF_3 + Me_3SiI \quad (c)$$

but there are more convenient ways for making Me_3SiI.

The same can be said of the formation of Me_3SiI in the cleavage of the Sb—Si bond by organoiodides[3], e.g.:

$$(Me_3Si)_3Sb + 3\ RI \rightarrow R_3Sb + 3\ Me_3SiI \quad (d)$$

In reactions involving $ClGeMe_2CH_2Cl$, the cleavage of the Ge—P or Ge—As bond with Me_2PCl or Me_2AsCl is utilized to obtain a Cl—Ge-containing intermediate:

$$ClGeMe_2CH_2Cl + LiMMe_2 \rightarrow Me_2MGeMe_2CH_2MMe_2 + 2\ LiCl \quad (e)$$

$$Me_2MGeMe_2CH_2MMe_2 + Me_2MCl \rightarrow ClGeMe_2CH_2MMe_2 + Me_2MMMe_2 \quad (f)$$

where $ClGeMe_2CH_2P(orAs)Me_2$ can then be reacted with $LiNMe_2$ or $Me_3SiAs(orP)Me_2$ to produce the mixed chelating agents $Me_2MGeMe_2CH_2M'Me_2$ (M ≠ M′ = N,P,As) which is the real purpose of the procedure[4].

(J. E. DRAKE)

1. H. Schaefer, *Z. Naturforsch., Teil B*, *33*, 351 (1978).
2. J. Apel, J. Grobe, *Z. Anorg. Allg. Chem.*, *453*, 28 (1979).

3. H. J. Brennig, V. Brennig-Lyriti, *Z. Naturforsch., Teil B, 34*, 926 (1979).
4. J. Grobe, J. Hendriock, *J. Organomet. Chem., 132*, 77 (1977).

2.5.12. Group-IVB Halides by Halide–Halide Exchange Reactions (Metathesis)

2.5.12.1. by Metal Halides

2.5.12.1.1. of the Alkali or Alkaline Earths.

Aliphatic chlorides or bromides are converted to the corresponding iodo derivatives by reaction with dry NaI, or less often KI or CaI_2, in a solvent[1]:

$$RX + I^- \rightleftharpoons RI + X^- \tag{a}$$

where X = Cl, Br. Dry acetone is used because of the solubility of the inorganic halides and the speed of establishing the equilibrium. Sodium iodide is soluble in acetone (1.29M, ca. 20 times that of KI), while NaBr, and particularly NaCl (5.5×10^{-6}M), are only slightly soluble[2]. The organic halide is stirred with NaI in acetone at RT or under reflux until the precipitation of NaBr or NaCl is complete. The yield of iodo product is 60–100%.

The ease of conversion depends on the organic halide. Allyl and benzyl halides and α-halo carbonyl compounds are reactive, with Br being more easily replaced than Cl. Primary alkyl halides exchange quicker than secondary or tertiary ones[1,3]. Higher boiling solvents, such as methyl ethyl ketone or acetyl acetone, are employed for unreactive alkyl halides. The exchange is also achieved in MeOH or EtOH (2-bromo to 2-iodoheptane using NaI, 6-h reflux)[4] or H_2O (chloroacetic acid to iodoacetic acid using KI, 50%, 2 h)[5].

Simple aromatic halides cannot be converted to the corresponding iodo derivative by this procedure unless an electron-withdrawing substituent (such as o- or p-NO_2) is present. 1-Chloro-2,4-dinitrobenzene can be converted to the corresponding 1-iodo derivative in ca. 70% yield by refluxing with 5:1 xs NaI in dimethylformamide (DMF)[6]. Other procedures for halogen–iodine exchange involving aromatic halides are available (see §2.5.13).

Treating aliphatic dibromides and dichlorides with NaI in acetone leads to the corresponding diiodo derivatives, but the reaction may also produce unsaturated or cyclic products; i.e., olefins may accompany the halogen exchange of $\alpha - \beta$-dihalides:

$$RCHXCHXR' + 2\ I^- \rightarrow RCH{=}CHR' + 2\ X^- + I_2 \tag{b}$$

Methylene diiodide is prepared by reacting CH_2Cl_2 with NaI in $PhCH_2OH$[7]. Partial iodination of some systems can be achieved if equimolar reactants are used[8]:

$$ClCH_2CH_2CH_2Cl \xrightarrow[\text{24 h, reflux}]{\text{NaI-acetone}} ICH_2CH_2CH_2Cl \tag{c}$$

Other halogen-replacement reactions that use group IA or IIA bromides are less important from a preparative point of view than the iodo conversion; e.g., LiBr (in refluxing MeOH or acetone), NaBr (in refluxing MeOH or EtOH) or $CaBr_2$ can be used to substitute Br for halogen next to a C=C or C≡C group[2].

Reaction of NaI or KI in acetone with organic halides gives iodoalkyl organosilanes from the corresponding Cl derivative, e.g., refluxing Me_3SiCH_2Cl with NaI in dry acetone produces Me_3SiCH_2I in 70% yield[9]:

$$Me_3SiCH_2Cl + NaI \xrightarrow{\text{acetone}} Me_3SiCH_2I + NaCl \tag{d}$$

Other iodoalkylsilanes prepared by this procedure include ICH_2SiH_3[10], $ICH_2Si_2H_5$[11], $ICH_2Si_2Me_5$[12], $ICH_2\overline{SiMe[OSiMe_2]_3O}$[13], $ICH_2SiMe_2OSiMe_3$[14], $ICH_2SiMe_2CH_2SiMe_3$[15], $CH_3CHISiMe_3$[16], and the (iodoalkyl)trialkoxysilanes, $I(CH_2)_nSi(OR)_3$ (R = Me, Et; n = 1–3)[17]. The iodination of these silanes proceeds faster than with similar organic chlorides[18]. Thus, Me_3SiCH_2Cl reacts[19] with KI in acetone 25 times faster than n-BuCl at RT. The chloromethyl disilane, $ClCH_2SiMe_2SiMe_3$, is more reactive than $ClCH_2SiMe_3$ in its conversion to the corresponding iodomethyl derivative[12].

Silicon halides undergo rapid metathesis reactions with certain alkali-metal and alkaline-earth halides in solvents such as acetone, nitromethane, xylene, MeCN and Et_2O; e.g., Et_3SiI is formed[20] from Et_3SiCl with NaI in MeCN, and Me_3SiCl with anhyd $MgBr_2$ or MgI_2 in xylene produces Me_3SiBr and Me_3SiI in 57 and 61% yields, respectively[21]:

$$2\ Me_3SiCl + MgX_2 \xrightarrow{\text{xylene}} 2\ Me_3SiX + MgCl_2 \tag{e}$$

where X = I, Br. Partial exchange of halogen may also occur[21]:

$$Me_2SiCl_2 + MgBr_2 \rightarrow Me_2SiClBr + MgBrCl \tag{f}$$

Chlorosilanes are used in the conversion, although fluorosilanes may also be used; e.g., Et_3SiF with $MgBr_2$ in Et_2O produces Et_3SiBr in 50% yield[22].

Although it is possible to recover the bromo- and iodoalkylsilanes by distillation, metathesis can be used for the in situ generation[23] of Me_3SiI and Me_3SiBr from Me_3SiCl, e.g., when Me_3SiCl is added to anhyd NaI in MeCN, NaCl is precipitated and Me_3SiI forms, which subsequently can be used in organic transformations[23–25]. Proton- and ^{13}C-NMR in acetone suggest[24]:

$$CH_3CN + Me_3SiCl + NaI \rightleftharpoons [CH_3C{\equiv}NSiMe_3]^+I^- + NaCl \tag{g}$$

Reagents used for the in situ generation of Me_3SiI and Me_3SiBr by metathesis are given in Table 1. These halosilanes are versatile synthetic organic chemistry, and their in situ generation and subsequent use offer advantages over the pure halosilane in many instances[23].

Germanium and tin halides also undergo exchange with alkali-metal halides in solvent. Acetone commonly is used, and the pure halogermane or -stannane prepared by this method can be isolated from the medium; e.g., the reaction of $(n\text{-hexyl})_3GeCl$ with NaI in acetone at RT produces $(n\text{-hexyl})_3GeI$ in 56% yield[34]. Treating Me_5Ge_2Cl with NaBr and NaI in acetone or tetrahydrofuran (THF) produces the corresponding bromo- and iodopentamethyldigermanes in 47 and 9% yields, respectively[35]. The reaction of Me_2SnCl_2 with NaI in acetone produces Me_2SnI_2 in 77.5% yield[36]. The reaction of

TABLE 1. IN SITU GENERATION OF BROMO- AND IODOSILANES FROM Me_3SiCl WITH HALIDES OF THE ALKALI AND ALKALINE-EARTH METALS

Me_3SiCl + metal halide reagent	Trimethylsilyl halide or in situ equivalent formed	Solvent	Ref.
Me_3SiCl-NaI	Me_3SiI	MeCN	24, 25, 26
		CH_2Cl_2	27
		Neat or in xs substrate	27, 28, 29
Me_3SiCl-KI	Me_3SiI	MeCN	30
Me_3SiCl-LiI	Me_3SiI	CH_2Cl_2	27
		CCl_4	31
Me_3SiCl-NaBr	Me_3SiI	Xylene	21
Me_3SiCl-NaBr	Me_3SiBr	MeCN	22
Me_3SiCl-LiBr	Me_3SiBr	MeCN	32
		Substrate	33
Me_3SiCl-KBr	Me_3SiBr	Substrate	33
Me_3SiCl-$MgBr_2$	Me_3SiBr	Substrate	33
Me_3SiCl-$MgBr_2$	Me_3SiBr	Et_2O	21

$ClSn[Mn(CO)_5]_3$ with NaI in THF gives[37] $ISn[Mn(CO)_5]_3$. As expected, Cl on Sn undergoes halogen exchange with NaI in halophenylchlorotins[38,39]:

$$(XC_6H_4)_2SnCl_2 + 2\ NaI \xrightarrow{\text{alcohol}} (XC_6H_4)_2SnI_2 + 2\ NaCl \qquad \text{(h)}$$

where X = I, Br, Cl.

Halogen exchange involving Mg metal or organomagnesium halides has been noted in reactions of Br- or I-containing reagents (RMghal) with chloro derivatives of the group-IVB elements. When it occurs, exchange leads to some unexpected products, e.g.:

$$ClCH_2SiMe_2Cl + CH_3MgI \rightarrow Me_3SiCH_2I \qquad \text{(i)}^{40}$$

$$3\ Me_2SiCl_2 + 4\ EtMgBr \rightarrow Et_2Si(CH_3)_2 + EtMe_2SiCl + EtMe_2SiBr \qquad \text{(j)}^{41}$$

$$Ph_2SiCl_2 + PhMgBr \rightarrow Ph_3SiCl + Ph_3SiBr \qquad \text{(k)}^{42}$$

$$GeCl_4 + 4\ MeMgI \rightarrow GeI_4 + 4\ MeMgCl \qquad \text{(l)}^{43}$$

At 610°C, $MgCl_2$ with GeF_4 produces $GeCl_4$, whereas under similar conditions (620–630°C) it reacts with SiF_4 to produce $SiCl_4$ and chlorofluorosilanes $F_{4-y}SiCl_y$ (y = 1–3).[44,45]

(C. H. VAN DYKE)

1. H. Finkelstein, *Chem. Ber. 43*, 1528 (1910).
2. C. Weygan, G. Hilgetag, in *Preparative Organic Chemistry*, G. Hilgetag, A. Martini, eds.; Wiley-Interscience, New York, 1972.
3. J. B. Conant, R. E. Hussey, *J. Am. Chem. Soc.*, *47*, 476 (1925).
4. M. Schirm, H. Besendorf, *Arch. Pharm.* (Weinheim, Ger.), *280*, 64 (1942).
5. E. Abderhalden, M. Guggenheim, *Chem. Ber.*, *41*, 2853 (1908).
6. J. F. Bunnett, R. M. Conner, *J. Org. Chem.*, *23*, 305 (1958); *Org. Synth.*, *40*, 34 (1960).
7. E. D. Laskina, *Zhur. Priklad, Khim.*, *32*, 878 (1959); *Chem. Abstr.*, *53*, 17039 (1959).
8. H. B. Hass, H. C. Huffman, *J. Am. Chem. Soc.*, *63*, 1233 (1941).

9. F. C. Whitmore, L. H. Sommer, *J. Am. Chem. Soc.*, *68*, 481 (1946).
10. J. M. Bellama, A. G. MacDiarmid, *J. Organomet. Chem.*, *18*, 275 (1969).
11. J. Morrison, J. M. Bellama, *Inorg. Chem.*, *14*, 1614 (1975).
12. Data from M. Kumada, quoted by H. Sakurai, *J. Organomet. Chem.*, *200*, 261 (1980).
13. J. E. Bulkowski, N. D. Miro, D. Sepelak, C. H. Van Dyke, *J. Organomet. Chem.*, *101*, 267 (1975).
14. G. F. Roedel, *J. Am. Chem. Soc.*, *71*, 269 (1949).
15. L. H. Sommer, G. M. Goldberg, G. H. Barnes, L. S. Stone Jr., *J. Am. Chem. Soc.*, *76*, 1609 (1954).
16. L. H. Sommer, D. L. Bailey, G. M. Goldberg, C. E. Buck, T. S. Bye, F. J. Evans, F. C. Whitmore, *J. Am. Chem. Soc.*, *76*, 1613 (1954).
17. M. G. Voroknov, V. M. D'yakov, Yu. A. Lukina, G. A. Sansonova, N. M. Kudyakov, *J. Gen. Chem., USSR* (*Engl. Trans.*), *45*, 1974 (1975).
18. C. Eaborn, J. C. Jeffrey, *J. Chem. Soc.*, 4266 (1954).
19. G. D. Cooper, M. Prober, *J. Am. Chem. Soc.*, *76*, 3943 (1954).
20. C. Eaborn, *J. Chem. Soc.*, 3077 (1980).
21. U. Kruerke, *Chem. Ber.*, *95*, 174 (1962).
22. Y. Etienne, *C.R. Hebd. Séances Acad. Sci.*, *234*, 1985 (1952).
23. A. H. Schmidt, *Chem. Z.*, *104*, 253 (1980).
24. G. Olah, S. C. Narang, B. G. B. Gupta, R. Malhotra, *J. Org. Chem.*, *44*, 1247 (1979).
25. G. A. Olah, S. C. Narang, B. G. B. Gupta, R. Malhotra, *Synthesis*, 61 (1979).
26. T. Morita, Y. Okamoto, H. Sakurai, *Tetrahedron Lett.*, 2523 (1978).
27. M. Russ, *Ingenieurarbeit*, Fachhochschule Fresenius, Wiesbaden, Ger. 1976.
28. A. H. Schmidt, M. Russ, *Chem. Z.*, *102*, 26 (1978).
29. A. H. Schmidt, M. Russ, *Chem. Z.*, *102*, 65 (1978).
30. S. Scheibye, I. Thompsen, S.-O. Lawesson, *Bull. Soc. Chim. Belg.*, *88*, 1043 (1979).
31. Y. Machida, S. Nomomoto, I. Saito, *Synth. Commun.*, *9*, 97 (1979).
32. G. A. Olah, B. G. B. Gupta, R. Malhotra, S. C. Narang, *J. Org. Chem.*, *45*, 1638 (1980).
33. A. H. Schmidt, M. Russ, *Chem. Ber.*, *114*, 1099 (1981).
34. R. Fuchs, H. Gilman, *J. Org. Chem.*, *23*, 911 (1958).
35. A. J. Andy, *Diss. Abstr. Int.*, *1335*, 2537 (1974); *Chem. Abstr.*, *82*, 98,083 (1975).
36. D. Seyferth, E. G. Rochow, *J. Am. Chem. Soc.*, *77*, 1302 (1955).
37. J. A. J. Thompson, W. A. G. Graham, *Inorg. Chem.*, *6*, 1365 (1967).
38. K. A. Kocheshkov, A. N. Nesmeyanov, *Chem. Ber.*, *64*, 628 (1931).
39. A. N. Nesmeyanov, D. A. Kocheshkov, *Zh. Obshch. Khim.*, *1*, 219 (1931); *Chem. Abstr.*, *26*, 2182 (1932).
40. A. P. Petrov, V. M. Vdovin, *J. Gen. Chem. USSR* (*Engl. Transl.*), *29*, 2870 (1959); *Chem. Abstr.*, *50*, 11,981 (1960).
41. R. N. Lewis, *J. Am. Chem. Soc.*, *69*, 717 (1947).
42. U. Wannagat, H. Bürger, E. Ringel, *Monatsh. Chem.*, *93*, 1363 (1962).
43. R. Zablotna, *Bull. Acad. Polon. Sci. Ser. Chim.*, *12*, 475 (1966); *Chem. Abstr.*, *62*, 1311 (1965).
44. W. C. Schumb, D. W. Breck, *J. Am. Chem. Soc.*, *74*, 1754 (1952).
45. M. Schmeisser, H. Jenker, *Z. Naturforsch, Teil B*, 191 (1952).

2.5.12.1.2. by Silver and Other Transition-Metal Halides.

The use of silver halides in converting one carbon–halogen bond to another is limited in synthetic organic chemistry, but the method does have an important application in converting available iodoalkylmetal derivatives to the corresponding Cl or Br compound; e.g., the reaction of iodomethyltin compounds with AgCl and AgBr in MeCN at RT yields chloro- and bromomethyl organotins difficult to prepare otherwise[1]:

$$\geq SnCH_2I + AgX \xrightarrow[7\ days]{CH_3CN,\ 25^{\circ}C} \geq SnCH_2X + AgI \qquad (a)$$

where X = Cl, Br. The reactions are attributed to solubilities of the AgI vs. AgCl and AgBr. Haloamethyltins prepared by this procedure are given in Table 1.

TABLE 1. IODOMETHYLTIN COMPOUNDS WITH AgCl AND AgBr[1]

$SnCH_2I$ Compound	AgX	Product	Yield, %
Me_3SnCH_2I	AgCl	Me_3SnCH_2Cl	73
$PhMe_2SnCH_2I$	AgCl	$PhMe_2SnCH_2Cl$	93
$Me_2Sn(CH_2I)_2$	AgCl	$Me_2Sn(CH_2Cl)_2$	68
Me_3SnCH_2I	AgBr	Me_3SnCH_2Br	67
$Me_2Sn(CH_2I)_2$	AgBr	$Me_2Sn(CH_2Br)_2$	75

Halogen derivatives of Si, Ge and Sn readily undergo exchange with silver halides as long as the order is in the direction[2-4]:

$$M{-}I \rightarrow M{-}Br \rightarrow M{-}Cl \rightarrow M{-}F \quad \text{(b)}$$

where M = Si, Ge, Sn. Silicon, Ge or Sn halides can be converted to any corresponding derivative on its right, but not on its left, by heating the compound with the appropriate Ag salt; e.g., the reaction between Et_3SiBr and xs AgCl produces Et_3SiCl in 90% yield, whereas no reaction occurs between Et_3SiCl and AgBr[2]. Some of the conversions are carried out in nonpolar solvents, while others employ no solvent at all. Yields exceed 90%. The procedure works with silyl, methylsilyl and ethylsilyl derivatives, and the reaction should be applicable to most other organic and inorganic Si, Ge and Sn systems as well. The reaction converts iodogermanes to other halogermanes, e.g., prepares CF_3GeCl_3 and CF_3GeF_3 from CF_3GeI_3[5]:

$$CF_3GeI_3 + 3\ AgX \rightarrow CF_3GeX_3 + 3\ AgI \quad \text{(c)}$$

where X = Cl, Br. Other examples include:

$$C_5H_{10}GeI_2 + 2\ AgCl \rightarrow C_5H_{10}GeCl_2 + 2\ AgI \quad \text{(d)}^{6}$$

$$\text{(Me,R-cyclohexenyl)}GeI_2 + 2\ AgCl \rightarrow \text{(Me,R-cyclohexenyl)}GeCl_2 + 2\ AgI \quad \text{(e)}^{7}$$

where R = H, Me;

$$Ge_2H_5I + AgX \rightarrow Ge_2H_5X + AgI \quad \text{(f)}^{8}$$

where X = Br, Cl; and

$$GeH_3Br + AgCl \rightarrow GeH_3Cl + AgBr \quad \text{(g)}^{9}$$

Other metal halides are used in some of the conversions, but these are not used as extensively as the Ag salts[2]. Mercury(II) halides convert iodosilanes to the corresponding Cl or Br derivatives[10]:

$$Si_2I_6 + 3\ HgCl_2 \rightarrow Si_2Cl_6 + 3\ HgI_2 \quad \text{(h)}^{6}$$

and bromogermane can be converted[9] to GeH_3Cl by its reaction with $HgCl_2$:

$$2\ GeH_3Br + HgCl_2 \rightarrow 2\ GeH_3Cl + HgBr_2 \qquad \text{(i)}$$

The sterically hindered $(Me_3Si)_3CSiMe_2I$ derivative undergoes only partial conversion to the corresponding Cl (22%) and Br (20%) derivative by its reaction with $HgCl_2$ or $HgBr_2$, respectively, in acetic acid[11]. No exchange occurs in MeOH.

Iron(III) chloride converts GeF_4 to $GeCl_4$ at 350°C, but it does not react with SiF_4 (at 500°C) or CF_4 (at 720°C)[12,13].

Copper(I) chloride containing a small amount of $CuCl_2$ is capable of exchanging Cl for Br attached to an aromatic ring, particularly in 2-picoline or dimethylformamide (DMF)[13]:

$$PhBr + CuCl \rightarrow PhCl + CuBr \qquad \text{(j)}$$

(C. H. VAN DYKE)

1. D. Seyferth, S. B. Andrews, *J. Organomet. Chem.*, *30*, 151 (1971).
2. C. Eaborn, *J. Chem. Soc.*, 3077 (1950).
3. H. H. Anderson, *J. Am. Chem. Soc.*, *73*, 5439 (1951).
4. H. H. Anderson, J. A. Basta, *J. Org. Chem.*, *19*, 1300 (1954).
5. H. C. Clark, C. J. Willis, *J. Am. Chem. Soc.*, *84*, 898 (1962).
6. P. Mazerolles, *Bull. Soc. Chem. Fr.*, 1907 (1962).
7. P. Mazerolles, G. Manuel, *Bull. Soc. Chim. Fr.*, 327 (1966).
8. D. M. Mackay, P. Robinson, E. J. Spanier, A. G. MacDiarmid, *J. Inorg. Nucl. Chem.*, *28*, 1125 (1966).
9. S. Cradock, E. A. V. Ebsworth, *J. Chem. Soc.*, *A*, 12 (1967).
10. C. Friedel, *C.R. Hebd. Seances. Acad. Sci.*, *73*, 1011 (1871).
11. C. Eaborn, D. A. R. Hopper, S. P. Hopper, K. D. Safa, *J. Organomet. Chem.*, *188*, 179 (1980).
12. W. C. Schumb, D. W. Breck, *J. Am. Chem. Soc.*, *74*, 1754 (1952).
13. W. B. Hardy, R. B. Fortenbaugh, *J. Am. Chem. Soc.*, *80*, 1716 (1958).

2.5.12.1.3. by Halides.

Anhydrous Al halides undergo halogen–halogen exchange reactions with some specific carbon and Si halides. Aluminum bromide is a particularly reactive halogen-exchange reagent when used with organic halides and can be used to introduce Br into chlorinated hydrocarbons, e.g., $BrCCl_3$ can be obtained in 51.4% yield by refluxing anhyd $AlBr_3$ with CCl_4 for 45 min[1]. Carbon tetrachloride is converted to CBr_4 by this procedure if the reaction is carried out in EtBr at −10 to 0°C.[2]:

$$3\ CCl_4 + 4\ AlBr_3 \xrightarrow{C_2H_5Br} 3\ CBr_4 + 4\ AlCl_3 \qquad \text{(a)}$$

Chloroform likewise undergoes[3] halogen exchange with $AlBr_3$:

$$CHCl_3 + AlBr_3 \rightarrow CHBr_3 + AlCl_3 \qquad \text{(b)}$$

Polyfluoroalkanes can be converted to polychloroalkanes by their reaction with Al and other metal chlorides[4]:

$$3\ CF_4 + 4\ AlCl_3 \xrightarrow{190°C} 3\ CCl_4 + 4\ AlBr_3 \qquad \text{(c)}$$

Pure $PhCCl_3$ is prepared[5] by the reaction of $PhCF_3$ with $AlCl_3$ in acetyl chloride; partial exchange[6] occurs between F_3CCCl_3 and $AlCl_3$ yielding $F_2ClCCCl_3$. Heating a mixture of CCl_4 and CS_2 with AlI_3 produces[7] CI_4.

Halomethylorganosilicon compounds, important precursors for preparing silanes that have substituents at the a-carbon atom, are prepared by a halogen-exchange reaction that involves an Al halide; e.g., chloromethylorganosilicon compounds can be converted to bromomethyl or iodomethyl derivatives by their reaction[8] with $AlBr_3$ or AlI_3, respectively, in a low-bp solvent such as EtBr. The Cl on carbon of chloromethylchlorosilanes undergoes a selective halogen exchange with $AlBr_3$ producing the corresponding bromomethylchlorosilane derivative[9].

$$3\ ClCH_2SiCl_3 + AlBr_3 \rightarrow 3\ BrCH_2SiCl_3 + AlCl_3 \quad \text{(d)}$$

$$3\ Cl_2CHSiCl_3 + 2\ AlBr_3 \rightarrow 3\ Br_2CHSiCl_3 + 2\ AlCl_3 \quad \text{(e)}$$

$$3\ ClCH_2SiMeCl_2 + AlBr_3 \rightarrow 3\ BrCH_2SiMeCl_2 + AlCl_3 \quad \text{(f)}$$

Organofluorosilanes undergo exchange with $AlCl_3$, $AlBr_3$ and AlI_3 producing the corresponding organohalosilane[10–12].

$$3\ Et_3SiF + AlX_3 \rightarrow 3\ Et_3SiX + AlF_3 \quad \text{(g)}$$

where X = Cl, Br, I. Silicon tetrafluoride undergoes a similar exchange with Al halides producing[4,13] mixed halofluorosilanes and the corresponding SiX_4:

$$SiF_4 + AlX_3 \rightarrow F_{4-y}SiX \quad \text{(h)}$$

where X = Cl (180–190°C), Br (260°C), I (280–300°C); y = 1–4;

$$SiF_4 + AlI_3 \xrightarrow{400°C} SiI_4 \quad \text{(i)}$$

The reaction prepares organohalosilanes where the corresponding organofluorosilane is easy to prepare. The exchange may be carried out in the absence of a solvent. However, where there is danger of disproportionation occurring, a solvent such as Et_2O or benzene is recommended to moderate the T. Catalytic effects of the Al halide limit the value of this procedure where high distillation T are involved. However, by removing any dissolved Al halide by adding an alkylfluorosilane of appropriate bp before fractionation reduces the possibility of catalyzed side reactions. The method's greatest preparative significance is the synthesis of dialkylchlorofluorosilanes from the corresponding difluoro compound:

$$Et_2SiF_2 \xrightarrow{AlCl_3} Et_2SiFCl \quad \text{(j)}$$

(C. H. VAN DYKE)

1. G. Lehmann, B. Lucke, *J. Prakt. Chem.*, *22*, 230 (1963).
2. H. S. Nutting, P. S. Petrie, U.S. Pat. 2,120,675 (1938); *Chem. Abstr.*, *32*, 5851 (1938).
3. M. Meslans, *C.R. Hebd. Seances Acad. Sci.*, *110*, 717 (1890).
4. W. C. Schumb, D. W. Breck, *J. Am. Chem. Soc.*, *74*, 1754 (1952).
5. A. L. Henne, M. S. Newman, *J. Am. Chem. Soc.*, *60*, 1697 (1938).
6. W. T. Miller, *J. Am. Chem. Soc.*, *62*, 993 (1960).
7. J. F. Durand, *Bull. Soc. Chim. Fr.*, *41*, 1251 (1977).
8. M. G. Voronkov, V. M. D'yakov, L. I. Gubanova, Yu. A. Lukina, USSR, Pat. 523,100, 1976; *Chem. Abstr.*, *85*, 177607 (1976).
9. M. G. Voronkov, V. M. D'yakov, L. I. Guvanova, *Akad. Nauk USSR Bulletin Div. Chem. Sci.*, *(Engl. Transl.)* 1576 (1975).
10. C. Eaborn, *J. Chem. Soc.*, 3077 (1950).
11. C. Eaborn, *J. Chem. Soc.*, 2755 (1949).
12. C. Eaborn, *J. Chem. Soc.*, 494 (1953).
13. M. Schmeisser, H. Jenkner, *Z. Naturforsch. Teil B*, *7*, 191 (1952).

2.5.12.2. by Metathesis Using Nonmetal Halides

2.5.12.2.1. Exchange with Halogens or Interhalogen Compounds.

The action of Cl_2 or Br_2 on organic halogen compounds can lead to the formation of carbon–Cl or carbon–Br bonds. Alkyl iodides at low T with Cl_2 form unstable iodochlorides that decompose to RCl; e.g., CH_3I reacts with Cl_2 to form CH_3ICl_2, which decomposes to CH_3Cl and ICl[1]:

$$CH_3I + Cl_2 \rightarrow CH_3ICl_2 \rightarrow CH_3Cl + ICl \qquad \text{(a)}$$

2-Iodooctane with Cl_2 or Br_2 at −78°C in light pet. ether results in the formation of 2-chloro- and 2-bromooctane, respectively[2]. Bromo- and iodobenzene undergo a photochemical interchange of halogens with Cl_2 or ICl as illustrated below[3–5]:

$$C_6H_5Br + 0.5Cl_2 \xrightarrow{h\nu} C_6H_5Cl + 0.5Br_2 \qquad \text{(b)}$$

$$C_6H_5I + ICl \xrightarrow{h\nu} C_6H_5Cl + I_2 \qquad \text{(c)}$$

$$C_6H_5Br + ICl \xrightarrow{h\nu} C_6H_5Cl + IBr \qquad \text{(d)}$$

$$C_6H_5I + Br_2 \xrightarrow{h\nu} C_6H_5Br + IBr \qquad \text{(e)}$$

Molecular iodine converts aromatic and heterocyclic Cl and Br to the corresponding iodide derivative. The method is not a direct halogen exchange but requires the conversion of the Br or Cl to an organometallic derivative, usually an organomagnesium halide reagent:

$$2\ RX + 2\ Mg \rightarrow 2\ [RMgX] \xrightarrow{I_2} 2\ RI + MgX_2 + MgI_2 \qquad \text{(f)}$$

Preparing iodobenzene[6] calls for adding an Et_2O solution prepared from Mg metal and C_6H_5Br dropwise to I_2 dissolved in xs Et_2O.

Silicon tetrahalides undergo exchange with halogens in which either complete or partial exchange occurs. Some of the reactions have preparative value. Silicon tetrabromide is produced[7] from SiI_4 with Br_2, although the reaction is not of preparative significance for $SiBr_4$. In some reactions, Al metal promotes the exchange by acting as a halogen-transfer agent. Thus, passing a mixture of SiF_4 and Br_2 over Al chips at

500–600°C produces $SiBr_4$ quantitatively[8]. At lower T only partial exchange is achieved; e.g., at 400°C, the SiF_4–Br_2 reaction produces mostly SiF_3Br (80%), together with SiF_2Br_2 and $SiFBr_3$. At 500–600°C, SiF_4 reacts with xs Cl_2 in the presence of Al metal to produce $SiCl_4$ and some mixed fluorochlorosilanes[8] (mostly SiF_3Cl):

$$SiF_4 + X_2 \xrightarrow{Al(500-600°C)} SiX_4 \tag{g}$$

where X = Cl, Br. The mixed chloro-bromides of Si are prepared[9] by allowing $SiBr_4$ to reflux in an atmosphere of Cl_2 for 48 h:

$$SiBr_4 + Cl_2 \rightarrow SiCl_3Br + SiCl_2Br_2 + SiClBr_3 \tag{h}$$

The mixture can be separated by fractional distillation. Higher molecular weight chlorosilanes and Si_3Cl_8 with Br_2 also produce mixed chlorobromosilanes. The reaction of $SiFCl_3$ with Cl_2 produces fluorochlorobromosilanes:

$$4\ SiFBr_3 + 3\ Cl_2 \rightarrow 2\ SiFClBr_2 + 2\ SiFCl_2Br + 3\ Br_2 \tag{i}$$

while SiF_2Br_2 with Cl_2 proceeds with complete Br–Cl exchange:

$$SiF_2Br_2 + Cl_2 \rightarrow SiF_2Cl_2 + Br_2 \tag{j}$$

The reaction of Me_2SiCl_2 with Br_2 in naphthalene produces $Me_2SiBrCl$[10]. The interhalogen I–Cl in $CFCl_3$ or CCl_4 is used to convert carbon–iodine bonds to carbon–chlorine bonds at bridgehead positions in certain cyclic compounds[11]. This interhalogen also reacts with organosilicon iodides to produce the corresponding chlorosilanes, e.g., Et_3SiI with ICl in CCl_4 produces[12] Et_3SiCl and I_2:

$$Et_3SiI + ICl \rightarrow Et_3SiCl + I_2 \tag{k}$$

The sterically hindered tris(trimethylsilyl)methyl silicon derivative, $(Me_3Si)_3CSiMe_2I$, trisyl, also reacts rapidly with ICl in CCl_4 to produce $(Me_3Si)_3CSiMe_2Cl$ quantitatively[13]:

$$(Me_3Si)_3CSiMe_2I + ICl \rightarrow (Me_3Si)_3CSiMe_2Cl + I_2 \tag{l}$$

However, under the same conditions other related, sterically hindered trisyl derivatives with ICl may produce rearranged Cl products rather than the corresponding Cl product[14], e.g., both $(Me_3Si)_3CSiPh_2I$ and $(Me_3Si)_3CSiMeClI$ with ICl give the rearrangement products $(Me_3Si)_2C(SiMe_2Cl)(SiPh_2Me)$ and $(Me_3Si)_2C(SiMe_2Cl)_2$, respectively. The compounds $(Me_3Si)_3CSiEt_2I$ and $(Me_3Si)_3CSiEtMeI$ react with ICl to produce ca. 25 and 35%, respectively, of the rearranged Cl, $(Me_3Si)_2C(SiMe_2Cl)(SiEt_2Me)$ and $(Me_3Si)_2C(SiMe_2Cl)(SiMe_2Et)$, together with unrearranged Cl, $(Me_3Si)_3CSiEt_2Cl$ and $(Me_3Si)_3CSiEtMeCl$.

(C. H. VAN DYKE)

1. J. Thiele, W. Peter, *Chem. Ber.*, *38*, 2842 (1905).
2. F. M. Beringer, H. S. Schultz, *J. Am. Chem. Soc.*, *77*, 5533 (1955).
3. B. Milligan, R. L. Bradow, J. E. Rose, H. E. Hubbert, A. Roe, *J. Phys. Chem.*, *84*, 158 (1962).
4. B. Milligan, R. L. Bradow, *J. Phys. Chem.*, *66*, 2118 (1962).
5. J. T. Echols, V. T-C. Chuang, C. S. Parrish, J. E. Rose, B. Milligan, *J. Am. Chem. Soc.*, *89*, 4081 (1967).
6. R. L. Datta, H. K. Mitter, *J. Am. Chem. Soc.*, *41*, 287 (1919).
7. E. Hengge, in *Halogen Chemistry*, Vol. 2, V. Gutmann, Academic Press, New York, 1967.

8. M. Schmeisser, H. Jenkner, Ger. Pat. 912,330 (1954); *Chem. Abstr.*, *52*, 12,342 (1958).
9. W. C. Schumb, H. H. Anderson, *J. Am. Chem. Soc.*, *59*, 651 (1937).
10. K. A. Andrianov, G. A. Kurakov, V. M. Kopylov, L. M. Khananashvili, *J. Gen. Chem. USSR* (*Engl. Transl.*), *34*, 1695 (1964).
11. J. C. Kauer, Abstr. 159th National Meeting of the American Chemical Society, Houston, TX, Feb. 22–27, 1970, No. B14.
12. D. R. Deans, C. Eaborn, *J. Chem. Soc.*, 3169 (1954).
13. S. D. Dua, C. Eaborn, D. A. R. Happer, S. P. Happer, K. D. Safa, D. R. M. Walton, *J. Organomet. Chem.*, *178*, 75 (1979).
14. C. Eaborn, S. P. Hopper, *J. Organomet. Chem.*, *192*, 27 (1980).

(C. H. VAN DYKE)

2.5.12.2.2. by Hydrogen Halides.

Hydrogen halides exchange with group-IVB halogen derivatives, particularly Si and Ge. Exchange may accompany cleavage in the preparation of dihalosilanes by the hydrogen halide cleavage of the Si—Ph bond of phenylhalosilanes, e.g.[1]:

$$PhSiH_2Cl + HBr \rightarrow PhH + SiH_2ClBr \quad (a)$$

$$PhSiH_2Cl + HBr \rightarrow PhSiH_2Br + HCl \quad (b)$$

Both HCl and HBr (but not HI) are involved in the exchange, e.g.:

$$PhSiH_2I + HCl \rightarrow PhSiH_2Cl + HI \quad (c)$$

$$PhSiH_2I + HBr \rightarrow PhSiH_2Br + HI \quad (d)$$

and the reaction may also occur with mixed dihalosilanes.

$$SiH_2BrI + HBr \rightarrow SiH_2Br_2 + HI \quad (e)$$

$$SiH_2ClBr + HBr \rightarrow SiH_2Br_2 + HCl \quad (f)$$

$$SiH_2ClBr + HCl \rightarrow SiH_2Cl_2 + HBr \quad (g)$$

Halogen exchange reactions that utilized hydrogen halides are used in synthesis of Si and Ge halides. Silicon tetrachloride reacts with HI to produce iodochlorosilanes[26]:

$$SiCl_4 + x\ HI \rightarrow SiCl_{4-x}I_x + x\ HCl \quad (h)$$

and $SiICl_3$ reacts with HI to yield[3] SiI_2Cl_2:

$$SiICl_3 + HI \rightarrow SiI_2Cl_2 + HCl \quad (i)$$

The reaction of Me_3SiCl and Me_2SiCl_2 with HBr in the presence of an Fe or $FeCl_3$ catalyst produces Me_3SiBr and $Me_2SiClBr$ (40%), respectively[4]. No reaction occurs without the catalyst[5].

Chlorogermanes can be converted to the corresponding bromogermanes by their reaction[6,7] with HBr:

$$R_{4-x}GeCl_x + x\ HBr \rightarrow R_{4-x}GeBr_x + x\ HCl \quad (j)$$

where x = 1–4;

$$HGeCl_3 \xrightarrow[5\ h]{HBr} HGeBr_3 + GeBr_4 \quad (k)$$

A method for GeH_3Br and GeH_3I is based on the reaction of the available GeH_3Cl with HBr and HI, respectively[8]. Bromogermane also can be converted to GeH_3I in 70% yield by HI[8]:

$$GeH_3Cl + HX \rightarrow GeH_3X + HCl \quad (l)$$

where X = Br, I;

$$GeH_3Br + HI \rightarrow GeH_3I + HBr \quad (m)$$

Vinyltrichloro- and allyltrichlorogermane undergo halogen exchange with HBr without addition of the HBr to the double bond[9]:

$$CH_2{=}CHGeCl_3 + 3\ HBr \rightarrow CH_2{=}CHGeBr_3 + 3\ HCl \quad (n)$$

$$CH_2{=}CHCH_2GeCl_3 + 3\ HBr \rightarrow CH_2{=}CHCH_2GeBr_3 + 3\ HCl \quad (o)$$

As a comparison, HBr adds exclusively to the double bond of $CH_2{=}CHCH_2SiCl_3$:

$$CH_2{=}CHCH_2SiCl_3 + HBr \rightarrow BrCH_2CH_2CH_2SiCl_3 \quad (p)$$

Addition and exchange occur in the reaction of HBr with methallyltrichlorogermane:

$$CH_2{=}C(CH_3)CH_2GeCl_3 + 4\ HBr \rightarrow (CH_3)_2C(Br)CH_2GeBr_3 + 3\ HCl \quad (q)$$

Procedures for carrying out halogen exchange between Ge halides and hydrogen halides may not involve direct halogen exchange. It may be more convenient to hydrolyze the starting Ge halide and treat the product with the desired hydrogen halide. This procedure is used for Ph_2GeCl_2 from Ph_2GeBr_2 and Ph_2GeBr_2 and Ph_2GeF_2 from Ph_2GeCl_2[10]:

$$Ph_2GeBr_2 \xrightarrow{H_2O/HCl} Ph_2GeCl_2 \quad (r)$$

$$Ph_2GeCl_2 \xrightarrow{H_2O/HX} Ph_2GeX_2 \quad (s)$$

where X = Br, F.

(C. H. VAN DYKE)

1. G. Fritz, D. Kummer, *Z. Anorg. Allg. Chem.*, *310*, 327 (1961).
2. A. Besson, *C. R. Hebd. Seances Acad. Sci.*, *112*, 611, 1314 (1891).
3. R. West, E. G. Rochow, *Inorg. Synth.*, *4*, 41 (1953).
4. K. A. Andrianov, G. A. Kurakov, V. M. Kopylov, L. M. Khananashvili, *J. Gen. Chem. USSR (Engl. Trans.)*, *34*, 1695 (1964).
5. P. A. McCusker, E. L. Reilly, *J. Am. Chem. Soc.*, *75*, 1583 (1953).
6. V. F. Mironov, N. G. Dzhurinskaya, A. D. Petrov, *Bull. Acad. Sci. USSR Div. Chem. Sci*, 1956 (1961).
7. V. F. Mirinov, T. K. Gar, *Bull. Acad. Sci. USSR Div. Chem. Sci*, 740 (1965).
8. S. Cradock, E. A. V. Ebsworth, *J. Chem. Soc.*, *A*, 12 (1967).
9. V. F. Mironov, *Izv. Akad. Nauk SSSR*, 1826 (1959); *Chem. Abstr.*, *54*, 8607 (1960).
10. C. A. Kraus, C. L. Brown, *J. Am. Chem. Soc.*, *52*, 3690 (1930).

2.5.12.2.3. by Other Nonmetal Halides Including Exchange Redistribution.

Covalent halides undergo exchange with group-IV halides (for fluorine exchange reactions, see §2.5.12.3.). In carbon–halogen systems, the reactions often involve exchanging bromine or iodine for Cl in alkyl halide redistributions:

$$RCl + R'X \xrightarrow{AlCl_3} RX + R'Cl \quad (a)$$

where X = Br, I. The equilibrium is catalyzed by $AlCl_3$ to give high yields. Thus, CI_4 can be prepared[1] by the equilibration of CCl_4 and CHI_3, while HCI_3 and $HCBr_3$ result from the equilibration of $HCCl_3$ with EtI and EtBr, respectively[1,2]. Mixed chlorobromoalkanes can also be prepared[3,4] by this redistribution:

$$CHCl_3 + CHBr_3 \xrightarrow{AlCl_3} CHCl_2Br + CHClBr_2 \quad (b)$$

$$ClCH_2CH_2Cl + BrCH_2CH_2Br \xrightarrow{AlCl_3} 2\ ClCH_2CH_2Br \quad (c)$$

$$CCl_4 + CBr_4 \xrightarrow{AlCl_3} CCl_3Br + CCl_2Br_2 + CClBr_3 \quad (d)$$

A simple, fast, efficient and important procedure for halogen (Cl, Br, I) exchanges at bridgeheads of certain ring systems involves treating the bridgehead halide with an Al halide catalyst generated in situ by reaction of xs Al with Br_2 in a halogenated solvent[5]. Under these conditions, the exchange occurs with the halogenated solvent. Thus, 1-adamantyl bromide is converted to 1-adamantyl chloride (89%) of iodide (83%) by its reaction with the solvents $CHCl_3$ or CH_2I_2, respectively, in the presence of the Al halide catalyst generated in situ. Similarly, tetrabromadamantane is converted to the tetraiodo analogue by its catalyzed reaction with a CH_2I_2 solvent. Whereas the exchange of halogen between methyl- and Al halides proceeds only in the order[6]:

$$MeCl \xrightarrow{AlBr_3} MeBr \xrightarrow{AlI_3} CH_3I \quad (e)$$

the reaction promoted by the in situ generated Al halide catalyst does not have this restriction.

Tri-n-butyltin halides and certain alkyl halides undergo halogen exchange reactions; e.g., heating 1:1 $C_6H_5CH_2Br$ with n-Bu_3SnCl neat at 50°C in the presence of a R_4NBr catalyst proceeds[7,8] to an equilibrium containing 64% $C_6H_5CH_2Cl$:

$$n\text{-}Bu_3SnCl + C_6H_5CH_2Br \rightleftharpoons n\text{-}Bu_3SnBr + C_6H_5CH_2Cl \quad (f)$$

Benzyl bromide and iodide also undergo[9] similar catalyzed exchanges with R_3ECl where E = Si, Ge, Sn and Pb. The R_4MX catalyst is not required for Pb exchanges:

$$R_3ECl + C_6H_5CH_2X \rightleftharpoons R_3EX + C_6H_5CH_2Cl \quad (g)$$

where E = Si, Ge, Sn, Pb; X = Br, I. Tri-n-butyltin bromide or iodide and di-n-butyltin dibromide or diiodide are prepared[10] from the corresponding organotin chlorides by halogen redistribution with alkyl bromides or iodides. Tin tetrabromide takes part in a Br–Cl exchange[11] with $BrCl_2CSnBr_3$:

$$Br{-}\underset{\underset{Cl}{|}}{\overset{\overset{Cl}{|}}{C}}{-}SnBr_3 + SnBr_4 \rightarrow Br{-}\underset{\underset{Cl}{|}}{\overset{\overset{Br}{|}}{C}}{-}SnBr_3 + SnBr_3Cl \quad (h)$$

Primary, secondary, or tertiary alkyl fluorides and tertiary alkyl chlorides react with Me_3SiI in an inert solvent such as CH_2Cl_2 to produce[12,13] the corresponding alkyl iodides in high yields.

$$RF + Me_3SiI \rightarrow RI + Me_3SiF \quad (i)$$

Reagents that generate Me_3SiI in situ such as Me_3SiCl-NaI-CH_3CN and Me_6Si_2-I_2 can also be used to accomplish the fluoride-iodide halogen exchanges[12]. Alkyl chlorides and bromides react with Et_3SiI under both photochemical and thermal conditions to form[14] the corresponding alkyl iodide, although the reactions are slow and most of the yields low:

$$RX + Et_3SiI \xrightarrow{h\nu \text{ or } \Delta} RI + Et_3SiX \qquad (j)$$

where RX = n-$C_5H_{11}Cl$, i-C_3H_7Cl, t-C_4H_9Cl, CH_2Cl_2, $HCCl_3$, CCl_4.

Organosilicon and -germanium compounds undergo exchanges with many covalent halides. The reactions are not limited to halide–halide exchanges but include other organosilicon and -germaniums as well. The order of exchange is governed by distilling the most volatile compound from mixture. The reactants are refluxed, and the most volatile component is distilled out of the mixture. Thus, n-$PrSiCl_3$ can be obtained[15] in 82% yield by its distillation from a refluxing mixture of n-$PrSiI_3$ and $PhPCl_2$. Chlorine transfers from phosphorus to Si or Ge in the reaction of $SiBr_4$ or $GeBr_4$ with $PhPCl_2$:

$$SiBr_4 + 2\ PhPCl_2 \rightarrow SiCl_4 + 2\ PhPBr_2 \qquad (k)$$

$$GeBr_4 + 2\ PhPCl_2 \rightarrow GeCl_4 + 2\ PhPBr_2 \qquad (l)$$

Under distillation conditions, the reaction of $SbCl_3$ with proper proportions of $SiBr_4$ can be used to prepare any one of the three chlorobromides, $SiCl_3Br$, $SiCl_2Br_2$ or $SiClBr_3$ in 80% yield[16]. The reaction of $SbCl_3$ with $SiFBr_3$ yields[16] the fluorochlorobromosilanes $SiFClBr_2$ and $SiFCl_2Br$[16].

Mixtures of inorganic and organic Si halides containing different halogen substituents undergo exchanges that have preparative significance as well as theoretical interest; e.g., organofluorosilanes can be converted to the corresponding chloro-, bromo- or iodosilane by a Si–halide exchange (redistribution) that employs a tertiary amine, phosphine or silylamine catalyst[17]:

$$2\ Me_3SiCl + MePhSiF_2 \rightarrow Me_3SiF + MePhSiCl_2 \qquad (m)$$

$$SiBr_4 + 2\ MePhSiF_2 \rightarrow SiF_4 + 2\ MePhSiBr_2 \qquad (n)$$

$$SiI_4 + 2\ MePhSiF_2 \rightarrow SiF_4 + 2\ MePhSiI_2 \qquad (o)$$

The volatile fluoride derivative is removed as it is formed in the reaction. Halide–halide exchange occurs when Me_3SiCl and 1-bromo-1,2-dimethyl-1-silacyclobutane are combined in the presence of hexamethylphosphoramide (HMPA). No exchange occurs[18] without the HMPA:

$$\text{1-bromo-1,2-dimethylsilacyclobutane (Si–Br)} + Me_3SiCl \xrightarrow{\text{HMPA-}CCl_4} \text{1-chloro-1,2-dimethylsilacyclobutane (Si–Cl)} + Me_3SiBr \qquad (p)$$

Redistributions involving mixtures of halosilanes prepare derivatives that contain more than one type of halogen on Si. Thus, methyliodochlorosilanes can be prepared[19] by heating a mixture of $MeSiCl_3$ and $MeSiI_3$ at 300°C for 20 h and ethylchlorobromosilanes are produced by heating a mixture of $EtSiCl_3$ and $EtSiBr_3$ at 200°C with $AlCl_3$ present[20].

$$MeSiCl_3 + MeSiI_3 \rightarrow MeSiCl_2I + MeSiClI_2 \tag{q}$$

$$EtSiCl_3 + EtSiBr_3 \xrightarrow{AlCl_3} EtSiCl_2Br + EtSiClBr_2 \tag{r}$$

Mixed chloroiodo derivatives result[19] on heating dichloro- and diiodoorganosilanes at 400–450°C:

$$R_2SiCl_2 + R_2SiI_2 \rightleftharpoons 2\ R_2SiClI \tag{s}$$

where R = Me, Et. Heating mixtures of $SiCl_4$ and $SiBr_4$ to ca. 600°C without a catalyst produces[21] the silicon chlorobromides, $SiCl_3Br$, $SiCl_2Br_2$ and $SiClBr_3$. Likewise, fluorobromo- and fluoroiodosilanes result[22] from heating mixtures of SiF_4 with $SiBr_4$ or SiI_4. The halogen exchange that $HSiI_3$ undergoes with bromosilanes is used to prepare[23,24] mixed halosilanes, including one that has four different halogens attached to Si:

$$SiFClBr_2 + HSiI_3 \rightarrow SiFClBrI + HSiBrI_2 \tag{t}$$

$$SiFBr_3 + HSiI_3 \rightarrow SiFBr_2I + SiFBrI_2 + SiFI_3 \tag{u}$$

$$SiF_2Br_2 + HSiI_3 \rightarrow SiF_2BrI + SiF_2I_2 \tag{v}$$

Mixed Ge halides compounds also form in the equilibration of Ge halides. Thus, the mixed dihalide, GeH_2BrI, was detected[25] in the ^{1}H-NMR spectrum of 1:1 GeH_2Br_2 and GeH_2I_2.

Halide–halide exchange occurs[26] at asymmetric Si in optically active α-naphthylphenylmethylhalosilanes (R_3Si^*X); e.g., the exchange between R_3Si^*Cl and c-$C_6H_{11}NH_3F$ in $HCCl_3$ produces R_3Si^*F with at least 90% inversion of configuration, while the reaction of R_3Si^*Br with c-$C_6H_{11}NH_3Cl$ proceeds with predominant inversion of configuration and a minimum stereospecificity of 80%.

Halogen exchange involving covalent halides is important in preparing certain halotin compounds, e.g., a convenient preparation of Br_3Sn– and I_3Sn– complexes of the transition metals is based on the reaction of the corresponding trichlorotin compound with a 10:1 molar excess of anhyd tin(II)bromide or -iodide in MeOH or acetone[27]:

$$2\ h^5\text{-}C_5H_5Fe(CO)_2SnCl_3 + 3\ SnX_2 \rightleftharpoons 2\ h^5\text{-}C_5H_5Fe(CO)_2SnX_3 + 3\ SnCl_2 \tag{w}$$

where X = Br, I. The Sn(II) halide reactions are reversible, so it is possible to obtain[27] h^5-$C_5H_5Fe(CO)_2SnCl_3$ by xs $SnCl_2$ on h^5-$C_5H_5Fe(CO)_2SnI_3$. Trioganotin halides undergo exchange[28] with R_3SiX:

$$R_3SnX + Me_3SiY \rightarrow Me_3SiX + R_3SnY \tag{x}$$

where R = Me, n-Bu; X, Y = Cl, Br or I with Y heavier than X;

$$R_3SnF + Et_3SiX \rightarrow R_3SnX + Et_3SiF \tag{y}$$

where R = n-Bu, Ph; X = Cl, Br, I.

Organotin halides that contain more than one type of halogen are prepared by exchange[28,29]:

$$R_2SnX_2 + R_2SnY_2 \rightarrow 2\ R_2SnXY \quad \text{(z)}$$

$$R_2SnX_2 + Me_3SiY \rightarrow R_2SnXY + Me_3SiX \quad \text{(aa)}$$

where R = Me, Et, n-Bu; X, Y = ClBr, ClI, BrI.

NMR studies of halogen–halogen exchange equilibria between methyl Si, Ge and phosphorus halides show that the redistributions are not random; e.g., between dimethyl and methyl moieties, the lower at wt halogen at equilibrium associates with the dimethylSi grouping[30]. Other exchanges include those of halogens between MeSi and MeGe[31], Me_2Si and Me_2Ge[32,33]. MeP and Me_2Ge[34], MeP and Me_2Si[35] and MeP and MeGe[36,37]. Multicomponent scrambling equilibria in systems composed of $MeGeCl_3$, $MeGeBr_3$ and $MeGeI_3$ are also investigated[38].

Group-IVB compounds that contain mixed halogen substituents also undergo redistribution; e.g., although pure $MeSiBrCl_2$ can be refluxed at its boiling point (86.5°) without decomposition, it undergoes disproportionation to form $MeSiCl_3$ and $MeSiBr_3$ in the presence of $AlCl_3$, other metal halides or in basic solution[39].

$$3\ MeSiCl_2Br \xrightarrow{AlCl_3} 2\ MeSiCl_3 + MeSiBr_3 \quad \text{(ab)}$$

Other halogen redistributions include:

$$2\ PhSiHClBr \rightarrow PhSiHCl_2 + PhSiHBr_2 \quad \text{(ac)}^{40}$$

$$2\ SiHBrCl \rightarrow SiH_2Br_2 + SiH_2Cl_2 \quad \text{(ad)}^{40}$$

$$h\text{-}C_5H_5Fe(CO)_2SnBr_2Cl \xrightarrow{\text{refluxing THF}} n\text{-}C_5H_5Fe(CO)_2SnBr_3 \quad \text{(ae)}^{41}$$

Reviews on redistribution of the group-IV elements are available[42–45].

(C. H. VAN DYKE)

1. H. Soroos, J. B. Hinkamp, *J. Am. Chem. Soc., 67*, 1642 (1945).
2. J. W. Walker, *J. Chem. Soc., 85*, 1082 (1904).
3. M. S. Kharash, B. M. Kuderna, W. Urry, *J. Org. Chem., 13*, 895 (1948).
4. G. Calingaert, *J. Am. Chem. Soc., 62*, 1545 (1940).
5. J. W. McKinley, R. E. Pincock, W. S. Bruce, *J. Am. Chem. Soc., 95*, 2030 (1973).
6. H. C. Brown, W. J. Wallace, *J. Am. Chem. Soc., 75*, 6279 (1953).
7. E. C. Friedrich, P. F. Vartanian, R. L. Holmstead, *J. Organomet. Chem., 102*, 41 (1975).
8. E. C. Friedrich, P. F. Vartanian, *J. Organomet. Chem. 110*, 159 (1976).
9. E. C. Friedrich, C. B. Abma, P. F. Vartanian, *J. Organomet. Chem. 187*, 203 (1980).
10. E. C. Friedrich, C. B. Abma, G. Delucca, *J. Organomet. Chem., 228*, 217 (1982).
11. M. Weidenbruch, C. Pierrand, *J. Organomet. Chem. 71*, C29 (1974).
12. G. A. Olah, S. C. Narang, L. D. Field, *J. Org. Chem., 46*, 3727 (1981).
13. E. C. Friedrich, G. Delucca, *J. Organomet. Chem., 226*, 143 (1982).
14. Y. Nagai, H. Muramatsu, M.-A. Ohtsuki, H. Matsumoto, *J. Organomet. Chem., 17*, P19 (1969).
15. H. H. Anderson, *J. Am. Chem. Soc. 75*, 1576 (1953).
16. W. C. Schumb, H. H. Anderson, *J. Am. Chem. Soc., 59*, 651 (1937).
17. B. Kanner, D. L. Bailey, U.S. Pat. 3,128,297 (1964); *Chem. Abstr., 61*, 8340 (1964).
18. B. G. McKinnie, F. K. Cartledge, *J. Organomet. Chem., 104*, 407 (1976).
19. H. H. Anderson, *J. Am. Chem. Soc., 73*, 5804 (1951).
20. M. Kumada, *J. Chem. Soc. Jpn. (Ind. Chem. Sect.), 55*, 373 (1952); *Chem. Abstr., 48*, 10,543 (1954).

21. G. S. Forbes, H. H. Anderson, *J. Am. Chem. Soc.*, *66*, 931 (1944).
22. H. H. Anderson, *J. Am. Chem. Soc.*, *72*, 2091 (1950).
23. F. Höfler, W. Veigl, *Angew. Chem., Int. Ed., Engl.*, *10*, 919 (1971).
24. F. Höfler, H. D. Pletka, *Monatsh. Chem.*, *104*, 1 (1973).
25. S. Cradock, E. A. V. Ebsworth, *J. Chem. Soc., A*, 1226 (1967).
26. L. H. Sommer, F. O. Stark, K. W. Michael, *J. Am. Chem. Soc.*, *86*, 5683 (1964).
27. M. J. Mays, S. M. Pearson, *J. Chem. Soc., A*, 136 (1969).
28. D. A. Armitage, A. Tarassoli, *Inorg. Chem.*, *14*, 1210 (1975).
29. D. L. Alleston, A. G. Davies, *J. Chem. Soc.*, 2050 (1962).
30. K. Moedritzer, *Inorg. Chim. Acta*, *10*, 163 (1969).
31. K. Moedritzer, J. R. Van Wazer, *Inorg. Chem.*, *5*, 547 (1966).
32. J. R. Van Wazer, K. Moedritzer, L. C. D. Groenweghe, *J. Organomet. Chem.*, *5*, 420 (1966).
33. K. Moedritzer, J. R. Van Wazer, *J. Inorg. Nucl. Chem.*, *28*, 957 (1966).
34. K. Moedritzer, *Inorg. Chim. Acta*, *17*, 205 (1976).
35. K. Moedritzer, J. R. Van Wazer, *Inorg. Chem.*, *12*, 2856 (1973).
36. K. Moedritzer, J. R. Van Wazer, *Rev. Chim. Miner.*, *6*, 293 (1969).
37. K. Moedritzer, *J. Inorg. Nucl. Chem.*, *32*, 2529 (1970).
38. K. Moedritzer, J. R. Van Wazer, R. E. Muller, *Inorg. Chem.*, *7*, 1638 (1968).
39. R. S. Feinberg, E. G. Rochow, *J. Inorg. Nucl. Chem.*, *24*, 165 (1962).
40. G. Fritz, D. Kummer, *Z. Anorg. Allg. Chem.*, *310*, 327 (1961).
41. B. O'Dwyer, A. R. Manning, *Inorg. Chem. Acta*, *38*, 103 (1980).
42. J. C. Lockhart, *Chem. Rev.*, *65*, 131 (1965).
43. V. Chvalovský, *Organomet. React.*, *3*, 191 (1972).
44. K. Moedritzer, *Organomet. Chem. Rev.*, *1*, 179 (1966); *Adv. Organometal. Chem.*, *6*, 171.
45. D. R. Weyenberg, L. G. Mahone, W. H. Atwell, *Ann. N.Y. Acad. Sci.*, *159*, 38 (1969).

2.5.12.3. by Fluorinating Agents

Halogen exchange is used to prepare fluorine derivatives:

$$R{-}X + M{-}F \rightarrow R{-}F + M{-}X \tag{a}$$

where X = Cl, Br. Hydrogen fluoride and fluorine derivatives of metals such as Sb, Hg, Ag, Zn and K are used as exchange reagents. Although iodine is more easily replaced than Br or Cl, the latter two are employed because of their availability and the lack of side reactions[1]. Saturated organic halides exchange less readily than systems where the halogen is part of a carbonyl or sulfonyl halide or in the α position to a carbonyl or carboxyl group or to a C=C bond (allyl position). Vinyl halides generally do not undergo halogen-fluorine exchange. In aromatic systems exchange is only possible when an o- or p-NO_2 substituent is present.

(C. H. VAN DYKE)

1. C. Weygan, G. Hilgetag, in *Préparative Organic Chemistry*, G. Hilgetag, A. Martins, eds. Wiley-Interscience.

2.5.12.3.1. That Utilize SbF_3 or Antimony(V) Halides.

An old but still important method of preparing fluorine derivatives of carbon, Si and Ge is based on the halogen–fluorine exchange, which occurs when halides of these elements are treated with SbF_3 or a suitable modification thereof[1,2]. Antimony trifluoride itself is not effective; however, its activity can be increased significantly by having a pentavalent antimony compound present as a catalyst. Thus, combinations of SbF_3 with 2–5% $SbCl_5$, ca. 1% of Cl_2 or ca. 5% Br_2 are used[3]. The active fluorine exchange reagent in these cases is an antimony(V) halofluoride that is formed; e.g., for the $SbCl_5$, Cl_2 and Br_2 catalysts mentioned:

$$SbF_3 + SbCl_5 \rightarrow SbF_3Cl_2 + SbCl_3 \quad (a)$$

$$SbF_3 + X_2 \rightarrow SbF_3X_2 \quad (b)$$

where $X_2 = Cl_2$, Br_2. The fluorine–halogen exchange that occurs between the antimony(V) derivative and the group IV halide produces the corresponding group IV fluoride and regenerates the catalyst ($SbCl_5$) for further reaction:

$$3\ SiCl_4 + SbF_3Cl_2 \rightarrow 3\ SiCl_3F + SbCl_5 \quad (c)$$

Anhydrous SbF_3 is the usual fluorinating agent in these conversions, although aq SbF_3 can convert carbon–chlorine bonds to carbon–fluorine bonds[4].

In carbon–halogen exchanges by SbF_3/pentavalent Sb fluorination, the $—CCl_3$ group is the most reactive, producing $—CCl_2F$ and $—CClF_2$ groups, but rarely the $—CF_3$ group[2,3]. The $—CHCl_2$ groups undergo exchange slowly to produce the —CHClF group, and with greater difficulty the $—CHF_2$ group. The $—CH_2Cl$ and —CH(Cl)— groups generally are not affected by the treatment. A similar pattern exists for bromocarbon systems[5].

Antimony(III) fluoride with or without a catalyst converts Si and Ge halides to the corresponding fluoro derivatives:

$$3\ Me_3SiCl + SbF_3 \rightarrow Me_3SiF + SbCl_3 \quad (d)^{6}$$

$$CF_3SiCl_3 + SbF_3 \rightarrow CF_3SiF_3 + SbCl_3 \quad (e)^{7}$$

$$3\ (Me_3Si)_3SiBr + SbF_3 \xrightarrow{C_6H_6/CH_3CN} (Me_3Si)_3SiF + SbBr_3 \quad (f)^{8}$$

$$3\ Me_3GeX + SbF_3 \rightarrow 3\ Me_3GeF + SbX_3 \quad (g)^{9}$$

where X = Cl, Br;

$$3\ (n\text{-}Bu)_2GeI_2 + SbF_3 \rightarrow 3\ (n\text{-}Bu)_2GeF_2 + 2\ SbI_3 \quad (h)^{10}$$

The reaction is catalyzed by $SbCl_5$ and, in the Si systems, is inhibited by BF_3[11] and $AlCl_3$[12]. The ease of fluorination depends on the compound undergoing exchange, e.g., treatment of a mixture of i-Pr_2GeBr_2 and i-Pr_3GeBr with SbF_3 results in fluorine exchange involving the dibromo derivative, while a significant portion of the monobromo derivative can be recovered unchanged[13].

Partially fluorinated products are obtained on treating polychlorosilanes with SbF_3, although better yields of such products are obtained if other metal fluorides (e.g., CaF_2) are used[14]. In the SbF_3–Si halide exchange systems, the rate of fluorination increases as the fluorine content of the halide increases, thus the chlorofluoride in which all but one of the Cl atoms is replaced is obtained in smallest yield, e.g., the average yields of fluorinated products of i-$PrSiCl_3$ are i-$PrSiF_3$ (70–75%), i-$PrSiF_2Cl$ (8–10%) and i-$PrSiFCl_2$ (17–20%)[15]. Silicon tetrachloride and $GeCl_4$ react with a deficit of SbF_3 in the presence of $SbCl_5$ to produce mixtures of chlorofluorosilanes and germanes, respectively, from which the mixed halides can be isolated[16,17]. The Br of $SiCl_2Br_2$ can be selectively exchanged[18] by the fluorine of SbF_3:

$$3\ SiCl_2Br_2 + SbF_3 \rightarrow 3\ SiCl_2BrF + SbBr_3 \quad (i)$$

Complete fluorination of the methylchlorosilanes Me_nSiCl_{4-n}(n = 1–3) using SbF_3 is possible[19]. The reaction is general in organosilicon chemistry and is used in the synthesis of organic and inorganic fluorosilanes including fluorine derivatives of polysilanes[20–22]. In derivatives that contain both carbon–halogen and Si–halogen bonds, SbF_3 treatment results in fluorine substitution[23] at Si, although both carbon–halogen and Si–halogen bonds can undergo substitution:

$$Cl(CH_2)_3SiCl_3 \xrightarrow{SbCl_3} Cl(CH_2)_3SiF_3 \qquad (j)^{23}$$

$$Cl_3SiCH_{3-n}Cl_n \xrightarrow{SbF_3} F_3SiCH_{3-n}Cl_n \qquad (k)^{24}$$

where n = 1, 2;

$$Cl_3CC_6H_4SiCl_3 \xrightarrow{SbF_3\text{-}SbCl_5} F_3CC_6H_4SiF_3 \qquad (l)^{25}$$

Silicon systems fluorinated by this method are listed in Ref. 26.

Graphite intercalated SbF_5 is used as a mild fluorinating agent for organic and Si chlorides[27,28]:

$$Me_2PhSiCl \xrightarrow[10\ h;\ 25°C]{SbF_5-C} \underset{(85\%)}{Me_2PhSiF} \qquad (m)$$

The system cannot fluorinate Ge–X (Br, I) bonds[28].

(C. H. VAN DYKE)

1. S. Swarts, *Bull. Acad. R. Med. Belg., New York* (1972); *24*, 309, 474 (1892); *35*, 1533 (1924).
2. A. L. Henne, *Org. React.*, *2*, 49 (1944).
3. C. Weygan, G. Hilgetag, in *Preparative Organic Chemistry*, G. Hilgetag, A. Martins, eds., Wiley-Interscience, New York, 1972.
4. R. Muller, C. Dathe, *Z. Anorg. Allg. Chem.*, *330*, 195 (1964).
5. A. M. Lovelace, D. A. Rausch, W. Postelnek, *Aliphatic Fluorine Compounds*, Reinhold, New York, 1958.
6. H. S. Booth, J. F. Suttle, *J. Am. Chem. Soc.*, *68*, 2658 (1946).
7. H. Beckers, H. Burger, P. Bursch, I. Ruppert, *J. Organomet. Chem. 316*, 41 (1986).
8. H. Burger, W. Kilian, K. Burczyk, *J. Organomet. Chem.*, *21*, 291 (1970).
9. V. A. Ponomarenko, G. Ya. Vzenkova, *Izv. Akad. Nauk SSSR, Otd. Khim. Nauk*, 994.
10. H. H. Anderson, *J. Am. Chem. Soc.*, *83*, 547 (1961).
11. R. Müller, C. Dathe, *J. Prakt. Chem.*, *13*, 306 (1961).
12. R. Müller, C. Dathe, *Z. Anorg. Allg. Chem.*, *313*, 207 (1961).
13. H. H. Anderson, *J. Am. Chem. Soc.*, *75*, 814 (1953).
14. H. S. Booth, W. F. Martin, *J. Am. Chem. Soc.*, *68*, 2655 (1946).
15. H. S. Booth, D. R. Spessard, *J. Am. Chem. Soc.*, *68*, 2660 (1946).
16. H. S. Booth, C. F. Swinehart, *J. Am. Chem. Soc.*, *57*, 1333 (1935).
17. H. S. Booth, W. C. Morris, *J. Am. Chem. Soc.*, *58*, 90 (1936).
18. W. C. Schumb, H. H. Anderson, *J. Am. Chem. Soc.*, *59*, 651 (1937).
19. A. P. Hagen, L. L. McAmis, *Inorg. Synth.*, *16*, 139 (1976).
20. J. F. Bald, K. G. Sharp, A. G. MacDiarmid, *J. Fluorine Chem.*, *3*, 433 (1973).
21. J. E. Drake, N. Goddard, *J. Chem. Soc.*, *A*, 2587 (1970).
22. J. E. Drake, N. Goddard, N. P. C. Westwood, *J. Chem. Soc.*, *A*, 3305 (1971).
23. V. B. Puchnarevic, J. Vcelak, M. G. Voronkov, V. Shualovsky, *Coll. Czech. Chem. Commun.*, *39*, 2616 (1974).
24. R. Müller, S. Raichel, C. Dathe, *Chem. Ber.*, *97*, 1673 (1964).
25. L. W. Frost, U.S. Pat. 2,636,896 (1953); *Chem. Abstr.*, *48*, 4002 (1954).

26. C. H. Van Dyke, in *Organometallic Compounds of the Group IV Elements*, A. G. MacDiarmid, ed., Vol. 2, Part 1, Marcel Dekker, New York, 1972.
27. J. M. Lalancette, J. Lafontaine, *J. Chem. Soc., Chem. Commun.*, 815 (1973).
28. R. J. P. Corriu, J. M. Fernandez, G. Guerin, *J. Organomet. Chem.*, *192*, 347 (1980).

2.5.12.3.2. by HF.

Hydrogen fluoride alone undergoes halogen–fluorine exchange with only the reactive acyl or allyl halides. However, it converts organic halides to the fluoride if small amounts of SbF_3 and $SbCl_5$ are added[1,2]. The catalytic active Sb cleavage reagent is continually regenerated since the $SbCl_3$ formed reacts with HF to form SbF_3 and HCl:

$$3\ CCl_4 + 2\ SbF_3 \rightarrow 3\ CCl_2F_2 + 2\ SbCl_3 \qquad (a)$$

$$3\ HF + SbCl_3 \rightarrow SbF_3 + 3\ HCl \qquad (b)$$

Anhydrous, dil, conc or aq ROH HF convert halosilanes to the corresponding fluorosilanes[3,4], e.g., c-$C_6H_{11}SiF_3$ is produced[5] in 70% yield by shaking the corresponding trichloro derivative with 48% aq HF for 30 min at 65°C. Organohalosilanes undergo this conversion, including derivatives with Si—H and Si—Ph bonds:

$$Ph_nSiHCl_{3-n} \xrightarrow{\text{conc HF, } -30°\text{ to } -50°\text{C}} \underset{(70\text{–}90\%)}{Ph_nSiHF_{3-n}} \qquad (c)^6$$

However, the presence of HF may result in cleaving the Si—Ph bond in conversions involving phenylhalosilanes[7].

Aqueous HF is a very good reagent for the facile synthesis of trialkyltin(IV) fluorides and dialkyltin(IV)difluoride[8] from the corresponding chlorides.

$$R_2SnCl_2 + 2\ HF \rightarrow R_2SnF_2 + 2\ HCl \qquad (d)$$

$$(R = Me, Et, n\text{-}C_3H_7, n\text{-}C_4H_9, n\text{-}C_8H_{17})$$

It is also used to prepare[8] monofluoromethyltin chlorides from Me_2SnCl_2 and $MeSnCl_3$:

$$Me_2SnCl_2 + CHF \rightarrow Me_2SnClF + HCl \qquad (e)$$

$$MeSnCl_3 + HF \rightarrow MeSnCl_2F + HCl \qquad (f)$$

An attempt to prepare $SnCl_3F$ by reacting excess $SnCl_4$ with HF resulted[8] in the sole formation of "$SnCl_4 \cdot SnF_4$."

(C. H. VAN DYKE)

1. C. Wegan, G. Hilgetag, in *Preparative Organic Chemistry*, G. Hilgetag, A. Martins, eds., Wiley-Interscience, New York, 1972.
2. A. L. Henne, *Org. React.*, *2*, 49 (1944).
3. R. J. H. Voorhoeve, *Organohalosilanes-Precursors to Silicones*, American Elsevier, New York, 1967.
4. C. Eaborn, *Organosilicon Compounds*, Butterworths, London, 1960.
5. H. H. Anderson, *J. Am. Chem. Soc.*, *81*, 4785 (1959).
6. E. A. Chernyshev, M. E. Dolgaya, A. D. Petrov, *Akad. Nauk SSSR Bulletin, Div. Chem. Sci.* (*Engl. Transl.*) 1604 (1961).
7. G. Fritz, D. Kummer, *Z. Anorg. Allg. Chem.*, *308*, 105 (1961).
8. L. E. Levchuk, J. R. Sams, F. Aubke, *Inorg. Chem.*, *11*, 43 (1972).

2.5.12.3.3. by Other Metal–Fluorine Derivatives and NH_4F.

Metal fluorides can be used to carry out halogen–fluorine exchanges involving the group-IVB elements. Anhydrous ZnF_2 neat or in solution is a commonly used fluorinating agent in Si chemistry but is not widely used for the other elements, e.g.:

$$2\ \text{n-PrSiCl}_3 + 3\ ZnF_2 \rightarrow 2\ \text{n-PrSiF}_3 + 3\ ZnCl_2 \qquad \text{(a)}^1$$

$$Si_2Ph_{6-n}Cl_n + \frac{n}{2} ZnF_2 \xrightarrow{Ag} Si_2Ph_{6-n}F_n + \frac{n}{2} ZnCl_2 \qquad \text{(b)}^2$$

$$\text{(1-Me-2-Me-2-Cl-silacyclobutane)} + ZnF_2 \xrightarrow[\text{nonstereospecific}]{} \text{(1-Me-2-Me-2-F-silacyclobutane)} + ZnCl_2 \qquad \text{(c)}^3$$

$$Si_2Br_6 + 3\ ZnF_2 \rightarrow Si_2F_6 + 3\ ZnBr_2 \qquad \text{(d)}^4$$

The structure of the silane compound may influence the extent to which fluorine exchange can occur, e.g., the Si—Cl bonds in carbosilanes that have unsubstituted CH_2 bridges can be converted to Si—F bonds by ZnF_2 without skeletal cleavage:

$$(Cl_2SiCH_2)_3 \xrightarrow{ZnF_2} (F_2SiCH_2)_3 \qquad \text{(e)}$$

However, Cl–F exchange does not take place at Si of perchlorinated cyclic carbosilanes, e.g., $(Cl_2SiCCl_2)_3$ with either ZnF_2 or SbF_3–$SbCl_5$[5].

Fluorine derivatives of Ag are used as halogen–fluorine exchange reagents with the group-IV halides. (see conversion series, §2.5.12.1.2). Silver(I) fluorine is not used because it is deliquescent, light sensitive and easily reduced[6]. Furthermore, 2 mol of AgF must be used for each exchangable halogen, because the AgX salt produced in the reaction reacts with AgF to form the double salt AgXAgF, which has little fluorinating ability. Despite these drawbacks, AgF is employed in certain fluorination reactions, e.g., although adamantyl bromide can be converted to adamantyl fluoride in 61% yield by its reaction with anhyd ZnF_2, the conversion of the di-, tri- and tetrabromoadamantane to the respective fluoroadamantane requires[7] the use of AgF. It is also used to convert[8] CF_3GeI_3 to CF_3GeF_3:

$$CF_3GeI_3 \xrightarrow{AgF} CF_3GeF_3 \qquad \text{(f)}$$

The advantage of using AgF is that is reacts under mild conditions. It is not used in systems that contain Si—H bonds[9], although it converts[10] GeH_3Br to GeH_3F at low T:

$$GeH_3Br \xrightarrow[-22.9°C]{AgF} GeH_3F \qquad \text{(g)}$$

Chlorotin compounds can be converted[11,12] to the corresponding fluorotin derivative by using aq AgF:

$$h^5\text{-}C_5H_5Fe(CO)_2SnCl_3 \xrightarrow[H_2O]{AgF} h^5\text{-}C_5H_5Fe(CO)_2SnF_3 \qquad \text{(h)}$$

$$Me_2SnCl_2 \xrightarrow[H_2O]{AgF} Me_2SnF_2 \qquad \text{(i)}$$

Silver(I) fluoride cannot be used[13] to convert SiH_3CH_2I to SiH_3CH_2F. The reaction produces a mixture of SiF_4 and CH_2F_2.

Silver(II) fluoride (and CoF_3) replace both halogen and hydrogen by fluorine in halogenated hydrocarbons[14].

The alkali-metal fluorides NaF and in particular KF are used as halogen-exchange agents with group IV halides. Alkyl halides can be converted to alkyl fluorides by their reaction with KF in ethylene glycol or diethylene glycol[15], e.g., 1-bromohexane is converted[16] to 1-fluorohexane by KF. Potassium fluoride can also be used for halogen exchange in aromatic compounds, providing the halogen is activated by an o- or p-NO_2 substituent. The attempt to prepare a fluoromethyl derivative by Et_3SiCH_2X (X = Cl, I) with KF results in the formation of rearrangement and cleavage products[17].

$$2\ Et_3SiCH_2X + KF \xrightarrow[DMFA]{160^\circ} \underset{(25\%)}{Et_3SiF} + \underset{(34\%)}{Et_2Si(F)Pr\text{-}n} \qquad \text{(j)}$$

However, $F(CH_2)_3SiMe_3$ is prepared in a 61% yield by refluxing $Cl(CH_2)_3SiMe_3$ with anhyd KF in diethylene glycol[18]. Potassium fluoride with Crown-6 in benzene is used to fluorinate Si—halogen bonds.

$$(Me_3Si)_2C(H){-}Si(Br)(t\text{-}Bu)_2 \rightarrow (Me_3Si)_2C(H){-}Si(F)(t\text{-}Bu)_2 \qquad \text{(k)}^{19}$$

Potassium fluoride and KHF_2 either with or without a solvent, convert Si—Cl bonds to Si—F bonds[18,20]:

$$F(CH_2)_3SiCl_3 \xrightarrow{KHF_2} F(CH_2)_3SiF_3 \qquad \text{(l)}$$

With $F(CH_2)_3SiCl_3$, γ-Elimination reactions may accompany the exchange leading to the formation of cyclopropane and SiF_4. The elimination reaction is least important when KHF_2 is employed as the fluorinating reagent[18]. Refluxing a mixture of $Cl(CH_2)_3SiCl_3$ with dry KF in nitrobenzene (179°C) for 5 h produces the Si-fluorinated product $Cl(CH_2)_3SiF_3$ and the completely fluorinated product $F(CH_2)_3SiF_3$ (45% yield and 10% yield, respectively)[18]. Almost half of the starting silane is involved with a γ-elimination decomposition reaction.

Potassium fluoride can also be used as a fluorine exchange reagent with Sn and Pb chlorides; e.g., n-Bu_2SnF_2 can be prepared[21] by treating alcoholic n-Bu_2SnCl_2 with aq KF. Other alkyltin halides (Cl, Br, I) are converted to the corresponding fluorine derivatives in a similar manner[22]. Sodium iodide in $CHCl_3$ converts $MeSnCl_3$ to $MeSnI_3$ [23]. Triphenyllead fluoride precipitates when aq KF is shaken with Ph_3PbCl in benzene[24]:

$$Ph_3PbCl + KF \rightarrow Ph_3PbF + KCl \qquad \text{(m)}$$

$$Ph_2PbCl_2 + 2\ KF \rightarrow Ph_2PbF_2 + 2\ KCl \qquad \text{(n)}^{25}$$

Mixtures of fused alkali-metal fluorides (e.g., KF, LiF and NaF, mp. 454°C) fluorinate[26] CCl_4 and organochlorosilanes at high T.

Several other metal fluorides are used as fluorine-exchange reagents with group-IVB halides. Mercury(I) fluoride is a more effective exchange reagent than AgF for replacing iodine or Br (but not Cl) from alkyl halides. Mercury(II) chloride, fluoride and the fluoride iodide are also effective fluorinating agents. Mercury(II) fluoride is an effective fluorine-exchange reagent with alkyl halides and polyhalides[15].

$$2\ F_2CHBr + HgF_2 \rightarrow 2\ F_3CH + HgBr_2 \qquad \text{(o)}$$

Both fluorine atoms exchange rapidly, and the reactions are moderated by using a hydrocarbon or fluorocarbon solvent. The compound is also capable of fluorinating other group-IVB halides[27], such as SiH_3I:

$$2\ SiH_3I + HgF_2 \rightarrow 2\ SiH_3F + HgI_2 \qquad \text{(p)}$$

Other less widely used fluorides employed as fluorine-exchange agents with group-IVB halides include CoF_2[28], CaF_2[29], and PbF_2[30].

Ammonium fluoride acts as a mild fluorinating agent with certain group-IVB halides; e.g., it reacts with Et_2SiCl_2 to form[31] both Et_2SiF_2 and Et_2SiClF. It undergoes halogen exchange with certain chlorinated siloxanes such as $Et_2ClSiOSiEt_2Cl$ and $Et_3SiOSiEt_2Cl$ to produce good yields of the corresponding fluorosiloxanes[32]. Exchange can be accompanied by cleavage of the Si—O bond. Cleavage accompanies the exchange[33] in $(ClMe_2Si)_2NH$ with NH_4F:

$$(ClMe_2Si)_2NH + 4\ NH_4F \rightarrow 2\ Me_2SiF_2 + 3\ NH_3 + 2\ NH_4Cl \qquad \text{(q)}$$

Saturated aq $[NH_4]HF_2$ is also a fluorinating agent in organosilicon chemistry:

$$R_3SiCl \xrightarrow[\text{aq}]{[NH_4]HF_2} R_3SiF \qquad \text{(r)}^{34}$$

where R = Me, Et, n-Pr. The halogen exchange that takes place between α-naphthylphenylmethyl chlorosilane and $c\text{-}C_6H_{11}NF$ proceeds with predominant inversion of configuration[35]:

$$\underset{[\alpha]_D\ -5.7^\circ}{(-)\text{-}R_3Si^*Cl} + [c\text{-}C_6H_{11}NH_3]F \rightarrow \underset{[\alpha]_D\ -36^\circ}{(-)\text{-}R_3Si^*F} + [c\text{-}C_6H_{11}NH_3]Cl \qquad \text{(s)}$$

The reaction of $(CF_3)_3GeX$ with F^- in water or acetonitrile gives[36] trigonalbipyramidal $[(CF_3)_3GeF_2]^-$.

(C. H. VAN DYKE)

1. L. Tansjo, *Acta Chem. Scand.*, *18*, 456 (1964).
2. E. Hengge, F. Schrank, *J. Organometal. Chem.*, *299*, 1 (1986).
3. B. G. McKinnie, N. S. Bhacca, F. K. Cartledge, *J. Org. Chem.*, *41*, 1534 (1976).
4. M. Schmeisser, K.-P. Ehlers, *Angew. Chem., Int. Ed. Engl.*, *3*, 700 (1964).
5. G. Fritz, M. Berndt, *Angew. Chem., Int. Ed. Engl.*, *10*, 510 (1971).
6. D. R. Martin, *Inorg. Synth.*, *4*, 133 (1953).
7. K. S. Bhandari, R. E. Pincock, *Synthesis*, 655 (1974).
8. H. C. Clark, C. J. Willis, *J. Am. Chem. Soc.*, *84*, 898 (1962).
9. A. G. MacDiarmid, *Prep. Inorg. React.*, *1*, 165 (1964).
10. T. N. Srivastava, J. E. Griffiths, M. Onyszchuk, *Can. J. Chem.*, *40*, 579 (1962).
11. H. C. Clark, R. G. Geol, *J. Organomet. Chem.*, *7*, 263 (1967).

12. T. J. Marks, A. M. Seyam, *Inorg. Chem.*, *13*, 1624 (1974).
13. J. M. Bellama, A. G. MacDiarmid, *J. Organomet. Chem.*, *18*, 275 (1969).
14. E. T. McBee, B. W. Hotten, L. R. Evans, A. A. Alberts, Z. D. Welch, W. B. Ligett, R. C. Schreyer, K. W. Krantz, *Ind. Eng. Chem.*, *39*, 310 (1947).
15. A. L. Henne, *Org. React.*, *2*, 49 (1944).
16. A. I. Vogel, J. Leicester, W. A. T. Macey, *Org. Synth.*, *36*, 40 (1956).
17. M. G. Voronkov, S. V. Kirpichenko, V. V. Keiko, V. A. Pestunovich, E. O. Tsetina, V. Chvalovsky, V. Jaroslav, *Akad. Nauk SSSR Bulletin* (*Engl. Transl.*) 1932 (1975).
18. V. B. Puchnarevič, J. Včelak, M. G. Voronkov, V. Chvalovský, *Coll. Czech. Chem. Commun.*, *39*, 2616 (1974).
19. N. Wiberg, C. Wagner, G. Mueller, J. Riede, *J. Organometal. Chem.*, *271*, 381 (1984).
20. M. G. Voronkov, Yu. I. Skorik, *J. Gen. Chem. USSR* (*Engl. Transl.*), *33*, 3382 (1963).
21. D. L. Alleston, A. G. Davies, *J. Chem. Soc.*, 2050 (1962).
22. E. Krause, *Chem. Ber.*, *51*, 1447 (1918).
23. C. H. W. Jones, M. Dombsky, *Can. J. Chem.*, *59*, 1585 (1981).
24. R. J. H. Clark, A. G. Davies, R. J. Puddephatt, *Inorg. Chem.*, *8*, 457 (1969).
25. F. Huber, K. L. Schillings, *J. Fluorine Chem.*, *19*, 521 (1982).
26. W. Sundermeyer, *Z. Anorg. Allg. Chem.*, *310*, 100 (1962).
27. V. G. Noskov, A. A. Kirbichnikova, M. A. Sokai'skii, *J. Gen. Chem. USSR* (*Engl. Transl.*), *43*, 2090 (1973).
28. M. F. Shostakovskii, B. A. Sokolov, A. N. Grishko, K. F. Lavrova, G. J. Kagan, *J. Gen. Chem. USSR* (*Engl. Transl.*), *32*, 3882 (1962).
29. H. S. Booth, W. F. Martin, *J. Am. Chem. Soc.*, *68*, 2655 (1946).
30. S. Cradock, E. A. V. Ebsworth, *J. Chem. Soc.*, *A*, 1226 (1967).
31. C. J. Wilkins, *J. Chem. Soc.*, 2726 (1951).
32. D. A. Payne, *J. Chem. Soc.*, 2143 (1954).
33. J. Silbiger, J. Fuchs, *Inorg. Chem.*, *4*, 1371 (1965).
34. M. G. Voronkov, Yu. I. Skorik, *Akad. Nauk SSSR*, *Bulletin* (*Engl. Transl.*) 1127 (1964).
35. L. H. Sommer, F. O. Stark, K. W. Michael, *J. Am. Chem. Soc.*, *86*, 5683 (1964).
36. D. J. Brauer, J. Wilke, *J. Organometal Chem.*, *316*, 261 (1986).

2.5.12.3.4. by Fluorine Extraction from Fluorine-containing Anions.

Metal salts and related derivatives that contain fluoro anions such as $[BF_4]^-$, $[PF_6]^-$, $[SbF_6]^-$ and $[SiF_6]^-$ can convert Si, Ge and Sn halides to the corresponding fluorine derivatives. Silver salts or Na_2SiF_6, NH_4BF_4 and HBF_4 are used[1,2].

The exchange proceeds through the formation of fluroanion intermediate that decomposes via F^- abstraction[3], e.g., the isolation of Me_3SnF from Me_3SiBr with $AgPF_6$ in SO_2 can be attributed to the decomposition of Me_3SnPF_6 formed in the reaction:

$$Me_3SnBr + AgPF_6 \rightarrow Me_3SnPF_6 + AgCl \quad \text{(a)}$$

$$Me_3SnPF_6 \rightarrow Me_3SnF + PF_5 \quad \text{(b)}$$

In related reactions, Me_2SnF_2 results from Me_2SnCl_2 with Ag salts containing fluoroanions[4] such as Ag_2SiF_6 (in MeOH), $AgBF_4$ (in MeOH), $AgPF_6$ (in SO_2) and $AgSbF_6$ (in SO_2). Chloro-Si and -Ge transition-metal complexes react $AgBF_4$ (in acetone or benzene), $AgPF_6$ (in benzene) and $AgSbF_6$ (in CH_2Cl_2 or benzene) to form the corresponding fluoro derivatives in high yields and BF_3, PF_5 and SbF_5, respectively[5]. The reactions are straight forward, efficient and easy to perform and so represent one of the best procedures for preparing fluoro-group IVB metal-transition-metal compounds. Compounds prepared by this method include[5–8]. h^5-$C_5H_5M(CO)_3SiMe_{3-n}F_n$ (M = Cr, Mo, W, n = 1-3), h^5-$C_5H_5Fe(CO)_2SiF_2CH{=}CH_2$, h^5-$C_5H_5Fe(CO)_2SiF_3$, $(CO)_5MnGePh_2F$, $(CO)_4CoGePh_{3-n}F_n$ (n = 1, 3), h^5-$C_5H_5Mo(CO)_3GePh_{3-n}F_n$ (n = 1-3) h^7-$C_7H_7Mo(CO)_2GeF_3$. Treating

h^5-$C_5H_5Mo(CO)_3SnPh_2Cl$ with $AgBF_4$ in benzene produces a stable salt $[h^5\text{-}C_5H_5Mo(CO)_3SnPh_2]BF_4$ which on heating does not give the corresponding —$SnPh_2F$ derivative[5]. Treating $C(SiMe_2Br)_4$ with $AgBF_4$ produces[9] $C(SiMe_2F)_4$.

Sodium salts containing fluoroanions such as Na_2SiF_6 and $NaBF_4$ act as fluorinating agents in organosilicon chemistry:

$$MePhSiCl_2 + Na_2SiF_6 \rightarrow MePhSiF_2 + SiF_4 + 2\ NaCl \qquad (c)^1$$

$$Ph_3SiCl + NaBF_4 \xrightarrow{\text{acetone}} Ph_3SiF + NaCl + BF_3 \qquad (d)^{10}$$

Salts derived from organopentafluorosilicates also act as fluorinating agents[11]:

$$2\ Me_3SiCl + Na_2[MeSiF_5] \rightarrow 2\ Me_3SiF + MeSiF_3 + 2\ NaCl \qquad (e)$$

$$2\ MeSiCl_3 + 3\ [NH_4]_2[MeSiF_5] \rightarrow 5\ MeSiF_3 + 6\ NH_4Cl \qquad (f)$$

(C. H. VAN DYKE)

1. D. L. Bailey, R. M. Pike, U.S. Pat. 3,020,302 (1962); *Chem. Abstr., 56*, 15548 (1962).
2. R. Müller, C. Cathe, East Ger. Pat. 43,688 (1966); *Chem. Abstr., 65*, 7217 (1966).
3. H. C. Clark, R. J. O'Brien, *Inorg. Chem., 2*, 1020 (1963).
4. H. C. Clark, R. G. Geol, *J. Organomet. Chem., 7*, 263 (1967).
5. T. J. Marks, A. M. Seyam, *Inorg. Chem., 13*, 1624 (1974).
6. W. Malisch, P. Panster, *J. Organomet. Chem., 64*, C5 (1974).
7. W. Malisch, *Chem. Ber., 107*, 3835 (1974).
8. E. E. Isaacs, W. A. G. Graham, *Can. J. Chem., 53*, 975 (1975).
9. C. Eaborn, P. D. Lickiss, *J. Organometal. Chem., 294*, 305 (1985).
10. E. A. Lawton, A. Levy, *J. Am. Chem. Soc., 77*, 6083 (1955).
11. R. Müller, C. Dathe, H. J. Frey, *Chem. Ber., 99*, 1614 (1966).

2.5.12.3.5. with Covalent Fluorides.

Fluorine-halogen exchange occurs between covalent fluorides and group-IVB halogen derivatives, e.g., SF_4 undergoes halogen exchange[1] with organic polyhalides at elevated temperatures. Usually only partial exchange occurs, e.g.,

$$CCl_4 + SF_4 \xrightarrow{100-225^\circ C} CF_2Cl_2\ (15\%),\ CFCl_3\ (66\%) \qquad (a)$$

Replacement of Br by F is more facile than replacement of Cl by this method. The reaction of SF_4 with chloroalkenes and C_6Cl_6 produces[1] chlorofluoroalkanes and chlorofluorocyclohexenes respectively[6].

Covalent fluorides interact with chlorosilanes to produce fluorosilanes; e.g., Me_3SiCl reacts with SF_4 at 20°C to produce Me_3SiF in 90% yield[2]:

$$4\ Me_3SiCl + SF_4 \xrightarrow{20^\circ C} 4\ Me_3SiF + SCl_2 + Cl_2 \qquad (b)$$

Likewise, Me_3SiCl reacts with thionylfluoride to produce Me_3SiF, although the yield is low[3]:

$$Me_3SiCl + SOF_2 \rightarrow Me_3SiF\ (23\%) + SOCl_2 \qquad (c)$$

The reaction of Me_3SiCl with $CF_3\overline{CFCF_2O}$ also produces[4] Me_3SiF:

$$Me_3SiCl + CF_3\overline{CFCF_2O} \xrightarrow{-14^\circ C} Me_3SiF + CF_3CFClCOF \qquad (d)$$

The reaction of PF_5 with CH_3SiH_2Cl results in the substitution of one hydrogen and the Cl by fluorine[5]:

$$CH_3SiH_2Cl + PF_5 \rightarrow CH_3SiHF_2 + PF_3 + HCl \tag{e}$$

Iodine pentafluoride reacts with Me_3SiCl at $-45°C$ to produce Me_3SiF and an unstable yellow solid, presumably[6] IF_4Cl:

$$Me_3SiCl + IF_5 \rightarrow Me_3SiF + [IF_4Cl] \tag{f}$$

The reaction of $SnCl_4$ with ClF produces SnF_2Cl_2 [7].

$$SnCl_4 + 2\ ClF \xrightarrow{80°C} SnF_2Cl_2 + Cl_2 \tag{g}$$

(C. H. VAN DYKE)

1. C. W. Tullock, R. A. Carboni, R. J. Harder, W. C. Smith, D. D. Coffman, *J. Am. Chem. Soc.*, *82*, 5107 (1960).
2. R. Müller, D. Mross, *Z. Anorg. Allg. Chem.*, *324*, 78 (1963).
3. R. Müller, D. Mross, *Z. Anorg. Allg. Chem.*, *324*, 86 (1963).
4. L. Heinrich, *Z. Chem.*, *8*, 257 (1968).
5. S. K. Gondal, A. G. MacDiarmid, F. E. Saalfeld, M. V. McDowell, *Inorg. Nucl. Chem. Lett.*, *5*, 351 (1969).
6. G. Oates, J. M. Winfield, O. R. Chambers, *J. Chem. Soc., Dalton Trans.*, 1380 (1974).
7. K. Dehnicke, *Chem. Ber.*, *98*, 280 (1965).

2.5.13. Cleavage of Other Group-IVB Element Bonds (Cleavage of Group-IVB Element–Metal Bonds)

2.5.13.1. by Halogens.

2.5.13.1.1. Cleavage of Carbon–Metal Bonds.

Halogens (Cl_2, Br_2 and I_2) cleave carbon–metal bonds to produce organic halides and metal–halogen derivatives; e.g., the preparation of the organolead halides, R_3PbX and R_2PbX_2, can be achieved[1] by Cl_2, Br_2, I_2 with R_4Pb:

$$R_4Pb + X_2 \xrightarrow{-70\ °C} R_3PbX + RX \tag{a}$$

$$R_3PbX + X_2 \rightarrow R_2PbX_2 + RX \tag{b}$$

Groups other than simple alkyl or aryl substituents also participate; e.g., the carbon–metal bond of the carbamoyl complexes, $Hg(CONEt_2)_2$ and h^5-$C_5H_5(CO)FeCONHMe$, are cleaved by Br_2 to produce the metal bromide and the carbamoyl bromide in the initial cleavage step[2,3]:

$$Hg(CONEt_2) + 2\ Br_2 \rightarrow HgBr_2 + 2\ BrCONEt_2 \tag{c}$$

$$h^5\text{-}C_5H_5(CO)_2FeCONHMe + Br_2 \rightarrow h^5\text{-}C_5H_5(CO)_2FeBr + [BrCONHMe]$$

$$[BrCONHMe] \downarrow\ HBr + CH_3NCO \tag{d}$$

Preparative halogen cleavage chiefly involves C—Mg and —Hg bonds. To exchange halogens in an organic halide that does not exchange by other procedures (see §2.5.12.2.1) involves the reaction of a halogen with the Mg derivative of the organic halide:

$$RX + Mg \rightarrow RMgX \quad (e)$$

$$RMgX + I_2 \rightarrow RI + MgXI \quad (f)$$

The reaction converts aromatic or heterocyclic Cl or Br derivatives to the corresponding iodo derivatives[4]; e.g., iodobenzene is formed[5] in 90% yield by the addition of PhMgBr to I_2 in Et_2O:

$$PhMgBr + I_2 \rightarrow PhI + MgBrI \quad (g)$$

The yields of halogen-exchange products may be low and are influenced by the order in which the reagents are added[5].

Although neopentyl iodide can be prepared without rearrangement by treating neopentyl-MgCl with I_2, the product is contaminated with hydrocarbon byproducts that are difficult to remove[6]. Pure neopentyl iodide can be prepared in 92% yield by treating the corresponding Hg compound (prepared from neopentyl-MgCl) with I_2:

$$RMgCl + HgCl_2 \rightarrow RHgCl + MgCl_2 \quad (h)$$

$$RHgCl + I_2 \rightarrow RI + HgClI \quad (i)$$

where R = neopentyl. The reaction of Br_2 with neopentyl-HgCl produces pure neopentyl bromide in 82% yield. Organomercury compounds are less reactive than organomagnesium compounds in halogen cleavage of the carbon–metal bonds. Bromination can be used to characterize certain organomercury compounds, e.g., $PhHgCHCl_2$ is characterized by the analysis of $HCCl_2Br$ produced as its bromination product[7]:

$$PhHgCHCl_2 \xrightarrow{Br_2} HCCl_2Br \quad (j)$$

(C. H. VAN DYKE)

1. I. Ruidisch, H. Schmidbaur, H. Schumann, in *Halogen Chemistry*, Vol. 2, V. Gutmann, ed., Academic Press, New York, 1967, p. 233.
2. U. Schollkopf, F. Gerhart, *Angew. Chem., Int. Ed. Engl., 5*, 664 (1966).
3. W. Jetz, R. J. Angelici, *Inorg. Chem., 11*, 1960 (1972).
4. C. Weygand, G. Hilgetag, *Preparative Organic Chemistry*, G. Hilgetag, A. Martini, eds., Wiley-Interscience, New York, 1972.
5. R. L. Datta, H. K. Mitter, *J. Am. Chem. Soc., 41*, 287 (1919).
6. F. C. Whitmore, E. L. Wittle, B. R. Harriman, *J. Am. Chem. Soc., 61*, 1585 (1939).
7. D. Seyferth, M. E. Gordon, K. V. Darrah, *J. Organomet. Chem., 14*, 43 (1968).

2.5.13.1.2. with Compounds that Contain Lower Group-IVB Element–Transition-Metal Bonds

Halogens react with transition-metal derivatives of the lower group-IVB elements two ways. The reactions can proceed via cleavage of the metal–group-IVB element bond, producing the corresponding group-IVB element halide and a halogen derivative of the metal complex system; e.g., keeping 1:1 $Me_3SnMn(CO)_5$ and I_2 in pentane at 25°C for 12 h in a sealed tube produces Me_3SnI (75% recovered) and $IMn(CO)_5$ (100% recovered)[1]:

$$Me_3SnMn(CO)_5 + I_2 \rightarrow Me_3SnI + IMn(CO)_5 \quad (a)$$

In other systems the reaction proceeds by substituting halogen for an organic group (or hydrogen) attached to the group-IVB element; e.g., the reaction of $Ph_3SnMn(CO)_5$ with Cl_2, Br_2 or I_2 in CCl_4 at 20°C produces halotin manganese pentacarbonyl complexes[2]:

$$Ph_3SnMn(CO)_5 + n\,X_2 \xrightarrow{CCl_4} Ph_{3\text{-}n}X_nSnMn(CO)_5 + n\,PhX \qquad (b)$$

where X = Cl_2, Br_2, I_2; n = 1–3. The reaction can fail to occur, or substitution can occur at the transition-metal center:

$$Ph_3SiMn(CO)_5 + I_2(20\%\text{ in }CCl_4) \rightarrow \text{no reaction} \qquad (c)^2$$

$$(C_6F_5)_3SnMn(CO)_5 + xsCl_2 \rightarrow \text{no reaction} \qquad (d)^3$$

$$2\,RuH_3[Si(OEt)_3](PPh_3)_3 + I_2 \rightarrow 2\,RuH_2I[Si(OEt)_3](PPh_3)_3 + H_2 \qquad (e)^4$$

Both substitution and cleavage can occur on treating the group-IVB element-metal complexes with a halogen; e.g., the reaction of Cl_2 with $Ph_2Sn[Mn(CO)_5]_2$ produces[5] both $ClMn(CO)_5$ and $Cl_3SnMn(CO)_5$:

$$Ph_2Sn[Mn(CO)_5]_2 + 3\,Cl_2 \rightarrow Cl_3SnMn(CO)_5 + ClMn(CO)_5 + 2\,PhCl \qquad (f)$$

In the reaction of $MeGeH_2Mn(CO)_5$ with Br_2, both hydrogen substitution and metal cleavage occur, although the former predominates[6]:

$$MeGeH_2Mn(CO)_5 \xrightarrow{Br_2} MeGeBr_2Mn(CO)_5 + MeGeBr_3 \qquad (g)$$

The reaction of Br_2 with $Ph_3SnMn(CO)_5$ proceeds initially with preferential substitution at the tin–phenyl bond, although when all the $Ph_3SnMn(CO)_5$ is converted to the Br_3Sn derivative the Sn—Mn bond is cleaved[2], forming $SnBr_4$,

The course of cleavage-substitution is dependent on the group-IVB element, the transition metal, the substituents involved and other less well-defined factors. The cleavage reactions are electrophilic substitutions at the metal–metal bond. However, the reactions are often complex and may involve charge-transfer complexes; e.g., the proposed scheme involving $Me_3SnCr(CO)_3C_5H_5\text{-}h^5$ and is[7]:

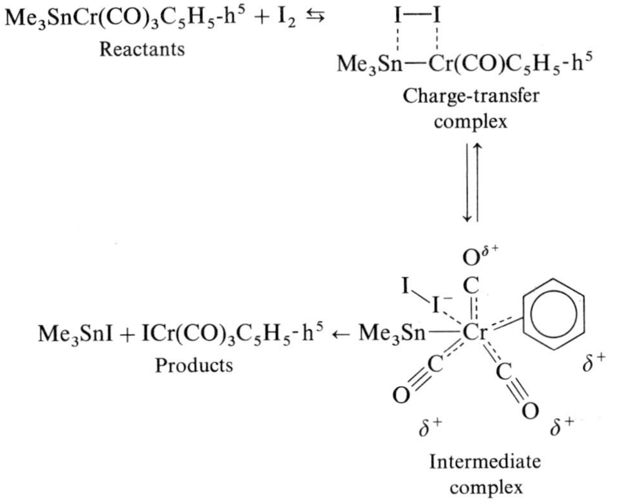

$$Me_3SnCr(CO)_3C_5H_5\text{-}h^5 + I_2 \leftrightarrows \underset{\text{Charge-transfer complex}}{\overset{I—I}{Me_3Sn—Cr(CO)C_5H_5\text{-}h^5}}$$

Reactants

$\updownarrow$ (h)

$$Me_3SnI + ICr(CO)_3C_5H_5\text{-}h^5 \leftarrow Me_3Sn—Cr(CO)_3(C_5H_5)\cdots I^-\cdots I$$

Products

Intermediate complex ($O^{\delta+}$, δ^+, δ^+, δ^+)

Kinetic information on the effects of changing the group-IVB element, the alkyl groups and the transition-metal ligands is available for iodine and bromine cleavages[2,7–12]. Bromine is more reactive[11] than I_2 and in:

$$Me_3EMn(CO)_5 + I_2 \xrightarrow[CCl_4]{30\ °C} Me_3EI + IMn(CO)_5 \tag{i}$$

where E = Si, Ge, Sn, Pb, the relative rates are Si < Ge < Sn < Pb[9,10]. Steric effects determine these relative rates of iodination[12].

The reaction of the interhalogens, ICl and IBr with $Me_3SnMn(CO)_5$ and $Me_3SnMo(CO)_3C_5H_5$-h^5 in CCl_4 are specific[11]. The tin–metal bond cleaves, producing the corresponding iodo-transition-metal complex:

$$IX + Me_3SnML_n \rightarrow Me_3SnX + IML_n \tag{j}$$

where X = Cl, Br; ML_n = $Mn(CO)_5$, $Mo(CO)_3C_5H_5$-h^5. The reaction of ICl with $Me_3EFe(CO)_2C_5H_5$-h^5 in cyclohexane is similar:

$$Me_3EFe(CO)_2C_5H_5\text{-h}^5 + ICl \rightarrow Me_3ECl + IFe(CO)_2C_5H_5\text{-h}^5 \tag{k}$$

where E = Si, Ge, Sn, although Me_3SnI forms as a side product from ICl with $Me_3SnMn(CO)_5$ in pentane[1].

Systems that undergo halogen cleavage of the group IVB—metal bond are given in Table 1.

Systems which undergo halogen substitution at the group-IVB element without appreciable cleavage of the transition-metal bond are listed in Table 2.

Compounds that contain group-IVB elements bonded to nontransition metals usually undergo halogen-metal exchange producing the metal halide and the halogenated group-IVB derivative. Subsequent reaction of the group IVB halide with the metal

TABLE 1. HALOGEN CLEAVAGES OF THE GROUP-IVB ELEMENT-TRANSITION-METAL BOND

Compound	Halogen	Solvent, T, °C	Ref.
Si			
$Ph_3SiMn(CO)_5$	Cl_2, Br_2	CCl_4, 20	2
$Me_3SiMn(CO)_5$	I_2	CCl_4	9, 10
$Me_3SiFe(CO)_2C_5H_5$-h^5	Cl_2	$CHCl_3$, 0	3
$Me_3SiFe(CO)_2C_5H_5$-h^5	I_2	CCl_4, 20	10
1,2-$C_6H_{10}(SiMe_2)_2Pt(PPh_3)_2$ (cyclohexane ring bearing two $SiMe_2$ groups bridged to $Pt(PPh_3)_2$)	Br_2	benzene	13
$(Et_3Si)_2Cd$	Br_2	CCl_4	14
Ge			
$Me_3GeMn(CO)_5$	I_2	CCl_4, 20	9, 10
$Me_3GeFe(CO)_2C_5H_5$-h^5	I_2	CCl_4, 20	10
		$CHCl_3$, 0	2
$(Me_3Ge)_2RhCOC_5H_5$-h^5	I_2[a]	hexane, RT	15

TABLE 1 (*continued*)

Compound	Halogen	Solvent, T, °C	Ref.
$R_3GeHgPt(PPh_3)_2GeR_3$ ($R = C_6F_5$)	Br_2[b]	—, 20	16
$Ph_3GeMn(CO)_5$	Br_2	heptane, RT	17
$Ph_2Ge[Mn(CO)_5]_2$	Br_2[c]	benzene, RT	17
$Pt(PEt_3)_2(GePh_3)_2$	I_2	benzene, 20	18
$Me_3GeW(CO)_3C_5H_5$-h[5]	I_2[d]	benzene, RT	19
Sn			
$Me_3SnCr(CO)_3C_5H_5$-h[5]	I_2	CCl_4, 20	7
		$CHCl_3$, RT	3
$Ph_3SnMn(CO)_5$	Br_2 (high conc)	CCl_4, 20	2
$R_3SnMn(CO)_5$	I_2	CCl_4, 20	10
R = Me, Et or C_6H_{11}			
$Ph_3SnFe(CO)_2C_5H_5$-h[5]	Cl_2	CCl_4	5
$R_3SnFe(CO)_2C_5H_5$-h[5]	I_2	CCl_4, 20	10
R = Me, Bu or Ph			
$Me_3SnFe(CO)_2C_5H_5$-h[5]	Cl_2	C_5H_{12}, −78	3
	I_2	C_5H_{12}, 60	3
$Me_3SnFe(CO)_2C_5H_5$-h[5]	I_2	$CDCl_3$, MeOD, $(CD_3)_2SO$	18
$Me_3SnFe(CO)_2C_5H_5$-h[5]	I_2	CCl_4	8
$Me_2Sn[Fe(CO)_2C_5H_5\text{-h}^5]_2$	I_2	$(CD_3)_2SO$	20
$Ph_3SnMo(CO)_3(C_5H_4Me\text{-h}^5)$	I_2	CCl_4, 20	10
$Me_3SnMo(CO)_3C_5H_5$-h[5]	I_2	$CDCl_3$, MeOD $(CD_3)_2SO$,	21
		$[MeOCH_2]_2$, RT	3
$R_3SnMo(CO)_3C_5H_5$-h[5]	I_2	CCl_4, 20	10
R = Me, Bu, Ph			
$Me_3SnMo(CO)_3C_5H_5$-h[5]	Cl_2	C_5H_{12}, RT	3
	I_2	CCl_4, RT	8
$Me_3SnW(CO)_3C_5H_5$-h[5]	I_2	CCl_4, 20	8
		$CHCl_3$, RT	3
$Ph_3SnW(CO)_3C_5H_5$-h[5]	I_2	$CHCl_3$, RT	3
$Me_3SnW(CO)_3C_5H_5$-h[5]	I_2	various solvents	21
$Me_3SnCo(CO)_4$	I_2	hexane, −10	22
		cyclohexane, 28.5	11
$Me_3SnRe(CO)_5$	I_2	CCl_4, 30	11
$(Me_3Sn)_2RhCOC_5H_5$-h[5]	I_2	hexane, RT	15
$(R_3Sn)_2M$	I_2	hexane, −20	23
M = Cd, Hg			
$R = Me_3SiCH_2$			
Pb			
$Et_3PbMn(CO)_5$	Cl_2, Br_2	—	1
$Me_3PbMn(CO)_5$	I_2	CCl_4, 20	9, 10
trans-$[Pt(PEt_3)_2(PbPh_3)_2]$	Br_2[e]	CCl_4	24

[a] One Ge—Rh bond cleaved.
[b] Products are $HgBr_2$, R_3GeBr and $R_3GePt(PPh_3)_2Br$.
[c] One $Mn(CO)_5$ group cleaved producing $Ph_2Ge(Br)Mn(CO)_5$.
[d] Products are R_3GeI, h[5]-$C_5H_5(CO)_2WI_3$ and CO.
[e] Products are trans-$[Pt(PEt_3)_2Br_2]$ and $PbPh_2Br_2$.

TABLE 2. HALOGENATION RESULTING IN SUBSTITUTION AT THE GROUP-IVB ELEMENT IN TRANSITION-METAL-GROUP-IVB ELEMENT COMPOUNDS

Compound	Halogen	Solvent, T, °C	Ref.
$MeGeH_2Mn(CO)_5$	Br_2[a]		6
$Ph_3GeMn(CO)_5$	Br_2	$Br_2C_2H_4$	17
$Ph_3SnMn(CO)_5$	Cl_2, Br_2, I_2[b]	CCl_4, 20	2, 5
		(CH_2Cl_2 for Cl_2)	5
$Ph_3SnMn(CO)_4PPh_3$	Cl_2	CH_2Cl_2, 25	5
$Ph_2(C_6F_5)SnMn(CO)_5$	Cl_2[c]	CH_2Cl_2, RT	3
$Ph(C_6F_5)_2SnMn(CO)_5$	Cl_2[c]	CH_2Cl_2, RT	3

[a] The main product is $MeGeBr_2Mn(CO)_5$; however some cleavage of the Ge—Mn bond occurs.
[b] Complete and partial replacement of Ph groups is possible, some cleavage of the Sn—Mn bond occurs with Br_2 and I_2.
[c] Only the Ph groups are replaced.

reagent forms a coupled product. Thus, the reaction of Ph_3GeLi with Br_2 produces Ph_3GeBr together with $Ph_3GeGePh_3$[25]:

$$Ph_3GeLi + Br_2 \xrightarrow{-20\ °C} LiBr + Ph_3GeBr \tag{l}$$

$$Ph_3GeBr + Ph_3GeLi \rightarrow Ph_3GeGePh_3 + LiBr \tag{m}$$

The reaction of Ph_3SiLi with I_2 involves, at least in part, a similar exchange followed by coupling[26].

(C. H. VAN DYKE)

1. M. R. Booth, D. J. Cardin, N. A. D. Carey, H. C. Clark, B. R. Sreenathan, *J. Organomet. Chem*, *21*, 171 (1970).
2. J. R. Chipperfield, J. Ford, D. E. Webster, *J. Organomet. Chem.*, *102*, 417 (1975).
3. R. E. J. Bichler, H. C. Clark, B. K. Hunter, A. T. Rake, *J. Organomet. Chem.*, *69*, 367 (1974).
4. R. N. Haszeldine, L. S. Malkin, R. V. Parish, *J. Organomet. Chem.*, *182*, 323 (1979).
5. R. D. Gorsich, *J. Am. Chem. Soc.*, *84*, 2486 (1962).
6. B. W. L. Graham, K. M. Mackey, S. R. Stobart, *J. Chem. Soc., Dalton Trans.*, 475 (1975).
7. J. R. Chipperfield, A. C. Hayter, D. E. Webster, *J. Chem. Soc., Dalton Trans.*, 2048 (1975).
8. J. R. Chipperfield, J. Ford, D. E. Webster, *J. Chem. Soc., Dalton Trans.*, 2042 (1975).
9. J. R. Chipperfield, A. C. Hayter, D. E. Webster, *J. Chem. Soc., Chem. Commun.*, 625 (1975).
10. J. R. Chipperfield, J. Ford, A. C. Hayter, D. E. Webster, *J. Chem. Soc. Dalton Trans.*, 360 (1976).
11. J. R. Chipperfield, J. Ford, A. C. Hayter, D. J. Lee, D. E. Webster, *J. Chem. Soc., Dalton Trans.*, 1024 (1976).
12. J. R. Chipperfield, A. C. Hayter, D. E. Webster, *J. Organomet. Chem.*, *121*, 185 (1976).
13. C. Eaborn, T. N. Metham, A. Pidcock, *J. Organomet. Chem.*, *63*, 107 (1973).
14. N. S. Vyazankin, G. A. Razuvsaev, *J. Gen. Chem. USSR (Engl. Transl.)*, *35*, 394 (1965).
15. R. Hill, S. A. R. Knox, *J. Organomet. Chem.*, *84*, C31 (1975).
16. G. A. Razuvaev, *J. Organomet. Chem.*, *200*, 243 (1980).
17. A. N. Nesmeyanov, K. N. Anisimov, N. E. Kolobova, A. B. Antonova, *Acad. Nauk SSSR, Bulletin, Div. Chem. Sci., (Engl.)* 139 (1966).
18. R. J. Cross, F. Glockling, *J. Chem. Soc.*, *1965*, 5422.
19. A. Carrick, F. Glockling, *J. Chem. Soc. (A)*, 913 (1968).

20. R. M. G. Roberts, *J. Organomet. Chem.*, *47*, 359 (1973).
21. R. M. G. Roberts, *J. Organomet. Chem.*, *40*, 359 (1972).
22. E. W. Abel, G. V. Hutson, *J. Inorg. Nucl. Chem.*, *30*, 2339 (1968).
23. G. S. Kalinina, O. A. Kruglaya, B. I. Petrov, E. A. Shchupak, N. J. Vayazankin, *J. Gen. Chem. USSR (Engl. Transl.)*, *43*, 2215 (1973).
24. G. Deganello, G. Carturan, U. Belluco, *J. Chem. Soc.*, *A*, 2873 (1968).
25. M. Lesbre, P. Mazerolles, J. Satgé, *The Organic Compounds of Germanium*, Interscience, New York, 1971, p. 650.
26. H. Gilman, R. L. Harrell, *J. Organomet. Chem.*, *9*, 67 (1967).

2.5.13.1.3. Hydrogen Halides with Lower Group-IVB Element–Metal Bonds

Hydrogen halides react with transition-metal derivatives of the lower group-IVB elements via cleavage of the group-IVB–metal bond to form either the halide or hydride:

$$R_3E{-}ML_n + HX \rightarrow R_3EX + HML_n \qquad \text{(a)}$$

$$R_3E{-}ML_n + HX \rightarrow R_3EH + XML_n \qquad \text{(b)}$$

where E is the group-IVB element, ML_n the metal-ligand system and X the halide. Both reactions can occur. The reaction can also proceed by substituting halogen for an organic group or hydrogen on the group-IVB element without cleaving the group-IV bond:

$$R_3E{-}ML_n + HX \rightarrow XR_2EML_n + RH \qquad \text{(c)}$$

The Sn derivatives undergo halogen substitution (element–carbon cleavage), whereas the Si and Ge analogues undergo metal cleavage; e.g., the reaction of $Me_3SiFe(CO)_2C_5H_5\text{-}h^5$ and $Me_3GeFe(CO)_2C_5H_5\text{-}h^5$ with HCl in CH_2Cl_2 at 50–60°C produces Me_3SiCl and Me_3GeCl, respectively, and $HFe(CO)_2C_5H_5\text{-}h^5$ or its decomposition product[1]:

$$Me_3EFe(CO)_2C_5H_5\text{-}h^5 + HCl \rightarrow Me_3ECl + HFe(CO)_2C_5H_5\text{-}h^5 \qquad \text{(d)}$$

where E = Si, Ge. The reaction of $Me_3SnFe(CO)_2C_5H_5\text{-}h^5$ with xs HCl in CH_2Cl_2 produces $ClMe_2SnFe(CO)_2C_5H_5\text{-}h^5$ and Me_3SnCl, while the reaction of HCl with $Ph_3SnFe(CO)_2C_5H_5\text{-}h^5$ produces $ClPh_2Sn{-}$, $Cl_2PhSn{-}$ or $Cl_3Sn{-}Fe(CO)_2C_5H_5\text{-}h^5$, depending on the reaction conditions[1]. The substitution reactions form a particular product unless forcing conditions are used. Thus, in the reaction of HCl with $Ph_3SnFe(CO)_2C_5H_5\text{-}h^5$ the equilibrium favors the formation of $Cl_2PhSnFe(CO)C_5H_5\text{-}h^5$ whereas in the analogous reaction with HBr, the formation of $BrPh_2SnFe(CO)_2C_5H_5\text{-}h^5$ is favored[1]. Halogen diphenyltin derivatives of $Mn(CO)_5$ cannot be made by this reaction because the equilibrium is in the favor of the dihalophenyl derivative[2]. The monohalo system can be prepared by redistribution[2]:

$$2\ Ph_3SnMn(CO)_5 + Cl_3SnMn(CO)_5 \rightarrow 3\ ClPh_2SnMn(CO)_5 \qquad \text{(e)}$$

Both $ClMe_2SnMn(CO)_5$ and $Cl_2MeSnMn(CO)_5$ can be prepared[3] by the reaction of HCl with $Me_3SnMn(CO)_5$; however, the chlorine substitution becomes progressively more difficult and does not occur beyond $Cl_2MeSnMn(CO)_5$.

The cleavage reaction which group-IVB–transition–metal bonds undergo with hydrogen halides can produce either the group-IVB halide or the hydride, depending on the polarity of the group-IVB–metal bond involved; e.g., the group-IVB chloride is formed[1] in the reaction of HCl with $Me_3EFe(CO)_2C_5H_5$-h^5:

$$Me_3EFe(CO)_2C_5H_5\text{-h}^5 + HCl \rightarrow Me_3ECl + HFe(CO)_2C_5H_5\text{-h}^5 \quad \text{(f)}$$

where E = Si, Ge, Sn. However, HCl with $Ph_3GeAuPPh_3$ produces[4] Ph_3GeH and $ClAuPPh_3$, in accord with a polarity $(\delta-)$Ge—Au$(\delta+)$:

$$Ph_3GeAuPPh_3 + HCl \rightarrow Ph_3GeH + ClAuPPh_3 \quad \text{(g)}$$

The reaction of HCl with $CH_2{=}CHSiMe_2Fe(CO)_2C_5H_5$-h^5 produces[5] $CH_2{=}CHSiMe_2H$ and $ClFe(CO)_2C_5H_5$-h^5:

$$CH_2{=}CHSiMe_2Fe(CO)_2C_5H_5\text{-h}^5 + HCl \rightarrow CH_2{=}CHSiMe_2H + ClFe(CO)_2C_5H_5\text{-h}^5 \quad \text{(h)}$$

Other systems that produce the group IVB hydride and the halotransition element include HBr with $Ph_3SiHf(Cl)(C_5H_5\text{-h}^5)_2$[6] and HCl with compounds containing Si—Zr[6], Si—Pt[7], Ge—Pt[7,8], Ge—Pd[9] and Pb—Pt[10] bonds. Both modes of cleavage may also occur, and the nature of the organic groups on the group-IVB element can influence where the cleavage occurs when there is a choice, e.g.:

$$(Et_3P)_2Pt(GePh_3)GeMe_3 + HCl \begin{cases} \uparrow \; (Et_3P)_2Pt(H)GePh_3 + Me_3GeCl \\ \downarrow \; (Et_3P)_2Pt(Cl)GePh_3 + Me_3GeH \; \text{(ca. 2\%)} \end{cases} \quad \text{(i)}^7$$

Hydrogen halides also cleave tin derivatives of the rare earths[11]:

$$(R_3Sn)_3Pr \cdot DME \xrightarrow{HCl} R_3SnCl + PrCl_3 + H_2 + CH_4 \quad \text{(j)}$$

where R = CH_2SiMe_3; DME = dimethyl formamide.

Representative group-IVB-metal compounds that undergo cleavage with hydrogen halides are summarized in Table 1; systems that result in substitution at the group-IVB element are summarized in Table 2.

TABLE 1. CLEAVAGE OF THE GROUP-IVB ELEMENT-TRANSITION-METAL BOND BY HYDROGEN HALIDES

Compound	Hydrogen halide	Products	Ref.
Si			
$RuH_3(SiR_3)(PPh_3)_3$ $R_3 = (OEt)_3$, $MeCl_2$	HCl	$RuCl_2(PPh_3)_2$, R_3SiH	12
trans-$(Et_3P)_2Pt(Cl)SiMe_3$	HCl	Me_3SiCl, trans-$(Et_3P)_2Pt(Cl)H$	7
$Me_3SiFe(CO)_2C_5H_5$-h^5	HCl (CH_2Cl_2)	Me_3SiCl, $HFe(CO)_2C_5H_5$-h^5, $[Fe(CO)_2C_5H_5$-$h^5]_2$	1
$R_3SiCo(CO)_4$ $R_3 = Cl_3$, $MeCl_2$, Me_2Cl	HX [a] (X = Cl, Br)	R_3SiX	13
$Me_3SiCo(CO)_4$	HX (X = Cl, Br)	R_3SiX	13
$SiH_3Co(CO)_4$ [b]	HX (X = F, Cl)	SiH_3X, $HCo(CO)_4$	14
$(Me_3Si)_2Hg$	HCl	Me_3SiCl, Me_3SiH, Hg	15
$CH_2{=}CHSiMe_2Fe(CO)_2C_5H_5$-$h^5$	HCl	$CH_2{=}CHSiMe_2H$, $ClFe(CO)_2C_5H_5$-h^5	5
Ge			
$R_3GeHgPt(PPh_3)_2GeR_3$ $R = C_6F_5$	HCl (C_6H_6, 80°C)	Hg, R_3GeCl, R_3GeH, $R_3GePt(PPh_3)_2H$	11
$GeH_3Co(CO)_4$	HCl	GeH_3Cl, GeH_4	16
$(Et_3P)_2Pd(GePh_3)_2$	HCl	Ph_3GeH, trans-$(Et_3P)_2PdCl$	9
trans-$(Ph_3Ge)_2Pt(PEt_3)_2$	HCl	Ph_3GeH, $(Et_3P)_2PtCl_2$, Ph_3GeCl, $(Et_3P)_2Pt(H)Cl$	8
trans-$(Et_3P)_2Pt(Cl)GeMe_3$	HCl (Et_2O or PhH)	Me_3GeCl, trans-$(Et_3P)_2Pt(Cl)H$	7
cis-$(Et_3P)_2Pt(Ph)GeMe_3$	HCl	Me_3GeH, cis-$(Et_3P)_2Pt(Ph)Cl$, trans-$(Et_3P)_2Pt(Ph)Cl$	7
trans-$(Et_3P)_2Pt(GePh_3)GeMe_3$	HCl	Me_3GeH, trans-$(Et_3P)_2Pt(H)GePh_3$, Me_3GeCl, trans-$(Et_3P)_2Pt(Cl)GePh_3$	7
trans-$(Et_3P)_2Pt(H)GePh_3$	HCl	Ph_3GeH, trans-$(Et_3P)_2Pt(H)Cl$	7
$Me_3GeFe(CO)_2C_5H_5$-h^5	HCl	Me_3GeCl, $HFe(CO)_2C_5H_5$-h^5, $[Fe(CO)_2C_5H_5$-$h^5]_2$	1
$R_3GeMo(CO)_3C_5H_5$-h^5 (R = Me, Et)	HCl	R_3GeCl, $HMo(CO)_3C_5H_5$-h^5	17
Sn			
$Me_3SnFe(CO)_2C_5H_5$-h^5	HCl (CH_2Cl_2)	Me_3SnCl, unidentified materials	1
Pb			
trans-$(Et_3P)_2Pt(PbPh_3)_2$	HCl[c]	(Ph_3PbH), trans-$(Et_3P)_2Pt(Ph_3Pb)Cl$	10

[a] High pressure reaction (90°C, 4000 atm).
[b] HCl reacts more readily than HF.
[c] With excess HCl, the products are trans-$(Et_3P)_2PtCl_2$ and $PbCl_2$.

TABLE 2. HYDROGEN HALIDE SUBSTITUTION AT THE GROUP-IVB ELEMENT IN TRANSITION-METAL–GROUP-IVB COMPOUNDS

Compound	Hydrogen halide	Products	Ref.
Si			
$SiH_3Mn(CO)_5$	HCl (75°C)	$SiH_{3-n}Cl_nMn(CO)_5$ (n = 1–3)	18
Ge			
$GeH_3Mn(CO)_5$	HX (X = Cl, Br)	$GeH_{3-n}X_nMn(CO)_5$ (n = 1–3)	19,20
$GeH_3Re(CO)_5$	HBr	$GeHBr_2Re(CO)_5$	21
Sn			
$Ph_2Sn[Mn(CO)_5]$	HCl	$Cl_2Sn[Mn(CO)_5]_2$	2
$Ph_3SnFe(CO)_2C_5H_5$-h^5	HCl (xs, 25°C) (CH_2Cl_2)	$Cl_3SnFe(CO)_2C_5H_5$-h^5	2
$Ph_3SnMn(CO)_5$	HCl (xs, 0–5°C)	$PhCl_2SnMn(CO)_5$	2
$Me_3SnMn(CO)_5$	HCl	$Me_2ClSnMn(CO)_5$, $MeCl_2SnMn(CO)_5$	3
$[Ph_3SnW(CO)_5]^-$	HCl	$[Cl_3SnW(CO)_5]^-$	22
$Ph_3SnFe(CO)_2C_5H_5$-h^5	HCl	$Ph_2ClSnFe(CO)_2C_5H_5$-h^5 $PhCl_2SnFe(CO)_2C_5H_5$-h^5	1
$Ph_3SnFe(CO)_2C_5H_5$-h^5	HBr	$Ph_2BrSnFe(CO)_2C_5H_5$-h^5 $PhBr_2SnFe(CO)_2C_5H_5$-h^5	1
$Ph_3SnFe(CO)(PR_3)C_5H_5$-h^5 R = Ph, Et, OPh	HCl	$Ph_{3-n}Cl_nSnFe(CO)(PR_3)C_5H_5$-h^5 (n = 1–3)	1
$Ph_3SnFe(C_5H_5$-h$^5)[P(OPh)_3]_2$	HCl	$Ph_{3-n}Cl_nSnFe[P(OPh)_3]_2C_5H_5$-h^5	1
$Ph_3SnFe(CO)P(OEt)_3C_5H_5$-h^5	HCl	$Cl_3SnFe(CO)P(OEt)_3C_5H_5$-h^5	1
$Me_3SnFe(CO)_2C_5H_5$-h^5	HCl	$Me_2ClSnFe(CO)_2C_5H_5$-h^5 $MeCl_2SnFe(CO)_2C_5H_5$-h^5	1
$Ph_3SnMo(CO)_3C_5H_5$-h^5	HCl	$Cl_3SnMo(CO)_3C_5H_5$-h^5	1
$Me_3SnW(CO)_3C_5H_5$-h^5	HCl	$Me_2ClW(CO)_3C_5H_5$-h^5 $MeCl_2W(CO)_3C_5H_5$-h^5	1
Pb			
$Et_3PbMn(CO)_5$	HCl	$Et_2ClPbMn(CO)_5$	3

(C. H. VAN DYKE)

1. R. E. J. Bichler, H. C. Clark, B. K. Hunter, A. T. Rabe, *J. Organomet Chem.*, *69*, 367 (1974).
2. R. D. Gorsich, *J. Am. Chem., Soc.*, *84*, 2486 (1962).
3. M. R. Booth, D. J. Cardin, N. A. D. Carey, H. C. Clark, B. R. Sreenathan, *J. Organomet. Chem.*, *21*, 171 (1970).
4. F. Glockling, M. D. Wilbey, *J. Chem. Soc.*, *A*, 2168 (1968).
5. W. Malisch, P. Panster, *J. Organomet. Chem.*, *64*, C5 (1974).
6. B. M. Kingston, M. F. Lappert, *J. Chem. Soc.*, *A*, 69 (1972).
7. F. Glockling, K. A. Hooton, *J. Chem. Soc.*, *A* 826 (1968).
8. R. J. Cross, F. Glockling, *J. Chem. Soc.*, 5422 (1965).
9. E. H. Brooks, F. Glockling, *J. Chem. Soc.*, *A*, 1241 (1966).
10. G. Deganello, G. Carturan, U. Belluco, *J. Chem. Soc.*, *A*, 2873 (1968).
11. G. A. Razuvaev, *J. Organomet. Chem.*, *200*, 243 (1980).
12. R. N. Haszeldine, L. S. Malkin, R. V. Parish, *J. Organomet. Chem.*, *182*, 323 (1979).

13. A. P. Hagen, L. McAmis, M. A. Stewart, *J. Organomet. Chem.*, *66*, 127 (1974).
14. B. J. Aylett, J. M. Campbell, *J. Chem. Soc.*, *A*, 1910 (1969).
15. E. Wiberg, O. Stecher, H. J. Andrascheck, L. Kreuzbichler, E. Stanue, *Angew. Chem., Int. Ed. Engl.*, *2*, 507 (1963).
16. R. D. George, K. M. Mackay, S. R. Stobart, *J. Chem. Soc., Dalton Trans.*, 974 (1972).
17. A. Carrick, F. Glockling, *J. Chem. Soc.*, *A*, 913 (1968).
18. B. J. Aylett, J. M. Campbell, *J. Chem. Soc.*, *A*, 1916 (1969).
19. B. W. L. Graham, K. M. Mackay, S. R. Stobart, *J. Chem. Soc., Dalton Trans.*, 475 (1975).
20. R. D. George, K. M. Mackay, S. R. Stobart, *J. Chem. Soc., Dalton Trans.*, 1505 (1972).
21. K. M. Mackay, S. R. Stobart, *J. Chem. Soc., Dalton Trans.*, 214 (1973).
22. E. E. Isaacs, W. A. G. Graham, *Can. J. Chem.*, *53*, 467 (1975).

2.5.13.1.4. Hg(II) Halides with Lower Group-IVB–Metal Bonds.

Mercury(II) halides cleave group-IVB-transition-metal bonds, although halogen substitution at the group IVB element may occur; e.g., $HgBr_2$ with $Me_3EFe(CO)_2C_5H_5$-h^5 (E = Si, Ge, Sn) produces[1] Me_3EBr and the Hg-transition-metal compound:

$$Me_3EFe(CO)_2 + HgBr_2 \rightarrow Me_3EBr + Fe(C_5H_5\text{-h}^5)(CO)_2HgBr \quad (a)$$

where E = Si, Ge, Sn. The rates in dioxan are Si, 1; Ge, 2.1 and Sn, 443. However, $HgBr_2$ cleaves not the Sn-transition-metal bond in $Ph_3SnFe(CO)_2C_5H_5$-h^5 but the Sn—Ph bond, forming $BrPh_2SnFe(CO)_2C_5H_5$-h^5. Substitution at Sn also occurs in $HgBr_2$ with $Ph_3SnMn(CO)_5$, where the favored product is $Br_2PhMn(CO)_5$:

$$Ph_3SnFe(CO)_2C_5H_5\text{-h}^5 + HgBr_2 \rightarrow BrPh_2SnFe(CO)_2C_5H_5\text{-h}^5 + PhHgBr \quad (b)$$

$$Ph_3SnMn(CO)_5C_5H_5\text{-h}^5 + 2\ HgBr_2 \rightarrow Br_2PhSnMn(CO)_5C_5H_5\text{-h}^5 + 2\ PhHgBr \quad (c)$$

Cleavage of the Sn—Mn bond does not occur with $HgBr_2$ in these Sn—Br systems; however, $HgBr_2$ cleaves the Si—Co bond of chlorosilyl-$Co(CO)_4$ derivatives and is a good procedure for preparing chlorobromides of Si [2]:

$$ClSiMe_2Co(CO)_4 \xrightarrow[90\ ^\circ C]{HgBr_2} Me_2Si(Cl)Br \quad (d)$$

$$Cl_2SiMeCo(CO)_4 \xrightarrow[90\ ^\circ C]{HgBr_2} MeSiCl_2Br \quad (e)$$

$$Cl_3SiCo(CO)_4 \xrightarrow[90\ ^\circ C]{HgBr_2} Cl_3SiBr \quad (f)$$

Organomercury halides also cleave the group IVB—transition-metal bond in an exchange[3]:

$$Me_3SnM(CO)_3C_5H_5\text{-h}^5 + RHgCl \rightarrow Me_3SnCl + RHgM(CO)_3C_5H_5\text{-h}^5 \quad (g)$$

$$2\ RHgM(CO)_3C_5H_5\text{-h}^5 \rightarrow R_2Hg + Hg[M(CO)_3C_5H_5\text{-h}^5]_2 \quad (h)$$

where M = Mo, W; R = Me, Ph, Allyl. Two modes of cleavage are possible, and the course of the reaction depends on the organomercury halide used. Thus, $Me_3SnMn(CO)_5$ reacts with MeHgBr:

$$2\ Me_3SnMn(CO)_5 + MeHgBr \rightarrow Hg[Mn(CO)_5]_2 + Me_4Sn + Me_3SnBr \quad (i)$$

while its reaction with PhHgCl proceeds[4]:

$$2\ Me_3SnMn(CO)_5 + 2\ PhHgCl \rightarrow Ph_2Hg + Hg[Mn(CO)_5]_2 + 2\ Me_3SnCl \quad (j)$$

The difference is attributed to variation in the polarity of the organomercury halides and the four-center intermediates involved.

$$\begin{array}{l} Me_3Sn{-}Mn(CO)_5 \\ \vdots \qquad \vdots \\ Cl{-}Hg{-}Ph \end{array} \rightarrow Me_3SnCl + PhHgMn(CO)_5 \rightleftharpoons Ph_2Hg + Hg[Mn(CO)_5]_2 \quad (k)$$

$$\begin{array}{l} Me_3Sn{-}Mn(CO)_5 \\ \vdots \qquad \vdots \\ Me{-}Hg{-}Br \end{array} \rightarrow Me_4Sn + BrHgMn(CO)_5 \rightleftharpoons HgBr_2 + Hg[Mn(CO)_5]_2 \xrightarrow{Me_3SnMn(CO)_5} Me_3SnBr + Hg[Mn(CO)_5]_2 \quad (l)$$

Dimethylmercury redistributes with Hg(II) halides to produce stable methylmercury halides[5]:

$$Me_2Hg + HgX_2 \rightarrow 2\ MeHgX \quad (m)$$

where X = Cl, Br, I. However, the reaction of $(Me_3Si)_2Hg$ with Hg(II) halides gives the corresponding trimethylsilyl halide and Hg metal without evidence of the silylmercury halide[6]:

$$(Me_3Si)_2Hg + HgX_2 \rightarrow 2\ Me_3SiX + 2\ Hg \quad (n)$$

where X = Cl, Br, I. Systems that undergo reaction with Hg(II) and organomercuric halides are given in Table 1.

(C. H. VAN DYKE)

1. J. R. Chipperfield, A. C. Hayter, D. E. Webster, *J. Chem. Soc., Dalton Trans.*, 921 (1977).
2. A. P. Hagen, L. McAmis, M. A. Stewart, *J. Organomet. Chem.*, *66*, 127 (1974).
3. R. M. G. Roberts, *J. Organomet. Chem.*, *40*, 359 (1972).
4. R. A. Burnham, F. Glockling, S. R. Stobart, *J. Chem. Soc., Dalton Trans.*, 1991 (1972).
5. M. D. Rausch, J. R. Van Wazer, *Inorg. Chem.*, *3*, 761 (1964).
6. A. G. Lee, *J. Organomet. Chem.*, *16*, 321 (1969).
7. B. J. Aylett, J. M. Campbell, *J. Chem. Soc., A*, 1910 (1969).
8. R. D. George, K. M. Mackay, S. R. Stobart, *J. Chem. Soc., Dalton Trans.*, 924 (1972).
9. F. Glockling, M. D. Wilbey, *J. Chem. Soc., A*, 2168 (1968).
10. B. W. L. Graham, K. M. Mackay, S. R. Stobart, *J. Chem. Soc., Dalton Trans.*, 475 (1975).
11. J. R. Chipperfield, A. C. Hayter, D. E. Webster, *J. Chem. Soc., Dalton Trans.*, 485 (1977).
12. R. M. G. Roberts, *J. Organomet. Chem.*, *47*, 359 (1973).

TABLE 1. CLEAVAGE PRODUCTS OF Hg(II) HALIDES WITH GROUP IVB–TRANSITION-METAL BONDS

Compound	Hg(II) Halide	Products	Ref.
Si			
$SiH_3Co(CO)_4$	HgX_2 (X = I, Cl)	SiH_3X, $Hg[Co(CO)_4]$, $XHg[Co(CO)_4]$	7
$Cl_nMe_{3-n}SiCo(CO)_4$, n = 0–3	$HgBr_2$	$Cl_nMe_{3-n}SiBr$	2
$Me_3SiMn(CO)_5$	$HgBr_2$	$Me_3SiBr + BrHgMn(CO)_5$	1
$Me_3SiFe(CO)_2C_5H_5$-h[5]	$HgBr_2$	$Me_3SiBr + BrHgFe(CO)_2C_5H_5$-h[5]	1
$(Me_3Si)_2Hg$	$HgCl_2$	Me_3SiCl, Hg	6
Ge			
$GeH_3Co(CO)_4$	$HgCl_2$	GeH_3Cl, $Hg[Co(CO)_4]_2$ $ClHg[Co(CO)_4]$	8
$Ph_3GeAuPPh_3$	$HgCl_2$	Ph_3GeCl, Ph_3PAuCl, Hg	9
$MeGeH_2Mn(CO)_5$	$HgCl_2$	$Cl_2MeGeMn(CO)_5$, Hg, H_2	10
$Me_3GeMn(CO)_5$	$HgBr_2$	Me_3GeBr, $BrHgMn(CO)_5$	1
$Me_3GeFe(CO)_2C_5H_5$-h[5]	$HgBr_2$	Me_3GeBr, $BrHgFe(CO)_2C_5H_5$-h[5]	1
Sn			
$Me_3SnCr(CO)_3C_5H_5$-h[5]	$HgBr_2$	Me_3SnBr, $BrHgCr(CO)_3C_5H_5$-h[5]	11
$Me_3SnMo(CO)_3C_5H_5$-h[5]	HgX_2 (X = Cl, Br, I)	Me_3SnX, $XHgMo(CO)_3C_5H_5$-h[5]	11
$Me_3SnW(CO)_3C_5H_5$-h[5]	$HgBr_2$	Me_3SnBr, $BrHgW(CO)_3C_5H_5$-h[5]	11
$Me_3SnMn(CO)_5$	HgX_2 (X = Cl, Br, I)	Me_3SnX, $XHgMn(CO)_5$	
$R_3SnMn(CO)_5$, R = Et, n-Bu, C_6H_{11}	$HgBr_2$	R_3SnBr, $BrHgMn(CO)_5$	1
$Me_3SnMn(CO)_5$	$HgCl_2$	Me_3SnCl, $Hg[Mn(CO)_5]_2$	4
$Me_3SnRe(CO)_5$	$HgBr_2$	Me_3SnBr, $BrHgRe(CO)_5$	11
$Me_3SnFe(CO)_2C_5H_5$-h[5]	HgX_2 (X = Cl, Br, I)	Me_3SnX, $XHgFe(CO)_2C_5H_5$-h[5]	11
$R_3SnFe(CO)_2C_5H_5$-h[5], R = n-Bu, C_6H_{11}	$HgBr_2$	R_3SnBr, $BrHgFe(CO)_2C_5H_5$-h[5]	1
$Me_3SnFe(CO)_2C_5H_5$-h[5]	$HgCl_2$	$Me_3SnCl + ClHgFe(CO)_2C_5H_5$-h[5]	12
$(allyl)_3SnM(CO)_3C_5H_5$-h[5], M = W, Mo	HgX_2 (X = Cl, Br, I)	$(allyl)_2Sn(X)M(CO)_3C_5H_5$-h[5] allyl-HgX	3
$Me_3SnM(CO)_3C_5H_5$-h[5], M = W, Mo	HgX_2 (X = Cl, Br, I)	Me_3SnX, $XHgM(CO)_3C_5H_5$-h[5]	3
$Me_3SnM(CO)_3C_5H_5$-h[5], M = W, Mo	RHgCl R = Me, Ph, allyl	Me_3SnCl, $RHgM(CO)_3C_5H_5$-h[5]	3
$Me_3SnFe(CO)_2C_5H_5$-h[5]	MeHgCl	Me_3SnCl, $MeHgFe(CO)_2C_5H_5$-h[5]	12
$Me_3SnMn(CO)_5$	MeHgCl	Me_3SnCl, $MeHgMn(CO)_5$	12

2.5.13.1.5. of Covalent Halides with Group-IVB–Metal Bonds.

(i) Coupling and Halogen–Metal Inverconversion Reactions. Reactions of covalent halides with group I or II metal derivatives of the group IVB elements yield organometallic compounds by straight forward coupling:

$$R_3E{-}M + R'X \rightarrow R_3E{-}R' + MX \qquad (a)$$

where E = group-IVB element, X = halogen, R = organic group, M = group I metal, although products arising from halogen-metal interconversion may also be present:

$$R_3E{-}M + R'X \rightarrow R_3E{-}X + R'M \quad \text{(b)}$$

$$R_3E{-}X + R_3E{-}M \rightarrow R_3E{-}ER_3 + MX \quad \text{(c)}$$

The coupling involving metal-organic derivatives, such as organomagnesium halide and organolithium reagents, with covalent halides is an important synthetic procedure:

$$E{-}X + M{-}\text{organic} \rightarrow E{-}\text{organic} + MX \quad \text{(d)}$$

where X = halogen, E = element less electropositive than M. Likewise, alkali-metal derivatives of Si, Ge and Sn are used in related couplings:

$$SiH_3K + Me_3SiCl \rightarrow Me_3SiSiH_3 + KCl \quad \text{(e)}^1$$

$$Et_3GeLi + EtMe_2SiCl \rightarrow Et_3GeSiMe_2Et + LiCl \quad \text{(f)}^2$$

$$Ph_3SnK + Ph_3SiCl \rightarrow Ph_3SnSiPh_3 + KCl \quad \text{(g)}^3$$

These couplings are discussed elsewhere under the formation of the particular element–element bond and are not considered here.

The reaction of organic and inorganic halides with metal–organic compounds may result in an exchange of the halogen and the metal in a halogen–metal interconversion[4]. The exchange is encountered with organolithium compounds:

$$RLi + R'X \rightarrow R'Li + RX \quad \text{(h)}$$

where the Li becomes attached to the more electronegative R group. With the proper choice of organolithium reagent, an organic halide (R′X) can be converted to the desired organolithium compound (R′Li) in high yield. Halogen–metal interconversion is important for the preparation of organolithium compounds that cannot be prepared from organic halides with Li. The most commonly used organolithium reagent in halogen–metal interconversion[4] is n-BuLi in Et_2O.

Halogen–metal exchange between metal–organic reagents and other metal halides can account for certain products; e.g., variable amounts of $Ph_3SiSiPh_3$ in Ph_3CNa with Ph_3SiBr can be accounted for by[5]:

$$Ph_3CNa + Ph_3SiBr \rightarrow Ph_3CBr + Ph_3SiNa \quad \text{(i)}$$

$$Ph_3SiNa + Ph_3SiBr \rightarrow Ph_3SiSiPh_3 + NaBr \quad \text{(j)}$$

The formation of a polymeric material containing Ph_3Ge, Ph_2Ge, PhGe and Ge groups in the reaction of GeI_2 with PhLi can be accounted for by the formation and subsequent reaction of Ge—Li intermediates produced in metal-halogen exchanges[5]:

$$PhGeI + PhLi \rightarrow PhGeLi + PhI \quad \text{(k)}$$

$$GeI_2 + PhLi \rightarrow LiGeI + PhI \quad \text{(l)}$$

Alkali-metal derivatives of the lower group-IVB elements may also undergo halogen–metal interchanges, e.g., Ph_3GeK with Ph_3SiCl produces Ph_6Ge_2, Ph_6Si_2 and $Ph_3GeSiPh_3$, which can be accounted for by halogen–metal interconversion followed by coupling[7]:

$$Ph_3GeK + Ph_3SiCl \rightarrow Ph_3GeCl + Ph_3SiLi \quad \text{(m)}$$

That the mixed species $Ph_3GeSiPh_3$ can be obtained in good yield from Ph_3SiK and Ph_3GeBr or Ph_3GeCl indicates that halogermanes do not undergo halogen–metal interconversion reactions as readily as halosilanes[7]. Thus the sole product[8] from Ph_3CNa and Ph_3GeBr is the simple coupling product Ph_3CGePh_3, whereas Ph_3CNa with Ph_3SiBr produces[5] $Ph_3SiSiPh_3$ as well as Ph_3CSiMe_3. Product analysis shows that halogen–metal exchange occurs in Ph_3SiK with Ph_3CCl[5], but not in Ph_3GeK with Ph_3SiCl[9]. The extent of halogen–metal exchange can be influenced by the order in which the reagents are combined. Thus, the addition of $Me_2PhSiLi$ to Ph_3SiCl produces the coupled product $Me_2PhSiSiPPh_3$ in 47% yield, whereas the addition of Ph_3SiCl to $Me_2PhSiLi$ produces $PhSiSiPh_3$ and $Me_2PhSiSiPhMe_2$ in 35 and 71% yields, respectively[10]:

$$Ph_3SiCl + Me_2PhSiLi \rightarrow Ph_3SiLi + Me_2PhSiCl \quad \text{(n)}$$

$$Ph_3SiLi + Ph_3SiCl \rightarrow Ph_3SiSiPh_3 + LiCl \quad \text{(o)}$$

$$Me_2PhSiLi + Me_2PhSiCl \rightarrow Me_2PhSiSiPhMe_2 + LiCl \quad \text{(p)}$$

In the interaction of Ph_3Si–metal derivatives with organic halides the extent of halogen–metal exchange is dependent on the particular metal and halogen involved[11]. Thus, Li, Na, and K derivatives undergo exchange; however, neither Ph_3SiRb nor Ph_3SiCs undergoes exchange with t-BuCl. Fluorine derivatives do not participate in halogen–metal exchange, and in reactions involving Ph_3SiLi and ethyl halides there is an increase in the extent of exchange $EtCl < EtBr < EtI$. This order is observed in other halogen–metal exchange reactions as well[12]. The framework of the organic halide will also influence the extent of and the actual products isolated in the exchange[11].

(ii) Cleavages Involving Inorganic Halogen Derivatives. Metal derivatives of group-IVB elements undergo reaction with metal halides to produce the corresponding group-IVB halide which can subsequently undergo a coupling reaction with the original group-IVB metal derivative. Thus, the addition of Ph_3SiLi to a solution of $HgCl_2$ in tetrahydrofuran (THF) at RT forms Ph_3SiCl (45% yield) as well as $Ph_3SiSiPh_3$ (18.1% yield)[13]:

$$Ph_3SiLi + HgCl_2 \rightarrow [Ph_3SiHgCl] + LiCl \quad \text{(q)}$$

$$[Ph_3SiHgCl] \rightarrow Ph_3SiCl + Hg \quad \text{(r)}$$

$$Ph_3SiCl + Ph_3SiLi \rightarrow Ph_3SiSiPh_3 + LiCl \quad \text{(s)}$$

Other metal halides, such as $PbCl_2$, $SnCl_4$ and Hg_2Cl_2, react similarly, although in the reaction of Ph_3SiLi with AgCl, $CuCl_2$, $FeCl_3$ and $SnCl_2$ the hydride derivative Ph_3SiH is isolated[14] with the $Ph_3SiSiPh_3$. The reaction of halogen derivatives of the group VB elements with Ph_3SiLi results[14] in $Ph_3SiSiPh_3$ in high yields, e.g., with $BiCl_3$:

$$Ph_3SiLi + BiCl_3 \rightarrow Ph_3SiBiCl_2 + LiCl \quad \text{(t)}$$

$$Ph_3SiBiCl_2 \rightarrow Ph_3SiCl + BiCl \quad \text{(u)}$$

$$Ph_3SiCl + Ph_3SiLi \rightarrow Ph_3SiSiPh_3 + LiCl \quad \text{(v)}$$

$$3\ BiCl \rightarrow BiCl_3 + Bi \quad \text{(w)}$$

(iii) Inorganic and Organic Halides with Transition-Metal Derivatives of the Lower Group-IVB Elements. Inorganic and organic halides react with transition-metal derivatives of the group IVB elements by cleaving the group IVB–metal bond; e.g., $TiCl_4$, $SiCl_4$ and $SnCl_4$ react with $Me_3SnFe(CO)_2C_5H_5$-h^5 to form Me_3SnCl and $TiCl_3Fe(CO)_2C_5H_5$-h^5, $SiCl_3Fe(CO)_2C_5H_5$-h^5 and $SnCl_3Fe(CO)_2C_5H_5$-h^5, respectively[15]:

$$Me_3SnFe(CO)_2C_5H_5\text{-h}^5 + ECl_4 \rightarrow Me_3SnCl + Cl_3EFe(CO)_2C_5H_5\text{-h}^5 \quad \text{(x)}$$

where E = Ti, Si, Sn. The order of reactivity in the cleavage is $TiCl_4 > SiCl_4 > SnCl_4$. The Ge—Au bond of $Ph_3GeAuPPh_3$ is cleaved[16] by $SnCl_4$, producing Ph_3GeCl, Ph_3PAuCl and $SnCl_2$. However, reactions may proceed by a substitution of the organic group by a halide at the group-IVB element; e.g. $SnCl_4$ with $Me_3SnMn(CO)_5$ eventually forms[17] $Cl_3SnMn(CO)_5$, by successive replacement of Me groups by Cl:

$$Me_3SnMn(CO)_5 + n\ SnCl_4 \rightarrow Me_{3\text{-}n}Cl_nMn(CO)_5 + n\ MeSnCl_3 \quad \text{(y)}$$

Likewise, under UV irradiation $HSiCl_3$ and CF_3COCl react with $Me_3SnMn(CO)_5$ under mild conditions at the Sn—C bond, replacing one Me group by Cl:

$$Me_3SnMn(CO)_5 + HSiCl_3 \rightarrow ClMe_2SnMn(CO)_5 + MeHSiCl_2 \quad \text{(z)}$$

$$Me_3SnMn(CO)_5 + CF_3COCl \xrightarrow{UV} ClMe_2SnMn(CO)_5 + CF_3COCH_3 \quad \text{(aa)}^{17}$$

The Sn—Me bond of $Me_3SnMn(CO)_5$ is cleaved[18] by BF_3; the products include $MeBF_2$, Me_2BF and $[Me_2SnMn(CO)_5]BF_4$. The latter is formed by the reaction of $FMe_2SnMn(CO)_5$ with xs BF_3 present.

The reactions of $Me_3SnFe(CO)_2C_5H_5$-h^5 and $Me_3SnMn(CO)_5$ with 10-fold xs Me_3SiCl result in Me_3SnCl in 88% and 80% yields, respectively[15]. The other expected products, $Me_3SiFe(CO)_2C_5H_5$-h^5 and $Me_3SiMn(CO)_5$, are not isolated but appreciable amounts of $ClFe(CO)_2C_5H_5$-h^5 and $ClMn(CO)_5$ are obtained.

Anhydrous $ZnCl_2$ reacts with $Me_3SnFe(CO)_2C_5H_5$-h^5 and $Me_3SnMn(CO)_5$ in dry acetone to produce Me_3SnCl in 25% and 19% yields, respectively[15]. Magnesium bromide with $Ph_3GeAuPPh_3$ forms Ph_3PAuBr and the germyl reagent, $Ph_3GeMgBr$[16]. Molybdenum and tungsten-magnesium halide reagents are formed in $MgBr_2$ with $R_3GeM(CO)_3C_5H_5$-h^5(M = Mo, W) compounds[19]:

$$R_3GeM(CO)_3C_5H_5\text{-h}^5 + MgBr_2 \xrightarrow{THF} h_5\text{-}C_5H_5(CO)_3MMgBr(THF) + R_3GeBr \quad \text{(ab)}$$

where M = Mo, W. The Ge—Pt bond in $(n\text{-}Pr_3P)_2Pt(GePh_3)_2$ undergoes stepwise cleavage[20] by MgI_2

$$(n\text{-}Pr_3P)_2Pt(GePh_3)_2 + MgI_2 \rightarrow (n\text{-}Pr_3P)_2Pt(I)GePh_3 + Ph_3GeMgI$$

$$(n\text{-}Pr_3P)_2Pt(I)GePh_3 \xrightarrow{MgI_2} (n\text{-}Pr_3P)_2PtI_2 \quad \text{(ac)}$$

The monoiodo product cannot be obtained[20] by $(n\text{-}Pr_3P)_2Pt(GePh_3)_2$ with 1 mol of I_2.

Organic halides cleave group IVB-transition metal bonds; e.g., CCl_4 cleaves[20] the Ge—Pt bonds of $(R_3P)_2Pt(GePh_3)_2$ derivatives forming Ph_3GeCl and the corresponding $(R_3P)_2PtCl_2$ (R = Et, n-Pr). The Ge—Au bond of $Ph_3GeAuPPh_3$ is cleaved[16] by CCl_4, producing Ph_3GeCl and Ph_3PAuCl. Chloroform quantitatively cleaves[21] the Zr—Si bond of $(C_5H_5\text{-}h^5)_2Zr(Cl)SiPh_3$. Methyl iodide with $(R_3P)_2Pt(GePh_3)_2$ at 110°C produces[20] Ph_3GeI, $(R_3P)_2PtI_2$ and $MeGePh_3$. The reaction of CH_3I with

$(PPh_3)_2Pt$ [chelate ring with two $SiMe_2$ groups bridged by o-phenylene]

produces $(PPh_3)_2Pt(Me)I$ in 73% yield[22]. The Si—Pt bonds of Pt(II) complexes that contain phosphinoethylsilyl bifunctional chelates are cleaved[23] by MeI and (+)$EtCH(Me)CH_2I(R^*I)$:

$$\text{Pt}[\text{SiRR}'\text{CH}_2\text{CH}_2\text{PPh}_2]_2 \xrightarrow{2\,CH_3I} \text{trans-}[PtI_2(PPh_2CH_2CH_2SiMe_3)_2]$$

$$\xrightarrow{2\,R^*I} \overset{R}{PtI_2(PPh_2CH_2CH_2SiMePhR^*)_2} \quad \overset{S}{PtI_2(PPh_2CH_2CH_2SiMePhR^*)_2} \qquad \text{(ad)}$$

where for the CH_3I reaction, R, R′ = Me, and for the R*I reaction, R = Me, R′ = Ph. Methyl iodide cleaves the Ge—Au bond of $Ph_3GeAuPPh_3$, producing[16] Ph_3GeMe, Ph_3PAuI and small amounts of CH_4, C_2H_6 and Au metal. Ethyl bromide also cleaves the Ge—Mo and —W bonds, yielding products that suggest radical processes[19]:

$$R_3GeM(CO)_3C_5H_5\text{-}h^5 + EtBr \rightarrow R_3GeBr + [C_5H_5\text{-}h^5(CO)_3M]_2 + C_2H_4 + C_2H_6 + C_4H_{10} \qquad \text{(ae)}$$

where M = Mo, R = Et; M = W, R = Me. Neither EtCl nor C_6F_5Br react[17] with $Me_3SnMn(CO)_5$.

Trifluoromethyl iodide reacts with $Me_3SiFe(CO)_2C_5H_5\text{-}h^5$ to produce Me_3SiF and $IFe(CO)_2C_5H_5\text{-}h^5$; however, with the Ge and Sn analogues the CF_3I reactions produce mixtures of Me_3MF, Me_3MI (M = Ge, Sn), $CF_3Fe(CO)_2C_5H_5\text{-}h^5$ and $IFe(CO)_2C_5H_5\text{-}h^5$ derivatives[24]. Group IVB-fluorine derivatives result from the elimination of CF_2 from Me_3MCF_3:

$$CF_3I + Me_3SiFe(CO)_2C_5H_5\text{-}h^5 \rightarrow Me_3SiCF_3 + IFe(CO)_2C_5H_5\text{-}h^5 \qquad \text{(af)}$$

$$Me_3SiCF_3 \rightarrow Me_3SiF + [CF_2] \qquad \text{(ag)}$$

The reaction of CF_3I with $Me_3SnMn(CO)_5$ produces the metal cleavage products[18] Me_3SnI and $IMn(CO)_5$ (but no CF_3-metal derivative), as well as the iodotin derivative, $IMe_2SnMn(CO)_5$.

1,2-Dibromoethane reacts with group IVB–transition-metal bonds to give C_2H_4 and the Br derivatives of the group IVB element and the metal complex, e.g.:

$$Me_3SnMn(CO)_5 + BrCH_2CH_2Br \rightarrow Me_3SnBr + BrMn(CO)_5 + C_2H_4 \quad \text{(ah)}^{17}$$

$$(R_3P)_2Pt(GePh_3)_2 + 2\ BrCH_2CH_2Br \rightarrow 2\ Ph_3GeBr + (R_3P)_2PtBr_2 + 2\ C_2H_4 \quad \text{(ai)}^{20}$$

where R = Et or n-Pr;

$$(Et_3P)_2Pd(GePh_3)_2 + 2\ BrCH_2CH_2Br \rightarrow 2\ Ph_3GeBr + (Et_3P)_2PdBr_2 + 2\ C_2H_4 \quad \text{(aj)}^{25}$$

$$Ph_3GeMPR_3 + BrCH_2CH_2Br \rightarrow Ph_3GeBr + (R_3P)_nMBr + C_2H_4 \quad \text{(ak)}^{26}$$

where M = Cu(I), Ag(I) or Au(I) and n = 1 for Au, 1 or 3 for Cu and Ag;

$$R_3GeMo(CO)_3C_5H_5\text{-}h^5 + BrCH_2CH_2Br \rightarrow R_3GeBr + BrMo(CO)_3C_5H_5\text{-}h^5 + C_2H_4 \quad \text{(al)}^{19}$$

$$(R_3Si^*)_2Hg + BrCH_2CH_2Br \rightarrow R_3SiBr \quad \text{(am)}^{27}$$

where R_3Si = Me(1-$C_{10}H_7$)PhSi. The reactions are quantitative and can be used as a diagnostic test for group IVB–transition-metal bonds[20]. 1,2-Dichloroethane is less reactive than the dibromo analogue[16].

(C. H. VAN DYKE)

1. E. Amberger, E. Mühlhofer, *J. Organomet. Chem.*, *12*, 55 (1968).
2. N. S. Vyazankin, G. A. Razuvaev, E. N. Gladyshev, S. P. Korneva, *J. Organomet. Chem.*, *7*, 353 (1967).
3. H. Gilman, S. D. Rosenberg, *J. Am. Chem. Soc.*, *74*, 531 (1952).
4. R. G. Jones, H. Gilman, *Org. React.*, *6*, 339 (1951).
5. A. G. Brook, H. Gilman, L. S. Miller, *J. Am. Chem. Soc.*, *75*, 4759 (1953).
6. F. Glockling, K. A. Hooton, *J. Chem. Soc.*, 1849 (1963).
7. H. Gilman, C. W. Gerow, *J. Am. Chem. Soc.*, *78*, 5823 (1956).
8. A. G. Brook, H. Gilman, *J. Am. Chem. Soc.*, *76*, 77 (1954).
9. H. Gilman, C. W. Gerow, *J. Am. Chem. Soc.*, *77*, 5509 (1955).
10. H. Gilman, G. D. Lichtenwalter, *J. Am. Chem. Soc.*, *80*, 608 (1958).
11. H. Gilman, H. J. S. Winkler, in *Organometallic Chemistry*, H. Zeiss, ed. Reinhold, New York, 1960, p. 270.
12. J.-P. Quintard, S. Hauvette-Frey, M. Pereyre, *J. Organomet. Chem.*, *159*, 147 (1978).
13. M. V. George, G. D. Lichtenwalter, H. Gilman, *J. Am. Chem. Soc.*, *81*, 978 (1959).
14. M. V. George, H. Gilman, *J. Am. Chem. Soc.*, *81*, 3288 (1959).
15. R. M. G. Roberts, *J. Organomet. Chem.*, *47*, 359 (1973).
16. F. Glockling, M. D. Wilbey, *J. Chem. Soc., A*, 2168 (1968).
17. R. A. Burnham, F. Glockling, S. R. Stobart, *J. Chem. Soc., Dalton Trans.*, 1991 (1972).
18. M. R. Booth, D. J. Cardin, N. A. D. Carey, H. C. Clark, B. R. Smeenathan, *J. Organomet. Chem.*, *21*, 171 (1970).
19. A. Carrick, F. Glocking, *J. Chem. Soc., A*, 913 (1968).
20. R. J. Cross, F. Glockling, *J. Chem. Soc.*, 5422 (1965).
21. B. M. Kingston, M. F. Lappert, *J. Chem. Soc., A*, 69 (1972).
22. C. Eaborn, T. N. Metham, A. Pidcock, *J. Organomet. Chem.*, *63*, 107 (1973).
23. R. D. Holmes-Smith, S. R. Stobart, T. S. Cameron, K. Jochem, *J. Chem. Soc., Chem Commun.* 937 (1981).
24. R. E. J. Bickler, H. C. Clark, B. K. Hunter, A. T. Rake, *J. Organomet. Chem.*, *69*, 367 (1974).
25. E. H. Brooks, F. Glockling, *J. Chem. Soc., A*, 5422 (1965).
26. F. Glockling, K. A. Hooton, *J. Chem. Soc.*, 2658 (1962).
27. C. Eaborn, R. A. Jackson, D. J. Tune, D. R. M. Walton, *J. Organomet. Chem.*, *63*, 85 (1973).

2.5.14. from Oxidative Halogenation of Compounds of Low-Valent Group-IVB Elements (Including Addition Reactions to C–C Multiple Bonds)

The formation of group-IV–halogen bonds by the oxidative addition of halogens and halogen compounds to divalent compounds of the group-IVB elements represents only a small part of the chemistry of the divalent carbenes[1–3] and their inorganic group-IVB analogues[4–11]. This section will also include addition reaction which halogens and hydrogen halides undergo with carbon–carbon multiple bonds.[12] Included for discussion are examples of the addition of halogens and hydrogen halides to organosilicon and -germanium compounds that contain unsaturated organic groups. Reviews are available[13,14] on the addition of halogens and halogen compounds to silenes and other unsaturated group-IVB compounds.

(C. H. VAN DYKE)

1. E. Chinoporos, *Chem. Rev.*, *63*, 235 (1963).
2. W. Kirmse, ed., *Carbene Chemistry*, 2nd ed., Academic Press, New York, 1971.
3. J. Hine, *Divalent Carbon*, Ronald Press, New York, 1964.
4. O. M. Nefedov, M. N. Manakov, *Angew, Chem.*, *Int. Ed. Engl.*, *5*, 1021 (1966).
5. W. H. Atwell, D. R. Weyenberg, *Angew, Chem.*, *Int. Ed. Engl.*, *8*, 469 (1969).
6. P. L. Timms, *Prep. Inorg. React.* *5*, 59 (1968).
7. P. P. Gaspar, B. J. Herold, in *Carbene Chemistry*, 2nd ed., W. Kirmse, ed., Academic Press, New York, 1977.
8. J. L. Margrave, P. W. Wilson, *Acc. Chem. Res.*, *4*, 145 (1971).
9. J. C. Thompson, J. L. Margrave, *Science*, *155*, 669 (1967).
10. D. L. Perry, J. L. Margrave, *J. Chem. Educ.*, *53*, 696 (1976).
11. J. Satgé, M. Massol, P. Rivière, *J. Organomet. Chem.*, *56*, 1 (1973).
12. C. Weygand, G. Hilgetag, in *Preparative Organic Chemistry*, G. Hilgetag, A. Martini, eds., Wiley Interscience, New York, 1972.
13. A. G. Brook, K. M. Baines, *Advances in Organometal. Chem.* *25*, 1 (1986).
14. N. Wiberg, *J. Organometal. Chem.*, *273*, 141 (1973).

2.5.14.1. by Halogens.

The interaction of group-IVB divalent derivatives with halogens can form two new group-IVB–halogen bonds by oxidative halogenation:

$$R_2E + X_2 \rightarrow R_2EX_2 \qquad \text{(a)}$$

where X is a halogen. The reaction works for tetravalent group-IVB–halogen derivatives; e.g., convenient preparations of CF_2I_2 and SiF_2I_2 are based on reacting I_2 with $[CF_2]$ and SiF_2, respectively,[1,2]:

$$[CF_2] + I_2 \rightarrow CF_2I_2 \qquad \text{(b)}$$

$$SiF_2 + I_2 \rightarrow SiF_2I_2 \qquad \text{(c)}$$

The $[CF_2]$ reaction is carried out by heating I_2 in the presence of the carbene-transfer reagent $(CF_3)_3PF_2$ at 120°C for 24 h, while the SiF_2 reaction takes place by cocondensing SiF_2 with I_2 at −196°C. The products of the $[CF_2]$–I_2 reaction include CF_2I_2 (30% yield) and the higher homologues ICF_2CF_2I and $ICF_2CF_2CF_2I$, while the only two products of the SiF_2–I_2 reaction are SiF_2I_2 and a small amount of SiF_3I. Silicon difluoride also reacts[3] with Br_2 at 350°C to form SiF_2Br_2, while the reaction of Cl_2 with the $[CF_2]$-transfer reagent $(CF_3)_3PF_2$ at 120°C results in the formation of CF_2Cl_2 (70%), CF_3Cl (20%) and $CFCl_3$ (10%)[1].

The halogenation of dialkytin compounds forms R_2SnX_2 derivatives. However, in most cases monomeric divalent tin compounds are not involved. The Sn–Sn bonds of the polymeric dialkytin compounds cleave[4], e.g.:

$$(Et_2Sn)_n + n\text{-}Cl_2 \rightarrow n\ Et_2SnCl_2 \tag{d}$$

The divalent tin compound bis[bis(trimethylsilyl)methyl]tin(II), SnR_2, is monomeric in benzene or cyclohexane and does undergo[5] straightforward halogenation with Cl_2 and Br_2:

$$SnR_2 + X_2 \rightarrow R_2SnX_2 \tag{e}$$

where R = $CH(SiMe_3)_2$; X = Cl, Br. The reaction of $(h^5\text{-}C_5H_5)_2Sn$ with halogens in benzene also produces the corresponding organotin(IV) dihalide[6]:

$$(h^5\text{-}C_5H_5)_2Sn + X_2 \xrightarrow{C_6H_6} (h^1\text{-}C_5H_5)_2SnX_2 \tag{f}$$

where X = Cl, Br, I. In the reaction of $(h^5\text{-}C_5H_5)_2Sn$ with I_2, the detection of $(h^1\text{-}C_5H_5)_3SnI$, $h^5\text{-}C_5H_5SnI$ and $h^1\text{-}C_5H_5SnI_3$ suggests a mechanism more complicated than simple addition[7]:

$$(h^5\text{-}C_5H_5)_2Sn \xrightarrow{I_2} (h^1\text{-}C_5H_5)_2SnI_2 \xrightarrow{(h^5\text{-}C_5H_5)_2Sn} (h^1\text{-}C_5H_5)_3SnI + h^5\text{-}C_5H_5SnI \tag{g}$$

$$h^5\text{-}C_5H_5SnI \xrightarrow{I_2} h^1\text{-}C_5H_5SnI_3; \quad h^1\text{-}C_5H_5SnI_3 \xrightarrow{(h^1\text{-}C_5H_5)_3SnI} (h^1\text{-}C_5H_5)_2SnI_2$$

Unsaturated organic compounds undergo addition with Cl_2, Br_2 and I_2 at low T. Chlorine and Br_2 react with alkenes at RT or below to produce the corresponding vicinal dihalide[8]:

$$\gt C{=}C\lt + X_2 \rightarrow -\overset{X}{\underset{|}{\overset{|}{C}}}-\overset{X}{\underset{|}{\overset{|}{C}}}- \tag{h}$$

Chlorine adds more readily than Br_2 while I_2 shows little tendency to react with most alkenes. Iodine monochloride adds to alkenes producing chloroiodo derivatives[9]. Fluorine reacts explosively with olefins, but PbF_4 can be used as a mild fluorine transfer agent[10].

The reaction of alkenyl-Si and -Ge compounds with halogens forms addition products, although in certain cases the group-IVB element–carbon bond is cleaved[11–15].

Thus, Cl_2 and Br_2 add to vinylsilanes, producing the corresponding dihaloalkyl derivatives in good yields:

$$CH_2{=}CHSiMe_3 + X_2 \rightarrow CH_2XCHXSiMe_3 \qquad (i)^{16}$$

where X = Cl, Br;

$$CH_2{=}CHSiCl_3 + Cl_2 \rightarrow CH_2ClCHClSiCl_3 \qquad (j)^{17}$$

$$CH_2{=}CHSi(OEt)_3 + Br_2 \rightarrow CH_2BrCHBrSi(OEt)_3 \qquad (k)^{17}$$

The reactions vary with conditions. Thus, the products obtained[17] in the reaction of Cl_2 with $CH_2{=}CHSiCl_3$ vary with T:

$$CH_2{=}CHSiCl_3 + Cl_2 \xrightarrow{100\text{–}250\ ^\circ C} CH_2ClCHClSiCl_3 \qquad (l)$$

$$CH_2{=}CHSiCl_3 + Cl_2 \xrightarrow{200\text{–}350\ ^\circ C} CH_2{=}C(Cl)SiCl_3 + HCl \qquad (m)$$

$$CH_2{=}CHSiCl_3 + Cl_2 \xrightarrow{300\text{–}400\ ^\circ C} SiCl_4 + \text{decomposition products} \qquad (n)$$

Bromination of trans-β-trimethylsilylstyrene at $-20^\circ C$ in CCl_4 cleaves[18] the Si—C bond; however, when the reaction is carried out at $-100^\circ C$ in CS_2, the dibromo adduct which is stable for several hours at RT, is formed[19].

The reaction of $CH_2{=}CHCH_2SiCl_3$ and $CH_2{=}CHCH_2SiMe_3$ with Cl_2 produces the correspnding vicinal dichloroalkyl addition product[20,21]

$$CH_2{=}CHCH_2SiR_3 + Cl_2 \rightarrow CH_2ClCHClCH_2SiR_3 \qquad (o)$$

where R = Me, Cl; however, the reaction with Br_2 results in cleavage of the Si—C bond[21]:

$$CH_2{=}CHCH_2SiMe_3 + Br_2 \rightarrow Me_3SiBr + CH_2{=}CHCH_2Br \qquad (p)$$

Similarly, Br_2 adds to vinylgermanes but cleaves the Ge—C bond of allylgermanes[14,22]:

$$CH_2{=}CHGe(Bu\text{-}n)_3 + Br_2 \rightarrow CH_2BrCH_2BrGe(Bu\text{-}n)_3 \qquad (q)$$

$$CH_2{=}CHCH_2Ge(Bu\text{-}n)_3 + Br_2 \rightarrow (n\text{-}Bu)_3GeBr + CH_2{=}CHCH_2Br \qquad (r)$$

Bromine undergoes normal addition[23] with the alkene derivatives $Et_3Ge(CH_2)_nCH{=}CH_2$ except for when n = 1 or 2, where the alkene group is cleaved from Ge.

Bromination of $Ph_3SiCH{=}CH_2$ and $Ph_3GeCH{=}CH_2$ is effected by a solution of Br_2 in CCl_4 in the presence of UV[24]. The reactions are slower in the dark, and yields are poorer. In systems where the double bond is activated by an aryl group, such as in the β-silylstyrenes, UV is not required when CH_2Cl_2 or $CHCl_3$ (but not CCl_4) is the solvent[24].

The stereochemistry of the bromination, debromination and debromosilylation of silylstyrene and other vinylsilanes is known[25]. For example, bromination and debromination of trans-triphenylsilylstyrene occur with *syn* stereochemistry, while debromosilylation occurs with *anti* stereochemistry in polar solvents[25]. With other vinylsilanes, different stereochemistries may be found.

Acetylenes also undergo addition with Cl_2 and Br_2. The reaction proceeds in two distinct steps, producing a dihaloalkene followed by the formation of a tetrahaloalkane:

$$—C{\equiv}C— + X_2 \rightarrow (X)C{=}C(X) \text{ (trans)} \quad (s)$$

$$(X)C{=}C(X) + X_2 \rightarrow —CX_2—CX_2— \quad (t)$$

The dihaloalkene produced in the first addition normally has a trans configuration and can be isolated if 1 mol of halogen is used. The reaction of I_2 with alkynes produces only the 1:1 addition product:

$$H—C{\equiv}C—H + I_2 \rightarrow (I)(H)C{=}C(H)(I) \text{ (trans)} \quad (u)$$

The rates of addition of halogens to alkynes are $Cl_2 > Br_2 > I_2$.

Reactions of halogens with alkynes are more exothermic than with alkenes, and explosive mixtures can result when chlorinations are attempted[26].

Organosilicon acetylenes react with Br_2, e.g.:

$$Me_3SiC{\equiv}CSiMe_3 + Br_2 \rightarrow Me_3SiCBr{=}CBrSiMe_3 \quad (v)^{27}$$

$$Et_3SiC{\equiv}CMe + Br_2 \rightarrow Et_3SiCBr{=}CBrMe \xrightarrow{Br_2} Et_3SiCBr_2CBr_2Me \quad (w)^{28}$$

$$Me_3SiC{\equiv}C—CH{=}CHR + Br_2 \xrightarrow[0°–5\ °C]{CCl_4} Me_3SiC(Br){=}C(Br)—CH{=}CHR \quad (x)^{29}$$

where R = CO_2Et, Ac, C_6F_5.

(C. H. VAN DYKE)

1. W. Mahler, *Inorg. Chem.*, *2*, 230 (1963).
2. J. L. Margrave, K. G. Sharp, P. W. Wilson, *J. Inorg. Nucl. Chem.*, *32*, 1813 (1970).
3. D. C. Pease, U.S. Pat. 3,026,173 (1962); *Chem. Abstr.*, *57*, 3081 (1962).
4. W. P. Neumann, *Angew. Chem., Int. Ed. Eng.*, *2*, 165 (1963).
5. J. D. Cotton, P. J. Davidson, M. F. Lappert, *J. Chem. Soc., Dalton Trans.*, 2275 (1976).
6. H.-J. Albert, U. Schröer, *J. Organomet. Chem.*, *60*, C6 (1973).
7. K. D. Bos, E. J. Bulten, J. G. Noltes, *J. Organomet. Chem.*, *67*, C13 (1974).
8. K. A. V'yunov, A. I. Ginak, *Russ. Chem. Rev.*, (*Engl. Transl.*), *50*, 151 (1981).
9. *Rodd's Chemistry of Carbon Compounds*, 2nd ed., S. Coffey, ed., Vol. 1, Part A, Elsevier, Amsterdam, 1964.
10. A. L. Henne, T. P. Walker, *J. Am. Chem. Soc.*, *67*, 1639 (1945).

11. C. Eaborn, *Organosilicon Compounds*, Butterworths, London, 1960.
12. C. Eaborn, R. W. Bott, in *Organometallic Compounds of the Group IV Elements*, A. G. MacDiarmid, ed., Vol. 1, Part I, Marcel Dekker, New York, 1968.
13. P. D. George, M. Prober, J. R. Elliot, *Chem. Rev.*, *56*, 1065 (1956).
14. F. Glockling, K. A. Hooton, *Organometallic Compounds of the Group IV Elements*, A. G. MacDiarmid, ed., Vol. 1, Part II, Marcel Dekker, New York, 1960.
15. M. Lesbre, P. Mazerolles, J. Satgé, *The Organic Compounds of Germanium*, Wiley, New York, 1971.
16. L. H. Sommer, D. L. Bailey, G. M. Goldberg, C. E. Buck, T. S. Bye, F. J. Evans, F. C. Whitmore, *J. Am. Chem. Soc.*, *76*, 1613 (1954).
17. G. H. Wagner, D. L. Bailey, A. N. Pines, M. L. Dunham, D. B. McIntire, *Ind. Eng. Chem.*, *45*, 367 (1953).
18. J. J. Eisch, M. W. Foxton, *J. Org. Chem.*, *36*, 3520 (1971).
19. K. E. Koenig, W. P. Weber, *Tetrahedron Lett.*, 2533 (1973).
20. D. L. Bailey, A. N. Pines, *Ind. Eng. Chem.*, *46*, 2363 (1954).
21. L. H. Sommer, L. J. Tyler, F. C. Whitmore, *J. Am. Chem. Soc.*, *70*, 2872 (1948).
22. P. Mazerolles, M. Lesbre, *C.R. Hebd. Séances Acad. Sci.*, *248*, 2018 (1959).
23. P. Mazerolles, M. Lesbre, S. Narre, *C.R. Hebd. Séances Acad. Sci.*, *261*, 4134 (1965).
24. A. G. Brook, J. M. Duff, G. E. Legrow, *J. Organomet. Chem.*, *122*, 31 (1976).
25. A. G. Brook, J. M. Duff, W. F. Reynolds, *J. Organomet. Chem.*, *121*, 293 (1976).
26. C. Weygand, G. Hilgetag, in *Preparative Organic Chemistry*, G. Hilgetag, A. Martin, eds., Wiley-Interscience, New York, 1972.
27. K. C. Frisch, R. B. Young, *J. Am. Chem. Soc.*, *74*, 4853 (1952).
28. A. D. Petrov, L. L. Shchukovskaya, *J. Gen. Chem. USSR* (*Engl. Transl.*), *25*, 1128 (1955).
29. T. R. Boronoeva, N. N. Belyaev, M. D. Stadnichuk, A. A. Petrov, *J. Gen. Chem. USSR* (*Engl. Transl*), *46*, 1514 (1976).

2.5.14.2. by Hydrogen Halides.

The reaction of hydrogen halides with group-IVB divalent derivatives results in new group-IVB–halogen and -hydrogen bonds:

$$R_2E + HX \rightarrow R_2E \begin{smallmatrix} \diagup H \\ \diagdown X \end{smallmatrix} \qquad \text{(a)}$$

Thus, $[CF_2]$ generated from $(CF_3)_3PF_2$ at 100°C reacts[1] with HCl to give HCF_2Cl. For Si, the low T reaction between SiF_2 and HBr results in a characteristic two atom insertion product, $BrSiF_2SiF_2H$, which undergoes redistribution[2] at RT to form Si_2F_5H, $SiF_3SiFHBr$, SiF_3SiHBr_2, $SiF_2BrSiFHBr$ and $SiF_2BrSiHBr_2$. The general reaction also accounts for the formation[3] of $HSiCl_3$ from HCl with elementary Si:

$$2\ HCl + Si \rightarrow [SiCl_2] + H_2 \qquad \text{(b)}$$

$$[SiCl_2] + HCl \rightarrow HSiCl_3 \qquad \text{(c)}$$

Divalent Ge compounds react with hydrogen halides to form a new Ge–halogen bond. The commonly cited equilibrium reaction observed with $HGeCl_3$ and its etherates illustrates the reaction[4]:

$$HGeCl_3 \rightarrow H[GeCl_3] \rightleftharpoons HCl + GeCl_2 \qquad \text{(d)}$$

$$2\ Et_2O{\cdot}HGeCl_3 \rightleftharpoons 2\ Et_2O{\cdot}HCl + GeCl_2 \qquad \text{(e)}$$

The reaction of HCl or its etherates with $(GeCl_2)_n$ polymer leads to the formation of $HGeCl_3$, or the trichlorogermane etherate[5,6]:

$$2\ Et_2O{\cdot}HCl + \frac{1}{n}(GeCl_2)_n \rightarrow 2\ Et_2O{\cdot}GeHCl_3 \qquad \text{(f)}$$

The mixed etherate $I_2ClHGe{\cdot}2\ OEt_2$ is prepared[5–7] from HCl etherate with GeI_2. In conc HBr or HI, $GeBr_2$ and GeI_2 exist in equilibrium with $HGeBr_3$[8,9] and $HGeI_3$[10,11], respectively:

$$HGeBr_3 \rightleftharpoons GeBr_2 + HBr \qquad \text{(g)}$$

$$HGeI_3 \rightleftharpoons GeI_2 + HI \qquad \text{(h)}$$

The associated form of the organohalogermylenes, EtGeCl and PhGeCl, insert[12,13] into the H—Cl bond in conc aq HCl:

$$RGeCl + HCl \xrightarrow{80\,°C} RGeCl_2H \qquad \text{(i)}$$

where R = Et, Ph. The reaction of HCl with dimethylgermylene polymer $(Me_2Ge)_n$ or telomer $Me(Me_2Ge)_nMe$ ($n \geq 2$) at 200–250°C forms[5,14] Me_2GeHCl and Me_2GeCl_2.

The divalent tin compound bis[bis(trimethylsilyl)methyl]tin undergoes oxidative addition with HCl or HF in Et_2O at RT[15].

$$R_2Sn + HX \rightarrow R_2Sn\begin{smallmatrix} \diagup H \\ \diagdown X \end{smallmatrix} \qquad \text{(j)}$$

where $R = CH(SiMe_3)_2$ and X = F, Cl. The reaction of $SnCl_2$ with HCl in Et_2O leads to[16,17] solvated $HSnCl_3$:

$$SnCl_2 + HCl \xrightleftharpoons{Et_2O} HSnCl_3 \qquad \text{(k)}$$

The active halogenostannanes generated in situ can undergo hydrostannation with, e.g., α-, β-unsaturated carbonyl compounds[18]:

$$\begin{array}{c} SnX_2 + HX \rightarrow [HSnX_3] \\ \downarrow RR'C{=}CR''{-}CR'''{=}O \\ X_3SnCRR'{-}CHR''{-}CR'''{=}O \end{array} \qquad \text{(l)}$$

when X = Cl, Br, I. These active halogenostannane intermediates are also produced from halogen acids and tin metal[18].

The reaction of $(h^5\text{-}C_5H_5)_2Pb$ with HCl results in the stepwise cleavage of the $h^5\text{-}C_5H_5$ groups[19]:

$$(h^5\text{-}C_5H_5)_2Pb \xrightarrow{HCl} C_5H_6 + h^5\text{-}C_5H_6PbCl \xrightarrow{HCl} C_5H_6 + PbCl_2 \qquad \text{(m)}$$

Hydrogen halides add to unsaturated organics to produce haloalkene derivatives, e.g.:

$$MeCH{=}CHMe + HCl \rightarrow MeCH_2CHClMe \quad (n)$$

$$\text{cyclohexene} + HBr \rightarrow \text{cyclohexyl-Br} \quad (o)$$

However, for HF, the addition may be complicated by polymerization of the alkene[20]. With unsymmetrical alkenes, the hydrogen halides undergo an addition in which the hydrogen adds to the carbon of the double bond which contains the greatest number of hydrogens, e.g.:

$$CH_2{=}CHMe + HCl \rightarrow MeCH(Cl)Me \quad (p)$$

In the presence of peroxides, HBr adds to unsymmetrical alkenes in the reverse manner. Mechanisms of these additions can be found in standard organic textbooks and reference works.

The addition of hydrogen halides to alkenylsilanes places the hydrogen on the carbon with more hydrogens, but exceptions are noted in addition reactions involving vinyl derivatives. Thus, the hydrogen halides add normally[21] to the double bond of $Me_3SiCH_2CH{=}CH_2$:

$$Me_3SiCH_2CH{=}CH_2 + HX \rightarrow Me_3SiCH_2{-}CHX{-}Me \quad (q)$$

where X = Cl, Br, I. The presence of benzoyl peroxide does not reverse the mode of addition[21] of HBr to $Me_3SiCH_2CH{=}CH_2$. The addition of HI to $Me_3SiCH{=}CH_2$ proceeds by a reverse addition, as does the addition of HBr to $Me_3SiCH{=}CH_2$ in the presence of benzoyl peroxide[22]. The latter addition does not take place during a reasonable time without the peroxide. The $AlCl_3$-catalyzed addition of HCl to $CH_2{=}CHSiCl_3$ produces β-chloroethyltrichlorosilane in a reversible equilibrium reaction[23]:

$$CH_2{=}CHSiCl_3 + HCl \xrightleftharpoons{AlCl_3} ClCH_2CH_2SiCl_3 \quad (r)$$

Hydrogen chloride does not add to $CH_2{=}CHCH_2SiCl_3$ at RT and its reaction with $CH_2{=}CHCH_2Si(OEt)_3$ cleaves the Si—C bond[24]:

$$CH_2{=}CHCH_2Si(OEt)_3 + HCl \rightarrow ClSi(OEt)_3 + CH_3CH{=}CH_2 \quad (s)$$

The addition of HBr to vinylgermanes places hydrogen on the α-carbon[25,26]:

$$n\text{-}Bu_3GeCH{=}CH_2 + HBr \rightarrow n\text{-}Bu_3GeCH_2CH_2Br \quad (t)$$

The reaction of HBr with $n\text{-}Bu_3GeCH_2CH{=}CH_2$ produces cleavage products[25,26]:

$$n\text{-}Bu_3GeCH_2CH{=}CH_2 + HBr \rightarrow n\text{-}Bu_3GeBr + CH_3CH{=}CH_2 \quad (u)$$

Hydrogen halides add to alkynes in two steps producing haloalkenes or geminal dihalides depending on whether 1 or 2 mol of hydrogen halide are used. Both additions place hydrogen on the carbon atom with more hydrogens. Thus, propyne reacts with

1 mol of HCl to form 2-chloropropene, while the reaction with 2 mol of HCl produces 2,2-dichloropropane:

$$\text{MeC}{\equiv}\text{CH} \xrightarrow{\text{HCl}} \text{MeClC}{=}\text{CH}_2 \xrightarrow{\text{HCl}} \text{Me}_2\text{CCl}_2 \qquad \text{(v)}$$

Hydrogen chloride does not react with $Me_2Si(C{\equiv}CPh)_2$, but it adds to[28] $Me_3SiC{\equiv}CCH{=}CH_2$:

$$\text{Me}_3\text{SiC}{\equiv}\text{CCH}{=}\text{CH}_2 + \text{HCl} \xrightarrow{-70^\circ\text{C}} \text{Me}_3\text{SiCH}{=}\text{CClCH}_2\text{CH}_2\text{Cl}, \quad \text{Me}_3\text{SiCH}_2\text{CCl}{=}\text{CHCH}_2\text{Cl} \qquad \text{(w)}$$

(C. H. VAN DYKE)

1. W. Mahler, *Inorg. Chem.*, *2*, 230 (1963).
2. K. G. Sharp, J. F. Bald, *Inorg. Chem.*, *14*, 2553 (1975).
3. J. Joklik, V. Bazant, *Coll. Czech. Chem. Commun.*, *29*, 603, 834 (1964).
4. V. F. Mironov, T. K. Gar, *Organomet. Chem. Rev.*, *A*, *3*, 311 (1968).
5. O. M. Nefedov, S. P. Kolesnikov, *Akad. Nauk SSSR, Bulletin Div. Chem. Sci. (Engl. Transl.)* 187 (1966).
6. O. M. Nefedov, S. P. Kolesnikov, V. I. Scheichenko, Yu. N. Sheinker, *Akad. Nauk SSSR, Proceedings, Chem. Sect. (Engl. Transl.) 162*, 510 (1965).
7. O. M. Nefedov, S. P. Kolesnikov, V. I. Scheichenko, *Angew. Chem., Int. Ed. Engl.*, *3*, 508 (1964).
8. V. F. Mironov, T. K. Gar, *Akad. Nauk SSSR, Bulletin, Div. Chem. Sci. (Engl. Transl.)* 740 (1965).
9. T. K. Gar, V. F. Mironov, *Akad. Nauk SSSR, Bulletin, Div. Chem. Sci. (Engl. Transl.)* 827 (1965).
10. P. Mazerolles, G. Manuel, *Bull. Soc. Chim. Fr.*, 2511 (1967).
11. V. F. Mironov, L. N. Kalinina, E. M. Berliner, T. K. Gar, *J. Gen. Chem. USSR* (*Engl. Transl.*), *40*, 2590 (1970).
12. J. Satgé, M. Massol, P. Rivière, *J. Organomet. Chem.*, *56*, 1 (1973).
13. M. Massol, J. Satgé, P. Rivière, and J. Barrau, *J. Organomet. Chem.*, *22*, 599 (1970).
14. O. M. Nefedov, S. P. Kolesnikov, *Akad. Nauk SSSR, Bulletin Div. Chem. Sci. (Engl. Transl.)* 725 (1964).
15. J. D. Cotton, P. J. Davidson, M. F. Lappert, *J. Chem. Soc., Dalton Trans.*, 2275 (1976).
16. O. M. Nefedov, S. P. Kolesnikov, *Izv. Nauk SSSR, Ser. Khim.*, *2*, 201 (1966).
17. O. M. Nefedov, S. P. Kolesnikov, V. I. Scheichenko, Yu. N. Sheinker, *Dokl. Akad. Nauk SSSR*, *162*, 589 (1965).
18. J. W. Burley, R. E. Hutton, V. Oakes, *J. Chem. Soc., Chem. Commun.*, 803 (1976).
19. A. K. Holliday, P. H. Makin, R. J. Puddephatt, J. D. Wilkins, *J. Organomet. Chem.*, *57*, C45 (1973).
20. A. von Grosse, C. B. Linn, *J. Organomet. Chem.*, *3*, 26 (1938).
21. L. H. Sommer, L. J. Tyler, F. C. Whitmore, *J. Am. Chem. Soc.*, *70*, 2872 (1948).
22. L. H. Sommer, D. L. Bailey, G. M. Goldberg, C. E. Buck, T. S. Bye, F. J. Evans, F. C. Whitmore, *J. Am. Chem. Soc.*, *76*, 1613 (1954).
23. G. H. Wagner, D. L. Bailey, A. N. Pines, M. L. Dunham, D. B. McIntire, *Ind. Eng. Chem.*, *45*, 367 (1953).
24. D. L. Bailey, A. N. Pines, *Ind. Eng. Chem.*, *46*, 2363 (1954).
25. F. Glockling, K. A. Hooten, *Organometalic Compounds of the Group IV Element*, A. G. MacDiarmid, ed., Vol. 1, Pt. 2, Marcel Dekker, New York, 1960.
26. P. Mazerolles, M. Lesbre, *C.R. Hebd. Séances Acad. Sci.*, *248*, 2018 (1959).
27. S. D. Ibekwe, M. J. Newlands, *J. Chem. Soc.*, 4608 (1965).
28. A. D. Petrov, S. I. Sadykh-Zade, Yu. P. Egorov, *Izv. Akad. Nauk SSSR, Otd, Khim. Nauk.*, 722 (1954); *Chem. Abstr.*, *49*, 10835 (1955).

2.5.14.3. by Other Covalent Halides.

Divalent group-IVB derivatives undergo oxidative addition (or insertion) reactions with covalent halides forming new tetravalent group IV–halogen derivatives:

$$R_2E + Z{-}X \rightarrow R_2E(X)(Z) \qquad (a)$$

where X is a halogen. For carbon compounds the reaction is illustrated by $[CH_2]$ with organic halides. Thus, the light-induced reaction of CH_2N_2 (acting as a source of $[CH_2]$) with polyhalomethanes and α-haloesters results in a methylenation at the carbon–halogen bonds[1]:

$$CCl_4 + 4\ CH_2N_2 \xrightarrow[60\ \%]{-4\ N_2} C(CH_2Cl)_4 \qquad (b)$$

$$BrCCl_3 + 4\ CH_2N_2 \xrightarrow[40\ \%]{-4\ N_2} BrCH_2C(CH_2Cl)_3 \qquad (c)$$

$$CCl_3COOMe + 3\ CH_2N_2 \xrightarrow[60\ \%]{-3\ N_2} (ClCH_2)_3CCOOMe \qquad (d)$$

The reaction of $[CH_2]$ with CF_3I occurs exclusively at the C—I bond[2]. The reaction of $[CH_2]$ with organic halides proceeds by insertion into C—H and C—C bonds as well as into C–halogen bonds, e.g.:

$$HCCl_3 + 3\ CH_2N_2 \xrightarrow[45\ \%]{-N_2} MeC(CH_2Cl)_3 \qquad (e)^1$$

$$MeCHClMe + [CH_2] \rightarrow MeCHClCH_2Me + Me_2CHCH_2Cl + Me_3CCl \qquad (f)^3$$

$$C_6H_5Cl + [CH_2] \rightarrow \text{(chlorocycloheptatriene, 54\%)} + C_6H_5CH_2Cl\ (30\%) + ClC_6H_4Me\ (16\%) \qquad (g)^4$$

The reaction of CH_2N_2 with acyl and many inorganic halides (boron halides excluded) also leads to methylenation products[5], e.g.:

$$MCl_4 + CH_2N_2 \rightarrow ClCH_2MCl_3 \qquad (h)$$

where M = Si, Ge, Sn; other systems are summarized in ref. 5. However, $[CH_2]$ is not required to be the reactive species. Indeed, a polar mechanism involving the nucleophilic attack of CH_2N_2 can give an addition complex, which then decomposes to methylenation products[5].

Halocarbenes insert into the metal–halogen bond of certain metal halides; e.g., a preparation of trichloromethyl- and dichloromethyl-Hg compounds is based on the insertion of dichloro- or monochlorocarbene into an Hg—Cl bond[6]. Phenyltrihalo- or dihalomethyl mercury compounds act as the carbene transfer reagent, e.g.;

$$PhHgCCl_2Br + PhHgCl \xrightarrow{\text{benzene, reflux}} PhHgBr + PhHgCCl_3 \quad (i)$$

$$PhHgCHClBr + PhHgCl \xrightarrow[\text{chlorobenzene}]{130°C} PhHgBr + PhHgCHCl_2 \quad (j)$$

Dichlorocarbene also inserts[6,9] into the tin–halogen bond of Me_3SnBr and Me_3SnCl but not the Si—Cl bond of Me_3SiCl or Me_2SiCl_2.

$$Me_3SnBr + PhHgCCl_2Br \xrightarrow{C_6H_6,\ 80°C} Me_3SnCCl_2Br + PhHgBr \quad (k)$$

Dichlorocarbene transfer occurs[9] between $PhHgCCl_3$ and the tetrabromides of Si, Ge and Sn, e.g.:

$$MBr_4 + PhHgCCl_3 \rightarrow BrCCl_2MBr_3 + PhHgCl \quad (l)$$

where M = Si, Ge;

$$SnBr_4 + PhHgCCl_3 \rightarrow PhHgCl + [BrCCl_2SnBr_3] \xrightarrow{SnBr_4} ClBr_2CSnBr_3 + SnBr_3Cl \quad (m)$$

The carbene, 1,2,2-trifluoroethylidene, inserts[10] into the Si—Cl bond of chlorosilanes and the Si—Br bond of Me_3SiBr:

$$Me_3SiX + [CHF_2{—}CF] \rightarrow Me_3SiCFXCHF_2 \quad (n)$$

where X = Cl, Br; yield is 2–15% based on the carbene available. Insertion of the carbene into the Si–halogen bond is more difficult than it is into the Si—H bond, and reactivity toward Si—Cl insertion decreases $Me_2SiCl_2 > MeSiCl_3 > Me_3SiCl, SiCl_4$.

Silicon difluoride and dichloride undergo oxidative addition (insertion) with covalent halides; e.g., SiF_2 reacts with CF_3I produces CF_3SiF_2I together with other products such as the two-atom insertion product, $CF_3SiF_2SiF_2I$, and other higher iodofluorosilanes[11,12]:

$$CF_3I + SiF_2 \rightarrow CF_3SiF_2I \quad (o)$$

The condensation of SiF_2 generated from SiF_4 at 1200–1800°C with BF_3 at −196°C forms[13] SiF_3BF_2 as well as $Si_2F_5BF_2$ and $Si_3F_7BF_2$:

$$SiF_2 + BF_3 \rightarrow F_2Si(BF_2)(F) \quad (p)$$

Under different conditions SiF_2 with BF_3 produces[14] $Si_2F_5BF_2$ and $SiF_3(SiF_2)_2BF_2$ together with smaller amounts of higher homologues in the series $SiF_3(SiF_2)_nBF_2$. Silicon difluoride undergoes an oxidative addition (insertion) reaction with the C—F bonds of trifluoroethylene and perfluorobenzene producing F_3Si derivatives[15,16]:

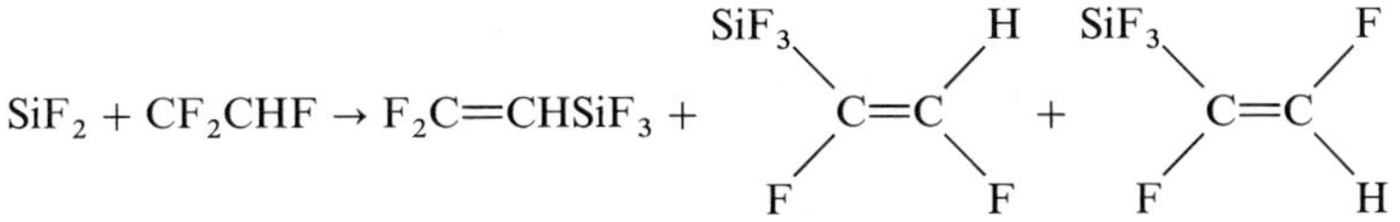

$$SiF_2 + CF_2CHF \rightarrow F_2C{=}CHSiF_3 + (SiF_3)(F)C{=}C(H)(F) + (SiF_3)(F)C{=}C(F)(H) \qquad (q)$$

$$2\ C_6F_6 + 3\ SiF_2 \rightarrow C_6F_5SiF_3 + C_6F_4(SiF_3)_2 \qquad (r)$$

Among the compounds formed[17] from SiF_2 and vinyl chloride is $\approx 1\%$ of the insertion product, $H_2C{=}C(H)SiF_2Cl$. Silicon difluoride does not react with SiF_4 at low T, and its reactions with GeF_4, NF_3, PF_3 and SiF_4 give[18] products that are not stable owing to halogen abstraction by Si.

Covalent halides oxidatively add to $SiCl_2$ which is useful in syntheses; e.g., $SiCl_3BCl_2$ and $SiCl_3PCl_2$ can be prepared in ca. 15% yield by cocondensing $SiCl_2$ with BCl_3 and PCl_3, respectively[18,19]:

$$SiCl_2 + BCl_3 \rightarrow SiCl_3BCl_2 \qquad (s)$$

$$SiCl_2 + PCl_3 \rightarrow SiCl_3PCl_2 \qquad (t)$$

The reaction of $SiCl_2$ with CCl_4 produces $SiCl_3CCl_3$, but the yield is low. Silicon dichloride inserts into the C—Cl of α-napthyl-CH_2Cl and a Si—Cl bond of:

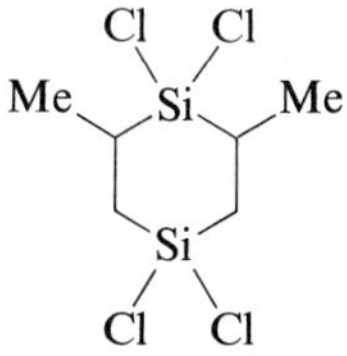

1,1,3,3-tetrachloro-2,4-dimethyl-1,3-disilacyclohexane (ring: Cl_2Si, CHMe, CH_2, CH_2, $SiCl_2$, CHMe)

to form the corresponding $—SiCl_3$ derivatives,[18,20], e.g.:

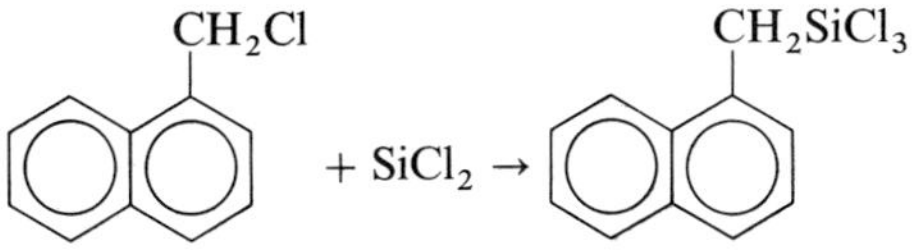

$$\alpha\text{-}C_{10}H_7CH_2Cl + SiCl_2 \rightarrow \alpha\text{-}C_{10}H_7CH_2SiCl_3 \qquad (u)$$

$$\text{(ring with } Si(Cl)_2 \text{ top, 2 Me, } SiCl_2 \text{ bottom)} + SiCl_2 \rightarrow \text{(same ring with } Si(Cl)(SiCl_3) \text{ top)} \qquad (v)$$

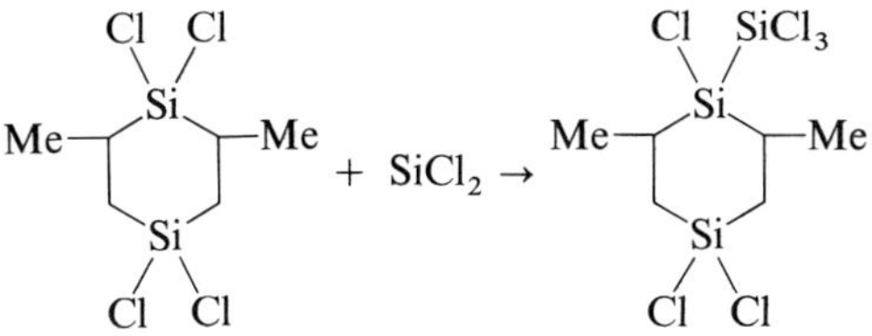

Germanium(II) halides undergo oxidative addition (insertion) with covalent halides. In particular, GeI_2 inserts into the C—I bond of organic iodides producing triiodogermane derivatives, e.g.:

$$RCH_2I + GeI_2 \xrightarrow{80-140°C} \underset{(80\text{–}100\%)}{RCH_2GeI_3} \qquad (w)^{21-23}$$

where R = H, Me n-Bu, MeO, EtOCO, I;

$$PhI + GeI_2 \xrightarrow{160°C} PhGeI_3, \qquad (x)^{23}$$

$$CF_3I + GeI_2 \xrightarrow{135°C} \underset{(43\%)}{CF_3GeI_3} \qquad (y)^{24}$$

The reaction of GeI_2 with CH_2I_2 results in the formation of the organofunctional derivative, ICH_2GeI_3, while its reaction with $Br(CH_2)_4Br$ occurs at one C—Br bond.

$$GeI_2 + Br(CH_2)_4Br \rightarrow X_3Ge(CH_2)_4Br \qquad (z)^{23}$$

where X = Br or I.

Germanium difluoride, dichloride and dibromide also undergo carbenoid insertion reactions with organic halides[25–27], e.g.:

$$ClCH_2OMe + GeF_2 \rightarrow ClGeF_2CH_2OMe \xrightarrow{EtMgBr} Et_3GeCH_2OMe \qquad (aa)$$

$$ClCH_2OMe + GeCl_2 \rightarrow Cl_3GeCH_2OMe \qquad (ab)$$

$$BrCH_2CH{=}CH_2 + GeBr_2 \rightarrow Br_3GeCHCH{=}CH_2 \qquad (ac)$$

The insertion reactions of the Ge dihalides are not limited to systems that contain carbon–halogen bonds, e.g., $GeCl_2$ inserts into the Ni—Cl bond of h^5-$C_5H_5(PR_3)NiCl$ to form the corresponding $GeCl_3$—Ni derivative[28]:

$$h^5\text{-}C_5H_5Ni(PR_3)Cl + GeCl_2{\cdot}THF \xrightarrow{benzene} h^5\text{-}C_5H_5Ni(PR_3)GeCl_3 \qquad (ad)$$

where THF = tetrahydrofuran, R = Ph, Bu. Chlorocarbonyl compounds of Mn, Fe, Mo, and W react with $GeCl_2$ (from $HGeCl_3$) in a similar manner.

$$ClM(CO)_3C_5H_5\text{-}h^5 + GeCl_2 \rightarrow Cl_3GeM(CO)_2C_5H_5\text{-}h^5 \qquad (ae)^{29}$$

$$(M, = Mo, W)$$

$$ClFe(CO)_2C_5H_5\text{-}h^5 + GeCl_2 \rightarrow Cl_3GeFe(CO)_2C_5H_5\text{-}h^5 \qquad (af)^{30}$$

$$ClMn(CO)_5 + GeCl_2 \rightarrow Cl_3GeMn(CO)_5 \qquad (ag)^{31}$$

$$ClMo(CO)_2(C_7H_7) + GeCl_2 \rightarrow Cl_3GeMo(CO)_2(C_7H_7) \qquad (ah)^{32}$$

Trihalogermyl and trihalostannyl pentacarbonyl metalates of Cr, Mo, and W can be prepared by a similar Ge or Sn dihalide insertion into a metal-halogen bond[33]

$$[Et_4N][ClM(CO)_5] + M'Cl_2 \rightarrow [Et_4N][Cl_3M'M(CO)_5] \qquad (ai)$$

$$(M' = Ge, Sn; M = Cr, Mo, W)$$

$$[Et_4N][BrW(CO)_5] + SnBr_2 \rightarrow [Et_4N][Br_3SnW(CO)_5] \qquad (aj)$$

Germanium dibromide inserts into the Ge—Br bond of $GeBr_4$ to give $Br_3GeGeBr_3$ in a reversible reaction[34] and GeF_2 reacts with chlorosilanes to form Ge—Si bonds[35]:

$$GeBr_2 + GeBr_4 \rightleftharpoons Br_3GeGeBr_3 \tag{ak}$$

$$Ph_{4-n}SiCl_n + GeF_2 \rightarrow Ph_{4-n}Si(GeF_2Cl)_n \tag{al}$$

where n = 1–3 the dibromide does not react with $GeCl_4$ or methylbromogermanes.[34] Phenylhalogermylenes, PhGeX (X = Cl, Br), insert into the Ge–halogen bond of phenylhalogermanes to form phenylhalopolygermanes[36]:

$$(PhGeCl) + PhGeCl_3 \xrightleftharpoons{20°C} \begin{cases} Ph(Cl)_2Ge{-}Ge(Cl)_2Ph \quad (20\%) \\ Ph(Cl)_2Ge{-}Ge(Ph)(Cl){-}Ge(Cl)_2Ph \quad (6\%) \\ [Ph(Cl)_2Ge]_3GePh \quad (12\%) \end{cases} \tag{am}$$

The oxidative addition (insertion) which divalent tin compounds undergo with reactive covalent halides is used to synthesize tin(IV) halogen derivatives, e.g., although $RSnX_3$ can be prepared by redistribution reactions involving, e.g., $SnCl_4$ with R_4Sn or R_2SnCl_2 compounds[37], a more direct and highly recommended procedure involves the reaction of SnX_2 with alkyl halides[38]:

$$SnX_2 + RX' \rightarrow RSnX_2X' \tag{an}$$

where R = alkyl; X, X' = halogen. The reaction can proceed without a catalyst, although the yields are low[38,39]:

$$SnI_2 + MeI \xrightarrow{160°C} \underset{(25\%)}{MeSnI_3} \tag{ao}$$

Trialkylantimonys are effective catalysts for the production of short- and long-chain $(C_1\text{-}C_{18})RSnX_3$[38]. No solvent is employed, since the procedure calls for a 200% xs of the alkyl halide to provide a suitable reaction medium. Tin(II) bromide does not insert[38] into the C—Br bond of bromobenzene, even after 15 h at 160°C in the presence of Et_3Sb. The direct reaction of tin(II) halides with α-, ω-dihaloalkanes in the presence of R_3Sb yields ω-haloalkyltins[40]:

$$SnX_2 + X(CH_2)_nX \xrightarrow{R_3Sb} X_3Sn(CH_2)_nX \tag{ap}$$

where n ≥ 3, X = halogen, mostly Br. Bis insertion occurs[40] with $SnBr_2$ and CH_2I_2:

$$SnBr_2 + 3\ CH_2I_2 \xrightarrow[120°C]{Et_3Sb} \underset{(10\%)}{ICH_2SnBr_2I} + \underset{(90\%)}{(IBr_2Sn)_2CH_2} \tag{aq}$$

Tin(II) chloride inserts into the C—Cl bond of CCl_4, producing Cl_3CSnCl_3, which then eliminates $SnCl_4$, producing perchlorinated hydrocarbons[41]. Carbon-functional chloromethylsilanes react with $SnCl_2$ in the presence of $[R_4N]^+$ salts to produce[42,43] silylmethyltin trichlorides, e.g.:

$$R_3SiCH_2Cl + SnCl_2 \rightarrow R_3SiCH_2SnCl_3 \quad \text{(ar)}$$

where $R_3 = EtOMe_2$, $(EtO)_2Me$, $(EtO)_3$. The reaction of Me_3SiCH_2I with $SnCl_2$ produces organotin chloride iodides, which can be methylated to MeMgI to produce $Me_3SiCH_2SnMe_3$:

$$Me_3SiCH_2I + SnCl_2 \rightarrow Me_3SiCH_2SnCl_mI_{3\text{-}m} \xrightarrow{MeMgI} Me_3SiCH_2SnMe_3 \quad \text{(as)}$$

where m = 0–3.

Organotin(II) compounds also undergo oxidative addition with halides, e.g.:

$$SnR_2 + MeI \rightarrow MeSnR_2I \quad \text{(at)}^{44}$$

where $R = CH(SiMe_3)_2$;

$$SnR_2 + RCl \rightarrow R_3SnCl \quad \text{(au)}^{44}$$

where $R = CH(SiMe_3)_2$;

$$(h^5\text{-}C_5H_5)_2Sn + MeI \rightarrow Me(h^1\text{-}C_5H_5)_2SnI \quad \text{(av)}^{45-47}$$

$$Sn(acac)_2 + RX \rightarrow RSn(acac)_2X \quad \text{(aw)}$$

where R = MeI, PrI, CH_2I_2, $CH_2{=}CHCH_2Br$, $PhCH_2Br$, Ph_3CBr, $BrCH_2COOEt$. The reaction of $(h^5\text{-}C_5H_5)_2Sn$ with MeI in benzene produces the insertion product $Me(h^1\text{-}C_5H_5)_2SnI$ in 90% yield; however, the precipitation of $h^5\text{-}C_5H_5SnI$ during the reaction suggests the following scheme[46,47]:

$$(h^5\text{-}C_5H_5)_2Sn + MeI \rightarrow Me(h^1\text{-}C_5H_5)_2SnI \underset{}{\overset{(h^5\text{-}C_5H_5)_2Sn}{\rightleftharpoons}} Me(h^1\text{-}C_5H_5)_3Sn + h^5\text{-}C_5H_5SnI$$

$$h^5\text{-}C_5H_5SnI \xrightarrow{MeI} Me(h^1\text{-}C_5H_5)SnI_2 \xrightarrow{Me(h^1\text{-}C_5H_5)_3Sn} Me(h^1\text{-}C_5H_5)_2SnI \quad \text{(ax)}$$

The interaction of $(h^5\text{-}C_5H_5)_2Sn$ with covalent halides may not proceed by oxidative addition; e.g., $(h^5\text{-}C_5H_5)_2Sn$ with benzyl bromide and benzyl chloride produces $h^5\text{-}C_5H_5SnBr$ and $h^5\text{-}C_5H_5SnCl$, respectively[47]:

$$PhCH_2X + (h^5\text{-}C_5H_5)_2Sn \xrightarrow{benzene} PhCH_2\text{-}C_5H_5 + PhCH_2\text{-}C_5H_5 + h^5\text{-}C_5H_5SnX \quad \text{(ay)}$$

and the reaction of $(h^5\text{-}C_5H_5)_2Sn$ with Ph_3CBr produces $h^5\text{-}C_5H_5SnBr$ together with 1- and 2-triphenylmethylcyclopentadiene[47]:

$$Ph_3CBr + (h^5\text{-}C_5H_5)_2Sn \xrightarrow{benzene} Ph_3C\text{-}C_5H_5 + h^5\text{-}C_5H_5SnBr \quad \text{(az)}$$

Alkytin(IV) chlorides exchange with $(h^5\text{-}C_5H_5)_2Sn$, producing[46] alkylcyclopentadienyltin(IV) chlorides and $h^5\text{-}C_5H_5SnCl$ or $SnCl_2$:

$$(h^5\text{-}C_5H_5)_2Sn + Me_2SnCl_2 \xrightarrow{\text{benzene}} Me_2(h^1\text{-}C_5H_5)SnCl + h^5\text{-}C_5H_5SnCl \quad \text{(ba)}$$

$$(h^5\text{-}C_5H_5)_2Sn + 2\ MeSnCl_3 \xrightarrow{\text{benzene}} 2\ Me(h^1\text{-}C_5H_5)SnCl_2 + SnCl_2 \quad \text{(bb)}$$

Tin(II) compounds react with transition-metal and other metal halides to produce tin(IV) halogen derivatives, e.g., $h^5\text{-}C_5H_5(CO)_2FeCl$ with $SnCl_2\cdot 2\ H_2O$ forms[48] $h^5\text{-}C_5H_5(CO)_2FeSnCl_3$:

$$h^5\text{-}C_5H_5(CO)_2FeCl + SnCl_2\cdot 2\ H_2O \xrightarrow{\text{MeOH}} h^5\text{-}C_5H_5(CO)_2FeSnCl_3 + 2\ H_2O \quad \text{(bc)}$$

Tin(II) chloride also inserts into the Pt—Cl bond, forming Pt—$SnCl_3$ complexes[49–51]:

$$\text{trans-}[PtHCl(PPh_3)_2] + SnCl_2\cdot 2\ H_2O \xrightarrow{\text{benzene-ether}} \text{trans-}[PtH(SnCl_3)(PPh_3)_2] \quad \text{(bd)}$$

The latter are active and selective hydroformylation catalysts[51,52].

The reaction of $SnCl_2$ with $Pt(\pi\text{-}C_3H_5)(PPh_3)_2Cl$ produces the ionic and covalent complexes $[Pt(\pi\text{-}C_3H_5)(PPh_3)_2][SnCl_3]$ and $PtSnCl_3(\pi\text{-}C_3H_5)PPh_3$, respectively[53]. Numerous other transition metal complexes containing the trihalogenotin group are prepared in a similar manner[53–57].

A five coordinated Rh(I) complex containing the $SnClBr_2$ group is also prepared by an $SnBr_2$ insertion involving a Rh—Cl bond[58]:

$$\tfrac{1}{2}\ [RhCl(COD)]_2 + dppe + SnBr_2 \rightarrow Rh(SnClBr_2)COD(dppe) \quad \text{(be)}$$

where COD = cyclooctadiene and dppe = bis(1,2-diphenylphospino)ethane. The reaction of $SnCl_2$ with chlorobis(polyfluorophenyl)thallium(III) compounds forms neutral bis(polyfluorophenyl)tin(IV) dichloro compounds that react with QCl halides (Q = Et_4N or Ph_3BzP) to form stable solids[59]:

$$TlR_2Cl + SnCl_2 \xrightarrow{-TlCl} SnR_2Cl_2 \xrightarrow{QCl} Q[SnR_2Cl_3] \quad \text{(bf)}$$

where R = C_6F_5, 2,3,5,6-C_6F_4H, 2,4,6-$C_6F_3H_2$; Q = Ph_3BzP, Et_4N.

Anhydrous Fe_2Cl_6 in Et_2O with 1:3 $(h^5\text{-}C_5H_5)_2Sn$ leads[60] to rapid ligand exchange and redox, producing $(h^5\text{-}C_5H_5)_2Fe$, $SnCl_2$ and the tin(IV) derivative $(h^1C_5H_5)_2SnCl_2$:

$$3\ (h^5\text{-}C_5H_5)_2Sn + Fe_2Cl_6 \rightarrow 2\ SnCl_2 + (h^1\text{-}C_5H_5)_2SnCl_2 + 2\ (h^5\text{-}C_5H_5)_2Fe \quad \text{(bg)}$$

The reactions of dienylmetal carbonyl halides, $(h^4\text{-dienyl})M(CO)_nL_mY$, with SnX_2 compounds (M = Fe, Mo or W; Y = Cl, Br, I; X = Cl, Br, or Y; L = ligand) produces the corresponding trihalotin(IV) derivatives, $[(h^4\text{-dienyl})M(CO)_nL_mSnX_2Y]$, in most cases, although reactions involving SnI_2 do not follow this pattern[61]. Bis[bis(trimethylsilyl)methyl]tin undergoes reaction with $h^5\text{-}C_5H_5(CO)_2FeCl$ and $[PtCl_2(PEt_3)_2]_2$ to produce chlorotin derivatives[44]:

$$SnR_2 + h^5\text{-}C_5H_5(CO)_2FeCl \rightarrow h^5\text{-}C_5H_5(CO)_2FeSnR_2Cl \quad \text{(bh)}$$

$$4\ SnR_2 + [PtCl_2(PEt_3)_2]_2 \rightarrow 2\ PtClPEt_3(SnR_2)(SnR_2Cl) + PEt_3 \quad \text{(bi)}$$

where R = $CH(SiMe_3)_2$. However, $Sn[CH(SiMe_3)_2]_2$ with $RhCl(PPh_3)_3$ in refluxing toluene forms[44] $[RhCl(PPh_3)_2]_2$ and a small amount of $RhCl(PPh_3)_2SnR_2$ [R = $CH(SiMe_3)_2$].

A procedure for preparing alkyltriiodo lead compounds is based on the reaction of an alkyl iodide with PbI_2 in the presence of Me_3Sb[62].

$$RI + PbI_2 \xrightarrow{140°C} RPbI_3 \tag{bj}$$

(R = Et, n-Pr, n-Bu)

Polyhalomethanes add to unsaturated group IVB derivatives to form haloalkyl addition compounds; e.g., CCl_4 and CCl_3Br add to vinylsilanes and stannanes at >50°C in the presence of an organic peroxide[63–65], e.g.:

$$Et_3SiCH{=}CH_2 + CCl_4 \xrightarrow{Bz_2O_2} \underset{43.5\%}{Et_3SiCHClCH_2CCl_3} \tag{bk}$$

$$Cl_3SiCH{=}CH_2 + CCl_3Br \xrightarrow[136.5°C]{Bz_2O_2} \underset{70\%}{Cl_3SiCHBrCH_2CCl_3} \tag{bl}$$

$$Et_3SnCH{=}CH_2 + CCl_3X \xrightarrow[90°–95°C]{Bz_2O_2} Et_3SnCHXCH_2CCl_3 \tag{bm}$$

where X = Cl or Br.

Halosilanes add to organosilanes containing Si multiply bonded to other elements; e.g., the silaethylene $Me_2Si{=}CH_2$ (thermally produced from dimethylsilacyclobutane) reacts with halosilanes to produce halogen-substituted disilamethylene derivatives (halosilylation reaction) difficult to prepare otherwise[66]:

$$Me_2Si{=}CH_2 + SiX_4 \rightarrow X_3SiCH_2SiMe_2X \tag{bn}$$

where X = F (44%), X = Cl (20%);

$$Me_2Si{=}CH_2 + MeSiCl_3 \rightarrow \underset{16\%}{Cl_3SiCH_2SiMe_2Cl} \tag{bo}$$

$$Me_2Si{=}CH_2 + Me_2SiCl_2 \rightarrow \underset{13\%}{ClMe_2SiCH_2SiMe_2Cl} \tag{bp}$$

In the reaction of $HSiCl_3$ with thermally produced $Me_2Si{=}CH_2$, the Si—Cl adds (halosilylation) rather than the Si—H bond (hydrosilylation)[66]:

$$Me_2Si{=}CH_2 + HSiCl_3 \rightarrow \underset{(22\%)}{HCl_2SiCH_2SiMe_2Cl} \tag{bq}$$

A Si—Cl bond results[67] from the addition of Me_3SiCl to the double-bonded intermediate, $Me_2Si{=}NSiMe_3$:

$$Me_2Si{=}NSiMe_3 + Me_3SiCl \rightarrow Me_2\underset{\substack{|\\ Cl}}{Si}N(SiMe_3)_2 \tag{br}$$

The silene $Me_2Si{=}C(SiMe_3)_2$ undergoes insertion[68] into group-IVB–chlorine bonds of Me_3ECl (E = Si, Ge, Sn), although the reaction is fast and quantitative only for Me_3SnCl:

$$Me_2Si{=}C(SiMe_3)_2 + Me_3ECl \rightarrow \underset{\displaystyle Cl}{Me_2Si}{-}\underset{\displaystyle EMe_3}{C(SiMe_3)_2} \qquad \text{(bs)}$$

where E = Si, Ge, Sn.

(C. H. VAN DYKE)

1. W. H. Urry, J. R. Eiszner, *J. Am. Chem. Soc.*, *74*, 5882 (1952).
2. M. Hudlický, *Coll. Czeck. Chem. Comm.*, *28*, 2824 (1963).
3. V. Franzen, *Justus Liebigs Ann. Chem.*, *627*, 22 (1959).
4. H. Meerwein, H. Disselnkötter, F. Rappen, H. v. Rintelen, H. von de Vloed, *Justus Liebigs Ann. Chem.*, *604*, 151 (1957).
5. D. Seyferth, *Chem. Rev.*, *55*, 1155 (1955).
6. D. Seyferth, M. E. Gordon, K. V. Darragh, *J. Organomet. Chem.*, *14*, 43 (1968).
7. M. E. Gordon, K. V. Darragh, D. Seyferth, *J. Am. Chem. Soc.*, *88*, 1831 (1966).
8. D. Seyferth, F. M. Armbrecht, *J. Organomet. Chem.*, *16*, 249 (1969).
9. M. Wiedenbruch, C. Pierrard, *J. Organomet. Chem.*, *71*, C29 (1974).
10. R. N. Haszeldine, A. E. Tipping, R. O'B. Watts, *J. Chem. Soc., Chem. Commun.*, 1364 (1969).
11. J. L. Margraave, K. G. Sharp, P. W. Wilson, *J. Inorg. Nucl. Chem.*, *32*, 1817 (1970).
12. K. G. Sharp, T. D. Coyle, *J. Fluorine Chem.*, *1*, 249 (1971/72).
13. D. L. Smith, R. Kirk, P. L. Timms, *J. Chem. Soc., Chem. Commun.*, 295 (1972).
14. P. L. Timms, T. C. Ehlert, J. L. Margrave, F. E. Brinckmann, F. E. Farrar, T. D. Coyle, *J. Am. Chem. Soc.*, *87*, 3819 (1965).
15. J. C. Thompson, J. L. Margrave, P. L. Timms, *J. Chem. Soc., Chem. Commun.*, 566. (1966).
16. P. L. Timms, D. D. Stump, R. A. Kent, J. L. Margrave, *J. Am. Chem. Soc.*, *88*, 940 (1966).
17. C. Liu, H. Hwang, *J. Am. Chem. Soc.*, *101*, 2996 (1979).
18. P. L. Timms, *Adv. Inorg. Chem. Radiochem.*, *14*, 121 (1972).
19. P. L. Timms, *Inorg. Chem.*, *7*, 387 (1968).
20. E. A. Chernyshev, N. G. Komalenkova, *J. Gen. Chem. USSR* (*Engl. Transl.*), *46*, 1286 (1976).
21. E. A. Flood, *J. Am. Chem. Soc.*, *55*, 4935 (1933).
22. E. A. Flood, L. S. Foster, K. L. Godfrey, *Inorg. Synth.*, *3*, 64 (1950).
23. M. Lesbre, P. Mazerolles, G. Manuel, *C.R. Hebd. Séances Acad. Sci.*, *257*, 2303 (1963).
24. H. C. Clark, C. J. Willis, *J. Am. Chem. Soc.*, *84*, 898 (1962).
25. P. Rivière, J. Satgé, A. Boy, *J. Organomet. Chem.*, *96*, 25 (1975).
26. V. F. Mironov, T. K. Gar, V. Z. Anisimova, E. M. Berliner, *J. Gen. Chem. USSR* (*Engl. Transl.*), *37*, 2323 (1967).
27. V. F. Mironov, T. K. Gar, *Organomet. Chem. Rev.*, *A*, *3*, 311 (1968).
28. F. S. Denisov, A. V. Gur'ev, *Vestn. Mosk. Univ. Khim.*, *17*, 624 (1976); *Chem. Abstr.*, *86*, 140223 (1977).
29. A. N. Nesmeyanov, K. N. Anisimov, N. E. Kolobova, M. Ya. Zakharova, *Akad. Nauk SSSR, Bulletin, Div. Chem. Sci.* (*Engl. Transl.*), 1813 (1967).
30. A. N. Nesmeyanov, K. N. Anisimov, N. E. Kolobova, F. S. Denisov, *Akad. Nauk SSSR, Bulletin, Div. Chem. Sci.* (*Engl. Transl.*), 2185 (1966).
31. A. N. Nesmeyanov, K. N. Anisimov, N. E. Kolobova, A. B. Antonova, *Akad. Nauk SSSR, Bulletin, Div. Chem. Sci.* (*Engl. Transl.*), 1284 (1965).
32. E. E. Isaacs, W. A. G. Graham, *Can. J. Chem.*, *53*, 975 (1975).
33. E. E. Isaacs, W. A. G. Graham, *Can. J. Chem.*, *53*, 467 (1975).
34. M. D. Curtis, P. Wolber, *Inorg. Chem.*, *11*, 431 (1972).
35. J. Satgé, P. Rivière, A. Boy, *C.R. Hebd. Séances, Acad. Sci., Ser. C.*, *278*, 1309 (1974).
36. P. Rivière, J. Satgé, *Synth. Inorg. Metal-Org. Chem.*, *1*, 13 (1971).
37. J. G. A. Luijten, G. J. M. Van der Kerk, in *Organometallic Compounds of the Group IV Elements*, A. G. MacDiarmid, ed., Vol. 1, Pt. 1, Marcel Dekker, New York, 1968.
38. E. J. Bulten, *J. Organomet. Chem.*, *97*, 167 (1975).
39. P. Pfeiffer, I. Heller, *Chem. Ber.*, *37*, 4618 (1904).
40. E. J. Bulten, H. F. M. Gruter, H. F. Martens, *J. Organomet. Chem.*, *117*, 329 (1976).

41. V. F. Mironov, V. I. Shiryaev, V. P. Kochergin, *J. Gen. Chem. USSR (Engl. Transl.)*, *46*, 715 (1976).
42. V. F. Mironov, V. I. Shiryaev, E. M. Stepina, V. V. Yankov, V. P. Kochergin, *J. Gen. Chem. USSR (Engl. Transl.)* *45*, 2404 (1976).
43. V. I. Shiryaev, V. P. Kochergin, E. M. Stepina, T. S. Kuptsova, E. M. Protasov, A. V. Kisin, V. F. Mironov, *J. Gen. Chem. USSR (Engl. Transl.)* *47*, 1601 (1977).
44. J. D. Cotton, P. J. Davidson, M. F. Lappert, *J. Am. Chem. Soc., Dalton Trans.*, 2275 (1976).
45. H.-J. Albert, U. Schröer, *J. Organomet. Chem.*, *60*, C6 (1973).
46. K. D. Bos, E. J. Bulten, J. G. Noltes, *J. Organomet. Chem.*, *67*, C13 (1974).
47. K. D. Bos, E. J. Bulten, J. G. Noltes, *J. Organomet. Chem.*, *99*, 397 (1975).
48. F. Bonati, G. Wilkinson, *J. Chem. Soc.*, 179 (1964).
49. J. C. Bailar, H. Itatani, *Inorg. Chem.*, *4*, 1618 (1965).
50. M. C. Baird, *J. Inorg. Nucl. Chem.*, *29*, 367 (1967).
51. C.-Y. Hsu, M. Orchin, *J. Am. Chem. Soc.*, *97*, 3553 (1975).
52. I. Schwager, J. F. Knifton, *J. Catal.*, *45*, 256 (1976).
53. J. N. Crosby, R. D. W. Kemmitt, *J. Organomet. Chem.*, *26*, 277 (1971).
54. P. A. McArdle, A. R. Manning, *J. Chem. Soc., Chem. Commun.*, 417 (1967).
55. M. van der Akker, F. Jellinek, *J. Organomet. Chem.*, *10*, P37 (1967).
56. J. V. Kingston, G. R. Scollary, *J. Chem. Soc. (A)*, 3399 (1971).
57. H.-E. Sasse, G. Hoch, M. L. Ziegler, *Z. Anorg. Allgem. Chem.*, *406*, 263 (1974).
58. R. Uson, L. A. Oro, M. T. Pinillos, A. Arruebo, K. A. Ostoja Starzewski, P. S. Pregosin, *J. Organomet. Chem.*, *192*, 227 (1980).
59. R. Uson, A. Laguna, T. Cuenca, *J. Organomet. Chem.*, *194*, 271 (1980).
60. P. G. Harrison, J. A. Richards, *J. Organomet. Chem.*, *108*, 35 (1976).
61. B. O'Dwyer, A. R. Manning, *Inorg. Chim. Acta*, *38*, 103 (1980).
62. G. Chobert, M. Devaud, *J. Organomet. Chem.*, *153*, C23 (1978).
63. A. F. Gordon, U.S. Pat. 2,715,113 (1955); *Chem. Abstr.*, *50*, 7131 (1956).
64. A. D. Petrov, E. A. Chernyshev, M. Bisku, *Izvest. Akad. Nauk SSSR, Otd. Khim. Nauk*, 1445 (1956).
65. D. Seyferth, *J. Org. Chem.*, *22*, 1252 (1957).
66. R. D. Bush, C. M. Golino, L. H. Sommer, *J. Am. Chem. Soc.*, *96*, 7105 (1974).
67. N. Wiberg, G. Preiner, *Angew. Chem. Int. Ed. Engl.*, *17*, 362 (1978).
68. N. Wiberg, *J. Organometal. Chem.*, *273*, 141 (1984).

2.5.15. Halides of Low-Valent Group-IVB Elements

2.5.15.1. Divalent Carbon Halides

2.5.15.1.1. Mono- and Di-halocarbenes

Halocarbenes can be generated as reactive intermediates by [1–3]:

(i) Base with a Haloform or Related Derivative. Mechanistic studies identify halocarbenes in the basic hydrolysis of haloforms. When the reaction is used to add the dihalocarbene to nucleophiles, such as olefins, aprotic solvents are required to produce good yields of product. Inert solvents, such as benzene, can be used or the reactions involving olefins are carried out in xs olefin as solvent. The base commonly used is t-BuOK, e.g.[4–7]:

$$CHCl_3 + [OH]^- \rightarrow H_2O + [CCl_3]^- \rightarrow CCl_2 + Cl^- \quad \text{(a)}$$

$$CHCl_3 + [\text{t-BuO}]^- \rightarrow \text{t-BuOH} + [CCl_3]^- \rightarrow CCl_2 + Cl^- \quad \text{(b)}$$

$$CHClF_2 + [\text{i-PrO}]^- \rightarrow \text{i-PrOH} + [CClF_2]^- \rightarrow CF_2 + Cl^- \quad \text{(c)}$$

$$CHBr_3 + [\text{t-BuO}]^- \rightarrow \text{t-BuOH} + [CBr_3]^- \rightarrow CBr_2 + Br^- \quad \text{(d)}$$

Dihalocarbenes can be generated from haloforms and aq base in a two-phase system[8].

An unusually simple procedure for generating CCl_2 on a small scale is by the action of ultrasound derived from a common laboratory cleaner on a stirred $NaOH/CHCl_3$ two phase system[9].

At 150°–200°C, ethylene oxide reacts with the haloforms $HCCl_3$, $HCClF_2$ and $HCCl_2F$ in the presence of tetraalkylammonium halides to produce CCl_2, CF_2 and CFCl respectively[10].

(ii) Thermal Decarboxylation of Trihaloacetic Acid Derivatives. Salts of trichloroacetic acid undergo thermal decarboxylation in aprotic solvents to produce CCl_2 under neutral conditions[11,12]:

$$CCl_3CO_2Na \rightarrow CO_2 + [NaCCl_3] \rightarrow NaCl + CCl_2 \tag{e}$$

The sodium salt is used since it is easily prepared anhydrous. 1,2-Dimethoxyethane (DME) is a suitable solvent. Difluorocarbene can be prepared by the thermal decomposition of CF_2ClCO_2Na in diglyme[13]:

$$CF_2ClCO_2Na \rightarrow CO_2 + [CF_2ClNa] \rightarrow CF_2 \tag{f}$$

and by the decomposition[14] of $CF_3CO_2SiMe_3$ at 450°C:

$$CF_3CO_2SiMe_3 \xrightarrow{450\ °C} Me_3SiF + CO_2 + CF_2 \tag{g}$$

(iii) Alkoxides with Esters of Trihaloacetic Acid and Related Derivatives. A synthesis of CCl_2 is based on esters of trichloroacetic acid with alkoxides[15,16] such as t-BuOK, NaOEt or NaOMe:

$$Cl_3C{-}\overset{\overset{O}{\|}}{C}{-}OR + [RO]K \rightarrow RO\overset{\overset{O}{\|}}{C}{-}OR + KCl + CCl_2 \tag{h}$$

where R = t-Bu.

$$Cl_3C{-}\overset{\overset{O}{\|}}{C}{-}OMe + [MeO]^- \rightarrow MeO{-}\overset{\overset{O}{\|}}{C}{-}OMe + [CCl_3]^- \rightarrow CCl_2 + Cl^- \tag{i}$$

Excellent yields of CCl_2 are obtained as determined by the amount of dichlorocyclopropanes produced when the reaction is carried out with an olefin present. Pentane or xs olefin is a suitable solvent. Dichlorocarbene is also produced in the basic cleavage of t-butyldichloroacetate[15,17], although the yield is much lower than when the trichloroacetate is used.

Dichlorocarbene can also be generated by treating $(Cl_3C)_2CO$ with a base such as NaOMe in an aprotic solvent[18,19]:

$$Cl_3C{-}\overset{\overset{O}{\|}}{C}{-}CCl_3 + [MeO]^- \rightarrow Cl_3C{-}\overset{\overset{O}{\|}}{C}{-}OMe + [CCl_3]^- \rightarrow CCl_2 + Cl^- \tag{j}$$

Chlorofluorocarbene can be prepared[20] by sym-difluorotetrachloroacetone with t-BuOK:

$$FCl_2C\overset{\overset{O}{\|}}{C}{-}CCl_2F + 2\,[RO]^- \rightarrow RO{-}\overset{\overset{O}{\|}}{C}{-}OR + 2\,[CCl_2F]^- \rightarrow 2\ CFCl + 2\ Cl^- \tag{k}$$

(iv) Lithium Alkyls or Lithium and Tetrahalomethanes (via Halogen–Metal Exchange Products). Organolithium compounds and polyhalogenated methanes result in halogen–metal exchange that forms dihalocarbenes; e.g., the exchange product formed from n-BuLi with Br_2CF_2 decomposes via elimination[21–23], producing CF_2:

$$Br_2CF_2 + \text{n-BuLi} \rightarrow \text{n-BuBr} + [CBrF_2Li] \tag{l}$$

$$[CBrF_2Li] \rightarrow LiBr + CF_2 \tag{m}$$

Low yields of CCl_2 result from MeLi or n-BuLi with CCl_4; however, $CBrCl_3$ or $CIBr_3$ results[24] in CCl_2:

$$CBrCl_3 + \text{n-BuLi} \rightarrow \text{n-BuBr} + CCl_3Li \rightarrow LiCl + CCl_2 \tag{n}$$

$$CCl_3I + MeLi \rightarrow MeI + CClI_3Li \rightarrow LiCl + CCl_2 \tag{o}$$

The action of Li alkyls with CH_2Cl_2 produces monochlorocarbene[25,26].

$$CH_2Cl_2 + RLi \rightarrow RH + [CHCl_2Li] \rightarrow LiCl + CHCl \tag{p}$$

The action of CCl_4 with Li metal also forms[27] CCl_2. The action of Li metal with $HCCl_3$ produces both CCl_2 and CClH, while its reaction with $HCBr_3$ produces mainly CBrH.

(v) Thermal or Photochemical Decomposition of Polyhalomethyl Derivatives. Several polyhalomethyl metal or metalloid derivatives produce halocarbenes during thermal decomposition; e.g., trihalo(trichloromethyl)silanes produce CCl_2 in the gas phase or in solution upon thermolysis, although utility is limited by the elevated T required[28–31]:

$$X_3SiCCl_3 \rightarrow X_3SiCl + CCl_2 \tag{q}$$

where X = Cl, 250°C; X = F, 100–120°C. The conversion can be achieved at RT by trimethyl(trihalomethyl)silanes with finely powdered dry KF, preferably in the presence of a catalytic amount of 18-crown-6 ether[32]:

$$Me_3SiCX_3 + KF \xrightarrow{25\,°C} Me_3SiF + KX + CX_2 \tag{r}$$

where X = Cl, Br. The formation of CF_2 also results from the thermal decomposition of Me_3SnCF_3 [33] and trifluoromethyl derivatives of pentavalent phosphorus[34], e.g.:

$$Me_3SnCF_3 \xrightarrow{150\,°C} Me_3SnF + CF_2 \tag{s}$$

$$(CF_3)_3PF_2 \xrightarrow{120\,°C} (CF_3)_2PF_3 + CF_2 \tag{t}$$

$$(CF_3)_2PF_3 \rightarrow CF_3PF_4 + CF_2 \tag{u}$$

$$CF_3PF_4 \rightarrow PF_5 + CF_2 \tag{v}$$

The reaction of $SnCl_2$ with CCl_4 produces Cl_3CSnCl_3, which decomposes by eliminating $SnCl_4$, providing a source[35] of CCl_2:

$$SnCl_2 + CCl_4 \rightarrow Cl_3CSnCl_3 \rightarrow SnCl_4 + CCl_2 \tag{w}$$

The thermolysis or I^- induced decomposition of phenyl (trihalomethyl)mercury reagents generates dihalocarbenes under mild conditions[36–38]:

$$PhHgCX_3 \xrightarrow[NaI + 30\,°C]{80\,°C\ or} PhHgX + CX_2 \tag{x}$$

This is a good procedure for generating dihalocarbenes in anhyd systems. The compound $PhHgCCl_2Br$ is an effective CCl_2-transfer agent even at RT. Sodium iodide is used to aid the generation of dihalocarbene from the more stable phenyl(trihalomethyl)mercury compounds, such as $PhHgCF_3$ and $PhHgCCl_2F$, e.g.:

$$PhHgCF_3 + I^- \rightarrow PhHgI + [CF_3]^- \rightarrow CF_2 + F^- \tag{y}$$

Among the photolysis reactions that yield dihalocarbenes is the irradiation[39] of C_2F_4 with CF_3I or C_2F_5I and the irradiation of C_2F_4 in the presence of Hg metal[40] or NOF[41]. Difluorocarbene is generated from halogenated ketones by high-intensity flash-photolysis techniques[42].

The photochemical or thermal decomposition of difluorodiazirine, CF_2N_2 forms difluorocarbene[43]:

$$CF_2N_2 \xrightarrow[UV]{120\,°C\ or} CF_2 + N_2 \tag{z}$$

Dichlorocarbene is a product of the pyrolysis[44] of $HCCl_3$ and the cracking of CCl_4 on a hot tungsten wire[45].

Electrical discharge generates dihalocarbenes; e.g., subjecting perfluorocyclopropane, CF_4, C_2F_3Cl or $(CF_3)_2CO$ to an electrical discharge forms[46,47] CF_2. The electrochemical preparation of CF_2 is known[48].

(C. H. VAN DYKE)

1. E. Chinoporos, *Chem. Rev., 63,* 235 (1963).
2. W. Kirmse, *Carbene Chemistry,* Academic Press, New York, 1964.
3. J. Hine, *Divalent Carbon,* Ronald Press, New York, 1964.
4. W. von E. Doering, A. K. Hoffman, *J. Am. Chem. Soc., 76,* 6162 (1954).
5. J. Hine, A. M. Dowell, J. E. Singly, *J. Am. Chem. Soc., 78,* 479 (1956).
6. J. Hine, K. Tanabe, *J. Am. Chem. Soc., 80,* 3002 (1958).
7. P. S. Skelly, A. U. Garner, *J. Am. Chem. Soc., 78,* 5430 (1956).
8. M. Makosza, *Pure Appl. Chem., 43,* 439 (1975).
9. S. L. Regen, A. Singh, *J. Org. Chem., 47,* 1587 (1982).
10. P. Weyerstahl, D. Klamann, C. Finger, F. Nerdel, J. Buddrus, *Chem. Ber., 100,* 1858 (1967).
11. W. M. Wagner, H. Kloosterziel, S. van der Ven, *Recl. Trav. Chim. Pays-Bas, 80,* 740 (1961).
12. W. M. Wagner, H. Kloosterziel, A. F. Bickel, *Recl. Trav. Chim. Pays-Bas, 81,* 925, 933 (1962).
13. J. M. Birchall, G. W. Cross, R. N. Haszeldine, *Proc. Chem. Soc.,* 81, (1960).
14. V. F. Mironov, V. D. Sheludyakov, V. V. Shcherbinin, N. A. Viktorov, O. M. Rad'kova, *J. Gen. Chem.* USSR (*Engl. Transl.*), 44, 2394 (1974).
15. W. E. Parham, F. C. Loew, *J. Org. Chem., 23,* 1705 (1958).
16. W. E. Parham, E. E. Schweizer, *J. Org. Chem., 24,* 1733 (1959).
17. W. E. Parham, F. C. Loew, E. E. Schweizer, *J. Org. Chem., 24,* 1900 (1959).
18. P. K. Kadaba, J. O. Edwards, *J. Org. Chem., 25,* 1431 (1960).
19. F. W. Grant, W. B. Cassic, *J. Org. Chem., 25,* 1433 (1960).
20. B. Farah, S. Horensky, *J. Org. Chem., 28,* 2494 (1963).
21. V. Franzen, *Angew. Chem., 72,* 566 (1960).
22. V. Franzen, L. Firkentscher, *Chem. Ber., 95,* 1958 (1962).
23. V. Franzen, *Chem. Ber., 95,* 1964 (1962).

24. W. Miller, C. S. Y. Kim, *J. Am. Chem. Soc.*, *81*, 5008 (1959).
25. G. L. Closs, L. E. Closs, *J. Am. Chem. Soc.*, *81*, 4996 (1959).
26. G. L. Closs, L. E. Closs, *J. Am. Chem. Soc.*, *82*, 5723 (1960).
27. O. M. Nefedov, A. A. Ivaschenko, M. N. Manakov, W. I. Sherjajev, A. D. Petrov, *Akad. Nauk SSSR Bulletin, Div. Chem. Sci.*, (*Engl. Transl.*), 343 (1962).
28. W. I. Bevan, R. N. Haszeldine, J. C. Young, *Chem. Ind.* (*London*), 789 (1961).
29. R. N. Haszeldine, W. I. Bevan, US Pat. 3,233,000 (1966); *Chem. Abstr.*, *64*, 15,770 (1960).
30. R. Fields, R. N. Haszeldine, D. Peters, *J. Chem. Soc.*, C, 167 (1969).
31. J. M. Birchall, G. N. Gilmore, R. N. Haszeldine, *J. Chem. Soc. Perkins Trans.*, 2530 (1974).
32. R. F. Cunico, B. B. Chou, *J. Organomet. Chem.*, *154*, C45 (1978).
33. H. C. Clark, C. J. Willis, *J. Am. Chem. Soc.*, *82*, 1888 (1960).
34. W. Mahler, *Inorg. Chem.*, *2*, 230 (1963).
35. V. F. Mironov, V. I. Shiryaev, V. P. Kochergin, *J. Gen. Chem. USSR* (*Engl. Transl.*), 46, 715 (1976).
36. D. Seyferth, J. M. Burlitch, J. K. Herren, *J. Org. Chem.*, *27*, 1491 (1962).
37. D. Seyferth, *Acc. Chem. Res.*, *5*, 65 (1972).
38. D. Seyferth, in *Carbenes*, Vol. 2, R. A. Moss, M. Jones, Jr. eds., Wiley, New York, 1975.
39. R. N. Haszeldine, *J. Chem. Soc.*, 3761 (1953).
40. E. Atkinson, *J. Chem. Soc.*, 2684 (1952).
41. S. Andreades, *Chem. Ind.* (*London*), 782 (1962).
42. J. P. Simons, A. J. Yarwood, *Nature* (*London*), *187*, 316 (1960).
43. R. A. Mitsch, *J. Heterocycl. Chem.*, *1*, 59, 225 (1964); *J. Am. Chem. Soc.*, *87*, 758 (1965).
44. G. P. Semeluk, R. B. Bernstein, *J. Am. Chem. Soc.*, *79*, 46 (1957).
45. L. P. Blanchard, P. LeGoff, *Can. J. Chem.*, *35*, 89 (1957).
46. S. V. R. Mastrangelo, US. Pat. 3,196,114 (1965); *Chem. Abst.*, *63*, 9807 (1965).
47. F. X. Powell, D. R. Lide, *J. Chem. Phys.*, *45*, 1067 (1966).
48. H. P. Fritz, W. Kornrumpf, Abstr., Meeting, Int. Soc. Electrochem., 1979; *Chem. Abstr.*, *92*, 197, 955 (1980).

2.5.15.1.2. Alkyl and Aryl Halocarbenes.

Alkylhalocarbenes [R(X)C] are prepared by 1,1-dihaloalkanes with Li alkyls in Et_2O or pentane[1], e.g.:

$$RCHCl_2 + R'Li \rightarrow R'H + [RCCl_2Li] \rightarrow LiCl + [R(Cl)C] \qquad (a)$$

where R = Me, Et, n-Pr. They can also be prepared by dihalocarbenes with Li alkyls or organomagnesium halide[1,2] reagents:

$$R\text{-}Li + [CCl_2] \rightarrow [RCCl_2Li] \rightarrow LiCl + [R(Cl)C] \qquad (b)$$

The source of the dihalocarbene is either a haloform:

$$RM + CHCl_3 \rightarrow RH + MCl + CCl_2 \xrightarrow{RM} [R(Cl)C] + MCl \qquad (c)$$

or a tetrahalomethane:

$$RM + CF_2Br_2 \rightarrow RBr + MBr + CF_2 \xrightarrow{RM} MF + [R(F)C] \qquad (d)$$

The thermal decomposition of fluoroethylhalosilanes produces carbon-functional alkylfluorocarbenes via α-elimination[3,4]:

$$CHFClCF_2SiCl_3 \rightarrow SiFCl_3 + [CHFCl\text{-}\ddot{C}F] \qquad (e)$$

$$CFCl_2CF_2SiCl_3 \rightarrow SiFCl_3 + [CFCl_2\text{-}\ddot{C}F] \qquad (f)$$

$$CHF_2CF_2SiF_3 \rightarrow SiF_4 + [CHF_2\text{-}\ddot{C}F] \qquad (g)$$

(C. H. VAN DYKE)

1. W. Kirmse, B.-G. V. Bülow, *Chem. Ber.*, *96*, 3316 (1963).
2. V. Franzen, L. Fikentscher, *Chem. Ber.*, *95*, 1958 (1962).
3. R. N. Haszeldine, P. J. Robinson, R. F. Simmons, *J. Chem. Soc.*, 1890 (1964).
4. R. N. Haszeldine, A. E. Tipping, R. O'B. Watts, *J. Chem. Soc.*, *Perkin Trans.*, 2391 (1974).

2.5.15.2. Divalent Silicon Halides (Dihalosilylenes, Silicon Dihalides)

2.5.15.2.1. by Reduction.

The most straightforward synthesis of dihalosilylenes is[1–4] reduction of the corresponding tetrahalide with elemental Si:

$$SiX_4\ (g) + Si\ (s) \xrightarrow{\text{heat}} 2\ SiX_2\ (g) \qquad (a)$$

where X = F, Cl, Br, I. The preparation of SiF_2 by this method[3, 5–9] calls for passing SiF_4 gas (freed from SO_2) over lumps of Si heated to ca. 1200°C. The conversion of SiF_4 to SiF_2 is ca. 50%. At 1400°C, the conversion increases to 85%. To prevent disproportionation, the SiF_2 is pumped out of the reaction zone. Ferrosilicon or Si carbide can be used in place of pure Si as the reducing agent for SiF_4.With Si carbide, SiF_4 is quantitatively reduced to SiF_2 at 1800°C.

The synthesis of $SiCl_2$ from $SiCl_4$ by reduction[10–14] requires $T > 900$°C. A 95% conversion of $SiCl_4$ to $SiCl_2$ can be achieved[3] at 1300–1350°C. Gaseous $SiCl_2$ is unstable outside of the reaction hot zone, with lifetimes of only a few milliseconds at pressures of 1 μ.

The synthesis of $SiBr_2$ can be accomplished by passing $SiBr_4$ over heated Si at 1200°C in vacuum[15]. The dibromide is isolated in a polymeric form. Similarly, polymeric $(SiI_2)_n$ results when SiI_4 is passed ovcr Si heated to 800–900°C in vacuum[16]. The yield of product is only ca. 1%. Silicon diodide undergoes dissocation at low pressures and high T making its isolation and study difficult.

(C. H. VAN DYKE)

1. O. M. Nefedov, M. N. Manakov, *Angew. Chem., Int. Ed. Engl.*, *5*, 1021 (1966).
2. W. H. Atwell, D. R. Weyenberg, *Angew. Chem., Int. Ed. Engl.*, *8*, 469 (1969).
3. P. L. Timms, *Prep. Inorg. React.*, *4*, 59 (1968).
4. P. P. Gaspar, B. J. Herold, in *Carbene Chemistry*, 2nd ed., W. Kirmse, ed., Academic Press, New York, 1971.
5. P. L. Timms, R. A. Kent, T. C. Ehlert, J. L. Margrave, *J. Am. Chem. Soc.*, *87*, 2824 (1965).
6. J. L. Margrave, P. W. Wilson, *Acc. Chem. Res.*, *4*, 145 (1971).
7. J. C. Thompson, J. L. Margrave, *Science*, *155*, 669 (1967).
8. D. L. Perry, J. L. Margrave, *J. Chem. Educ.*, *53*, 696 (1976).
9. D. C. Pease, US Pat. 2,840,588 (1958); *Chem. Abstr.*, *52*, 19,245 (1958).
10. H. Schäfer, J. Nickel, *Z. Anorg. Allg. Chem.*, *274*, 250 (1953).
11. O. Alstrup, C. O. Thomas, *J. Electrochem. Soc.*, *112*, 319 (1965).
12. K. Wieland, M. Heise, *Angew. Chem.*, *63*, 438 (1951).
13. R. Teichmann, E. Wolf, *Z. Anorg. Allg. Chem.*, *347*, 145 (1966).
14. P. Timms, *Inorg. Chem.*, *7*, 387 (1968).
15. M. Schmeisser, M. Schwarzmann, *Z. Naturforsch, Teil B*, *11*, 278 (1956).
16. M. Schmeisser, K. Friederich, *Angew. Chem., Int. Ed. Engl.*, *3*, 699 (1964).

2.5.15.2.2. by Thermal and Other Methods.

The thermal decomposition of certain polyhalosilanes generates divalent Si halides; e.g., SiF_2 can be prepared free from O_2 by the pyrolysis of Si_2F_6 at 700 °C in vacuum,[1] and $SiCl_2$ is generated from the thermal decomposition of perchlorinated polysilanes[2]:

$$Si_2F_6 \xrightarrow{700\,°C} SiF_4 + SiF_2 \quad (a)$$

$$Cl_3Si(SiCl_2)_nSiCl_3 \xrightarrow{\Delta} SiCl_2 \quad (b)$$

The decomposition of certain halomonosilanes and halodisilanes[3] induced by pyrolysis or by other energy input also proceeds by the formation of a silylene derivative:

$$HSiCl_3 \xrightarrow{>1000°C} HCl + SiCl_2 \quad (c)^{4}$$

$$SiX_4 \xrightarrow{>900°C} X_2 + SiX_2 \quad (d)^{5-8}$$

where X = Cl, I;

$$SiH_3X \xrightarrow[\text{photolysis}]{\text{flash}} H_2 + SiHX \quad (e)^{9}$$

where X = Cl, Br;

$$SiF_4 \xrightarrow[\text{discharge}]{\text{glow}} F_2 + SiF_2 \quad (f)^{10,11}$$

The energetic ^{31}Si atoms produced in the transmutation, $^{31}P(n,p)^{31}Si$, abstract fluorine atoms from PF_3 producing[12] both singlet and triplet $^{31}SiF_2$ in the ratio 1.0:3.3:

$$^{31}Si + PF_3 \rightarrow {}^{31}SiF_2 \quad \text{(singlet and triplet)} \quad (g)$$

(C. H. VAN DYKE)

1. M. Schmeisser, K.-P. Ehlers, *Angew. Chem., Int. Ed. Engl., 3*, 700 (1964).
2. E. A. Chernyshev, N. G. Komalenkova, S. A. Bashkirova, *Nov. Khim. Karbenoy. Mater. Vses. Soveshch. Khim. Karbenov. Ikh. Analogov*, 1st 243 (1972); *Chem. Abstr., 82*, 57,793 (1975).
3. R. L. Jenkins, A. J. Vanderwielen, S. P. Ruis, S. R. Gird, M. A. Ring, *Inorg. Chem., 12*, 2968 (1973).
4. E. Sirtl, K. Reuschel, *Z. Anorg. Allg. Chem., 332*, 113.
5. K. Wieland, M. Heise, *Angew. Chem., 63*, 438 (1951).
6. M. Schmeisser, K. Friederich, *Angew. Chem. Int. Ed. Engl., 3*, 700 (1964).
7. H. Schäfer, *Z. Anorg. Allg. Chem., 274*, 2651 (1963).
8. E. Wolf, C. Herbst, *Z. Chem., 7*, 34 (1967).
9. G. Herzberg, R. D. Verma, *Can. J. Phys., 42*, 395 (1964).
10. J. W. C. Johns, A. W. Chantry, R. F. Barrow, *Trans. Faraday Soc., 54*, 1580 (1958).
11. D. R. Rao, P. Venkateswarlu, *J. Mol. Spectrosc., 7*, 287 (1967).
12. O. F. Zeck, Y. Y. Su, G. P. Gennaro, Y.-N. Tang, *J. Am. Chem. Soc., 98*, 3474 (1976).

2.5.15.3. Divalent Germanium Halides (Dihalogermylenes, Germanium Dihalides, Organohalogermylenes.

Germanium difluoride can be prepared in 93% yield by heating Ge powder with anhyd HF for 16 h at 225°C under pressure[1]:

$$Ge + 2\,HF \rightarrow GeF_2 + H_2 \quad (a)$$

or by heating GeF_4 with Ge at 150°C under reduced pressure[2]:

$$GeF_4 + Ge \rightarrow 2\ GeF_2 \tag{b}$$

Germanium difluoride is obtained as a stable solid, somewhat volatile, but decomposes at >160°C to form GeF_4 and $(GeF)_n$ polymers.

Germanium dichloride can be prepared by heating $GeCl_4$ with Ge metal at 350–430°C at low pressures[3]:

$$GeCl_4\ (g) + Ge\ (s) \xrightarrow{350\text{–}430\ °C} 2\ GeCl_2\ (g) \tag{c}$$

The dichloride gas deposits as a yellow-white, polymeric solid, $(GeCl_2)_n$, beyond the hot reaction zone. An improvement in the procedure[4] calls for 370–400°C in a flow system under a $GeCl_4$ pressure of 1 atm. The polymeric $(GeCl_2)_n$ can be a source of monomeric $GeCl_2$ molecules. Germanium dichloride is formed by the decomposition of $HGeCl_3$, in the pure state (by pumping HCl off in a vacuum at −22°C over several hours[5]) or in 1,4-dioxane[6] or Et_2O [7]:

$$GeHCl_3 + B \rightarrow GeCl_2{\cdot}B + HCl \tag{d}$$

where B = 1,4-dioxane or Et_2O. To stabilize pure $GeHCl_3$ prior to its decomposition, the liq is saturated with HCl. Germanium dibromide also can be generated[8] from an Et_2O complex of $HGeBr_3$.

$$Ge(OH)_2 \xrightarrow[Et_2O]{HBr\ in} Et_2O{\cdot}GeHBr_3 \xrightarrow[-Et_2O]{distillation} GeBr_2 + HBr \tag{e}$$

Germanium diiodide, the most stable of the Ge dihalides can be prepared by several good methods. It can be prepared from the reaction of HI with GeS[9]:

$$GeS + 2\ HI \rightarrow GeI_2 + H_2S \tag{f}$$

or GeO[10]:

$$GeO + 2\ HI \rightarrow GeI_2 + H_2O \tag{g}$$

or by the partial reduction[10] of GeI_4 with H_3PO_2:

$$GeI_4 + H_2O + H_3PO_2 \rightarrow GeI_2 + H_3PO_3 + 2\ HI \tag{h}$$

or by the decomposition[7] of unstable $HGeI_3$ produced by HI with $HGeCl_3$ or $Ge(OH)_2$:

$$\left.\begin{array}{r} HGeCl_3 \\ H_2O \\ Ge(OH)_2 \end{array}\right] \xrightarrow{HI} [HGeI_3] \rightarrow GeI_2 + HI \tag{i}$$

The reaction of EtGeCl or PhGeCl with GeX_4, X = Cl, Br, I, forms unstable organopentahalodigermanes that decompose, producing divalent Ge halides in a state of high purity[11,12]:

$$RGeCl + GeX_4 \xrightarrow{20\ °C} R(Cl)XGeGeX_3 \xrightarrow{20\ °C} R(X)_2GeCl + GeX_2 \tag{j}$$

where R = Et, Ph; X = Cl, Br, I.

The organohalogermylenes EtGeCl and PhGeX (X = F, Cl, Br, I) are generated by the decomposition of organomethoxyhalogenogermanes[13]:

$$\text{Ph(H)Ge(X)(OMe)} \xrightarrow{\alpha\text{-elimination}} \text{Ph—Ge—X} + \text{MeOH} \quad \text{(k)}$$

where X = F, Cl, Br, I.

(C. H. VAN DYKE)

1. E. L. Muetterties, *Inorg. Chem.*, *1*, 342 (1962).
2. N. Bartlett, K. C. Yu, *Can. J. Chem.*, *39*, 80 (1961).
3. L. M. Dennis, H. L. Hunter, *J. Am. Chem. Soc.*, *51*, 2703 (1956).
4. E. Vajda, L. Hargittai, M. Kolonits, K. Ujazászy, J. Tamás, A. K. Maltsev, R. G. Mikaelian, O. M. Nefedov, *J. Organomet. Chem.*, *105*, 33 (1976).
5. L. Andrews, D. L. Frederick, *J. Am. Chem. Soc.*, *92*, 775 (1970).
6. S. P. Kolesnickov, V. I. Shirya, O. M. Nefedov, *Izv. Akad. Nauk. SSSR, Ser. Chim.*, 584 1966; *Chem. Abstr.*, *65*, 6705 (1966).
7. O. M. Nefedov, S. P. Kolesnikov, V. I. Sheichenko, *Angew. Chem., Int. Ed. Engl.*, *3*, 508 (1964).
8. T. K. Gar, V. F. Mironov, *Izv. Akad. Nauk. SSSR, Ser. Khim.* 855 (1965); *Chem. Abstr.*, *63*, 5666 (1965).
9. E. A. Flood, L. S. Foster, E. W. Pietrusza, *Inorg. Synth.*, *2*, 106 (1946).
10. G. Brauer, ed., *Handbook of Preparative Inorganic Chemistry*, Vol. 1, Academic Press, New York, 1963.
11. M. Massol, J. Barrau, J. Satgé, *Inorg. Nucl. Chem. Lett.*, *7*, 895 (1971).
12. P. Rivière, J. Satgé, *Synth. React. Inorg. Metal Org. Chem.*, *1*, 13 (1971).
13. M. Massol, J. Satgé, P. Rivière, J. Barrau, *J. Organomet. Chem.*, *22*, 599 (1970).

2.5.15.4. Divalent Sn Halides.

Tin(II) fluoride is prepared[1] from SnO with 40% HF at 60°C:

$$\text{SnO} + 2\ \text{HF} \rightarrow \text{SnF}_2 + \text{H}_2\text{O} \quad \text{(a)}$$

The reaction of anhyd HF with Sn metal at 200°C also forms[2] SnF_2. The synthesis of the other analogues is also based on dissolving Sn metal in the corresponding hydrohalic acid[3]. Anhydrous $SnCl_2$ can be prepared[1] by the acetic anhydride dehydration of commercially available $SnCl_2 \cdot 2\ H_2O$:

$$\text{SnCl}_2 \cdot 2\ \text{H}_2\text{O} + 2\ (\text{MeCO})_2\text{O} \rightarrow \text{SnCl}_2 + 4\ \text{MeCOOH} \quad \text{(b)}$$

The diiodide can also be prepared[1] from $SnCl_2$ with KI.

Divalent organotin compounds of the type h^5-C_5H_5SnX, where X = Cl, Br, can be prepared as white crystalline solids by mixing conc $(h^5\text{-}C_5H_5)_2Sn$ in tetrahydrofuran (THF) and the Sn dihalide[4].

$$(\text{h}^5\text{-C}_5\text{H}_5)_2\text{Sn} + \text{SnX}_2 \xrightarrow{\text{THF}} 2\ \text{h}^5\text{-C}_5\text{H}_5\text{SnX} \quad \text{(c)}$$

where X = Cl, Br. The preparation of h^5-C_5H_5SnCl can also be achieved[4] from anhyd HCl with $(h^5\text{-}C_5H_5)_2Sn$ in THF.

The formation of h^5-C_5H_5SnX derivatives is also noted from $(h^5\text{-}C_5H_5)_2Sn$ with organic and organometallic halides[5,6]:

$$PhCH_2X + (h^5\text{-}C_5H_5)_2Sn \xrightarrow{\text{benzene}} PhCH_2\text{—}C_5H_5 + PhCH_2\text{—}C_5H_5 + h^5\text{-}C_5H_5SnX \quad \text{(d)}$$

where X = Cl, Br;

$$Ph_3CBr + (h^5\text{-}C_5H_5)_2Sn \xrightarrow{\text{benzene}} Ph_3C\text{—}C_5H_5 + h^5\text{-}C_5H_5SnBr \quad \text{(e)}$$

$$Me_2SnCl_2 + (h^5\text{-}C_5H_5)_2Sn \rightarrow Me_2(h^1\text{-}C_5H_5)SnCl + h^5\text{-}C_5H_5SnCl \quad \text{(f)}$$

Some covalent halides, such as BCl_3, $SnBr_4$ and $FeCl_3$, undergo halogen–ligand exchange with $(h^5\text{-}C_5H_5)_2Sn$, producing[6–9] the corresponding tin(II) dihalide:

$$(h^5\text{-}C_5H_5)_2Sn + BX_3 \rightarrow SnX_2 + (R_2BX) \quad \text{(g)}$$

where X = Cl, Br;

$$3\ (h^5\text{-}C_5H_5)_2Sn + Fe_2Cl_6 \rightarrow 2\ SnCl_2 + (h^1\text{-}C_5H_5)_2SnCl_2 + 2\ (h^5\text{-}C_5H_5)_2Fe \quad \text{(h)}$$

while $(h^5\text{-}C_5H_5)_2Sn$ forms addition compounds with BF_3, BBr_3, BI_3, $AlCl_3$ and $AlBr_3$:

$$2\ (h^5\text{-}C_5H_5)_2Sn + Al_2X_6 \xrightarrow{C_6H_6} 2\ (h^5\text{-}C_5H_5)_2Sn\text{:}AlX_3 \quad \text{(i)}$$

where X = Cl, Br;

$$2\ (h^5\text{-}C_5H_5)_2Sn + BI_3 \rightarrow h^5\text{-}C_5H_5SnI\cdot SnI_2 + [B(C_5H_5\text{-}h^1)_3] \quad \text{(j)}$$

(C. H. VAN DYKE)

1. G. Brauer, ed., *Handbook of Preparative Inorganic Chemistry*, Vol. 1, Academic Press, New York, N.Y., 1963.
2. E. L. Muetterties, *Inorg. Chem., 1*, 342 (1962).
3. O. M. Nefedov, M. N. Manakov, *Angew. Chem., Int. Ed. Engl., 5*, 1021 (1966).
4. K. D. Bos, E. J. Bulten, J. G. Noltes, *J. Organomet. Chem., 39*, C52 (1972).
5. K. D. Bos, E. J. Bulten, J. G. Noltes, *J. Organomet. Chem., 99*, 397 (1975).
6. K. D. Bos, E. J. Bulten, J. G. Noltes, *J. Organomet. Chem., 67*, C13 (1974).
7. P. G. Harrison, J. A. Richards, *J. Organomet. Chem., 108*, 35 (1976).
8. W. Siebert, K. Kinberger, *J. Organomet. Chem., 116*, C7 (1976).
9. P. G. Harrison, J. J. Zuckerman, *J. Am. Chem. Soc., 92*, 2577 (1970).

2.5.16. Lead(II) Halides, PbX_2 (X = F, Cl, Br, I) and Organolead(II) Halides.

The Pb(II) halides (PbX_2, X = Cl, Br, I) are precipitated[1] from solutions of soluble Pb salts. For example, $PbCl_2$ is prepared by dissolving granulated Pb in dil HNO_3 followed by addition of dil HCl.

$$Pb + 8\ HNO_3 \longrightarrow 3\ Pb[NO_3]_2 + 2\ NO + 4\ H_2O \quad \text{(a)}$$

$$Pb[NO]_3 + 2\ HCl \longrightarrow PbCl_2 + 2\ HNO_3 \quad \text{(b)}$$

The bromide is prepared[1] similarly by adding HBr to $Pb[NO_3]_2$. The synthesis and purification of $PbCl_2$ and $PbBr_2$ is achieved by this procedure[2,3]. Lead(II) iodide is also prepared[4] by the addition of KI to $Pb(NO_3)_2$:

$$Pb[NO_3]_2 + 2\ KI \longrightarrow PbI_2 + 2\ KNO_3 \qquad \text{(c)}$$

A $PbI_2 \cdot HI \cdot 5\ H_2O$ adduct crystallizes[1] from HI[1]. The Pb(II) halides crystallize[4] from hot aq solutions according to their solubilities at 0°, 20° and 100 °C.

TABLE 1. SOLUBILITIES OF LEAD(II) HALIDES

	Solubility (g per 100 ml H_2O)		
Salt	100 °C	20 °C	0 °C
$PbCl_2$	3.34	0.99	0.67
$PbBr_2$	4.71	0.84	0.46
PbI_2	0.41	0.063	0.004

Pure $PbCl_2$ crystallizes in the form of white, silky needles or lamellae, the dibromide as white, silky, rhombic needles, while PbI_2 crystallizes from H_2O or dilute acetic acid in the form of crystal flakes with a golden luster. Although solid PbI_2 is yellow, its aq solutions are colorless[1].

Lead(II) chloride is also prepared[1] by treating PbO or $Pb_3(OH)_2CO_3$ with HCl:

$$PbO + 2\ HCl \longrightarrow PbCl_2 + H_2O \qquad \text{(d)}$$

$$Pb_3(OH)_2CO_3 + 6\ HCl \longrightarrow 3\ PbCl_2 + 2\ CO_2 + 4\ H_2O \qquad \text{(e)}$$

Lead(II) fluoride is prepared[5] by heating nitrate- and acetate-free $PbCO_3$ or $Pb(OH)_2$ with xs HF, e.g.:

$$PbCO_3 + 2\ HF \longrightarrow PbF_2 + H_2O + CO_2 \qquad \text{(f)}$$

Stable mixed Pb(II) dihalides are also known, e.g., PbFX (X = Cl, Br, I) are prepared[6] from stoichiometric aq PbX_2 and KF at 25 °C:

$$PbX_2 + KF \longrightarrow PbFX + KX \qquad \text{(g)}$$

where X = Cl, Br, I.

The organolead(II) halides h^5-C_5H_5PbX, X = Cl, Br, are prepared[7] anhydr hydrogen halides (HCl and HBr) with $(h^5\text{-}C_5H_5)_2Pb$. With xs protic reagent, both cyclopentadienyl groups are cleaved:

$$(h^5\text{-}C_5H_5)_2Pb + HX \longrightarrow h^5\text{-}C_5H_5PbX + C_5H_6 \qquad \text{(h)}$$

where X = Cl, Br.

$$(h^5\text{-}C_5H_5)_2Pb + 2\ HX \longrightarrow PbX_2 + 2\ C_5H_6 \qquad \text{(i)}$$

where X = Cl, Br.

Transition-metal halides cleave one cyclopentadienyl group.

$$(h^5\text{-}C_5H_5)_2Pb + TiCl_4 \longrightarrow h^5\text{-}C_5H_5PbCl + h^1\text{-}C_5H_5TiCl_3 + (h^1\text{-}C_5H_5)_2TiCl_2 \qquad \text{(j)}$$

(C. H. VAN DYKE)

1. H. Remy, *Treatise on Inorganic Chemistry*, Vol. 1, Elsevier Pub. Co. (1956).
2. G. P. Baxter, F. L. Grover, *J. Am. Chem. Soc., 37*, 1027 (1915).
3. G. P. Baxter, T. Thorvaldson, *J. Am. Chem. Soc., 37*, 1020 (1915).
4. G. G. Schlesinger, *Inorganic Laboratory Preparations*, Chemical Publishing Co., Inc., New York, (1962).
5. W. Kwasnik in *Handbook of Preparative Inorganic Reactions*, G. Brauer, ed. Vol. 1, Academic Press, New York (1963).
6. A. Rulmont, *C. R. Hebd. Seances Acad. Sci., Ser. C, 276*, 775 (1973).
7. A. K. Holliday, P. H. Makin, R. J. Puddephatt, J. D. Wilkins, *J. Organometal. Chem., 57*, C45 (1973).

2.5.17. Preparation of Astatine–Group-IVB Element Bonds

2.5.17.1. Preparation of Astatine–Carbon Bonds

The preparation of organoastatine compounds has been studied extensively in recent years, particularly in connection with the synthesis of astatine-labelled proteins[1,2].

(W. D. LEE)

1. K. Berei, L. Vasáros, *Organic Chemistry of Astatine*. Hungarian Acad. Sciences, Central Research Institute for Physics, Budapest, 1981. Rep. KFKI-1981-10.
2. K. Berei, L. Vasarós, *Organic Chemistry of Astatine* in The Chemistry of halides, pseudo-halides, and azides. S. Patai, Z. Rappoport, eds. John Wiley & Sons Ltd., NY 1983, pp. 405–440.

2.5.17.1.1. by Homogeneous and Heterogeneous Halogen Exchange

Reaction of At^- with ICH_2CO_2H at 40 °C in acidic aq iodide solution produces $AtCH_2CO_2H$[1]. Astatobenzene has been made at RT by the reaction of C_6H_5I with AtI dissolved in it[2] or by reaction of AtI and I_2 with $C_6H_5NHNH_2$ in aq KI[2]:

$$C_6H_5NHNH_2 + AtI + I_2 \longrightarrow C_6H_5At + N_2 + 3\,HI \quad \text{(a)}$$

Formation of C_6H_5At may also be brought about by the reaction sequence[2]:

$$(C_6H_5)_2I{\cdot}I + At^- \xrightarrow[100\,°C]{C_2H_5OH} (C_6H_5)_2I{\cdot}At + I^- \quad \text{(b)}$$

$$(C_6H_5)_2I{\cdot}At \xrightarrow{175\,°C} C_6H_5I + C_6H_5At \quad \text{(c)}$$

The starting material can also be $(C_6H_5)_2ICl$[3].

Alkyl and arylastatides can also be prepared by heterogeneous exchange reactions in which RI vapor at 130–200 °C is passed through a Kieselguhr column on which At^- has been adsorbed[2,4]. Arylastatides are prepared by exchange between NaAt (with or without NaI carrier) and the aryl bromides or iodides at 50 – 250 °C[5–8]. Alkyl amines are obtained by heating iodo- or 3,5-diiodotyrosine with At^- in vacuo for 30 min at 120 °C[9].

(W. D. LEE)

1. G. Samson, A. H. W. Aten, Jr. *Radiochim. Acta, 9*, 53 (1968). See also G. Samson, Ph.D., Univ. Amsterdam, 1971.
2. G. Samson, A. H. W. Aten, Jr., *Radiochim. Acta, 13*, 220 (1970).

3. V. D. Nefedov, Yu. V. Norseev, H. Savlevich, E. N. Sinotova, M. A. Toropova, V. A. Khalkin, *Bull. Acad. Sci. USSR, Div. Chem. Sci, 144*, 507 (1962).
4. G. Sampson, A. H. W. Aten, Jr., *Radiochim. Acta, 12*, 55 (1969).
5. L. Vasáros, K. Berei, Yu. V. Norseev, V. A. Khalkin, *Radiochem. Radioanal. Lett., 27*, 329 (1976).
6. L. Vasáros, Yu. V. Norseev, D. D. Nhan, V. A. Khalkin, *Radiochem. Radioanal. Lett., 47*, 313, 403 (1981) and *50*, 275 (1982).
7. L. Vasáros, Yu. V. Norseev, V. I. Fominykh, V. A. Khalkin, *Soviet Radiochem., 24*, 84 (1982).
8. C. Y. Shiue, G.-J. Meyer, T. J. Ruth, A. P. Wolf, *J. Labelled Compd. Radiopharm., 18*, 1039 (1980).
9. G. W. M. Visser, E. L. Diemer, F. M. Kaspersen, *Int. J. Appl. Radiat. Isot., 30*, 743 (1979).

2.5.17.1.2. by Decomposition of Diazonium Salts.

Arylastatides may be prepared by reaction of At^- with the corresponding diazonium chloride[1–4]:

$$RC_6H_4N_2Cl \xrightarrow[-5\,°C]{At^-} RC_6H_4N_2At \xrightarrow[80\,°C]{-N_2} RC_6H_4At \qquad (a)$$

A similar procedure is used to prepare 5-astatouracil[5] and 5-astatodeoxyuridine[6], and astatine-substituted derivatives of methylene blue[7]. Para-astatobenzoic acid prepared in this way is used to incorporate At into bovine serum albumen[5], and similar procedures result in other At-labelled proteins[8].

(W. D. LEE)

1. G. W. M. Visser, E. L. Diemer, *Radiochem. Radioanal. Lett., 51*, 135 (1982).
2. G.-J. Meyer, K. Rössler, G. Stöcklin, *Radiochem. Radioanal. Lett., 21*, 247 (1975).
3. G. W. M. Visser, E. L. Diemer, F. M. Kaspersen, *Recl. Trav. Chim. Pays-Bas, 99*, 93 (1980).
4. M. R. Zalutsky, A. M. Friedman, F. C. Buckingham, W. Wung, F. P. Stuart, S. J. Simonian, *J. Labelled Compd. Radiopharm., 13*, 181 (1977).
5. G.-J. Meyer, K. Rössler, G. Stöcklin, *J. Labelled Compd. Radiopharm., 12*, 449 (1976).
6. K. Rössler, G.-J. Meyer, G. Stöcklin, *J. Labelled Compd. Radiopharm., 13*, 271 (1977).
7. I. Brown, R. N. Carpenter, E. Link, J. S. Mitchell, *J. Radioanal. Nucl. Chem., 107*, 337 (1986).
8. A. T. M. Vaughn, *int. J. Appl. Radiat. Isot., 30*, 576 (1979).

2.5.17.1.3. from Organometallic Compounds

Aromatic astatine compounds are prepared in good yield by treatment of the arylmercury(II) chloride with At^- in sulfite soln followed by addition of KI_3[1]:

$$RC_6H_4HgCl \xrightarrow[I_3^-]{At^-} RC_6H_4At + HgI_2 \qquad (a)$$

The method is applied to the introduction of At into tyrosine[2] and steroids[3]. A similar method is used to make 6-At-2-methyl-1,4-naphthoquinone[4].

Astatine-substituted benzoic acid and anisole are made in similar fashion from the arylthallium(III) trifluoroacetates[5]:

$$RC_6H_5 \xrightarrow[CF_3CO_2H]{Tl(CF_3CO_2)_3} RC_6H_4Tl(CF_3CO_2)_2 \xrightarrow[I^-]{At^-} RC_6H_4At + TlI \qquad (b)$$

This method is only applicable to substrates that are resistant to oxidation by Tl(III).

Astatinated derivatives of benzoic acid may be produced via electrophilic destannylation of the corresponding trialkylstannane by astatine in the presence of aq. H_2O_2, followed by hydrolysis[6].

$$\text{Bu}_3\text{Sn-C}_6\text{H}_4\text{-oxazoline} \xrightarrow[\text{H}_2\text{O}_2]{\text{Na}^{211}\text{At–NaI}} {}^{211}\text{Al-C}_6\text{H}_4\text{-oxazoline} \xrightarrow[\text{1 N NaOH, 100°C, 10 min; acidify to pH 3}]{\text{(a) IN HCl, 100°C, 10 min}} {}^{211}\text{At-C}_6\text{H}_4\text{-COOH} \quad \text{(c)}$$

A similar method is used to produce astatinated vinylsteroid hormones[7] and 3-astatotamoxifen[6].

(W. D. LEE)

1. G. W. M. Visser, E. L. Diemer, F. M. Kaspersen, *J. Labelled Compd. Radiopharm.*, *17*, 657 (1980).
2. G. W. M. Visser, E. L. Diemer, F. M. Kaspersen, *Int. J. Appl. Radiat. Isot.*, *30*, 743 (1979).
3. G. W. M. Visser, E. L. Diemer, F. M. Kaspersen, *J. Labelled Compd. Radiopharm.*, *18*, 799 (1981).
4. I. Brown, *Int. J. Appl. Radiat. Isot.*, *33*, 75 (1982).
5. G. W. M. Visser, E. L. Diemer, *Int. J. Appl. Rad. Isot.*, *33*, 389 (1982).
6. R. A. Milius, W. H. McLaughlin, R. M. Lambrecht, A. P. Wolf, J. J. Carroll, S. J. Adelstein, W. D. Bloomer, *Appl. Radiat. Isot.*, *37*, 799 (1986).
7. K. M. R. Pillai, W. H. McLaughlin, R. M. Lambrecht, W. D. Bloomer, *J. Labelled Compd. Radiopharm.*, *24*, 1117 (1987).

2.5.17.1.4. by Electrophilic Substitution After Chemical or Electrochemical Oxidation.

Astatine can be substituted for hydrogen in aromatic compounds by reaction at 80–120 °C with At solutions in acetic acid containing $H_2Cr_2O_7$[1], or by reaction at 60 °C with AtCl or AtBr (see section 2.2.4.7.)[2,3].

Astatine is introduced into tyrosine and proteins from neutral or weakly alkaline solutions containing H_2O_2 and a small amount of KI[4–6], but the nature of the bonding remains uncertain. The chloramin-T method that can be used to iodinate proteins does not work for astatine[4].

Proteins and lymphocytes are labelled with At via electrochemical oxidation[4,7]. The success of the labelling is dependent on the value of the applied potential, but the nature of the bonding is unknown.

(W. D. LEE)

1. L. Vasáros, Yu. V. Norseev, V. A. Khalkin, *Bull. Acad. Sci. USSR (Engl. Transl.)*, *266*, 297 (1982).
2. K. Rössler, G.-J. Meyer, G. Stöcklin, *J. Labelled Compd. Radiopharm.*, *13*, 271 (1977).
3. G.-J. Meyer, K. Rössler, G. Stöcklin, *Radiochim. Acta*, *24*, 81 (1977).
4. C. Aaij, W. R. J. M. Tschroots, L. Lindner, T. E. W. Feltkamp, *Int. J. Appl. Radiat. Isot.*, *26*, 25 (1975).
5. A. T. M. Vaughan, J. H. Fremlin, *Int. J. Appl. Rad. Isot.*, *28*, 595 (1977).

6. A. T. M. Vaughan, J. H. Fremlin, *Int. J. Nucl. Med. Biol.*, *5*, 229 (1978).
7. J. A. Smith, J. A. A. Myburgh, R. D. Neirinckx, *Clin. Exp. Immunol.*, *14*, 107 (1973).

2.5.17.1.5. Hot-Atom Reactions.

Astatobenzene is prepared by the α-particle bombardment of $Bi(C_6H_5)_3$[1]. Astatine-211 can be incorporated into both aromatic and aliphatic hydrocarbons when it is formed by the electron-capture decay of ^{211}Rn in the presence of the hydrocarbon[2–6]. The hydrocarbon can be present as gas, liquid, or solid.

(W. D. LEE)

1. G. Sampson, A. H. W. Aten, Jr., *Radiochim. Acta*, *13*, 220 (1970).
2. V. D. Nefedov, M. A. Toropova, V. A. Khalkin, Yu. V. Norseev, V. I. Kuzin, *Soviet Radiochem*, *12*, 176 (1970).
3. V. I. Kuzin, V. D. Nefedov, Yu. V. Norseev, M. A. Toropova, E. S. Filatov, V. A. Khalkin, *Khim. Vys. Energ.*, *6*, 181 (1972); *High Energy Chemistry*, *6*, 161 (1972).
4. K. Berei, L. Vasáros, Yu. V. Norseev, V. A. Khalkin, *Radiochem. Radioanal. Lett.*, *26*, 177 (1976).
5. L. Vasáros, Yu. V. Norseev, G.-J. Meyer, K. Berei, V. A. Khalkin, *Radiochim. Acta*, *26*, 171 (1979).
6. L. Vasáros, Yu. V. Norseev, V. I. Fominkyh, V. A. Khalkin, *Radiokhimiya*, *24*, 95 (1982).

2.5.17.2. Preparation of Astatine-Lead Bonds.

Astatine in the −1 state coprecipitates from H_2O with PbI_2, presumably as PbIAt[1]. Addition of Pb^{2+} to aqueous I_2-I^- containing At precipitates $PtI_2 \cdot x\,I_2$, with which the At coprecipitates. The I_2 is presumably present as I_3^-, and the At as $[AtI_2]^-$ [2].

(W. D. LEE)

1. E. H. Appelman, *J. Am. Chem. Soc.*, *83*, 805 (1961).
2. E. H. Appelman, U. S. Atomic Energy Commission Report UCRL-9025 (1960); *Chem. Abst.*, *54*, 16998 (1960).

Abbreviations

abs	absolute
a.c.	alternating current
Ac	acetyl, CH_3CO
acac	acetylacetonate anion
acacH	acetylacetone, $CH_3C(O)CH_2C(O)CH_3$
AcO	acetate anion, $CH_3C(O)O$
Ad	adamantyl
ads	adsorbed
AIBN	2,2′-azobis(isobutyronitrile), 2,2′-$[(CH_3)_2CCN]_2N_2$
Alk	alkyl
am	amine
amt	amount
Am	amyl, C_5H_{11}
amu	atomic mass unit
anhyd	anhydrous
aq	aqueous
Ar	aryl
asym	asymmetrical, asymmetric
at	atom (not atomic, except in atomic weight)
atm	atmosphere (not atmospheric)
av	average
BBN	9-Borabicyclo[3.3.1]nonane
bcc	body-centered cubic
BD	butadiene
bipy	2,2′-bipyridyl
bipyH	protonated 2,2′-bipyridyl
bp	boiling point
Bu	butyl, C_4H_9
Bz	benzyl, $C_6H_5CH_2$
ca.	circa, about, approximately
catal	catalyst (not catalyzing, catalysis, catalyzed, etc.)
CDT	cyclododecatriene
cf.	compare
Ch.	chapter
CHD	1,3-cycloheptadiene
Chx	cyclohexyl
ChxD	1,3-cyclohexadiene
COD	cyclooctadiene
conc	concentrated (not concentration)
const.	constant
COT	cyclooctatriene
Cp	cyclopentadienyl, C_5H_5
CPE	controlled-potential electrolysis
cpm	counts per minute
CT	charge-transfer
CV	cyclic voltammetry
CVD	chemical vapor deposition
CW	continuous wave

d	day, days
DABIP	N,N′-diisopropyl-1,4-diazabutadiene
DBA	dibenzylideneacetone
d.c.	direct current
DCM	dicyclopentadienylmethane
DCME	$Cl_2CHC(O)CH_3$
DCP	1,3-dicyclopentadienylpropane
DDT	dichlorodiphenyltrichloroethane, 1,1,1′-trichloro-2,2-bis-(4-chlorophenyl)ethane
dec	decomposed
DED	1,1-bis(ethoxycarbonyl)ethene-2,2-dithiolate, $[[(H_5C_2OC(O)]_2C{=}CS_2]^{2-}$
depe	1,2-bis(diphenylphosphino)ethene, $(C_6H_5)_2PCH{=}CHP(C_6H_5)_2$
DIAD	diindenylanthracenyl
diars	1,2-bis(dimethylarsino)benzene, o-phenylenebis(dimethylarsine), 1,2-$(CH_3)_2AsC_6H_4As(CH_3)_2$
dien	diethylenetriamine, $[H_2N(CH_2)_2]_2NH$
diglyme	diethyleneglycol dimethylether, $CH_3O(CH_2CH_2O)_2CH_3$
dil	dilute
diop	2,3-O-isopropylidene-2,3-dihydroxy-1,4-bis(diphenylphosphino)butane, $(C_6H_5)_2PCH_2CH[OCH(CH_3){=}CH_2]CH[OCH(CH_3){=}CH_2]CH_2P(C_6H_5)_2$
diphos	1,2-bis(diphenylphosphino)benzene, 1,2-$(C_6H_5)_2PC_6H_4P(C_6H_5)_2$
Div.	division
DMA	dimethylacetamide
dme	dropping mercury electrode
DME	1,2-dimethoxyethane, glyme, $CH_3O(CH_2)_2OCH_3$
DMF	N,N-dimethylformamide, $HC(O)N(CH_3)_2$
DMG	dimethylglyoxime, $CH_3C({=}NOH)C({=}NOH)CH_3$
DMP	1,2-dimethoxybenzene, 1,2-$(CH_3O)_2C_6H_4$
dmpe	1,2-bis(dimethylphosphino)ethane, $(CH_3)_2P(CH_2)_2P(CH_3)_2$
DMSO	dimethylsulfoxide, $(CH_3)_2SO$
dpam	bis(diphenylarsino)methane, $[(C_6H_5)_2As]_2CH_2$
dpic	dipicolinate ion
DPP	differential pulse polarography
dppb	1,4-bis(diphenylphosphino)butane, 1,4-$(C_6H_5)_2P(CH_2)_4P(C_6H_5)_2$
dppe	1,2-bis(diphenylphosphino)ethane, 1,2-$(C_6H_5)_2P(CH_2)_2P(C_6H_5)_2$
dppm	bis(diphenylphosphino)methane, $[(C_6H_5)_2P]_2CH_2$
dppoe	bis(diphenylphosphoryl)ethane
dppp	1,3-bis(diphenylphosphino)propane, 1,3-$(C_6H_5)_2P(CH_2)_3P(C_6H_5)_2$
dptpe	1,2-bis(di-p-tolylphosphino)ethane, 1,2-$(4\text{-}CH_3C_6H_4)_2P(CH_2)_2P\text{-}(C_6H_4CH_3\text{-}4)_2$
DTA	differential thermal analysis
DTBQ	3,5-di-t-butyl-o-benzoquinone
DTH	1,6-dithiahexane, butane-1,4-dithiol, 1,4-$HS(CH_2)_4SH$
DTS	dithiosquarate
ed.	edition, editor
eds.	editors
EDTA	ethylenediaminetetraacetic acid, $[HOC(O)]_2N(CH_2)_2N[C(O)OH]_2$
e.g.	exempli gratia, for example
emf	electromotive force

en	ethylenediamine, $H_2N(CH_2)_2NH_2$
enH	protonated ethylenediamine
EPR	electron paramagnetic resonance
equimol	equimolar
equiv	equivalent
EPR	electron paramagnetic resonance
Eq.	equation
ERF	effective reduction factor
ES	excited state
ESR	electron-spin resonance
esu	electrostatic unit
Et	ethyl, CH_2CH_2
etc.	et cetera, and so forth
Et_2O	diethyl ether, $(C_2H_5)_2O$
EtOH	ethanol, C_2H_5OH
et seq.	et sequentes, and the following
eu	entropy unit
fac	facial
fcc	face-centered cubic
ff.	following
Fig.	figure
Fl	fluorenyl
fp	freezing point
g	gas
g-at	gram-atom
glyme	1,2-dimethoxyethane, $CH_3O(CH_2)_2OCH_3$
graph	graphite
GS	ground state
h	hour, hours
HD	1,5-hexadiene
Hex	hexyl
HMDB	hexamethyl(Dewar benzene)
hmde	hanging mercury drop electrode
HMI	heptamethylindenyl
HMPA	hexamethylphosphoramide, $[(CH_3)_2N]_3PO$
HOMO	highest occupied molecular orbital
i.e.	id est, that is
Im	imidazole
inter alia	among other things
IPC	isopinocamphylborane
IR	infrared
irrev	irreversible
ISC	intersystem crossing
isn	isonicotinamide
l	liquid
L	ligand
LC	ligand centered
LF	ligand field
LFER	linear free-energy relationship
liq	liquid
LMCT	ligand-to-metal charge transfer
Ln	lanthanides, rare earths
LSV	linear-scan voltammetry

LUMO	lowest unoccupied molecular orbital
m	meta
max	maximum
M	metal
MC	metal centered
Me	methyl, CH_3
Men	menthyl
MeOH	methanol, CH_3OH
mer	meridional; the repeating unit of an oligomer or polymer
mhp	2-hydroxy-6-methylpyridine, 2-HO, 6-$CH_3C_5H_3N$
min	minimum, minute, minutes
MLCT	metal-to-ligand charge transfer
MO	molecular orbital
mol	molar
mp	melting point
MV	methyl viologen, 1,1′-dimethyl-4,4′-bipyridinium dichloride
n.a.	not available
napy	naphthyridine
NBD	norbornadiene, [2.2.1]bicyclohepta-2,5-diene
neg	negative
nhe	normal hydrogen electrode
NMR	nuclear magnetic resonance
No.	number
np	tris-[2-(diphenylphosphino)ethyl]amine, $N[CH_2CH_2P(C_6H_5)_2]_3$
Np	naphthyl
NPP	normal pulse polarography
NQR	nuclear quadrupole resonance
NTA	nitrilotriacetate
o	ortho
obs	observed
Oct	octyl
oep	octaethylporphyrin
OF	oxidation factor
O_h	octahedral
Oq	oxyquinolate
p	para
p.	page
P	pressure
Pat.	patent
pet.	petroleum
Ph	phenyl, C_6H_5
phen	1,10-phenanthroline
Ph_2PPy	2-(diphenylphosphino)pyridine, 2-$(C_6H_5)_2PC_5H_4N$
pip	piperidine, $C_5H_{10}N$
PMDT	pentamethyldiethylenetriamine, $(CH_3)_2N(CH_2)_2N(CH_3)(CH_2)_2N(CH_3)_2$
PMR	proton magnetic resonance
pn	propylene-1,3-diamine, 1,3-$H_2NCH_2CH_2CH_2NH_2$
pos	positive
pp.	pages
ppb	parts per billion
ppm	parts per million

ppn	bis(diphenylphosphino)amine, $[(C_6H_5)_2P]_2NH$
ppt	precipitate
Pr	propyl, C_3H_7
PSS	photostationary state
PVC	poly(vinyl chloride)
PY	pyridine, C_5H_5N
pyr	pyrazine
PZE	potential of zero charge
rac	racemic mixture, racemate
R	organic group; universal gas constant
RDE	rotated disk electrode
RE	rare earths, lanthanides
ref.	reference
rev	reversible
rf	radiofrequency
RF	reduction factor
rh	rhombohedral
rms	root mean square
rpm	revolutions per minute
RT	room temperature
s	second, seconds; solid
sce	saturated calomel electrode
SCE	standard calomel electrode
sec	secondary
Sep	sepulcrate, 1,3,6,8,10,13,16,19-octaazabicyclo[6.6.6]eicosane
Sia	Diisamyl
soln	solution
solv	solvated
sp	specific
STP	standard temperature and pressure
subl	sublimes
Suppl.	supplement
sym	symmetrical, symmetric
t	time; tertiary
T	temperature
T_d	tetrahedral
TCNE	tetracyanoethylene
TEA	tetraethylammonium ion, $[(C_2H_5)_4N]^+$
terpy	2,2′2″-terpyridyl
tetraphos	$Ph_2PCH_2CH_2PPhCH_2CH_2PPhCH_2CH_2PPh_2$
TGA	thermogravimetric analysis
TGL	triethyleneglycol dimethylether
THF	tetrahydrofuran
THP	tetrahydropyran
THT	tetrahydrothiophene
Thx	thexyl
TLC	thin-layer chromatography
TMED	N,N,N′,N′-tetramethylethylenediamine, $(CH_3)_2N(CH_2)_2N(CH_3)_2$
TMPH	2,2,6,6-tetramethylpiperidine, 2,2,6,6-$(CH_3)_4C_5H_6N$
Tos	tosyl, tolylsulfonyl, 4-$CH_3C_6H_4SO_2$
TPA	tetraphenylarsonium ion, $[(C_6H_5)_4As]^+$

Author Index

The entries of this index were derived directly by computer program from the lists of references. The accuracy of the references was the sole responsibility of the authors. No editorial check, except for format and journal-title abbreviation, was applied. Consequently, errors occurring in authors' names in the references will recur in this index.
Each entry in the index refers to the appropriate section number.

B

D

G

L

O

P

Q

R

S

T

U

V

Compound Index

This index lists individual, fully specified compositions of matter that are mentioned in the text. It is an index of empirical formulas, ordered according to the following system: the elements within a given formula occur in alphabetical sequence except for C, or C and H if present, which always come first. The formulas are ordered alphanumerically without exception.

The index is augmented by successively permuted versions of all empirical formulas. As an example, $C_3H_3AlO_9$ will appear as such and, at the appropriate positions in the alphanumeric sequence, as $H_3AlO_9{*}C_3$, $AlO_9{*}C_3H_3$ and $O_9{*}C_3H_3Al$. The asterisk identifies a permuted formula and allows the original formula to be reconstructed by shifting to the front the elements that follow the asterisk.

Whenever an empirical formula does not show how the elements are combined in groups, it is followed by a linearized structural formula, which reveals the connectivity of the compound(s) underlying the empirical formula and serves to distinguish substances which are identical in composition but differ in the arrangement of elements.

The nonpermuted empirical formulas are followed by keywords. They describe the context in which the compounds represented by the empirical formulas are discussed. Section numbers direct the reader to relevant positions in the book.

Subject Index

This index supplements the compound index and the table of contents by providing access to the text by way of methods, techniques, reaction conditions, properties, effects and other phenomena. Reactions of specific bonds and compound classes are noted when they are not accessed by the heading of the section in which they appear.
For multiple entries, additional keywords indicate contexts and thereby avoid the retrieval of information that is irrelevant to the user's need.
Section numbers are used to direct the reader to those positions in the volume where substantial information is to be found.

B

E

F

H

I

K

L

M

N

R

S

T

W

Date Due

BJJH

Periodic Tabl

Period	Group IA	Group IIA	Group IIIA	Group IVA	Group VA	Group VIA	Group VIIA	
1 1*s*	1 **H**							
2 2*s*2*p*	3 **Li**	4 **Be**						
3 3*s*3*p*	11 **Na**	12 **Mg**						
4 4*s*3*d* 4*p*	19 **K**	20 **Ca**	21 **Sc**	22 **Ti**	23 **V**	24 **Cr**	25 **Mn**	26 **Fe**
5 5*s*4*d* 5*p*	37 **Rb**	38 **Sr**	39 **Y**	40 **Zr**	41 **Nb**	42 **Mo**	43 **Tc**	44 **Ru**
6 6*s* (4*f*) 5*d* 6*p*	55 **Cs**	56 **Ba**	57* **La**	72 **Hf**	73 **Ta**	74 **W**	75 **Re**	76 **Os**
7 7*s* (5*f*) 6*d*	87 **Fr**	88 **Ra**	89** **Ac**	104	105	106	107	108

*Lanthanide series 4*f*	58 **Ce**	59 **Pr**	60 **Nd**	61 **Pm**	62 **Sm**	63 **Eu**
**Actinide series 5*f*	90 **Th**	91 **Pa**	92 **U**	93 **Np**	94 **Pu**	95 **Am**

Adapted from F.A. Cotton, G. Wilkinson, *Advanced Inorganic Che*